CELLULAR

AND

MOLECULAR

IMMUNOLOGY

SAUNDERS TEXT AND REVIEW SERIES

CELLULAR
AND
MOLECULAR
IMMUNOLOGY

THIRD EDITION

ABUL K. ABBAS, M.B., B.S.
Professor of Pathology
Harvard Medical School
Brigham and Women's Hospital
Boston, Massachusetts

ANDREW H. LICHTMAN, M.D., Ph.D.
Associate Professor of Pathology
Harvard Medical School
Brigham and Women's Hospital
Boston, Massachusetts

JORDAN S. POBER, M.D., Ph.D.
Director, Molecular Cardiobiology
Boyer Center for Molecular Medicine
Professor of Pathology and Immunobiology
Yale University School of Medicine
New Haven, Connecticut

W.B. SAUNDERS COMPANY
A Division of Harcourt Brace & Company
Philadelphia   London   Toronto   Montreal   Sydney   Tokyo

W.B. SAUNDERS COMPANY
A Division of Harcourt Brace & Company

The Curtis Center
Independence Square West
Philadelphia, Pennsylvania 19106

Library of Congress Cataloging-in-Publication

Abbas, Abul K.
Cellular and molecular immunology / Abul K. Abbas, Andrew H. Lichtman,
Jordan S. Pober.—3rd ed.

p.   cm.

Includes bibliographical references and index.

ISBN 0-7216-4024-9

1. Cellular immunity.   2. Molecular immunology.   I. Lichtman, Andrew H.
    II. Pober, Jordan S.   III. Title.
    [DNLM: 1. Immunity, Cellular.   2. Lymphocytes—immunology.   QW
    568 A122c 1997]

QR185.5.A23   1997   616.07′9—dc21

DNLM/DLC   96-49579

CELLULAR AND MOLECULAR IMMUNOLOGY   ISBN 0-7216-4024-9

Last digit is the print number: 9   8   7   6   5   4   3   2   1

To

Ann, Jonathan, Rehana

Sheila, Eben, Ariella, Amos, Ezra

Barbara, Jeremy, Jonathan

PREFACE

In the 3 years since the last edition of *Cellular and Molecular Immunology,* a wealth of new information in the field of immunology has emerged. Some of the areas in which the increase in knowledge is particularly impressive are the structures of molecules involved in immune responses, signal transduction, and lymphocyte apoptosis. Furthermore, the application of transgenic and knockout mouse models to analyses of the immune system has provided many new insights into the development and functions of immune cells. The challenge in writing this third edition has been to incorporate the important new information while maintaining our focus on the principles of immunology without greatly increasing the length of the book. To accomplish this, we have reduced much of the discussion of ideas that are largely of historical value. Where possible, we have tried to reorganize the information to highlight the conclusions of key experiments and the main concepts derived from such observations. Many of the illustrations are new or significantly revised, and color plates are used to illustrate the three-dimensional structures of important molecules. Our fundamental approach to immunology as an experiment-based discipline remains unchanged, and we have continued to emphasize the relevance of the science of immunology to human disease.

Many colleagues and students have helped our task by offering constructive criticisms and suggestions for improvements. At W.B. Saunders, three individuals have been instrumental in putting together this edition—William Schmitt, our editor; Risa Clow, our indispensable illustrator; and Janice Gaillard, development editor. To all of them we owe our sincere appreciation and gratitude.

ABUL K. ABBAS
ANDREW H. LICHTMAN
JORDAN S. POBER

PREFACE

Cellular and Molecular Immunology is intended as an introductory textbook primarily for students of medicine and related disciplines. This book evolved from a course we teach to first-year medical students at Harvard Medical School who are enrolled in the joint Harvard–MIT M.D. program in Health Sciences and Technology. The impetus for writing this book is the remarkable development of immunology as a science and the equally remarkable effect that this science has had upon clinical medicine. Over the past 20 years, the field of immunology has undergone radical changes. We can now identify the specific cells and molecules that are the essential components of the immune system, and we can predict the functions of these components based on a limited number of general principles. Students and practitioners of medicine need to be conversant with these advances in order to understand the immunologic diseases and to use the rapidly emerging methods of diagnosis and therapy that are based on immunologic approaches.

Accordingly, we believe that there is a need for a new textbook that should meet two major goals. The first and foremost is to convey an accurate and up-to-date understanding of the immune system. This book emphasizes the organizing principles of immunology and is not intended to be simply a compendium of facts. The principles of immunology are derived from the shared interpretations of key experiments. To enable the student to appreciate the basis of modern immunology, these key experiments and their interpretations are described, usually in summary or schematic form. The discussion of experimental studies also serves to illustrate the evolution of immunology as a science and, we hope, to convey some of the excitement that accompanies scientific discoveries. The most important methods used in experimental analyses are described in "boxes," which are separated from the main text. As an aid to students, each chapter concludes with a list of recent review articles that may serve as a bridge to the primary scientific reports. In addition, we have listed selected research papers that present experimental studies described in the text.

The second goal is to provide students of medicine with an appreciation of how immunologic principles are being applied to understand human diseases. The importance of the immune system in clinical medicine is greatest in two broad areas—

defense against infections, and diseases due to abnormal immune responses. These and other connections between immunology and medicine are highlighted throughout the book. More detailed descriptions of selected clinical disorders that illustrate important points have also been included in the boxes.

The book is organized into four sections, each focusing on different aspects of immunology. Section I presents an introduction to the cells and tissues of the immune system. Section II examines the molecular mechanisms used by the immune system to recognize antigens and the process of activation of the immune system that results from antigen recognition. Section III describes the means by which the stimulated immune system eliminates foreign molecules, cells, and organisms. Section IV is specifically devoted to clinical problems that are primarily immunologic or in which modern immunology has made a major contribution.

This book would not have been possible without the help and support of many individuals. Foremost among these were colleagues who provided invaluable constructive criticisms. Dr. Geoffrey Sunshine, of Tufts University School of Medicine, and Hal Burstein, an M.D.–Ph.D. candidate at Harvard Medical School, read the entire book and guided us through many problems of clarity and consistency. Individual chapters or topics were reviewed by the following immunologists, who are listed here in alphabetical order; Drs. Hugh Auchincloss, J. Latham Claflin, Robert Colvin, George Eisenbarth, Vic Engelhard, Frank Fitch, Steven Galli, Richard Hodes, Keith James, Stephanie James, Anne Marshak-Rothstein, Rick Mitchell, Harry Orr, David Parker, Jose Quintans, Ray Redline, Alan Sher, Richard Titus, and Janis Weis. We consider ourselves fortunate that we have been able to draw upon such a wealth of expertise. Valuable input also came from the first-year medical students on whom we first tried out the approach that is the cornerstone of this book.

We owe a great debt to many members of the staff of W.B. Saunders Company. In particular, Marty Wonsiewicz and Rosanne Hallowell, editors, and Risa Clow, illustrator, have shown extraordinary dedication and have been very much a part of the planning and writing of this book. Important contributions have also been made by Carol Robins, copy editor; Pat Morrison, Assistant Manager of Illustration and Design; Paul Fry, designer; and Pete Faber, production manager.

Many thanks are also due to Mary Jane Tawa, David Lence, and Jim Throp, who typed most of the manuscript, and to Pam Battaglino for hand-drawn illustrations.

Finally, we are grateful to the people who faithfully supported us even when we were not available for them—the members of our laboratories, who kept our research projects alive and well; Dr. Ramzi Cotran, our department chairman, whose indulgence was more than we could have asked for; and, above all, our families, who were tolerant of our many demands and awaited the completion of the book with an eagerness that matched our own.

ABUL K. ABBAS
ANDREW H. LICHTMAN
JORDAN S. POBER

CONTENTS

COLOR PLATES

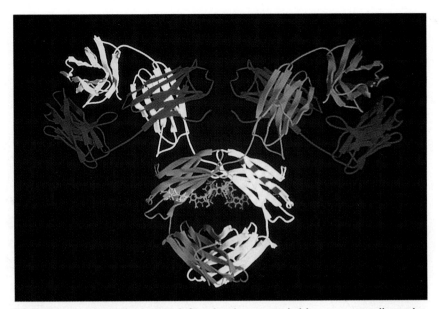

PLATE I. Structure of a human IgG molecule as revealed by x-ray crystallography. In this ribbon diagram of a secreted IgG molecule, the light chain polypeptides are depicted in red, the heavy chain polypeptides in magenta and yellow, and the carbohydrates in blue and green. Note that each light chain is folded into two Ig domains and each heavy chain into four Ig domains. The first Ig domain of each light chain (V_L) pairs with the first Ig domain of each heavy chain (V_H) to form an antigen-combining site. The carbohydrates are attached to the second constant region of each heavy chain ($C\gamma 2$) and occupy a space between these domains. The red squares on the yellow $C\gamma 2$ indicate positions involved in the activation of complement. (Courtesy of Dr. A. Edmundson, Oklahoma Medical Research Foundation, Oklahoma City, OK; reproduced with permission from the cover of Immunology Today, Vol. 16, February 1995. Copyright Elsevier Science, Ltd.)

COLOR PLATE II

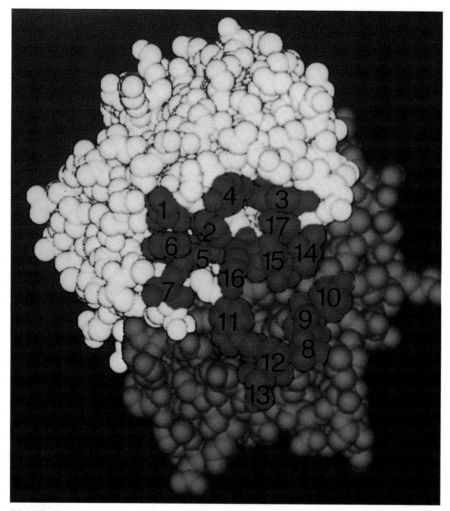

PLATE II. *En face* view of the antigen-binding site of an anti-lysozyme antibody, revealed by x-ray crystallography of an antigen-antibody complex. In this space-filling model, the variable regions of the heavy and light chains are shown in blue and yellow, respectively. The numbered side chains in red indicate amino acid residues in the complementarity-determining regions (CDRs) of the light chain (1–7) and heavy chain (8–17) that contact the protein antigen on this planar surface. (Courtesy of Dr. R. J. Poljak, Pasteur Insitute, Paris, France; reprinted with permission from Amit, A. G., Mariuzza, R. A., Phillips, S. E. V., and Poljak, R. J. Three-dimensional structure of an antigen-antibody complex at 2.8Å resolution. Science 233:747–753, 1986. Copyright 1986 American Association for the Advancement of Science.)

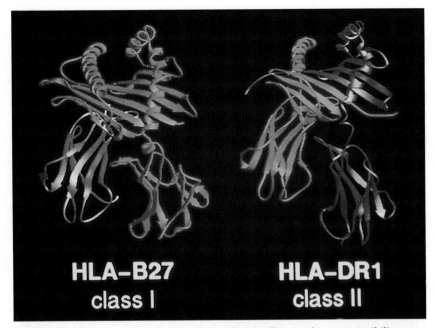

PLATE III. Structures of human class I and class II major histocompatibility complex molecules as revealed by x-ray crystallography. In these ribbon diagrams of the extracellular portions of the HLA-B27 (class I) and HLA-DR1 (class II) molecules, the α chains are depicted in blue and the β chains in red. Note that the peptide-binding domain of the class I molecule (left) is formed entirely from the α chain, whereas the homologous region in the class II molecule is formed from a complex of the α and β chains. Despite this difference, the overall structures of both molecules are strikingly similar. (Courtesy of Dr. L. Stern, Massachusetts Institute of Technology, Cambridge, MA.)

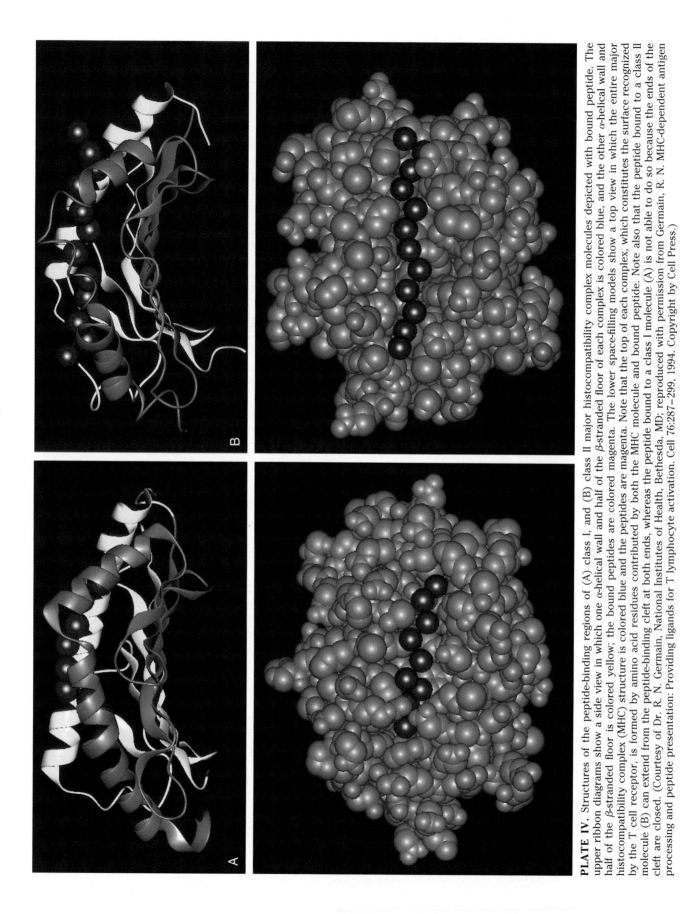

PLATE IV. Structures of the peptide-binding regions of (A) class I, and (B) class II major histocompatibility complex molecules depicted with bound peptide. The upper ribbon diagrams show a side view in which one α-helical wall and half of the β-stranded floor of each complex is colored blue, and the other α-helical wall and half of the β-stranded floor is colored yellow; the bound peptides are colored magenta. The lower space-filling models show a top view in which the entire major histocompatibility complex (MHC) structure is colored blue and the peptides are magenta. Note that the top of each complex, which constitutes the surface recognized by the T cell receptor, is formed by amino acid residues contributed by both the MHC molecule and bound peptide. Note also that the peptide bound to a class II molecule (B) can extend from the peptide-binding cleft at both ends, whereas the peptide bound to a class I molecule (A) is not able to do so because the ends of the cleft are closed. (Courtesy of Dr. R. N. Germain, National Institutes of Health, Bethesda, MD; reproduced with permission from Germain, R. N. MHC-dependent antigen processing and peptide presentation: Providing ligands for T lymphocyte activation. Cell 76:287–299, 1994. Copyright by Cell Press.)

COLOR PLATE V

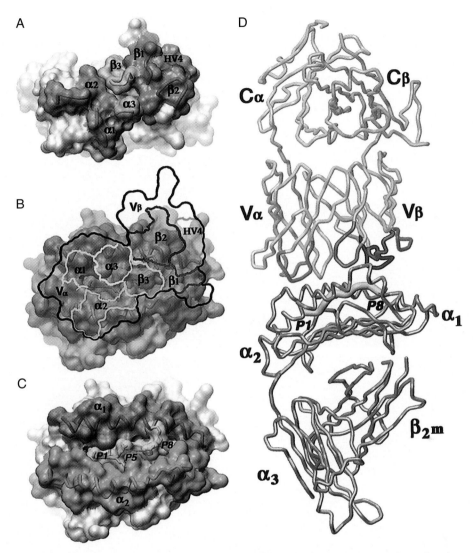

PLATE V. Structure of the complex formed between a T cell receptor, peptide, and a class I MHC molecule, as revealed by x-ray crystallography. In panels A, B and D, the TCR CDRs α1 and α2 are depicted in magenta, CDRs β1 and β2 in blue, CDRs α3 and β3 in green, and the hypervariable region found within the β chain constant region, sometimes called HV4 in orange. The binding site surface of a mouse TCR, specific for a complex of self-peptide bound to the H-2K^b class I MHC molecule, is shown in the first panel (A). A footprint of the TCR on the surface of an H-2K^b-peptide complex is shown in the next panel (B), in which the outline of the footprint is depicted by the black line, the bound peptide in solid green, and the outlines of the CDRs in the colors described above. The surface of the H-2K^b-peptide complex, which is depicted in (B), is shown more clearly in the next panel (C). A representation of the polypeptide backbone of the TCR-peptide-MHC complex is shown in the righthand panel (D), with the TCR on top, the peptide depicted as a thick green tube, and the MHC molecule on the bottom. Note that all CDRs 1–3 of the α and β chains of the TCR contact both peptide and MHC residues, with CDR3s straddling the middle of the peptide. (Courtesy of Dr. Ian A. Wilson and colleagues, Scripps Research Institute, La Jolla, CA; reprinted with permission from Garcia, K. C., Degano, M., Stanfield, R. L., Brunmark, A., Jackson M. R., Peterson, P. A., Teyton, L., and Wilson, I. A. Science 274:209–219, 1996. Copyright 1996 by the American Association for the Advancement of Science.)

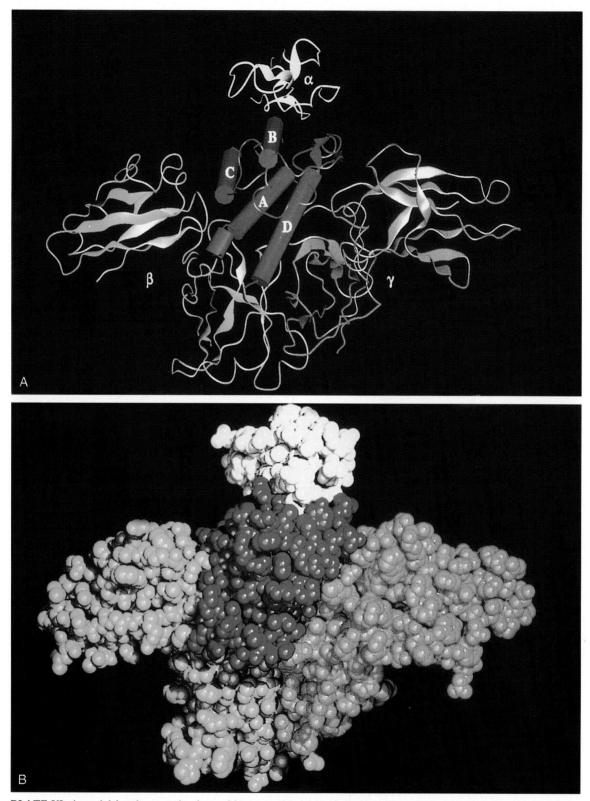

PLATE VI. A model for the complex formed between interleukin 2 (IL-2) and its trimeric receptor as deduced from the structure of growth hormone with its receptor. In both the ribbon model (A) and the space-filling model (B), IL-2 is colored red, and the α, β, and γ chains of the receptor are colored yellow, green, and blue, respectively. The four α-helices of the IL-2 molecule are labeled A, B, C, and D. In these top views, note that IL-2 interacts with all three chains of the heterotrimeric receptor, and the β and γ chains also interact with each other. Despite these contacts, isolated γ chain has no appreciable affinity for IL-2. (Courtesy of Dr. J. Thèze, Pasteur Institute, Paris, France; reproduced with permission from Thèze, J., Alzari, P. M., and Bertoglio, J. Interleukin 2 and its receptors: Recent advances and new immunological functions. Immunology Today 17:481–486, 1996.)

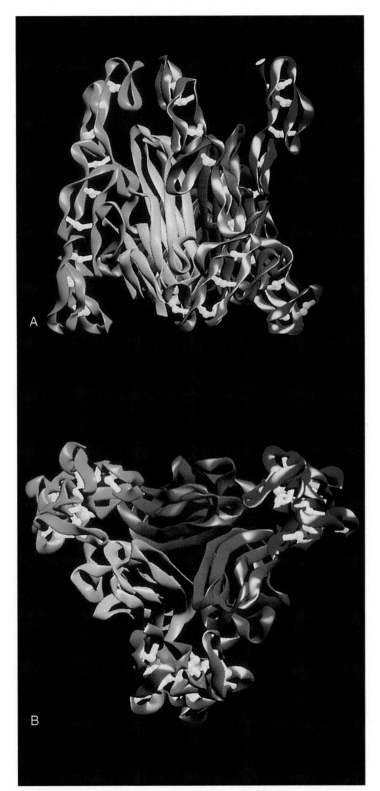

PLATE VII. Structure of the complex formed between lymphotoxin (LT) and the extracellular portions of the tumor necrosis factor receptor I (TNF-RI), as revealed by x-ray crystallography. In these ribbon structures depicting (A) a side view and (B) a top view, LT forms a homotrimer colored green, indigo, and dark blue. Three separate TNF-RIs, colored magenta, cyan, and red, associate with the trimer. Disulfide bonds in the receptor chains are colored yellow. The LT-homotrimer forms an inverted three-sided pyramid with its base at the top and apex at the bottom. The receptor, which is a type II membrane protein, has its carboxy terminus at the top. Note that each receptor interacts with two different LT monomers in the homotrimer complex. (Courtesy of Dr. W. Lesslauer, F. Hoffmann-LaRoche, Basel, Switzerland; reproduced with permission from Banner, D. W., D'Arcy, A., Jones, W., Gentz, R., Schoenfeld, H-J, Broger, C., Loetscher, C., and Lesslauer, W. Crystal structure of the soluble human 55kd TNF receptor–human TNFβ complex: Implications for TNF receptor activation. Cell 73:431–445, 1993. Copyright by Cell Press.)

COLOR PLATE VIII

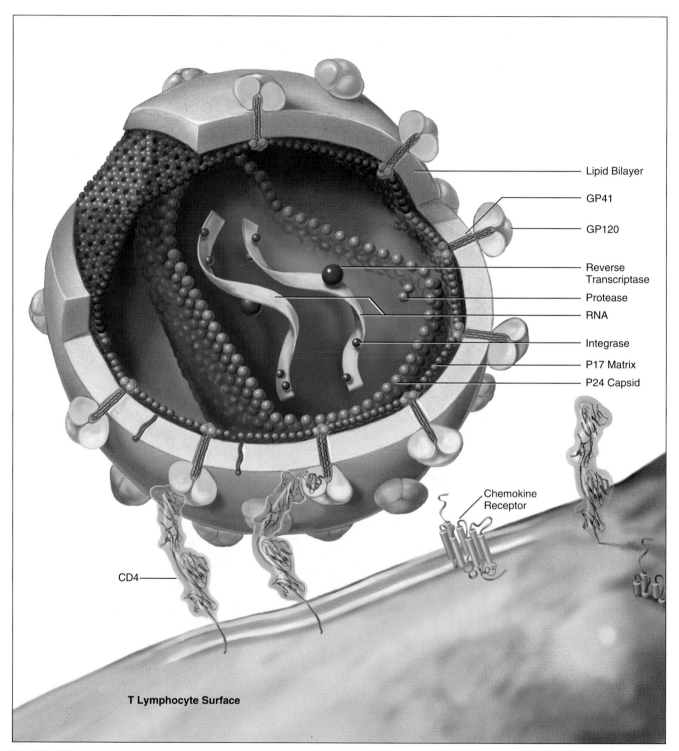

Lipid Bilayer

GP41

GP120

Reverse
Transcriptase

Protease

RNA

Integrase

P17 Matrix

P24 Capsid

Chemokine
Receptor

CD4

T Lymphocyte Surface

PLATE VIII. Structure of human immunodeficiency virus-1 (HIV-1). An HIV-1 virion is shown next to a T cell surface. HIV-1 consists of two identical strands of RNA (the viral genome) and associated enzymes, including reverse transcriptase, integrase, and protease, packaged in a cone-shaped core composed of p24 capsid protein with a surrounding p17 protein matrix, all surrounded by a phospholipid membrane envelope derived from the host cell. Virally encoded membrane proteins (gp41 and gp120) are bound to the envelope. CD4 and chemokine receptors on the T cell surface function as HIV-1 receptors. (From front cover, "The New Face of AIDS." Science 272:1841–2012, 1996. Copyright Terese Winslow.)

SECTION I

INTRODUCTION TO

IMMUNOLOGY

The first two chapters introduce the nomenclature of immunology and the components of the immune system. Chapter 1 describes different categories of immune responses and their general properties, and introduces the fundamental principles that govern all immune responses. Chapter 2 is devoted to a description of the cells and tissues of the immune system, with an emphasis on their anatomic organization and structure-function relationships. This will set the stage for more thorough discussions of the individual cells that participate in immune responses and how the immune system recognizes and responds to antigens.

CHAPTER ONE

GENERAL PROPERTIES OF

IMMUNE RESPONSES

The term immunity is derived from the Latin word *immunitas,* which referred to the exemption from various civic duties and legal prosecution offered to Roman senators during their tenures in office. Historically, immunity meant protection from disease, and, more specifically, infectious disease. The cells and molecules responsible for immunity constitute the **immune system,** and their collective and coordinated response to the introduction of foreign substances is the **immune response.** We now know that many of the mechanisms of resistance to infections are also involved in the individual's response to noninfectious foreign substances. Furthermore, mechanisms that normally protect individuals from infections and eliminate foreign substances are themselves capable of causing tissue injury and disease in some situations. Therefore, a more inclusive and modern definition of immunity is a reaction to foreign substances, including microbes, as well as macromolecules such as proteins and polysaccharides, without implying a physiologic or pathologic consequence of such a reaction. Immunology is the study of immunity in this broader sense and of the cellular and molecular events that occur after an organism encounters microbes and other foreign macromolecules.

Historians often credit Thucydides, who lived in Athens during the fifth century B.C., as having first mentioned immunity to an infection that he called "plague" (but that was probably not the bubonic plague we recognize today). The concept of immunity may have existed long before, as suggested by the ancient Chinese custom of making children inhale powders made from the crusts of skin lesions of patients recovering from smallpox. Immunology, in its modern form, is an experimental science, in which explanations of immunologic phenomena are based on experimental observations and the conclusions drawn from them. The evolution of immunology as an experimental discipline has depended on our ability to manipulate the function of the immune system under controlled conditions. Historically, the first clear example of this, and one that remains among the most dramatic ever recorded, was Edward Jenner's successful vaccination against smallpox. Jenner, an English physician, noticed that milkmaids who had recovered from cowpox never contracted the more serious smallpox. Based on this observation, he injected the material from a cowpox pustule into the arm of an 8-year-old boy. When this boy was later intentionally inoculated with smallpox, the disease did not develop. Jenner's landmark treatise on **vaccination** (Latin *vaccinus,* of or from cows) was published in 1798. It led to the widespread acceptance of this method for inducing immunity to infectious diseases. An eloquent testament to the importance and progress of immunology was the announcement by the World Health Organization in 1980 that smallpox was the first infectious disease that had been eradicated worldwide by a program of vaccination.

In the past 30 years, there has been a remarkable transformation in our understanding of the immune system and its functions. Advances in cell culture techniques, recombinant DNA methodology, and protein biochemistry have changed immunology from a largely descriptive science into one in which diverse immune phenomena can be tied together coherently and explained in quite precise structural and biochemical terms. This chapter outlines the general features of immune responses and introduces the concepts that form the cornerstones of modern immunology and that recur throughout the remainder of this book.

INNATE AND SPECIFIC (ADAPTIVE) IMMUNITY

Healthy individuals protect themselves against microbes by means of many different mechanisms (Table 1–1). Some of these protective mechanisms comprise **innate** (also called **natural,** or **native**) **immunity.** The characteristics of innate immunity are a limited capacity to distinguish one microbe from another and its fairly stereotypic nature, in that it functions in much the same way against most infectious agents. The principal components of innate immunity are (1) physical and chemical barriers, such as epithelia and antimicrobial substances produced at epithelial surfaces; (2) blood proteins, including members of the complement system and other mediators of inflammation; and (3) phagocytic cells (neutrophils, macrophages) and other leukocytes, such as natural killer cells. Innate immunity provides the early lines of defense against microbes. The pathogenicity of microbes is, in part, related to their ability to resist the mechanisms of innate immunity.

TABLE 1–1. Features of Innate and Specific (Adaptive) Immunity

Feature	Innate	Specific (Adaptive)
Characteristics		
Specificity for microbes	Relatively low	High
Diversity	Limited	Large
Specialization	Relatively stereotypic	Highly specialized
Memory	No	Yes
Components		
Physical and chemical barriers	Skin, mucosal epithelia; anti-microbial chemicals (e.g., defensins)	Cutaneous and mucosal immune systems; secreted antibodies
Blood proteins	Complement	Antibodies
Cells	Phagocytes (macrophages, neutrophils), natural killer cells	Lymphocytes

In contrast to innate immunity, there are more highly evolved defense mechanisms that are stimulated by exposure to infectious agents, and increase in magnitude and defensive capabilities with each successive exposure to a particular microbe. Since this form of immunity evolves as a response to infection, it is called **adaptive immunity.** The characteristics of adaptive immunity are: exquisite specificity for distinct molecules; specialization, which enables them to respond in particular ways to different types of microbes; and their ability to "remember" and respond more vigorously to repeated exposures to the same microbe. Because of its capacity to distinguish between different microbes, adaptive immunity is also called **specific immunity;** in this book these terms are used interchangeably. The elements of specific immunity are present prior to exposure to microbes, but these elements are highly responsive to, and rapidly stimulated by, microbes. The components of specific immunity are **lymphocytes** and their products, such as **antibodies.** Foreign substances that induce specific immune responses or are the targets of such responses are called **antigens.** By convention, immunology is the study of specific immunity, and "immune responses" refer to responses that are specific for different antigens, which may be microbial antigens or noninfectious substances.

The mechanisms of both innate and specific immune responses make up an integrated system of host defense in which numerous cells and molecules function cooperatively. *Innate immunity not only provides early defense against microbes, but also plays several important roles in the induction of specific immune responses.* For instance, the inflammation that is associated with many infections provides a "warning" signal that triggers specific immune responses. One mechanism by which this occurs is the response of macrophages to inflammatory stimuli, leading to the secretion of protein hormones, called cytokines, that promote the activation of lymphocytes specific for microbial antigens. Another mechanism is the activation of complement, also a component of innate immunity, which enhances the production of specific antibodies. In addition, the nature of the early innate response influences the type of specific immune response that develops subsequently. For example, macrophages containing ingested microbes often secrete a particular cytokine, which, in turn, stimulates the development of T lymphocytes that are particularly effective activators of macrophages. Thus, the interactions between innate and specific immunity are bidirectional.

The specific immune system has retained many of the effector mechanisms of innate immunity that function to eliminate foreign invaders, and has added to them three important additional properties:

First, *the specific immune response enhances the protective mechanisms of innate immunity, and thus makes them better able to eliminate foreign antigens.* Therefore, specific immune responses are able to combat microbes that have evolved to successfully resist innate immunity. For example, complement, a system of plasma proteins, is activated by some bacteria and participates in phagocytic clearance and lysis of these bacteria, a form of innate immunity. Binding of antibodies (a component of specific immunity) to the bacteria markedly enhances complement activation, and leads to bacterial elimination even if the bacteria do not activate complement by themselves. As another example, macrophages ingest and destroy many microbes, but some bacteria have learned to resist this type of innate immunity. T lymphocytes activated by the bacteria as part of the specific immune response enhance the microbicidal activities of macrophages and eliminate the bacteria. We will return to these examples of the cooperation between innate and specific immunity later in this chapter and throughout the book.

Second, *the specific immune system has superimposed upon the relatively stereotyped mechanisms of innate immunity a high degree of specialization.* Whereas innate immunity functions similarly against most microbes, the nature of the specific immune response varies according to the type of microbe, and is designed to most effectively eliminate that microbe. For example, the specific immune system responds to blood-borne microbes by producing antibodies, which are designed to eliminate circulating particles. In contrast, the specific response to phagocytosed microbes consists of T lymphocytes, which enhance the functions of phagocytes.

Third, *the specific immune system "remembers" each encounter with a microbe or foreign antigen, so that subsequent encounters stimulate increasingly effective defense mechanisms.* This is called **immunologic memory,** and is the basis of protective vaccination against infectious diseases.

The concept that specific immune responses enhance and "improve" upon innate immunity is also reflected in the phylogeny of defense mechanisms (Box 1–1). In invertebrates, host defense against foreign invaders was mediated largely by the mechanisms of innate immunity, including phagocytes and circulating molecules that resemble the proteins of the complement system. Specific immunity, consisting of lymphocytes and antibodies, appeared in vertebrates, and became increasingly specialized with evolution.

TYPES OF SPECIFIC IMMUNE RESPONSES

Specific immune responses are normally stimulated when an individual is exposed to a foreign antigen. The form of immunity that is induced by this process of immunization is called **active immunity** because the immunized individual plays an

BOX 1–1.   Evolution of the Immune System

Mechanisms for defending the host against foreign invaders and for healing injured self tissues are present in some form in all members of the enormously diverse and large numbers of phyla of invertebrates. These mechanisms constitute innate immunity. The more discriminating and specialized defense mechanisms that constitute specific or adaptive immunity are generally found in vertebrates only. Various cells in invertebrates respond to microbes by enclosing these infectious agents within aggregates and destroying them. These responding cells resemble phagocytes and have been called phagocytic amebocytes in acelomates, hemocytes in mollusks and arthropods, coelomocytes in annelids, and blood leukocytes in tunicates. Invertebrates do not contain antigen-specific lymphocytes and do not produce immunoglobulin molecules or complement proteins. However, they contain a number of soluble molecules that bind to and lyse microbes. These molecules include lectin-like proteins, which bind to carbohydrates on microbial cell walls and agglutinate the microbes, and numerous lytic and antimicrobial factors such as lysozyme, which is also produced by neutrophils in higher organisms. Phagocytes in some invertebrates may be capable of secreting cytokines that resemble macrophage-derived cytokines in the vertebrates. Thus, host defense in invertebrates is mediated by the cells and molecules that resemble the effector mechanisms of innate immunity in higher organisms.

Many studies have shown that invertebrates are capable of rejecting foreign tissue transplants, or allografts. If sponges (Porifera) from two different colonies are parabiosed by being mechanically held together, they become necrotic in 1 to 2 weeks, whereas sponges from the same colony become fused and continue to grow. Earthworms (annelids) and starfish (echinoderms) also reject tissue grafts from other species of the phyla. These rejection reactions are mediated mainly by phagocyte-like cells. They differ from graft rejection in vertebrates in that specific memory for the grafted tissue either is not generated or is difficult to demonstrate. Nevertheless, such results indicate that even invertebrates must express cell surface molecules that distinguish self from non-self, and such molecules may be the precursors of histocompatibility molecules in vertebrates.

The various components of the mammalian immune system appear to have arisen virtually together in phylogeny and have become increasingly specialized with evolution (see Table). Thus, of the cardinal features of specific immune responses, specificity, specialization, memory, self/non-self discrimination, and a capacity for self-limitation are present in the lowest vertebrates, and diversity of antigen recognition increases progressively in the higher species. All vertebrates contain antibody molecules. Fishes have only one type of antibody, called immunoglobulin M; this number increases to two types in anuran amphibians like *Xenopus* and to seven or eight in mammals. The diversity of antibodies is much lower in *Xenopus* than in mammals, even though the genes coding for antibodies are structurally similar. Lymphocytes that have some characteristics of both B and T cells are probably present in the earliest vertebrates, such as lampreys, and become specialized into functionally and phenotypically distinct subsets in amphibians and most clearly in birds and mammals. The major histocompatibility complex, which is a genetic locus that controls graft rejection and T lymphocyte antigen recognition, is present in some of the more advanced species of amphibians and fishes and in all birds and mammals. Its absence from some amphibians and fishes and all reptiles suggests that these histocompatibility genes have evolved independently on several occasions during vertebrate phylogeny. The earliest organized lymphoid tissues detected during evolution are the gut-associated lymphoid tissues; spleen, thymus, and lymph nodes (see Chapter 2) are found in higher vertebrates.

	Innate Immunity		Specific (Adaptive) Immunity		
	Phagocytosis	*NK Cells*	*Antibodies*	*T and B Lymphocytes*	*Lymph Nodes*
Invertebrates					
Protozoa	+	—	—	—	—
Sponges	+	—	—	—	—
Annelids	+	+	—	—	—
Arthropods	+	—	—	—	—
Vertebrates					
Elasmobranchs (sharks, skates, rays)	+	+	+ (IgM only)	+	—
Teleosts (common fish)	+	+	+ (IgM, ? other)	+	—
Amphibians	+	+	+ (2 or 3 classes)	+	—
Reptiles	+	+	+ (3 classes)	+	—
Birds	+	+	+ (3 classes)	+	+ (some species)
Mammals	+	+	+ (7 or 8 classes)	+	+

Abbreviations: NK, natural killer; IgM, immunoglobulin M.
Key: +, present; −, absent.

active role in responding to the antigen. Specific immunity can also be conferred upon an individual by transferring cells or serum from a specifically immunized individual. The recipient of such an **adoptive transfer** becomes immune to the particular antigen without ever having been exposed to or having responded to that antigen. Therefore, this form of immunity is called **passive immunity.** Passive immunization is a useful method for conferring resistance rapidly, without having to wait for an active immune response to develop. For instance, passive immunization against snake venoms by the administration of antibodies from immunized individuals is a life-saving treatment for potentially lethal snake bites. The technique of adoptive transfer has also made it possible to define the various cells and molecules that are responsible for mediating specific immunity.

Clinically, immunity cannot be measured by transferring cells or antibodies, or by testing an individual's resistance to an infection. Therefore, immunity is actually assayed by determining whether individuals who have been previously exposed to a foreign substance manifest a detectable reaction when re-exposed to, or challenged with, that substance. Such a reaction is an indication of "sensitivity" to challenge, and individuals who have been exposed to a foreign substance are said

to be "sensitized." Diseases caused by abnormal or excessive immune reactions are called **hypersensitivity diseases.**

Specific immune responses are classified into two types, based on the components of the immune system that mediate the response (Fig. 1–1):

1. **Humoral immunity** is mediated by molecules in the blood that are responsible for specific recognition and elimination of antigens; these are called **antibodies.** It can be transferred to unimmunized (also called "naive") individuals by cell-free portions of the blood (i.e., plasma or serum).

2. **Cell-mediated immunity,** also called **cellular immunity,** is mediated by cells called **T lymphocytes.** It can be transferred to naive individuals with T lymphocytes from an immunized individual but not with plasma or serum.

The first definitive experimental demonstration of humoral immunity was provided by Emil von Behring and Shibasaburo Kitasato in 1890. They showed that if serum from animals who had recovered from diphtheria infection was transferred to naive animals, the recipients became specifically resistant to diphtheria infection. The active components of the serum were called **antitoxins** because they neutralized the pathologic effects of the bacterial toxin. In the early 1900s, Karl Landsteiner

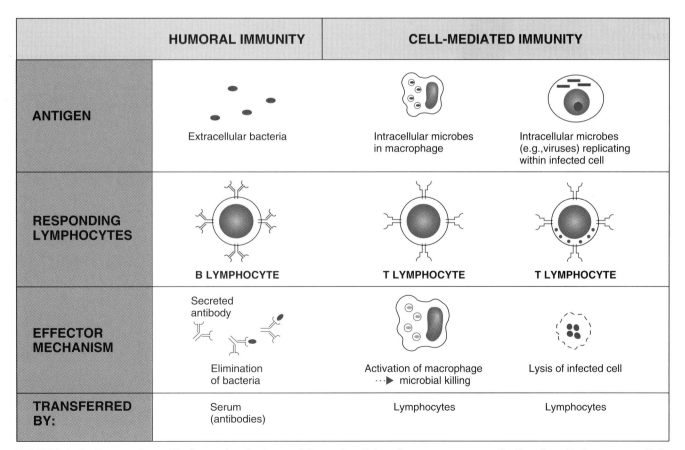

	HUMORAL IMMUNITY	CELL-MEDIATED IMMUNITY	
ANTIGEN	Extracellular bacteria	Intracellular microbes in macrophage	Intracellular microbes (e.g.,viruses) replicating within infected cell
RESPONDING LYMPHOCYTES	**B LYMPHOCYTE**	**T LYMPHOCYTE**	**T LYMPHOCYTE**
EFFECTOR MECHANISM	Secreted antibody · Elimination of bacteria	Activation of macrophage ···▶ microbial killing	Lysis of infected cell
TRANSFERRED BY:	Serum (antibodies)	Lymphocytes	Lymphocytes

FIGURE 1–1. Forms of specific immunity. In humoral immunity, B lymphocytes secrete antibodies that eliminate extracellular microbes. In cell-mediated immunity, T lymphocytes activate macrophages to kill intracellular microbes or destroy infected cells (e.g., virus-infected cells).

and other investigators showed that not only toxins but also other, non-microbial substances could induce humoral immunity. From such studies arose the more general term **antibodies** for the serum proteins that mediate humoral immunity. Substances that bound antibodies and generated the production of antibodies were then called **antigens.** (The properties of antibodies and antigens are described in Chapter 3.) In 1900, Paul Ehrlich provided a theoretical framework for the specificity of antigen-antibody reactions, the experimental proof for which came over the next 50 years from the work of Landsteiner and others using simple chemicals as antigens. Ehrlich's theories of the physicochemical complementarity of antigens and antibodies are remarkable for their prescience. This early emphasis on antibodies led to the general acceptance of the **humoral theory of immunity,** according to which immunity is mediated by substances present in body fluids (humors).

The **cellular theory of immunity,** which stated that host cells were the principal mediators of immunity, was championed initially by Elie Metchnikoff. His demonstration of phagocytes surrounding a thorn stuck into a translucent starfish larva, published in 1893, was perhaps the first experimental evidence that cells responded to foreign invaders. Sir Almroth Wright's observation in the early 1900s that factors in immune serum enhanced the phagocytosis of bacteria, a process known as **opsonization,** lent support to the belief that antibodies merely prepared microbes for ingestion by phagocytes. These early "cellularists" were unable to prove that specific protective immunity could be mediated by cells. In 1942, Landsteiner and Merrill Chase reported that skin reactions to various chemicals (a type of hypersensitivity) could be transferred to naive animals with cells but not with serum from specifically immunized animals. The cellular theory of immunity became firmly established in the 1950s, when George Mackaness showed that resistance to an intracellular bacterium, *Listeria monocytogenes,* could also be adoptively transferred with cells but not with serum. We now know that the specificity of cell-mediated immunity is due to lymphocytes, which often function in concert with other cells such as phagocytes to control or eliminate microbes.

Adoptive transfer of specific immunity is one of the principal techniques for analyzing immune responses. It is now complemented by *in vitro* experiments, in which the cells of the immune system can be stimulated by defined antigens and the development of specific immune responses can be examined. As we shall discuss in subsequent chapters, such studies have shown that *humoral immunity and cell-mediated immunity are mediated by responses of distinct types of lymphocytes.* The cells of humoral immunity are **B lymphocytes,** which respond to foreign antigens by developing into antibody-producing cells, whereas **T lymphocytes** are the mediators of cellular immunity. Humoral immunity is the principal defense mechanism against extracellular microbes and their secreted toxins because antibodies can bind to these and assist in their elimination. In contrast, intracellular microbes, such as viruses and some bacteria, survive and proliferate inside phagocytes and other host cells, where they are inaccessible to circulating antibodies. Defense against such infections is due to cell-mediated immunity, which functions by promoting the destruction of microbes residing in phagocytes, or the lysis of infected cells (see Fig. 1–1).

CARDINAL FEATURES OF IMMUNE RESPONSES

Humoral and cell-mediated immune responses to all antigens have a number of fundamental properties. The experimental analysis of the immune response is, in fact, an attempt to provide molecular and mechanistic explanations for these cardinal features of specific immunity.

1. *Specificity.* Immune responses are specific for distinct antigens and, in fact, for different structural components of a single complex protein, polysaccharide, or other antigen (Fig. 1–2). The portions of such antigens that are specifically recognized by individual lymphocytes are called **determinants,** or **epitopes.** This fine specificity exists because individual B and T lymphocytes that respond to foreign antigens express membrane receptors that distinguish subtle differences between distinct antigens. Antigen-specific lymphocytes develop without antigenic stimulation, so that clones of cells with different antigen receptors and specificities are available in unimmunized individuals to recognize and respond to many foreign antigens.

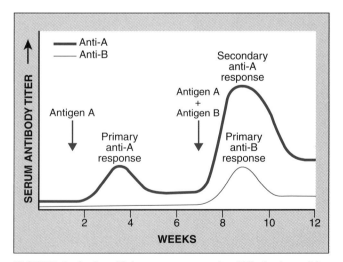

FIGURE 1–2. Specificity, memory, and self-limitation of immune responses. Antigens A and B induce the production of different antibodies (specificity). The secondary response to antigen A is more rapid and larger than the primary response (memory). Antibody levels, or titers, decline with time after each immunization (self-limitation).

This concept is the basic tenet of the **clonal selection hypothesis,** which is discussed in more detail later in this chapter.

2. *Diversity.* The total number of antigenic specificities of the lymphocytes in an individual, called the **lymphocyte repertoire,** is extremely large. It is estimated that the mammalian immune system can discriminate at least 10^9 distinct antigenic determinants. This extraordinary diversity of the repertoire is a result of variability in the structures of the antigen-binding sites of lymphocyte receptors for antigens. In other words, different clones of lymphocytes differ in the structures of their antigen receptors and, therefore, in their specificity for antigens, creating a total repertoire that is extremely diverse. One of the most important advances in immunology has been the elucidation of the molecular mechanisms that produce such structural diversity. These mechanisms are discussed in Chapters 4 and 8.

3. *Memory.* Exposure of the immune system to a foreign antigen enhances its ability to respond again to that antigen. Thus, responses to second and subsequent exposures to the same antigen, called **secondary immune responses,** are usually more rapid, larger, and often qualitatively different from the first, or primary, immune responses to that antigen (Fig. 1–2). This property of specific immunity is called **immunologic memory.** Several features of lymphocytes are responsible for memory:

 a. Lymphocytes proliferate when stimulated by antigens and the progeny of a particular antigen-responsive lymphocyte has the same antigen receptors and, hence, specificity as the original cell. Therefore, each exposure to antigen expands the clone(s) of lymphocytes specific for that antigen.

 b. Memory cells, which develop from the progeny of antigen-stimulated lymphocytes, survive for long periods and are prepared to respond rapidly to antigenic challenge. Moreover, the responses of memory cells are often more efficient at eliminating the antigen than are primary responses. For instance, memory B lymphocytes produce antibodies that bind antigen with higher affinities than do previously unstimulated B cells, and memory T cells home to sites of antigen entry more rapidly than do naive T cells. The mechanisms and physiologic importance of such differences will be discussed in later chapters.

4. *Specialization.* The immune system responds in distinct and special ways to different microbes. Such adaptations have developed to maximize the efficiency of anti-microbial defense mechanisms. Understanding the mechanisms of this specialization and learning how to exploit it for preventing and treating diseases are major themes of modern immunology and will be discussed throughout this book.

5. *Self-limitation.* All normal immune responses wane with time after antigen stimulation (Fig. 1–2). This is largely because immune responses function to eliminate antigens and thus eliminate the essential stimulus for lymphocyte activation. Antigen-stimulated lymphocytes also perform their functions for brief periods after stimulation and then die or differentiate into functionally quiescent memory cells. In addition, antigens and the immune responses to them stimulate numerous mechanisms whose main role is feedback regulation of the response itself. These regulatory mechanisms are discussed in Chapter 10.

6. *Discrimination of self from non-self.* One of the most remarkable properties of every normal individual's immune system is its ability to recognize, respond to, and eliminate many foreign (non-self) antigens while not reacting harmfully to that individual's own (self) antigenic substances. Immunologic unresponsiveness is also called **tolerance.** Self-tolerance is maintained partly by the elimination of lymphocytes that may express receptors specific for self antigens and partly by functional inactivation of self-reactive lymphocytes after their encounter with self antigens. A great deal is now known about the processes that are responsible for self-tolerance, and these will be discussed in Chapter 19. Abnormalities in the induction or maintenance of self-tolerance lead to immune responses against self (autologous) antigens, often resulting in **autoimmune diseases.** The development and pathologic consequences of autoimmunity are also described in Chapter 19.

These cardinal features of specific immunity are necessary if the immune system is to perform its normal function of host defense. Specificity and memory enable the immune system to mount heightened responses to persistent or recurring stimulation with the same antigen and thus to combat infections that are prolonged or occur repeatedly. Diversity is essential if the immune system is to defend individuals against the many potential pathogens in the environment. Specialization enables the host to "custom design" responses to best combat the many different types of microbes that infect humans. Self-limitation allows the system to return to a state of rest after it eliminates each foreign antigen, thus enabling it to respond optimally to other antigens that the individual encounters. Self-tolerance and the ability to distinguish between self and non-self are vital for preventing reactions against one's own cells and tissues while maintaining a diverse repertoire of lymphocytes specific for foreign antigens.

PHASES OF IMMUNE RESPONSES

All immune responses are initiated by the recognition of foreign antigens. This leads to activation of the lymphocytes that specifically recognize the antigen and culminates in the development of mechanisms that mediate the physiologic function of the response, namely elimination of the antigen. Thus, specific immune responses may be divided into (1) the **recognition phase,** (2) the **activation phase,** and (3) the **effector phase** (Fig. 1–3). Throughout this book, we will discuss the mechanisms of specific immunity in the context of these three phases.

Recognition Phase

The recognition phase of immune responses consists of the binding of foreign antigens to specific receptors on mature lymphocytes, which exist prior to antigen exposure. B lymphocytes, the cells of humoral immunity, express antibody molecules on their surfaces that can bind foreign proteins, polysaccharides, lipids, or other chemicals in extracellular or cell-associated forms. T lymphocytes, which are responsible for cell-mediated immunity, express receptors that recognize only short peptide sequences in protein antigens. Moreover, T lymphocytes have the unique property of recognizing and responding only to peptide antigens that are present on the surfaces of other cells. The structural basis of antigen recognition by T cells and its physiologic implications are discussed in Chapters 6 and 7.

Activation Phase

The activation phase of immune responses is the sequence of events induced in lymphocytes as a consequence of specific antigen recognition. All lymphocytes undergo two major changes in re-

sponse to antigens. First, they *proliferate,* leading to expansion of the clones of antigen-specific lymphocytes and amplification of the protective response. Second, the progeny of antigen-stimulated lymphocytes *differentiate* either into effector cells that function to eliminate the antigen or into memory cells that survive ready to respond to re-exposure to the antigen. (Some of the progeny of antigen-stimulated lymphocytes die and are lost from the responding pool.) We will return to a more detailed discussion of the fates of antigen-stimulated lymphocytes in Chapter 2. Different classes of lymphocytes differentiate into distinct effector cells. Antigen-stimulated B lymphocytes differentiate into antibody-secreting cells, and the secreted antibody binds the antigen and triggers the mechanisms that eliminate the antigen. Some T lymphocytes differentiate into cells that activate phagocytes to kill intracellular microbes, and other T lymphocytes directly lyse cells that are producing foreign antigens such as viral proteins. The ability of T cells to recognize cell-bound antigens focuses T cell responses in such a way that cell-mediated immunity is effective against intracellular microbes. A general feature of lymphocyte activation is that it usually requires two types of signals: the first is provided by the antigen, and the second by other cells, which may be **helper cells** or **accessory cells.** The nature of these stimuli and the sequence of T and B cell activation are discussed in Chapters 7 and 9.

Two aspects of lymphocyte activation are important in order to allow the small number of cells that respond to any one antigen to perform the many functions that lead to elimination of the anti-

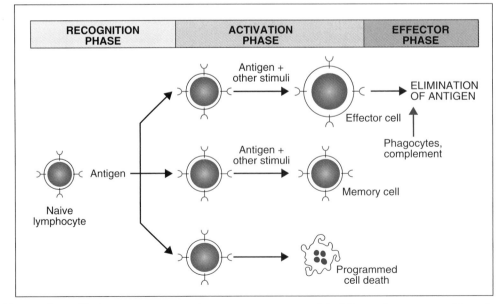

RECOGNITION PHASE	ACTIVATION PHASE	EFFECTOR PHASE

Antigen + other stimuli → Effector cell → ELIMINATION OF ANTIGEN

Phagocytes, complement

Naive lymphocyte → Antigen

Antigen + other stimuli → Memory cell

Programmed cell death

FIGURE 1–3. Phases of specific immune responses. Immune responses consist of three phases: recognition of antigen, activation (proliferation and differentiation of lymphocytes), and effector (elimination of antigen). Some antigen-stimulated lymphocytes die by a process called programmed cell death. Since this applies to both B and T lymphocytes, the lymphocytes shown can be of either class.

gen. First, immunization and antigen recognition trigger numerous amplification mechanisms that rapidly increase the number of cells that respond to the antigen. Second, lymphocytes efficiently home to sites of antigen entry and persistence. The cellular and biochemical mechanisms of amplification and lymphocyte migration are discussed in later chapters.

Effector Phase

The effector phase of immune responses is the stage at which lymphocytes that have been specifically activated by antigens perform the functions that lead to elimination of the antigen. Lymphocytes that function in the effector phase of immune responses are called **effector cells.** Many effector functions require the participation of other, non-lymphoid cells (which are also often referred to as "effector cells") and defense mechanisms that are also mediators of innate immunity. For instance, antibodies bind to foreign antigens and enhance their phagocytosis by blood neutrophils and mononuclear phagocytes. Antibodies also activate a system of plasma proteins termed **complement,** which

participates in the lysis and phagocytosis of microbes (see Chapter 15). Activated T lymphocytes secrete protein hormones, called **cytokines,** which enhance the functions of phagocytes and stimulate inflammatory responses (see Chapters 12 and 13). Phagocytes, complement, mast cells, cytokines, and the leukocytes that mediate inflammation are all components of innate immunity because they do not specifically recognize or distinguish between different foreign antigens, and they are all involved in defense against microbes even without specific immune responses. Thus, the effector phase of specific immunity illustrates a fundamental concept that was emphasized earlier in this chapter—that *specific immune responses serve to amplify and focus onto foreign antigens a variety of effector mechanisms that are also functional in the absence of lymphocyte activation* (Fig. 1–4).

THE CLONAL SELECTION HYPOTHESIS

From the initial demonstration that the immune system could respond specifically to a vast number of foreign antigens, the problem of explaining how such a diverse repertoire could be gener-

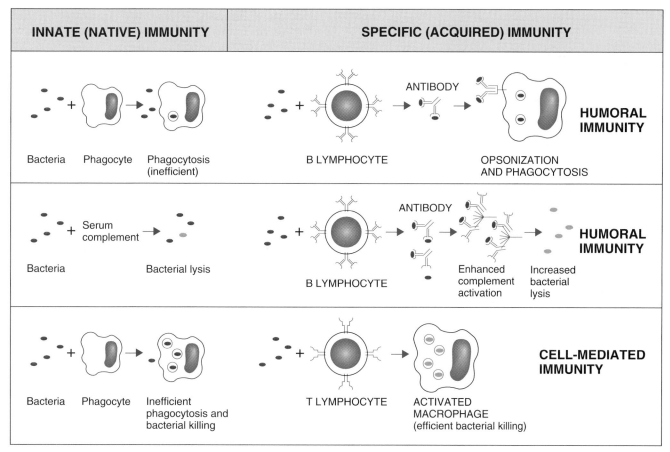

FIGURE 1–4. Cooperation of innate and specific immunity in host defense against infections. In the examples shown, antibodies promote phagocytosis or activate serum complement to kill microbes, and T lymphocytes enhance the phagocytic and microbicidal functions of macrophages.

ated and maintained was appreciated by immunologists. Two mutually exclusive hypotheses were proposed to explain the specificity and diversity of immune responses, even before there was a clear understanding of the importance of lymphocytes in antigen recognition. According to the instructional theory, immunocompetent cells and antibodies acquired their specificities after the introduction of antigen by changing the conformation of antigen-binding receptors, so that these became capable of recognizing the antigen. The alternative view, which we now know is correct, was first suggested by Niels Jerne in 1955, modified by David Talmadge and Macfarlane Burnet, and most clearly enunciated by Burnet in 1957. The key postulates of this theory, called the **clonal selection hypothesis,** have been convincingly proved by a variety of experiments and form the cornerstone of the current concepts of lymphocyte specificity and anti-

gen recognition. In essence, the clonal selection hypothesis states the following (Fig. 1–5):

1. *Every individual contains numerous clonally derived lymphocytes, each clone having arisen from a single precursor and being capable of recognizing and responding to a distinct antigenic determinant.* The development of antigen-specific clones of lymphocytes occurs prior to and independent of exposure to antigen. The cells constituting each clone have identical antigen receptors, which are different from the receptors on the cells of all other clones. Although it is difficult to place an upper limit on the number of antigenic determinants that can be recognized by the mammalian immune system, a frequently used estimate is in the order of 10^9 to 10^{11}. This is a reasonable approximation of the potential number of antigen receptor proteins that can be produced and may, therefore, reflect

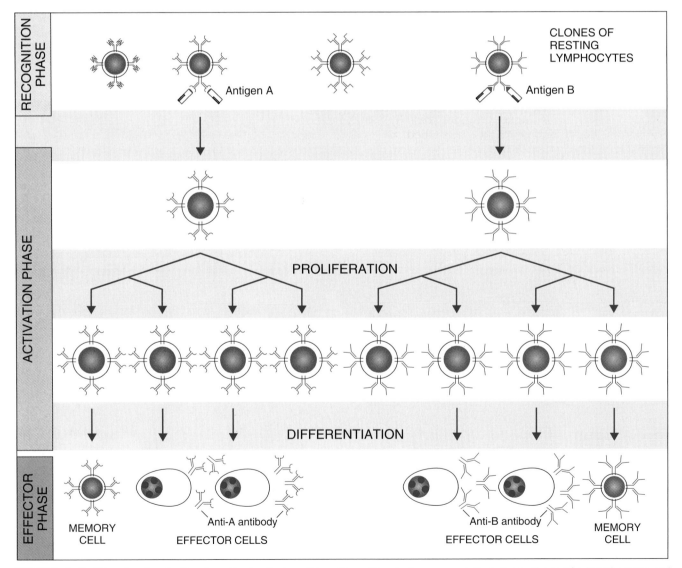

FIGURE 1–5. The clonal selection hypothesis. Each antigen (A or B) selects a pre-existing clone of specific lymphocytes and stimulates its proliferation and differentiation. The diagram shows only B lymphocytes giving rise to antibody-secreting cells and memory cells, but the same principle applies to T lymphocytes as well.

the number of distinct clones of lymphocytes present in each individual.

2. *Antigen selects a specific pre-existing clone and activates it,* leading to its proliferation and its differentiation into effector and memory cells. The observation that a secondary immune response is more rapid and larger than the primary response is explained by the clonal expansion of antigen-specific lymphocytes as a result of priming (first immunization) with antigen.

Many lines of evidence prove that both B and T lymphocytes with diverse receptors and specificities exist prior to the introduction of antigen, and clones with distinct specificities are selectively activated by different antigens.

1. If an individual is immunized with multiple antigens, each activated B lymphocyte isolated from that individual will produce antibody molecules specific for only one of the antigens. This has been shown by culturing lymphocytes from immunized individuals at "limiting dilution," so that each culture well contains the clonal progeny of one lymphocyte.

2. Different antigens bind to different lymphocytes, and no two structurally distinct foreign antigens bind to the same cell.

3. If an animal is injected with an antigen to which a highly radioactive tag is attached, that antigen binds to specific lymphocytes. The lymphocytes, being radiosensitive, are killed. Injection of such an antigen makes the animal incapable of responding to that antigen (until new lymphocytes develop), but the animal responds normally to all other structurally different antigens.

4. In lymphocytes with distinct specificities, the antigen receptors have structurally different combining sites. This has been established by nucleotide and amino acid sequencing of receptors isolated from lymphocytes stimulated with different antigens and from cloned cell lines derived from normal lymphocytes.

The specific immune system is remarkable for its complexity and diversity. Immune responses require the coordinated and precisely regulated interplay of many different cells and secreted molecules. Perhaps the greatest achievement of modern immunology is the application of a reductionist approach to analyzing the complexity of the immune system. As we shall see throughout this book, it is now possible to separate the interacting components of the immune system and to dissect their properties and functions individually. Such analyses have led to a clear understanding of the molecular basis of specific antigen recognition, so that the essential features of the recognition phase of humoral and cell-mediated immune responses are known. In particular, the structure of B and T lymphocyte receptors for antigens is now understood, and the molecular genetic basis of the expression of diverse antigen receptors is known. The antigenic epitopes of numerous model protein antigens

that are recognized by lymphocytes have been completely defined. The selection of the lymphocyte repertoire and the mechanisms responsible for discrimination between self and non-self are being elucidated, so that approaches for analyzing autoimmune diseases are becoming feasible. We have also learned a great deal about lymphocyte surface proteins and biochemical signals involved in the proliferative and differentiative responses of these cells to foreign antigens. Among the most impressive advances is the identification of the cytokines that mediate many of the effector functions of lymphocytes and are responsible for the communications among cells of the immune system that serve to amplify and regulate immune responses. The potential use of cytokines as biological response modifiers for treating human diseases is one of the exciting applications of this basic research. Thus, immunology has progressed from a science of phenomena to one of defined genes, molecules, and cells. The challenge in the years to come is to apply our knowledge of these molecules and cells to understand how physiologic immune responses are initiated and regulated in normal individuals and how responses become deficient or aberrant, leading to pathologic tissue injury and clinical diseases.

SUMMARY

The specific immune response is initiated by the recognition of foreign antigens by specific lymphocytes, which respond by proliferating and differentiating into effector cells whose function is to eliminate the antigen. The effector phase of specific immunity requires the participation of various defense mechanisms, including the complement system, phagocytes, inflammatory cells, and cytokines, that are also operative in innate immunity. The specific immune response amplifies the mechanisms of innate immunity and enhances their function, particularly upon repeated exposures to the same foreign antigen. The immune system possesses several properties that are of fundamental importance for its normal functions. These include specificity for distinct antigens, diversity of antigen recognition, memory for antigen exposure, specialized responses to different microbes, self-limitation, and the ability to discriminate between self and foreign antigens.

In the remainder of this book we describe the biology of lymphocytes, in particular, the structural basis of antigen recognition by lymphocytes, their stimulation leading to the development of effector cells, their regulation, the nature of effector mechanisms, and the abnormalities that lead to diseases of deficient or excessive immunity.

Selected Readings

Burnet, F. M. A modification of Jerne's theory of antibody production using the concept of clonal selection. Australian Journal of Science 20:67–69, 1957.

du Pasquier, L. Origin and evolution of the vertebrate immune system. Acta Pathologica Microbiologica et Immunologica Scandinavica 100:383–392, 1992.

Fearon, D. T., and R. M. Locksley. The instructive role of innate immunity in the acquired immune response. Science 272:50–54, 1996.

Jerne, N. K. The natural-selection theory of antibody formation. Proceedings of the National Academy of Sciences USA 41: 849–857, 1955.

Silverstein, A. M. A History of Immunology. Academic Press, San Diego, 1989.

CHAPTER TWO

CELLS AND TISSUES OF THE

IMMUNE SYSTEM

The cells of the immune system are normally present as circulating cells in the blood and lymph, as anatomically defined collections in lymphoid organs, and as scattered cells in virtually all tissues except the central nervous system. The anatomic organization of these cells and their ability to circulate and exchange among the blood, lymph, and tissues are of critical importance for the generation of immune responses. The immune system has to be able to respond to a very large number of foreign antigens introduced at any site in the body, and only a small number of lymphocytes specifically recognize and respond to any one antigen. These lymphocytes not only must locate foreign antigens but also have to activate the many effector mechanisms that are required to eliminate these antigens. The ability of the immune system to optimally perform its protective functions is dependent on several properties of its constituent cells and tissues. These include the following:

1. The immune system has developed specialized tissues, called peripheral lymphoid organs, which efficiently concentrate antigens that are introduced through the common portals of entry (skin and gastrointestinal and respiratory tracts). The same tissues attract lymphocytes and other cells that are needed to initiate specific immune responses.

2. To ensure that immunity is systemic (i.e., protective mechanisms can act anywhere in the body), effector and memory lymphocytes are able to circulate and home to sites of antigen entry and are efficiently retained at these sites.

3. Bidirectional interactions between antigen-specific lymphocytes and other cells that are involved in the activation and effector phases of immune responses serve to optimize these responses.

4. Multiple amplification loops magnify the effects of stimulating the few lymphocytes that are specific for any one antigen.

This chapter describes the morphology of the cells and tissues of the immune system, with an emphasis on the ways in which their structural characteristics reflect or contribute to their functions. The circulation of lymphocytes and the functional anatomy of immune responses are discussed in Chapter 11, after the description of the cellular and biochemical bases of antigen recognition and lymphocyte activation.

Lymphocytes are the cells that specifically recognize and respond to foreign antigens. However, both the recognition and activation phases of specific immune responses depend on non-lymphoid cells, called **accessory cells,** which are not specific for different antigens and whose functions will be described in greater detail in later chapters. Mononuclear phagocytes, dendritic cells, and several other cell populations function as accessory cells in the induction of immune responses. Lympho-

TABLE 2–1. Normal Blood Cell Counts

Cells	Number per mm³ (Mean ± S.D.)	Per Cent of Leukocytes	
		Mean	*95 Per Cent Range*
White blood cells (×10³) (leukocytes)	7.25 ± 1.7		
Neutrophils		55	34.6–71.4
Eosinophils		3	0–7.8
Basophils		0.5	0–1.8
Lymphocytes		35	19.6–52.7
Monocytes		6.5	2.4–11.8
Red blood cells (×10⁶) (erythrocytes)	5.0 ± 0.35		
Platelets (×10³)	248 ± 50		

Abbreviation: S.D., standard deviation.

cytes and accessory cells are organized in anatomically discrete lymphoid organs, where they interact with one another to initiate and amplify immune responses. The activation of lymphocytes leads to the generation of numerous effector mechanisms. Many of these effector mechanisms require the participation of effector cells, such as activated T lymphocytes, mononuclear phagocytes, and other leukocytes (white blood cells). Lymphocytes and effector cells are also present in the blood, from where they can migrate to peripheral sites of antigen exposure and function to eliminate the antigen. The cellular constituents of the blood are listed in Table 2–1. We will describe first the properties of the individual cell types and then the functional anatomy of lymphoid organs.

LYMPHOCYTES

Specific immune responses are mediated by lymphocytes, which are the only cells in the body capable of specifically recognizing and distinguishing different antigenic determinants. This has been established by several lines of evidence:

1. Protective immunity to microbes can be adoptively transferred from immunized to naive individuals only by lymphocytes or their secreted products.

2. Some congenital and acquired immunodeficiencies are associated with reduction of lymphocytes in the peripheral circulation and in lymphoid tissues. Furthermore, selective depletion of lymphocytes with drugs, irradiation, or cell type–specific antibodies leads to impaired immune responses.

3. Lymphocytes are often found in increased numbers at sites of immunization and/or in lymphoid tissues that drain these sites.

4. Stimulation of lymphocytes by antigens in culture leads to responses *in vitro* that show many

of the characteristics of immune responses induced under more physiologic conditions *in vivo.*

5. Most importantly, specific high-affinity receptors for antigens are produced by lymphocytes and no other cells.

Lymphocyte Development and Heterogeneity

The **small lymphocyte** is 8 to 10 micrometers (μm) in diameter and has a large nucleus with dense heterochromatin. There is a thin rim of cytoplasm that contains a few mitochondria, ribosomes, and lysosomes but no specialized organelles (Fig. 2–1). This bland morphologic pattern provides no clues to the remarkable functional capabilities of lymphocytes. Like all blood cells, lymphocytes originate in the bone marrow. This was first demonstrated by experiments with radiation-induced bone marrow chimeras. Lymphocytes and bone marrow stem cells (described later) are radiosensitive and are killed by high doses of γ-irradiation. If an irradiated mouse of one inbred strain is injected with bone marrow cells of another strain that can be distinguished from the host, all the lymphocytes that develop subsequently are derived from the bone marrow cells of the donor (Fig. 2–2). Such approaches have proved useful for defining the maturation of lymphocytes and other blood cells. In the initial stages of their development, lymphocytes do not produce surface receptors for antigens and are, therefore, unresponsive to antigens. As they mature, they begin to express antigen receptors, become responsive to antigenic stimulation, and develop into different functional classes.

Lymphocytes consist of distinct subsets that are quite different in their functions and protein products, even though they all appear morphologically similar (Table 2–2 and Fig. 2–3). One class of lymphocytes consists of **B lymphocytes,** so called because in birds they were first shown to mature in an organ called the bursa of Fabricius. In mammals, there is no anatomic equivalent of the bursa, and the early stages of B cell maturation occur in the bone marrow. Thus, "B" lymphocyte refers to bursa- or bone marrow–derived. *B lymphocytes are the only cells capable of producing antibodies.* The antigen receptors of B lymphocytes are membrane-bound forms of antibodies. Interaction of antigens with these membrane antibody molecules initiates the sequence of B cell activation, which culminates in the development of effector cells that actively secrete antibody molecules (see Chapter 9).

The second major class of lymphocytes consists of **T lymphocytes,** whose precursors arise in the bone marrow and then migrate to and mature in the thymus (the name "T" lymphocyte referring to thymus-derived). T lymphocytes are further subdivided into functionally distinct populations, the best defined of which are **helper T cells** and **cytolytic (or cytotoxic) T cells** (Fig. 2–3). *The principal functions of T lymphocytes are to regulate all immune responses to protein antigens and to serve as effector cells for the elimination of intracellular microbes.* T cells do not produce antibody molecules. Their antigen receptors are membrane molecules distinct from but structurally related to antibodies (see Chapter 7). Helper and cytolytic T lymphocytes have an unusual specificity for antigens: they recognize only peptide antigens attached to proteins that are encoded by genes in the major histo-

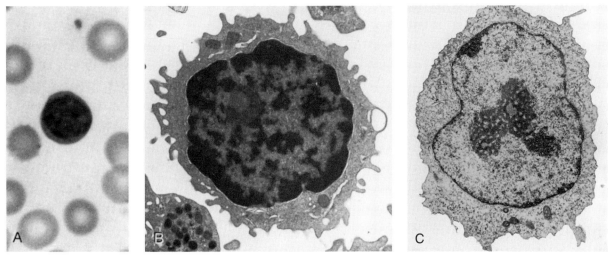

FIGURE 2–1. Morphology of lymphocytes.
A. Light micrograph of a small lymphocyte in a peripheral blood smear.
B. Electron micrograph of a small lymphocyte. (Courtesy of Dr. Noel Weidner, Department of Pathology, Brigham and Women's Hospital, Boston.)
C. Electron micrograph of a large lymphocyte (lymphoblast). (Reproduced with permission from W. Bloom and D. W. Fawcett. Textbook of Histology, 12th ed. W. B. Saunders, Philadelphia, 1994.)

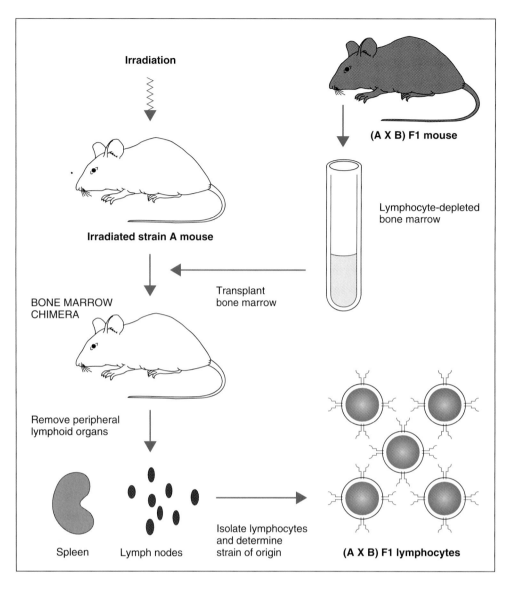

Irradiation

Irradiated strain A mouse

(A X B) F1 mouse

Lymphocyte-depleted
bone marrow

BONE MARROW
CHIMERA

Transplant
bone marrow

Remove peripheral
lymphoid organs

Spleen   Lymph nodes   Isolate lymphocytes
and determine
strain of origin

(A X B) F1 lymphocytes

FIGURE 2–2. Origin of lymphocytes defined in bone marrow chimeras. In an irradiated strain A mouse reconstituted with bone marrow cells from an (A × B)F1 hybrid (i.e., the offspring of a mating between a strain A and a strain B mouse), the mature lymphocytes that are recovered from peripheral lymphoid tissues are derived from the donor strain.

compatibility complex and expressed on the surfaces of other cells. As a result, these T cells recognize and respond to cell surface–associated but not soluble antigens (see Chapter 6). In response to antigenic stimulation, helper T cells secrete protein hormones called **cytokines,** whose function is to promote the proliferation and differentiation of the T cells as well as other cells, including B cells and macrophages. Cytokines also recruit and activate inflammatory leukocytes, including macrophages and granulocytes, providing important links between specific T cell immunity and the effector mechanisms of innate immunity. Cytolytic T lymphocytes (CTLs) lyse cells that produce foreign antigens, such as cells infected by viruses and other intracellular microbes. The mechanisms of action and physiologic functions of these T cell populations are described in detail in later chapters. In addition to providing helper and cytolytic functions, T cells may also inhibit im-

mune responses. There is, at present, considerable controversy about the nature and physiologic roles of these so-called "suppressor T cells." In fact, it is not clear whether suppression of immune responses is mediated by a distinct T cell subset or by cells that, under various situations, can also function as helper or cytolytic cells (see Chapter 10).

An important advance in the identification and analysis of these T cell subsets was the discovery that *functionally distinct populations express different membrane proteins.* These proteins serve as phenotypic markers of different lymphocyte populations. For instance, most helper T cells express a surface protein called CD4, and most CTLs express a different marker called CD8. Antibodies against such markers can, therefore, be used to identify and isolate various lymphocyte populations. Many of the surface proteins that were initially recognized as phenotypic markers for various lymphocyte subpopulations turned out, upon further anal-

TABLE 2–2. Lymphocyte Classes

Class	Functions	Antigen Receptor	Selected Phenotypic Markers	Per Cent of Total Lymphocytes		
				Blood	*Lymph Node*	*Spleen*
B lymphocytes	Antibody production (humoral immunity)	Surface antibody (immunoglobulin)	Fc receptors; class II MHC; CD19; CD21	10–15	20–25	40–45
T lymphocytes Helper T lympho- cytes	Stimuli for B cell growth and differen- tiation (humoral im- munity) Macrophage activation by secreted cyto- kines (cell-mediated immunity)	$\alpha\beta$ heterodimers	CD3$^+$, CD4$^+$, CD8$^-$	50–60*	50–60	50–60
Cytolytic T lympho- cytes	Lysis of virus-infected cells, tumor cells, allografts (cell-medi- ated immunity) Macrophage activation by secreted cyto- kines (cell-mediated immunity)	$\alpha\beta$ heterodimers	CD3$^+$, CD4$^-$, CD8$^+$	20–25	15–20	10–15
Natural killer cells	Lysis of virus-infected cells, tumor cells; antibody-dependent cellular cytotoxicity	Ig superfamily member	Fc receptor for IgG (CD16)	~10	Rare	~10

* In most tissues, the ratio of CD4$^+$CD8$^-$ to CD8$^+$CD4$^-$ cells is about 2:1.
Some T lymphocytes function to inhibit immune responses; these are called "suppressor cells," but it is unclear if they are a distinct subpopulation of T lymphocytes.
Abbreviation: MHC, major histocompatibility complex; Ig, immunoglobulin.

ysis, to play important roles in the activation and functions of these cells. In this book, we use the unified CD nomenclature for lymphocyte markers. CD stands for "cluster of differentiation" and refers to a molecule recognized by a "cluster" of mono- clonal antibodies that can be used to identify the lineage or stage of differentiation of lymphocytes and thus to distinguish one class of lymphocyte from another (Box 2–1). CD molecules recognized by specific antibodies now serve as useful markers of lymphocytes and other leukocytes that partici- pate in immune and inflammatory responses. Ex- amples of some CD proteins are mentioned in Ta- ble 2–2, and the biochemistry and functions of the most important ones are described in later chap- ters. A current list of known CD markers for leuko- cytes is provided in the Appendix (p. 462).

The third major class of lymphocytes does not express markers of either T or B cells and was, therefore, initially called the null cell population. It is now apparent that most null cells are large lym- phocytes with numerous cytoplasmic granules that are capable of lysing a variety of virus-infected and tumor cells without overt antigenic stimulation. As a result, these lymphocytes are called **large granu- lar lymphocytes** or **natural killer (NK) cells.** The properties and functions of NK cells are discussed in more detail in Chapter 13.

Lymphocyte Responses to Antigens: Morphologic Changes and Fates of Activated Lymphocytes

Lymphocytes undergo a well-defined pattern of changes upon activation. In the past, it was techni- cally difficult to study the responses of lympho- cytes to antigens because only a very small frac- tion of the total population is specific for any one antigen. Immunologists overcame this problem by using **polyclonal activators** (e.g., antibodies against antigen receptors) that stimulate many B or T lymphocytes irrespective of antigenic specific- ity. More recently, such analyses have been done with lymphocytes from transgenic mice expressing single T or B cell antigen receptors of defined specificities. (Transgenic mice are described in Chapter 4, Box 4–1.) In these mice, many lympho- cytes express the same receptor and respond to the same known antigen. From such analyses, lym- phocytes are known to respond in the following ways to antigenic stimulation (see Fig. 1–3, Chap- ter 1).

BLAST TRANSFORMATION AND PROLIFERATION

Prior to antigenic or polyclonal stimulation, small lymphocytes are in a state of rest, or in the G0 stage of the cell cycle. It is believed that if

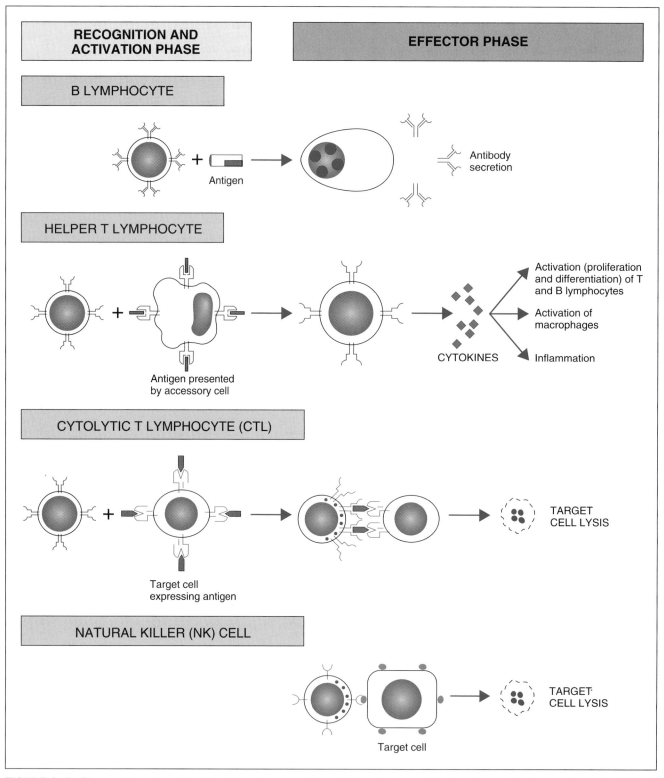

FIGURE 2–3. Classes of lymphocytes. B lymphocytes recognize soluble antigens and develop into antibody-secreting cells. Helper T lymphocytes recognize antigens on the surfaces of accessory cells and secrete cytokines, which stimulate different components of immunity and inflammation. Cytolytic T lymphocytes recognize antigens on target cells and lyse these targets. Natural killer cells use receptors that are not fully identified to recognize and lyse targets.

BOX 2–1.   Lymphocyte Markers: The CD Nomenclature

From the time that functionally and developmentally distinct classes of lymphocytes were recognized, immunologists have attempted to develop methods for distinguishing them. The basic approach was to produce antibodies that would selectively recognize different subpopulations. This was initially done by raising "alloantibodies" (i.e., antibodies that might recognize allelic forms of cell surface proteins) by immunizing inbred strains of mice with lymphocytes from other strains. Such techniques were remarkably successful and led to the development of antibodies that reacted with murine T cells (anti-Thy-1 antibodies) and even against functionally different subsets of T lymphocytes (anti-Lyt-1 and -Lyt-2 antibodies). The limitations of this approach, however, are obvious, since it is only useful for cell surface proteins that exist in allelic forms. Other approaches that met with some success but also have major limitations included searching for lymphocyte-specific autoantibodies in patients with autoimmune diseases. The advent of hybridoma technology gave such analyses a tremendous boost, and the most dramatic development was the production of monoclonal antibodies that reacted specifically and selectively with defined populations of lymphocytes, first human and subsequently in many other species. (Alloantibodies and monoclonal antibodies are described in Chapter 3.)

The cell surface molecules recognized by monoclonal antibodies are called "antigens," since antibodies can be raised against them, or "markers," since they identify and discriminate between ("mark") different cell populations. These markers can be grouped into several categories: some are specific for cells of a particular lineage or maturational pathway, and the expression of others varies according to the state of activation or differentiation of the same cells. Biochemical analyses of cell surface proteins recognized by different monoclonal antibodies in the same species or even in different species demonstrated that in many instances these antibodies were specific for the same evolutionarily conserved cellular proteins. Considerable confusion was created because these surface markers were initially named according to the antibodies that reacted with them. In order to resolve this, a uniform nomenclature system was adopted, initially for human leukocytes. According to this system, a surface marker that identifies a particular lineage or differentiation stage, that has a defined structure, and that is recognized by a group ("cluster") of monoclonal antibodies is called a member of a cluster of differentiation. Thus, all leukocyte surface antigens whose structures are defined are given a "CD" designation (e.g., CD1, CD2). Although this nomenclature was originally used for human leukocyte antigens, it is now common practice to refer to homologous markers in other species and on cells other than leukocytes by the same CD designation. Newly developed monoclonal antibodies are periodically exchanged among laboratories, and the antigens recognized are assigned to existing CD structures or introduced as new "workshop" candidates ("CDw").

The value of CD antigens in classifying lymphocytes is enormous. For instance, most helper T lymphocytes are $CD3^+CD4^+CD8^-$, and most CTLs are $CD3^+CD4^-CD8^+$. This has allowed immunologists to identify the cells participating in various immune responses, isolate them, and individually analyze their specificities, response patterns, and effector functions. Such antibodies have also been used to define specific alterations in particular subsets of lymphocytes that might be occurring in various diseases. Further investigations of the effects of monoclonal antibodies on lymphocyte function have shown that these surface proteins are not merely phenotypic markers but are themselves involved in a variety of lymphocyte responses. The two most frequent functions attributed to various CD antigens are (1) to promote cell–cell interactions and adhesion and (2) to transduce signals that lead to lymphocyte activation. Examples of both types of functions are described in Chapter 7.

---

naive lymphocytes do not encounter antigen, they eventually die, and the population is maintained at a steady-state level by the development of new cells from precursors in the bone marrow. The life span of naive (unstimulated) lymphocytes is not definitively established and may be months or even years in humans. In response to antigenic (or polyclonal) stimulation, resting small lymphocytes enter the G1 stage of the cell cycle. They become larger and are called large lymphocytes, or **lymphoblasts** (Fig. 2–1). These cells are 10 to 12 μm in diameter and have a wider rim of cytoplasm, more organelles, and increased amounts of cytoplasmic ribonucleic acid (RNA) compared with unstimulated small lymphocytes. Progression to the S phase of the cell cycle continues, and the activated large lymphocytes divide. This sequence of events is called **blast transformation.** Mitotic division is responsible for proliferation of the antigen-responsive clones of lymphocytes.

DIFFERENTIATION INTO EFFECTOR CELLS

Some of the progeny of the antigen-stimulated lymphocytes differentiate from cognitive cells, which recognize antigen, to effector cells, which function to eliminate the antigen. Differentiated helper T cells have essentially the same morphologic appearance as small or large lymphocytes. Differentiated CTLs may contain increased numbers of cytoplasmic granules whose contents include proteins that lyse target cells. Antibody-producing B cells often differentiate into specialized forms called **plasma cells.** Plasma cells are found only in lymphoid organs and at sites of immune responses and normally do not circulate in the blood or lymph. They have a characteristic morphology with eccentric nuclei, abundant cytoplasm, and distinct perinuclear haloes (Fig. 2–4). Under the electron microscope, the cytoplasm can be seen to contain dense, rough endoplasmic reticulum, which is the site where antibodies (and other secreted and membrane proteins) are synthesized. There is also a large Golgi complex, which stains poorly with routinely used histologic stains and is responsible for the perinuclear halo; antibody molecules are converted to their final forms and packaged for secretion in this organelle. Plasma cells are believed to be terminally differentiated cells with little or no capacity for mitotic

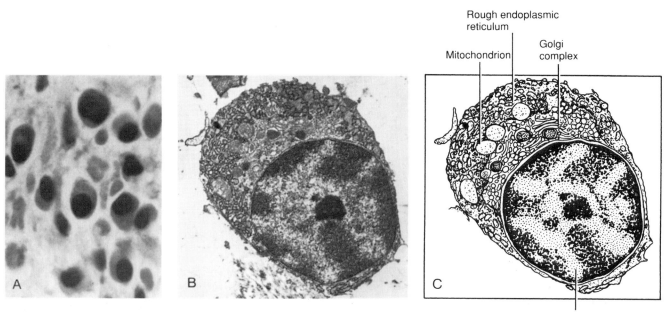

FIGURE 2–4. Morphology of plasma cells.
A. Light micrograph of plasma cells in tissues.
B. Electron micrograph of a plasma cell. (Courtesy of Dr. Noel Weidner, Department of Pathology, Brigham and Women's Hospital, Boston.)
C. Schematic diagram of the plasma cell depicted in *B*, illustrating the eccentric nucleus with a "cartwheel" pattern of chromatin, the abundant cisternae of the rough endoplasmic reticulum, and the prominent Golgi complex.

division and are, in essence, factories for the synthesis and secretion of antibody molecules. It is estimated that half or more of the messenger RNA in plasma cells codes for antibody proteins.

DIFFERENTIATION INTO MEMORY CELLS

Other progeny of antigen-stimulated B and T lymphocytes do not differentiate into effector cells. Instead, they become **memory lymphocytes** (Table 2–3), which are capable of surviving for long periods, perhaps 20 years or more, apparently in the absence of antigenic stimulation. It is still not established if memory cells truly survive without any stimulation, or if they are maintained by chronic low-level stimulation by persisting antigen or by environmental antigens that cross-react with the antigen that initiated the response. Memory cells are functionally quiescent (i.e., they do not produce effector molecules unless they are stimulated by antigens). The factors that determine whether a progeny of an activated T or B lymphocyte will become an effector cell or a memory cell are not known. The development of memory cells is crucial to the success of vaccination as a method of providing long-lived immunity against infections. It is believed that memory cells are morphologically similar to small lymphocytes. However, the characteristics of memory cells are not fully known because these cells are defined by their survival, and there are virtually no phenotypic markers to definitively distinguish them from naive and activated or effector lymphocytes. In humans, most naive T cells express a 200 kD isoform of a surface molecule called CD45 that contains a segment encoded by an exon designated "A." This CD45 isoform can be recognized by antibodies specific for the A en-

TABLE 2–3. Characteristics of Naive and Memory Lymphocytes

Lymphocyte	Naive Cells	Memory Cells
T Lymphocytes		
Migration pattern	Preferentially to lymph nodes	To peripheral tissues, sites of inflammation, mucosal tissues
Frequency of cells responsive to particular antigen	Relatively low	High
Surface proteins		
Peripheral lymph node homing receptor (L-selectin, CD62L)	High level of expression	Low level of expression
Cell-cell adhesion molecules: integrins, CD44	Low expression	High expression
Major CD45 isoform (humans only)	CD45RA	CD45RO
B Lymphocytes		
Membrane immunoglobulin (Ig) isotype	IgM and IgD	Often others (e.g., IgGs)
Affinity of Ig produced	Relatively low	Relatively high

coded segment and is therefore called CD45RA (for "restricted A"). In contrast, most activated and memory T cells express a 180 kD isoform of CD45 in which the A exon has been spliced out; this isoform is called CD45RO. However, this distinction is not absolute, and interconversion between CD45RA⁺ and CD45RO⁺ populations has been documented. Most memory T cells also express low levels of the peripheral lymph node homing receptor but higher levels of other adhesion molecules compared to naive T cells. The significance of these differences will be discussed in Chapter 11, when we describe lymphocyte recirculation and homing to various tissues.

PROGRAMMED CELL DEATH

Some of the progeny of antigen-stimulated lymphocytes proliferate but, instead of differentiating into effector or memory cells, they die by a process known as **apoptosis.** Apoptosis is a form of regulated, physiologic cell death in which the nucleus undergoes condensation and fragmentation, the plasma membrane shows blebbing and vesiculation, and the dead cell is rapidly phagocytosed without its contents being released. (This is contrasted with necrosis, in which the nuclear and plasma membranes break down, and cellular contents often spill out, causing a local inflammatory reaction.) In many physiologic situations where homeostasis is maintained by cell death, death occurs by apoptosis. For instance, elimination of unwanted cells during embryogenesis and atrophy and involution of hormone-responsive cells as a result of hormone depletion are all due to apoptosis. These are examples of **programmed cell death,** which refers to the concept that many cell types are "programmed" to die unless saved by specific growth and/or survival factors. (Although the terms "apoptosis" and "programmed cell death" are sometimes used interchangeably, the former refers to the morphologic pattern of cell death and the latter to a mechanism.) Programmed cell death is an important homeostatic mechanism in the immune system, whose main function is to maintain the size of the lymphoid pool fairly constant throughout life. It is responsible for (1) the elimination of precursor lymphocytes that fail to express functional antigen receptors and are not selected to survive in the generative lymphoid organs; (2) the death of naive lymphocytes that do not encounter their cognate antigen; and (3) the death of a fraction of activated lymphocytes that are not continuously exposed to antigen or sufficient concentrations of growth factors. There are no good estimates of what fraction of antigen-stimulated lymphocytes die by programmed cell death *in vivo,* but it has been proposed that this may be the fate of the majority of the clonal progeny of antigen-stimulated cells. Antigen-stimulated lymphocytes undergo apoptosis probably because the local concentrations of antigen and/or growth factors are limiting, and the cells become deprived of essential survival stimuli. In later chapters, we will return to discussions of many of these physiologic situations of programmed cell death. A considerable amount of information has accumulated about the mechanisms of apoptosis, and its regulation, in lymphoid cells (Box 2–2). The immune system utilizes another pathway of apoptosis, which is triggered not by deficient stimulation or growth factors but by repeated activation. This is called **activation-induced cell death.** It is particularly important for the maintenance of immunologic self-tolerance and will be discussed in more detail in Chapters 10 and 19.

MONONUCLEAR PHAGOCYTES

The **mononuclear phagocyte system** constitutes the second major cell population of the immune system and consists of cells that have a common lineage whose primary function is phagocytosis. In the early 20th century, morphologists observed that certain cells took up dyes injected intravenously (called "vital dyes," since they stained live cells). Aschoff identified these cells as macrophages in connective tissues, microglia in the central nervous system, endothelial cells lining vascular sinusoids, and reticular cells of lymphoid organs. He suggested that these varied cell types functioned in host defense by phagocytosis of foreign invaders such as microbes, and grouped them collectively into the **reticuloendothelial system.** It is now clear that the pinocytosis of which the endothelial and reticular cells are capable is fundamentally different from the active phagocytosis of macrophages. It is, therefore, more appropriate to classify monocytes and macrophages as members of the mononuclear phagocyte system.

Development

All the cells of the mononuclear phagocyte system originate in the bone marrow and, after maturation and subsequent activation, can achieve varied morphologic forms (Fig. 2–5). The first cell type that enters the peripheral blood after leaving the marrow is incompletely differentiated and is called the **monocyte.** Monocytes are 10 to 15 μm in diameter, and they have bean-shaped nuclei and finely granular cytoplasm containing lysosomes, phagocytic vacuoles, and cytoskeletal filaments (Fig. 2–6). Once they settle in tissues, these cells mature and become **macrophages,** which are also called "histiocytes." Macrophages can be activated by a variety of stimuli and may assume different forms. Some develop abundant cytoplasm and are called epithelioid cells because of their resemblance to epithelial cells of the skin. Macrophages can fuse to form polykaryons, also termed multinucleate giant cells. Macrophages are found in all organs and connective tissues and have been given special names to designate specific locations. For instance, in the central nervous system they are the "microglial cells"; when lining the vascular sinusoids of the liver, they are called "Kupffer cells"; in pulmonary airways, they are the "alveolar macrophages"; and multinucleate phagocytes in bone are called "osteoclasts." Some mononuclear phago-

cytes may differentiate into another cell type, called dendritic cells, which are described later.

Activation and Function

Mononuclear phagocytes are important participants in the bidirectional interactions between innate and specific immunity. Macrophages that re-

---

BOX 2-2. Programmed Cell Death

There are many situations in biology where cells need to be eliminated without eliciting harmful inflammatory reactions. For instance, during embryogenesis excess numbers of developing cells die, and in hormone-responsive tissues such as the uterus, cyclical depletion of particular hormones leads to death of hormone-dependent cells. In all these situations, cell death occurs by the process of **apoptosis,** which is characterized by nuclear condensation and fragmentation, membrane blebbing, cell shrinkage, and elimination of dead cells by phagocytosis. The examples mentioned above are examples of **programmed cell death.** This concept implies that many cell populations have a tendency to undergo apoptosis (i.e., are "programmed" to die unless specifically protected). In other words, cells must constitutively express, or try to express, molecules that function as effectors of apoptosis, and these effectors are actively suppressed in order to allow the cells to survive. Some of the most elegant analyses of proteins that induce or block apoptosis have been done in the worm, *Caenorhabditis elegans.* Particular cells die in a precisely defined sequence during the development of this worm, so that the consequences of genetic manipulations on death or survival of these cells can be accurately defined. By now, several "ced" genes (for "*C. elegans* death" genes) have been identified. Remarkably, homologues of these genes are found in mammalian cells and serve essentially the same functions, implying that the process of programmed cell death and its control have been preserved through evolution. For instance, an important effector of apoptosis is a protein called Ced-3 in the worm. Its mammalian homologue is an aspartate-directed cysteine protease that belongs to a family of proteases, one prototype of which is an enzyme, called interleukin-1 converting enzyme (ICE), that converts the precursor form of the cytokine, interleukin-1β, to its active form. (Interleukin-1 and other cytokines will be described in Chapter 12.) At least a dozen Ced 3/ICE–like proteases have been identified, and the number continues to grow. These enzymes are normally present in the cytoplasm in an inactive form. They are catalytically activated by a variety of stimuli, including glucocorticoids and p53 (activated in response to DNA damage), but how these proteases are activated spontaneously (i.e., in cells "programmed" to die) is still not known. The targets of these proteases include nucleoproteins and matrix proteins whose degradation results in nuclear DNA fragmentation and apoptosis.

Normally, the activation of "pro-apoptotic" proteases is blocked by a family of proteins, the prototype of which is Bcl-2. The *bcl-2* gene was first identified as part of a chromosomal translocation in a B lymphocyte–derived follicular lymphoma. By preventing cell death, Bcl-2 promotes cell viability and effectively functions like an oncogene. A second member of this family is called Bcl-x and is present in "long" and "short" forms, called Bcl-x$_L$ and Bcl-x$_S$. The former functions like Bcl-2 to prevent cell death, whereas Bcl-x$_S$ and a related protein called Bax promote cell death. Thus, the Bcl-2 family consists of several proteins that regulate apoptosis, especially in lymphocytes. The mechanisms of action of these proteins are not yet defined. Bcl-2 and Bcl-x$_L$ are activated by growth factors, such as interleukin-2 and -3, and by

survival signals, such as "costimulators" for T cells. It is believed that in growth factor–dependent cells programmed death is prevented by growth factor–mediated induction of one or more of these survival proteins. Withdrawal of the growth factor leads to depletion of these proteins, activation of the constitutive death pathway, and apoptosis.

Programmed cell death plays a key role in controlling the size of the lymphocyte pool at many stages of lymphocyte maturation and activation. Immature lymphocytes that do not express functional antigen receptors or are not positively selected in the thymus (for T cells) or in the bone marrow (for B cells) undergo programmed death. After their maturation, if lymphocytes never encounter antigen they die by apoptosis. Even if cells are activated by antigen, a fraction of the progeny do not receive adequate growth factors or continued stimulation and also die. Fluctuations in the levels of expression of Bcl-2 or Bcl-x$_L$ during lymphocyte maturation and activation appear to correlate inversely with susceptibility to apoptosis. Overexpression of *bcl-2* or *bcl-x$_L$* as a transgene (e.g., in B cells) results in enhanced survival of immature lymphocytes and prolonged antibody responses. Conversely, knockout of *bcl-2* or *bcl-x$_L$* leads to reduced survival of mature or immature lymphocytes. It has also been postulated that the long life span of memory lymphocytes may be due to constitutive expression of Bcl-2 and/or Bcl-x$_L$, but this hypothesis remains unproved.

Lymphocytes undergo a second form of apoptotic death, which occurs not because of deficient stimulation or lack of growth factors but as a result of receptor-mediated activation. This is called **activation-induced cell death** and is thought to be induced and regulated by mechanisms that are distinct from those that control Bcl-2–regulated programmed cell death. Activation-induced cell death is especially important for the death of lymphocytes that recognize self antigens and also plays a role in the induction of tolerance to some foreign antigens. This pathway of apoptosis in lymphocytes will be discussed in Chapter 10.

Apoptosis is measured by a number of assays. Electron microscopic analysis of tissues shows apoptotic cells as cells with membrane blebbing and nuclear condensation. These cells may be within phagocytes or in the extracellular space and are called "apoptotic bodies" by morphologists. *In vitro* assays for apoptosis are designed to detect the fragmentation of nuclear DNA. This may be done by labeling permeabilized cells with dyes that intercalate into DNA, such as propidium iodide, and flow cytometry to detect subnormal amounts of DNA in individual cells. A more precise assay is terminal end-labeling of broken DNA fragments with labeled nucleotides; this reaction is catalyzed by the enzyme terminal deoxyribonucleotidyl transferase (TdT) and is called the TUNEL assay. It is used in cell suspensions and for staining tissue sections. Fragmented DNA is also detected as characteristic nucleosome-sized pieces (often multimers of ~80 base pairs) by gel electrophoresis of extracted DNA. All assays other than electron microscopy tend to underestimate the frequency of apoptotic cells, because these cells are rapidly phagocytosed and are undetectable by staining or gel electrophoresis once inside phagocytes.

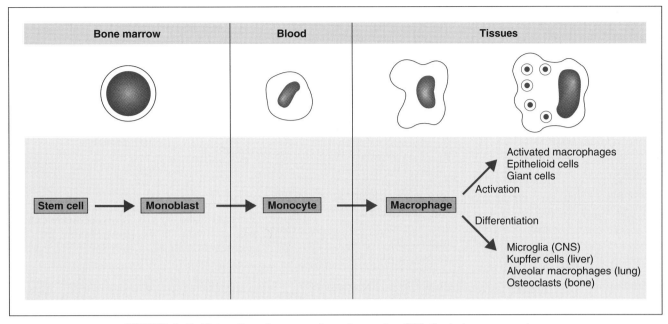

Bone marrow	Blood	Tissues

Stem cell → Monoblast → Monocyte → Macrophage

Activation
→ Activated macrophages
Epithelioid cells
Giant cells

Differentiation
→ Microglia (CNS)
Kupffer cells (liver)
Alveolar macrophages (lung)
Osteoclasts (bone)

FIGURE 2–5. Maturation of mononuclear phagocytes. CNS, Central nervous system.

spond to microbes as a reaction of innate immunity also function to trigger microbe-specific lymphocyte responses. Conversely, effector lymphocytes and their products enhance the antimicrobial functions of macrophages. These responses illustrate concepts that were introduced in Chapter 1, namely that innate immunity participates in the initiation and regulation of specific immunity, and specific immune responses augment the defense mechanisms of innate immunity.

The principal functions of mononuclear phagocytes in innate immunity include the following:

1. Macrophages phagocytose foreign particles, such as microbes, macromolecules, including antigens, and even self tissues that are injured or dead, such as senescent erythrocytes. Macrophage recognition of foreign substances and injured tissues may involve receptors for phospholipids and sugars, such as mannose, but the precise mecha-

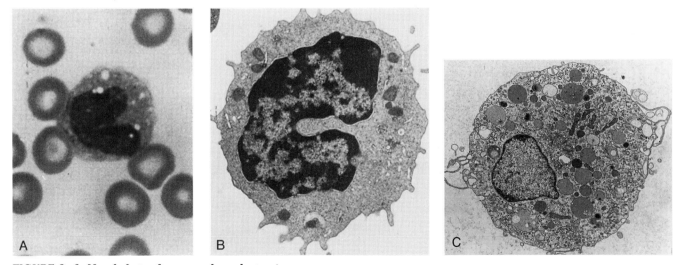

FIGURE 2–6. Morphology of mononuclear phagocytes.
A. Light micrograph of a monocyte in a peripheral blood smear.
B. Electron micrograph of a peripheral blood monocyte. (Courtesy of Dr. Noel Weidner, Department of Pathology, Brigham and Women's Hospital, Boston.)
C. Electron micrograph of an activated tissue macrophage. (Reproduced with permission from W. Bloom and D. W. Fawcett. Textbook of Histology, 12th ed. W. B. Saunders, Philadelphia, 1994.)

nisms are not well understood. Macrophages also actively phagocytose particles coated with complement proteins, which can be generated in innate and specific immune responses. Phagocytosed substances are degraded within macrophages by lysosomal enzymes. Macrophages function as the principal "scavenger cells" of the body. In addition, the cells secrete enzymes, reactive oxygen species, and nitric oxide (in mice), all of which serve to kill microbes and control the spread of infections, and can injure even normal tissues in the immediate vicinity.

2. Macrophages produce cytokines that recruit other inflammatory cells, especially neutrophils, and are responsible for many of the systemic effects of inflammation, such as fever. One of the most potent inducers of these macrophage responses is a bacterial cell wall component called endotoxin (or lipopolysaccharide), which binds to a macrophage surface molecule called CD14. Macrophages also produce growth factors for fibroblasts and vascular endothelium that promote the repair of injured tissues.

Mononuclear phagocytes play the following important roles in the recognition, activation, and effector phases of specific immunity.

1. Macrophages display foreign antigens on their surface in a form that can be recognized by antigen-specific T lymphocytes. This function of macrophages as **antigen-presenting cells** (APCs) is described in Chapter 6. Macrophages also produce soluble proteins (cytokines) that stimulate T cell proliferation and differentiation. One such cytokine, called interleukin-12, is especially important for the development of cell-mediated immunity (see Chapters 12 and 13). In addition, activated macrophages express surface proteins, called co-stimulators, that augment T cell responses at sites of infection and inflammation (see Chapter 7). Thus, *macrophages function as accessory cells in lymphocyte activation.*

2. In the effector phase of certain cell-mediated immune responses, antigen-stimulated T cells secrete cytokines that activate macrophages. Such activated macrophages are more efficient at performing microbicidal functions than are unstimulated cells and are thus better able to destroy phagocytosed microbes. Thus, *macrophages are among the principal effector cells of cell-mediated immunity* (see Chapter 13).

3. In the effector phase of humoral immune responses, foreign antigens, such as microbes, become coated, or opsonized, by antibody molecules and complement proteins. Because macrophages express surface receptors for antibodies and for certain complement proteins, they bind and phagocytose opsonized particles much more avidly than uncoated particles (see Chapter 3). Thus, *macro-*

phages participate in the elimination of foreign antigens by humoral immune responses.

The ability of macrophages and lymphocytes to stimulate each other's functions provides an important amplification mechanism for specific immunity. The mechanisms and physiologic consequences of the bidirectional interactions between immunocompetent lymphocytes and non-lymphoid accessory and effector cells will be referred to in many sections of this book.

DENDRITIC CELLS

Dendritic cells are accessory cells that play important roles in the induction of immune responses. These cells are identified morphologically as cells with membranous or spine-like projections. There are two types of dendritic cells that have different properties and functions. **Interdigitating dendritic cells,** which are usually called simply "dendritic cells," are present in the interstitium of most organs, are abundant in T cell–rich areas of lymph nodes and spleen, and are scattered throughout the epidermis of the skin, where they are called **Langerhans cells.** Langerhans cells contain an unusual cytoplasmic organelle called the Birbeck granule, whose function is unknown. These interdigitating dendritic cells are thought to arise from marrow precursors and are related in lineage to mononuclear phagocytes. They are extremely efficient at presenting protein antigens to CD4[+] helper T cells and may be the principal APCs for activating naive T cells (i.e., for initiating T cell responses; see Chapter 6). Langerhans cells are capable of picking up antigens that enter via the skin and transporting these antigens to draining lymph nodes, where immune responses are initiated. In fact, many of the dendritic cells in lymphoid organs and interstitial tissues may have arisen from Langerhans cells, which migrate from the skin into tissues after picking up antigens (see Chapter 11).

The second type of dendritic cells are called **follicular dendritic cells** because they are present in the germinal centers of lymphoid follicles in the lymph nodes, spleen, and mucosa-associated lymphoid tissues. Most follicular dendritic cells are not derived from precursors in the bone marrow and are unrelated to interdigitating dendritic cells. Follicular dendritic cells trap antigens complexed to antibodies or complement products and display these antigens on their surfaces for recognition by B lymphocytes. The functions of interdigitating and follicular dendritic cells in immune responses are discussed in more detail in Chapters 6 and 9.

GRANULOCYTES

In addition to lymphocytes and mononuclear phagocytes, other blood leukocytes, which are called **granulocytes** because they contain abundant cytoplasmic granules, participate in the effector phase of specific immune responses. The details of the morphology, biochemistry, and functions of granulocytes are beyond the scope of this book. These leukocytes are often referred to as **inflammatory cells** because they play important roles in inflammation and innate immunity and function to eliminate microbes and dead tissues. However, like macrophages, granulocytes are stimulated by T cell–derived cytokines and phagocytose opsonized particles, so that these cells serve important effector functions in specific immune responses as well.

Peripheral blood contains three types of granulocytes, which are classified according to the staining characteristics of their predominant granules.

Neutrophils, also called **polymorphonuclear leukocytes** because of their multilobed, morphologically diverse nuclei, are the most numerous. They respond rapidly to chemotactic stimuli; phagocytose and destroy foreign particles, such as microbes; can be activated by cytokines produced primarily by macrophages and endothelial cells; and are the major cell population in the acute inflammatory response. Neutrophils also possess receptors for a type of antibody called immunoglobulin (Ig)G and for complement proteins, and they migrate to and accumulate at sites of complement activation. Therefore, they phagocytose opsonized particles and function as effector cells of humoral immunity.

Eosinophils are thought to function mainly in defense against certain types of infectious agents. Eosinophils express receptors for a class of antibody called IgE and are able to bind avidly to IgE-coated particles. They are particularly effective at destroying infectious agents that stimulate the production of IgE, such as helminthic parasites. In fact, helminths may be relatively resistant to the lysosomal enzymes of neutrophils and macrophages but are often killed by the specialized granule proteins of eosinophils. Eosinophils are also abundant at sites of immediate hypersensitivity (allergic) reactions; in this setting, eosinophils contribute to tissue injury and inflammation (see Chapter 14). The growth and differentiation of eosinophils are stimulated by a helper T cell–derived cytokine called interleukin-5, and T cell activation contributes to eosinophil accumulation at sites of parasitic infestation and allergic reactions.

Basophils are circulating cells whose functions are similar to those of tissue **mast cells**. Both basophils and mast cells express high-affinity receptors for IgE and, therefore, avidly bind free IgE antibodies. Subsequent interaction of antigens with these bound IgE molecules stimulates basophils and mast cells to secrete their granule contents, which are the chemical mediators of immediate hypersensitivity (see Chapter 14). Thus, these granulocytes are effector cells of IgE-mediated immediate hypersensitivity.

FUNCTIONAL ANATOMY OF LYMPHOID TISSUES

In order to optimize the cellular interactions necessary for the recognition and activation phases of specific immune responses, the majority of lymphocytes, mononuclear phagocytes, and other accessory cells are localized and concentrated in anatomically defined tissues or organs, which are also the sites where foreign antigens are transported and concentrated. Such anatomic compartmentalization is not fixed because, as discussed in Chapter 11, many lymphocytes recirculate and constantly exchange between the circulation and tissues. *Lymphoid tissues can be classified into two groups: (1) the **generative organs,** also called primary lymphoid tissues, are the tissues where lymphocytes first express antigen receptors and attain phenotypic and functional maturity, and (2) the **peripheral organs,** also called secondary lymphoid tissues, are the sites where lymphocyte responses to foreign antigens are initiated and develop* (Fig. 2–7). Included in the generative lymphoid organs of mammals are the bone marrow, where all the lymphocytes arise, and the thymus, where T cells mature and reach a stage of functional competence. In birds, another generative organ is the bursa of Fabricius, the site of B cell maturation; the bursal equivalent in mammals is the bone marrow itself. The peripheral lymphoid tissues include the lymph nodes, spleen, mucosa-associated lymphoid tissues, and the cutaneous immune system. In addition, poorly defined aggregates of lymphocytes are found in connective tissues and in virtually all organs except the central nervous system.

Bone Marrow

During fetal life, the generation of all blood cells, called **hematopoiesis,** occurs initially in blood islands and then in the liver and spleen. This function is gradually taken over by the bone marrow and increasingly by the marrow of the flat bones, so that by puberty hematopoiesis occurs mostly in the sternum, vertebrae, iliac bones, and ribs. The red marrow that is found in these bones consists of a sponge-like reticular framework located between long trabeculae. The spaces in this framework are filled with fat cells and the precursors of blood cells, which mature and exit via the dense network of vascular sinuses to become part of the circulatory system.

All the blood cells originate from a common **stem cell** that becomes committed to differentiate along particular lineages (i.e., erythroid, megakaryocytic, granulocytic, monocytic, and lymphocytic; Fig. 2–8). The proliferation and maturation of precursor cells in the bone marrow are stimulated by

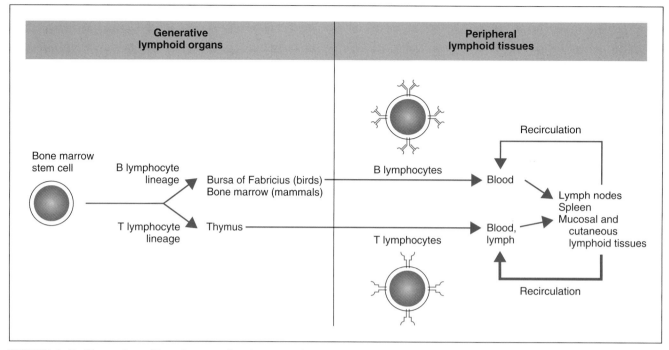

FIGURE 2–7. Maturation of lymphocytes. Development of mature lymphocytes prior to antigen exposure occurs in the generative lymphoid organs and immune responses to foreign antigens occur in the peripheral lymphoid tissues.

cytokines. Many of these cytokines are also called "colony-stimulating factors" because they are assayed by their ability to stimulate the growth and development of various leukocyte colonies from marrow cells. Hematopoietic cytokines are produced by stromal cells and macrophages in the bone marrow, thus providing the local environment for hematopoiesis, and by antigen-stimulated T lymphocytes, providing a mechanism for replenishing leukocytes that may be consumed during immune reactions. Different cytokines promote the proliferation and maturation of different lineages of bone marrow precursor cells (see Chapter 12). Little is known about the nature of the self-renewing pluripotential stem cell or the mechanisms that regulate its commitment to specific lineages. Stem cells express a surface protein called CD34, which is often used as a basis for purifying these cells.

The marrow contains mature B lymphocytes, which have developed from progenitor cells (see Chapter 4). There are also numerous antibody-secreting plasma cells, which develop in peripheral lymphoid tissues as a consequence of antigenic stimulation of B cells and then migrate to the marrow. The maturation of T lymphocytes occurs not in the bone marrow but in the thymus.

Thymus

The thymus is a bilobed organ situated in the anterior mediastinum. Each lobe is divided into multiple lobules by fibrous septa, and each lobule consists of an outer cortex and an inner medulla (Fig. 2–9). The cortex contains a dense collection of T lymphocytes, and the lighter-staining medulla is more sparsely populated with lymphocytes. Scattered throughout the thymus are non-lymphoid epithelial cells, which have abundant cytoplasm, as well as bone marrow-derived dendritic cells and macrophages. In the medulla are structures called Hassall's corpuscles, which are composed of tightly packed whorls of epithelial cells that may be remnants of degenerating cells. The thymus has a rich vascular supply and efferent lymphatic vessels that drain into mediastinal lymph nodes.

The lymphocytes in the thymus, also called **thymocytes,** are T lymphocytes at various stages of maturation. Precursors that are committed to the T cell lineage enter the thymic cortex via blood vessels. It is possible that other lymphocyte progenitors also enter the thymus but fail to survive. The most immature thymocytes do not express receptors for antigens or surface markers, including CD4 and CD8, that are characteristic of the mature phenotype. These immature cells migrate from the cortex toward the medulla and come into contact with epithelial cells, macrophages, and dendritic cells. En route to the medulla, thymocytes begin to express receptors for antigens and surface markers that are present on mature, peripheral T lymphocytes. Thus, the medulla contains mostly mature T cells, and only mature CD4$^+$ or CD8$^+$ T cells, committed to helper or cytolytic function, exit the thymus and enter the blood and peripheral lymphoid tissues.

From the large number of T cell precursors that enter the thymus, the majority die in the thymus, and only cells capable of recognizing and re-

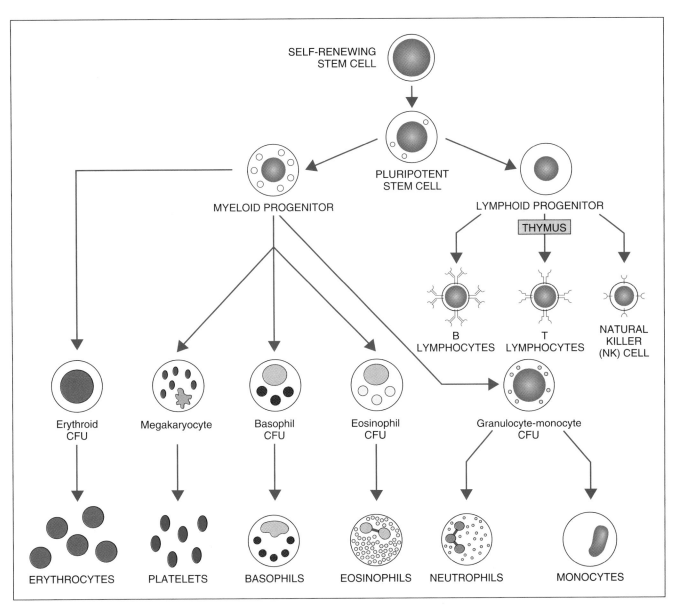

FIGURE 2–8. Maturation of blood cells: the hematopoietic "tree." CFU, colony-forming unit.

sponding to foreign antigens are selected to survive, proliferate, and mature. These selection processes are described in considerable detail in Chapter 8.

Lymph Nodes

Lymph nodes are small, nodular aggregates of lymphoid tissue situated along lymphatic channels throughout the body. Epithelia, such as the skin and the mucosa of the gastrointestinal and respiratory tracts, as well as connective tissues and most organs have a lymphatic drainage. Lymphatic vessels collect fluid and cells from the interstitial spaces and return this fluid, now called **lymph,** via lymph nodes into the blood. Antigens that enter through the skin and gastrointestinal and respiratory tracts end up in lymphatic vessels, largely

bound to accessory cells, and are transported to lymph nodes (see Chapter 11). Thus, the lymphatic system provides a mechanism for antigen collection, and the lymph nodes, which are strategically located along lymphatic vessels, "sample" the lymph for the presence of foreign antigenic material.

Each lymph node is surrounded by a fibrous capsule that is pierced by numerous afferent lymphatics, which empty the lymph into a subcapsular sinus (Fig. 2–10). The node consists of an outer cortex in which there are aggregates of cells constituting the **follicles.** Some follicles contain central areas called **germinal centers,** which stain lightly with commonly used histologic stains. Follicles without germinal centers are called primary follicles, and those with germinal centers are secondary follicles. The inner medulla contains less dense

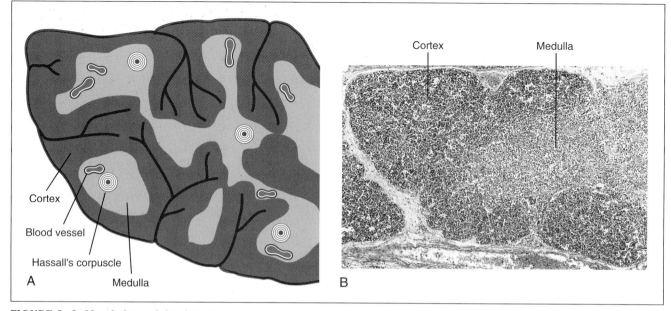

FIGURE 2-9. Morphology of the thymus.
A. Schematic diagram of the thymus, illustrating a portion of a lobe divided into multiple lobules by fibrous trabecula.
B. Light micrograph of a lobe of the thymus, showing the cortex and medulla.

lymphocytes and mononuclear phagocytes scattered among lymphatic and vascular sinusoids. Lymphocytes and accessory cells are often found in close proximity but do not form inter-cellular junctions, which is important for maintaining the ability of the lymphocytes to migrate and recirculate between the lymph, blood, and tissues. The lymph that enters the subcapsular sinus percolates through the cortex and medulla and exits via a single efferent lymphatic located in the hilum of the node. In addition, each node has a vascular supply with afferent and efferent vessels at the hilum.

Different classes of lymphocytes and non-lymphoid accessory cells are sequestered in particular areas of the node. Follicles are the B cell–rich

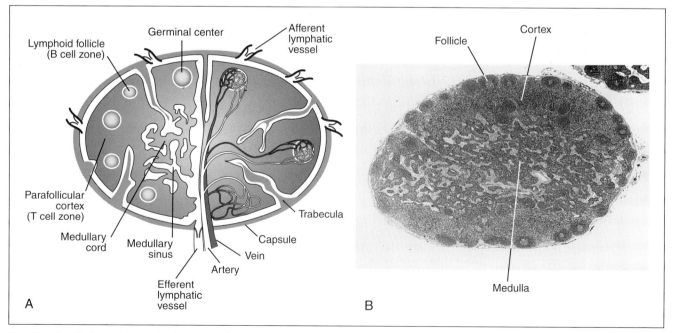

FIGURE 2-10. Morphology of a lymph node.
A. Schematic diagram of a lymph node illustrating the blood supply and the T cell– and B cell–rich zones.
B. Light micrograph of a lymph node.

areas of lymph nodes (Fig. 2–10). Primary follicles contain mostly mature, naive B lymphocytes that have apparently not been stimulated previously by antigens. Germinal centers develop in response to antigen stimulation. They are sites of remarkable B cell proliferation, selection of B cells producing high-affinity antibodies, and generation of memory B cells. Follicular dendritic cells, which reside in germinal centers, display antigens on their surfaces and function to selectively activate B cells that bind the antigen with high affinities. Fully differentiated plasma cells develop outside the germinal centers and may migrate out of lymph nodes to other tissues. The processes of B lymphocyte stimulation and antibody production are described in more detail in Chapter 9.

The T lymphocytes are located predominantly between the follicles and in the deep cortex, called the parafollicular areas (Fig. 2–10). Most of these T cells are CD4$^+$ helper T cells, intermingled with relatively sparse CD8$^+$ cells. Naive T lymphocytes, which have not been stimulated by their specific foreign antigens, enter each lymph node either via the lymph or through specialized venules lined by cuboidal endothelium, called **high endothelial venules,** that are abundant in T cell–rich zones. Here the T cells encounter foreign antigens that have been transported to the node in the lymph. Interdigitating dendritic cells, which are also abundant in the T cell areas, as well as other accessory cells, are located along the connective tissue scaffold of the node and present antigens they have picked up to naive helper T cells. Thus, the *lymph nodes are the sites where T cell responses to lymphborne protein antigens are initiated.*

The medulla contains scattered lymphocytes, large numbers of macrophages and dendritic cells, and, in nodes draining sites of immunization, numerous plasma cells, all of which are interspersed with lymphatic channels.

The mechanisms responsible for the anatomic sequestration of different classes of lymphocytes in distinct areas of the node are unclear. One possibility is that compartmentalization is maintained by specific adhesion of different lymphocytes to stromal cells or extracellular matrix proteins. In mice, knockout of the gene encoding a cytokine called lymphotoxin results in a failure of lymph node development (and abnormal spleen anatomy). The precise function of cytokines in lymphoid organogenesis is not yet known; it is postulated that they may induce the expression of adhesion molecules that regulate lymphocyte migration during the development of the lymphoid system (see Chapter 12 for a discussion of lymphotoxin). The anatomic organization of lymph nodes provides multiple sites for interactions between accessory cells and lymphocytes and between different classes of lymphocytes. Dendritic cells located in T cell–rich areas present antigens to CD4$^+$ helper T cells. Follicular dendritic cells in germinal centers present antigens to activated B cells. B

lymphocytes that enter the node from the blood must traverse helper T cell–rich zones en route to follicles, thus maximizing the chances for cooperative T cell–B cell interactions (see Chapter 9). It is, therefore, not surprising that the organization of various cell populations in lymph nodes is critical for the generation of immune responses (see Chapter 11). The structure of lymph nodes is not fixed but changes with antigen exposure. For instance, germinal centers develop within about 1 week after immunization and gradually regress after the antigenic stimulus is eliminated.

Spleen

The spleen is an organ weighing about 150 gm in adults, located in the left upper quadrant of the abdomen. It is supplied by a single splenic artery, which pierces the capsule at the hilum and divides into progressively smaller branches that remain surrounded by protective and supporting fibrous trabeculae (Fig. 2–11). Small arterioles are surrounded by cuffs of lymphocytes, called **periarteriolar lymphoid sheaths,** to which are attached the lymphoid follicles, some of which contain germinal centers. The periarteriolar lymphoid sheaths and follicles are surrounded by a rim of lymphocytes and macrophages, called the **marginal zone.** These dense lymphoid tissues constitute the **white pulp** of the spleen. The arterioles ultimately end in vascular sinusoids, scattered among which are large numbers of erythrocytes, macrophages, dendritic cells, sparse lymphocytes, and plasma cells; these constitute the **red pulp.** The sinusoids end in venules that drain into the splenic vein, which carries blood out of the spleen and into the portal circulation.

Lymphocytes and accessory cells are anatomically segregated in the spleen as they are in lymph nodes (Fig. 2–11). The periarteriolar sheaths contain mainly T lymphocytes, about two thirds of which are of the CD4$^+$ helper class and one third are CD8$^+$. The follicles and germinal centers are the predominantly B cell zones, having the same anatomic features and functions as in lymph nodes. The marginal zones contain macrophages, B lymphocytes, and CD4$^+$ helper T cells. Antigens and lymphocytes enter the spleen through the vascular sinusoids; the spleen lacks high endothelial venules. Activation of B cells is initiated in the marginal zones, which are adjacent to helper T cells in lymphoid sheaths. Activated B cells subsequently migrate into germinal centers or into the red pulp (see Chapter 9).

In principle, the function of the spleen and its responses to antigens are much like those of lymph nodes, the essential difference being that *the spleen is the major site of immune responses to blood-borne antigens, whereas lymph nodes are involved in responses to antigens in the lymph.* The spleen is also an important "filter" for the blood; its red pulp macrophages clear the blood of un-

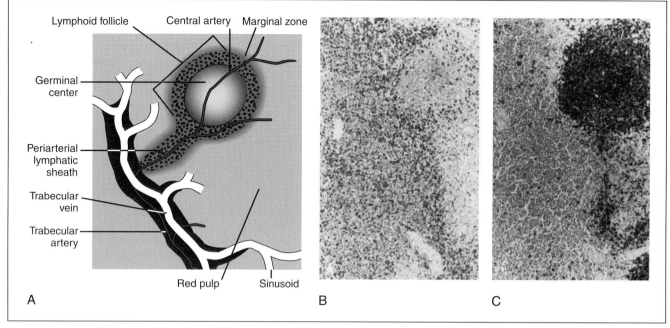

FIGURE 2–11. Morphology of the spleen.
A. Schematic diagram of the spleen. Note that the white pulp, made up of dense lymphoid tissues in periarteriolar sheaths and follicles, is intermingled with the red pulp, composed of vascular sinusoids and scattered cells.
B. T lymphocytes in the periarteriolar lymphoid sheath, detected by an immunoperoxidase stain with an antibody specific for T cells. (Courtesy of Dr. G. S. Pinkus, Department of Pathology, Brigham and Women's Hospital, Boston.)
C. B lymphocytes in a lymphoid follicle, detected by an immunoperoxidase stain with an antibody specific for B cells. (Courtesy of Dr. G. S. Pinkus, Department of Pathology, Brigham and Women's Hospital, Boston.)

wanted foreign substances and senescent erythrocytes even in the absence of specific immunity.

Other Peripheral Lymphoid Tissues

In addition to the lymph nodes and spleen, lymphocytes are found either scattered or in aggregates in many tissues. Some of these collections are anatomically well organized and have unique properties. Located beneath the mucosa of the gastrointestinal and respiratory tracts are aggregates of lymphocytes and accessory cells that resemble lymph nodes in structure and function. These aggregates include Peyer's patches in the lamina propria of the small intestine, tonsils in the pharynx, and submucosal lymphoid follicles in the appendix and throughout the upper airways. The lymphoid tissues in these sites constitute the **mucosal immune system**. The **cutaneous immune system** consists of lymphocytes and accessory cells in the epidermis and dermis. The specialized structural features and functions of the cutaneous and mucosal immune systems are described in Chapter 11.

In addition to these normal lymphoid organs, **ectopic lymphoid tissues** can develop at sites of strong immune responses. A striking example of this is the disease rheumatoid arthritis, in which an immune response in the synovium ultimately leads to destruction of the cartilage and bone in joints. In severe cases, synovial tissues contain well-developed lymphoid follicles with prominent germinal centers. Local production of cytokines such as lymphotoxin may play a role in the development of ectopic lymphoid tissues, because in mice overexpression of lymphotoxin at abnormal sites stimulates the local development of tissues resembling lymph nodes.

SUMMARY

The principal cellular constituents of the immune system are lymphocytes, mononuclear phagocytes, and related accessory cells. Lymphocytes are the only immunocompetent cells capable of specific recognition of antigens. They are morphologically homogeneous but consist of distinct subsets that perform different functions and can be distinguished phenotypically. B lymphocytes are the cells that produce antibodies and are thus the mediators of humoral immune responses. Some T lymphocytes express the CD4 surface marker, and function as helper cells to stimulate antibody production by B lymphocytes and to activate macrophages to destroy phagocytosed microbes. Other T lymphocytes express the CD8 marker, and function as cytotoxic cells to kill target cells expressing foreign antigens. NK cells are neither B nor T lymphocytes; they lyse virus-infected and certain tumor cells. Mononuclear phagocytes are critical for host defense in the absence of specific immunity and have also evolved into key participants in the recognition, activation, and effector phases of specific immune responses. Dendritic cells are accessory cells that are involved in the initiation of T lymphocyte responses to protein antigens.

Lymphocytes originate in the bone marrow and mature in different generative organs, the bone marrow itself for B cells and the thymus for T lymphocytes. Mature lymphocytes and accessory cells are located in anatomically defined peripheral lymphoid tissues. Lymph nodes are the sites where protein antigens are transported in the lymph and concentrated, and immune responses to these antigens are initiated and develop. The spleen is the organ where immune responses to blood-borne antigens are initiated. The structural organization of lymphoid tissues optimizes intimate contact and short-range interactions between the cell populations that cooperate in the generation of immune responses.

Selected Readings

Bhan, A. K., and I. Bhan. In situ characterization of human lymphoid cells using monoclonal antibodies. *In* M. Miyasaka and Z. Trnka (eds.). Differentiation Antigens in Lymphohemopoietic Tissues. M. Dekker, New York and Basel, 1988, pp. 13–46.

Brekelmans, P., and W. van Ewijk. Phenotypic characterization of murine thymic microenvironments. Seminars in Immunology 2:13–24, 1990.

Cory, S. Regulation of lymphocyte survival by the *BCL-2* gene. Annual Review of Immunology 13:513–543, 1995.

Heinen, E., N. Corwann, and C. Kinet-Denoel. The lymph follicle: a hard nut to crack. Immunology Today 9:240–243, 1988.

Mackay, C. R. Immunological memory. Advances in Immunology 53:217–265, 1993.

MacLennan, I. C. M. Germinal centers. Annual Review of Immunology 12:117–139, 1994.

Osmond, D. G. The turnover of B-cell populations. Immunology Today 14:34–37, 1993.

Sprent, J. T and B memory cells. Cell 76:315–322, 1994.

Vaux, D., and A. Strasser. The molecular biology of apoptosis. Proceedings of the National Academy of Sciences USA 93: 2239–2244, 1996.

Weissman, I. L. Developmental switches in the immune system. Cell 76:207–218, 1994.

ANTIGEN RECOGNITION

AND

LYMPHOCYTE ACTIVATION

The initial phases of specific immune responses are the specific recognition of antigen by lymphocytes and the responses of lymphocytes to antigen. This section is devoted to a discussion of the cellular and molecular basis of antigen recognition and lymphocyte activation.

We will begin with antibodies, which are the antigen receptors and effector molecules of B lymphocytes, because our understanding of the structural basis of antigen recognition has developed from studies of these molecules. Chapter 3 describes the molecular structure of antibodies and how these proteins recognize antigens and perform their effector functions. Chapter 4 deals with the structure and expression of antibody genes, the generation of antibody diversity, and the development of the B cell repertoire.

Before continuing with a discussion of antibody production, the next four chapters will consider antigen recognition by T cells, which play a central role in all immune responses to protein antigens, including antibody responses. In Chapter 5 we will describe the genetics and biochemistry of the major histocompatibility complex (MHC), whose products are integral components of the ligands that T cells specifically recognize. Chapter 6 discusses the association of foreign antigens with MHC molecules and the cell biology and physiologic significance of antigen presentation. Chapter 7 deals with the structure of the T cell antigen receptor, the role of other T cell surface proteins in responses to antigens, and the mechanisms of T lymphocyte activation. The expression of T cell receptor genes and the development of mature T lymphocytes in the thymus are described in Chapter 8.

We will return to humoral immunity in Chapter 9 and discuss the responses of B lymphocytes to antigens and to stimuli provided by helper T cells. Chapter 10 describes how these immune responses are regulated and the phenomenon of immunologic tolerance to foreign antigens. Finally, in Chapter 11 we will describe the development of immune responses at the organismal level and the special features of immune responses at different anatomic sites.

CHAPTER THREE

ANTIBODIES AND ANTIGENS

One of the earliest experimental demonstrations of specific immunity was the induction of humoral immunity to microbial toxins. The protective effects of humoral immunity are now known to be mediated by a family of structurally related glycoproteins called **antibodies.** *Antibodies always initiate their biologic effects by binding to antigens.* Antibodies are not enzymes and, except in unusual circumstances, do not modify the covalent structure of antigens. Antibody binding to antigen, although entirely non-covalent, is nevertheless exquisitely specific for one antigen versus another and often very strong. Antibodies, major histocompatibility complex (MHC) molecules (see Chapter 5), and T cell antigen receptors (see Chapter 7) constitute the three classes of molecules used by the specific immune system to recognize antigens. Of these three, antibodies show by far the widest range of antigenic structures they can recognize, the greatest ability to discriminate among different antigens, and the greatest strength of binding to antigen. Antibodies are also the best studied of these antigen-binding molecules. Therefore, we will begin our discussion of how the specific immune system recognizes antigens by describing in molecular terms how antibodies perform this function.

Antibodies are produced in a membrane-bound form by B lymphocytes, and these membrane molecules function as B cell receptors for antigens. *The interaction of antigen with membrane antibodies on B cells constitutes the recognition phase of humoral immunity.* Antibodies are also produced in a secreted form by the progeny of B cells that differentiate in response to antigenic stimulation. *These secreted antibodies bind to antigens and trigger several of the effector functions of the immune system.* The specificity of the effector phase is due to the antigen-antibody interaction, but the effector functions themselves are usually not specific for the eliciting antigen. In fact, these functions are often mediated by portions of the antibody molecule that are spatially distinct from the site of antigen binding. In this chapter, we will also describe the structural features of antibody molecules that underlie their effector functions. Finally, we will describe how antibodies can be used as laboratory reagents to analyze biologic systems, including the immune system itself.

MOLECULAR STRUCTURE OF ANTIBODIES

Structural analyses of antibody molecules involving the efforts of many laboratories have been in progress for over 70 years. Early studies were performed with naturally occurring mixtures of antibodies present in the blood of immunized individuals. Blood contains many different antibodies, each derived from a particular clone of B cells and each having a distinct structure and specificity for antigen. Nevertheless, antibodies are sufficiently similar to each other that Michael Heidelberger

and colleagues were able to purify mixtures of antibodies from other components of the blood, laying the groundwork for subsequent structural studies. Working with these mixtures, immunologists were able to deduce the overall structure of antibody molecules. However, the molecular heterogeneity of these polyclonal antibodies (i.e., antibodies produced by multiple clones of B cells) interfered with more detailed analysis of antibody structure, such as amino acid sequence determination. The key methodological breakthrough in this endeavor was the discovery that patients or animals with multiple myeloma, a monoclonal tumor of antibody-secreting plasma cells, often have high levels of biochemically identical antibodies or portions of antibodies in their blood or urine, providing a source of individual antibody molecules of a single (albeit usually unknown) specificity. In 1975, Georges Kohler and Cesar Milstein described a method for immortalizing individual antibody-secreting cells from an immunized animal, permitting the selection of individual **monoclonal antibodies** of predetermined specificity (Box 3–1). The availability of homogeneous populations of antibodies and antibody-producing cells permitted complete amino acid sequence determination of several individual antibody molecules and, eventually, molecular cloning and genetic analysis of antibodies. These studies have culminated in the x-ray crystallographic determinations of the three-dimensional structure of several antibody molecules and, in several cases, of antibody with bound antigen. As a result, we now know more about the structure of antibody molecules than about any other element of the immune system. This portion of the chapter describes the purification of antibody molecules and their general structural features.

Natural Distribution and Purification of Antibody Molecules

Although antibodies were first isolated from the fluid portion of the blood, they are found in several different anatomic locations:

1. Antibodies are present within cytoplasmic membrane-bound compartments (endoplasmic reticulum and Golgi apparatus) and on the surface of B lymphocytes, which are the only cells that synthesize antibody molecules.
2. Antibodies are present in the plasma (fluid portion) of the blood and, to a lesser extent, in the interstitial fluid of the tissues where secreted antibody from B cells accumulates.
3. Antibodies are bound to the surface of certain immune effector cells, such as mononuclear phagocytes, natural killer (NK) cells, and mast cells, which do not synthesize antibody but have specific receptors for binding antibody molecules.
4. Antibodies are present in secretory fluids such as mucus and milk, into which certain types of antibody molecules are specifically transported.

BOX 3–1.   Hybridomas and Monoclonal Antibodies

The technique of producing virtually unlimited quantities of a single antibody specific for a particular antigenic determinant has revolutionized immunology and has had a far-reaching impact on research in diverse fields as well as in clinical medicine. This technique is based on the fact that each B lymphocyte produces antibody of a single specificity. Therefore, each monoclonal tumor derived from a B lymphocyte, called a myeloma, produces only one antibody. Such tumors occur spontaneously in humans and can be induced experimentally by various treatments in mice. Myeloma-derived homogeneous antibodies have proved invaluable for elucidating the structure of Ig proteins, and Ig genes were first isolated from myelomas. However, most myelomas secrete antibodies of unknown antigenic specificities, because the transformation process that gives rise to these tumors affects B lymphocytes randomly and it is not possible to predict the specificity of any randomly transformed clone of B cells. Many attempts have been made to produce homogeneous or monoclonal antibodies of known specificity. Since normal B lymphocytes cannot grow indefinitely, such attempts have focused on immortalizing B cells that produce a specific antibody. The first and now generally used technique for doing this was described by Georges Kohler and Cesar Milstein in 1975. The method involves cell fusion or somatic cell hybridization between a normal antibody-producing B cell and a myeloma line, and selection of fused cells that secrete antibody of the desired specificity derived from the normal B cell. Such fusion-derived immortalized antibody-producing cell lines are called hybridomas, and the antibodies they produce are monoclonal antibodies.

The success of this technique depended on the development of cultured myeloma lines that would grow in normal culture medium but would not grow in a defined "selection" medium because they lacked a functional gene(s) required for DNA synthesis in this selection medium. Fusing normal cells to these defective myeloma fusion partners would provide the necessary gene(s) from the normal cells, so that only the somatic cell hybrids would continue to grow in the selection medium. Moreover, genes from the myeloma cell make such hybrids immortal. Cell lines that can be used as fusion partners are created by inducing defects in nucleotide synthesis pathways (Fig. A). Normal animal cells synthesize purine nucleotides and thymidylate de novo from phosphoribosyl pyrophosphate and uridylate, respectively, in several steps, one of which involves the transfer of a methyl or formyl group from activated tetrahydrofolate. Anti-folate

drugs, such as aminopterin, block the reactivation of tetrahydrofolate, thereby inhibiting the synthesis of purine and thymidylate. Since these are necessary components of DNA, aminopterin blocks DNA synthesis via the de novo pathway. Aminopterin-treated cells can use a salvage pathway in which purine is synthesized from exogenously supplied hypoxanthine using the enzyme hypoxanthine-guanine phosphoribosyltransferase (HGPRT) and thymidylate is synthesized from thymidine using the enzyme thymidine kinase (TK). Therefore, cells grow normally in the presence of aminopterin if the culture medium is also supplemented with hypoxanthine and thymidine (called HAT medium). Cell lines, however, can be made defective in HGPRT if they are mutagenized and selected in thioguanine or azaguanine, which are analogs of normal metabolites that function as substrates for HGPRT but give rise to nonfunctional purines. Similarly, cells can be made defective in TK by mutagenesis and selection in bromodeoxyuridine, which is metabolized by TK to form a light-sensitive, lethal product. Such HGPRT- or TK-negative cells cannot use the salvage pathway and will, therefore, die in HAT medium. If normal cells are fused to HGPRT-negative or TK-negative cells, the normal cells provide the necessary enzymes(s), so that the hybrids synthesize DNA and grow in HAT medium.

This principle was applied to the generation of antibody-producing hybridomas by first developing HGPRT-negative and/or TK-negative myeloma lines. Myeloma lines are the best fusion partners for B cells, since like cells tend to fuse and give rise to stable hybrids more efficiently than unlike cells. Kohler and Milstein fused an HGPRT-defective mouse myeloma line to normal B cells from mice immunized with a known antigen, using Sendai virus, which expresses an envelope protein ("fusion protein") that fuses cells together. Hybrids were selected for growth in HAT medium; under these conditions, unfused myeloma cells die because they cannot use the salvage pathway and the B cells cannot survive for more than 1 to 2 weeks because they are not immortalized, so that only hybrids will grow (Fig. B). More recent advances in this basic technique include the use of myeloma lines that do not produce their own Ig and the use of polyethylene glycol instead of Sendai virus as the fusing agent because of technical ease. The fused cells are cultured at a concentration at which each culture well is expected to contain only one hybridoma cell. The culture supernatant from each well in which growing cells are detected is then tested for the presence of antibody reactive with the antigen used

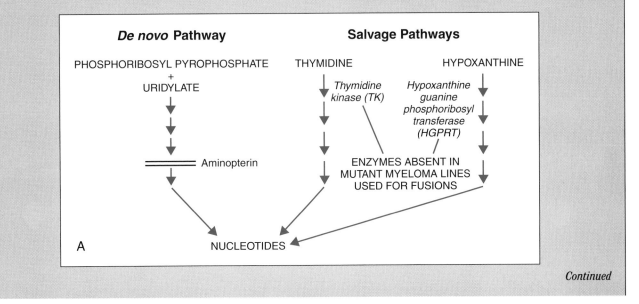

Continued

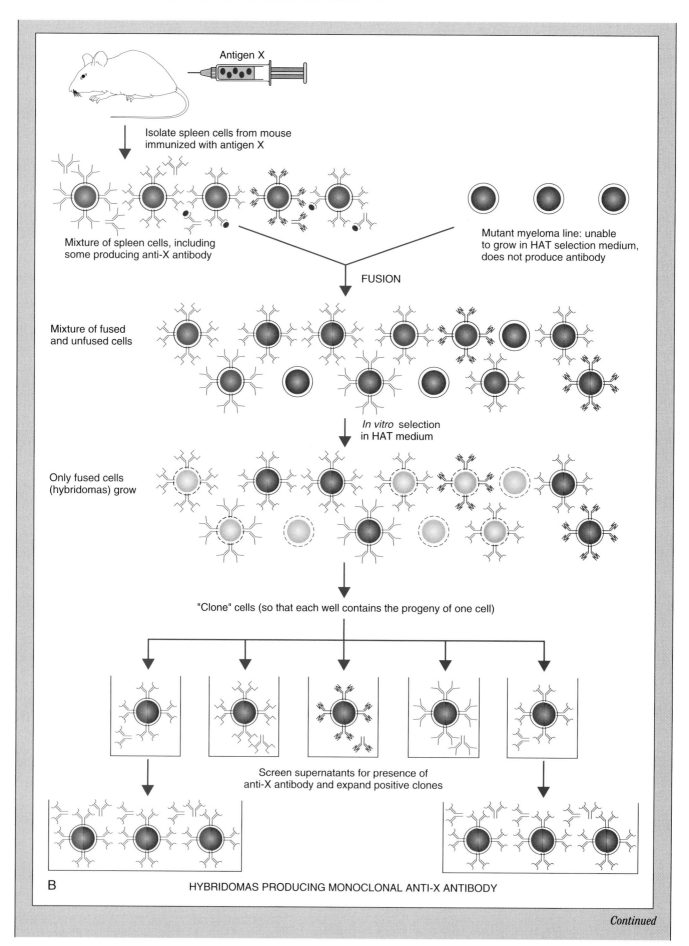

Antigen X

Isolate spleen cells from mouse immunized with antigen X

Mixture of spleen cells, including some producing anti-X antibody

Mutant myeloma line: unable to grow in HAT selection medium, does not produce antibody

FUSION

Mixture of fused and unfused cells

In vitro selection in HAT medium

Only fused cells (hybridomas) grow

"Clone" cells (so that each well contains the progeny of one cell)

Screen supernatants for presence of anti-X antibody and expand positive clones

B   HYBRIDOMAS PRODUCING MONOCLONAL ANTI-X ANTIBODY

Continued

for immunization. The screening method depends on the antigen being used. For soluble antigens, the usual technique is RIA or ELISA; and for cell surface antigens, a variety of assays for antibody binding to viable cells can be used (see Laboratory Uses of Antibodies later in this chapter). Once positive wells are identified (i.e., wells containing hybridomas producing the desired antibody), the cells are cloned in semisolid agar or by limiting dilution, and clones producing the antibody are isolated by another round of screening. These cloned hybridomas produce monoclonal antibodies of a desired specificity. Hybridomas can be grown in large volumes or as ascitic tumors in syngeneic mice in order to produce large quantities of monoclonal antibodies.

Two features of this somatic cell hybridization make it extremely valuable. First, it is the best method for producing a monoclonal antibody against a known antigenic determinant. Second, it can be used to identify unknown antigens present in a mixture because each hybridoma is specific for only one antigenic determinant. For instance, if several hybridomas are produced that secrete antibodies that bind to the surface of a particular cell, each hybridoma clone will secrete an antibody specific for only one surface antigenic determinant. These monoclonal antibodies can then be used to purify different cell surface molecules, some of which may be known molecules and others that may not have been identified previously. Some of the commonest applications of hybridomas and monoclonal antibodies include the following:

(1) Identification of phenotypic markers unique to particular cell types. The basis for the modern classification of lymphocytes and mononuclear phagocytes is the binding of population-specific monoclonal antibodies. These have been used to define "clusters of differentiation" for various cell types (see Chapter 2).
(2) Immunodiagnosis. The diagnosis of many infectious and systemic diseases relies upon the detection of specific antigens and/or antibodies in the circulation or in tissues, using monoclonal antibodies in immunoassays.

(3) Tumor diagnosis and therapy. Tumor-specific monoclonal antibodies are used for detection of tumors by imaging techniques and for immunotherapy of tumors *in vivo*.
(4) Functional analysis of cell surface and secreted molecules. In immunologic research, monoclonal antibodies that bind to cell surface molecules and either stimulate or inhibit particular cellular functions are invaluable tools for defining the functions of surface molecules, including receptors for antigens. Antibodies that neutralize cytokines are routinely used for detecting the presence and functional roles of these protein hormones *in vitro* and *in vivo*.

At present, hybridomas are most often produced by fusing HAT-sensitive mouse myelomas with B cells from immunized mice, rats, or hamsters. The same principle is used to generate mouse T cell hybridomas, by fusing T cells with a HAT-sensitive, T cell–derived tumor line; uses of such monoclonal T cell populations are described in Chapter 7. Attempts are being made to generate human monoclonal antibodies, primarily for administration to patients, by developing human myeloma lines as fusion partners. (It is a general rule that the stability of hybrids is low if cells from species that are far apart in evolution are fused, and this is presumably why human B cells do not form hybridomas with mouse myeloma lines at high efficiency.) As we shall discuss later in the chapter, only small portions of the antibody molecule are responsible for binding to antigen; the remainder of the antibody molecule can be thought of as a "framework." This structural organization allows the DNA segments encoding the antigen-binding sites from a murine monoclonal antibody to be "stitched" into a complementary DNA encoding a human myeloma protein, creating a hybrid gene. When expressed, the resultant hybrid protein, which retains antigen specificity, is referred to as a "humanized antibody." Humanized antibodies offer an alternative strategy for generating monoclonal antibodies that may be safely administered to patients.

When blood or plasma forms a clot, antibodies remain in the residual fluid, called **serum.** A sample of serum that contains a large number of antibody molecules that bind to a particular antigen is commonly called an **antiserum.** (The study of antibodies and their reactions with antigens is therefore classically called **serology.**) The number of antibody molecules in a serum specific for a particular antigen is often measured by serially diluting the serum until binding can no longer be observed; sera with a large number of antibody molecules specific for a particular antigen are said to be "strong" or have a "high titer."

Plasma or serum glycoproteins are traditionally separated by solubility characteristics into albumins and globulins and may be further separated by migration in an electric field, a process called electrophoresis. Elvin Kabat and colleagues demonstrated that most antibodies are found in the third fastest migrating group of globulins, named **gamma globulins** for the third letter of the Greek alphabet. Another common name for antibody is **immunoglobulin** (Ig), referring to the immunity-conferring portion of the gamma globulin fraction. The terms immunoglobulin and antibody are used interchangeably throughout this book.

Currently, antibody molecules are generally purified from plasma or other natural fluids by a two-step procedure. The first step is to precipitate antibodies from the biologic fluid by adding a concentration of ammonium sulfate that ranges from 40 to 50 per cent of saturation. Under these conditions, albumin and most small molecules remain in solution, so that partially purified antibody can be collected in a pellet by centrifugation. The antibody-containing pellet is redissolved in buffer and then purified by various forms of chromatography (the second step). When the antibody of interest in the biologic fluid is specific for a known antigen, the antigen can be immobilized on a column matrix and used to bind the antibody, a method called *affinity chromatography*. Antibody can be recovered from the column matrix by a change in pH.

Overview of Antibody Structure

A number of the structural and functional features of antibodies were determined from the early studies of these molecules:

1. *All antibody molecules are similar in overall structure, accounting for certain common physico-*

chemical features, such as charge and solubility. These common properties may be exploited as a basis for the purification of antibody molecules from fluids such as blood. *All antibodies have a common core structure of two identical light chains (each about 24 kilodaltons [kD]) and two identical heavy chains (about 55 or 70 kD)* (Fig. 3–1). One light chain is attached to each heavy chain, and the two heavy chains are attached to each other. Both the light chains and the heavy chains contain a series of repeating, homologous units, each about 110 amino acid residues in length, which fold independently in a common globular motif, called an **immunoglobulin domain** (Fig. 3–2). All Ig domains contain two layers of β-pleated sheet with three or four strands of antiparallel polypeptide chain. Certain Ig domains, such as those comprising variable regions (see later), have an extra strand in each of the two layers. As will be discussed in Chapter 7, many other proteins of importance in the immune system contain regions that use the same folding motif and show structural relatedness to Ig amino acid sequences. All molecules that contain this motif are said to belong to the **Ig superfamily,** and all of the gene segments encoding the Ig-like domains are believed to have evolved from the same common ancestral gene (see Chapter 7, Box 7–2).

2. Despite their overall similarity, *antibody molecules can be readily divided into a small number of distinct classes and subclasses, based on minor differences in physicochemical characteristics such as size, charge, and solubility and on their behavior as antigens* (Box 3–2). The classes of antibody molecules are also called **isotypes** and in humans are named IgA, IgD, IgE, IgG, and IgM (Table 3–1). IgA and IgG isotypes can be further subdivided into closely related subclasses, or subtypes, called IgA1 and IgA2, and IgG1, IgG2, IgG3, and IgG4, respectively. In certain instances, it will be convenient to refer to studies of mouse antibody. Mice have the same general isotypes as humans, but the IgG isotype is divided into the IgG1, IgG2a, IgG2b, and IgG3 subclasses. The heavy chains of all antibody molecules of an isotype or subtype share extensive regions of amino acid sequence identity but differ from antibodies belonging to other isotypes or subtypes. Heavy chains are designated by the letter of the Greek alphabet corresponding to the overall isotype of the antibody: IgA1 contains α1 heavy chains; IgA2, α2; IgD, δ; IgE, ε; IgG1, γ1; IgG2, γ2; IgG3, γ3; IgG4, γ4; and

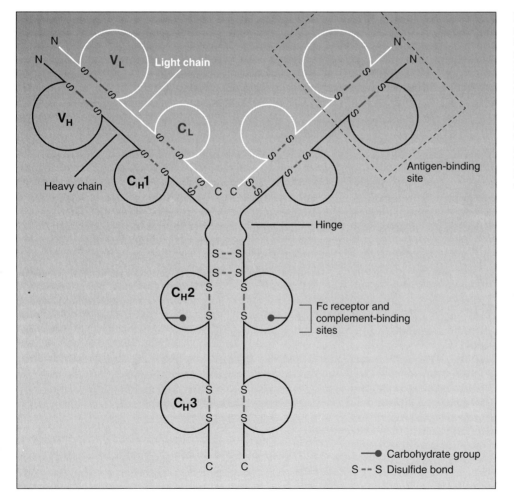

FIGURE 3–1. Schematic diagram of an immunoglobulin (Ig) molecule. In this drawing of an IgG molecule, the antigen-binding sites are formed by the juxtaposition of V_L and V_H domains. The locations of complement and Fc receptor-binding sites within the heavy chain constant regions are approximations. S––S refers to intrachain and interchain disulfide bonds; N and C refer to amino and carboxy termini of the polypeptide chains, respectively.

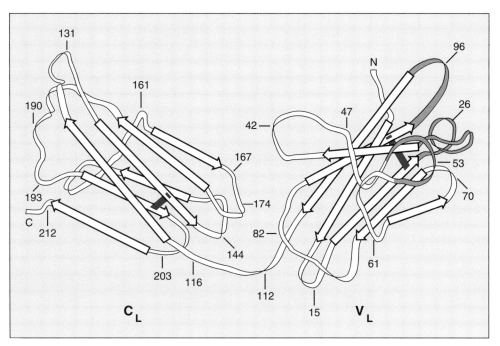

FIGURE 3–2. Polypeptide folding into immunoglobulin (Ig) domains in a human antibody light chain. The V and C regions each independently fold into Ig domains. The white arrows represent polypeptide arranged in β-pleated sheets, the dark blue bars are intrachain disulfide bonds, and the numbers indicate the positions of amino acid residues counting from the amino (N) terminus. The CDR1, CDR2, and CDR3 loops of the V region, colored in light blue, are brought together to form the antigen-binding surface of the light chain. (Adapted with permission from Edmundson, A. B., K. R. Ely, E. E. Abola, M. Schiffer, and N. Panagiotopoulos. Rotational allostery and divergent evolution of domains in immunoglobulin light chains. Biochemistry 14:3953–3961, 1975. Copyright 1975, American Chemical Society.)

IgM, μ. The shared regions of heavy chain amino acid sequences are responsible for both the common physicochemical properties and the common antigenic properties of antibodies of the same isotype. In addition, the shared regions of the heavy chains provide members of each isotype with common abilities to bind to certain cell surface receptors or to other macromolecules like complement and thereby activate particular immune effector functions. Thus, the separation of antibody molecules into isotypes and subtypes on the basis of common structural features also separates antibodies according to which set of effector functions they commonly activate. In other words, *different effector functions of antibodies are mediated by distinct isotypes and subtypes.* As we shall see later, there are two isotypes of antibody light chains, called κ and λ. The light chains do not mediate or influence the effector functions of antibodies. However, as we shall discuss shortly, both the heavy and light chains contribute to specific antigen recognition.

3. *There are more than 1×10^7, and perhaps as many as 10^9, structurally different antibody molecules in every individual, each with unique amino acid*

TABLE 3–1. Human Antibody Isotypes*

Antibody	Subtypes	H Chain (Designation)	H Chain Domains (Number)	Hinge	Tail Piece	Serum Concentration (mg/ml)	Secretory Form	Molecular Size of Secretory Form (kD)
IgA	IgA1	$\alpha1$	4	Yes	Yes	3	Monomer, dimer, trimer	150, 300, or 400
	IgA2	$\alpha2$	4	Yes	Yes	0.5	Monomer, dimer, trimer	150, 300, or 400
IgD	None	δ	4	Yes	Yes	Trace	—	180
IgE	None	ϵ	5	No	No	Trace	Monomer	190
IgG	IgG1	$\gamma1$	4	Yes	No	9	Monomer	150
	IgG2	$\gamma2$	4	Yes	No	3	Monomer	150
	IgG3	$\gamma3$	4	Yes	No	1	Monomer	150
	IgG4	$\gamma4$	4	Yes	No	0.5	Monomer	150
IgM	None	μ	5	No	Yes	1.5	Pentamer	950

* Multimeric forms of IgA and IgM are associated with J chain via the tail piece region of the heavy chain. IgA in mucus is also associated with secretory piece.
Abbreviations: Ig, immunoglobulin; kD, kilodalton.

BOX 3–2. Anti-Immunoglobulin Antibodies

Antibody molecules are proteins and therefore can be antigenic. Immunologists have exploited this fact to produce antibodies specific for Ig molecules that can be used as reagents to analyze the structure and function of Ig molecules. In order to obtain an anti-antibody response, it is necessary that the Ig molecules used to immunize an animal be recognized in whole or in part as foreign. The simplest approach is to immunize one species (e.g., rabbit) with Ig molecules of a second species (e.g., mouse). Populations of antibodies generated by such cross-species immunizations are largely specific for epitopes present in the constant regions of light or heavy chains. Such sera can be used to define the isotype of an antibody.

When an animal is immunized with Ig molecules derived from another animal of the same species, the immune response is confined to epitopes of the immunizing Ig that are absent or uncommon on the Ig molecules of the responder animal. Two types of determinants have been defined by this approach. First, determinants may be formed by minor structural differences (polymorphisms) in amino acid sequences located in the conserved portions of Ig molecules. Ig genes that encode such polymorphic structures are inherited as mendelian alleles. (The concepts of polymorphism and allelic genes are discussed more fully in Chapter 5.) Determinants on Ig molecules that differ among individuals of the same species that have inherited different alleles are called **allotopes**. All antibody molecules that share a particular allotope are said to belong to the same allotype. Most allotopes are located in the constant regions of light or heavy chains, but some are found in the framework portions of variable regions. Allotypic differences have no functional significance, but they have been important in the study of Ig genetics. For example, allotypes detected by anti-Ig antibodies were initially used to locate the position of Ig genes by linkage analysis. In addition, the remarkable observation that, in homozygous animals, all of the heavy chains of a particular isotype (e.g., IgM) share the same allotype even though the V regions of these antibodies have different amino acid sequences provided the first evidence that the constant portions of all Ig molecules of a particular isotype are encoded by a single gene that is separate from the genes encoding V regions. As will be discussed in Chapter 4, we now know that this surprising conclusion is correct.

The second type of determinant on antibody molecules that can be recognized as foreign by other individuals of the same species is that formed largely or entirely by the hypervariable regions of an Ig variable domain. When a homogeneous population of antibody molecules (e.g., a myeloma protein or a monoclonal antibody) is used as an immunogen, antibodies are produced that react with the hypervariable loops. These determinants are recognized as "foreign" because they are usually present in very small quantities in any given animal (i.e., at too low a level to induce self-tolerance; see Chapter 19). Such determinants on individual antibody molecules are called **idiotopes**, and all antibody molecules that share an idiotope are said to belong to the same idiotype. The term idiotype is also used to describe the collection of idiotopes expressed by an Ig molecule. As will be discussed in Chapter 4, hypervariable sequences that form idiotopes arise both from inherited germline diversity and from somatic events. Individual idiotopes that arise from somatic events are rare and may define the products of one or a few clones of antibody-producing B cells. Idiotopes that arise from the germline are less rare and, in some cases, may be present on the majority of antibody molecules that recognize a particular antigen (dominant idiotypes). Unlike allotopes, idiotopes may be functionally significant because they may be involved in regulation of B cell functions. The theory of lymphocyte regulation through antibody-binding idiotopes expressed on membrane Ig molecules, called the network hypothesis, is discussed further in Chapter 10.

In addition to experimentally elicited anti-Ig antibodies, immunologists have also been interested in naturally occurring antibodies reactive with self Ig molecules. Small quantities of anti-idiotypic antibodies may be found in normal individuals. Anti-Ig antibodies are particularly prevalent in an autoimmune disease called rheumatoid arthritis (see Chapter 20), in which setting they are known as rheumatoid factor. Rheumatoid factor is usually an IgM antibody that reacts with the constant regions of self IgG, but some rheumatoid factors cross-react with other molecules such as the cell surface proteins encoded by MHC genes (see Chapter 5). The significance of rheumatoid factor in the pathogenesis of rheumatoid arthritis is unknown.

---

sequences in their antigen-combining sites. This extraordinary diversity of structure (whose generation is explained in Chapter 4) accounts for the extraordinary specificity of antibodies for antigens, because each amino acid difference may produce a difference in antigen binding. In theory, such extensive sequence diversity poses a structural problem because the three-dimensional structure of any protein is completely determined by its amino acid sequence and certain sequences are incapable of folding into soluble, stable proteins. In the evolution of antibody molecules, this problem was avoided by confining the sequence diversity to three short stretches within the amino terminal Ig domains of the heavy and light chains. These amino terminal domains are called **variable (V) regions** to distinguish them from the more conserved **constant (C) regions** of the remainder of each chain. The three highly divergent stretches within the V regions are called **hypervariable**

regions, and they are held in place by more conserved **framework regions.** In an intact immunoglobulin, the three hypervariable regions of a light chain and the three hypervariable regions of a heavy chain can be brought together in three-dimensional space to form an antigen-binding surface. Because these sequences form a surface complementary to the three-dimensional surface of a bound antigen, the hypervariable regions are also called **complementarity-determining regions** (CDRs).

In summary, the V regions of the heavy and light chains account for antigen recognition and the C regions of the heavy chain are responsible for initiating effector functions. In other words, *Ig molecules spatially segregate recognition and effector functions.* As we shall see in Chapter 4, the immune system has evolved genetic mechanisms for recombining the portions of the Ig molecule involved in antigen recognition (i.e., V regions) with different effector function (i.e., C regions).

With this overview of antibody structure and function in mind, we will now consider antibody structure in greater detail.

Detailed View of Antibody Structure

LIGHT CHAIN STRUCTURE

All antibody light chains fall into one of two classes or isotypes, κ and λ. Each member of a light chain isotype shares complete amino acid sequence identity of the carboxy terminal C region with all other members of that isotype. In humans, antibodies with κ and λ light chains are present in about equal number. In mice, κ-containing antibodies are about 10 times more frequent than λ-containing antibodies. There are no known differences in function between κ-containing and λ-containing antibodies.

Each light chain, whether κ or λ, is folded into separate V and C domains corresponding to the amino terminal and carboxy terminal halves of the polypeptide, respectively (Fig. 3–2). Each domain is about 110 amino acids long. As noted above, most of the amino acid sequence variation among different light chains is confined to three separate locations in the V region. These three hypervariable segments, or CDRs, are each about ten amino acids long (Fig. 3–3). Proceeding from the amino terminus, these regions are the CDR1, CDR2, and CDR3, respectively. CDR3 is the most variable of the CDRs, and, as will be discussed in Chapter 4, there are more genetic mechanisms for generating sequence diversity in this region than in CDR1 and CDR2. V region folding into an Ig domain is mostly determined by the sequence of the framework regions adjacent to the CDRs. Within the framework regions, certain amino acid residues and certain structural features are very highly conserved. For example, all V region sequences contain an internal disulfide loop of about 90 amino acid residues. Other portions of the framework regions differ between κ and λ chains. When V_κ or V_λ regions fold into an Ig domain, the CDRs are positioned on the surface as projecting loops (see Fig. 3–2). Recent studies reveal that each CDR (except CDR3 of the heavy chain) folds into similarly shaped loops, regardless of the precise amino acid sequence, suggesting that there are conserved ("canonical") structural features within the hypervariable segments of antibodies. Sequence differences among the CDRs of different antibody molecules result in unique chemical structures being displayed by the projecting loops. As we shall discuss shortly, these *variations in surface structure account for specificity for binding to antigens.*

The carboxy terminus of the C region of the light chain also folds into an Ig domain. Although C_κ and C_λ differ in exact amino acid sequence, they are structurally related, or homologous, to each other and, to a lesser extent, to V_κ and V_λ.

HEAVY CHAIN STRUCTURE

All heavy chain polypeptides, regardless of antibody isotype, contain a tandem series of segments, each approximately 110 amino acid residues in length. These segments are homologous to each other, and all undergo characteristic folding into 12 kD Ig domains. As in light chains, the amino terminal variable, or V_H, domain displays the greatest sequence variation among heavy chains, and the most variable residues are concentrated into three short (up to ten amino acid residue) stretches called CDR1, CDR2, and CDR3 (Fig. 3–3). Also similar to light chains, the heavy chain CDR3 shows greater variability in sequence and folding pattern than CDR1 or CDR2.

The remainder of the heavy chain, which forms the constant (C) region, differs among isotypes; however, it is invariant among the member antibodies within a particular isotype. In IgM and IgE antibodies, the constant region folds to form four tandem Ig domains. In IgG, IgA, and IgD antibodies, the shorter constant regions form three Ig domains. (In the mouse, the δ chain gene has undergone a deletion such that the constant region forms only two Ig domains.) These domains are designated as C_H and numbered sequentially from amino terminus to carboxy terminus (i.e., C_H1, C_H2, etc). In particular isotypes (e.g., for IgG1), these regions may be designated more specifically (e.g., as Cγ1, Cγ2, etc).

Since every molecule contains at least two light and two heavy chains, every Ig molecule contains at least two antigen binding sites, each formed by a pair of V_H and V_L domains. Many Ig molecules can orient these binding sites so that two antigen molecules on a planar (e.g., cell) surface may be engaged at once. This flexibility is conferred by a **hinge region** located between C_H1 and C_H2 in certain isotypes. The hinge may contain from about ten (in α1, α2, γ1, γ2, and γ4) to over 60 (in γ3 and δ) amino acid residues. Although portions of this sequence form rodlike helical structures, other portions assume a random and flexible conformation, permitting an extensive range of molecular motion between C_H1 and C_H2. Some of the greatest differences between the constant regions of the IgG subclasses are concentrated in the hinge. As will be discussed later in this chapter, the ability to bind to two antigens at once may increase the strength of the attachment.

All heavy chains may be expressed in secreted or cell membrane–associated forms that differ in amino acid sequence on the carboxy terminal side of the last C_H domain. The secretory form, found in blood plasma, terminates with a sequence containing charged and hydrophilic amino acid residues. The membrane form, found only on the plasma membrane of the B lymphocyte that synthesized the antibody, has distinct carboxy terminal sequences that include approximately 26 uncharged, hydrophobic side chains followed by variable num-

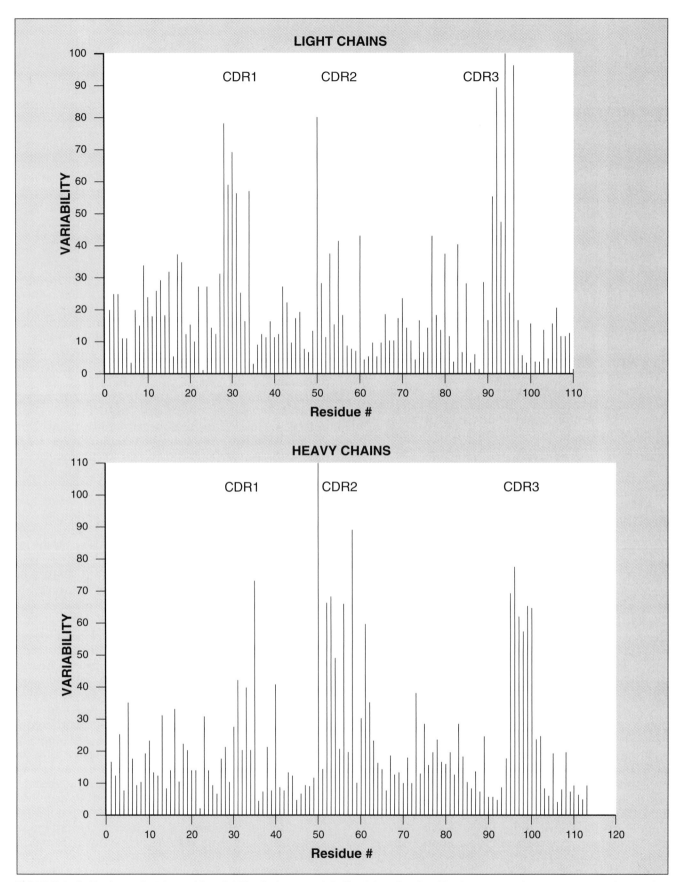

FIGURE 3–3. Regions of amino acid variability in immunoglobulin (Ig) molecules. The histograms depict the extent of variability, defined as the number of differences in each amino acid residue among various independently sequenced Ig heavy and light chains, plotted against amino acid residue number, measured from the amino terminus. This method of analysis, developed by Elvin Kabat and Tai Te Wu, indicates that the most variable residues are clustered in three "hypervariable" regions, colored in blue, corresponding to the three CDRs shown in Figure 3–2. (Courtesy of Dr. E. A. Kabat, Department of Microbiology, Columbia University College of Physicians and Surgeons, New York.)

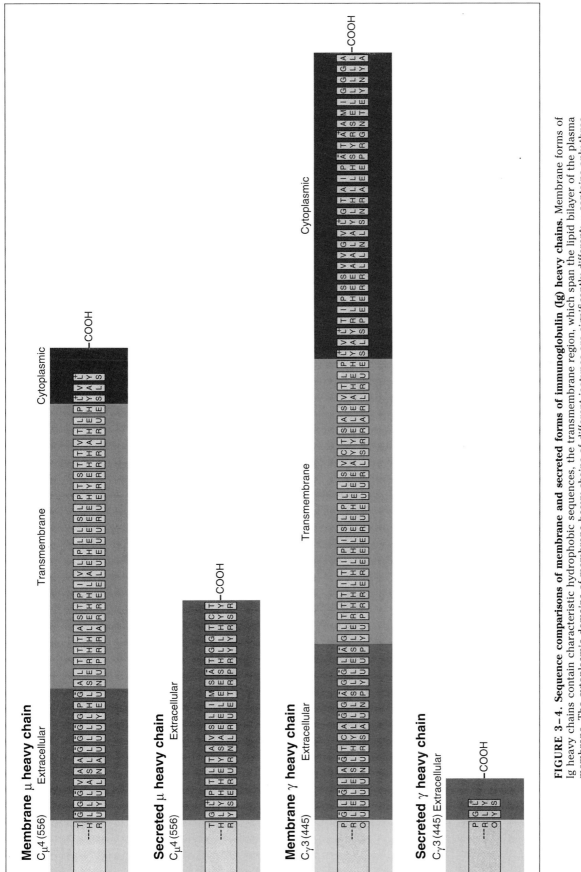

FIGURE 3-4. Sequence comparisons of membrane and secreted forms of immunoglobulin (Ig) heavy chains. Membrane forms of Ig heavy chains contain characteristic hydrophobic sequences, the transmembrane region, which span the lipid bilayer of the plasma membrane. The cytoplasmic domains of membrane heavy chains of different isotypes are significantly different: μ contains only three residues, whereas γ3 contains 28. The carboxy termini of secreted forms also differ among isotypes: μ has a long tail piece involved in pentamer formation, whereas γ3 does not. Amino acids are shown in the three-letter code, and charged residues are marked + or −; the numbers in parentheses mark the amino acid residue number of the carboxy terminus of the last Ig domain (i.e., Cμ4 or Cγ3).

bers of charged (usually basic) amino acid residues (Fig. 3–4). This structural motif is characteristic of transmembrane proteins. The hydrophobic residues are believed to form an α-helix, which extends across the hydrophobic portion of the membrane lipid bilayer; the basic side chains of the cytoplasmic amino acids interact with the phospholipid head groups on the cytoplasmic surface of the membrane. In membrane IgM or IgD, the extreme carboxy terminus or cytoplasmic portion of the heavy chain is very short, only three amino acid residues; in membrane IgG or IgE, it is somewhat longer, up to about 30 amino acid residues in length.

The secretory forms of μ, α, and δ heavy chains, but not γ or ϵ, have additional extended nonglobular sequences on the carboxy terminal side of the last C_H domain. These extensions are called **tail pieces.** In secreted IgM and IgA molecules, the tail pieces contribute toward intermolecular interactions that result in the formation of multimeric Ig molecules. Specifically, IgM forms a pentamer, containing ten heavy chains and ten light chains, and IgA can form dimers containing four heavy chains and four light chains, or trimers, containing six heavy chains and six light chains (Fig. 3–5). Little is known about the usual form of circulating IgD because it is normally present in only trace amounts in the blood. Multimeric IgM and IgA also contain an additional 15 kD polypep-

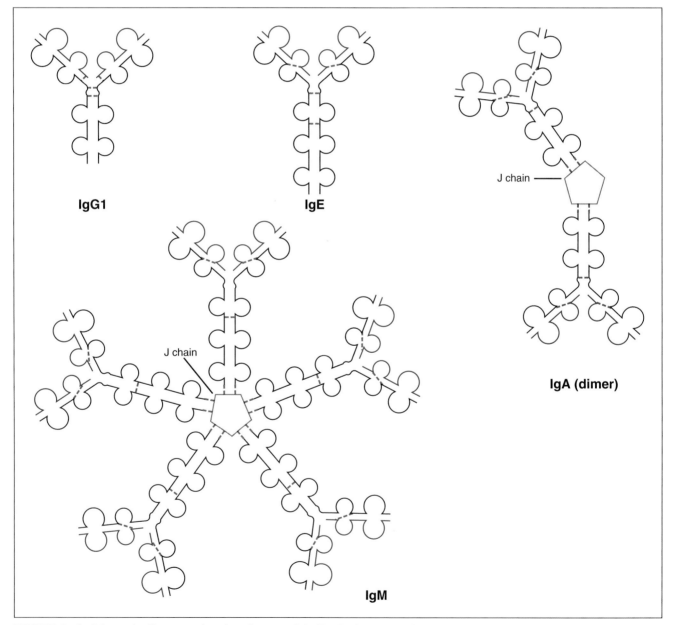

FIGURE 3–5. Schematic diagrams of various immunoglobulin (Ig) isotypes. IgG and IgE circulate as monomers, whereas secreted forms of IgA and IgM are dimers and pentamers, respectively, stabilized by the J chain. (Some IgA molecules are trimers, not shown.)

tide, encoded by a separate gene, called the **joining (J) chain,** which is disulfide-bonded to the tail pieces, stabilizing the multimer. All membrane Ig molecules, regardless of isotype, are believed to be monomeric, containing two heavy and two light chains.

All heavy chains are characteristically N-glycosylated; that is, the polypeptide contains N-linked oligosaccharide groups attached to asparagine side chains. The location of oligosaccharides may vary in different Ig isotypes. The precise composition of the oligosaccharides is not fully determined by the polypeptide sequence and may vary with the physiologic state of the host at the time of antibody synthesis.

ASSOCIATION OF LIGHT AND HEAVY CHAINS

The basic pattern of chain association in all antibody molecules is that each light chain is attached to a heavy chain and each heavy chain pairs with another heavy chain (see Plate I, preceding Section I). The association between light and heavy chains involves both covalent and non-covalent interactions (see Fig. 3–2). The covalent interactions are disulfide bonds formed between the cysteine residues in the carboxy terminus of the light chain and the C_H1 domain of the heavy chain. The exact position of the heavy chain cysteine that participates in disulfide bond formation varies with the isotype. Non-covalent interactions consist primarily of hydrophobic interactions between V_L and V_H domains and between the C_L domain and the C_H1 domain. This association of V_L and V_H domains

produces a spatial apposition such that the juxtaposed V domains can each contribute to the binding of antigen (see Plate II, preceding Section I).

The pairing of heavy chains is best understood from studies of IgG molecules. As in the case of light and heavy chain association, both covalent and non-covalent interactions are involved. Heavy chains form interchain disulfide bonds in the region near the carboxy terminus of the hinge. Extensive non-covalent interactions occur between the $C\gamma3$ domains. In contrast, there is little favorable interaction between the polypeptides of the $C\gamma2$ domains. Some of the N-linked oligosaccharides are located in a physical gap formed between these portions of the chain and may positively interact with each other, contributing to interchain associations. The length and flexibility of the hinge regions differ significantly among IgG subclasses because, as noted above, most of the amino acid sequence differences among the four subclasses are located in the hinge region. These sequence differences lead to very different overall shapes among the IgG subtypes, as depicted in Figure 3–6.

These structure/function relationships of Ig molecules were first suggested by the results of the classical limited proteolysis studies of rabbit IgG conducted by Rodney Porter and colleagues. The theory of limited proteolysis is that globular or rodlike domains of folded proteins are more resistant to the peptide-bond cleaving actions of proteolytic enzymes than are extended, flexible regions of polypeptide. In IgG molecules, the most susceptible region is therefore the hinge located between $C\gamma1$ and $C\gamma2$ of the heavy chain. The pro-

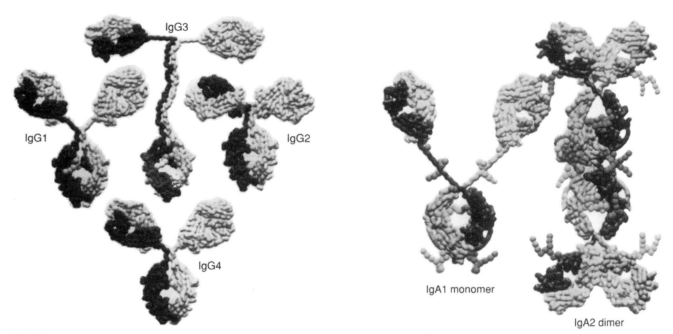

FIGURE 3–6. Three-dimensional shapes of various immunoglobulin (Ig) isotypes. These computer-generated space-filling models of different Ig isotypes illustrate that the shapes of antibody molecules are quite distinct largely owing to differences in the lengths of the hinge regions. (Courtesy of Dr. R. S. H. Pumphrey, Regional Immunology Service, St. Mary's Hospital, Manchester.)

teolytic enzyme papain preferentially cleaves rabbit IgG molecules into three separate pieces (Fig. 3–7). Two of the pieces are identical to each other and consist of an intact light chain associated with a V_H–$C\gamma1$ fragment of the heavy chain. These fragments each retain the ability to bind antigen, a function of the V_L and V_H domains, and are therefore called **Fab** (fragment, antigen-binding). The third piece contains identical fragments of the γ heavy chain composed of the $C\gamma2$ and $C\gamma3$ domains. This piece of IgG has a propensity to self-associate and to crystallize into a lattice. It is therefore called **Fc** (fragment, crystalline). Lattice formation depends upon a uniformity of structure. The propensity of Fc regions to form a lattice reflects the presence of common amino acid sequences of the $C\gamma2$ and $C\gamma3$ domains shared by all antibodies of the same subtype. As we shall discuss later in this chapter, many of the effector functions of immunoglobulins are mediated by the Fc portions of the molecule. These proteolysis experiments provided the first evidence that the anti-

gen recognition functions and the effector functions of Ig molecules are spatially segregated.

Different results are obtained when the proteolytic enzyme pepsin is used instead of papain to cleave rabbit IgG molecules (Fig. 3–7). In this case, under limiting conditions of enzyme concentrations and time, proteolysis is restricted to the carboxy terminus of the hinge region near the $C\gamma2$ domain such that the antigen-binding fragment of IgG retains the hinge and the interchain disulfide bonds. Fab fragments retaining the heavy chain hinge are called Fab'; when the interchain disulfide bonds are intact, the two Fab' fragments remain associated in a form called **F(ab')$_2$**. The Fc fragment is often extensively degraded and does not survive proteolysis by pepsin. Fab and F(ab')$_2$ are often useful as experimental tools because they can bind to antigens without activating Fc-dependent effector mechanisms.

These proteolysis experiments are not readily extended to other antibody isotypes such as IgM. In fact, they are not even applicable to all IgG

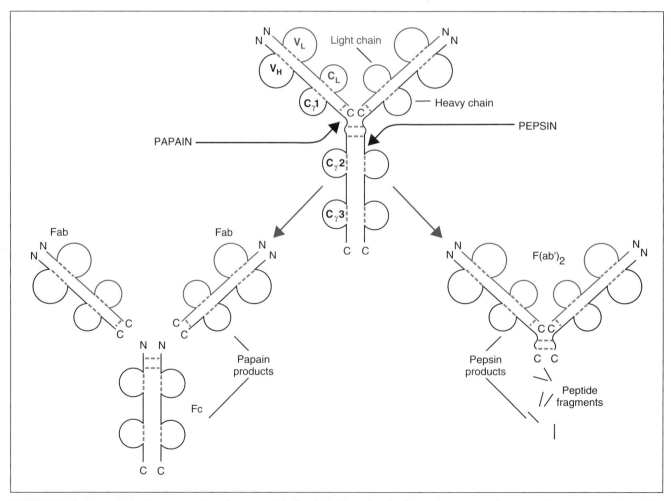

FIGURE 3–7. Proteolytic fragments of an immunoglobulin G (IgG) molecule. Sites of papain and pepsin cleavage are indicated by arrows. Papain digestion allows separation of two antigen-binding regions (the Fab fragments) from the portion of the IgG molecule that activates complement and binds to Fc receptors (the Fc fragment). Pepsin generates a single bivalent antigen-binding fragment [F(ab')$_2$] with higher avidity for antigen than the two monovalent Fab fragments produced by papain cleavage.

molecules in many species other than rabbit. However, the basic organization of the Ig molecule that Porter deduced from his studies of rabbit IgG is common to all Ig molecules of all isotypes and of all species.

ANTIBODY BINDING OF ANTIGENS

In the preceding sections, we have developed a general description of the structure of antibody molecules. Now we will turn to a more detailed discussion of the structural basis and physico-chemical characteristics of antigen binding.

Structural Aspects of Biologic Antigens

An **antigen** can be defined as any substance that may be specifically bound by an antibody molecule or T cell receptor. This differs from the original (historical) definition of antigen as a molecule that generates an antibody. We now know that almost every kind of biologic molecule, including simple intermediary metabolites, sugars, lipids, autacoids, and hormones as well as macromolecules such as complex carbohydrates, phospholipids, nucleic acids, and proteins, can serve as antigens. However, only macromolecules can initiate lymphocyte activation necessary for an antibody response. Molecules that generate immune responses are called **immunogens.** (Although technically less precise, the more inclusive term "antigens" is still commonly used to refer to "immunogens.") In order to generate antibodies specific for small molecules, immunologists commonly attach such small molecules to macromolecules before immunization. In this system, the small molecule is called a **hapten** and the macromolecule, usually a foreign protein, is called a **carrier.** The hapten-carrier complex, unlike free hapten, can act as an immunogen.

In general, macromolecules are much larger than the antigen-binding region of an antibody molecule. Therefore, an antibody binds to only a specific portion of the macromolecule, called a **determinant,** or **epitope.** These two words are synonymous and are used interchangeably throughout the book. A hapten may be thought of as an exogenous determinant that is attached to a macromolecule.

Macromolecules typically contain multiple determinants, each of which, by definition, can be bound by an antibody. The presence of multiple determinants is referred to as multivalency (or, in the case of many determinants, polyvalency). In the case of polysaccharides and nucleic acids, identical epitopes may be regularly spaced. Multimeric proteins may also be polyvalent.

Proteins commonly display different determinants that are spatially well separated, and two separate antibody molecules can be bound to the same antigen molecule without influencing each other; such determinants are said to be non-overlapping. In other cases, the first antibody bound to an antigen may sterically interfere with the binding of the second, and the determinants of the antigen are said to be overlapping. In rarer cases, binding of the first antibody may cause a conformational change in the structure of the antigen, influencing the binding of the second antibody by means other than steric hindrance. Such interactions are called allosteric effects.

In the case of phospholipids or of complex carbohydrates, the antigenic determinants are entirely a function of the covalent structure of the macromolecule. However, in the case of nucleic acids, and even more so in the case of proteins, the non-covalent folding of the macromolecule may also contribute to the formation of determinants. In proteins, epitopes formed by adjacent amino acid residues in the covalent sequence are called **linear determinants** (Fig. 3–8). It is estimated that, in a protein antigen, the size of the linear determinant that forms contacts with specific antibody is about six amino acids long. Linear determinants may be accessible to antibodies in the native folded protein if they appear on the surface or in a region of extended conformation. More often, linear determinants may be inaccessible in the native conformation and appear only when the protein is denatured. In contrast, **conformational determinants** are formed by amino acid residues from separated portions of the linear amino acid sequence that are spatially juxtaposed only upon folding (see Fig. 3–8). In theory, denatured proteins could transiently give rise to conformational determinants; however, such determinants are too short-lived unless they are maintained by energetically favorable interactions such as those found in native proteins. Thus, antibodies specific for certain linear determinants and antibodies specific for conformational determinants can be used to ascertain whether a protein is denatured or in its native conformation, respectively. When there is more than one stable conformation, antibodies may be specific for one or the other. The energy of antibody binding may actually alter the relative stability of the two conformations, shifting the dynamic equilibrium.

Proteins may be subjected to covalent modifications such as phosphorylation or specific proteolysis. These modifications, by altering the covalent structure, can produce new antigenic epitopes. Such epitopes, called **neoantigenic determinants,** may be recognized by specific antibodies (see Fig. 3–8).

At the beginning of this chapter, we made the general assertion that antibodies are not enzymes. However, certain antigen-binding sites may coincidentally resemble the active sites of enzymes and function to catalyze reactions. This is an area of intense research activity as a means of designing enzymes but is unlikely to play much of a role in normal immunity.

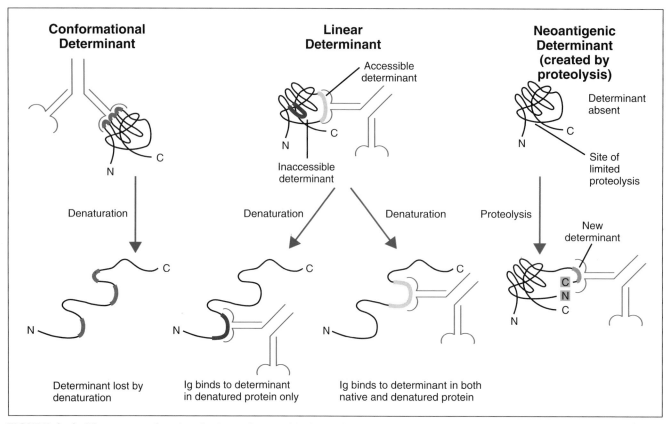

Conformational Determinant

Denaturation

Determinant lost by denaturation

Linear Determinant

Accessible determinant

Inaccessible determinant

Denaturation

Denaturation

Ig binds to determinant in denatured protein only

Ig binds to determinant in both native and denatured protein

Neoantigenic Determinant (created by proteolysis)

Determinant absent

Site of limited proteolysis

Proteolysis

New determinant

FIGURE 3–8. The nature of antigenic determinants. Antigenic determinants (shown in gray) may depend upon protein folding (conformation) as well as upon covalent structure. Some linear determinants are accessible in native proteins, whereas others are exposed only upon protein unfolding. Neodeterminants arise from covalent modifications such as peptide bond cleavage. Ig, immunoglobulin.

Structural Basis of Antigen Binding

The limited proteolysis of antibody molecules described above indicated that the antigen-binding region of antibody is contained within the Fab fragment. Several lines of evidence provided a more precise localization of this function to the hypervariable regions of V_L and V_H.

1. V_L and V_H vary among antibodies of different antigenic specificity. By recombinant deoxyribonucleic acid (DNA) technology, V_L and V_H can be combined into a single polypeptide, and such recombinant single chain antibodies display the antigen specificity of the original antibody in the complete absence of C_L and C_H domains.

2. Changes in the hypervariable regions within the V regions, either by spontaneous mutation or by specifically directed mutagenesis, can alter antigen-binding specificity. By recombinant DNA technology, it is possible to replace the hypervariable sequences of one antibody with those of a second antibody. In this case, specificity for antigen is entirely determined by the hypervariable sequences and is independent of the rest of the V region sequences.

3. Crystallographic analyses of many antibody structures reveal that the hypervariable regions form extended loops that are exposed on the surface of the antibody and are thus available to

interact with antigen. More significantly, crystallographic analyses of antigen-antibody complexes show that the amino acid residues of the hypervariable regions form extensive contact with bound antigen. The most extensive contact is with the third hypervariable region, the most variable of the three (see Plate II, preceding Section I).

The assignment of antigen-binding specificity to the hypervariable regions is what led to the alternative name for these sequences as CDRs described earlier in the chapter. *The amino acid sequences of the CDRs are primarily responsible for the specificity of antigen binding.* However, it should be noted that antigen binding is not solely a function of the CDRs. Some framework region residues also may contact the antigen. Moreover, in binding of some antigens, one or more of the CDRs may be outside the region of contact with antigen, thus not participating in antigen binding.

The original models for antigen binding, based on analogy to enzymes, proposed that antibodies contained clefts for binding antigens. Indeed, some of the earliest characterized antibody-antigen complexes involved small carbohydrate antigens, and these small molecules were bound in a cleft between V_L and V_H that were partly formed by amino acid residues of the hypervariable regions. However, more recently analyzed antibody-antigen complexes *have revealed that native protein antigens may interact with a more planar antibody-combining site.* This is critical for recognition of native proteins, as globular protein antigens are unlikely to fit into clefts. As will be discussed in Chapter 5, this is a key difference between the antigen-binding sites of antibody molecules and those of certain other antigen-binding molecules of the immune

system, namely MHC molecules, which cannot bind to native globular proteins.

The recognition of antigen by antibody involves non-covalent reversible binding. The strength of the binding between a single antibody combining site and a monovalent antigen, which can be determined by experiment (Box 3–3), is called the **affinity** of the interaction. The affinity is

---

BOX 3–3.   Measurement of Antibody-Antigen Interactions

To describe the physicochemical characteristics of antigen binding to antibody, we will first consider a simplified system consisting of an antigen that has only one determinant per molecule and a population of identical antibody molecules specific for this determinant. When this antigen and antibody are mixed in solution, antigen-antibody complexes constantly form and then spontaneously dissociate. After a period of time, the rate of complex formation will exactly equal the rate of complex dissociation and a state of **dynamic equilibrium** will have been reached. If antibody is present at a much lower concentration than antigen, the proportion of antibody molecules that have bound antigen at equilibrium is determined by two factors: the concentration of antigen molecules and the strength of the binding interaction. (The strength of the binding interaction is also influenced by other factors such as temperature and solvent conditions, but to simplify the analysis, we will hold these constant.) Under such conditions, the concentration of antigen that allows one half of the antibodies to be in complex with antigen and leaves one half free is a measure of the strength, or **affinity,** of the binding interaction. This concentration of antigen, measured in molarity, is called the **dissociation constant** (K_d) of the interaction. *A smaller K_d means a greater affinity* (i.e., a lower concentration of antigen is needed to reach half maximal occupancy). Affinity may also be represented by the reciprocal of the dissociation constant (i.e., $1/K_d$), which is called the **association constant** (K_a); since K_a is the reciprocal of K_d, a larger K_a signifies greater affinity. It is important to note that affinity depends on both the antibody and the antigen. A given antibody molecule can have different affinities for different related antigens.

The K_d of antigen binding can be measured directly for small antigens (e.g., haptens) by means of equilibrium dialysis (see Figure). In this method, a solution of antibody is confined within a "semipermeable" membrane of porous cellulose and is immersed in a solution containing the antigen.

(Semipermeable in this context means that small molecules, like antigen, can pass freely through the membrane pores but that macromolecules, like antibody, cannot.) If no antibody were present within the membrane, the antigen in the bathing solution would enter the membrane-bound compartment until the concentration of antigen within the membrane-bound compartment became exactly the same as that outside. Another way to view the system is that, at dynamic equilibrium, antigen enters and leaves the membrane-bound compartment at exactly the same rate. However, when antibody is present inside the membrane, the net amount of antigen inside the membrane at equilibrium increases by the quantity that is bound to antibody. This occurs because only unbound antigen can diffuse across the membrane, and, at equilibrium, it is the unbound concentration of antigen that must be identical inside and outside the membrane. The extent of the increase in antigen inside the membrane depends on the antigen concentration, on the antibody concentration, and on the K_d of the binding interaction. By measuring the antigen and antibody concentrations, by spectroscopy or by other means, the K_d can be calculated.

An alternative way to determine the K_d is by measuring the rates of antigen-antibody complex formation and dissociation. These rates depend on the concentrations of antibody and antigen, on the affinity of the interaction, and on certain geometric parameters that equally influence the rate in both directions. All parameters except the concentrations can be summarized as rate constants, and both the **on rate constant** (k_{on}) and the **off rate constant** (k_{off}) can be calculated experimentally by determining the concentrations and the actual rates of association or dissociation, respectively. The ratio of k_{off}/k_{on} allows one to cancel out all of the parameters not related to affinity and is exactly equal to the dissociation constant K_d. Thus, one can measure K_d at equilibrium by equilibrium dialysis or calculate K_d from rate constants measured under non-equilibrium conditions.

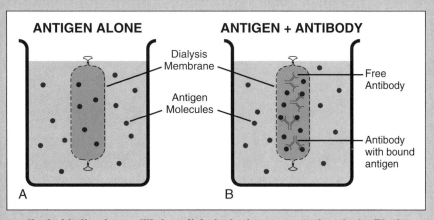

Analysis of antigen-antibody binding by equilibrium dialysis. In the presence of antibody (B), the amount of antigen within the dialysis membrane is increased compared with the absence of antibody (A). As described in the text, this difference, caused by antibody binding of antigen, can be used to measure the affinity of the antibody for the antigen. This experiment can only be performed when the antigen is a small molecule (e.g., a hapten) capable of freely crossing the dialysis membrane.

commonly represented by a dissociation constant (K_d), which describes the concentration of antigen that is required to occupy one half of the antibody-combining sites present in a solution of antibody molecules. A smaller K_d indicates a stronger or higher affinity interaction, since a lower concentration of antigen is needed to occupy the sites. For antibodies specific for natural antigens, the K_d usually varies from about 10^{-7} M to 10^{-10} M. In a natural serum, there will be a mixture of such antibodies with different affinities for the specific antigen, depending primarily upon the precise amino acid sequences of the CDRs. (The average affinity of the antibody molecules in a population will increase with repeated immunization, a phenomenon called **affinity maturation;** this is discussed in Chapter 4.)

Natural antigens, as opposed to small haptens, usually contain more than one determinant and therefore bind more than one antibody molecule. Moreover, some multivalent antigens will have more than one copy of a particular determinant. (Such multivalency may arise because a macromolecule may be multimeric or may have an internal repeating structure. A cell surface will also be multivalent by virtue of having multiple copies of a particular surface antigen.) Unless inhibited by steric constraints, a single antibody may attach to a single multivalent antigen by more than one binding site. For IgG or IgE, this attachment can involve, at most, two binding sites because there are only two combining regions per antibody molecule, one on each Fab. For IgM, however, a single antibody may bind at up to ten different sites! Although the affinity of any one site will be unchanged, the overall strength of attachment must take into account binding at all of the sites. This overall strength of attachment is called the **avidity** and will be much stronger than the affinity of any given site. Mathematically, the strength of the avidity increases almost geometrically (rather than additively) for each occupied site. Thus, a low-affinity IgM molecule can still bind very tightly to a multivalent antigen because many low-affinity interactions can produce a single high-avidity interaction.

Multivalent interactions between antigen and antibody are of biologic significance. If a multivalent antigen is mixed with a specific antibody in a test tube, the two will associate to form **immune complexes** (Fig. 3–9). At the correct concentrations, called a "zone of equivalence," antibody and antigen form an extensively cross-linked network of non-covalently attached molecules such that most or all of the antigen and antibody molecules are complexed into large masses. Immune complexes can be dissociated into smaller aggregates either by increasing the concentration of antigen so that free antigen molecules will displace cross-linked antigen from antibody-combining sites ("zone of antigen excess") or by increasing antibody so that free antibody molecules will displace cross-linked antibody from antigen determinants ("zone of antibody excess"). If a "zone of equivalence" is reached *in vivo,* large immune complexes can form in the circulation. Immune complexes (small or large) that are trapped or formed in tissue can initiate an inflammatory reaction resulting in "immune complex diseases" (see Chapter 20).

FUNCTIONS OF ANTIBODIES

The effector function of an antibody is triggered by binding of antigen. Once antigen is bound, different consequences may ensue, depend-

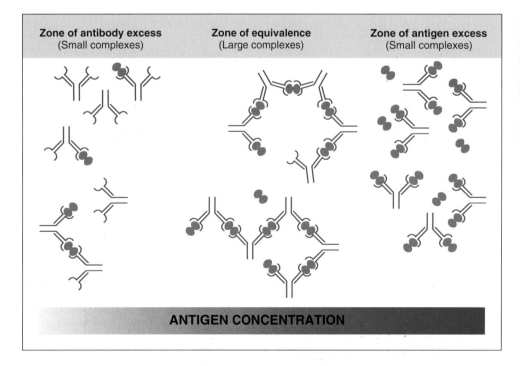

| Zone of antibody excess (Small complexes) | Zone of equivalence (Large complexes) | Zone of antigen excess (Small complexes) |

ANTIGEN CONCENTRATION

FIGURE 3–9. Sizes of antigen-antibody (immune) complexes as a function of relative concentrations of antigen and antibody. Large complexes are formed at concentrations of multivalent antigens and antibodies that are termed the "zone of equivalence"; the complexes are smaller in relative antigen or antibody excess.

ing on the structure, anatomic location, and isotype of the antibody. The following sections consider the various biologic effects of antibody binding to antigen.

Membrane Antibody as the B Cell Antigen Receptor

The binding of antigen to specific B cells is the signal that initiates proliferation and secretion of antibody by the B cells. Recognition of antigen is provided by the membrane forms of antibody expressed on the B cell surface. At different points in the history of a B cell, different heavy chain isotypes may be expressed. For example, immature B cells may express IgM, previously unstimulated mature B cells may express IgM and IgD, and previously stimulated memory B cells may express any isotype or subtype. As noted earlier, membrane Ig differs structurally from secreted Ig of the same isotype in that membrane Ig contains an extra hydrophobic sequence of about 26 amino acids near the carboxy terminus (see Fig. 3–4). This sequence spans the hydrophobic region of the plasma membrane lipid bilayer so that the extreme carboxy terminal amino acids are located in the cytoplasm. Much evidence suggests that cross-linking of cell surface antibody by polyvalent antigen may be an important signal that contributes to B cell activation. As will be discussed in Chapter 9, membrane Ig molecules associate with two other proteins (called Igα and Igβ) that are involved in signal generation.

Neutralization of Antigen by Secreted Antibody

The initial discovery of antibodies arose by analyzing the humoral factors that protect immunized hosts against microbial toxins. Many injurious agents, such as toxins, drugs, viruses, bacteria, and other parasites, initiate cell injury by binding to specific cell surface receptors. Secreted antibodies can sterically hinder this interaction by binding to antigenic determinants on the agent (or, less commonly, on the cell receptor), thereby **neutralizing** the toxic or infectious process. This action may be mediated by antibodies of any isotype and experimentally can also be mediated by Fab or F(ab')$_2$ fragments.

Isotype-Specific Functions of Antibodies

Many functions of antibody molecules are mediated by their Fc portions and are, therefore, specific for particular isotypes or subtypes. As will be seen in Chapter 4, B cells may undergo **heavy chain isotype switching**, allowing the progeny of a single B cell to express different isotypes of Ig molecules without changing the antigen-binding regions. As a consequence, the same antibody specificity for antigen can be utilized to activate different effector functions. Although the V regions

of antibodies are responsible for the spec and diversity of humoral immunity, the availability of multiple C regions in different heavy chain isotypes provides an additional measure of adaptability. The production of various heavy chain isotypes serves to direct the humoral immune response along different functional and anatomic pathways, involving diverse interactions with the body's mechanisms of innate immunity and inflammation.

ACTIVATION OF COMPLEMENT BY IgG AND IgM

The **complement system** consists of a family of serum proteins that can be activated by a proteolytic cascade to generate effector molecules. The name complement refers to a heat-labile serum component that was needed to "complement" the function of heat-stable antibody in order to produce lysis of certain target cells. The complement system mediates many of the cytolytic and inflammatory effects of humoral immunity (Chapter 15). Here we note that certain human isotypes of antibody have the ability to activate the complement system through the **classical complement pathway**. This response is triggered when a complement protein called C1q binds to the Cγ2 region of IgG1 or IgG3 or the Cμ3 region of IgM antibodies that are involved in immune complexes or bound to a cell surface. (In mice, the IgG subtypes that activate complement are IgG2a and IgG2b.)

OPSONIZATION BY IgG FOR ENHANCED PHAGOCYTOSIS

Both mononuclear phagocytes and granulocytes have the ability to ingest particulate matter as a prelude to intracellular killing and degradation. The ingestion process of particulate matter, called **phagocytosis,** involves attachment of surface membrane to the foreign material and then "zipping up" of membrane around it. The efficiency of the process is markedly improved if the membrane of the phagocytic cell can attach itself with specificity to the object undergoing phagocytosis (Fig. 3–10).

Both mononuclear phagocytes and neutrophils express receptors for the Fc portions of IgG molecules. In fact, at least three distinct types of Fcγ receptors are expressed, each with different affinities for different IgG subtypes and distinct functions (Box 3–4). It is interesting that all three classes of IgG Fc receptors on leukocytes (written FcγR) contain Ig-like domains and are thus members of the Ig superfamily. When IgG molecules bind to and coat antigenic particles, a process called **opsonization,** the bound IgG is recognized by the FcγR molecules on the leukocyte, serving to enhance the efficiency of phagocytosis. Both high- and low-affinity FcγRs contribute to phagocytosis, and the IgG subtypes that bind best to these receptors (IgG1 and IgG3) are most efficient for promoting phagocytosis.

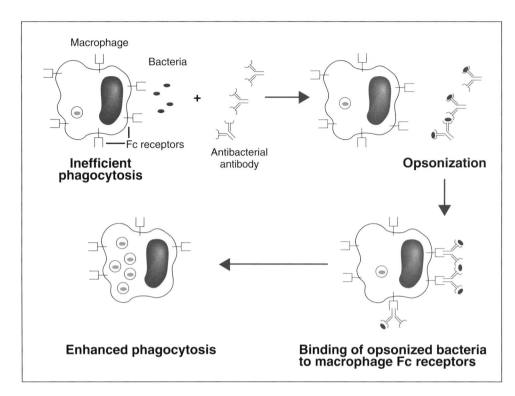

Macrophage

Bacteria

Fc receptors

Inefficient phagocytosis

Antibacterial antibody

Opsonization

Enhanced phagocytosis

Binding of opsonized bacteria to macrophage Fc receptors

FIGURE 3–10. Antibody-dependent opsonization and phagocytosis of bacteria. Antibody binding to particles such as bacteria can markedly enhance the efficiency of phagocytosis. The enhancement of phagocytosis in macrophages involves both increased attachment to the cell surface and activation of the phagocyte, both mediated through occupancy of $Fc\gamma$ receptors.

A fragment derived from the third component of complement (C3b) can also opsonize particles for phagocytosis through binding to a leukocyte receptor for C3b. Since C3b can be generated and attached to cells as a consequence of the classical pathway of complement activation, IgG or IgM binding can indirectly lead to opsonization and enhanced phagocytosis.

Opsonization may lead to more than merely increasing the binding of the particulate matter to phagocytes. Specifically, the high-affinity $Fc\gamma$Rs also appear to be involved in metabolic activation

---

BOX 3–4 Leukocyte Receptors for Immunoglobulins

Cell surface receptors for the Fc portions of Ig molecules fall into two general classes: specific and non-specific. Non-specific receptors include the lectin-like molecules CD23 and galectin 3 that bind IgE; the function of these molecules is not known and they will not be discussed further. Specific receptors include the transporters of Ig across epithelium, such as secretory component (also called the polyimmunoglobulin receptor) and the "neonatal" Fc receptor (FcRN) that are described in the text. Here we will discuss the specific Fc receptors of leukocytes that mediate functional responses, namely $Fc\gamma$RI, $Fc\gamma$RII, $Fc\gamma$RIII, $Fc\epsilon$RI, and $Fc\alpha$R. All of these receptors have a single Fc-binding polypeptide chain, which is folded into two or three Ig motifs. Many of these receptors contain additional polypeptide chains involved in signal transduction.

The Ig-binding subunit of $Fc\gamma$RI (CD64) is a 40 kD protein that binds human IgG1 and IgG3 with a sufficiently high affinity (K_d of 10^{-8} to 10^{-9} M) to allow occupancy by monomeric IgG at physiologic concentrations of antibody. Interactions with other IgG subtypes (i.e., IgG2 and IgG4) are much weaker. (In mice, high-affinity interactions involve IgG2a and IgG2b). The large extracellular amino terminal region of the receptor polypeptide folds into three tandem Ig-like domains. Expression of $Fc\gamma$RI is dependent upon association of the Ig-binding subunit with a disulfide-linked homodimeric signaling complex, initially identified as the γ chains of a different Fc receptor, namely $Fc\epsilon$RI (see below); the polypeptide is now commonly called the FcR γ chain. FcR γ chain is a 7 kD

polypeptide with a short extracellular amino terminus, a transmembrane sequence, and a large intracellular carboxy terminus. As we will discuss in Chapter 7, the FcR γ chain is structurally homologous to the signaling components (ζ chains) of the T cell receptor (TCR) complex. Both the FcR γ chain and the TCR ζ chain contain specific signaling regions, called immunoreceptor tyrosine activation motifs (ITAMs), that couple ligand-induced receptor clustering to non-receptor protein tyrosine kinases. The $Fc\gamma$RI complex is expressed on activated macrophages but not resting monocytes; transcription of the $Fc\gamma$RI gene is regulated by the macrophage-activating cytokine interferon-γ (IFN-γ).

Both humans and mice express two different classes of low-affinity receptors of IgG, called $Fc\gamma$RII and $Fc\gamma$RIII. $Fc\gamma$RII (CD32) is a 30 kD polypeptide whose large extracellular amino terminus folds into two tandem Ig domains. $Fc\gamma$RII binds certain human IgG subtypes (IgG1 and IgG3) with sufficiently low affinity (K_d of 10^{-6} M) that monomeric IgG molecules are unable to occupy this receptor at physiologic antibody concentrations. Consequently, $Fc\gamma$RII binding of Ig is largely restricted to immune complexes or to IgG-coated (opsonized) microbes or to cells displaying aggregates of Fc regions. In humans there are three different genes encoding $Fc\gamma$RII molecules (called A, B, and C), and each gene encodes proteins with structural variations arising from messenger ribonucleic acid processing (e.g., alternative splicing). $Fc\gamma$RIIA is expressed by phagocytic cells (neutrophils and

Continued

mononuclear phagocytes) and probably participates in the process of phagocytosis. FcγRIIA is also expressed by certain vascular endothelia and other cell types where its function is unknown, but may play a role in the binding of molecules other than Ig, such as lipoproteins. The carboxy terminus of FcγRIIA and C contain ITAMs and, upon clustering by IgG1- and or IgG3-coated particles or cells, can deliver an activation signal to phagocytes. FcγRIIB is expressed exclusively by lymphocytes, especially B cells. The intracellular carboxy terminus of this isoform contains a short amino acid sequence that can deliver inhibitory signals, counteracting the positive signals delivered to B cells by the clustering of membrane Ig. Thus, FcγRIIB clustering, induced by occupancy with polyvalent IgG1 or IgG3, may shut off immunoglobulin synthesis (see Chapter 9). Knockout mice lacking FcγRII on their B cells may develop autoimmunity (see Chapter 20).

FcγRIII (CD16) is similar to FcγRII in size, structure, and affinity for IgG. Two separate human genes code for this molecule. The FcγRIIIA isoform, the product of the A gene, is a conventional transmembrane protein expressed largely on NK cells. This form of the molecule associates, like FcγRI, with homodimers of the FcR γ chain, or with heterodimers composed of the FcR γ chain and the ζ chain of the TCR complex. This association is necessary for cell surface expression and, through the ITAMs in these signaling chains, can deliver intracellular activating signals. Receptor clustering induced by occupancy with IgG1 or IgG3 bound to a target cell surface activates NK cells to cause target cell death (see Chapter 13), thereby mediating antibody-dependent cell-mediated cytotoxicity. The FcRγIIIB isoform, a product of the B gene, is a phosphatidylinositol (PI)–linked protein expressed largely on neutrophils. It may participate in phagocytosis or in neutrophil activation. Clustering of PI-linked proteins has been observed to cause activation of

certain non-receptor protein tyrosine kinases in a variety of cells, but the precise downstream signals activated by clustering of FcRγIIIB are not known.

The IgE-binding subunit of FcεRI is a 25 kD protein whose extracellular amino terminus folds into two tandem Ig-like domains. The protein is the product of a single gene and is of such high affinity (K_d of 10^{-10} M) that it readily binds monomeric IgE molecules even at very low physiologic concentrations of this antibody isotype. The full receptor complex involves, in addition to the IgE-binding α chain, a 27 kD β chain and two FcR γ chains. The β chain spans the plasma membrane four times; both its amino terminus and carboxy terminus are intracellular, and its carboxy terminus contains an ITAM. FcεRI is constitutively expressed on mast cells and basophils, where, in association with bound IgE, it serves as an antigen receptor for initiation of antigen-specific activation. FcεRI can be induced on eosinophils by interleukin-5; once induced, it can mediate IgE-directed ADCC (see Chapter 14).

FcαR (CD89), the most recently described member of the FcR family, is encoded by a single gene in humans. The IgA binding polypeptide is a 30 kD core protein which may be heavily glycosylated. The IgA binding chain of FcαR forms two external tandem Ig domains and is coexpressed with two FcR γ signaling chains. FcαR is expressed on neutrophils, monocytes, and eosinophils, where it may mediate IgA-directed ADCC.

As noted above, FcγRI, FcγRIIIA, FcεRI, and FcαR each depend on association with a homodimeric (or heterodimeric) complex involving the FcR γ chain for expression and signaling. Knockout of this polypeptide in mice results in loss of expression of all of these molecules. Such animals are impaired in their ability to mount inflammatory reactions in response to immune complexes, establishing the role of Fc receptors as important mediators of leukocyte function.

Human Leukocyte Receptors for Immunoglobulin Fc Regions				
Family	**CD**	**Affinity for Ig**	**Cell distribution**	**Function**
FcγRI	CD64	High ($K_d = 10^9$ M)	Activated phagocytes	Cell activation, phagocytosis
FcγRIIA	CD32	Low ($K_d > 10^{-7}$ M)	Phagocytes	Cell activation, phagocytosis
FcγRIIB	CD32	Low ($K_d > 10^{-7}$ M)	B cells	Antibody feedback
FcγRIIIA	CD16	Low ($K_d > 10^{-7}$ M)	NK cells	Cell activation, ADCC
FcγRIIIB	CD16	Low ($K_d > 10^{-7}$ M)	Neutrophils	Phagocytosis
FcεRI		High ($K_d = 10^{-10}$ M)	Mast cells, basophils, eosinophils, Langerhans cells	Cell activation, ADCC
FcαR	CD89	Low ($K_d = 10^{-6}$ M)	Eosinophils, monocytes, neutrophils	Cell activation, ADCC

of phagocytes, increasing the efficiency of the subsequent intracellular degradation of ingested particles.

ANTIBODY-DEPENDENT CELL-MEDIATED CYTOTOXICITY TARGETED BY IgG, IgE, AND IgA

Several different leukocyte populations other than cytolytic T lymphocytes, including neutrophils, eosinophils, mononuclear phagocytes, and especially NK cells, are capable of lysing various target cell types. In many cases, killing of target cells requires that the target cell be precoated with specific IgG, and the lytic process is called

antibody-dependent cell-mediated cytotoxicity (ADCC) (Fig. 3–11A). Recognition of bound antibody occurs through a low-affinity receptor for Fcγ on the leukocyte, called FcγRIII or CD16. In the case of NK cells, the predominant cellular mediators of ADCC, it is now appreciated that IgG serves two distinct functions. First, it provides a recognition function (i.e., those target cells that have bound IgG will be preferentially killed compared with those not displaying IgG). Second, the occupancy of FcγRIII serves to activate the NK cell to synthesize and secrete cytokines such as tumor necrosis factor and interferon-γ as well as to discharge their granules. The released cytokines me-

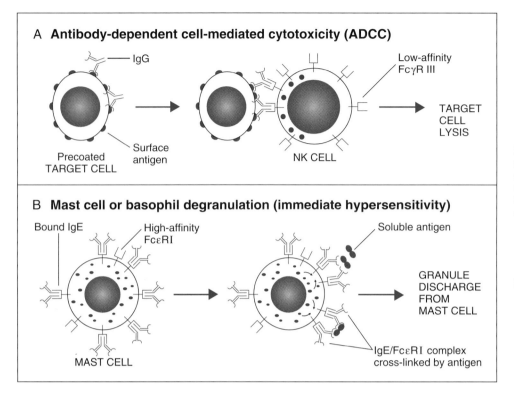

A Antibody-dependent cell-mediated cytotoxicity (ADCC)

IgG

Surface
antigen

Precoated
TARGET CELL

Low-affinity
FcγR III

NK CELL

TARGET
CELL
LYSIS

B Mast cell or basophil degranulation (immediate hypersensitivity)

Bound IgE

High-affinity
FcεRI

MAST CELL

Soluble antigen

GRANULE
DISCHARGE
FROM
MAST CELL

IgE/FcεRI complex
cross-linked by antigen

FIGURE 3–11. Functions of antibodies in natural killer (NK) cell–mediated cytolysis (A) and in degranulation of mast cells (B).
A. NK cells express low-affinity Fcγ receptors that recognize clustered immunoglobulin (Ig)G molecules prebound to antigens on the surface of a target cell.
B. Monomeric IgE antibodies attach to high-affinity Fcε receptors on mast cells or basophils in the absence of antigen; subsequent exposure to antigen causes cross-linking of the IgE/FcεRI complex, leading to mast cell or basophil degranulation.

diate inflammatory functions, and the released granule proteins mediate many of the cytolytic functions of this cell type (see Chapters 12 and 13). Since FcγRIII is a low-affinity receptor and more efficiently binds aggregated IgG than monomeric IgG, monomeric IgG in plasma neither activates NK cells nor competes effectively with cell-bound IgG for recognition. *Thus, ADCC occurs only when the target cell is precoated with antibody.*

Eosinophils mediate a special type of ADCC directed against parasites such as helminths. Helminths are relatively resistant to lysis by neutrophils and mononuclear phagocytes, but they can be killed by a basic protein present in the granules of eosinophils. In this case, IgE rather than IgG serves as the principal isotype that provides recognition and effector cell activation because eosinophils express high-affinity Fc receptors for IgE antibodies, called FcεRI. Eosinophils may also use IgA to direct ADCC through Fcα receptors (Box 3–4).

IMMEDIATE HYPERSENSITIVITY TRIGGERED BY IgE

Mast cells and basophils express high-affinity receptors for the Fc portion of IgE molecules (Box 3–4 and Chapter 14). Because of their high affinity, FcεRI receptors are occupied by IgE monomer in the absence of antigen, a key difference from FcγRIII involved in ADCC (Fig. 3–11B). The introduction of specific antigen causes aggregation of the bound IgE molecules and their receptors. This clustering, in turn, causes the mast cell or basophil to release inflammatory and vasoactive mediators (e.g., histamine) from preformed storage granules

and to synthesize and secrete lipid-derived mediators (e.g., leukotrienes, prostaglandins, and platelet-activating factor) and cytokines *de novo*. The consequence of the release of these mediators is a vascular and inflammatory response called **immediate hypersensitivity,** which is discussed in Chapter 14.

MUCOSAL IMMUNITY MEDIATED BY IgA

Although IgA is a relatively minor component of systemic humoral immunity, it plays a key role in mucosal immunity. This is because IgA alone of the various isotypes can be selectively transported across mucosal barriers into the lumens of mucosa-lined organs (see Chapter 11). More IgA is synthesized by a normal individual than any other isotype, but because synthesis occurs mainly in mucosal lymphoid tissues and transport into the mucosal lumen is so efficient, IgA constitutes less than one quarter of the antibody in plasma. Epithelial cells of organs such as the intestine express specific Fc receptors for dimeric IgA molecules. This IgA receptor is commonly called the **poly-Ig receptor** or **secretory component.** Initially, secretory component binds IgA on the basal surface of the epithelial cell that is facing the blood. Bound IgA is passaged through the cell to the mucosal surface by vesicular transport. Surprisingly, IgA is not simply released from its receptor. Rather, secretory component itself is specifically cleaved, leaving a bound secretory component peptide (called **secretory piece**) attached to the dimeric IgA molecule. Once in mucosal secretions, IgA functions to neutralize injurious agents and to direct ADCC by effector cells bearing FcαR (see Chapter 11).

NEONATAL IMMUNITY MEDIATED BY MATERNAL IgG

Neonatal mammals often lack the ability to mount effective immune responses against microbes. However, maternally produced antibodies can provide a protective action. Maternal IgG is transported across the placenta and enters the fetal circulation. In addition, maternal IgA is secreted into breast milk and ingested by the infant, so that it can neutralize pathogenic organisms that attempt to colonize the infant's gut. Maternal IgG is also present in breast milk. IgG is specifically taken up from the gut lumen into the blood of the neonate, the opposite direction of IgA secretion. A distinct receptor for IgG has been identified that mediates this transport function. Interestingly, the gut receptor for IgG (called FcR neonatal or FcRN) is unique among Fc receptors in that it structurally resembles a class I MHC molecule containing a transmembrane heavy chain that is non-covalently associated with β_2 microglobulin (see Chapter 5).

Knockout mice lacking β_2 microglobulin fail to express both conventional class I MHC molecules and the FcRN. As expected, such animals cannot take up maternal IgG and neonatal animals have depressed IgG levels. Surprisingly, adult animals also have depressed IgG levels, apparently due to a shortened half-life of IgG in the blood stream. Thus, it seems that FcRN has a second role: protecting IgG from degradation by cells. The mechanism of this effect is not known.

FEEDBACK INHIBITION OF IMMUNE RESPONSES MEDIATED BY IgG

Various lymphocyte populations express Fc receptors for different Ig isotypes. These receptors are believed to modulate lymphocyte function independent of antigenic specificity. The best example of this phenomenon is the binding of aggregated IgG or IgG-containing antibody-antigen complexes to a low-affinity FcγR expressed on B cells (FcγRIIB), which may inhibit activation of these B cells, a process called **antibody feedback** (see Chapter 9).

LABORATORY USES OF ANTIBODIES

This portion of the chapter describes how antibodies may be used as tools in research and in clinical diagnosis. Historically, many of the uses of antibody depended upon the ability of antibody and specific antigen to form large immune complexes. Immunochemists could detect antigen by observing the formation of such complexes in solution by light scattering. In addition, specific antigens could be purified from solutions containing mixtures of molecules by collecting the specific immune complexes by centrifugation or by precipitation of antibody with chemical agents. The presence of antigen could be detected by allowing antibody-antigen precipitates to form in gels. These methods were of great importance in early studies, but now have been almost entirely replaced by simpler methods based on immobilized antibodies or antigens. In the remainder of this chapter, we describe four common applications of antibodies in widespread current use.

Quantitation of Antigen

Immunologic methods of quantifying antigen concentration provide exquisite sensitivity and specificity and have become standard techniques for both research and clinical applications. All modern immunochemical methods of quantitation are based upon having a simple and accurate method for measuring the quantity of indicator molecules that bind to a solid phase surface. When the indicator molecule is labeled with a radioisotope, as first introduced by Rosalyn Yalow and colleagues, it may be quantified by counting radioactive decay events in a scintillation counter; the assay is called a **radioimmunoassay** (RIA). When the indicator molecule is covalently coupled to an enzyme, it may be quantified by determining with a spectrophotometer the initial rate at which the enzyme converts a clear substrate to a colored product; the assay is called an **enzyme-linked immunosorbent assay** (ELISA). Several variations of RIA and ELISA exist for quantifying antigen in a test solution, but the most commonly used and most sensitive version is the sandwich assay (Fig. 3-12).

The sandwich assay uses two different antibodies reactive with the antigen whose concentration is to be measured. A fixed quantity of antibody one is attached to a series of replicate solid supports, such as plastic microtiter wells. Test solutions containing antigen at unknown concentrations or a series of standard solutions with known concentrations of antigen are added to the wells and allowed to bind. Unbound antigen is removed, and antibody two, which is enzyme-linked or radiolabeled, is allowed to bind. The antigen serves as a bridge, so the more antigen in the test or standard solutions, the more enzyme-linked or radiolabeled second antibody will bind. The results from the standard solutions are used to construct a binding curve for second antibody as a function of antigen concentration, from which the quantities of antigen in the test solutions may be inferred. When this test is performed with two monoclonal antibodies, it is essential that these antibodies see non-overlapping determinants on the antigen; otherwise, the second antibody cannot bind.

In an important clinical variant of immunobinding assays, patient samples may be tested for the presence of antibodies to a specific microbial antigen (e.g., antibodies reactive with proteins from human immunodeficiency virus or hepatitis B virus) as indicators of infection. In this case, a

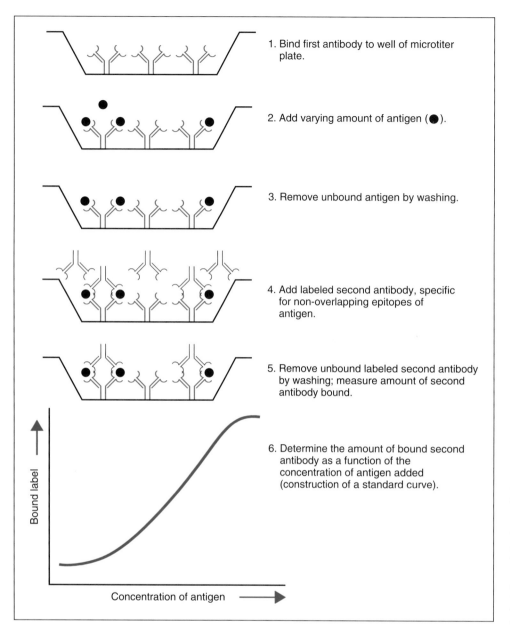

1. Bind first antibody to well of microtiter plate.

2. Add varying amount of antigen (●).

3. Remove unbound antigen by washing.

4. Add labeled second antibody, specific for non-overlapping epitopes of antigen.

5. Remove unbound labeled second antibody by washing; measure amount of second antibody bound.

6. Determine the amount of bound second antibody as a function of the concentration of antigen added (construction of a standard curve).

Bound label

Concentration of antigen

FIGURE 3–12. Sandwich enzyme-linked immunosorbent assay or radioimmunoassay. A fixed amount of one immobilized antibody is used to capture an antigen. The binding of a second, labeled antibody, which recognizes a nonoverlapping determinant on the antigen, will increase as the concentration of antigen increases, allowing quantification of antigen.

saturating quantity of antigen is added to replicate wells containing plate-bound antibody from a non-human source (or antigen is attached directly to the plate) and serial dilutions of the patient's serum are added to replicate wells. After washing to remove unbound antibody, the presence of patient antibody bound to the immobilized antigen is determined by use of an enzyme-linked or radiolabeled second anti-human Ig antibody. The results are often reported as the greatest dilution of serum at which specific antibody could still be detected.

Identification and Characterization of Protein Antigens

The two major approaches used by immunochemists to identify and characterize protein antigens are immunoprecipitation and Western blotting.

IMMUNOPRECIPITATION

An antibody directed against one protein antigen in a mixture of proteins is used to isolate the specific antigen from the mixture (Fig. 3–13). In most modern procedures, the antibody is attached to a solid phase particle (e.g., an agarose bead) either by direct chemical coupling or indirectly. Indirect coupling may be achieved by means of an attached "second antibody" such as rabbit anti-mouse Ig antibody or by means of some other protein with specific affinity for the Fc portion of Ig molecules, such as protein A or protein G from staphylococcal bacteria. After the antibody-coated beads are incubated with the solution of antigen,

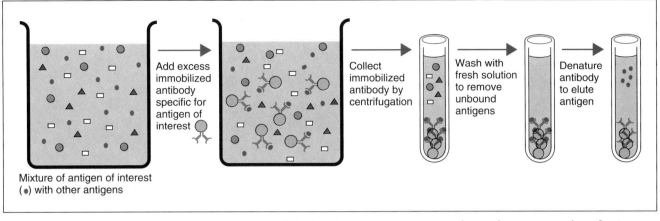

FIGURE 3–13. Isolation of an antigen by immunoprecipitation. Immunoprecipitation can be used as a means of purification, as a means of quantification, or as a means of identification of an antigen. Antigens purified by immunoprecipitation are often analyzed by sodium dodecyl sulfate–polyacrylamide gel electrophoresis.

unbound molecules are separated from the bead-antibody-antigen complex by washing. Specific antigen is then released (eluted) from the antibody by changing pH or by other solvent conditions that reduce the affinity of binding. If the beads are packed into a column, large quantities of antigen can be purified by this procedure, called affinity chromatography. (Recall that affinity chromatography is also used as a method for purifying antibody molecules reactive with bound antigen; see p. 41.) The purified antigen can then be analyzed by conventional protein chemical techniques. Alternatively, a small amount of radiolabeled protein can be purified, and the characteristics of the macromolecule can be inferred from the behavior of the radioactive label in analytical separation techniques such as sodium dodecyl sulfate–polyacrylamide gel electrophoresis (SDS-PAGE) or isoelectric focusing.

WESTERN BLOTTING

If the protein antigen to be characterized is in a mixture with other antigens, it may be first subjected to analytical separation, typically by SDS-PAGE, so that the positions of different proteins in the gel are a function of their molecular sizes. The array of separated proteins is then transferred from the separating polyacrylamide gel to a support membrane by capillary action (blotting) or by electrophoresis, such that the membrane acquires a replica of the array of separated macromolecules present in the gel (Fig. 3–14). SDS is displaced from the protein during the transfer process, and native antigenic determinants are often regained as the protein refolds. The position of the antigen on the membrane can then be detected by binding of labeled antibody, thus providing information about antigen size and quantity. The same technique can be used to determine if a protein has undergone covalent modification, e.g., whether it has been phosphorylated on tyrosine residues by probing

the membrane with antibody reactive with phosphotyrosine. Both enzyme-linked and radiolabeled antibodies are commonly used, most recently with enzymes that generate chemiluminescent signals. The sensitivity and specificity of this technique can be increased by starting with immunoprecipitated proteins instead of total protein mixtures, such as cell lysates. This sequential technique is especially useful for detecting protein-protein interactions (e.g., by immunoprecipitating with antibody directed against one member of a putative complex and staining the blot with antibody to a second protein that may be co-immunoprecipitated).

The technique of transferring proteins from a gel to a membrane is called Western blotting as a biochemist's joke. Southern is the last name of the scientist who first blotted DNA from a separating gel to a membrane, a technique since called Southern blotting. By analogy, "Northern blotting" was applied to the technique of transferring ribonucleic acid from a gel to a membrane, and "Western blotting" was applied to protein transfer. A more detailed description of both Southern and Northern blotting will be presented in Box 4–2, Chapter 4. Since most Western blots are analyzed by binding of antibodies, the term immunoblot is sometimes used to describe the procedure.

Cell Surface Labeling and Separation

Antibodies are commonly used to characterize, identify, and separate cell populations. In these methods, the antibody can be radiolabeled, enzyme-linked, or, most commonly, fluorescently labeled. In the case of fluorescent labels, the amount of bound antibody on every individual cell in a population is measured by passing suspended cells one at a time through a fluorimeter. This ability to analyze individual cells more than compensates for the fact that fluorescence intensity is always an arbitrary quantity and, unlike radioactivity or ab-

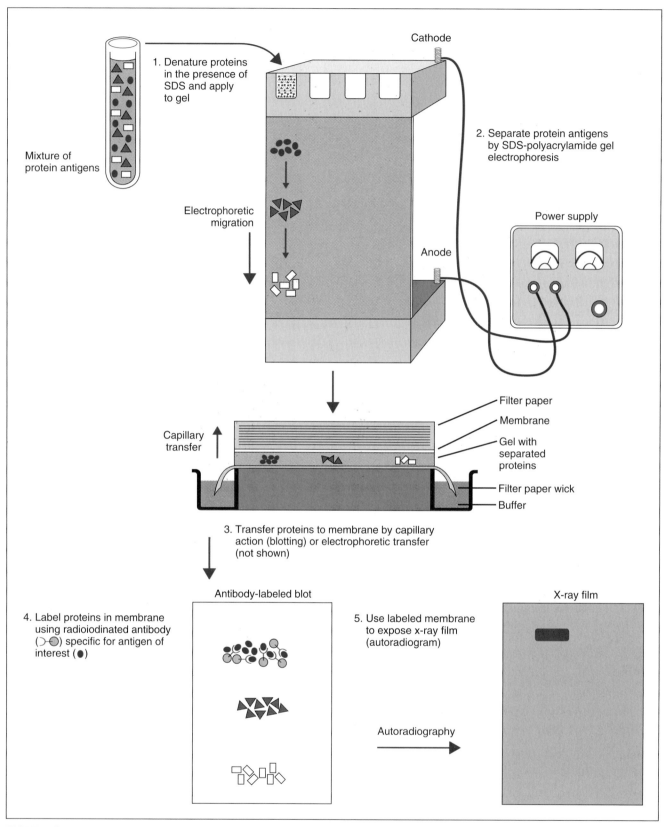

FIGURE 3–14. Characterization of antigens by Western blotting. Protein antigens, separated by sodium dodecyl sulfate (SDS)–polyacrylamide gel electrophoresis and transferred to a membrane, can be labeled with radioactive or (not shown) enzyme-coupled antibodies. Analysis of an antigen by Western blotting provides information similar to that obtained from immunoprecipitation followed by polyacrylamide gel electrophoresis. Some antibodies work only in one technique or the other.

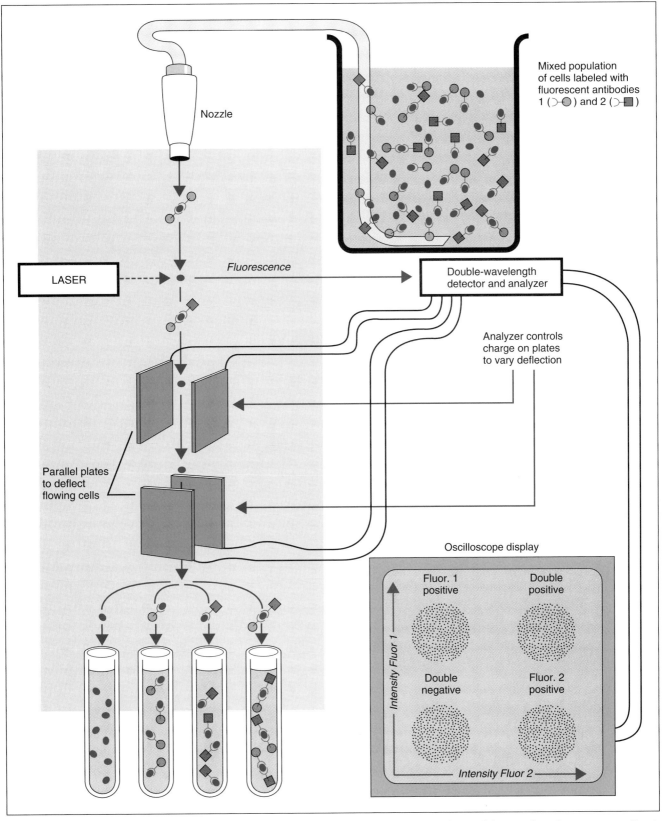

FIGURE 3–15. The principle of fluorescence-activated cell sorting. The separation depicted here is based upon two antigenic markers ("two-color sorting"). Modern instruments can routinely separate cell populations based upon three or more markers.

sorbance, cannot be related to an absolute number of molecules except by comparison with known standards. The flowing cells can also be differentially deflected by electromagnetic fields whose strength and direction are varied according to the measured intensity of the fluorescence signal, thereby allowing one to separate cell populations according to surface antibody binding (Fig. 3–15). The instrument that performs this task is called a *fluorescence-activated cell sorter* (FACS). Modern FACS instruments routinely allow simultaneous detection of three or more different fluorescent signals, each attached to a different antibody, permitting analysis or separation of cells according to any specified combination of surface antibody-binding patterns. A more rapid but less rigorous separation can be accomplished by allowing cells to attach to antibodies bound to plates ("panning") or to magnetic beads.

Localization of Antigen Within Tissues or Cells

Antibodies can be used to identify the anatomic distribution of an antigen within a tissue or within compartments of a cell. The common principle behind these techniques is that a label is attached to the specific antibody and that the position of the label in the tissue or cell, determined with a suitable microscope, is used to infer the position of the antigen. In the earliest version of this method, called **immunofluorescence,** the antibody was labeled with a fluorescent moiety and allowed to bind to a monolayer of cells or to a frozen section of a tissue. The stained cells or tissues were examined with a fluorescence microscope to locate the antibody. Although extremely sensitive, the fluorescence microscope is not an ideal tool for identifying the normal unstained structures of the cell or tissue. Thus, it is often difficult to interpret precisely where the label has been localized. This problem may be overcome by use of a confocal fluorescence microscope, which in addition to providing higher resolution, allows the fluorescent image to be superimposed on a phase contrast image. Alternatively, antibodies may be coupled to enzymes that convert colorless substrates to colored insoluble substances that precipitate at the position of the enzyme. A conventional light microscope may then be used to localize the antibody in a stained cell or tissue. The most common variant of this method utilizes the enzyme horseradish peroxidase, and the method is commonly referred to as the **immunoperoxidase technique.** Examples of this method have been shown in Chapter 2. Another commonly used enzyme is alkaline phosphatase. Different antibodies, coupled to different enzymes, may be used in conjunction to produce simultaneous "two-color" localizations of different antigens. In other variations, antibody can be coupled to an electron-dense probe, such as colloidal gold, and the loca-

tion of antibody can be determined subcellularly by means of an electron microscope, a technique called **immunoelectron microscopy.** Different sized gold particles have been used for simultaneous localization of different antigens at the ultrastructural level.

In all immunomicroscopic methods, signals may be enhanced by using "sandwich" techniques. For example, instead of attaching horseradish peroxidase to a specific mouse antibody directed against the antigen of interest, it can be attached to a second antibody (e.g., rabbit anti-mouse Ig antibody) that is used to bind to the first, unlabeled antibody. When the label is attached directly to the specific, primary antibody, the method is referred to as **direct;** when the label is attached to a secondary or even tertiary antibody, the method is **indirect.** In some cases, molecules other than antibody can be used in indirect methods. For example, staphylococcal protein A, which binds to IgG, or avidin, which binds to primary antibodies labeled with biotin, can be coupled to enzymes.

SUMMARY

Antibodies, or immunoglobulins, are a family of structurally related glycoproteins produced by B lymphocytes that function as the mediators of specific humoral immunity. All antibodies have a common core structure of two identical light chains and two identical heavy chains. Each chain consists of multiple independently folded domains of about 110 amino acids containing conserved intrachain disulfide bonds. The N-terminal domains of heavy and light chains form the variable regions of antibody molecules, which differ among antibodies of different specificities. The variable (V) regions of heavy and light chains each contain three separate hypervariable regions (also called complementarity-determining regions) of about ten amino acids that are located on one planar surface of the folded V region to form the antigen-combining site of the antibody molecule. Light chains contain one constant (C) region domain, and heavy chains contain three or four, which are similar in antibodies of the same class (isotype) and subclass but differ among antibodies of different classes and subclasses. Most of the effector functions of antibodies are mediated by the C regions of the heavy chains, but these functions are triggered by binding of antigens to the spatially distant combining site in the variable region.

Macromolecular antigens contain multiple epitopes, or determinants, each of which may be recognized by an antibody. The affinity of the interaction between the combining site of a single antibody molecule and a single epitope is measured as a dissociation constant (K_d). Macromolecules may contain multiple identical epitopes to which identical antibody molecules can bind. When such multivalent antigens react with antibodies, a zone of equivalence may be reached, where the relative concentrations of antigen and antibody favor the formation of large immune complexes. In zones of antigen excess or antibody excess, these extensively cross-linked aggregates dissociate into smaller complexes.

Antibodies are produced in membrane-associated and secreted forms. Membrane Ig, on the surface of a B cell, is the B cell receptor for antigen. Secreted antibody molecules neutralize antigens, activate the complement system, and opsonize antigens and enhance their phagocytosis by neutrophils and macrophages, which express receptors for the Fc portion of certain IgG isotypes. In addition, different antibody isotypes bind to Fc receptors on eosinophils, mast cells, and NK cells and stimulate the functions of these cells as a consequence of antigen binding. Other Fc receptors on epithelial cells mediate transepithelial transport of IgA and IgG antibodies, respectively.

Antibodies are also invaluable tools in the laboratory. RIA and ELISA are used to quantify antigens in solution. Immunoprecipitation and Western blotting are used for the purification and structural analysis of protein antigens. Fluorescence-activated cell sorting is used to identify, characterize, and purify cell populations that express particular surface antigens. Immunomicroscopic techniques such as immunofluorescence and immunoperoxidase are used to identify and localize antigens in cells and tissues.

Selected Readings

Alzari, P. M., M. Lascombe, and R. J. Poljak. Three-dimensional structure of antibodies. Annual Review of Immunology 6: 555–580, 1988.

Amit, A. G., R. A. Mariuzza, S. E. V. Phillips, and R. J. Poljak. Three-dimensional structure of an antigen-antibody complex at 2.8 A resolution. Science 233:747–753, 1986.

Burton, D. R. Structure and function of antibodies. *In* F. Calabi and M. S. Neuberger (eds.). Molecular Genetics of Immunoglobulins. Elsevier Science Publishers, Amsterdam, 1987.

Burton, D. R., and J. M. Woof. Human antibody effector function. Advances in Immunology 51:1–84, 1992.

Davies, D. R., and E. A. Padlan. Antibody-antigen complexes. Annual Review of Biochemistry 59:439–473, 1990.

Johnstone, A., and R. Thorpe. Immunochemistry in Practice, 2nd ed. Blackwell Scientific Publishers, Oxford, 1987.

Kohler, G., and C. Milstein. Continuous cultures of fused cells secreting antibody of predefined specificity. Nature 256: 495–497, 1975.

Porter, R. R. The hydrolysis of rabbit γ-globulin and antibodies by crystalline papain. Biochemical Journal 73:119, 1959.

Ravetch, J. V., and J.-P. Kinet. Fc receptors. Annual Review of Immunology 9:457–492, 1991.

Ravetch, J. V. Fc receptors: rubor redux. Cell 78:553–560, 1994.

Wu, T. T., and E. A. Kabat. An analysis of the sequences of the variable regions of Bence Jones proteins and myeloma light chains and their implications for antibody complementarity. Journal of Experimental Medicine 132:211–250, 1970.

CHAPTER FOUR

MATURATION OF B LYMPHOCYTES

AND EXPRESSION OF

IMMUNOGLOBULIN GENES

Antibodies, or immunoglobulins (Igs), are synthesized exclusively by B lymphocytes. Therefore, the humoral immune response to a foreign antigen reflects the types of Ig produced by the B cells that are stimulated by that antigen. At different stages of their maturation, B cells provide both recognition and effector functions in the humoral immune response. Membrane Ig-expressing B cells specifically recognize antigens. Following antigenic stimulation, they differentiate into effector cells that secrete Ig. Elucidation of the mechanisms of antibody synthesis has been among the major advances in modern immunology and remains one of the foundations for understanding the generation of repertoire diversity and the differentiation of lymphocytes. This chapter describes the molecular genetic basis of humoral immune responses, with particular reference to the organization and expression of Ig genes. The cellular interactions and extrinsic stimuli that lead to B cell proliferation and differentiation, including the functions of helper T cells in specific antibody responses, will be discussed in Chapter 9.

GENERAL FEATURES OF B LYMPHOCYTE MATURATION

The first analyses of specific antibody responses to foreign antigens were initiated in the early 1900s and focused on the types of antibodies produced in a person or experimental animal exposed to bacteria or microbial toxins. The total population of antibody specificities that an individual can produce is called the **antibody repertoire** and is a reflection of all the B cell clones capable of Ig synthesis and secretion in response to antigenic stimulation. During its life, each B lymphocyte and its clonal progeny go through a series of well-defined maturational or differentiative stages, each of which has a characteristic pattern of Ig production. Attempts to understand the development of B lymphocytes ultimately led to the isolation of Ig genes and to analyses of their expression and regulation. Our current concepts of B cell ontogeny are based on analyses of tumor lines, fetal and neonatal B cells, and bone marrow cultures in which B cell progenitors develop to maturity. More recently, a variety of gene knockout mice have illustrated some of the molecular requirements for normal B cell maturation *in vivo*.

Diversity of the Antibody Repertoire

The **primary antibody repertoire** consists of all the antibodies that an individual can produce in response to the first (primary) immunization with different antigens. It is determined by the number of B cell clones (estimated to be $>10^9$ in each individual) that exist prior to immunization and express membrane Ig molecules with distinct specificities for antigens. As the clonal selection hypothesis predicted, lymphocytes specific for different antigens develop before exposure to these antigens. It follows, therefore, that *the information needed to generate the enormously diverse repertoire of antibodies is present in the deoxyribonucleic acid (DNA) of each individual.* If, however, each Ig heavy and light chain were produced by an individual gene, more than one third of the genome that can code for functional proteins would be required to generate 10^9 antibody specificities. This is clearly not the case, and as we shall see later in this chapter, each Ig heavy chain and light chain polypeptide is not encoded by a discrete DNA sequence in the germline. Instead, B lymphocytes have developed remarkably effective genetic mechanisms for generating a large and highly diverse repertoire from a pool of Ig genes that is of more limited size.

Stages of B Lymphocyte Maturation

All B lymphocytes arise in the bone marrow from a stem cell that does not produce Ig but is committed to the B cell lineage (Fig. 4–1). The earliest cell type that synthesizes a detectable Ig gene product contains cytoplasmic μ heavy chains composed of variable (V) and constant (C) regions. This cell is called the **pre-B lymphocyte** and is found only in hematopoietic tissues, such as the bone marrow and fetal liver. It does not express functional, fully assembled membrane IgM, since surface expression requires synthesis of both heavy and light chains. Therefore, pre–B cells cannot recognize or respond to antigen. Some of the μ heavy chains in pre–B cells associate with proteins called **surrogate light chains,** which are structurally homologous to κ and λ light chains but are invariant (i.e., they do not have V regions and are identical in all B cells). Complexes of μ and surrogate light chains may be expressed on the cell surface at low levels and are postulated to play a role in stimulating the proliferation of immature B cells, the subsequent production of κ or λ light chains, and the continued maturation of B cells. The importance of surrogate light chains is underscored by studies of mice deficient in these proteins. For instance, knockout of the gene encoding one of the surrogate light chains, called λ5, or disruption of the μ gene by homologous recombination (Box 4–1), results in markedly reduced development of mature B cells. This result suggests that the μ-λ5 complex delivers signals required for B cell maturation, and if either gene is disrupted, maturation does not proceed. It is not known what the μ-λ5 complex recognizes, or what biochemical signals it delivers. In developing T lymphocytes, a similar role is played by a protein called pre-Tα, which associates with one chain of the T cell antigen receptor before the second chain is expressed (see Chapter 8).

At the next identifiable stage in B cell maturation, κ or λ light chains are also produced. These complex with μ heavy chains, and then the assem-

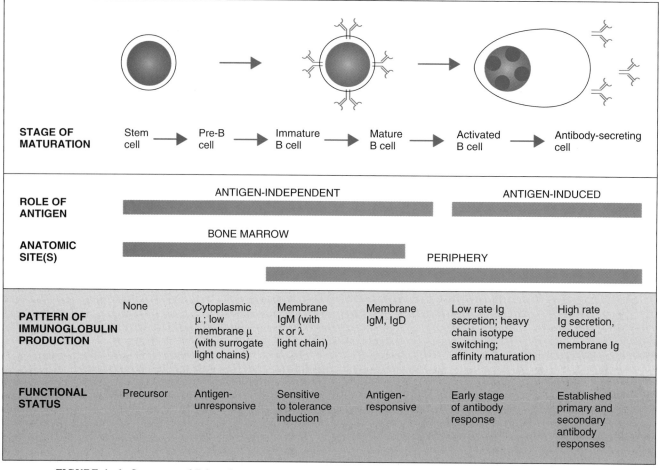

FIGURE 4–1. Sequence of B lymphocyte maturation and antigen-induced differentiation. Ig, immunoglobulin.

bled IgM molecules are expressed on the cell surface, where they function as specific receptors for antigens. The expression of surface Ig requires its association with other proteins, called Igα and Igβ, which also function in the transduction of signals generated by antigen binding to membrane Ig (see Chapter 9). IgM-bearing B cells that are recently derived from marrow precursors are called **immature B lymphocytes** because they do not proliferate and differentiate in response to antigens. In fact, their encounter with antigens, such as self antigens in the marrow, may lead to cell death or functional unresponsiveness (tolerance) rather than activation. This property is important for maintaining tolerance to self antigens that are present in the bone marrow (see Chapter 19). Once a B cell expresses a complete heavy or light chain, it cannot produce another heavy or light chain containing a different V region.

The stimuli that drive maturation of pre–B cells to immature and then mature B cells are incompletely defined. The role of the μ-λ5 complex was mentioned above. Interleukin (IL)-7, a cytokine produced by bone marrow stromal cells, is a growth factor for pre–B cells. Adhesive interactions between developing B cells and marrow stromal cells are also important in the maturation

process. A number of biochemical intermediates and transcription factors that are important in B cell maturation are being defined by gene knockouts. For instance, knockout of a transcription factor called E2A leads to severely impaired B cell development. An immunodeficiency disease called X-linked agammaglobulinemia, in which cells fail to develop beyond the pre-B stage (see Chapter 21), is due to a mutation in a tyrosine kinase of the src family, called B cell tyrosine kinase (Btk), suggesting that this enzyme plays a key role in B cell maturation.

Having acquired the ability to produce complete Ig molecules and, therefore, a specificity, B cells migrate out of the bone marrow and enter the peripheral circulation and lymphoid tissues. They continue to mature, still without antigenic stimulation. **Mature B cells** coexpress μ and δ heavy chains in association with the original κ or λ light chain and, therefore, produce both membrane IgM and IgD. Both classes of membrane Ig have the same V region and hence the same antigen specificity. Such cells are responsive to antigens. It is possible that some B cells may acquire functional responsiveness to antigenic stimulation without expressing IgD. It is believed that unless these mature B lymphocytes encounter antigen, they die in

BOX 4–1. Transgenic Mice and Targeted Gene Knockouts

Two important methods for studying the functional effects of specific gene products *in vivo* are the creation of transgenic mice, which overexpress a particular gene in a defined tissue(s), and gene knockout mice, in which a targeted disruption is used to ablate the function of a particular gene. Both techniques have been widely employed to analyze many biologic phenomena. As we shall see throughout this book, transgenic and knockout mice are providing valuable information about the development of the immune system and the functions of a variety of proteins in immune responses.

To create transgenic mice, foreign DNA sequences, called transgenes, are introduced into the pronuclei of fertilized mouse eggs, and the eggs are implanted into the oviducts of pseudopregnant females. Usually, if a few hundred copies of a gene are injected into pronuclei, about 25 per cent of the mice that are born are transgenic. One to 50 copies of the transgene insert in tandem into a random site of breakage in a chromosome and are subsequently inherited as a simple mendelian trait. Because integration usually occurs before DNA replication, most (about 75 per cent) of the transgenic pups carry the transgene in all their cells, including germ cells. In most cases, integration of the foreign DNA does not disrupt normal function. Also, each founder mouse carrying the transgene is a heterozygote, from which homozygous lines can be bred.

The great value of transgenic technology is that it can be used to express genes in particular tissues by attaching coding sequences of the gene to regulatory sequences that normally drive the expression of genes selectively in that tissue. For instance, lymphoid promoters and enhancers can be used to overexpress genes, such as rearranged antigen receptor genes, in lymphocytes, and the insulin promoter can be used to express genes in the β cells of pancreatic islets. Examples of the utility of these methods for studying the immune system will be mentioned in many chapters of this book. Transgenes can also be expressed under the control of inducible promoters, such as the tetracycline-responsive promoter, which responds to the antibiotic. In these cases, transcription of the transgene can be controlled at will by administering the inducing agent.

A powerful method for developing animal models of single gene disorders, and the most definitive way of establishing the obligatory function of a gene *in vivo*, is the creation of knockout mice by a targeted mutation or disruption of the gene. This technique relies on the phenomenon of homologous recombination. If an exogenous gene is inserted into a cell (e.g., by electroporation), it can integrate randomly into the cell's genome. However, if the gene contains sequences that are homologous to an endogenous gene, it will preferentially recombine with and replace endogenous sequences. To select for cells that have undergone homologous recombination, a drug-based selection strategy is employed. The fragment of homologous DNA to be inserted into a cell is placed in a vector typically containing a neomycin resistance gene and a viral thymidine kinase (TK) gene (see Figure). This "targeting vector" is constructed in such a way that the neomycin resistance gene is always inserted, but the TK gene is lost only if homologous recombination occurs. The vector is introduced into cells, and the cells are grown in neomycin and in ganciclovir, a drug that is metabolized by TK to generate a lethal product. Cells in which the gene is

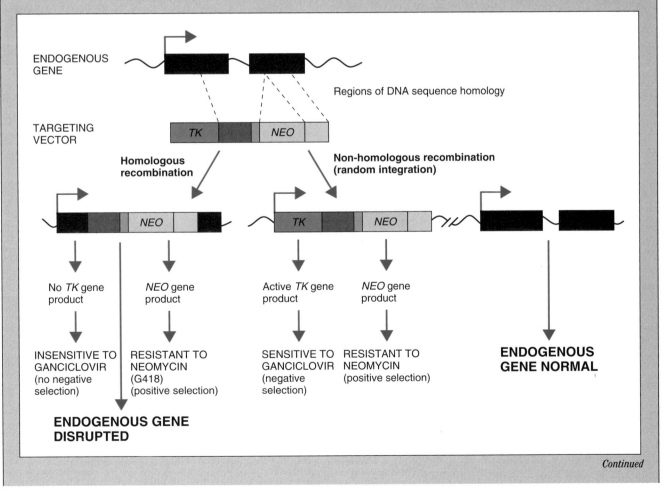

Continued

integrated randomly will be resistant to neomycin but will be killed by ganciclovir, whereas cells in which homologous recombination has occurred will be resistant to both drugs because the TK gene will not be incorporated. This "positive-negative selection" ensures that the inserted gene in surviving cells has undergone homologous recombination with endogenous sequences. The presence of the inserted DNA in the middle of an endogenous gene usually disrupts the coding sequences and ablates expression and/or function of that gene. In addition, targeting vectors can be designed such that homologous recombination will lead to deletion of one or more exons of the endogenous gene.

To generate a mouse carrying a targeted gene disruption or mutation, a targeting vector is used to first disrupt the gene in a murine embryonic stem (ES) cell line. ES cells are pluripotent cells, derived from mouse embryos, that can be propagated and induced to differentiate in culture or can be incorporated into a mouse blastocyst, which may be implanted in a pseudopregnant mother and carried to term. Importantly, the progeny of the ES cells develop normally into mature tissues, which will express exogenous genes that have been transfected into the ES cells. Thus, the targeting vector designed to disrupt a particular gene is inserted into ES cells, and colonies in which homologous recombination has occurred (on one chromosome) are selected with drugs, as described above. The presence of the desired recombination is verified by analysis of DNA, using techniques such as Southern blot hybridization or the polymerase chain reaction (PCR). The selected ES cells are injected into blastocysts, which are implanted into pseudopregnant females. Mice that develop will be chimeric for a heterozygous disruption or mutation (i.e., some of the tissues will be derived from the ES cells and others from the remainder of the normal blastocyst). Usually, the germ cells are also chimeric, but since these are haploid, only some will contain the chromosome copy with the disrupted (mutated) gene. If chimeric mice are mated with normal (wild-type) animals, and either sperm or eggs containing the chromosome with the mutation fuse with the wild-type partner, all cells in the offspring derived from such a zygote will be heterozygous for the mutation (so-called germline transmission). Such heterozygous mice can be mated to yield animals that will be homozygous for the mutation with a frequency that is predictable based on simple mendelian segregation. Such "knockout mice" are deficient in the expression of the targeted gene.

Knockout mice have opened new approaches for analyzing the functions of a wide variety of molecules. They have also led to an appreciation that many important gene products are "redundant," in that lack of expression of one gene may be compensated for by other gene products.

a few days or weeks. It is estimated that an adult mouse produces 5×10^7 B cells daily from progenitors in the bone marrow. Of these, only 1 to 2×10^6 survive and enter the circulation and peripheral lymphoid tissues, serving to maintain the peripheral pool of $\sim 5 \times 10^8$ mature B cells.

Once the mature B cells are stimulated by antigens (and other signals that will be described in Chapter 9), they are called **activated B lymphocytes.** Activated B cells proliferate and differentiate, producing an increasing proportion of their Ig in a secreted form and progressively less in a membrane-bound form. Some of the progeny of activated B cells undergo **heavy chain class (isotype) switching** and begin to express Ig heavy chain classes other than μ and δ (e.g., γ, α, or ϵ). Other activated B lymphocytes may not secrete antibody but instead persist as membrane Ig-expressing **memory cells.** Memory cells survive for weeks or months and actively recirculate among the blood, lymph, and lymphoid organs. The stimulation of memory B cells by antigens leads to secondary antibody responses. Memory B cells generally express Ig molecules whose affinities for antigens are higher than those of their unstimulated clonal precursors. Antigen-induced differentiation of naive or memory B lymphocytes culminates in the development of **antibody-secreting cells,** some of which can be morphologically identified as plasma cells. In the blood or lymphoid tissues of normal individuals, the majority of B cells are IgM$^+$ or IgM$^+$IgD$^+$.

Throughout their lives, the members of each B cell clone express the same V region and maintain essentially the same antigen specificity. However, subtle changes in antibody V regions do occur during responses to antigens. In particular, the average affinity of the secreted antibody and of membrane Ig on antigen-specific B cells increases after antigenic stimulation and is substantially higher in secondary than in primary antibody responses. This is called **affinity maturation** and is a property of humoral immune responses to protein antigens. Affinity maturation arises from mutations in the DNA coding for Ig V regions in individual antigen-stimulated B lymphocytes, followed by a selection process that ensures survival of the cells that recognize the antigen with high affinity. The mechanisms and consequences of somatic mutations in Ig genes are discussed later in this chapter.

Two other features of Ig production by B cells are noteworthy. First, each B cell clone and its progeny are specific for only one antigenic determinant. It is, therefore, necessary for each B cell to express only one set of Ig heavy and light chain V genes throughout its life, even though all heterozygous individuals inherit two sets of Ig genes, one from each parent. A single specificity is maintained because only one of the two parental alleles of Ig is expressed by each B cell clone and its progeny. This phenomenon is called **allelic exclusion** and is characteristic of both B and T lymphocyte receptors for antigens. Second, each B cell clone produces either a κ or a λ light chain but not both, and although heavy chain class switching occurs following activation, switches from one light chain class to the other do not occur throughout the life of each clone. This is called **light chain isotype exclusion.**

In addition to the antigen receptor, B lymphocytes express a variety of cell surface proteins at different stages of maturation (Table 4–1). These have been identified by numerous techniques, the most useful of which is the production of monoclonal antibodies specific for B cells. These surface

TABLE 4-1. Selected Surface Molecules of B lymphocytes

Surface Molecule	Function/Significance	Stage of B Cell Maturation
Immunoglobulin (Ig)	Antigen receptor	Immature, mature, and activated B cells
Class II MHC molecules	Role in helper T cell–B cell interactions	Immature, mature, and activated B cells
CD40	Role in helper T cell–B cell interactions	Mature and activated B cells
CD5 (Ly-1)	Marker for distinct subset of B cells	Distinct subset; mature cells
CD9	Marker for acute leukemias	Precursors, pre–B cells; some plasma cells
CD10 (CALLA)	Marker for acute leukemias	Precursors and pre–B cells
CD19 (B4)	? Role in B cell activation	All B cells
CD20 (B1)	? Role in B cell activation	Pre-B, immature, mature, and activated B cells
B7-1 (CD80), B7-2 (CD86) (also present on macrophages and dendritic cells)	Costimulator for T cell activation; bind to CD28 and CTLA-4 on T cells	Mature and activated B cells
CD45R (B220)	Form of leukocyte common antigen	All B cells
Complement receptor type 2 (CR2, CD21)	Enhancement of antibody responses Receptor for Epstein-Barr virus	Mature and activated B cells
Fc receptor, FcγRIIB (receptor for IgG)	Negative feedback control of B cell activation	Mature and activated B cells
Low-affinity FcεRII (CD23)	Unknown; induced by IL-4	Activated B cells

Abbreviations: MHC, major histocompatibility complex; CALLA, common acute lymphoblastic leukemia antigen; IL-4, interleukin-4.

proteins on B cells are important for several reasons. Some are lineage- and maturation-specific phenotypic markers for B cells. Various B cell surface molecules may function in the activation or regulation of these cells; these are described in Chapter 9. Antibodies against B cell–specific surface proteins can also be used to localize and treat tumors derived from B lymphocytes (see Chapter 18). One small subpopulation of B cells expresses a marker called CD5 (Ly-1), which was originally identified on a subset of T lymphocytes. Only 5 to 10 per cent of B cells in the blood and lymphoid organs are CD5+, and these may express a quite limited repertoire of V genes. Surprisingly, virtually all B cell–derived chronic lymphocytic leukemias are CD5+. Moreover, CD5+ B cells spontaneously secrete IgM antibodies that often react with self antigens, and these cells may be significantly expanded in autoimmune diseases. CD5+ B cells do not develop in the bone marrow, and in mice large numbers of these cells are found as a self-renewing population in the peritoneum. It is, therefore, likely that this small subset of B lymphocytes has unique properties in terms of ontogeny, function, and role in disease.

In summary, analyses of the antibody repertoire and B lymphocyte maturation and differentiation have led to the following general conclusions about the expression of Ig genes.

1. Each clone of B cells produces one set of Ig heavy chain and light chain V regions, and different clones produce Igs with different V regions.

2. During the life of each clone, the same heavy chain V region is expressed in association with membrane or secreted forms of C regions and with C regions of different isotypes. In contrast, the light chain C region remains the same.

3. The specificities of V regions in each B cell and its progeny remain essentially unaltered but are "fine-tuned" after antigenic stimulation, leading to affinity maturation.

It is now clear that the generation of antibody diversity, allelic exclusion, Ig secretion, heavy chain isotype switching, and affinity maturation are regulated by changes in the organization of Ig genes and by the transcription and translation of these genes. In the remainder of this chapter, we will focus on the expression of Ig genes, beginning with the earliest events in the B lymphocyte lineage and proceeding through the various stages of B cell maturation and differentiation.

REARRANGEMENT OF IMMUNOGLOBULIN GENES

Discovery of Immunoglobulin Gene Rearrangement

The first hypothesis about the organization of the genes that coded for different portions of Ig molecules was proposed even before B lymphocytes were recognized as antibody-producing cells. It was known in the 1950s that each chain of antibody molecules consisted of highly diverse V regions, which accounted for the specificities of antibodies, and relatively invariant C regions. Moreover, each C region had to be the product of a single allelic gene locus because in any individual the C regions present in all the antibodies of a particular isotype, irrespective of their antigenic specificities, contained sequences that were inherited in mendelian fashion. (These sequences constitute the allotypes of antibodies and have been described in Box 3–2, Chapter 3.) This apparent diversity of V regions and constancy of C regions of antibody molecules created a paradox, since the V and C regions made up one polypeptide chain and it was believed at that time that one protein was always encoded by one gene. Therefore, one would have to propose that only the variable portion of this putative gene was subject to extensive alterations or mutations. Realizing this paradox, Dreyer and Bennett postulated in 1965 that each antibody chain was actually encoded by at least two genes, one variable and the other constant, and the two became joined at the level of the DNA or messenger ribonucleic acid (mRNA) to give rise to functional Ig proteins.

Formal proof of this hypothesis came over a decade later. In a landmark study, Susumu Tonegawa and his colleagues demonstrated that the structure of Ig genes in the cells of an antibody-producing tumor, called a myeloma or plasmacytoma, is different from that in embryonic tissues or in non-lymphoid tissues not committed to Ig production. This observation is best demonstrated by a technique called Southern blot hybridization, which is used to examine the sizes of DNA fragments produced by enzymatic digestion of genomic DNA (Box 4–2). By this method, it has been shown that *the sizes of DNA fragments containing Ig genes are different in cells that do and do not make antibody* (Fig. 4–2). The explanation for these different sizes is that V and C regions for any Ig light (L) or heavy (H) chain are encoded by different gene segments that are located far apart in embryonic cells and are brought close together in cells committed to antibody synthesis (i.e., B lymphocytes). *Thus, Ig genes undergo a process of somatic DNA recombination or rearrangement during B cell ontogeny.* These findings provided the first clear molecular explanation for the production of antibodies. The current concepts are best understood by describing the unrearranged, or germline, organization of Ig genes and then the pattern and mechanisms of their rearrangement during B cell development.

BOX 4–2.   Southern Blot Hybridization

This technique, introduced by E. M. Southern, is used to characterize the organization of DNA surrounding a specific nucleic acid sequence (e.g., a particular gene). In a typical experiment (see figure), genomic DNA is chemically extracted from the nuclei of isolated cells or from whole tissues. At this point, the DNA will be present in extremely long segments and must be broken down into small fragments for analysis. This is accomplished by enzymatic cleavage with restriction endonucleases. (These bacterial enzymes are so named because they function to restrict the survival of foreign bacterial DNA in a host strain and because they cleave DNA at specific internal sites as opposed to digesting DNA from the ends as an exonuclease would). Restriction endonucleases cleave double-stranded DNA only at positions of particular symmetric nucleotide sequences, usually about 6 bp in length. Such sequences are called restriction sites, and each site is uniquely recognized by a particular restriction endonuclease. Within the genome of an individual, restriction sites are present at the same positions in every cell, barring somatic mutation or DNA rearrangement. Complete digestion with a particular enzyme leads to cleavage of the genomic DNA, giving rise to an array of DNA fragments ranging from about 0.5 to 10 kb in size. Different restriction endonucleases produce different arrays of restriction fragments, but each enzyme generally produces the same fragments from the DNA of every cell from an individual. This, of course, is not the case when one examines genes that rearrange in different ways in different clones of cells, such as Ig and T cell receptor genes.

To analyze the fragments, the digested DNA is separated according to size by electrophoresis in an agarose gel. The separated fragments are transferred by capillary action (blotting) from the gel to a membrane of nitrocellulose or nylon. Each fragment is then attached in place to the membrane by heating or ultraviolet irradiation. The net result is that the array of fragments is arranged by size on the membrane. Any particular nucleic acid sequence, such as a gene of interest, will be present on one or a few unique-sized fragments, depending only on the choice of restriction endonuclease and the distances between the relevant restriction sites that flank the gene or are located within the gene.

The analysis is completed by ascertaining the size of the restriction fragment(s) containing the gene of interest. This is accomplished by nucleic acid hybridization. Double-stranded DNA can be "melted" into single-stranded DNA by changing temperature and solvent conditions. When the temperature is lowered, single-stranded DNA will reanneal to form double-stranded DNA. The rate at which reannealing of a sequence occurs is determined by the concentration of DNA containing complementary nucleic acid sequences present in the system. In the Southern blot hybridization technique, the DNA on the membrane is melted and then allowed to reanneal in the presence of a solution containing a large excess of single-stranded DNA (probe) with a sequence complementary to the gene of interest. The probe is typically labeled with radioactive phosphorus. It preferentially anneals to the melted DNA on the membrane only at the position of the restriction fragment that contains the complementary sequence. After excess probe is removed by washing, the location of the bound probe is determined by autoradiography and compared with the positions of DNA fragments of known size. By varying the conditions of the hybridization and of the subsequent washing steps, one can control the quantity of probe that remains bound as a function of sequence complementarity. Practically, this means that at "high stringency" the probe may need to be an exact complement of the gene of interest, whereas at "low stringency" one can identify related but non-identical sequences.

Restriction sites are generally located at the same positions in the genomes of all individuals of a species. Occasionally, this is not true and a particular restriction site is present only in or near some allelic forms of a gene. This variability in the presence of a particular restriction site leads to variability in the length of the restriction fragment that is detected by Southern blot hybridization using a probe that hybridizes near the variable restriction site. The length of the fragment is inherited as a mendelian allele, and its variation among individuals is described as a restriction fragment length polymorphism (RFLP). Southern blotting for RFLPs is now commonly used to study the inheritance of nearby ("linked") genes.

Nucleic acid electrophoresis, blotting, and hybridization

Continued

with DNA probes have also been used to analyze mRNA molecules. In this case, mRNA is isolated from the cell of interest and subjected to electrophoresis without digestion; mRNA is already single-stranded, and the gel electrophoresis is run under denaturing conditions to prevent internal hybridization. The position of probe binding can indicate the size of the mRNA (rather than of a restriction fragment of DNA), and the extent of probe binding correlates with the abundance of mRNA present in the cell. This technique for RNA analysis is now universally referred to as Northern blotting, a biochemist's joke alluding to the DNA blotting technique introduced by Southern that now bears his name.

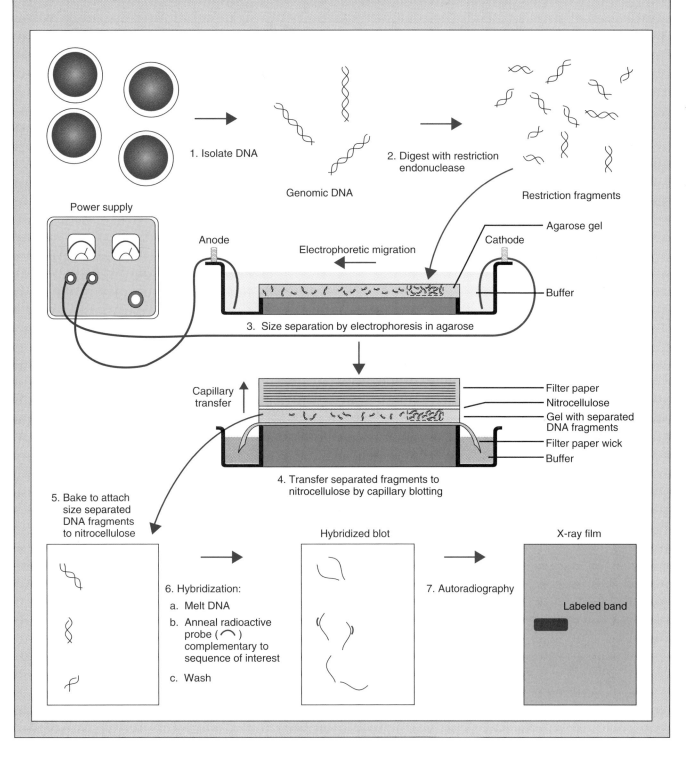

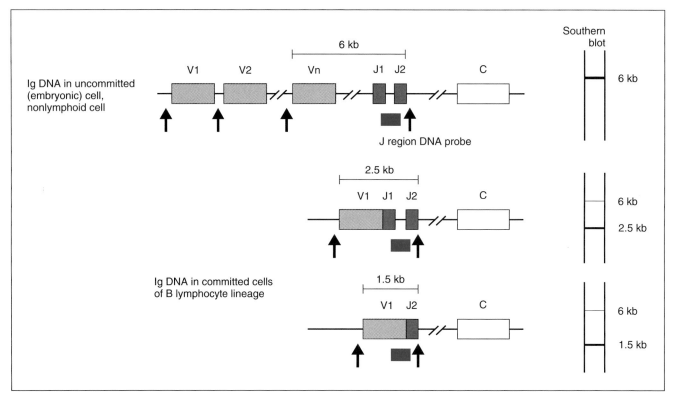

FIGURE 4–2. Detection of immunoglobulin (Ig) gene rearrangements by Southern blot hybridization. In this hypothetical example, genomic DNA from an embryonic or nonlymphoid cell is cut by a restriction enzyme at sites shown by bold arrows, producing a 6 kb fragment that is detected by Southern blot hybridization using a Jκ segment-specific DNA probe. In two different clones of B cells, the same enzyme may generate a 2.5 kb fragment or a 1.5 kb fragment because VJ rearrangement has led to the deletion of the DNA between the rearranged gene segments. (Note that a 6 kb band persists in the B cells; this band is derived from the unrearranged allelic locus.)

Organization of Immunoglobulin Genes in the Germline

The organization of Ig genes* in the germline is fundamentally similar in all species studied. Genes encoding the two light chains, κ and λ, and the single locus containing the various heavy chain genes are located on three different chromosomes.

---

* In Ig (and T cell receptor) loci, the term "gene" usually refers to the DNA encoding the complete H or L chain polypeptide. As we shall discuss presently, each H or L chain gene consists of multiple "gene segments" that code for V, C, and other regions and are separated from one another in the genome by large stretches of DNA that are never transcribed. Each V and C gene segment is further composed of coding sequences that are present in the mature mRNA; these are called **exons.** For instance, each C_H gene segment that gives rise to a heavy chain C region is composed of five to six exons. Exons are separated by pieces of DNA, called **introns,** that are transcribed and present in the primary (nuclear) RNA but are absent from the mRNA. The removal of introns from the primary transcript is the process of RNA **splicing.** Sometimes, the terms "gene," "gene segment," and "exon" are used interchangeably. For instance, "V gene" might refer to the gene segment coding for the complete V region of an Ig heavy or light chain, which actually consists of a V region exon and additional segments (J and D, as we shall see later), or "V gene" might refer to a V region exon only (as in "V gene families," discussed in this chapter).

Each locus has a similar basic organization, which is illustrated for mouse and human Ig genes in Figures 4–3 and 4–4, respectively.

The Ig heavy and light chain loci are composed of multiple genes that give rise to the V and C regions of the proteins, separated by stretches of non-coding DNA. At the 5′ end of each Ig locus are the **V region exons,** each about 300 base pairs (bp) long, separated from one another by non-coding DNA of varying lengths. About 90 bp 5′ of each V region exon is a small (60 to 90 bp long) exon that encodes the translation initiation signal and 20 to 30 amino terminal residues of the translated protein. These residues are moderately hydrophobic and make up the **signal (or leader) sequences.** Signal sequences are found in secreted and transmembrane proteins and are involved in guiding the emerging polypeptides during their synthesis on ribosomes into the lumen of the endoplasmic reticulum. Here, the signal sequences are rapidly cleaved and lost from the mature proteins, probably before translation is complete. The numbers of V genes (used here synonymously with V region exons) in the mouse vary from two for λ chains to about 1000 for heavy chains and are present in the genome over large stretches of DNA, at least 1000 to 2000 kilobases (kb) long (Figs. 4–3 and 4–4). In

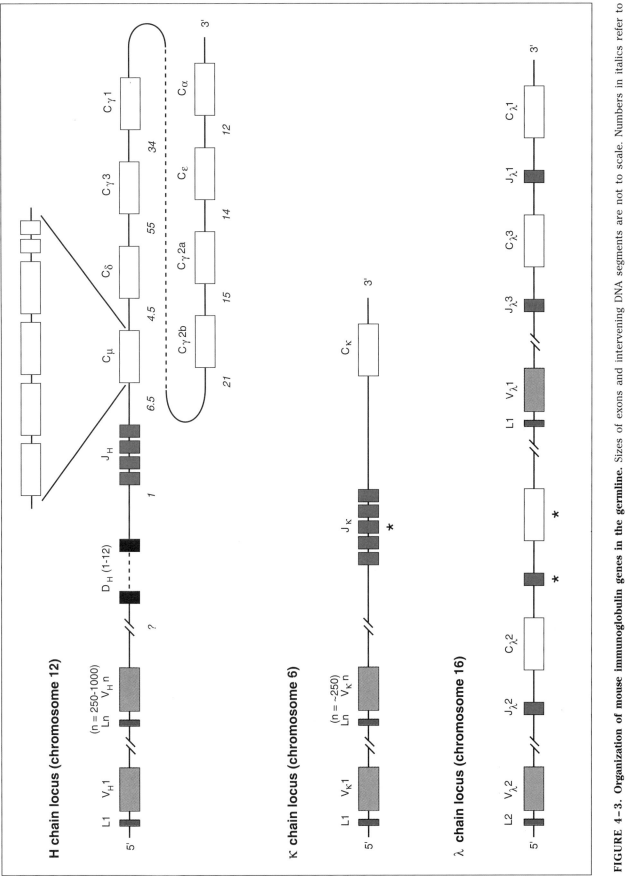

H chain locus (chromosome 12)

κ chain locus (chromosome 6)

λ chain locus (chromosome 16)

FIGURE 4–3. Organization of mouse immunoglobulin genes in the germline. Sizes of exons and intervening DNA segments are not to scale. Numbers in italics refer to approximate lengths of DNA segments in kilobases. Asterisks indicate nonfunctional pseudogenes. Each C_H gene is shown as a single box but is composed of several exons; the actual exon composition of C_H genes is shown for μ only. Gene segments are indicated as follows: L, leader (usually called "signal" sequence); V, variable; D, diversity; J, joining; C, constant.

75

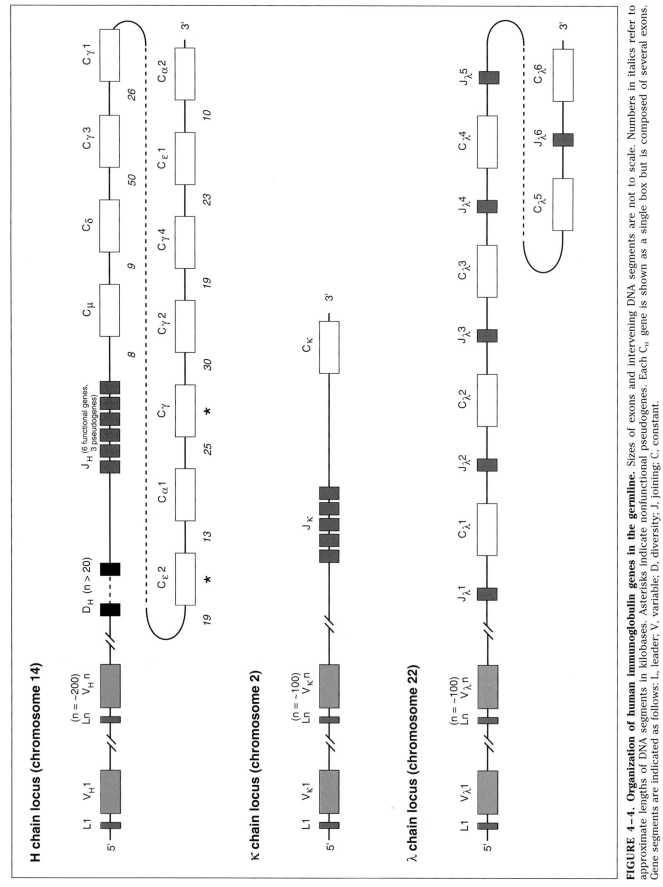

FIGURE 4–4. Organization of human immunoglobulin genes in the germline. Sizes of exons and intervening DNA segments are not to scale. Numbers in italics refer to approximate lengths of DNA segments in kilobases. Asterisks indicate nonfunctional pseudogenes. Each C_H gene is shown as a single box but is composed of several exons. Gene segments are indicated as follows: L, leader; V, variable; D, diversity; J, joining; C, constant.

humans, each H or L chain locus contains ~100 to 200 V genes.

In all species examined to date, V genes are organized into multiple families. The members of each family are identical at 70 to 80 per cent of their nucleotide sequences. Each family is thought to have originated by duplication of a single V gene exon. In the mouse H chain there are at least nine V gene (V_H) families, each consisting of two to 60 members. The functional significance of these V gene families remains uncertain; members of each family do not appear to encode antibody V regions with a similar specificity.

At varying distances 3' of the V genes are the **C region genes.** In both mice and humans, the κ light chain locus has a single C gene, λ has three to six, and the genes for heavy chain C regions (C_H) of different isotypes are arranged in a tandem array whose order is characteristic for each species. Each heavy chain C region gene actually consists of three to four exons (each similar in size to a V region exon) that make up the complete C region, and smaller exons that code for the carboxy terminal ends, including the transmembrane and cytoplasmic domains of the heavy chains. Between the V and C genes, and separated by introns of varying lengths, are additional coding sequences, 30 to 50 bp long, which make up the **joining (J) segments** and, in the H chain locus only, the **diversity (D) segments.** The J and D gene segments code for the carboxy terminal ends of the V regions, including the third hypervariable (complementarity-determining) regions of antibody molecules. Thus, in an Ig light chain protein (κ or

λ), the variable region is encoded by the V and J exons and the constant region by a C exon (which does not have transmembrane or cytoplasmic segments). In the heavy chain protein, the variable region is encoded by the V, D, and J exons. The constant region of the protein is derived from the multiple C exons and, for membrane-associated heavy chains, the exons encoding the transmembrane and cytoplasmic domains (Fig. 4–5).

Based on the tandem organization of V and C genes in each Ig locus and on the structural homologies between them, it is likely that these genes evolved from repeated duplication of a primordial gene. Each V and C exon codes for an individual domain of an antibody molecule. Other proteins that contain Ig-like domains are considered to be members of the **Ig gene superfamily** (see Box 7–2, Chapter 7) and are encoded by genes that are homologous to Ig V and C genes.

Although the introns and the untranscribed DNA sequences between exons are not expressed in mature mRNA, they play an important role in the production of antibodies. As we shall see later, recognition sequences that dictate rearrangement of various exons, and nucleotide sequences that regulate transcription and RNA splicing, are located between the exons.

Sequence of Immunoglobulin Gene Rearrangement

Cells other than B lymphocytes contain Ig genes in the germline configuration, and only B lymphocytes express these genes in properly rear-

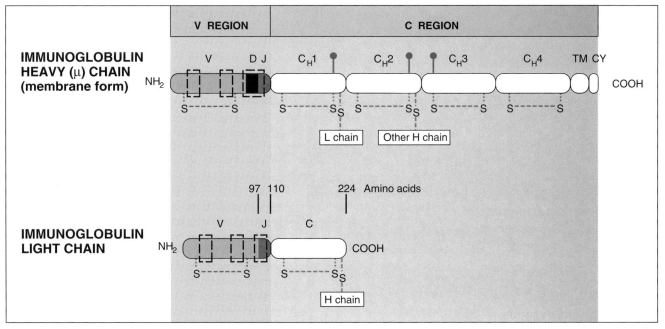

FIGURE 4–5. Relation of immunoglobulin (Ig) gene segments to domains of Ig polypeptide chains. The V and C regions of Ig polypeptides are encoded by different gene segments. The locations of intrachain and interchain disulfide bonds (S—S) and of carbohydrates (↑) as shown are approximate. Areas in dashed boxes indicate hypervariable (complementarity determining) regions. In the μ chain, transmembrane (TM) and cytoplasmic (CY) domains are encoded by separate exons. In the light chain, numbers refer to positions of amino acids; see Figure 3–2 for the locations of these residues in the three-dimensional structure.

ranged forms, capable of giving rise to functional proteins. *DNA rearrangements in Ig gene loci occur in a precise order*, which explains the pattern of B cell maturation described earlier (see Fig. 4–1). In a developing B cell, the first rearrangement involves the heavy chain locus and leads to the joining of one D and one J segment with deletion of the intervening DNA (Fig. 4–6). The D segments 5′

of the rearranged D and the J segments 3′ of the rearranged J are not affected by this recombination (D1 and J2-4 in Fig. 4–6). DJ rearrangement may actually occur prior to commitment of a lymphoid precursor to the B cell lineage, because about 20 per cent of T cell–derived tumors contain DJ rearrangements in Ig heavy chain loci. Following the DJ rearrangement, one of the many V

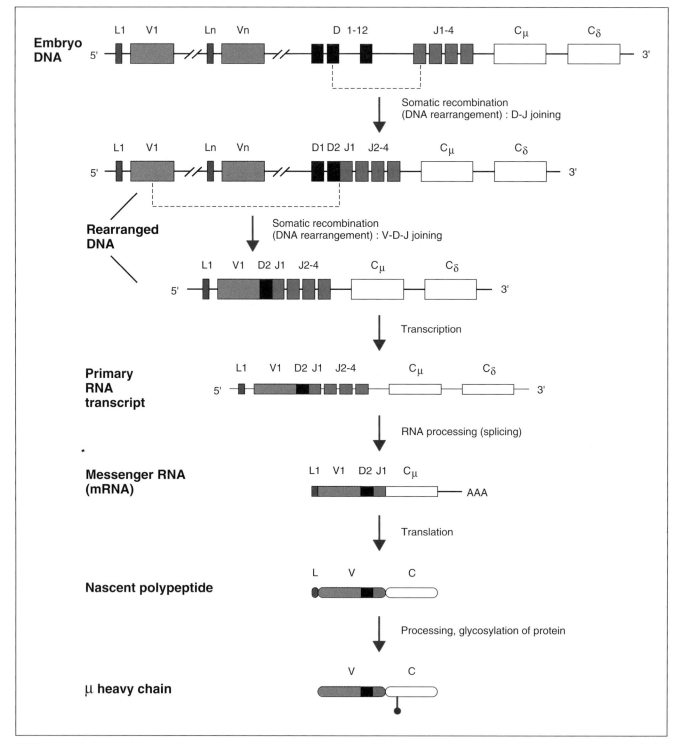

FIGURE 4–6. Sequence of gene rearrangement, transcription, and synthesis of the mouse immunoglobulin μ heavy chain. In this example, the V region of the μ chain is encoded by the exons V1, D2, and J1. V genes are indicated as V1 to Vn; C_H genes 3′ of Cδ are not shown; the location of carbohydrates is schematic; and gene segments and distances between them are not shown to scale.

genes is joined to the DJ complex, giving rise to a rearranged VDJ gene. At this stage, all D segments 5′ of the rearranged D are also deleted. This VDJ recombination occurs only in cells committed to become B lymphocytes and is the critical control point in Ig expression because only the rearranged V gene is subsequently transcribed. The C region genes remain separated from this VDJ complex by an intron (presumably containing the unrearranged J segments), and the primary (nuclear) RNA transcript has the same organization. It is not known whether all the C regions are expressed in the primary transcript.

Subsequent processing of the RNA leads to splicing out of the intron between the VDJ complex and the most proximal C region gene, which is Cμ, giving rise to a functional mRNA for the μ heavy chain. Multiple adenine nucleotides, called poly-A

tails, are added to one of several consensus polyadenylation sites located 3′ of the Cμ RNA. Genes coding for other C_H classes also have 3′ polyadenylation sites, which are utilized when these C regions are expressed (see below). It is thought that cleavage of the primary RNA transcript and splicing are tightly coupled to polyadenylation, since most functional, complete mRNAs have poly-A tails. Translation of the μ heavy chain mRNA leads to production of the μ protein, giving rise to the "cytoplasmic μ only" phenotype of the pre–B lymphocyte.

The next somatic DNA recombination involves a light chain locus (usually κ and then λ; see below) and follows an essentially similar sequence (Fig. 4–7). One V segment is joined to one J segment, forming a VJ complex, which remains separated from the C region by an intron, and this

FIGURE 4–7. Sequence of gene rearrangement, transcription, and synthesis of mouse Igκ light chain. In this example, the V region of the κ chain is encoded by the exons V1 and J1. V genes are indicated as V1 to Vn; the location of carbohydrates is schematic; and gene segments and distances between them are not shown to scale.

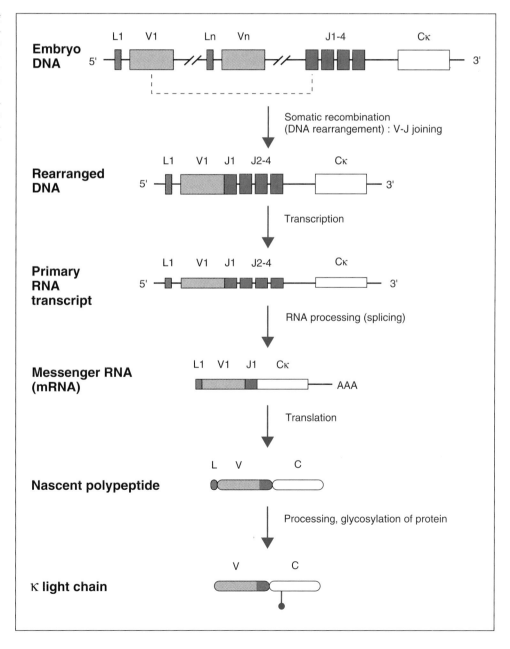

gives rise to the primary RNA transcript. Splicing of the intron from the primary transcript joins the C region transcript to the VJ complex, forming an mRNA that is translated to produce the κ or λ protein. The light chain assembles with the previously synthesized μ in the endoplasmic reticulum to form the complete membrane IgM molecule, which is expressed on the cell surface, and the cell is now an immature B lymphocyte.

Rearrangements of Ig genes are the essential first steps in the production of antibodies. *In addition, the μ heavy chain protein that is synthesized in a developing B cell itself regulates the somatic recombination of Ig genes in two ways* (Fig. 4–8):

1. First, the μ protein produced by the rearranged gene on one chromosome irreversibly inhibits rearrangement on the other chromosome, accounting for *allelic exclusion* of Ig heavy chains. Heavy chain genes on the second allelic chromosome will rearrange only if the first rearrangement is nonproductive. Such nonproductive DNA rearrangements may be due to deletions, mutations, or frameshifts during recombination that generate stop codons. Thus, in any B cell, one heavy chain allele is productively rearranged and expressed, and the other is in the germline configuration or is aberrantly rearranged. If both alleles undergo nonproductive recombinations, the cell cannot produce Ig, will be unable to recognize and respond to antigens, and apparently dies. This occurs frequently and is one of the reasons why only a small

fraction of the cells that arise from B cell progenitors develop into mature B lymphocytes. Studies with transgenic mice have shown that the membrane but not the secreted form of the μ protein suppresses heavy chain gene rearrangement and is responsible for allelic exclusion; however, the precise mechanism of this suppression is unknown.

2. A second effect of the production of a μ protein in a pre–B cell is *stimulation of light chain gene rearrangement*. This was first suggested by analyses of pre–B tumor lines in which VJ_{κ} rearrangements spontaneously occur in culture. Such light chain DNA recombinations are seen only in those daughter cells of the original clone that synthesize μ protein. Again, the molecular signals by which the μ protein stimulates light chain gene rearrangement are not known. However, some light chain gene rearrangement does occur in mice in which the μ gene is knocked out, indicating that the μ protein is not an obligatory stimulus for the rearrangement of light chain genes.

It was initially thought that rearrangement of light chain genes occurs first in the κ locus, and if the κ rearrangement is productive, giving rise to a κ protein, subsequent rearrangement of the λ light chain is blocked. However, more recent studies indicate that there is not a precise sequence of light chain gene rearrangement, and either light chain isotype may rearrange first. Production of either light chain inhibits rearrangement at the other locus, explaining why an individual B cell clone can

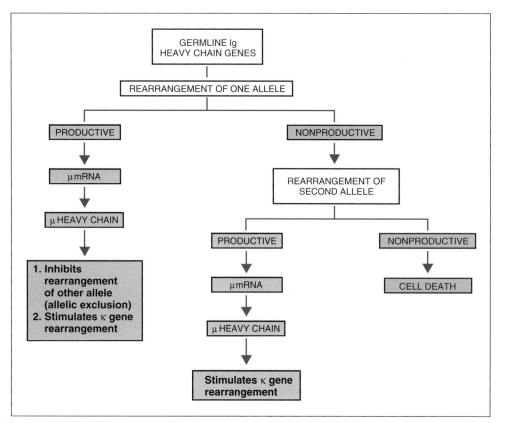

FIGURE 4–8. Order of rearrangement and expression of immunoglobulin (Ig) heavy chain (μ) genes. The consequences of productive (functional) and nonproductive (aberrant) rearrangements of Ig heavy chain genes on the two allelic chromosomes are shown. The μ chain is the first Ig protein produced in a developing B lymphocyte.

produce only one of the two types of light chains during its life (light chain isotype exclusion). As in the heavy chain locus, functional rearrangement involving either light chain locus on one of the two parental chromosomes actively prevents rearrangement at the other allele, accounting for allelic exclusion of light chains in individual B cells. Also, as for heavy chains, if one allele undergoes nonfunctional rearrangement, DNA recombination can occur on the other allele; however, if both alleles of both κ and λ chains are nonfunctional, that cell apparently dies.

Mechanisms of Immunoglobulin Gene Rearrangement

Rearrangement of Ig (and, as we shall discuss in Chapter 8, T cell receptor) genes is a special kind of recombination involving non-homologous gene segments. It is mediated by the coordinated activities of lymphocyte-specific **recombinases,** and more ubiquitous proteins, such as DNA repair enzymes, which are present in all cells. The lymphocyte-specific recombinases recognize specific DNA **recognition sequences** located in the intervening DNA 3' of each V exon and 5' of each J segment and flanking both sides of each D segment (Fig. 4–9A). The recognition sequences are highly conserved stretches of seven or nine nucleotides separated by non-conserved 12 or 23 nucleotide spacers. Recombination of V and J exons may occur by a process of deletion or inversion (Fig. 4–9B). If the recognition sequences face in the opposite direction, recombinases bring the V and J exons into apposition, forming a loop of intervening DNA. The intervening DNA is then excised, and the ends of the V and J exons are annealed to complete the rearrangement process. If the recognition sequences face the same direction, the intervening DNA is inverted and the V and J exons fuse at one end. Ig gene rearrangement usually occurs by deletion, and inversion appears to be important only in the κ locus, where some V exons are in a 3' to 5' orientation. The same process of deletion and ligation is responsible for DJ and VDJ recombinations in the H chain locus.

Two genes that stimulate Ig gene recombination, called recombination activating genes 1 and 2 (*RAG-1* and *RAG-2*), have been identified in pre–B and immature T cells. *The RAG-1 and RAG-2 proteins form a dimer that is responsible for lymphocyte-specific recombinase activity.* The *RAG-1–* and *RAG-2–*encoded recombinase has two important properties. First, it is cell type–specific, being active only in cells of the B and T lymphocyte lineages. Interestingly, the same recombinase mediates recombination at both Ig and T cell receptor loci, yet complete functional rearrangement of Ig genes normally occurs only in B cells and of T cell receptor genes in T cells. This suggests that some other as yet undefined mechanism controls the cell type specificity of recombination. Second, the recombi-

nase is active in immature B and T lymphocytes but not in mature cells, explaining why Ig and T cell receptor gene rearrangements do not continue in cells that express functional antigen receptors. Knockout of *RAG-1* or *RAG-2* genes from mice leads to a failure to produce both Ig and T cell receptor proteins and a complete lack of mature B and T lymphocytes. Other enzymes also participate in Ig and T cell receptor gene recombination. One is a double-stranded DNA repair enzyme that is defective in mice carrying the severe combined immunodeficiency mutation (see Chapter 21). There is no known human homolog of this mutation.

Rearrangement of Ig genes during B cell development is also controlled by accessibility of the DNA to recombinases. In eukaryotic cells, most regions of DNA are covered by proteins, condensed into a structure called chromatin, which prevent the uncontrolled expression of genes. Early in B cell maturation, the chromatin structure of Ig genes opens, so that the genes become accessible to recombinases. This allows the functional rearrangement and subsequent transcription of complete Ig genes.

Generation of the Antibody Repertoire

Several different genetic mechanisms contribute to the diversity of membrane-bound and secreted antibodies in each individual. The estimated contribution of each of these mechanisms to the total repertoire of antibodies is summarized in Table 4–2.

1. *Combinatorial diversity: somatic recombinations involving multiple germline gene segments.* Each clone of B lymphocytes and its progeny express a unique combination of V, D, and J gene segments in functionally rearranged Ig heavy and

TABLE 4–2. Mechanisms Contributing to the Generation of Diversity in the Antibody Repertoire*

	H	κ	λ
Germline genes*			
V gene segments	250–1000	250	2
J segments	4	4	3
D segments	12	0	0
Combinatorial joining			
V × J (× D)	10,000–40,000	1000	6
H-L chain associations			
H × κ		$1–4 \times 10^7$	
H × λ		$5–10 \times 10^4$	
Total potential repertoire with junctional diversity		$10^9–10^{11}$	

* The results are from mouse immunoglobulin loci. Numbers of gene segments and the contribution of junctional diversity are estimates. The mouse is unusual among all species examined because of the low number of V_λ genes and the limited diversity in antibodies with λ light chains. Apart from this, other mammals (including humans) are similar.

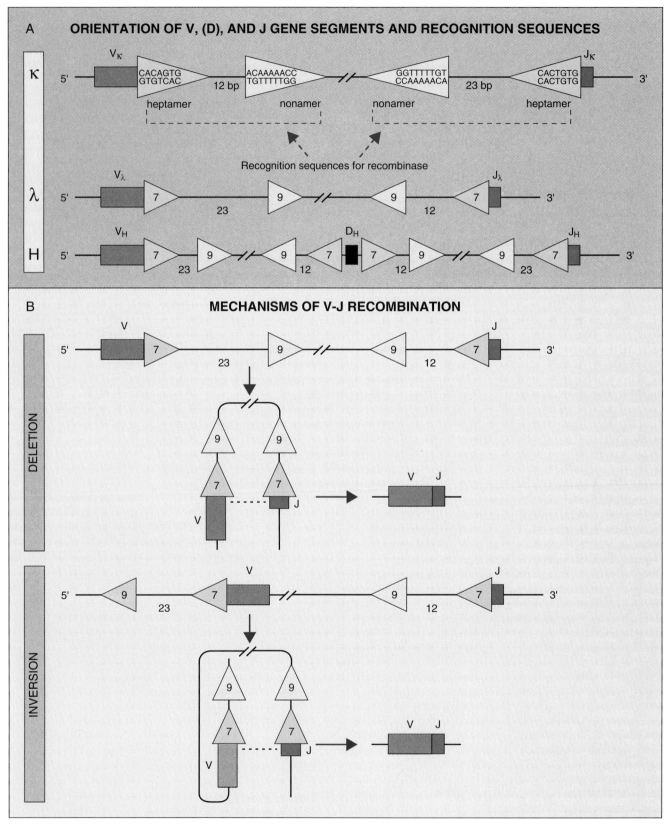

FIGURE 4–9. Mechanisms of immunoglobulin (Ig) gene recombination.

A. DNA recognition sequences for recombinases that mediate Ig gene rearrangement. Conserved heptamer (7 bp) and nonamer (9 bp) sequences, separated by 12 or 23 base pair spacers, are located adjacent to V and J exons (for κ and λ loci), or V, D, and J exons (in the H chain locus). Recombinases presumably recognize these regions and bring the exons together.

B. Recombination of V and J exons may occur by deletion of intervening DNA and ligation of the V and J segments or by inversion of the DNA followed by ligation of adjacent gene segments.

light chains. In the total population of B cells, the combinatorial associations of different V, D, and J genes lead to a large potential for generating different specificities, accounting for the diversity of the B cell repertoire. The maximum possible number of combinations is the product of the numbers of V, J, and D (if present) exons at each Ig locus. Therefore, the amount of diversity generated at each locus reflects the number of V, J, and D gene segments at that locus. For instance, in mice, antibodies containing the λ light chain comprise only 5 to 10 per cent of total antibodies, and they recognize a quite restricted set of antigens, because the λ locus contains only two functional V genes. In humans, on the other hand, the κ and λ loci contain roughly equal numbers of V genes, are equally represented in antibodies produced, and are similar in their diversity.

2. *Junctional diversity.* Even the same set of germline V, D, and J gene segments can generate different amino acid sequences at the junctions. This additional junctional diversity arises from two mechanisms (Fig. 4–10):

a. The first is inaccurate or **imprecise DNA rearrangement** that occurs because the 3′ end of a V gene and the 5′ end of a J seg-

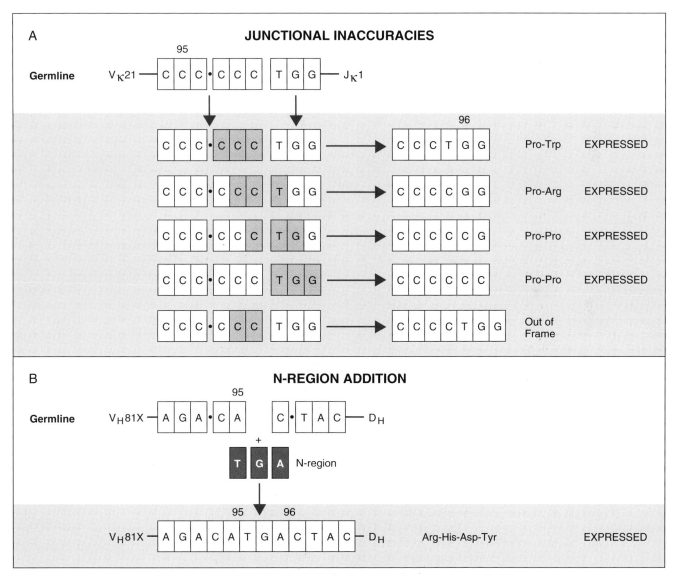

FIGURE 4–10. Generation of junctional diversity.
A. During the joining of different V and J exons, deletion of nucleotides (indicated as shaded boxes) may lead to the generation of novel nucleotide and amino acid sequences or to out-of-frame nonfunctional recombinations. All the indicated amino acid sequences have been detected in different V$_\kappa$21 containing mouse antibodies. (Adapted with permission from Weigert, M., R. Perry, D. Kelley, T. Hunkapiller, J. Schilling, and L. Hood. The joining of V and J gene segments creates antibody diversity. Nature 283: 497–499, 1980. Copyright © 1980, Macmillan Magazines Ltd.)
B. Random nucleotides (N regions) may be added to the site of VD recombination (in this example), generating new sequences. (Adapted with permission from Keyna, U., G. B. Beck-Engeser, J. Jongstra, S. E. Applequist, and H.-M. Jack. Surrogate light chain-dependent selection of Ig heavy chain regions. Journal of Immunology 155:5536–5542, 1995. Copyright © 1995, The American Association of Immunologists.)

ment in a light chain, or the ends of V, J, and D gene segments in a heavy chain, can each recombine at any of several nucleotides in the germline sequence. As long as the recombination does not generate non-functional DNA sequences, such as non-sense or stop codons, different nucleotide and, subsequently, amino acid sequences can arise (Fig. 4–10A). Imprecise joining can also lead to recombinations that are out of frame so that the DNA cannot be transcribed. This inefficiency may be a price that is paid for generating diversity.

b. The second mechanism for junctional diversity is **N region diversification.** Nucleotides that are not present in the Ig loci can be added to the junctions of rearranged VJ or VDJ genes (Fig. 4–10B), generating short sequences called N regions. N region diversification is more common in Ig heavy chains (and in T cell antigen receptors) than in Ig κ or λ chains. This addition of new nucleotides is a random process mediated by an enzyme called terminal deoxyribonucleotidyl transferase (TdT). In mice rendered deficient in TdT by gene knockout, the diversity of B (and T cell) repertoires is substantially less than in normal mice.

Because of combinatorial and junctional diversity, antibody molecules show the greatest variability at the junctions of V and C regions in heavy and light chains. These junctions form the third hypervariable region, or complementarity-determining region (CDR)3, which is the most important portion of the Ig molecule for binding antigen and for defining the specificity of the antibody (see Chapter 3). In fact, the number of different amino acid sequences that are present in the third hypervariable regions of antibody molecules is actually greater than the number that can be encoded by germline J and D segments present. As we shall see in Chapter 8, both imprecise joining and N region addition are even more important for generating diversity in T cell antigen receptor genes than in Ig genes. Also, the sequence of nucleotides at the VDJ recombination site is a marker for clonality of B (and T) cell–derived lymphomas and leukemias. Since each tumor arises from a single clone, it contains a unique sequence at the site of recombination (Box 4–3).

3. *Combinations of H and L chain proteins.* In addition to these mechanisms operative at the level of Ig genes, the combination of different H and L chain proteins in different B cells also contributes to diversity of the repertoire, because the V region of each chain participates in antigen recognition.

A practical application of the elucidation of the molecular basis of antibody diversity is the construction of antibodies by *in vitro* techniques. In theory, the mechanisms that generated a di-

BOX 4–3. Immunoglobulin Genes in B Lymphocyte–Derived Tumors

Since the presence of functionally rearranged Ig heavy and light chain genes is the *sine qua non* of a B lymphocyte, examination of tumors for the presence of such rearrangements has proved to be a useful diagnostic and analytical method. Perhaps the earliest significant result of such studies was that tumors that did not express phenotypic markers characteristic of a particular cell lineage could be classified unambiguously. For instance, the cell of origin of hairy cell leukemias was an issue of great debate until the demonstration that virtually all these tumors contained rearranged Ig genes, establishing their derivation from B lymphocytes. The expression of Ig proteins can also be used as a marker for the clonality of B cell tumors. This was initially done by analyzing the frequency of cells producing κ or λ light chains within a proliferative lesion of B lymphocytes. In humans, approximately half of all antibody molecules contain κ and half contain λ light chains. Therefore, if all the B cells in a lesion express either κ or λ, it is likely that this lesion is a monoclonal tumor. Conversely, if equal numbers of cells express κ and λ, the lesion is a polyclonal or non-neoplastic proliferation. This approach has been refined and tremendously improved by our ability to analyze Ig gene rearrangements by Southern blot hybridization, because every clone of B lymphocytes contains unique rearranged VJ and VDJ complexes in light chain and heavy chain loci, respectively. Thus, if genomic DNA from B cell tumors is isolated and digested with a panel of restriction enzymes, and then Southern blots are probed with cDNA probes for J or C regions, the pattern of restriction fragments containing these genes is unique for each clonally derived tumor. This pattern is reflected in the bands detected by Southern blot hybridization. Provided that a sufficiently large panel of restriction enzymes is used, no two tumors will exhibit identical bands. If the lesion being studied is not a tumor but a polyclonal hyperplasia of B lymphocytes, no one rearrangement is present in enough cells to give a discernible band, and a smear of DNA fragments will be visible in Southern blots.

Such a distinction between hyperplastic and neoplastic lesions has important therapeutic implications. Moreover, the identification of a unique restriction fragment pattern of a B cell tumor is useful for determining if recurrences in patients are due to new tumors or to growth of tumor cells that resisted treatment. Similarly, if circulating tumor cells develop in the blood of patients with lymphoma initially confined to lymphoid organs, Southern blot analysis can establish whether or not this reflects conversion of the original lymphoma to a leukemic growth phase. A more recent and much more sensitive variation of this approach employs the polymerase chain reaction (PCR; see Box 5–2, Chapter 5). If the N region of a particular B cell tumor is known, PCR analysis using primers spanning these N sequences can be done to identify very small numbers of residual or recurrent tumor cells after therapy, because the N region of each B cell clone has a unique sequence.

verse antibody repertoire can be replicated in a genetically engineered microorganism that expresses recombinases. This will allow an antibody repertoire to be generated from germline genes in the laboratory, without immunization of individuals. Such an approach could replace hybridoma technology as a means of producing monoclonal antibodies, and it is not limited to a particular species of animal. Although this is not yet feasible, some molecular approaches for generating antibodies have been initiated. In one example, the CDR3 of an Ig heavy chain complementary DNA (cDNA) was replaced with random DNA sequences, generating $\sim 5 \times 10^7$ potential CDRs. These were expressed on the surface of bacteriophage; phage that bound an antigen of interest, by virtue of expressing a specific CDR, were isolated by affinity purification. Thus, a randomly generated "combinatorial phage library" may be used to produce antibodies of any desired specificity.

EXPRESSION OF DIFFERENT CLASSES AND TYPES OF IMMUNOGLOBULINS

The molecular genetic events that we have described so far lead to the appearance of B cells that express membrane-associated IgM molecules. This IgM consists of μ heavy chains derived from one of the two parental chromosomes, complexed with κ or λ light chains also derived from one allele. The variable regions of both the heavy and light chains differ among different B cell clones.

The next series of changes in Ig gene expression occur in the constant regions of heavy chains and result in the production of membrane or secreted forms of heavy chains of various isotypes. Many of these changes lead to alterations in the functions of the antibody produced by the B cell, but the antigenic specificity of each Ig-producing B cell clone is retained. For instance, the switch in Ig production from a membrane to a secreted form converts the B cell from a cognitive to an effector cell. Because different Ig heavy chain isotypes perform distinct effector functions, heavy chain class switching is important for the elimination of mi-

crobes that are susceptible to different effector mechanisms (see Chapter 3). In addition, in the progeny of activated B cells, the V regions of Ig molecules may progressively accumulate somatic mutations that alter the affinities of these antibodies for their cognate antigens. We now know in considerable detail how the clonal progeny of a membrane IgM-producing B cell can express different types of Ig molecules with the same antigen-binding regions as the original, or parent, cell and how subtle changes in Ig genes after antigenic stimulation lead to increased affinity of antibodies.

Co-expression of IgM and IgD

The mature B lymphocyte is the functionally responsive stage in B cell maturation at which membrane-associated μ and δ heavy chains are co-expressed on the surface of each cell in association with κ or λ light chains. Both classes of Ig heavy chains on each cell have the same V region and, therefore, the same antigen specificity. Simultaneous expression of a single V_H with both C_μ and C_δ to form the two heavy chains occurs by alternative RNA splicing. A long primary RNA transcript is produced containing the rearranged VDJ complex as well as sequences encoded by both C_μ and C_δ (Fig. 4–11). If the introns are spliced out such that the VDJ complex is attached to the C_μ RNA, it gives rise to a μ mRNA. If, however, the C_μ RNA is spliced out as well so that the VDJ complex becomes contiguous with C_δ, a δ mRNA is produced. Subsequent translation results in the synthesis of a complete μ or δ heavy chain protein. Thus, *alternative splicing allows a B cell to simultaneously produce mature mRNAs of two different heavy chain isotypes.* The precise mechanisms that regulate the choice of polyadenylation and/or splice acceptor sites by which the rearranged VDJ is joined to C_μ

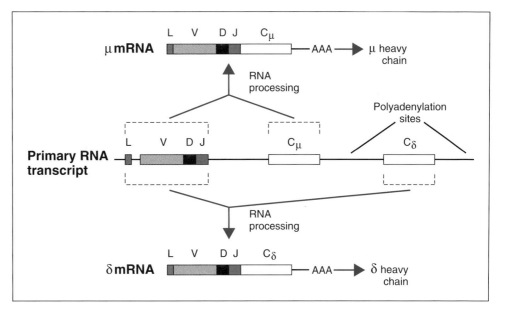

FIGURE 4–11. Co-expression of immunoglobulin (Ig)M and IgD in a B lymphocyte. Alternative processing of a primary RNA transcript results in the formation of a μ or δ mRNA. Dashed lines indicate the H chain segments that are joined by RNA splicing.

or C_δ are not known, nor are the signals that determine when and why a B cell expresses both IgM and IgD rather than IgM alone. The correlation between expression of IgD and acquisition of functional competence has led to the suggestion that IgD is the essential activating receptor of B cells. However, as we shall discuss in Chapter 9, there is no clear evidence for a functional difference between membrane IgM and IgD. Moreover, knockout of the δ gene in mice does not have a major impact on the maturation or antigen-induced responses of B cells.

Secretion of Immunoglobulin

Binding of antigen to the membrane Ig on specific B cells initiates the process that culminates in differentiation of the cells and secretion of Ig by the progeny of the activated cells. Membrane and secreted Ig molecules differ in their carboxy termini (see Fig. 3–4, Chapter 3). For instance, in secreted μ, the $C_\mu 4$ domain is followed by a tail piece containing charged amino acids. In membrane μ, on the other hand, $C_\mu 4$ is followed by a short spacer, 26 hydrophobic transmembrane residues, and a cytoplasmic tail of three amino acids (lysine, valine, and lysine). *The transition from membrane to secreted heavy chain occurs at the level of mRNA processing.* The primary RNA transcript in all IgM-producing B cells contains the rearranged VDJ, the four C_μ exons coding for the C region domains, a small exon encoding the tail piece immediately 3' of the fourth C_μ exon, and the two exons encoding the transmembrane and cytoplasmic domains. Alternative processing of this transcript, which is regulated by RNA cleavage and the choice of polyadenylation sites, determines whether or not the transmembrane and cytoplasmic exons are included in the mature mRNA (Fig. 4–12). If they are, the μ chain produced contains the amino acids that make up the transmembrane and cytoplasmic segments and is, therefore, anchored in the lipid bilayer of the plasma membrane. If, on the other hand, the transmembrane segment is excluded from the μ chain, the carboxy terminus consists of about 20 amino acids constituting the tail piece. Since this protein does not have a stretch of hydrophobic amino acids or a positively charged cytoplasmic domain, it cannot remain anchored in the cell membrane and is secreted. Thus, each B cell can synthesize both membrane and secreted Ig. As differentiation proceeds, more and more of the Ig mRNA is the secreted form. The biochemical signals that regulate this process of alternative RNA splicing are not known. All C_H genes contain similar membrane exons, and all heavy chains can apparently be expressed in membrane-bound and secreted forms. The secretory form of the δ heavy chain is rarely made, however, so that IgD is usually present as a membrane-bound protein.

Heavy Chain Class (Isotype) Switching

After antigenic stimulation, the progeny of mature IgM- and IgD-expressing B cells also undergo the process of heavy chain class (isotype) switching, leading to the production of antibodies with heavy chains of different classes, such as γ, α, and ϵ. The principal mechanism of isotype switching is a process called **switch recombination,** in which the rearranged VDJ gene segment recombines with a downstream C region gene and the intervening DNA is deleted. This phenomenon was first observed in myelomas, where it was found that cells producing one Ig isotype have deleted all genes encoding C_H segments that are located 5' of the expressed C_H. Thus, if the C_μ and C_δ genes are deleted, the rearranged VDJ will be attached to the next C_H complex, giving rise to γ_3 heavy chains (in the mouse); subsequent deletion of all γ subclass–encoding genes will result in the production of ϵ heavy chains; and so on (Fig. 4–13).

These DNA recombination events involve nucleotide sequences called switch regions, which are located in the introns at the 5' end of each C_H

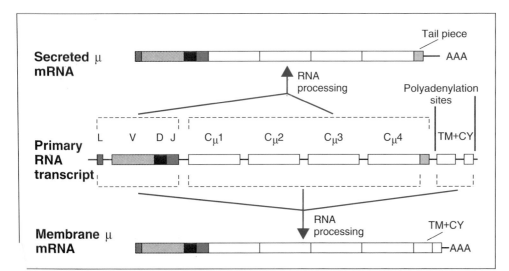

FIGURE 4–12. Expression of membrane and secreted μ chains by B lymphocytes. Alternative processing of a primary RNA transcript results in the formation of a membrane or secreted μ mRNA. Dashed lines indicate the μ chain segments that are joined by RNA splicing. TM and CY refer to transmembrane and cytoplasmic segments, respectively. $C_\mu 1$, $C_\mu 2$, $C_\mu 3$, and $C_\mu 4$ are four exons of the C_μ gene.

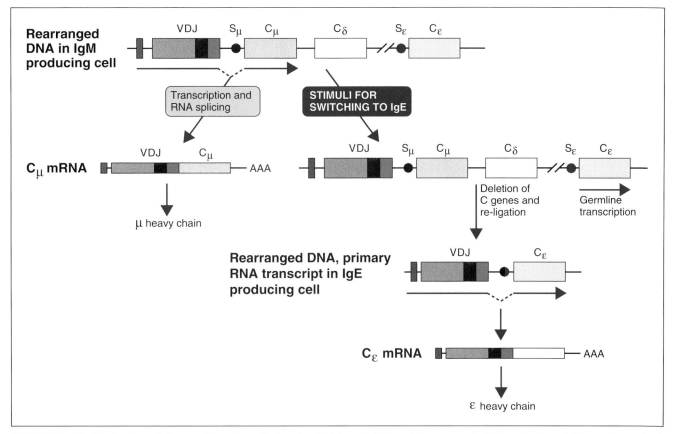

FIGURE 4–13. Mechanism of heavy chain class (isotype) switching. Deletion of C_H genes, of which only C_μ and C_δ are shown, leads to recombination of the VDJ complex with the next, i.e., 3′, C_H gene and expression of this gene, which is C_ϵ in the example shown. Switch regions are indicated by dark circles. (Note that the C_γ genes are located between C_δ and C_ϵ but are not shown.) The stimuli for switching to immunoglobulin (Ig)E include ligation of the CD40 surface molecule on B cells and the cytokine interleukin-4 (see Chapter 9).

locus. Switch regions occupy distances of 1 to 10 kb and contain numerous tandem repeats of highly conserved DNA sequences, each of which is up to 52 bp long. The stimuli that trigger class switching first induce transcription through particular switch regions and C_H genes, resulting in the production of incomplete (germline) transcripts, and this is followed by switch recombination of the rearranged VDJ complex to that C region. Knockout of the switch region for any heavy chain class leads to an inability to switch to that class. The enzymes that mediate switch recombination are not yet identified; as expected from the nature of the target sequences, the *RAG-1–* and *RAG-2–*encoded recombinase is not involved in heavy chain class switching.

Rarely, one observes co-expression of two or more heavy chain isotypes in single B cells from immunized animals or antigen-stimulated cultures. This may result from alternative splicing of a long transcript that contains all the C_H genes, much like the process that leads to co-expression of μ and δ or membrane and secreted Ig. However, alternative splicing does not appear to be a frequent or important mechanism of class switching.

Heavy chain class switching is not a random process but is regulated by helper T cells and

their secreted cytokines (see Chapter 9). For instance, the production of IgE is stimulated by T cell signals delivered via a B cell molecule called CD40 and a T cell–derived cytokine called IL-4, which is a necessary "IgE switch factor." Similarly, the cytokine interferon-gamma selectively induces switching to IgG2a in mice. The C_δ gene lacks a switch region, because of which switching to the IgD isotype does not occur after antigenic stimulation.

Somatic Mutations in Immunoglobulin Genes

After antigenic stimulation, the Ig heavy and light chain genes undergo another type of structural alteration, namely somatic mutations. *Somatic mutations in expressed Ig genes involve primarily V gene segments and are responsible for the affinity maturation of antibodies.*

Comparisons of amino acid and nucleotide sequences of V regions of IgM and IgG antibodies specific for the same antigen first revealed the existence of large numbers of point mutations in the IgG antibodies. Several features of the somatic mutations that occur in Ig genes have been established:

1. In the clonal progeny of B cells responding to a protein antigen or a hapten-protein conjugate,

the numbers of mutations in heavy and light chain V genes are higher in IgG than in IgM antibodies, increase with time after the first immunization, and are even more frequent after secondary and tertiary immunizations (Fig. 4–14). These conclusions are based largely on comparisons of antigen-specific hybridomas derived from B cells at different times after immunizations with an antigen.

2. Point mutations tend to be clustered in V region exons and adjacent flanking sequences of both H and L chains and are most numerous in the hypervariable regions (Fig. 4–14). One possible explanation for this localization of somatic mutations

is that particular areas of the gene are mutation-sensitive. Alternatively, and more likely, mutations may occur in rapidly proliferating B cells randomly throughout rearranged V genes, but only the ones involving antigen-combining sites lead to increased affinities of antibodies. B cells that produce antibodies of higher affinities are positively selected by antigen. This is because as the immune response develops, more antibody is produced and the concentration of available antigen decreases. Under these conditions, B cells that bind the antigen with higher affinity are preferentially stimulated. The germinal centers of lymphoid follicles in

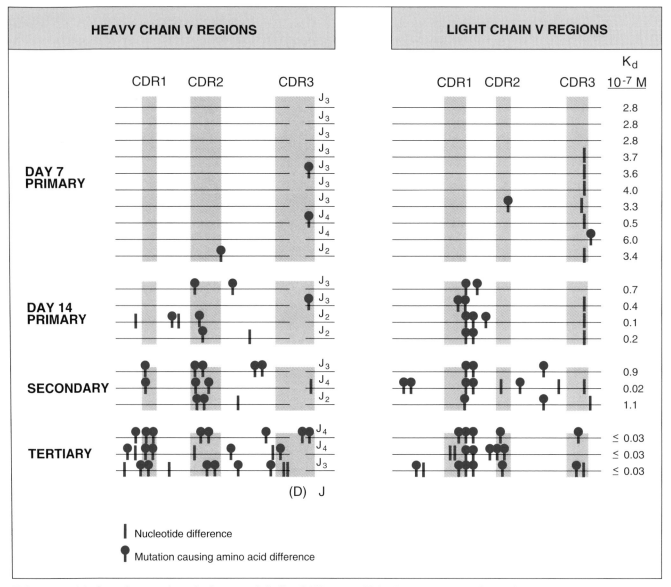

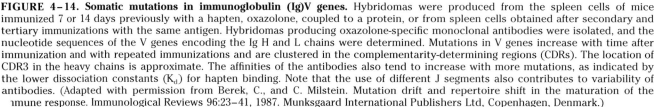

FIGURE 4–14. Somatic mutations in immunoglobulin (Ig)V genes. Hybridomas were produced from the spleen cells of mice immunized 7 or 14 days previously with a hapten, oxazolone, coupled to a protein, or from spleen cells obtained after secondary and tertiary immunizations with the same antigen. Hybridomas producing oxazolone-specific monoclonal antibodies were isolated, and the nucleotide sequences of the V genes encoding the Ig H and L chains were determined. Mutations in V genes increase with time after immunization and with repeated immunizations and are clustered in the complementarity-determining regions (CDRs). The location of CDR3 in the heavy chains is approximate. The affinities of the antibodies also tend to increase with more mutations, as indicated by the lower dissociation constants (K_d) for hapten binding. Note that the use of different J segments also contributes to variability of antibodies. (Adapted with permission from Berek, C., and C. Milstein. Mutation drift and repertoire shift in the maturation of the immune response. Immunological Reviews 96:23–41, 1987. Munksgaard International Publishers Ltd, Copenhagen, Denmark.)

peripheral lymphoid tissues are the principal sites of somatic mutation in Ig V genes and subsequent selection of high-affinity antigen-specific B cells (see Chapter 9). Since mutations occur after antigenic stimulation, they tend to be more numerous in isotype-switched antibodies and in memory B cells. Therefore, memory B cells have higher average affinities for antigen than do naive cells.

3. Mutations may lead to loss of antigen-binding activity, and B cells in which this happens become antigen-unresponsive and undergo programmed cell death. Alternatively, as a result of somatic mutations, the Ig produced by a particular B cell may no longer be specific for the immunizing antigen but may acquire a new specificity for a different antigen. This is a potential mechanism for increasing the diversity of antibodies, but it probably plays a minor role in the generation of diversity.

4. The somatic mutation rate is estimated to be 10^{-3} per V gene base pair per cell division, which is 10^3 to 10^4 times higher than the spontaneous rate of mutation in other mammalian genes. (For this reason, mutation in Ig V genes is also called "hypermutation.") Because the V genes of expressed heavy and light chains in each B cell contain a total of about 700 nucleotides, this implies that mutations will accumulate in expressed V regions at an average rate of almost one per cell division. It is estimated that as a result of somatic mutations, the nucleotide sequences of IgG antibodies derived from one clone of B cells can diverge 1 to 5 per cent from the original germline sequence. This usually translates to less than 10 amino acid substitutions because some mutations may not change the V region amino acid sequence.

5. Affinity maturation is observed only in antibody responses to helper T cell–dependent protein antigens. Therefore, it is likely that T cells or their products are involved in stimulating muta-

tional mechanisms or in selecting high affinity B cells (see Chapter 9).

The precise mechanisms of somatic mutations in Ig genes are poorly understood. Immunologists are actively searching for cell populations in which mutations can be induced *in vitro* by antigens and helper T cells or by T cell–derived cytokines. Such models should prove valuable for analyzing the extrinsic signals, enzymes, and regulatory mechanisms responsible for the extraordinarily high mutation rate in Ig V genes.

TRANSCRIPTIONAL AND TRANSLATIONAL CONTROL OF ANTIBODY PRODUCTION

The elucidation of the structure and expression of Ig genes during B cell maturation has been a remarkable milestone in the progress of immunology. More recently, attention has been devoted to a better understanding of the regulation of Ig gene expression. There are several reasons why this is an important area of scientific investigation. First, the production of Ig is an excellent example of the expression of cell type–specific, developmentally regulated genes. Second, changes in the pattern of Ig biosynthesis and secretion can be induced by well-defined external stimuli such as antigens, polyclonal activators, and helper T cells or T cell–derived cytokines (see Chapter 9). Finally, abnormalities in antibody production contribute to the pathogenesis of many diseases associated with deficient or excessive immune responses.

The general principles of transcriptional regulation of Ig genes are similar to those for other genes. Transcription is controlled primarily by *cis*-acting nucleotide sequences and by proteins that bind to these sequences. The production of antibodies is also influenced by the rate of turnover of

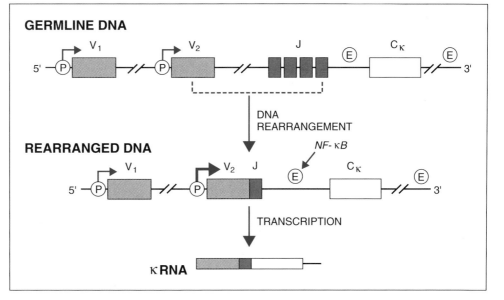

FIGURE 4–15. Transcriptional regulation of immunoglobulin genes. VDJ rearrangement brings promoter sequences (shown as P) close to the enhancer (E) located between the J and C loci, resulting in increased transcription of the rearranged V gene (V2, whose active promoter is indicated by a bold arrow).

mRNAs, their translation, and the post-translational modification and assembly of heavy and light chains.

Cell Type–Specific Regulation of Immunoglobulin Gene Transcription

The transcription of most eukaryotic genes, including Ig genes, is regulated by two types of DNA sequences, called **promoters** and **enhancers.** In Ig loci, promoters are located 5' of V, D, and C exons and are responsible for accurate and efficient transcription of these genes. Enhancer elements are located in the introns between J and C gene segments and at considerable distances 3' of the C genes (Fig. 4–15). It is believed that one function of VDJ recombination is to bring the intronic enhancer close to the V gene promoter, stimulating

high-level transcription of this gene. Recent analyses of knockout mice lacking the distant 3' heavy chain enhancer suggest that this element plays a role in heavy chain isotype switching and perhaps also in somatic hypermutation of V genes. It is not known how these effects may be brought about.

Multiple transcription factors (DNA-binding proteins) have been shown to regulate Ig gene transcription. A prototype of these transcription factors is NFκB, so called because it was discovered as a nuclear factor that bound to the κ promoter in B cells. NFκB stimulates the transcription of the κ gene in mature B cells but is absent in pre–B cells (which do not express a light chain). NFκB is now known to belong to a family of structurally related transcription factors that function to control the expression of many genes in response to diverse external stimuli (Box 4–4). We

---

BOX 4–4.   Transcriptional Regulation and NFκB

A common theme in immunology is that cellular activation, such as the lymphocyte response to antigen or a target cell response to cytokines, results in the initiation of transcription of genes that were previously silent or transcribed at very low rates. The set of genes that can be transcribed in any given differentiated cell type is predetermined by poorly understood mechanisms, including the accessibility of the DNA encoding a gene to soluble protein factors (i.e., the "openness of the chromatin") and chemical modifications of the DNA in the regulatory regions of a gene (e.g., by methylation of particular nucleotides), and by other signals, including tissue-specific protein factors that may promote or repress transcription. When a gene is available for active transcription, a complex of proteins, including the RNA polymerase enzyme, assemble near the site where transcription begins. The DNA sequence that binds this **basal transcription complex** of proteins is called the **promoter.** However, binding of the basal transcription complex to the promoter is not sufficient by itself to efficiently initiate transcription, and the necessary additional signals are provided by interactions with additional proteins called **transcription factors.** Transcription factors have dual functions: they must recognize and bind to specific DNA sequences, and they must interact with proteins of the basal transcription complex. The specific DNA sequences that bind these transcription factors must be located on the same segment of DNA as the promoter and are therefore called **cis-acting sequences.** In some cases, these sequences must be near and in a particular orientation to the transcriptional start site and are often considered to be part of the promoter. In other cases, these cis-acting regulatory sequences can be as much as several thousand nucleotides away and will work in either orientation. Such movable regulatory sequences are called **enhancers.** Although most enhancer sequences are located 5' of the transcriptional start site, the immunoglobulin κ gene enhancer that operates in B cells is located 3' of the start site and is contained within an intron. The protein transcription factors that bind to enhancer sequences are usually encoded by genes that are distant from the promoter or enhancer sequences (e.g., on a different chromosome), and transcription factors are therefore sometimes called **transacting factors,** meaning they need not be encoded by the same piece of DNA as the regulated gene. In fact, many transcription factors are multimeric proteins and their subunits may be encoded by widely separated genes on separate chromosomes.

The means by which transcription factors act provide several possible mechanisms for regulation of transcription.

(1) Transcription factors may not be present in the resting cell, and activation signals can result in their *de novo* synthesis. An example of this kind of regulation is the initiation of synthesis of c-Fos, a component of transcription factor AP-1, in response to growth factors or to antigenic stimulation of lymphocytes (see Chapter 7).

(2) Transcription factors may require assembly into multimeric complexes in order to bind to DNA, and this step may be regulated. An example of this kind of regulation is the activation of the STAT proteins by Janus kinases in response to cytokines (see Chapter 12, Box 12–2).

(3) Transcription factors may require covalent modification in order to bind DNA, and this step may be regulated. In the case of c-Jun, another component of transcription factor AP-1, regulation involves removal of a phosphate moiety from an amino acid side chain located in the carboxy terminal DNA-binding region of the protein by a protein phosphatase.

(4) Transcription factors may require covalent modification in order to interact with the proteins of the basal transcription complex, and this step may be regulated. Again, in the case of c-Jun, regulation involves phosphorylation of an amino acid side chain in the amino terminal trans-activating region of the protein by a c-Jun N-terminal kinase (see Chapter 7).

(5) Transcription factors may pre-exist in a form that is capable of binding DNA and of mediating transactivation, but the factor may not have access to the cis-acting regulatory sequences of the target gene. In the case of nuclear factor of activated T cells (NFAT), regulation involves removal of a phosphate moiety from an amino acid side chain, thereby exposing an amino acid sequence required for the NFAT protein to enter the nucleus. The protein phosphatase that mediates this reaction, calcineurin, is activated in the T cell by antigen recognition (see Chapter 7).

We will discuss the transcription factor called **nuclear factor κB (NFκB)** here because of its central role in the activation of Ig synthesis. NFκB is also important in many other cellular responses in the immune system, such as antigen-induced cytokine gene transcription in T lymphocytes and cytokine-mediated activation of mononuclear phago-

Continued

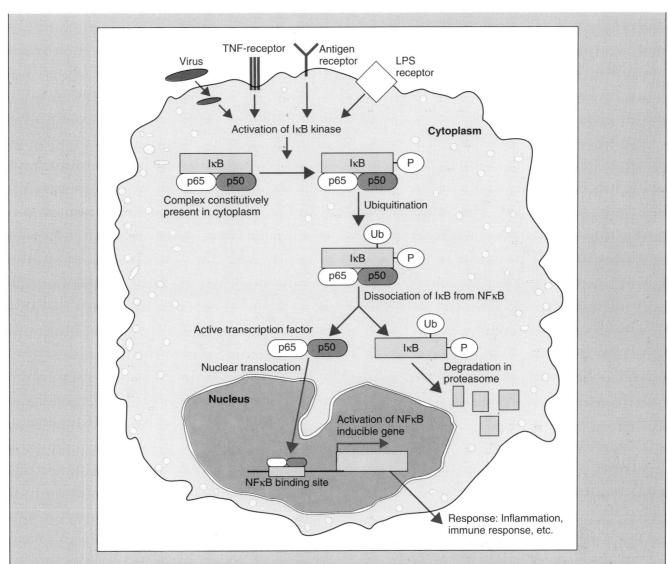

cytes. Each functional NFκB transcription factor is a dimer of either identical or structurally homologous protein subunits of about 50 to 65 kD (see Figure). The common structural motif shared by these proteins is called a rel homology domain because this sequence was first identified in the retroviral transforming gene, *v-rel*. The mammalian proteins that share this motif are: p50 (also called NFκB1), p52 (NFκB2), p65 (RelA), c-Rel, and RelB. Drosophila express a member of this family called *dorsal*. The mammalian forms of this protein are expressed in almost every cell type except for RelB, which is selectively expressed in lymphocytes. All of these proteins can participate in either homodimer or heterodimer formation, except for RelB, which only associates with p50 or p52. Minor variations in the κB enhancer-binding *cis*-acting DNA sequence may favor interactions with one form of NFκB or another.

In the resting cell, functional NFκB dimers are bound in the cytoplasm to one or more inhibitory proteins, called **inhibitors of κB (IκBs).** The "activation" of NFκB involves the signal-induced degradation of IκB proteins (see Figure). The details of this process are not fully known, but it involves the hyperphosphorylation of IκB by an as yet uncharacterized protein kinase(s), the subsequent attachment of ubiquitin to hyperphosphorylated IκB, and, finally, degradation of ubiquitinated IκB by the cytoplasmic proteasome complex (see Chapter 6). Once IκB is degraded, NFκB is free to enter the nucleus, where it binds to specific κB enhancer sequences in the target gene, initiating transcription. The IκBα isoform is one of the genes whose transcription is initiated by binding of NFκB, so that the degradation of IκBα

initiates its own synthesis, resulting in negative feedback of the signal. In contrast, IκBβ is not replenished, so that degradation of this protein leads to sustained NFκB activity. In some pre–B cells, tumor necrosis factor (TNF) triggers transient NFκB activity (IκBα degradation only), whereas interleukin (IL)-1 leads to sustained NFκB activity (IκBα and β degradation) and Ig production.

NFκB may interact with other proteins to optimally activate transcription. For example, NFκB may interact with interferon regulatory factor I to increase transcription of class I major histocompatibility complex genes (see Chapter 5, Box 5–3). The interactions between transcription factors may involve bending of DNA caused by the binding of constitutively expressed nuclear proteins (e.g., high-mobility group protein IY). The functional complex of different transcription factors bound to a contiguous segment of DNA has been called an **enhanceosome.**

NFκB was discovered as a regulatory factor of pre–B cells, but it is involved in transcriptional control of Ig throughout B cell maturation until the plasma cell stage. However, the NFκB proteins that mediate Ig gene transcription may change with developmental stage. NFκB is also involved in T cell activation, contributing to IL-2 transcription (see Chapter 7) and in the response of many diverse cell types to pro-inflammatory cytokines such as TNF and IL-1 (see Chapter 12). Although it was not thought to play a role in embryonic development, knock out of the p50 subunit (NFκB1) in mice proved to be a developmental lethal mutation. The basis of this effect is unknown. Studies of knockout mice lacking other NFκB or IκB proteins are in progress.

BOX 4–5. Chromosomal Translocations in Tumors of B Lymphocytes

Reciprocal chromosomal translocations in many tumors, including lymphomas and leukemias, were first noted by cytogeneticists in the 1960s, but their significance remained unknown until almost 20 years later. At this time, sequencing of switch regions 5′ of C_H genes in B cell tumors revealed the presence of DNA segments that were not derived from Ig genes. This was first observed in two tumors derived from B lymphocytes, human Burkitt's lymphoma, and murine myelomas. The "foreign" DNA was identified as a portion of the *c-myc* proto-oncogene, which is normally present on chromosome 8 in humans. Proto-oncogenes are normal cellular genes that often code for proteins nvolved in cell growth and regulation, such as growth factors, receptors for growth factors, or transcription-activating factors. In normal cells, their expression is tightly regulated. When these genes are altered by mutations, inappropriately expressed, or incorporated into and reintroduced in cells by RNA retroviruses, they can function as oncogenes. Such deregulation is postulated to be one of the mechanisms that lead to enhanced cellular growth and, ultimately, neoplastic transformation.

The most common translocation in Burkitt's lymphoma is t(8;14), involving the Ig heavy chain locus on chromosome 14; less commonly, t(2;8) or t(8;22) translocations are found, involving the κ or λ light chain loci, respectively. In all cases of Burkitt's lymphoma, *c-myc* is translocated from chromosome 8 to one of the Ig loci, which explains the reciprocal 8;14, 2;8, or 8;22 translocations detectable in these tumors. Because the Ig loci normally undergo several genetic rearrangements during B cell differentiation, they are likely sites for accidental translocations of distant genes. The *myc* gene product is a transcription factor, and translocation of the gene leads to its dysregulated expression. Increased *myc* activity is believed to lead to uncontrolled cell proliferation at the expense of differentiation, but the precise mechanism of

this effect is not known. Thus, myc is an example of a transcription factor that serves as an oncoprotein.

Since the discovery of *myc* translocations in B cell lymphomas, the genes involved in translocations in many other lymphoid (and non-lymphoid) tumors have been identified (see Table). In most cases, these translocations result in dysregulation of transcription factors, cell survival proteins, or signal-transducing molecules such as kinases. About 25 per cent of childhood acute lymphoblastic leukemias show a balanced t(1;19) (q23;p13) translocation, which produces a fusion gene consisting of the DNA-binding portion of PBX, a member of the Hox family of transcription factors, and the transcriptional activating domain of an unrelated transcription factor, E2A. It is postulated that the translocation converts PBX from a transcriptional repressor to an activator. About 40 per cent of acute myelogenous leukemias show t(12;21) (p12;q22) translocations, producing a fusion gene encoding portions of two unrelated transcription factors. The "Philadelphia chromosome," found in chronic myelogenous leukemia and less frequently in acute lymphoblastic leukemias, is a translocation yielding a fusion gene consisting of the *c-abl* and *bcr* genes, producing a novel tyrosine kinase. The t(14;18) translocation found in follicular B cell lymphomas leads to overexpression of the survival gene, *bcl-2*, and prevention of programmed cell death. The most frequent translocation in T cell–derived acute leukemias involves rearrangement of a transcription factor, TAL-1, to the TCRα locus. Less common in T cell leukemias is the translocation of a transmembrane receptor, NOTCH1, to the TCRβ locus. NOTCH1 plays a role in T cell development, but its mechanism of oncogenesis is not known. In some B cell lymphomas, the *bcl-1* gene is translocated to the IgH locus; *bcl-1* codes for cyclin D, a protein that regulates cell cycle progression.

This box was written with the generous assistance of Dr. Jon Aster, Department of Pathology, Brigham & Women's Hospital and Harvard Medical School, Boston.

Type of Tumor	Chromosomal Translocation	Genes Involved in Translocation
B Cell–Derived		
Burkitt's lymphoma	t(8;**14**) (q24.2;**q32.3**) most common	*c-myc*, IgH
Acute lymphoblastic leukemia (B-ALL)	t(1;19) (q23;p13) (25% of childhood cases)	PBX, E2A
B-ALL; also chronic myelogenous leukemia	t(9;22) (q23;q11) (25% of adult ALL, 3% of childhood ALL)	*bcr*, *abl* ("Philadelphia chromosome")
Follicular lymphoma	t(**14**;18) (**q23.3** q21.3)	*bcl-2*, IgH
Diffuse B cell lymphoma, large cell type	t(3;**14**) (q27;**q32.3**)	*bcl-6* (DNA-binding protein), IgH
T Cell–Derived		
Pre–T cell acute lymphoblastic leukemia	t(1;**14**) (p32;**q11**) (25% of cases)	TAL-1, TCRα
T cell ALL	t(**7**;9) (**q34**;q34.3) (3% of cases)	NOTCH1, TCRβ
Diffuse T cell lymphoma, large cell type	t(2;5) (p23;q35)	ALK (tyrosine kinase), NPM (unknown function)

The above are some samples of examples of chromosomal translocations that have been molecularly cloned in human lymphoid tumors. Each translocation is indicated by the letter t. The first pair of numbers refers to the chromosomes involved, for example, (8;14), and the second pair to the bands of each chromosome, for example, (q24.1;q32.3). The normal chromosomal locations of antigen receptor genes (indicated in bold) are: IgH, 14q32.3; Igκ, 2p12; Igλ, 22q11.2; TCRα/δ, 14q11.2; TCRβ, 7q34; TCRγ, 7p15.

Abbreviations: Ig, immunoglobulin; TCR, T cell receptor.

will return to discussions of this and other transcription factors throughout the book.

Since Ig genes are transcriptionally active in B cells and are sites for multiple DNA recombinational events, foreign genes can be abnormally translocated to Ig loci and, as a result, may become transcriptionally active. For instance, in certain tumors of B cells, oncogenes are translocated to switch sites located 5′ of heavy or light chain C region genes. Such chromosomal translocations are frequently accompanied by enhanced transcription of the oncogenes and are believed to be one of the factors causing the development of tumors of B lymphocytes (Box 4–5).

Regulation of Messenger RNA Turnover and Translation

The turnover rate of Ig mRNA changes during B cell differentiation and is another factor that determines the quantity of Ig synthesized. For instance, the half-life of Ig mRNA in myelomas, which actively synthesize and secrete large amounts of antibodies, is as long as 20 to 40 hours; in B cell lymphomas, which produce much lower quantities of Ig only in the membrane-bound form, the half-life is less than 6 hours. This may be one of the reasons why plasma cells contain 100 to 500 times more cytoplasmic Ig mRNA than do nonsecreting B lymphocytes. Moreover, if B lymphoma cell lines are stimulated to secrete more Ig, the level of steady-state Ig mRNA increases five to ten times more than the transcription rate. This implies that Ig mRNA is stabilized after stimulation and its half-life is prolonged. The mechanisms that control turnover of mRNA are largely unknown.

The importance of translational control in regulating Ig production is suggested by several observations. Mature B cells, despite containing about ten times more μ mRNA (for the membrane form) than δ mRNA, express more membrane IgD than IgM. Antibody-secreting cells often contain mRNAs for the membrane forms of heavy chains even when they do not express detectable membrane Ig. Such observations indicate that translation of Ig mRNA is a regulated process that varies according to the stage of B lymphocyte maturation and differentiation, but the control mechanisms are not yet identified.

Synthesis and Assembly of Immunoglobulin Molecules

Immunoglobulin heavy and light chains, like most secreted and membrane proteins, are synthesized on membrane-bound ribosomes in the rough endoplasmic reticulum. The 5′ signal peptides are cleaved co-translationally in the endoplasmic reticulum. The covalent association of heavy and light chains, created by the formation of disulfide bonds, probably occurs in the endoplasmic reticu-

lum as well, where high-mannose oligosaccharides may be added to asparagine residues (N-linked glycosylation). Following synthesis, the polypeptide chains are directed into the cisternae of the Golgi complex, where the high-mannose carbohydrates are converted to their mature forms by the trimming of terminal mannose and the addition of other sugar side chains. Assembled antibody molecules are transported to the plasma membrane in vesicles, where they become anchored into the cell membrane or are secreted by a process of reverse pinocytosis. Moreover, other proteins that bind to Ig are coordinately regulated. For instance, the "J chains" attached to IgA and IgM (see Chapter 3) are believed to be important for the polymerization of the secreted forms of these antibodies (and are not related to J gene segments in Ig loci). In cells producing such antibodies, transcriptional activation of Ig heavy and light chains is accompanied by coordinate stimulation of J chain gene transcription and biosynthesis.

Intracellular Transport, Surface Expression, and Secretion of Immunoglobulins

The expression of membrane and secreted forms of Ig is also controlled by mechanisms that determine the intracellular fate of newly synthesized proteins. For instance, pre–B cells synthesize both the membrane and secreted forms of μ heavy chains but do not express significant quantities of either (i.e., most of the μ in pre–B cells is in the cytoplasm). It is thought that the secreted form of μ is retained intracellularly bound to proteins that reside within the lumen of the endoplasmic reticulum. These retention proteins include calnexin, and one that has been given the rather uninformative designation "binding protein," or Bip. Bip is a member of the heat shock family of proteins and acts as a chaperonin, the name given to molecules that conduct newly synthesized proteins among various intracellular compartments. Bip complexed to μ heavy chains may shuttle between the endoplasmic reticulum and a pre-Golgi compartment. This is one of the mechanisms that ensures that μ heavy chains of the secretory form are not secreted in pre–B cells and probably also not in immature and mature B cells prior to activation by antigen. These retention mechanisms are evidently turned off after antigen exposure, allowing Ig to be secreted. Although most of the studies of retention and secretion have focused on μ heavy chains, similar mechanisms are probably operative for other heavy chain isotypes.

Most of the membrane-type μ heavy chains in pre–B cells are rapidly degraded intracellularly. As the cells mature, κ or λ light chains are produced, and their covalent assembly with μ proteins apparently protects the heavy chains from intracellular degradation. The result is expression of IgM on the surface of B cells. *In mature B lymphocytes, IgM and*

IgD are present on the cell surface non-covalently associated with two other proteins, called Igα and Igβ. Thus, the antigen receptor of B cells is actually a complex of proteins, analogous to the T cell antigen receptor complex in T lymphocytes (see Chapter 7). Igα and Igβ serve the same two functions for Ig in B cells as the CD3 and ζ proteins do for T cell antigen receptors. These functions are assembly and expression of the complex in the plasma membrane, and the transduction of signals generated by binding of antigen to membrane Ig (see Chapter 9).

SUMMARY

The maturation of B lymphocytes is accompanied by specific changes in Ig gene structure and mRNA expression, which correspond to changes in the production of Ig molecules in different forms (Fig. 4–16). The earliest detectable change in Ig genes during B cell ontogeny is a somatic recombination of variable (V), joining (J), and, for heavy chains, diversity (D) gene segments so that one of each is assembled together on one chromosome. The recombination of Ig genes occurs first in the H chain locus, followed by κ or λ. Functional rearrangement at each heavy or light chain locus inhibits rearrangement on the other allele (allelic exclusion). The constant (C) region heavy or light chain gene is attached to the VDJ or VJ complex by RNA splicing to generate a mature mRNA that is translated into heavy or light chain protein. The diverse repertoire of antibody specificities is generated by the combinatorial associations of multiple germline V, D, and J genes, and junctional diversity generated by imprecise VDJ recombination and the addition of random nucleotides to the sites of recombination. Co-expression of μ and δ heavy chains on the same B cell and the change from membrane to secreted Ig are mediated by alternative splicing of primary heavy chain RNA transcripts. Heavy chain class (isotype) switching results from a process of switch recombination and deletion of intervening C$_H$ genes. Somatic mutations in V genes give rise to affinity maturation in antibody responses. Transcription and translation of Ig are controlled by numerous processes that are incompletely elucidated; many of these processes are believed to be specific for B lymphocytes.

Analysis of these molecular events has provided a clear picture of the differentiation of B lymphocytes and the molecular basis of Ig expression. In addition, such studies form the basis of our rapidly increasing understanding of related fields, such as the evolution of B cell tumors, and are providing increasingly sophisticated tools for the diagnosis of diseases of B lymphocytes.

Selected Readings

Baldwin, A. S. The NF-κB and I-κB proteins. Annual Review of Immunology 14:649–683, 1996.

Kantor, A. B., and L. A. Herzenberg. Origin of murine B cell lineages. Annual Review of Immunology 11:501–538, 1993.

Koller, B. H., and O. Smithies. Altering genes in animals by gene targeting. Annual Review of Immunology 10:785–807, 1992.

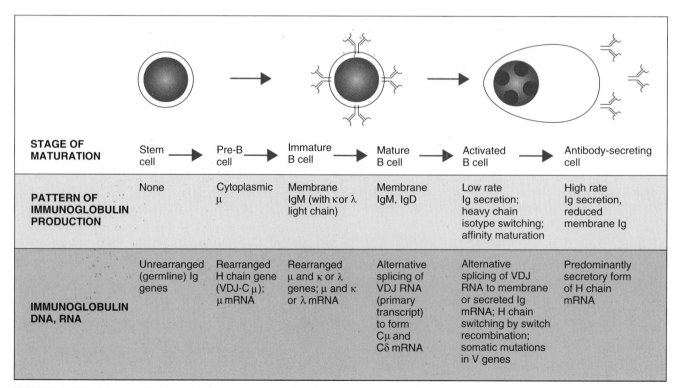

STAGE OF MATURATION	Stem cell →	Pre-B cell →	Immature B cell →	Mature B cell →	Activated B cell →	Antibody-secreting cell
PATTERN OF IMMUNOGLOBULIN PRODUCTION	None	Cytoplasmic μ	Membrane IgM (with κ or λ light chain)	Membrane IgM, IgD	Low rate Ig secretion; heavy chain isotype switching; affinity maturation	High rate Ig secretion, reduced membrane Ig
IMMUNOGLOBULIN DNA, RNA	Unrearranged (germline) Ig genes	Rearranged H chain gene (VDJ-Cμ); μ mRNA	Rearranged μ and κ or λ genes; μ and κ or λ mRNA	Alternative splicing of VDJ RNA (primary transcript) to form Cμ and Cδ mRNA	Alternative splicing of VDJ RNA to membrane or secreted Ig mRNA; H chain switching by switch recombination; somatic mutations in V genes	Predominantly secretory form of H chain mRNA

FIGURE 4–16. Scheme of B lymphocyte maturation, showing the patterns of immunoglobulin (Ig) gene expression at different stages of maturation.

Korsmeyer, S. J. Chromosomal translocations in lymphoid malignancies reveal novel proto-oncogenes. Annual Review of Immunology 10:785–807, 1992.

Matsuda, F., and T. Honjo. Organization of the human heavy-chain locus. Advances in Immunology 62:1–29, 1996.

Okada, A., and F. W. Alt. The variable region gene assembly mechanism. *In* T. Honjo and F. W. Alt (ed.). Immunoglobulin Genes, 2nd edition. Academic Press, New York, pp. 205–234, 1995.

Pillai, S. Immunoglobulin transport in B cell development. International Review of Cytology 130:1–36, 1990.

Rajewsky, K. Clonal selection and learning in the antibody system. Nature 381:751–758, 1996.

Rolink, A., and F. Melchers. B lymphopoiesis in the mouse. Advances in Immunology 53:123–156, 1993.

Schatz, D. G., M. A. Oettinger, and M. S. Schlissel. V(D)J recombination: molecular biology and regulation. Annual Review of Immunology 10:359–383, 1992.

Staudt, L. M., and M. J. Lenardo. Immunoglobulin gene transcription. Annual Review of Immunology 9:373–398, 1991.

Stavnezer, J. Antibody class switching. Advances in Immunology 61:79–146, 1996.

Tonegawa, S. Somatic generation of antibody diversity. Nature 302:575–581, 1983.

Wagner, S. D., and M. S. Neuberger. Somatic hypermutation of immunoglobulin genes. Annual Review of Immunology 14:441–457, 1996.

CHAPTER FIVE

THE MAJOR

HISTOCOMPATIBILITY COMPLEX

The **major histocompatibility complex (MHC)** is a region of highly polymorphic genes whose products are expressed on the surfaces of a variety of cells. MHC genes play a central role in immune responses to protein antigens. This is because *antigen-specific T lymphocytes do not recognize antigens in free or soluble form but instead recognize portions of protein antigens (i.e., peptides) that are non-covalently bound to MHC gene products.* In other words, MHC molecules provide a system for displaying antigenic peptides to T cells. This allows T cells to survey the body for the presence of peptides derived from foreign proteins. There are two different types of MHC gene products, called **class I** and **class II MHC molecules,** and any given T cell recognizes foreign antigens bound to only one class I or class II MHC molecule. Furthermore, as we shall discuss in Chapter 6, in each individual, the antigen receptor of mature T cells is actually specific for the complex of a foreign protein antigen plus a self MHC molecule. Thus, MHC molecules not only display peptides but are also integral components of the ligands that T cells recognize.

Before we discuss the discovery and structure of MHC molecules, it is useful to summarize the principal physiologic consequences of the specificity of T lymphocytes for self MHC–associated antigens.

1. The immune response to a foreign protein is determined by the expression of specific MHC molecules that can bind and present peptide fragments of that protein to T cells. MHC genes are very polymorphic; i.e., many different alleles exist within the population, and these alleles differ in their ability to bind and present different antigenic determinants of proteins. If a peptide is formed that does not bind to any of the allelic MHC molecules expressed by an individual, T cells cannot respond to that peptide. In this way, *MHC genes influence immune responses to protein antigens.*

2. Because MHC molecules are membrane-associated and not secreted, T lymphocytes can recognize foreign antigens only when bound to the surfaces of other cells. This limits T cell activation such that *T cells interact only with other cells that bear MHC-associated antigens and not with soluble antigens.* In fact, as we shall discuss in Chapter 7, *the T cell receptor for antigen binds to amino acid side chains of both the bound peptide and the MHC molecule itself.* The cell that bears the MHC molecules is said to *present* the antigen to the T lymphocyte. The recognition of antigen on a cell surface localizes T cell activation to the sites where antigen is concentrated, e.g., to the surface of an infected cell or to a specialized secondary lymphoid organ that functions to collect antigen (see Chapter 11). The same process also serves to localize the effector functions of the activated T cell to the anatomic site of antigen presentation. In contrast, antibodies can function in the circulation by binding to and neutralizing soluble antigens.

3. *The patterns of antigen association with class I or class II MHC molecules determine the kinds of T cells that are stimulated by different forms of antigens.* As we shall discuss in Chapter 6, the distinct pathways of biosynthesis and assembly of class I and class II MHC proteins determine the source of peptides associated with each class of MHC molecules. Peptide fragments generated in endosomes or lysosomes from endocytosed (usually exogenous) proteins bind to class II MHC molecules. In contrast, peptides generated in the cytosol (generally from endogenously synthesized proteins) associate with class I molecules. As a consequence, exogenously and endogenously synthesized proteins are, for the most part, recognized by functionally distinct T cell populations.

The functions of MHC gene products in T cell antigen recognition and in the development of the T cell repertoire are described in more detail in Chapters 6 to 8. We will begin our discussion of recognition of antigen by T cells with a discussion of the structure of MHC molecules. The terminology and genetics of the MHC are best understood from a historical perspective, and we will start with a description of how the MHC was discovered.

DISCOVERY OF THE MAJOR HISTOCOMPATIBILITY COMPLEX
Transplantation in Mice

The initial discovery of the murine MHC was made by George Snell and his colleagues, using classical genetic techniques to analyze the rejection of transplanted tumors and other tissues. In their simplest form, these experiments examined the outcome of skin grafts between individual animals. The key to the analysis was the use of inbred strains of laboratory mice.

In an animal population, some genes are represented by only one normal nucleic acid sequence; every variant nucleic acid sequence is an uncommon mutation and may result in a disease state. Such genes are said to be nonpolymorphic, and the normal, or wild type, gene sequence will usually be present on both chromosomes of a pair in each individual member of the species. (Recall that all chromosomes, except sex chromosomes in the male, are present in pairs in a normal diploid animal.) In other genes, the nucleic acid sequences may vary at a relatively high frequency among normal individuals in the population; i.e., at least 1 per cent of individuals may express a gene that differs from the homologous gene in remaining members of the population. Such genes are said to be **polymorphic.** Each common variant of a polymorphic gene present in the population is called an **allele.** For polymorphic genes, any individual animal can have the same allele at a genetic locus on both chromosomes of the pair (i.e., can be homozygous) or two different alleles, one on each chromosome (i.e., can be heterozygous).

Inbred mouse strains are produced by repetitive matings of siblings. After about 20 generations, every individual animal of a given inbred mouse strain will have identical nucleic acid sequences at all locations on both members of each pair of chromosomes. In other words, *inbred mice are completely homozygous at every genetic locus. In addition, every mouse of an inbred strain is genetically identical* (**syngeneic**) *to every other mouse of the same strain.* In the case of polymorphic genes, each inbred strain, because it is completely homozygous, expresses only one allele from the original population. Different strains may express different alleles and are said to be **allogeneic** to one another.

When a tissue or an organ, such as a patch of skin, is grafted from one animal to another, two possible outcomes may ensue. In some cases, the grafted skin survives and functions as normal skin. In others, the immune system destroys the graft, a process called **rejection.** By determining whether or not grafts exchanged among various inbred strains of mice were destroyed, several key observations were made about the genetic basis of graft rejection.

1. Grafts of skin from one animal to itself (autologous grafts, or autografts) or grafts between animals of the same inbred strain (syngeneic grafts, or syngrafts) are usually not rejected.

2. Grafts between animals of different inbred strains or between outbred mice (allogeneic grafts, or allografts) are almost always rejected.

We will return to these experiments in Chapter 17, when we discuss transplantation. Suffice it to say here that the different outcomes of grafts between syngeneic animals and grafts between allogeneic animals established that there is a genetic basis for recognizing a graft as foreign. The genes responsible for causing a grafted tissue to be perceived as similar to one's own tissues or as foreign were called **histocompatibility genes,** and the differences between foreign and self were attributed to genetic polymorphisms among different histocompatibility gene alleles.

The tools of classical genetics, namely breeding and analysis of the offspring, were then applied to identify the relevant genes (Box 5–1). The critical strategy in this effort was the breeding of **congenic mouse strains** that differed only by genes responsible for causing graft rejection. These studies indicated that although several different genes could contribute to rejection, *a single genetic region is responsible for rapid rejection.* The particular region identified in mice by George Snell's group was linked to a gene encoding a polymorphic blood group antigen called Antigen II, and this region was subsequently called histocompatibility-2 or, simply, H-2. Initially, H-2 congenic strains were

thought to differ at a single gene locus. However, occasional recombination events occurred within the H-2 locus during interbreeding of different strains, suggesting that the H-2 locus actually contained several different but closely linked genes, each involved in graft rejection. The H-2 region in mice (Fig. 5–1) is now known to be homologous to genes that determine the fate of grafted tissues in other species, and all of these are grouped under the generic name the **major histocompatibility complex.**

The genetics of graft rejection indicated that the products of MHC genes are co-dominantly expressed; i.e., the alleles on both chromosomes of a pair are expressed. As a consequence, each parent of a genetic cross between two different strains can reject a graft from the offspring by recognizing MHC alleles inherited from the other parent.

The more general importance of the MHC to immunology was first appreciated from studies of the strength of antibody production to synthetic polypeptides (e.g., random polymers of one or more amino acids) in guinea pigs and mice. Individual mendelian dominant genes, called **immune response (Ir) genes,** appeared to determine whether or not antibodies could be made to a particular amino acid polymer. Using the inbred and congenic mouse strains that were developed for transplantation studies, Baruj Benacerraf, Hugh McDevitt, and their colleagues discovered that the Ir genes all mapped to the MHC. We now know that each Ir gene encodes a particular allelic form of an MHC molecule that either can or cannot bind peptides derived from the amino acid polymer being tested. Ir genes that encode alleles of MHC molecules that can bind such peptides allow T cells specific for these peptides to be activated and provide "help" for antibody production by B cells. This results in a strong antibody response. Ir genes that encode alleles of MHC molecules that cannot bind the peptides derived from a specific polymer do not allow T cells specific for those peptides to be activated and, in the absence of T cell help, only a weak antibody response is produced. The mechanism of T cell recognition of such peptides is discussed in Chapter 6, and the basis of T cell help for B cell production of antibody is discussed in Chapter 9. Suffice it to say here that these mouse experiments moved the study of MHC genes and their products to the forefront of immunology research.

Serological Studies in Humans

The kinds of experiments used to discover and define MHC genes in mice, namely intentional inbreeding and skin graft rejection, obviously cannot be performed in humans. However, the development of allogeneic blood transfusion and especially allogeneic organ transplantation as methods of treatment in clinical medicine provided a strong

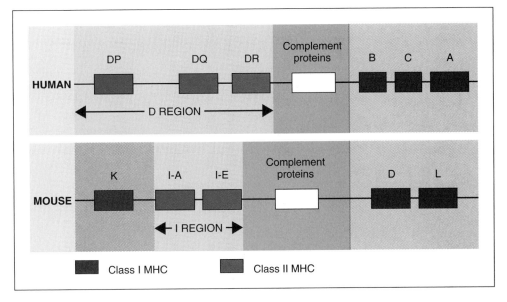

FIGURE 5–1. Schematic maps of human and mouse MHC loci. Sizes of genes and intervening distances are not shown to scale. Each class II locus (e.g., DP and DQ) consists of multiple genes. HLA-DQ is much closer to HLA-DR than to HLA-DP; in fact, HLA-DQ is in strong "linkage disequilibrium" with HLA-DR; that is, certain DQ alleles are inherited together with particular DR alleles, well out of proportion to their frequency in the population. MHC, major histocompatibility complex.

BOX 5–1.   Identification of MHC Genes in Mice

A key development in defining the genes responsible for causing graft rejection was the development of congenic mouse strains that differ only by the relevant genes, now known as the MHC. The strategy for such breeding depends upon two circumstances. First, there must already exist a homozygous inbred mouse strain that will reproducibly reject transplanted grafts of another strain. Second, there must be a simple assay for graft rejection. Both of these conditions are met by transplanting skin between inbred strains that differ in alleles of MHC genes. The procedure is outlined in the accompanying figure.

A mouse of one inbred strain A is mated with a mouse from a second strain B. The MHC of a strain A mouse is referred to as being *aa* homozygous (with italics representing alleles) and that of a strain B mouse as being *bb* homozygous. All of the first filial (F1) generation will be *ab* heterozygotes. Next, an F1 mouse is mated with a strain A mouse. This is called a backcross. Half of the offspring will be *aa* and half will be *ab*. Those mice that are *ab* can be identified by the fact that their skin will be rapidly rejected by strain A mice; in contrast, the skin of *aa* offspring will not be rapidly rejected by strain A mice. Step three is again to backcross one of the heterozygous *ab* mice with a strain A mouse. Once again, half of the offspring will be *aa* and half will be *ab*, and the heterozygous *ab* mice can be identified by the fact that strain A mice will rapidly reject their skin.

By continuing to carry out such backcrosses for multiple generations, two things occur. First, the *b* allele of the MHC will be indefinitely maintained in the offspring because its expression is necessary for skin graft rejection, and this is the positive selection being imposed by the experimenter. Second, all other genetic loci from the strain B mice will disappear as a result of random backcrosses into strain A mice. This second effect can be thought of as serial dilution. The F1 mice will contain 50 per cent strain B genes; the offspring of the first backcross will contain, on average, 25 per cent; the offspring of the next backcross, on average, 12.5 per cent, and so on, until by about 20 backcrosses no strain B genes except the MHC alleles will persist. At this point, a mouse strain has been bred that is identical to the strain A except that the MHC is heterozygous *ab*. If the new strain is allowed to interbreed, 25 per cent of the first gener-

ation offspring will be *aa* homozygous, 50 per cent will be *ab* heterozygous, and 25 per cent will be *bb* homozygous at the MHC. The *bb* MHC homozygotes can be identified as being the only mice that will rapidly reject strain A skin grafts. If the *bb* mice interbreed, a strain is produced that is identical to strain A at every locus except the MHC and is identical to strain B only at the MHC. These mice are congenic to strain A and are said to have the b MHC on an A background.

In mice, the MHC alleles of particular inbred strains are designated by lower case letters (e.g., a, b, c). The individual genes within the MHC are named for the MHC type of the mouse strain in which they were first identified. The two independent MHC loci known to be most important for graft rejection in mice are called H-2K and H-2D. The K gene was first discovered in a strain whose MHC had been designated k, and the D gene was first discovered in a strain whose MHC had been designated d. In the parlance of mouse geneticists, the allele of the K gene in a strain with the k-type MHC is called K^k (pronounced K of k), whereas the allele of the K gene in a strain of MHC d is called K^d (pronounced K of d). A third locus similar to K and D was discovered later and called L.

Several other genes were subsequently mapped to the region between the K and D genes responsible for skin rejection. For example, S genes that code for polymorphic serum proteins, now known to be components of the complement system, were identified. Most importantly, *the polymorphic Ir genes described in this chapter were assigned to a region within the MHC called I* (the letter, not the Roman numeral). The I region, in turn, was further subdivided into I-A and I-E subregions on the basis of recombination events during breeding between congenic strains. The I region was also found to code for certain cell surface antigens against which antibodies could be produced by interstrain immunizations. These antigens were called I region–associated or **Ia molecules.** As we shall discuss in Chapters 6 and 8, we now know that the I-A and I-E Ir genes are the structural genes that code for Ia antigens, which are called I-A and I-E molecules, respectively. The I-A molecule found in the inbred mouse strain with the K^k and D^k alleles is called I-A^k (pronounced I big A of k). Similar terminology is used for I-E molecules.

Continued

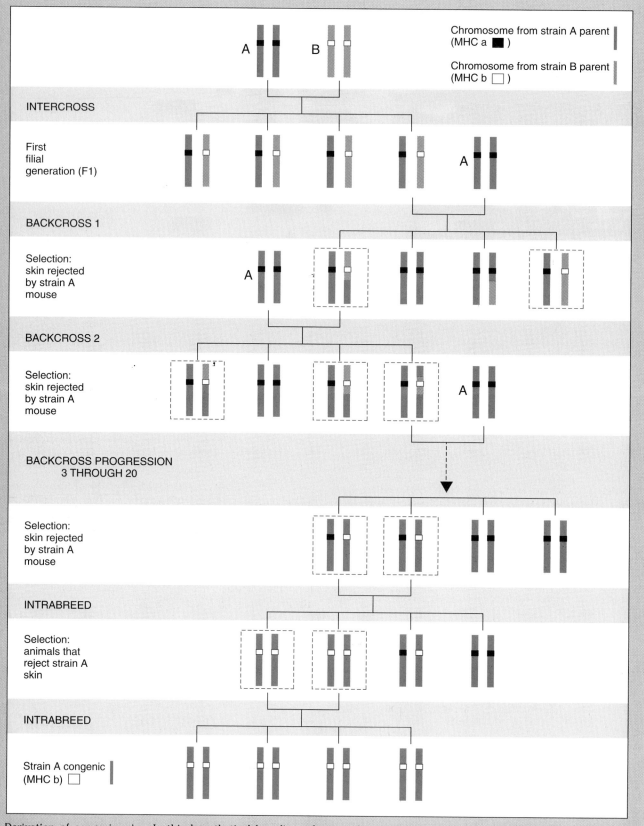

Derivation of congenic mice. In this hypothetical breeding scheme, selection for skin graft rejection is used to derive mice congenic for major histocompatibility complex (MHC) loci. To simplify the diagram, each animal is depicted as one chromosome pair, and the MHC is represented as a single locus on this chromosome. Animals that are positively selected are indicated by dashed blue borders around the chromosome pair. As described in the box, all non-MHC strain B genes are lost during progressive backcrosses with strain A animals; MHC b genes are preserved by selection, eventually allowing the breeding of a congenic strain that differs only at the MHC from parental strain A.

Continued

MHC genes were initially identified by graft rejection and Ir gene phenomena, which are T cell–mediated immune responses. This is not unexpected because, as stated earlier and discussed in detail in Chapter 6, MHC molecules are crucial for T cell recognition of foreign antigens. Nevertheless, immunization of one congenic mouse strain with cells of another can be used to produce antibodies (i.e., B cell products) specific for MHC gene products. Such antibodies were important in the biochemical analysis of murine MHC molecules and played a central role in the discovery of the MHC in humans.

impetus to detect and define genes that control rejection reactions in humans. Jean Dausset and others noted that patients who rejected kidneys or had transfusion reactions to white blood cells often developed circulating antibodies reactive with antigens on the white blood cells of the blood or organ donor. In the presence of complement, the recipient's serum would lyse lymphocytes from the donor and also lyse lymphocytes obtained from some, but not all, third parties (i.e., individuals other than the blood or organ donor or the recipient). These sera, which react against the cells of allogeneic individuals, are called **alloantisera** (or allosera for short) and are said to contain **alloantibodies,** whose molecular targets are called **alloantigens.**

It was presumed that these alloantigens are the products of polymorphic genes that distinguish foreign tissues from self. Panels of allosera from immunized donors, including multiparous women (who are immunized by paternal alloantigens expressed by the fetus during pregnancy) and actively immunized volunteers as well as transfusion or transplant recipients, were collected and compared for their ability to lyse panels of lymphocytes from different donors. Efforts at several international workshops, involving free exchanges of reagents among laboratories, led to the definition of at least six separate polymorphic genetic loci, clustered together in a single area of the genome, that can help to predict the strength of graft rejection. Because they are expressed on human leukocytes, the alloantigens recognized by these sera became known as **human leukocyte antigens (HLAs).** Family studies were then used to construct the map of these human genes (Fig. 5–1). The first three genes defined by purely serologic approaches were called HLA-A, HLA-B, and HLA-C. The second three to be identified mapped to an adjacent region, called HLA-D, which was originally detected by induction of proliferation of foreign T cells in the mixed leukocyte reaction (see below). The first gene product detected by alloantibodies that mapped to the HLA-D region was called "HLA-D related" or HLA-DR. The final two genes were subsequently called HLA-DQ and HLA-DP, with Q and P chosen for their proximity in the alphabet to R. *The HLA region is now known as the human MHC and is equivalent to the H-2 region of mice (Fig. 5–1).*

The various HLA and H-2 loci and their gene products are structurally and functionally homologous. *Specifically, human HLA-A, -B, and -C resemble mouse H-2K, D, and L (see Box 5–1) and are called class I MHC molecules, whereas human HLA-DP, -DQ, and -DR resemble mouse I-A and I-E and are called class II MHC molecules.* Indeed, similar polymorphic genes and protein products have been found in every vertebrate species examined!

The studies of the mouse MHC were accomplished with a limited number of inbred and congenic strains. Although it was appreciated that mouse MHC genes were polymorphic, only 10 to 20 alleles were identified at each locus. The human serological studies were conducted on outbred human populations. A remarkable feature to emerge from the studies of the human MHC genes is the unprecedented and unanticipated extent of their polymorphism. More than 150 separate alleles have been identified to date for some of the HLA loci, and this is undoubtedly an underestimate resulting from the limited resolution of serology. *MHC genes are by far the most polymorphic genes present in the genome of every species analyzed.* The significance of this polymorphism will become evident when we turn to the structure and functions of MHC molecules.

The use of antibodies to study alloantigenic differences between donors and recipients in human transplantation was complemented by the mixed leukocyte reaction (MLR), a test for T cell recognition of foreign MHC molecules. The MLR is also an *in vitro* model for allograft rejection and will be discussed more fully in the context of transplantation (see Chapter 17). Analysis of the allogeneic MLR led to the conclusion that two distinct classes of T lymphocytes recognize and respond to different types of MHC gene products. CD4+ T cells, most of which are cytokine-producing helper cells, are specific for class II MHC molecules, i.e., HLA-DR, -DQ, and -DP in human and I-A and I-E in mice. CD8+ T cells, most of which are cytolytic T lymphocytes (CTLs) or their precursors, are specific for class I MHC molecules, namely HLA-A, -B, and -C or H-2K, D, and L. As we shall see in Chapter 6, these MHC recognition specificities of CD4+ and CD8+ T cells apply not only to recognition of allogeneic MHC molecules but also to recognition of foreign proteins, e.g., microbial antigens, in association with self MHC molecules. In other words, *CD4+ T cells recognize foreign antigens bound to self class II MHC molecules, and CD8+ T cells recognize foreign antigens bound to self class I molecules.*

The total set of MHC alleles present on each chromosome is also called an **MHC haplotype.** In humans, each HLA allele is given a numerical designation. For instance, an HLA haplotype of an individual could be HLA-A2, -B5, -DR3, and so on. All heterozygous individuals, of course, have two HLA haplotypes. In mice, each H-2 allele is given a letter designation. Inbred mice, being homozygous, have a single haplotype. Thus, the haplotype of an H-2^k mouse is H-2K^k I-A^k I-E^k D^k L^k. In humans, certain HLA alleles at different loci are inherited together more frequently than would be predicted by random assortment, a phenomenon called linkage disequilibrium. Such extended haplotypes, in which multiple HLA genes remain linked, are associated with certain autoimmune diseases (see Chapter 19).

STRUCTURE OF MHC MOLECULES

The realization that MHC molecules play an essential part in the recognition of all protein antigens by T cells led to an enormous effort by many laboratories to elucidate the structure of these molecules. The biochemical analysis of MHC molecules has been highly successful, especially with the solution of the crystal structures for the extracellular portions of human class I and class II molecules by Don Wiley, Jack Strominger, and their colleagues. On the basis of this new knowledge, we now largely understand the structural basis of MHC molecule function.

In this section of the chapter, we will initially discuss the structure of class I and class II molecules separately, but, as we shall see, many features of these molecules now point to their fundamental similarity. We will conclude this section

with a more detailed description of peptide binding to both class I and class II molecules.

Class I MHC Molecules

All class I molecules contain two separate polypeptide chains: an MHC-encoded α or heavy chain of about 44 kD in humans, or about 47 kD in mice, and a non–MHC-encoded β chain of 12 kD in both species (Fig. 5–2). The α chain is formed by a core polypeptide of about 40 kD and contains one (human) or two (mouse) N-linked oligosaccharides. Each α chain is oriented so that about three quarters of the complete polypeptide, including the amino terminus and oligosaccharide group(s), extends into the extracellular milieu, a short hydrophobic segment spans the membrane, and the carboxy terminal 30 amino acid residues are located in the cytoplasm. The β chain interacts noncovalently with the extracellular portion of the α chain and has no direct attachment to the cell. Based upon primary amino acid sequences of many class I molecules and upon the crystal structure of the extracellular portions of several class I molecules, we can divide class I molecules into four separate domains (Fig. 5–2): an amino terminal extracellular peptide-binding domain; an extracellular immunoglobulin (Ig)-like domain; a transmembrane domain; and a cytoplasmic domain.

As we noted in the introductory section of this chapter, the principal function of MHC molecules is to bind fragments of foreign proteins, thereby

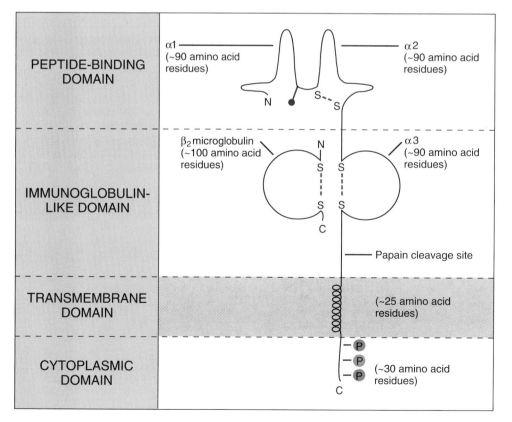

PEPTIDE-BINDING DOMAIN	α1 (~90 amino acid residues) — α2 (~90 amino acid residues) — N — S--S
IMMUNOGLOBULIN-LIKE DOMAIN	β_2 microglobulin (~100 amino acid residues) — N — α3 (~90 amino acid residues) — S-S-S-S — C — Papain cleavage site
TRANSMEMBRANE DOMAIN	(~25 amino acid residues)
CYTOPLASMIC DOMAIN	P P P (~30 amino acid residues) — C

FIGURE 5–2. Schematic diagram of a class I major histocompatibility complex molecule. Different segments are not shown to scale. N and C refer to amino and carboxy termini of the polypeptide chains, respectively; S--S, to intrachain disulfide bonds; ↑ to carbohydrate; and P, to phosphorylation sites. The papain cleavage site was used to prepare the extracellular portion of class I molecules for x-ray crystallography.

forming complexes that can be recognized by T lymphocytes. The portion of class I molecules that interacts with peptide antigen is formed from approximately 180 amino acid residues at the amino terminus of the class I α chain. Analysis of amino acid sequences has indicated that this region is formed of two homologous segments of about 90 amino acid residues each, referred to as α1 and α2. The α2 segment contains a disulfide bond, forming a loop of about 63 amino acid residues. The α1 and α2 segments interact to form a platform of an eight-stranded, antiparallel β-pleated sheet supporting two parallel strands of α-helix (Fig. 5–3 and Color Plate III, preceding Section I). Four strands of the β-pleated sheet and one of the α-helices are formed from amino acid residues of α1; the remaining four strands of the β-pleated sheet and the other α-helix are formed from amino acid residues of α2. The two α-helices form the sides of a cleft whose floor is formed by the strands of the β-pleated sheet. The cleft is of appropriate size (25Å × 10Å × 11Å) to bind a 9 to 11 amino acid fragment of a protein in a flexible, extended conformation and, as we now know, is in fact the site where foreign peptides bind to MHC molecules for presentation to T cells. Significantly, the cleft is too small to bind an intact globular protein and thus differs from the more planar binding site of antibody molecules (see Chapter 3). The small size of the cleft in MHC molecules requires that native globular proteins be "processed" to produce smaller fragments that can bind to MHC molecules and be recognized by T lymphocytes (see Chapter 6).

The α3 segment of the heavy chain and the β chain form the Ig-like domain of the class I molecule. The α3 segment contains about 90 extracellular amino acid residues located between the carboxy terminal end of the α2 segment and the insertion into the plasma membrane. The amino acid sequence in this region is highly conserved among all class I molecules examined and by sequence analysis is homologous to Ig constant domains and contains an Ig-like disulfide-linked loop. The β chain of class I molecules, which is encoded by a gene outside the MHC, is invariant among all human class I molecules examined. (In the mouse, there are two common alleles.) This polypeptide is

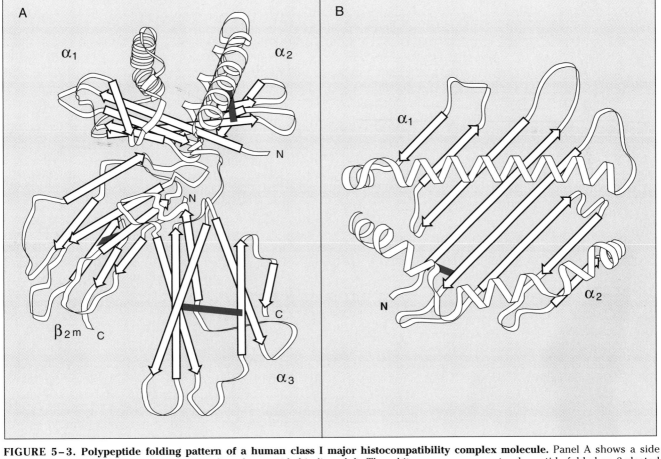

FIGURE 5–3. Polypeptide folding pattern of a human class I major histocompatibility complex molecule. Panel A shows a side view and Panel B depicts a top view, revealing the peptide-binding cleft. The white arrows represent polypeptide folded as β-pleated sheet, the white coils represent polypeptide folded as α-helix, and the blue bars represent disulfide bonds. N and C refer to the amino and carboxy termini of the polypeptide chains, respectively. (Adapted with permission from Bjorkman, P. J., M. A. Saper, B. Samraoui, W. S. Bennet, J. L. Strominger, and D. C. Wiley. Structure of the human class I histocompatibility antigen HLA-A2. Nature 329:506–512, 1987. Copyright 1987, Macmillan Magazines Ltd.)

identical to a protein previously identified in human urine, called **β_2 microglobulin** for its electrophoretic mobility (β_2), size (micro), and solubility (globulin). The β chain of class I molecules is usually called β_2 microglobulin even when attached to the cell surface. Like the $\alpha 3$ segment, β_2 microglobulin is structurally homologous to an Ig constant domain and contains an Ig-like disulfide-linked loop.

Solution of the class I MHC crystal structure has confirmed that both $\alpha 3$ and β_2 microglobulin are folded into Ig-like domains (see Chapter 3), and thus class I MHC molecules are considered to be part of the Ig superfamily (see Chapter 7, Box 7-2). These two domains interact with each other, and β_2 microglobulin also interacts with the β-pleated sheet platform of the peptide-binding domain, forming extensive contacts with amino acid residues in $\alpha 1$ and $\alpha 2$. The interaction of β_2 microglobulin with heavy chain $\alpha 1$ and $\alpha 2$ is strengthened when $\alpha 1$ and $\alpha 2$ are interacting with peptide; similarly, the energetically favorable interaction of β_2 microglobulin with class I α chain stabilizes the binding of peptide. Therefore, *the native class I molecule is best thought of as a heterotrimer, consisting of an α chain, a β_2 microglobulin, and a peptide.*

The strong correlation of CD8 expression on T cells with specificity for class I MHC–associated peptides led to the simple proposal that CD8 might function by binding to a nonpolymorphic portion of a class I molecule. This hypothesis has now been proved. Mutational and structural analyses of class I molecules have demonstrated that the nonpolymorphic $\alpha 3$ Ig domain contains a projecting loop that is responsible for binding of CD8.

The polypeptide of the α chain continues from the end of the $\alpha 3$ segment into a short connecting region and then into a stretch of approximately 25 hydrophobic amino acid residues. This segment is believed to form an α-helix that passes through the hydrophobic region of the plasma membrane lipid bilayer and anchors the MHC molecule in the membrane. As with all known transmembrane proteins, the hydrophobic sequence is immediately terminated at its carboxy terminal end with a cluster of basic amino acid residues that are believed to interact with the phospholipid head groups of the inner leaflet of the membrane bilayer. Some, but not all, class I α chains contain a cysteine residue within the hydrophobic sequence that may be modified by esterification with myristic acid. The significance of this covalent attachment of a fatty acid is unknown.

The extreme carboxy terminal portion of class I α chains contains approximately 30 amino acids that extend into the cytoplasm. The overall sequence of this intracytoplasmic domain is not conserved among different class I MHC molecules, although specific features (e.g., consensus phosphorylation sites) are conserved. Deletion of portions of the carboxy terminus has been found to inhibit internalization of class I molecules, directly implicating the carboxy terminal region in intracellular trafficking.

Class II MHC Molecules

All class II MHC molecules are composed of two non-covalently associated polypeptide chains (Fig. 5–4). In general, the two class II chains are similar to each other in overall structure. The α chain (32 to 34 kD) is slightly larger than the β chain (29 to 32 kD) as a result of more extensive glycosylation. In class II molecules, both polypeptide chains contain N-linked oligosaccharide groups, both polypeptide chains have extracellular amino termini and intracellular carboxy termini, and over two thirds of each chain is located in the extracellular space. The two chains of class II molecules are encoded by different MHC genes, and, with few exceptions, both class II chains are polymorphic.

The three-dimensional structure of class II molecules has also been solved by x-ray crystallography (see Color Plate III, preceding Section I). In addition, the nucleotide and amino acid sequences of many class II molecules are known. Both types of analysis reveal fundamental structural similarities between class I and class II molecules, especially in the peptide-binding cleft (see Color Plate IV, preceding Section I). In parallel with the structural features of the class I molecules, it is useful to divide class II MHC molecules into a peptide-binding domain, an Ig-like domain, a transmembrane domain, and a cytoplasmic domain.

The extracellular portions of both the α and β chains have been subdivided into two segments of about 90 amino acid residues each, called $\alpha 1$ and $\alpha 2$ or $\beta 1$ and $\beta 2$, respectively. The peptide-binding region of the class II molecule is formed by an interaction of both chains involving the $\alpha 1$ and $\beta 1$ segments. This is different from class I MHC molecules, in which only the α chain is involved in forming the peptide-binding cleft. The class II $\alpha 1$ and $\beta 1$ fold to form an eight-stranded, antiparallel β-pleated sheet platform supporting two α-helices; four strands of the β-pleated sheet and one of the α-helices are formed by $\alpha 1$, whereas the other four strands and the other α-helix are formed by $\beta 1$. Class II $\alpha 1$ (like class I $\alpha 1$) does not contain a disulfide-linked loop, whereas class II $\beta 1$ (like class I $\alpha 2$) does; the class II $\beta 1$ disulfide-linked loop is in the same position as the class I $\alpha 2$ disulfide-linked loop. As in the class I structure, the class II structure uses α-helices and β strands to form the sides and floor, respectively, of a peptide-binding cleft.

Both the $\alpha 2$ and $\beta 2$ segments of class II molecules contain internal disulfide bonds and, on the basis of amino acid sequence analysis, belong to the Ig superfamily. These segments, like class I $\alpha 3$ and β_2 microglobulin, are now known to be folded into Ig domains within the native MHC molecule.

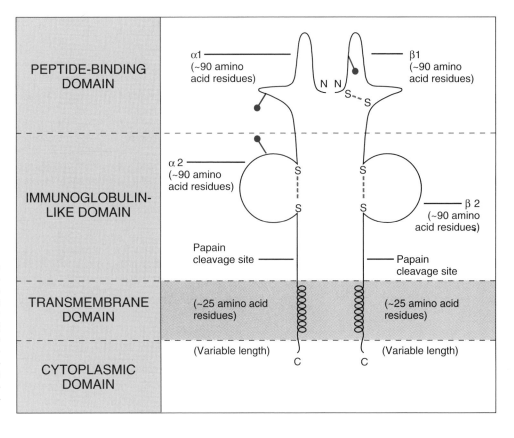

FIGURE 5–4. Schematic diagram of a class II major histocompatibility complex molecule. Different segments are not shown to scale. N and C refer to amino and carboxy termini of the polypeptide chains, respectively; S––S, intrachain disulfide bonds; and ↑ carbohydrate. The papain cleavage sites were used to prepare the extracellular portion of class II molecules for x-ray crystallography.

The class II $\alpha 2$ and $\beta 2$ segments are essentially nonpolymorphic among various alleles of a particular class II gene but show some differences among the different genetic loci. Thus, the $\alpha 2$ regions of all DR alleles are similar, but DR$\alpha 2$ differs from DP$\alpha 2$ and DQ$\alpha 2$. The correlation of CD4 expression on T cells with specificity for class II MHC molecules arises from binding of the CD4 molecule with a projecting loop of the Ig-like nonpolymorphic $\beta 2$ domain of the class II molecules, similar to the interaction of CD8 with $\alpha 3$ of the class I heavy chain.

The Ig-like regions of the class II molecules are probably important for non-covalent interactions between the two chains, although other portions of the polypeptide chains no doubt contribute as well. These interactions are quite strong and can be disrupted only by harsh denaturing conditions. In general, α chains of one locus (e.g., DR) pair best with β chains of the same locus and less commonly with β chains of other loci (e.g., DQ or DP).

The carboxy terminal ends of the $\alpha 2$ and $\beta 2$ segments continue into short connecting regions followed by approximately 25 amino acid stretches of hydrophobic residues that span the membrane. In both chains, the hydrophobic transmembrane region ends with a cluster of basic amino acid residues; these are followed by the carboxy terminal ends of the polypeptides, which form short, hydrophilic cytoplasmic tails.

The Structural Basis of Peptide Binding to MHC Molecules

Before considering the structural features of the binding of peptides to class I and class II MHC molecules, we will summarize the key features of this interaction that have been deduced from biochemical studies.

1. *The association of antigenic peptides and MHC molecules is a saturable, low-affinity interaction ($K_d \approx 10^{-6}$ M) with a slow "on rate" and a very slow "off rate."* These features were determined first by the techniques of equilibrium dialysis (see Chapter 3, Box 3–3) and gel filtration using purified class II MHC molecules and fluorescently or radioactively labeled peptides. The affinity of peptide-MHC interaction is much lower than that of antigen-antibody binding, which usually has a K_d of 10^{-7} to 10^{-11} M. In a solution, saturation of peptide binding to class II MHC molecules takes 15 to 30 minutes. Once bound, peptides may stay associated for hours to many weeks! The slow on rate of association of peptides with class II MHC molecules suggests that conformational changes in both peptide and MHC molecule are required before stable binding occurs. Dissociation of peptides from class I molecules is even slower than from class II molecules and usually requires separation of the α chain from β_2 microglobulin to occur at all. The extraordinarily slow off rates of peptide dissociation from MHC molecules allow peptide-MHC complexes to

persist long enough to interact with T cells despite the low affinity of the interaction.

2. *Each class I or class II MHC molecule binds only one peptide at a time.* This was apparent from the analysis of peptide binding to MHC molecules in solution and was confirmed by the solution of the x-ray crystallographic structure of both class I and class II MHC molecules, which show peptide occupying a single binding cleft.

3. *Multiple different peptides can bind to the same MHC molecule, albeit at different times.* This was first suggested by functional assays in which recognition of one peptide-MHC complex by a T cell could be inhibited by the addition of another structurally similar peptide. In these experiments, the MHC molecule apparently could bind different peptides, but the T cell recognized only one peptide-MHC complex. Definitive evidence for the ability of a single MHC molecule to bind different peptides came from direct binding studies with purified MHC molecules in solution as well as the analyses of peptides eluted from MHC molecules derived from intact cells. Although a wide variety of peptides with diverse amino acid sequences are capable of binding to each MHC molecule, there are certain structural constraints (discussed below) that prohibit all peptides from binding to any individual MHC molecule indiscriminately. These observations, together with the limited number of MHC alleles expressed in each individual, support the hypothesis that *MHC molecules show a broad specificity for peptide binding and that the fine specificity of antigen recognition must reside largely in the antigen receptors of T lymphocytes.*

4. *All of the peptides that bind to a particular allelic form of an MHC molecule show certain common features that may not be shared by peptides that bind to other allelic MHC molecules.* Examples of shared features are a hydrophobic residue at position 2 or a positively charged residue at position 7. Mutagenesis studies have confirmed that such motifs are crucial for peptide binding to particular allelic forms of MHC molecules.

5. *There are distinct differences in the nature of peptides that bind to class I or class II MHC molecules.* Most significantly, peptides that are eluted from class I molecules are typically 9 to 11 amino acid residues in length, whereas those eluted from class II molecules can range from 10 to 30 residues or more.

6. *The amino acid residues that vary among different alleles of class I and class II MHC molecules are largely confined to the amino terminal peptide binding domains.* Mutational analyses of MHC molecules confirm that *many of these polymorphic residues define the peptide binding specificity of the molecule encoded by a particular MHC allele. Other polymorphic residues, also located within the amino terminal peptide binding domains, do not affect peptide binding but do affect T cell recognition of the peptide-MHC molecule complex.*

These features of the peptide-MHC interaction can now be explained in precise structural terms. For example, the α-helical sides of the cleft of class I MHC molecules converge at the ends of the cleft, limiting the size of peptides that can be accommodated within the cleft to nine or ten residues. The binding of an 11-residue peptide is possible, but it requires that the peptide bow upward in the center in order to be accommodated. Twelve residues is simply too large to fit into a class I cleft. In contrast, the α-helical sides of the cleft of class II MHC molecules do not converge, allowing bound peptides to extend outward from the ends of the cleft. Thus, peptides that bind to class II molecules have no maximum length. This structural difference accounts for the observed difference in the size of peptides eluted from class I versus class II molecules. In class I MHC molecules, the charged amino terminal and carboxy terminal ends of the peptide interact electrostatically with counter-charges on the MHC molecule. Such interactions do not occur in class II molecules.

In any given class I or class II MHC molecule, the characteristically conserved features of the peptides that bind to that allelic form of the molecule are complemented by the presence of specific structural features of the MHC molecule such as the presence of pockets in the floor of the cleft. These pockets are actually spaces between the peptide backbones of the β-pleated strands. The presence or absence of a pocket is determined by the amino acid sequence of the β strands, and when a pocket is formed, the polymorphic residues of the MHC molecule that form the pocket determine the nature of the peptide side chain that can fit into the pocket (e.g., hydrophobic, charged, etc.). Conserved peptide residues that fit into the pockets of the MHC molecules are called "anchor residues" because they are critical for attaching the peptide to the MHC molecule. In the initial structures that were solved for class I molecules, the anchor residues were located near the ends of the peptide, placing little constraint upon the peptide sequence except at the ends. Some more recent structures indicate that this feature is not universal (i.e., anchor residues can be located in the middle of the peptide and interact with pockets in the middle of the cleft). Anchor residues make a strong contribution to peptide binding but are not the sole basis of attachment to MHC molecules. Some polymorphic residues in the α-helices of the MHC molecule make contacts with the peptide and may also contribute to specificity of binding. Finally, some of the contacts between the MHC molecule and the peptide involve non-polymorphic amino acid residues of the α-helices; these residues typically interact with conserved features of the peptide, such as its peptide backbone, and do not contribute to specificity but do stabilize binding.

Some peptide amino acid side chains and

some polymorphic residues of the α-helices point upward (i.e., away from the floor of the cleft). These residues do not contribute to peptide-MHC interactions but instead form the antigenic surface recognized by the T cell receptor. In other words, *amino acid side chains from both peptide and MHC molecules contribute to T cell recognition.*

These structural studies have established the significance of polymorphisms within the MHC, namely that *the polymorphic residues of MHC molecules contribute to determining the specificity of peptide binding and to determining the structure recognized by T cell antigen receptors.* We conclude our discussion of peptide binding to MHC molecules with a consideration of the genetic basis of MHC polymorphism. On a population level, there is an advantage to having multiple alleles, namely that the presence of polymorphism decreases the likelihood that any particular microbe can escape detection by the immune systems of all individuals in the population by encoding proteins that cannot be digested into peptides capable of binding to some host MHC molecule. It is hard to calculate how significant an advantage to the population MHC polymorphism actually confers because, as we have described, the structural requirements for peptide binding by particular MHC alleles are fairly broad. Furthermore, many different allelic forms of MHC molecules may have very similar binding specificities for peptides, an observation that has led some investigators to divide MHC molecules into a limited number of "supertypes." Despite the obvious advantage to the population of having widely polymorphic MHC molecules, it has not

been possible to demonstrate that infectious microbes have exerted selective pressure on generating or maintaining specific polymorphisms. Nevertheless, specific mechanisms have evolved for generating new MHC molecule polymorphisms, which we will discuss when we consider the genomic organization of the MHC.

GENOMIC ORGANIZATION OF THE MHC
Organization of the MHC Gene Loci

In humans, the MHC is located on the short arm of chromosome 6. β_2 microglobulin is encoded by a gene on chromosome 15. The human MHC occupies a large segment of DNA, extending about 3500 kilobases (kb). (For comparison, a large human gene may extend up to 50 to 100 kb, and 3500 kb is the size of the entire *Escherichia coli* genome!) In classical genetic terms, it extends about 4 centimorgans, meaning crossovers within the MHC occur with a frequency of over 4 per cent at each meiosis. A recent molecular map of the human MHC is shown in Figure 5–5. Many of the genes found within the MHC code for proteins whose function is not yet known. In addition, there are many as yet unidentified genes, especially within the class I region. Remarkably, expression of almost all of the genes located within the MHC is responsive to the cytokine, interferon-γ (IFN-γ). The class II genes are located closest to the centromere.

A surprise from these gene-mapping studies is that there may be two or three functional β chain genes for some class II loci but usually only one

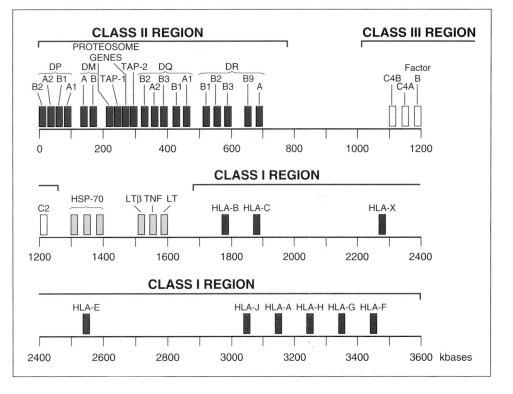

FIGURE 5–5. Molecular map of the human major histocompatibility complex. HLA-F, G, H, J, and X are class I–like molecules. This map is simplified to exclude other class I–and class II–like genes, genes not of immunologic interest, and numerous genes of unknown function. The pattern of class II genes may vary with the inherited allele. DM, TAP, and proteosome genes contribute to MHC molecule assembly; C2, C4A, C4B, and Factor B are complement proteins; HSP-70 is a heat shock protein; lymphotoxin (LT), lymphotoxin β (LT-β), and tumor necrosis factor (TNF) are cytokine genes.

functional α chain gene. The use of more than one β chain gene allows some class II gene products, especially HLA-DR, to be expressed in more than two "allelic" forms on a single cell. This contributes to an important difference between class I and class II loci. For class I, *a heterozygous individual expresses six different polymorphic alleles (three from each parent) and six different class I MHC molecules per cell.* For class II, although the individual also inherits only six different polymorphic alleles, more than six different class II MHC $\alpha\beta$ heterodimers can be expressed per cell. Some class II molecules are formed by more than one functional polymorphic β chain within the same allelic locus. Additional class II heterodimers are produced by the combination of the α chain from one allele and the β chain of the other allele. *Usually, individuals can express 10 to 20 different class II gene products per cell; the variation in number depends upon which alleles have been inherited.* This increases the potential number of foreign peptide antigens that can bind to and be presented in association with class II molecules.

An important recent discovery is that the class II region of the MHC also contains genes whose expression is necessary for the biosynthesis of class I and class II molecules. Two of these genes, called **transporter in antigen processing (TAP) 1 and 2,** encode subunits of a heterodimeric protein that transports peptides from the cytosol into the endoplasmic reticulum, where they can associate with newly translated class I MHC heavy chains. Other genes in this cluster encode subunits of a cytosolic protease complex (the **proteasome**) that is involved in generating the peptides from cytosolic proteins that are incorporated into class I MHC molecules. Another pair of genes, called **HLA-DMA and B,** encode a heterodimeric class II–like molecule that facilitates peptide binding to the polymorphic class II molecules. The functions of these proteins will be discussed in more detail in Chapter 6.

The complement or so-called class III region is telomeric to the class II region and encodes components of the complement system (C2, C4A, C4B, and Factor B, see Chapter 15), as well as the enzyme steroid 21-hydroxylase. The most telomeric portion of the MHC contains the class I α chain genes in the sequence HLA-B, C, and A. The genes for the cytokines, tumor necrosis factor (TNF), and lymphotoxin (LT), as well as for the structurally related protein lymphotoxin β (LT β), map between the complement and class I regions (see Chapter 12). Some heat shock proteins also map to this region; these molecules normally bind to denatured proteins and may be involved in delivery of denatured cytosolic proteins to the proteasome for degradation.

The large space located between the HLA-C and A genes contains additional genes that are class I–like. Many more class I–like genes have been found telomeric to HLA-A outside of the true MHC; these genes occupy another 11 centimorgans! Some of these class I–like sequences are pseudogenes, but some encode nonpolymorphic proteins that are expressed in association with β_2 microglobulin. Such gene products have been called class IB proteins to distinguish them from the conventional polymorphic class IA MHC molecules we have discussed in this chapter. In mice, one such gene, called Tla, is expressed on thymocytes and leukemic T cells. A recently described human class IB gene, called HLA-H, appears to participate in iron metabolism. Mutations in the HLA-H gene cause an inherited condition of iron overloading called hemochromatosis. Other β_2 microglobulin–associated proteins, such as the neonatal Fc receptor (FcRN, see Chapter 3) and the CD1 molecules, which may be involved in presenting non-peptidic antigens to unusual populations of T cells (see Chapter 6), are encoded outside of the MHC on different chromosomes. These molecules also show some homology to class IA MHC molecules and are sometimes included in the list of class IB molecules.

The function of the vast majority of the nonpolymorphic class IB molecules encoded within the MHC is not known. An attractive proposal for the function of these genes and the many pseudogenes is that they serve as a repository of alternative nucleic acid sequences to be used for generating polymorphic sequences in conventional class I and II MHC molecules by the process of **gene conversion.** Simply stated, gene conversion involves replacing a portion of the sequence of one gene with the portion of another gene without a reciprocal recombination event. Gene conversion is a far more efficient mechanism for genetic mutation without loss of function than point mutations because several changes can be introduced at once, and amino acids necessary for maintaining protein structure can remain unchanged if identical amino acids at those positions are encoded by both the genes involved in the conversion event. It is clear from population studies that gene conversion, rather than point mutations, has been responsible for the generation of the extraordinary polymorphism of the MHC molecules.

The murine MHC, located on chromosome 17, occupies a somewhat smaller region than the human MHC (about 2000 kb), and the genes are organized in a slightly different order. Specifically, one of the class I genes (H2-K) is located centromeric to the class II region, but the other class I genes and the nonpolymorphic class IB genes are telomeric to the class II region. A possible interpretation of this variant arrangement is that the basic form of the MHC—namely class II, complement, cytokines, and class I—arose prior to speciation between mouse and human. Subsequent to speciation, the murine MHC may have undergone a rearrangement that resulted in dividing the class I genes. The molecular structure of the murine class II region has also revealed some surprises not fully

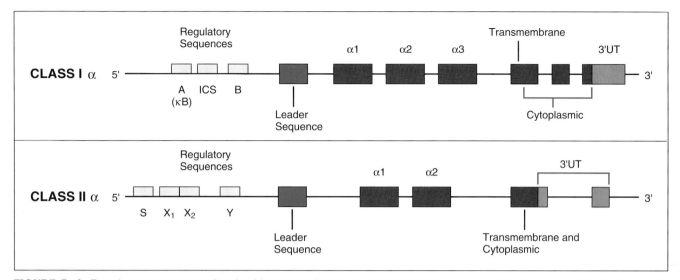

FIGURE 5–6. Exon-intron structures of major histocompatibility complex genes. Exons are depicted as boxes. The first exon also includes short 5′ untranslated sequences. The 5′ regulatory sequences include sequences described in Box 5–3. 3′UT indicates the 3′ untranslated sequence. Note that exons and introns are not shown to scale.

anticipated by classical genetics. The I-A subregion, originally defined by classical genetics, codes for the α and β chains of the I-A molecule as well as the highly polymorphic β chain of the I-E molecule. The I-E subregion of classical genetics codes only for the less polymorphic α chain of the I-E molecule. As in the human, β_2 microglobulin is not coded for by the MHC, but it is located on a separate chromosome (chromosome 2).

Organization of Individual Class I and Class II MHC Genes

The general patterns of intron-exon organization of the individual class I and class II genes are similar to each other. Schematic examples of class Iα and class IIα genes are depicted in Figure 5–6.

In all MHC genes, the first exon encodes the leader or signal sequences that target the nascent proteins to the endoplasmic reticulum. These amino acid residues are not found in mature cell surface MHC molecules. Each of the approximately 90 amino acid residue extracellular segments (e.g., class I α1, α2, and α3 or class II α1, α2, β1, and β2) is coded for by a separate large exon. Since polymorphisms are largely localized to the amino terminal domains of the proteins, the polymorphic regions are contained within one or two exons per gene. Consequently, the polymorphisms of different alleles can be studied by sequencing a portion of each gene, for example with the polymerase chain reaction (PCR) (Box 5–2). The transmembrane and cytoplasmic regions are encoded by several small exons.

BOX 5–2. Polymerase Chain Reaction

The PCR is a rapid and simple method for copying and amplifying specific DNA sequences up to about 1 kilobase in length. In order to use this method, it is necessary to know the sequence of a short region of DNA on each end of the larger sequence that is to be copied; these short sequences are used to specify oligonucleotide primers. The method consists of repetitive cycles of DNA melting, DNA annealing, and DNA synthesis (see figure). Double-stranded DNA containing the sequence to be copied and amplified is mixed with a large molar excess of two single-stranded DNA oligonucleotides (the primers). The first primer is identical to the 5′ end of the sense strand of the DNA to be copied, and the second primer is identical to the anti-sense strand at the 3′ end of the sequence. (Since double-stranded DNA is antiparallel, the second primer is also the inverted complement of the sense strand.)

The PCR is initiated by melting the double-stranded DNA at high temperature and cooling the mixture to allow DNA annealing. During annealing, the first primer (present in large molar excess) will hybridize to the 3′ end of the anti-sense strand, and the second primer (also present in large molar excess) will hybridize to the 3′ end of the sense strand. The annealed mixture is incubated with DNA polymerase I and all four deoxynucleotide triphosphates (A, T, G, and C), allowing new DNA to be synthesized. DNA polymerase I will extend the 3′ end of each bound primer, synthesizing the complement of the single-stranded DNA templates. Specifically, the original sense strand is used as a template to make a new anti-sense strand and the original anti-sense strand is used as a template to make a new sense strand. This ends cycle one. Cycle two is initiated when the reaction mixture is remelted and then allowed to reanneal with the primers. In the DNA synthetic step of the second cycle, each strand synthesized in the first cycle serves as an additional template, having hybridized with the appropriate primers. (It follows that the number of templates in the reaction doubles with each cycle, hence the name "chain reaction.") The second

Continued

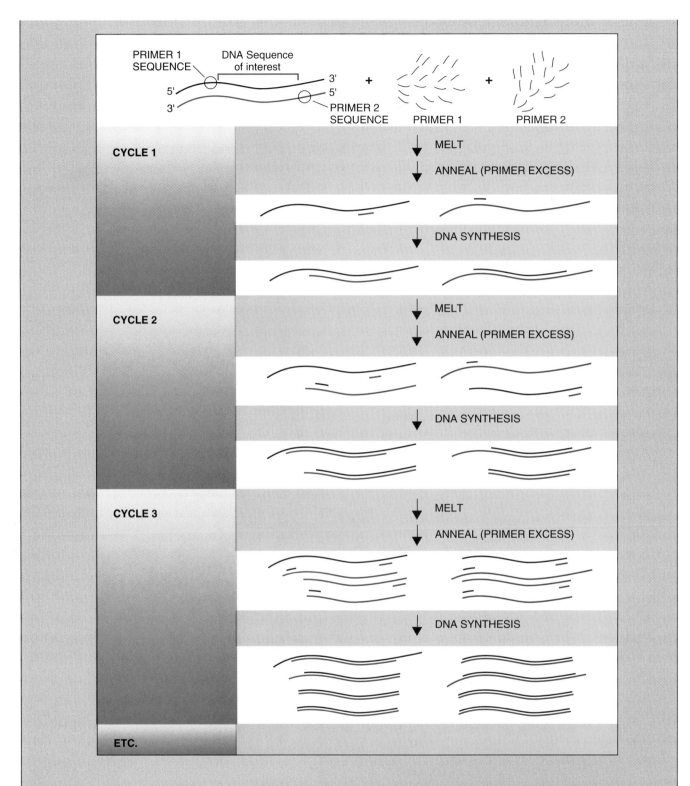

cycle is completed when DNA polymerase I extends the primers to synthesize the complement of the templates. PCR reactions can be conducted for 20 or more cycles, doubling the sequence flanked by the primers at each step. These reactions are routinely automated by using temperature controlled cyclers to regulate melting, annealing, and DNA synthesis and by using a DNA polymerase I enzyme isolated from thermostable bacteria that can withstand the temperatures used for melting DNA.

PCR has many uses. For example, by choosing suitable primers, exon 2 of an unknown HLA-DR β allele (i.e., the

exon that encodes the polymorphic $\beta 1$ protein domain) can be amplified, the amplified DNA ligated into a suitable vector, and the DNA sequence determined without ever isolating the original gene from genomic DNA. This was the original purpose for which PCR was invented. PCR is also widely used for the cloning of known genes or genes related to known genes (e.g., by choosing primers from highly conserved regions of a gene family). PCR can be used to detect, or in some cases even quantify, the presence of particular DNA sequences in a sample (e.g., the presence of viral se-

Continued

quences in a clinical specimen). Levels of messenger ribonucleic acid (mRNA) can also be quantified by using reverse transcriptase to make a complementary DNA copy of the mRNA prior to PCR amplification. Finally, since PCR can amplify DNA only when the two primers are near each other (i.e., within about 1 kb), PCR can be used to assay for specific gene rearrangements by choosing primers complementary to sequences that are brought together only when the gene is rearranged (e.g., with specific V and J segment sequences in the Ig genes; see Box 4–2).

REGULATION OF MHC MOLECULE EXPRESSION

The expression of MHC molecules on different cell types determines whether or not T lymphocytes can interact with foreign (e.g., microbial) antigens present within these cells. CD8$^+$ CTLs recognize foreign antigens such as viral peptides when they are bound to class I MHC molecules. The ability of CTLs to lyse a virally infected cell is a direct function of the quantity of class I molecules expressed. In contrast, CD4$^+$ helper T lymphocytes recognize antigens bound to class II MHC molecules. Class II MHC molecules are expressed by fewer cell types than class I MHC molecules and are under different regulation. As a consequence, fewer cell types can present antigens to helper T cells (see Chapter 6). Furthermore, the ability to present antigen to helper T cells is, in large part, a function of the level of class II MHC molecule expression. Thus, regulating the level of class II MHC molecule expression on an antigen-presenting cell may regulate the strength of the CD4$^+$ T cell response to foreign antigens presented by that cell. The concluding portion of this chapter describes the regulation of MHC gene expression.

The expression of class I and class II MHC molecules is regulated both by differentiation, in a cell- and tissue-specific manner, and by extrinsic immune and inflammatory stimuli. There are four important features of MHC molecule expression:

1. *The constitutive expression of class I molecules is distinct from that of class II molecules.* In general, class I molecules are present on virtually all nucleated cells, whereas class II molecules are normally expressed only on dendritic cells, B lymphocytes, macrophages, endothelial cells, and a few other cell types.

2. *The rate of transcription is the major determinant of MHC molecule expression on the cell surface.* Some post-transcriptional regulation may occur, but generally these effects are small.

3. *Transcription and expression of the various class I genes and molecules are coordinately regulated; similarly, the transcription and expression of different class II genes and their products are also coordinately regulated.* In many cells, β_2 microglobulin is coordinately regulated with class I α chains, despite the fact that the β_2 microglobulin gene is not located within the MHC. However, the transcription of class I and class II MHC molecules may be independently regulated on the same cell.

4. *Cytokines are the key modulators of the rate of transcription of class I and class II genes in a wide variety of cell types.* This is an important amplification mechanism for T cell responses because most of the cytokines that enhance MHC expression are secreted by T cells, and MHC molecules are components of the ligands that T cells recognize and respond to. On almost all cell types examined, IFN-α, -β, and -γ markedly increase the level of expression of class I molecules. TNF and LT can also increase class I molecule expression. (The biologic activities of cytokines are discussed in detail in Chapter 12.) These cytokine effects are mediated by increased levels of gene transcription that result from cytokine-activated transcription factors binding to regulatory deoxyribonucleic acid (DNA) sequences in class I genes (see Box 5–3).

Class II molecules not only show marked differences in expression among various cell types but also show cell type–specific differences in regulation by cytokines. Among the cells that commonly present antigens to T cells, mononuclear phagocytes express only low levels of class II molecules until stimulated to do so by IFN-γ or certain other cytokines. These cytokine responses involve increased transcription of the class II genes (see Box 5–3). This expression is antagonized by IFN-α, IFN-β, and IL-10. Vascular endothelial cells, like mononuclear phagocytes, increase class II expression in response to IFN-γ. Endothelial cell expression is antagonized by IFN-α and IFN-β but not by IL-10. Immature dendritic cells (e.g., Langerhans cells or circulating dendritic cells) express class II molecules in response to granulocyte-monocyte–colony stimulating factor, whereas mature dendritic cells in lymph nodes are constitutively positive for class II expression and do not appear to respond further to cytokines. B lymphocytes constitutively express class II molecules and can also respond to cytokines. However, resting B cells in mice increase expression of class II molecules in response to IL-4 and IL-10 and may decrease expression in response to IFN-γ. Although most class II–positive human cells express HLA-DR and HLA-DP at considerably higher levels than they express HLA-DQ molecules, B cells express HLA-DQ at greater levels than they express HLA-DP.

Most cell types express little if any class II MHC molecules unless exposed to high levels of IFN-γ. In some cells, such as pancreatic islet cells, additional cytokines like TNF may be necessary as co-signals for induction. These cells are unlikely to present antigens to CD4$^+$ T cells except in unusual circumstances. Some cells, such as neurons, do not respond to any known cytokine and remain

BOX 5-3. Transcriptional Regulation of MHC Genes

In most cell types, the class I MHC genes are constitutively transcribed. This response is dependent upon two short DNA sequences located 5' to the transcriptional start site of the gene, called enhancer A and enhancer B. In resting cells, enhancer A, which has a κB consensus sequence, is occupied by a transcription factor (KBF1) now known to be a homodimer of p50 (NFκB-1) subunits (see Chapter 4, Box 4-4). The factor that binds to enhancer B is unknown.

Cytokines increase the rate of class I gene transcription. Some but not all of the cytokine responsive elements are located 5' to the transcriptional start site. When cells are treated with TNF, p50 homodimeric forms of NFκB are displaced by p50/p65 (NFκB-1/RelA) heterodimers and transcription increases. Positioned 20 nucleotides downstream of enhancer A is a DNA sequence that conforms to the requirements of an IFN consensus sequence. In cells treated with IFN (α, β, or γ), this site becomes occupied by a transcription factor called IFN response factor-1 (IRF-1), and transcription of the gene increases. Co-treatment of cells with TNF and IFN causes a greater increase in transcriptional rate than treatment with either cytokine alone.

For reasons not currently understood, the onset in the cytokine-induced transcription of the class I genes is delayed by several hours, although activation of transcription factors is not. During this time, the cell is not idle. Both TNF and IFN cause the cell to increase transcription and translation of TAP proteins, which (as will be discussed in Chapter 6) are needed for peptide loading and assembly of newly synthesized class I molecules in the endoplasmic reticulum. The

IFN response of the TAP genes utilizes a Jak/STAT pathway (see Chapter 12, Box 12-2).

In many cell types the class II MHC genes are constitutively silent, whereas in others class II genes are constitutively expressed. The promoters of many class II genes contain several conserved DNA sequences commonly called "boxes." These include (5' to 3') a seven nucleotide S box (also called the H box) that is part of a larger (approximately 20 nucleotide) W or Z box; a pyrimidine-rich (i.e., A,G-rich) sequence, a 15 nucleotide X box, which overlaps by two nucleotides with an eight nucleotide X_2 box, and finally, about 10 to 15 nucleotides further downstream, the 10 nucleotide Y box. The S, X, X_2, and Y boxes, with appropriate spacing and alignment, are all required for class II gene expression in transfected cells and in transgenic mice. A similar array of *cis*-acting regulatory sequences is found in genes encoding proteins needed for class II molecule synthesis (i.e., in HLA-DM and in the non–MHC-encoded invariant chain; see Chapter 6). As predicted from this sharing of *cis* regulatory elements, the genes encoding proteins involved in class II synthesis are coordinately transcribed with the class II structural genes.

Multimeric proteins bind to all of the *cis* regulatory elements of the class II genes in cells that actively express class II molecules, but often not in cells that fail to express class II molecules. The X box is occupied by a protein complex called regulatory factor X (RFX), the X_2 box by a complex called X_2 binding protein (X_2BP), and the Y box by a complex called nuclear factor-Y (NF-Y). The binding of these

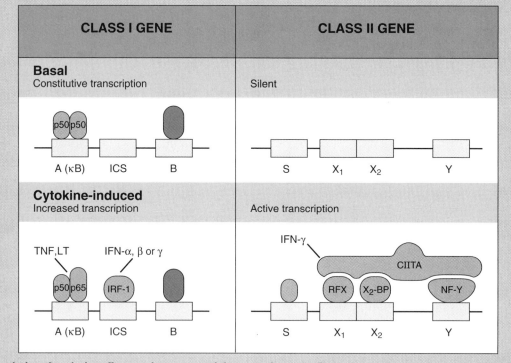

Occupancy of class I and class II major histocompatibility complex gene 5' regulatory regions by transcription factors. The upper panels depict the state of these genes in resting cells where class I molecules (left) are constitutively expressed and class II genes are not transcribed (right). The lower panels depict the changes in occupancy induced by cytokines, leading to increased rates of transcription of class I genes and the onset of transcription of class II genes. Abbreviations: A, enhancer A; B, enhancer B; ICS, interferon consensus sequence; IFN, interferon; IRF-1, interferon regulatory factor-1; p50 and p65, subunits of nuclear factor κB; TNF, tumor necrosis factor; LT, lymphotoxin; S, X_1, X_2, and Y are the names of conserved *cis* regulatory sequences described in the text of Box 5-3; CIITA, class II transcriptional activator; RFX, regulatory factor X box; X2-BP, X_2 box binding protein; NF-Y, nuclear factor of the Y box.

Continued

protein complexes to DNA is cooperative and requires that all three complexes be present in the cell for any one to bind efficiently. Not all of the protein subunits of these complexes have been fully characterized, and the precise subunit composition of these factors may vary with cell type and activation state of the cell. Proteins that bind to the S box are even less well characterized.

Mutations in the RFX protein complex have been identified in several families and result in a failure to express class II molecules (a form of the "bare lymphocyte syndrome"; see Chapter 21). Other families with a similar phenotype proved to have a mutation in a different protein, called class II transcription activator (CIITA). CIITA does not directly bind to the class II gene promoter but instead binds to the assembled RFX/X_2BP/NF-Y complex and is necessary for interaction of these transcription factors with the basal transcription complex. Synthesis of CIITA is induced by IFN-γ, and expression of this protein in IFN-γ–treated cells mediates the response of class II genes to this cytokine.

class II-negative. Finally, human but not mouse T cells express class II molecules upon activation; however, no cytokine has been identified in this response, and its functional significance is unknown.

SUMMARY

The MHC is a large genetic region coding for the class I and class II MHC molecules as well as other proteins. MHC molecules are extremely polymorphic, with up to 150 or more common alleles for each individual gene. Both class I and class II molecules were originally recognized for their role in triggering T cell responses that caused the rejection of transplanted tissue. It is now appreciated that MHC-encoded class I and class II molecules bind foreign peptide antigens and form complexes that are recognized by antigen-specific T lymphocytes. Antigens associated with class I molecules are recognized by CD8$^+$ cytolytic T lymphocytes, whereas class II–associated antigens are recognized by CD4$^+$ helper T cells. The class I molecules are composed of a 44 kD transmembrane glycoprotein in a non-covalent complex with a non-polymorphic 12 kD polypeptide (β_2 microglobulin). The class II molecules contain two MHC-encoded polymorphic chains (about 31 to 34 kD and 29 to 32 kD). The three-dimensional structures of both classes of MHC molecules are similar and may be divided into a polymorphic amino terminal extracellular peptide-binding domain, an extracellular Ig-like domain, a transmembrane domain, and a cytoplasmic domain. The peptide-binding region of class I molecules is formed by the α1 and the α2 segments of the heavy chain. It consists of a cleft measuring approximately 25Å $\times$ 10Å $\times$ 11Å with α-helical sides and an eight-strand anti-parallel β-pleated sheet floor. The analogous cleft of class II molecules is formed by the α1 and β1 domains of the two chains. The Ig-like domains of class I and class II molecules show only limited polymorphisms and contain the conserved interaction regions for CD8 and CD4 molecules, respectively.

MHC molecules bind only one peptide at a time, and all of the peptides that bind to a specific allelic form of an MHC molecule share common structural motifs. In general, peptide binding is low affinity ($K_d \approx 10^{-6}$ M), but the off rate is very slow so that complexes, once formed, persist for a sufficiently long time to be recognized by T cells. The peptide binding cleft of class I molecules is closed at its ends, limiting bound peptides to 11 amino acid residues in length, whereas that of class II molecules is open, allowing larger peptides (up to 30 amino acid residues or more) to bind. The polymorphic residues of MHC molecules are localized to the peptide-binding domain. Some polymorphic MHC residues determine the binding specificities for peptides by forming complementary structures (e.g., pockets) that interact with conserved features of the bound peptide (e.g., anchor residues). Some polymorphic MHC residues and some residues of the peptide are not involved in peptide–MHC molecule binding but instead form the antigenic determinant recognized by T cells.

The human MHC is very large (about 3500 kb) and is organized as follows: (1) class II genes (HLA-DP, HLA-DQ, HLA-DR), (2) complement genes, (3) heat shock protein and cytokine (TNF, LT, and LT-β) genes, and (4) class I genes (HLA-B, HLA-C, and HLA-A). The mouse MHC is smaller (about 2000 kb), and the sequence of genes is (1) class I (H-2K), (2) class II (I-A, I-E), (3) complement genes, (4) cytokine genes, and (5) class I (H-2D, H-2L). All MHC genes have similar exon-intron structure, and most regulatory sequences are located in the 5' flanking region. The expression of the MHC gene products is largely regulated at the level of transcription both by cell type–specific factors and by inflammatory and immune stimuli, including cytokines like IFN-γ. In general, class I genes are expressed more widely, i.e., on more diverse cell types, than are class II genes. Different cell types have distinct patterns of expression of class II MHC molecules. Some cells, such as mononuclear phagocytes, can be induced to express class II molecules by cytokines, especially IFN-γ. Other cells, such as dendritic cells and B lymphocytes, constitutively express class II molecules.

Selected Readings

Bjorkman, P. J., and P. Parham. Structure, function and diversity of class I major histocompatibility complex molecules. Annual Review of Biochemistry 59:253–288, 1990.

Bjorkman, P. J., M. A. Saper, B. Samraoui, W. S. Bennett, J. L. Strominger, and D. C. Wiley. Structure of the human class I histocompatibility antigen HLA-A2. Nature 329:506–512, 1987.

Brown, J. H., T. S. Jardetzky, J. C. Gorga, L. J. Stern, R. G. Urban, J. L. Strominger, and D. C. Wiley. Three-dimensional structure of the class II histocompatibility antigen HLA-DR1. Nature 364:33–39, 1993.

Campbell, R. D., and J. Trowsdale. Map of the human MHC. Immunology Today 14:349–352, 1993.

David-Watine, B., A. Israel, and P. Kourilsky. The regulation and expression of MHC class I genes. Immunology Today 11: 286–292, 1990.

Engelhard, V. H. Structure of peptides associated with class I and class II MHC molecules. Annual Review of Immunology 12:181–207, 1994.

Glimcher, L. H., and C. J. Kara. Sequences and factors: a guide to MHC class II transcription. Annual Review of Immunology 10:13–49, 1992.

Mach, B., V. Steimle, E. Martinez-Soria, and W. Reith. Regulation of MHC class II genes: lessons from a disease. Annual Review of Immunology 14:301–331, 1996.

Madden, D. The three dimensional structures of peptide-MHC complexes. Annual Review of Immunology 13:587–622, 1995.

Silver, M. L., H.-C. Guo, J. L. Strominger, and D. C. Wiley. Atomic structure of a human MHC molecule presenting an influenza virus peptide. Nature 360:367–369, 1992.

CHAPTER SIX

ANTIGEN PROCESSING AND

PRESENTATION TO

T LYMPHOCYTES

T lymphocytes play a central role in specific immune responses to protein antigens. In Chapter 5, we introduced the concept that the physical forms of antigens recognized by T cells are actually peptide fragments that are derived from protein antigens and are bound to cell surface proteins encoded by genes of the major histocompatibility complex (MHC). MHC molecules serve to display peptides to T cells. Peptides derived from cytosolic proteins are bound to class I MHC molecules and are recognized by CD8$^+$ T cells, which are usually cytolytic T lymphocytes (CTLs). CTLs provide the major host defense mechanism against intracellular microbes, which produce proteins in the host cell cytoplasm. In contrast, peptides derived from extracellular proteins that are endocytosed by specialized cell types are bound to class II MHC molecules and are recognized by CD4$^+$ T cells, which are usually helper T lymphocytes. Helper T cells are required for the induction of humoral and cell-mediated responses, which are most effective in eliminating extracellular and phagocytosed pathogens. In this chapter, we will describe how peptide antigens become associated with MHC molecules inside cells, and the characteristics of the cells that form and display these peptide-MHC complexes. The T cell receptors that recognize these complexes and the activation events that occur subsequent to their recognition are described in Chapter 7.

GENERAL FEATURES OF ANTIGEN PROCESSING AND PRESENTATION TO T LYMPHOCYTES

Our current understanding of T cell antigen recognition is the culmination of a vast amount of work that began with studies of the physicochemical forms of antigens that stimulated cell-mediated immunity. These studies led to the discovery that cells other than T lymphocytes play an obligatory role in T cell activation by foreign antigens and later to the elucidation of the function of MHC molecules in T cell antigen recognition.

The Forms of Antigens Recognized by T Lymphocytes

The specificity of T lymphocytes for complexes of peptide antigens and MHC molecules determines several characteristics of T cell antigen recognition, which differ in fundamental ways from antigen recognition by antibody molecules.

1. *Most T lymphocytes recognize only peptides,* whereas B cells can specifically recognize peptides, proteins, nucleic acids, polysaccharides, lipids, and small chemicals. As a result, T cell–mediated immune responses are induced only by protein antigens (the natural source of foreign peptides), whereas humoral immune responses are seen with protein and non-protein antigens. Some T cells are specific for chemically reactive forms of haptens

such as dinitrophenol. In these situations, it is likely that the haptens bind to cell surface proteins, including MHC molecules, and peptides containing these hapten conjugates are recognized by T cells. Rare examples of T cells that recognize non-peptide antigens, including lipids bound to nonpolymorphic MHC-like molecules, have been described, but the significance of these T cells is not known.

2. *T cells recognize only linear determinants of peptides defined predominantly by primary amino acid sequences.* Such linear peptides assume extended conformations within the peptide-binding clefts of MHC molecules (see Chapter 5 and Color Plate IV). In contrast, B cells specific for protein antigens may recognize conformational determinants that exist when proteins are in their native tertiary (folded) configuration or determinants that are exposed by denaturation or proteolysis. Thus, when an animal is immunized with a native protein, the antigen-specific T cells that are stimulated will respond to denatured or even proteolytically digested forms of that protein. In contrast, antibodies produced by B cells after immunization with the native protein react only with the native protein (Table 6–1). This has important implications for the development of vaccines, since it is theoretically possible to induce T cell immunity to pathogens using peptide immunogens.

3. *T cells recognize and respond to foreign peptide antigens only when the antigen is attached to the surfaces of other cells,* whereas B cells and secreted antibodies bind soluble antigens in body fluids or cell surface antigens. This is because MHC molecules form part of the complex that T cells recognize, and these molecules are cell surface–bound integral membrane proteins. The display of peptide-MHC complexes in a form that can be recognized by T cells is called **antigen presentation.** Cells that display antigens in this form are called **antigen-presenting cells (APCs).** The properties and functions of APCs are discussed later in the chapter. Although historically the term APC has most often been used to describe cells that present antigen to CD4$^+$ helper T lymphocytes, it is appropriate to describe target cells of CTL lysis as APCs as well, since CTLs also recognize peptide-MHC complexes on the surface of these target cells.

The Phenomenon of MHC-Restricted Antigen Recognition by T Lymphocytes

A fundamental aspect of antigen recognition by helper T cells and CTLs is that any one T lymphocyte is restricted to recognizing a peptide antigen only when it is complexed to a single allelic form of an MHC molecule. This phenomenon is called **MHC restriction.** It was first discovered in the 1970s when investigators mixed T cells and APCs from different inbred strains of animals and measured various kinds of T cell responses. In

TABLE 6–1. Qualitative Differences in Antigen Recognition by T and B Lymphocytes

		Secondary Immune Response	
Immunizing Antigen	Secondary Antigen Exposure	B Cell–Mediated (Antibody Production)	T Cell–Mediated (Delayed-Type Hypersensitivity)
Native protein	Native protein	+	+
Denatured protein	Native protein	–	+
Native protein	Denatured protein	–	+
Denatured protein	Denatured protein	+	+

Antigen recognition by T and B lymphocytes is qualitatively different. In an immunized animal, B cells are specific for conformational determinants of the immunogen and, therefore, distinguish between native and denatured protein antigens. T cells, however, do not distinguish between native and denatured protein antigens because T cells recognize linear epitopes on short peptides derived from the intact proteins by proteolysis.

these experiments, T cells from an animal immunized with an antigen would subsequently recognize and respond to that antigen in vitro only if the APCs came from the same animal (or from another that shared MHC alleles with the first animal). In other words, T cells are self-MHC restricted; they recognize and respond to antigen presented by an APC only if that APC expresses MHC molecules that the T cell recognizes as self. The MHC molecules that T cells recognize as self are those that the T cells encountered during their maturation from precursors (discussed in Chapter 8). Therefore, "self MHC" does not refer to MHC molecules expressed by the T cells themselves but to MHC molecules on the APCs or target cells. In the normal situation, T cells would only be exposed to self APCs, and therefore the concept of self MHC restriction may seem trivial. There are two reasons, however, why self-MHC restriction is important.

1. The discovery of self-MHC restriction provided the initial evidence that T cell recognition of antigen involves MHC molecules and, furthermore, that T cells recognize polymorphic residues of MHC molecules, i.e., residues that distinguish self and non-self MHC alleles.

2. The phenomenon of self-MHC restriction provided important insights into the process of T cell maturation, including the revelation that developing T cells must recognize self MHC molecules in order to mature into functional T lymphocytes (discussed in Chapter 8).

One of the earliest and clearest demonstrations of MHC restriction came from Rolf Zinkernagel and Peter Doherty's studies of virus-specific CTL-mediated lysis of virally infected target cells (Fig. 6–1). In most of these experiments, the virus-infected target cells are lysed by the CTLs only if they express allelic forms of MHC molecules that are expressed in the animal from which the CTLs were derived. Furthermore, by using congenic strains of mice, it was shown that the CTLs and the target cell must be derived from mice that share a particular class I MHC allele. Thus, CTL recognition of antigen is restricted by self class I MHC

alleles. Essentially similar experiments demonstrated that helper T lymphocyte responses to antigen are also self MHC–restricted, and usually class II MHC–restricted. For instance, helper T cells will respond to antigens presented by macrophages or B cells that express self class II MHC molecules, and this form of recognition is critical for cell-mediated and humoral immune reactions. In fact, the class I or class II MHC restriction of T cells correlates more strongly with their expression of CD8 or CD4 than with the functional capabilities of the cells. This is because the responses of both developing and mature T cells to antigen receptor stimuli are dependent on the binding of CD8 to class I MHC molecules or CD4 to class II MHC molecules (see Chapters 7 and 8). Therefore, all CD8+ T cells are class I–restricted. Most of these cells are CTLs, although some may function mainly as cytokine-producing cells. Similarly, all CD4+ T cells are restricted by class II MHC molecules, because CD4 binds to class II MHC molecules. Most CD4+ T cells are helper cells, although CD4+ CTLs (again class II MHC-restricted) also exist.

The molecular basis of MHC-restricted antigen recognition by T cells was elucidated by parallel studies of antigen presentation and the structure of the T cell receptor for antigen. As we shall see in Chapter 7, a T cell expresses antigen receptors that simultaneously interact with a peptide epitope of a protein antigen and with polymorphic residues of MHC molecules which bind that peptide (Fig. 6–2). Later in this chapter, we will address the question of how large, complex proteins are converted to peptides that bind to MHC molecules and get displayed on APC surfaces.

The Role of Accessory Cells in T Cell Activation

Since both CD4+ and CD8+ T cells recognize peptide-MHC complexes, and these are only present on the surface of other cells, it is clear that T cell activation by antigen requires the participation of other cells. Even before it was known what form of antigen T cells recognized, immunologists were aware that cells other than T lymphocytes, often

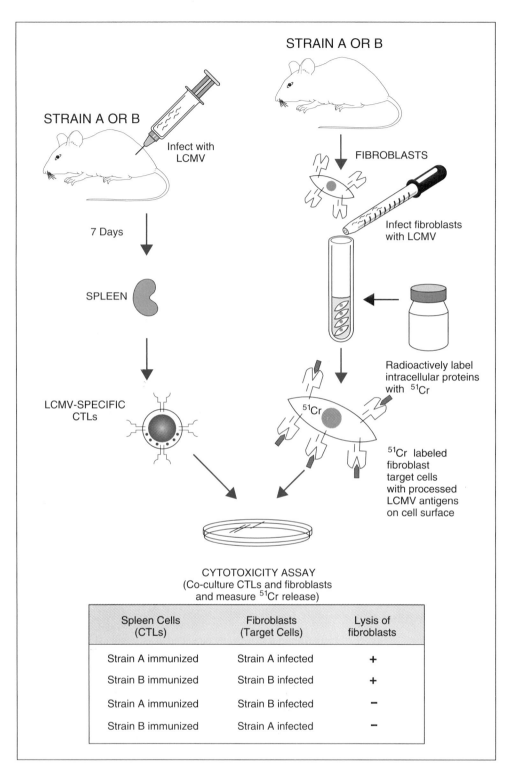

STRAIN A OR B

STRAIN A OR B

Infect with
LCMV

FIBROBLASTS

Infect fibroblasts
with LCMV

7 Days

SPLEEN

Radioactively label
intracellular proteins
with ^{51}Cr

LCMV-SPECIFIC
CTLs

^{51}Cr

^{51}Cr labeled
fibroblast
target cells
with processed
LCMV antigens
on cell surface

CYTOTOXICITY ASSAY
(Co-culture CTLs and fibroblasts
and measure ^{51}Cr release)

Spleen Cells (CTLs)	Fibroblasts (Target Cells)	Lysis of fibroblasts
Strain A immunized	Strain A infected	+
Strain B immunized	Strain B infected	+
Strain A immunized	Strain B infected	−
Strain B immunized	Strain A infected	−

FIGURE 6–1. Major histocompatibility complex restriction of cytolytic T lymphocytes (CTLs). Virus-specific CTLs from a strain A or strain B mouse lyse only syngeneic target cells infected with the specific virus. The CTLs do not lyse uninfected targets and are not alloreactive. Further analysis has shown that the CTLs and target cells must come from animals that share class I MHC alleles in order for the target cell to present viral antigens to the CTLs. Thus, CTL recognition of antigen is self class I MHC–restricted. LCMV, lymphocytic choriomeningitis virus.

called **accessory cells,** were required for the activation of T cells. For example, T cells isolated from the blood, spleen, or lymph nodes of individuals immunized with a protein antigen can be restimulated in tissue culture by that antigen. Stimulation may be measured by assaying the production of cytokines by the T cells or by the proliferation of the T cells. When contaminating macrophages, B cells, and dendritic cells are removed from the cultures, the purified T lymphocytes no longer respond to antigen, and responsiveness can be restored by adding back the macrophages or dendritic cells. Such experimental approaches provide the basis for defining the accessory functions of various cell types in T lymphocyte activation. The importance of accessory cells in immune re-

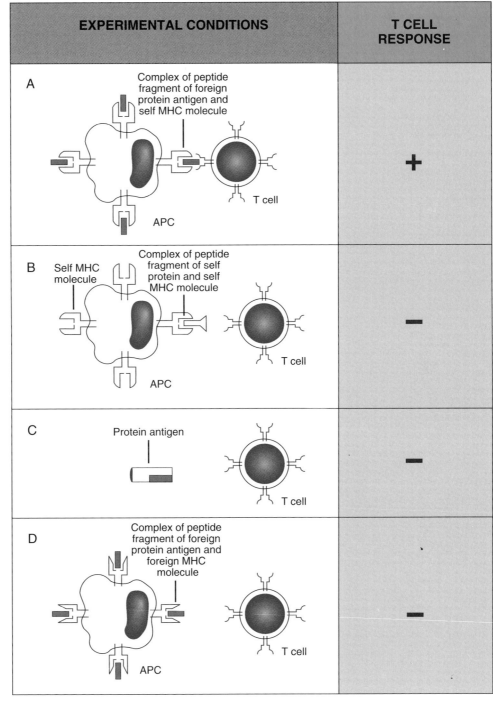

FIGURE 6-2. Specificity of major histocompatibility complex (MHC)-restricted T cells. Helper T cells and cytolytic T lymphocytes (CTLs) recognize complexes of self MHC molecules and peptide fragments of foreign antigens (A). MHC-restricted T cells do not recognize self MHC with self peptide (B), foreign antigens without MHC molecules (C), or complexes of foreign MHC molecules and peptide fragment of antigen (D). (T cells specific for foreign peptide antigen plus self MHC molecules often also respond to complexes of an unrelated peptide with non-self allelic forms of MHC molecules. This is called alloreactivity and is discussed in Chapter 17). APC, antigen-presenting cell. (Note that some MHC molecules in this and other figures are depicted without bound peptides for the sake of clarity. Most MHC molecules actually do have bound self peptides.)

sponses *in vivo* is suggested by the observation that adjuvants often need to be administered in addition to antigen in order to elicit an immune response to the antigen. These adjuvants are usually insoluble or undegradable substances that promote nonspecific inflammation, with recruitment and activation of mononuclear phagocytes and dendritic cells at the site of immunization.

Accessory cells serve two important functions in the activation of CD4+ T cells. First, accessory cells are APCs, i.e., they convert protein antigens to peptides and they present peptide-MHC complexes in a form that can be recognized by the T cells. The conversion of native proteins to MHC-associated peptide fragments by APCs is called **antigen processing.** As early as the 1950s, it was demonstrated that radioactively or fluorescently labeled antigens injected into animals were found in mononuclear phagocytes or follicular dendritic cells and not in lymphocytes. Later studies showed that an antigen that was taken up by macrophages *in vitro* and then injected into mice was up to 1000 times more immunogenic on a molar basis than the same antigen administered by itself, in a cell-free form. The explanation for this finding is that T cells respond only to antigen associated with APCs, and only a small fraction of an injected soluble antigen ends up in this processed, immunogenic cell-associated form.

The second function of accessory cells is to provide stimuli to the T cell, in addition to those initiated by peptide-MHC complexes binding to the T cell antigen receptor. These stimuli, referred to as **costimulators,** are required for full physiologic activation of the T cells. They are provided by membrane-bound or secreted products of accessory cells. In fact, adjuvants may enhance immune responses *in vivo* in part by inducing the expression of costimulator molecules on accessory cells. The costimulatory functions of APCs are discussed further in Chapter 7.

The Source of Antigens Presented to Class I and Class II MHC–Restricted T Cells

As we will discuss in later chapters, class II MHC–restricted CD4+ helper T cells and class I MHC–restricted CD8+ CTLs perform distinct effector functions and are, therefore, involved in eliminating different types of microbes. Specifically, CD4+ T cells enhance both the production of antibodies by B cells and the microbicidal activities of macrophages. These are the two main mechanisms for elimination of microbes that live in extracellular locations or within phagocytic vesicles. In contrast, CD8+ T cells lyse cells expressing foreign antigens, and are therefore useful for eliminating viruses and bacteria that infect and live within various cell types. The antigen receptors of CD4+ and CD8+ T cells cannot distinguish between extracellular and intracellular microbes. Therefore, specific antigen recognition cannot be the basis for the segregated responses of CD4+ and CD8+ T cells to different types of microbes. The problem of distinguishing extracellular from intracellular microbes has been solved by the segregation of the antigen presentation pathways, which are responsible for generating peptides from these microbes and displaying them in association with MHC molecules (Fig. 6–3). As mentioned previously, CD4+ helper T cells recognize peptides bound to class II MHC molecules. *The major source of peptides that bind to class II MHC molecules are extracellular proteins, including microbial proteins, which are endocytosed by APCs and enter the acidic vesicular pathway that cells use to break down internalized proteins.* In contrast, CD8+ CTLs recognize peptides bound to class I MHC molecules. *The major source of peptides that bind to class I MHC molecules are proteins found in the cytosol of APCs, including viral and intracellular bacterial proteins.* Therefore, the segregation of different types of antigens into the class I or class II MHC pathways of antigen presentation ensures that CD4+ and CD8+ T cells respond to the

types of microbes they are most effective at eliminating. The cellular mechanisms that account for the partitioning of peptides derived from endocytosed or cytosolic proteins to class II or class I MHC molecules, respectively, will be discussed in detail later in the chapter.

Common Features of the Cellular Pathways of Antigen Processing and Presentation

The generation of complexes of peptide antigen and MHC molecules on the surface of APCs involves intricate multistep pathways within the APC. Our understanding of these *antigen-processing pathways* has increased greatly over the past decade with the discovery and characterization of the molecules and subcellular organelles that mediate the production of peptides from intact proteins and promote the assembly of MHC molecules. It has become clear that there are distinct subcellular pathways that generate complexes of peptides with class I versus class II MHC molecules. Before we describe the details that distinguish these pathways, it is useful to consider the important general features that are common to both pathways.

1. *Peptide antigens are generated by proteolysis of intact proteins within subcellular organelles of APCs.*

2. *Peptide binding to MHC molecules occurs prior to cell surface expression.* In fact, peptide association is required for the stable assembly and surface expression of both class I and class II MHC molecules.

3. *Both the class I and class II MHC pathways of antigen processing and presentation utilize subcellular organelles and enzymes that have generalized protein degradation and recycling functions that are not exclusively used for antigen display to the immune system.* In other words, both class I and class II MHC antigen presentation pathways have evolved as adaptations of basic cellular functions.

4. *Both class I and class II MHC antigen-processing and presentation pathways do not distinguish normal self proteins from foreign antigens, but rather display peptides from a sampling of all endocytosed and cytoplasmic proteins for T cell surveillance.* The presentation of self peptides is integral to the antigen display mechanisms of APCs and the surveillance function of T cells. The specificity and sensitivity of cell-mediated immune responses reside in the capacity of T cells to recognize small numbers of foreign peptide–self MHC molecule complexes on APCs and to ignore the large numbers of self peptide–self MHC molecule complexes presented by normal self APCs.

The major features of the class I and class II MHC pathways of antigen presentation are compared in Table 6–2. We will return to specific examples of each of these features now, as we discuss the class I and class II MHC pathways in more detail.

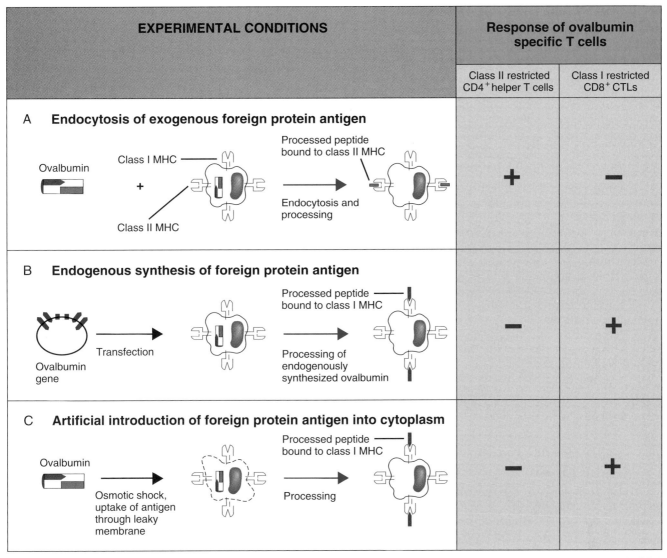

FIGURE 6–3. Presentation of exogenous and endogenous antigens. When ovalbumin is added to an antigen-presenting cell (APC) that expresses class I and class II major histocompatibility complex (MHC) molecules, ovalbumin-derived peptides are presented only in association with class II (A). When ovalbumin is synthesized intracellularly as a result of transfection of its gene (B) or when it is introduced into the cytoplasm through membranes made leaky by osmotic shock (C), ovalbumin-derived peptides are presented in association with class I MHC molecules. The measured response of class II–restricted helper T cells is cytokine secretion, and the measured response of class I–restricted cytolytic T lymphocytes (CTLs) is killing of the APCs.

TABLE 6–2. Comparative Features of Class I and Class II MHC Pathways of Antigen Processing and Presentation

Feature	Class I MHC Pathway	Class II MHC Pathway
Composition of peptide-MHC complex	Polymorphic α chain, β_2 microglobulin, peptide	Polymorphic α and β chains, peptide
Types of APCs	All nucleated cells	Mononuclear phagocytes, B lymphocytes, dendritic cells, endothelial cells, thymic epithelium
Responsive T cells	CD8$^+$ T cells (mostly CTLs)	CD4$^+$ cells (mostly helper T cells)
Source of protein antigens	Cytosolic proteins (mostly synthesized on polyribosomes in the cell)	Endosomal/lysosomal proteins (mostly from extracellular environment)
Site of peptide generation	Cytosolic proteasome	Endosome/lysosome
Site of peptide loading of MHC	ER	CIIV or MIIC (specialized endosomal/lysosomal compartment)
Proteins associated with MHC before peptide loading	Calnexin, BiP, TAP in ER	Calnexin in ER; invariant chain in ER, Golgi, and CIIV
Peptide transport to site of loading	TAP	Not necessary
Transport to cell surface	Secretory pathway	? Secretory pathway

Abbreviations: APC, antigen-presenting cell; MHC, major histocompatibility complex; ER, endoplasmic reticulum; CIIV, class II vesicle; MIIC, MHC class II compartment; CTLs, cytolytic T lymphocytes; TAP, transporter associated with antigen processing; BiP, binding protein.

MECHANISMS OF ANTIGEN PRESENTATION TO CLASS II MHC–RESTRICTED CD4+ T CELLS

The earliest studies of antigen processing and presentation examined the uptake of extracellular bacteria by macrophages *in vitro* and the subsequent activation of bacterial antigen-specific T cells. Only later was it appreciated that the events being examined in such experimental systems comprised the pathway of antigen processing and presentation to class II MHC–restricted CD4+ T cells. The principal steps in this pathway are

1. Endocytosis of native protein antigens from the extracellular environment into APCs
2. Processing of the antigen in acidic endosomes or lysosomes, leading to the generation of peptide fragments
3. Binding of peptides to class II MHC molecules within a specialized form of endocytic vesicles
4. Surface expression of peptide–class II MHC molecule complexes
5. Recognition of the complexes by CD4+ T cells that are specific for the foreign peptide and the self MHC molecule

We will now discuss the types of cells capable of performing these functions and then the details of the subcellular events.

Types of APCs for CD4+ Class II MHC–Restricted T Cells

The two requisite properties that allow a cell to function as an APC for class II MHC–restricted helper T lymphocytes are the ability to process endocytosed antigens and the expression of class II MHC gene products (Table 6–3). Most mammalian cells appear to be capable of endocytosing and proteolytically digesting protein antigens, but only a restricted group of cell types normally express class II MHC. The best-defined APCs for helper T lymphocytes include (1) dendritic cells, (2) mononuclear phagocytes, and (3) B lymphocytes. These

cell types are often called **professional APCs** (Table 6–4). In humans, microvascular endothelial cells also express class II MHC molecules and can perform antigen presentation functions.

Dendritic cells of the spleen and lymph nodes are irregularly shaped, nonphagocytic cells making up a small fraction (<1 per cent) of the total cell population of these organs. They are derived from the bone marrow and are related to the mononuclear phagocytic lineage (see Chapter 2). Dendritic cells are competent at presenting protein antigens to helper T cells, including naive T cells that have not previously been exposed to antigen. This is likely due to the fact that dendritic cells constitutively express high levels of certain molecules called costimulators, which are required in addition to peptide-MHC complexes to fully activate naive T cells (see Chapter 7). Therefore, the initiation of immune responses to extracellular protein antigens, sometimes called "priming," is dependent on antigen presentation by dendritic cells. It is also believed that dendritic cells are important for inducing T cell responses to foreign (allogeneic) MHC molecules in tissue allografts. Consistent with this hypothesis is the observation that dendritic cells are potent stimulators of mixed lymphocyte reactions (see Chapter 17). A subset of dendritic cells present in the epidermal layer of the skin are called *Langerhans cells*. They express the CD1 marker and contain an unusual cytoplasmic organelle called the Birbeck granule. Langerhans cells are capable of migrating from skin to lymph nodes and are likely to be the origin of dendritic cells in lymph nodes that drain cutaneous sites. Langerhans cells in the skin are not fully differentiated and are inefficient at stimulating naive T cells. Full antigen-presenting capabilities are attained during or after migration from the skin to lymph nodes. It is believed that Langerhans cells transport antigens from the skin to draining lymph nodes, where the antigens are presented to specific naive T cells (see Chapter 11).

Macrophages and other cells of the *mononuclear phagocyte system* are the only APCs that can

TABLE 6–3.   Requirement for Class II MHC Expression in Antigen Presentation to CD4+ Antigen-Specific T Cells

APCs	Genes Transfected Into APCs	Surface Class II MHC	Surface Class I MHC	Antigen	Response of Cytochrome c–Specific, I-E^{k-} Restricted T Cell Line (Cytokine Secretion)
3T3 (murine fibroblast)	None	None	K^k, D^k	Cytochrome c	−
3T3 (murine fibroblast)	Murine class II Eα^k and Eβ^k	I-E^k	K^k, D^k	None	−
3T3 (murine fibroblast)	Murine class II Eα^k and Eβ^k	I-E^k	K^k, D^k	Cytochrome c	+

Class II MHC expression is required for antigen presentation to CD4+ antigen-specific T cells. In this experiment, a murine fibroblast cell line, 3T3, derived from an H-2 mouse, which expresses class I, but not class II, MHC molecules, does not present cytochrome c to a cytochrome c–specific, I-E–restricted T cell hybridoma line. When functional genes encoding the α and β chains of the I-E molecule are transfected into 3T3 cells, they become competent at presenting antigen to the T cell line.
Abbreviations: MHC, major histocompatibility complex; APC, antigen-presenting cell.

TABLE 6–4. Properties and Functions of Antigen-Presenting Cells

| | Expression of: | | |
Cell Type	*Class II MHC*	*Costimulators*	Principal Function
Dendritic cells (Langerhans cells, lymphoid dendritic cells)	Constitutive	Constitutive	Initiation of CD4$^+$ T cell responses (priming); allograft rejection
Macrophages	Inducible by IFN-γ	Inducible by CD40 ligand, LPS, IFN-γ	Development of CD4$^+$ effector T cells
B lymphocytes	Constitutive; increased by IL-4	Inducible by T cells	Stimulation of CD4$^+$ helper T cells in humoral immune responses (cognate T cell–B cell interactions)
Vascular endothelial cells	Inducible by IFN-γ	Constitutive	? Enhanced recruitment of antigen-specific T cells to site of antigen exposure or inflammation; ? role in allograft rejection
Various epithelial and mesenchymal cells	Inducible by IFN-γ	Probably none	No known physiologic function; ? role in exacerbation of autoimmune reactions in tissues

Abbreviations: LPS, lipopolysaccharide; IFN-γ, interferon-γ; IL-4, interleukin-4; MHC, major histocompatibility complex.

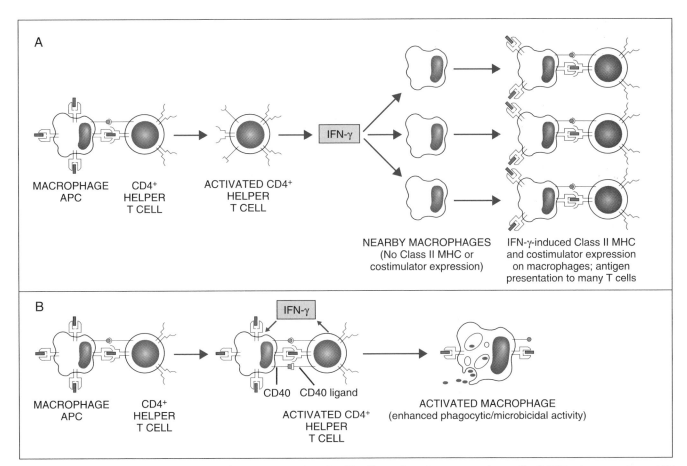

FIGURE 6–4. The consequences of antigen presentation for T cells and antigen-presenting cells (APCs). A macrophage APC expressing complexes of foreign peptide and class II major histocompatibility complex (MHC) molecules activates a CD4$^+$ T cell, leading to the expression of cytokines. Class II MHC and costimulator expression are enhanced on nearby macrophages in response to interferon (IFN)-γ produced by activated T cells, thereby amplifying the number of effective APCs in the area (A). Phagocytic and microbicidal functions of the macrophage APC are enhanced by IFN-γ and CD40 ligand expressed by the activated T cell (B).

actively phagocytose large particles. Therefore, they probably play an important role in presenting antigens derived from extracellular infectious organisms such as bacteria and parasites.

B lymphocytes utilize their membrane immunoglobulin (Ig) molecules to bind and internalize soluble protein and present processed peptides derived from these proteins to helper T cells. The antigen-presenting function of B cells is particularly important in helper T cell–dependent antibody production (see Chapter 9).

In humans, *microvascular endothelial cells* express class II MHC molecules and may present antigen to blood T cells that have become adherent to the vessel wall (see Chapter 13). This may be particularly important in presentation of alloantigens during graft rejection (see Chapter 17) and perhaps in cell-mediated immune reactions (see Chapter 13).

Thymic epithelial cells constitutively express class II MHC and play a specialized role in presenting peptide-MHC complexes to maturing T cells in the thymus as part of the selection processes that shape the repertoire of T cell specificities (see Chapter 8).

The T cell–APC interaction has a number of functional consequences (Fig. 6–4):

1. Professional APCs present peptide-MHC complexes to specific helper T cells and thereby induce helper T cells to express cytokines and new cell surface molecules, initiating the process of T cell clonal expansion and performance of effector functions.

2. The products of activated helper T cells enhance the antigen-presenting functions of nearby cells, thereby amplifying the number of effective APCs. For example, interferon-γ (IFN-γ) secreted by activated T cells upregulates the expression of class II MHC and costimulator molecules on the surface of APCs, both of which are required for activation of CD4$^+$ T cells.

3. The products of activated T cells also enhance the effector functions of B cell and macrophage APCs. For example, shortly after a B cell presents antigen to a CD4$^+$ T cell, the T cell expresses certain new surface molecules (e.g., CD40 ligand) and cytokines, which, in turn, bind to receptors on the B cell and activate antibody production and isotype switching (see Chapter 9). Similarly, after a macrophage presents antigen to a CD4$^+$ T cell, CD40 ligand and IFN-γ produced by activated T cells can enhance the microbicidal activities of the macrophage (see Chapter 13). Thus, bidirectional APC–T cell interactions serve to amplify immune responses.

Epithelial and mesenchymal cells also express class II MHC molecules in response to IFN-γ. The physiologic significance of antigen presentation by these "non-professional APCs" is unclear. Since they generally do not express costimulators, it is unlikely that they play an important role in the initiation of most T cell responses. However, since memory T cells usually have less stringent costimulator requirements than do naive T cells, it is possible that non-professional APCs play a role in some secondary cell-mediated immune responses.

Uptake of Extracellular Proteins Into Vesicular Compartments of APCs

The initial step in the processing and presentation of an extracellular protein antigen is the binding of the native antigen to an APC. Different APCs can bind protein antigens in several ways and with varying efficiencies and specificities. Macrophages and dendritic cells bind many different proteins, with little or no specificity, to surface molecules that are currently undefined, and internalize then by the process of phagocytosis or pinocytosis. There are, however, special cases in which well-characterized surface molecules on the APC mediate binding and subsequent internalization of the antigen by receptor-mediated endocytosis in clathrin-coated vesicles. For example, specific receptors for the Fc portions of immunoglobulins and receptors for the complement protein C3b, which are present on the surface of macrophages, can efficiently bind opsonized antigens and enhance their internalization. This may partially explain why secondary immune responses require lower doses of antigen than primary responses, since at the time of secondary immunization pre-existing specific antibody may augment binding of the antigen to APCs. Macrophages also express receptors that bind to mannose residues present on bacterial wall polysaccharides, and these mannose receptors can mediate phagocytosis of the bacteria. Because of differences in carbohydrate-modifying enzymes, mammalian cells usually do not have exposed mannose residues on their surfaces. Another example of specific receptors on APCs is the surface Ig on B cells, which, because of their high affinity for antigens, can effectively mediate the internalization of proteins present at very low concentrations in the extracellular fluid. Soluble protein antigens may also be internalized into APCs by fluid phase pinocytosis, without actually binding to the cell surface.

As mentioned above, internalized antigens become localized in intracellular membrane-bound vesicles called **endosomes** (Fig. 6–5). The precise ultrastructural and biochemical characteristics of endosomes are not well described. Rather, these organelles are defined mostly by their function, which is intracellular transport and degradation of internalized proteins. It is known that the endosomal pathway of protein traffic in the cell is continuous with and ends up at the lysosome, a more dense membrane-bound vesicular organelle with well-defined ultrastructural features.

Both endosomes and lysosomes may provide intracellular sites for processing of internalized antigens. Proteins found in endosomes yield peptides

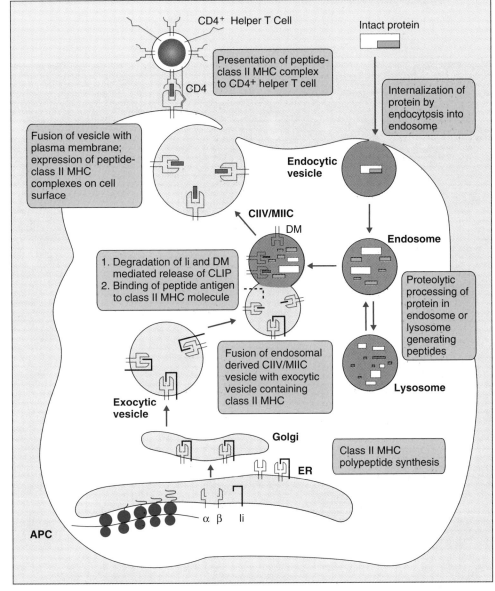

FIGURE 6-5. The class II major histocompatibility complex (MHC) pathway of antigen presentation. CLIP, class II–associated invariant chain peptide; Ii, invariant chain; ER, endoplasmic reticulum. Details of the functions of Ii and DM are shown in Figure 6–7.

that bind to class II MHC molecules, and most of these proteins are internalized from the extracellular environment. Thus, antigens made by extracellular bacteria, fungi, protozoa, and helminths are usually presented by the class II MHC pathway and activate CD4+ T cells. Additionally, some intact microorganisms can enter a cell by endocytosis or phagocytosis and survive within intracellular membrane-bound vesicles. Peptides derived from proteins made by these intracellular microorganisms may also be presented by class II MHC molecules.

Processing of Internalized Proteins in Endosomal/Lysosomal Vesicles

The next step in antigen presentation is the processing of the antigen that was internalized in its native form. Several characteristics of the pro-

cessing of extracellularly derived protein antigens are known:

1. *Antigen processing is a time- and metabolism-dependent phenomenon that takes place subsequent to internalization of antigen by APCs.* If macrophages (or other APCs) are incubated briefly ("pulsed") with a protein antigen such as ovalbumin, rendered metabolically inert by chemical fixation at various times thereafter, and tested for their ability to stimulate ovalbumin-specific T cells, functional antigen presentation occurs only if 1 to 3 hours elapse between the antigen pulse and fixation (Fig. 6–6). This time is required for the APCs to process the antigen and present it in association with class II MHC molecules on the cell surface. Processing of antigen is inhibited by maintaining the APCs below physiologic temperatures, by adding metabolic inhibitors such as azide, or by fixation earlier than 1 hour after the antigen pulse.

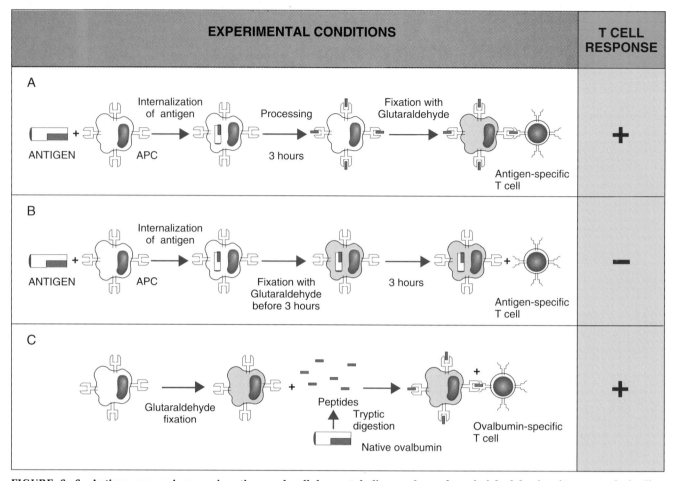

FIGURE 6–6. Antigen processing requires time and cellular metabolism and can be mimicked by in vitro proteolysis. If an antigen-presenting cell (APC) is allowed to process antigen and is then chemically fixed (rendered metabolically inert) 3 hours or more after antigen internalization, it is capable of presenting antigen to T cells (A). Antigen is not processed or presented if APCs are fixed less than 1 to 3 hours after antigen uptake (B). Fixed APCs bind and present proteolytic fragments of antigens to specific T cells (C). The artificial proteolysis, therefore, mimics physiologic antigen processing by APCs. Effective antigen presentation is assayed by measuring a T cell response, such as cytokine secretion. (Note that T cell hybridomas respond to processed antigens on fixed APCs, but growth factor–dependent T cells may require costimulators that are destroyed by fixation.)

2. *The endosomes and lysosomes where antigen processing takes place have an acidic pH, which is required for the processing.* Chemical agents that increase the pH of intracellular acid vesicles, such as chloroquine and ammonium chloride, are potent inhibitors of antigen processing.

3. *Cellular proteases are required for the processing of many protein antigens.* Several types of proteases, including cathepsin and leupeptin, are present in endosomes and lysosomes, and specific inhibitors of these enzymes block the presentation of protein antigens by APCs. The function of proteases is to cleave native protein antigens into small peptides. These proteases also probably act on the invariant chain, promoting its dissociation from class II MHC molecules, as discussed later. Most of these proteases function optimally at acid pH, and this is the likely reason why antigen processing occurs best in acidic compartments.

The processed forms of most protein antigens that T cells recognize can be artificially generated

by proteolysis in the test tube. Macrophages that are fixed or that are treated with chloroquine before exposure to antigen can effectively present pre-digested peptide fragments of that antigen, but not the intact protein, to specific T cells (Fig. 6–6). Peptides that bind to MHC molecules and stimulate T cells can be analyzed for amino acid sequence and secondary structure to determine the nature of the potential ligands for T cell antigen receptors. Immunogenic peptides derived from many complex globular proteins, such as cytochrome c, ovalbumin, myoglobin, and lysozyme, have been characterized in detail in this way. More recently, naturally generated peptides have been eluted from the class II MHC molecules of APCs and analyzed for common structural characteristics. The physicochemical features of peptides that permit their binding to MHC molecules were described in Chapter 5.

The net result of processing of a protein antigen is the generation of peptides, many of which are 10 to 30 amino acids long and capable of binding to the peptide-binding clefts of class II MHC molecules. The requirement for antigen processing prior to T cell stimulation explains why T cells recognize linear but not conformational determinants of proteins and why T cells cannot distinguish between native and denatured forms of a protein antigen (see Table 6–1). It is likely that most types of APCs, including macrophages, B cells, and dendritic cells, are qualitatively similar in their ability to process endocytosed antigens; however, there may be quantitative differences. For instance, macrophages contain many more proteases than do B cells and are more actively phagocytic, so that macrophages may be more efficient than B cells at internalizing and processing large particulate antigens and presenting peptide fragments of these antigens. It is also possible that different APCs generate distinct sets of peptides from the same native protein because of differences in their endosomal proteases. Furthermore, different APCs may present different peptides because the set of class II MHC molecules expressed by one APC may not be identical to those expressed by another. Therefore, it is possible that the APCs involved in presenting a particular protein antigen can influence which T cells are activated by that antigen.

Association of Processed Peptides With Newly Synthesized Class II MHC Molecules

Peptides generated by proteolysis of proteins in endosomes and lysosomes bind to newly synthesized class II MHC molecules within intracellular vesicles (see Fig. 6–5). The exact site of this association is not definitely known, but a variety of experimental data indicate that it occurs within an organelle of the endocytic pathway. An understanding of how peptide–class II MHC complexes are formed requires knowledge of the biosynthesis and subcellular transport of new class II MHC molecules. Several steps and key features of this process have been defined.

1. *The α and β chains of class II MHC molecules are coordinately synthesized and associate with each other in the endoplasmic reticulum (ER).* These chains are translated from messenger ribonucleic acid (mRNA) molecules on membrane-bound ribosomes and are co-translationally inserted into the membrane of the ER.

2. *Newly synthesized class II heterodimers temporarily associate with two other nonpolymorphic polypeptides, not encoded by the MHC, which are required for proper assembly and transport of the MHC molecule.* The first of these proteins is called **calnexin** and it functions as a molecular chaperone, ensuring that the α and β chains are properly folded during assembly of a class II MHC molecule. Calnexin is also involved in the assembly of other multichain molecules in the ER, including class I MHC molecules and the T cell antigen receptor

(see Chapter 7). The second nonpolymorphic protein associated with the class II MHC αβ heterodimers in the ER is called the **invariant chain (Ii).** This protein is a 30 kD Ig superfamily member which is a type II membrane protein, i.e., it has a reverse orientation to most transmembrane proteins, so that the amino terminus is intracytoplasmic and the carboxy terminus is intraluminal. The native invariant chain is a homotrimer. Each subunit binds one newly synthesized class II αβ heterodimer, forming a nine polypeptide chain complex (i.e., three αβ heterodimers bound to one invariant chain homotrimer). Only after the invariant chain binds the αβ heterodimer is calnexin released, and the class II–invariant chain complex is able to move out of the ER.

3. *The invariant chain prevents peptides or nascent unfolded polypeptides in the ER from binding to newly formed class II MHC αβ heterodimers.* The invariant chain binds to the class II MHC heterodimer in a way that interferes with peptide loading of the cleft formed by the α and β chains. There are, in fact, peptides within the ER derived from cytosolic proteins, as we will discuss later. Since the effector functions of class II–restricted T cells are best suited for dealing with extracellular microbes, it would be counterproductive to have class II MHC molecules loaded with peptides derived from cytosolic proteins. Furthermore, since the peptide binding cleft of class II MHC molecules has open ends, it can theoretically accommodate binding of newly translated polypeptides which have not yet folded into their tertiary structural conformation. Such polypeptides are abundant in the ER, but the presence of the invariant chain prevents their association with class II MHC molecules.

4. *The invariant chain also directs newly formed class II MHC molecules to specialized endosomal/lysosomal organelles where internalized proteins are proteolytically degraded into peptides.* In the ER, N-linked oligosaccharides are added to the newly translated class II MHC α and β chains, the two chains form heterodimers, and the heterodimers associate with invariant chains. Subsequent to these events, the class II MHC–invariant chain complexes pass through the Golgi apparatus, where the oligosaccharides are further modified. Then the invariant chain targets the movement of the mature class II MHC molecules to specialized membrane-bound organelles of the endocytic pathway that contain proteolytically degraded proteins derived from the extracellular milieu. The invariant chain performs this function by virtue of certain amino acid sequences in its amino terminal cytoplasmic tail. Immunoelectronmicroscopy and subcellular fractionation studies have been used to define specific characteristics of this subcellular compartment targeted by the invariant chain. In macrophages, it is called the MHC class II compartment or MIIC and has the properties of a vesicle in transition between endosome and lysosome, in-

cluding high density and a characteristic multivesiculated appearance. In some B cells, a similar but less dense organelle containing invariant chain and class II MHC has been identified and named the class II vesicle (CIIV). These organelles likely represent specialized branch points in the vesicular transport pathways that allow newly formed class II MHC molecules on their way to the cell surface to become exposed to endocytically derived peptides. Thus, *the invariant chain plays a key role in getting MHC molecules to the same place as peptides derived from extracellular protein antigens.*

5. *Within the MIIC/CIIV compartment, the invariant chain is removed from class II MHC molecules by the combined action of proteolytic enzymes and the HLA-DM molecule* (see Fig. 6–7). Since the invariant chain blocks access to the peptide-binding groove of a class II MHC molecule, it must be removed before complexes of peptide and class II MHC can form. The same proteolytic enzymes that generate peptides from internalized proteins also act on the invariant chain in a stepwise fashion, leaving only a 24 amino acid remnant called **class II–associated invariant chain peptide (CLIP).** X-ray crystallographic analysis has shown that the CLIP peptide sits in the peptide-binding cleft in the same way that other peptides bind to class II MHC molecules. Therefore, removal of CLIP is required before ac-

cess is provided to peptides from extracellular proteins. This is accomplished by the action of a molecule called HLA-DM (or H-2M in the mouse), which is encoded within the MHC and has a structure very similar to that of class II MHC molecules. HLA-DM molecules differ from class II MHC molecules in several respects: they are not polymorphic, they do not necessarily associate with invariant chain, they are not expressed on the cell surface, and their subcellular distribution is distinct from class II MHC molecules. Nonetheless, HLA-DM is found in the MIIC compartment. Mutant cell lines which lack DM are defective in presenting peptides from extracellularly derived proteins. When class II MHC molecules are isolated from these DM-mutant cell lines, they are found to have almost exclusively CLIP peptides in their peptide-binding clefts, consistent with a role for DM in removing CLIP. *In vitro* studies have confirmed that HLA-DM acts as a peptide exchange molecule, facilitating the removal of CLIP and the addition of other peptides to class II MHC molecules. Predictably, DM gene knockout mice have profound defects in class II MHC–restricted antigen presentation.

6. *Once CLIP peptides are removed, peptides generated by proteolytic cleavage of extracellularly derived protein antigens bind to class II MHC mole-*

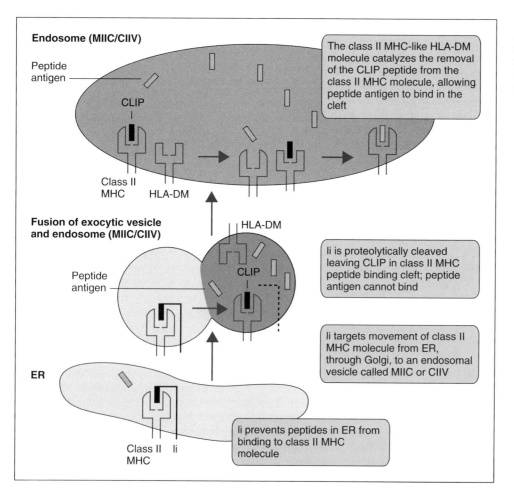

FIGURE 6–7. The functions of class II major histocompatibility complex (MHC)–associated invariant chains and HLA-DM. ER, endoplasmic reticulum; Ii, invariant chain.

Endosome (MIIC/CIIV)

Peptide antigen

The class II MHC-like HLA-DM molecule catalyzes the removal of the CLIP peptide from the class II MHC molecule, allowing peptide antigen to bind in the cleft

CLIP

Class II MHC   HLA-DM

Fusion of exocytic vesicle and endosome (MIIC/CIIV)

HLA-DM

Peptide antigen

CLIP

Ii is proteolytically cleaved leaving CLIP in class II MHC peptide binding cleft; peptide antigen cannot bind

Ii targets movement of class II MHC molecule from ER, through Golgi, to an endosomal vesicle called MIIC or CIIV

ER

Ii prevents peptides in ER from binding to class II MHC molecule

Class II   Ii
MHC

cules. Although initial studies of the physical interaction of peptides with class II MHC molecules indicated a very slow association rate, requiring up to 48 hours to achieve saturation, more recent analyses indicate that HLA-DM greatly enhances this process, so that peptides can form stable complexes with class II MHC molecules within 20 minutes. Since the ends of the class II MHC peptide-binding cleft are open, large peptides or even unfolded whole proteins may bind, yet the size of peptides eluted from cell surface class II MHC molecules is restricted to between 10 and 30 amino acids. It is possible, therefore, that proteolytic enzymes "trim" bound polypeptides to the appropriate size for T cell recognition after the polypeptides bind to class II MHC molecules.

7. *Peptide binding to class II MHC molecules stabilizes the αβ heterodimer, and the peptide dissociation rate is extremely slow.* The ability of peptide to increase the tightness of association of the class II MHC α and β chains serves to increase the likelihood that only properly loaded peptide-MHC complexes will survive long enough to get displayed on the cell surface. A similar phenomenon occurs in class I MHC assembly. The long life of a peptide-MHC complex increases the chance that a T cell specific for such a complex will make contact, bind, and be activated by that complex.

8. *Stable peptide-class II MHC complexes are delivered to the cell surface by membrane fusion with exocytic vesicles, and they are displayed there for surveillance by CD4+ T cells.*

Only a very small fraction of cell surface peptide-MHC complexes will contain the same peptide. Furthermore, most of the bound peptides will be derived from normal self proteins, since there is no mechanism to distinguish self from foreign proteins in the process that generates the peptide-MHC complexes. This has been demonstrated by amino acid analysis of peptides eluted from class II MHC molecules purified from B cells grown in tissue culture. Most of these peptides were derived from self proteins. These findings raise two important questions. First, if individuals process their own proteins and present them in association with their own class II MHC molecules, why do we normally not develop immune responses against self proteins? It is likely that self-tolerance is mainly due to the absence or inactivation of T cells capable of recognizing and responding to self antigens, and this is why self peptide-MHC complexes do not normally induce autoimmunity (see Chapters 8 and 19). Second, how can a T cell recognize and be activated by specific foreign antigen when it encounters an APC surface that is predominantly displaying self-peptide–MHC complexes? The answer lies in part with the extraordinary sensitivity of T cells for specific peptide-MHC complexes. It has been estimated that as few as 100 to 200 complexes of a particular peptide with a particular allelic form of class II MHC molecule on the surface of

an APC can lead to activation of a T cell. This represents less than 0.1 per cent of the total number of class II molecules likely to be present on the surface of the APC, most of which would be occupied with self peptides. In fact, the indiscriminate ability of the APC to internalize, process, and present the heterogeneous mix of self and foreign extracellular proteins ensures that the immune system will not miss transient or quantitatively small exposures to foreign antigens. Furthermore, there is evidence that a single T cell will sequentially engage multiple peptide-MHC complexes until achieving a sufficient threshold of activating signals (see Chapter 7).

Although the bulk of experimental evidence supports the model described above for the generation of most class II MHC–peptide complexes, there are potentially important alternate intracellular pathways for the generation of these complexes that may be immunologically significant. First, it is possible that cell surface class II molecules may be recycled by internalization into endosomes, where they bind newly generated peptide fragments of internalized protein. This process would likely require an exchange of previously bound peptides with the new ones. Second, there are exceptions to the general case that class II MHC molecules bind peptides derived from internalized exogenous proteins. Cell surface complexes of class II MHC molecules with peptides derived from endogenously synthesized proteins have been detected both by T cell responses to such proteins and by direct analysis of eluted peptides from cell surface–derived class II MHC molecules. In some cases, this may result from a normal cellular pathway for the turnover of cytoplasmic contents, referred to as autophagy. In this pathway, cytoplasmic contents are entrapped within ER-derived membrane vesicles called autophagosomes, these vesicles fuse with lysosomes, and the cytoplasmic proteins are proteolytically degraded. The association of the peptides generated by this route would require movement of the peptides to a class II–bearing vesicular compartment, as described previously for trafficking of exogenously derived peptides. In addition, some peptides that associate with class II MHC are derived from endogenously synthesized membrane proteins. Before they are expressed on the surface, these proteins may have ready access to class II MHC molecules because they would be synthesized and transported through the same ER-Golgi compartments as the membrane-bound class II MHC molecules themselves. How such membrane proteins are processed is currently unknown. It is also possible that after cell surface expression, membrane proteins may reenter the cell by the same endocytic pathway as exogenous proteins. In this way, peptides derived from virally encoded membrane proteins may enter the class II–MHC pathway of antigen presentation. This is a theoretically important way in which viral antigen-specific CD4+ helper T cells may be activated.

MECHANISMS OF ANTIGEN PRESENTATION TO CLASS I MHC–RESTRICTED CD8+ T CELLS

As we have mentioned previously, CD8+ T cells, most of which are CTLs, recognize peptides that are usually derived from protein antigens that are synthesized within APCs, processed, and subsequently expressed on the APC surface in association with class I MHC molecules. Examples of endogenously synthesized foreign proteins are viral proteins and the products of mutated or dysregulated genes in tumor cells. CTLs are the principal immunologic defense mechanisms against viruses and may be important in the immune destruction of tumors. In contrast to the restricted expression of class II MHC molecules, almost all cells express class I MHC molecules and have the ability to display peptide antigens in association with these MHC molecules on the cell surface. This ensures that any cell synthesizing viral or mutant proteins can be marked for recognition and killing by CD8+ CTLs. As is the case with class II MHC–associated antigen presentation, generation of peptide–class I MHC complexes is a continuous normal function of cells, which does not discriminate between foreign and self proteins. This portion of the chapter describes the known features of the generation of peptide–class I MHC complexes on the surface of cells. The principal steps in this pathway are as follows (Fig. 6–8):

1. Synthesis of protein antigens in the cytosol or delivery of protein antigens into the cytosol
2. Proteolytic degradation of cytosolic proteins into peptides
3. Transport of peptides into the ER
4. Assembly of peptide–class I MHC complexes within the ER
5. Expression of peptide–class I MHC complexes on the cell surface

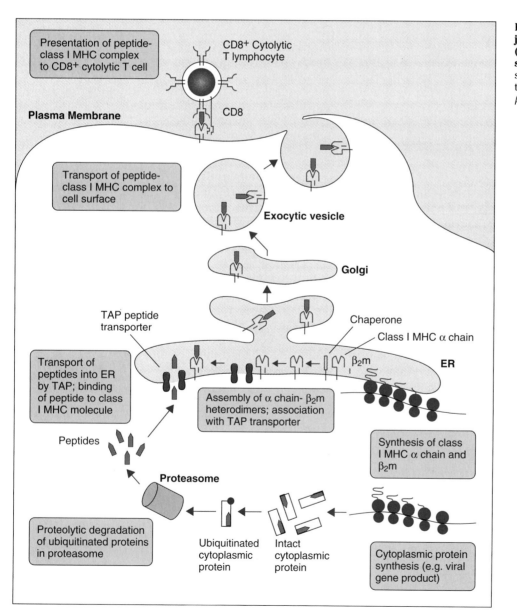

FIGURE 6–8. The class I major histocompatibility complex (MHC) pathway of antigen presentation. TAP, transporter associated with antigen presentation; ER, endoplasmic reticulum; β_1m, β_2 microglobulin.

Presentation of peptide-class I MHC complex to CD8+ cytolytic T cell

CD8+ Cytolytic T lymphocyte

Plasma Membrane

CD8

Transport of peptide-class I MHC complex to cell surface

Exocytic vesicle

Golgi

TAP peptide transporter

Chaperone

Class I MHC α chain

Transport of peptides into ER by TAP; binding of peptide to class I MHC molecule

β_2m

ER

Assembly of α chain- β_2m heterodimers; association with TAP transporter

Synthesis of class I MHC α chain and β_2m

Peptides

Proteasome

Proteolytic degradation of ubiquitinated proteins in proteasome

Ubiquitinated cytoplasmic protein

Intact cytoplasmic protein

Cytoplasmic protein synthesis (e.g. viral gene product)

Entry of Cytosolic Proteins Into the Class I–MHC Pathway of Antigen Presentation

The prerequisite for entry of a protein into the processing pathway leading to peptide–class I MHC association is simply location in the cytosol. Several lines of evidence support this.

1. If a viral protein, such as influenza nucleoprotein, or a protein like ovalbumin, is added in soluble form to a cell that expresses class I and class II MHC molecules, the antigen is internalized, processed, and presented only in association with class II MHC molecules. Such exogenously added antigens will be recognized by class II–restricted, antigen-specific CD4+ T cells but will not sensitize the APC to lysis by CD8+ T cells. On the other hand, if the gene encoding the viral protein or ovalbumin is transfected into the APCs so that the antigen is synthesized on polyribosomes in the cytosol, the cell becomes sensitive to lysis by specific class I–restricted CD8+ T cells (see Fig. 6–3).

2. If an antigen is introduced into the cytoplasm of a cell by making the plasma membrane transiently permeable to macromolecules or by membrane fusion of an APC with lipid vesicles containing the protein, the antigen is subsequently processed and peptides associate only with class I MHC molecules (see Fig. 6–3).

The significance of having cytosolic proteins enter the class I MHC pathway of antigen presentation lies in the fact that endogenously synthesized foreign or mutant proteins will be present in the cytosol, and therefore will target cells for lysis by CD8+ class I MHC–restricted CTL. For example, viruses encode RNA transcripts, which are translated into foreign proteins in the host cell cytoplasm. Therefore, peptides derived from viral protein antigens end up being displayed on class I MHC molecules on the surface of virally infected cells. This enables class I MHC–restricted CD8+ cytolytic T cells to recognize the virally infected cells and destroy them. Since virtually all nucleated cells express class I MHC, any virus-infected cell is susceptible to CTL-mediated lysis. Similarly, CTLs may be important in recognizing and killing cancer cells, which often express mutated genes or unmutated genes that are not expressed in normal adult cells (see Chapter 18). The products of such endogenous genes may be expressed in the cytosol. In addition, some intracellular microbes, such as mycobacteria, reside for long periods of time within phagocytic vesicles. It is possible that there will be some breakdown in the membrane barrier of these vesicles, resulting in the microbial proteins leaking into the cytoplasm, and thus gaining access to the class I MHC pathway of antigen presentation. Alternatively, there may be specific transport mechanisms that deliver proteins or peptides from these vesicles to the cytoplasm.

Processing of Cytosolic Antigens

The intracellular mechanisms that generate antigenic peptides which bind to class I MHC molecules are very different from the mechanisms described earlier for peptide–class II MHC molecule associations. This is evident from the observations that the agents that raise endosomal and lysosomal pH, or directly inhibit endosomal proteases, block class II– but not class I–associated antigen presentation.

Peptides that bind to class I MHC molecules are proteolytically generated in the cytoplasm prior to entry into the exocytic pathway that delivers the peptide–MHC protein complex to the cell surface. This conclusion is supported by a variety of experimental observations.

1. A cell infected with a virus becomes sensitized to lysis by virus-specific CTLs; this is because the cell displays peptides derived from viral proteins in association with class I MHC molecules on the cell surface. Some of these proteins, such as influenza nucleoprotein, are neither membrane bound nor secreted, i.e., they do not gain access to exocytic pathways in their intact form. Furthermore, the genes encoding viral membrane proteins can be altered to eliminate the membrane insertion sequences. When these genes are transfected into cells, the encoded proteins cannot gain access to the ER and exocytic pathway, yet peptides from these proteins are still presented to CD8+ CTLs.

2. When peptide epitopes for CTL recognition are synthesized directly in the cytoplasm of a cell as products of transfected minigenes, the cell becomes sensitized for lysis. This implies that peptides generated in the cytoplasm have direct access to the exocytic pathway for cell surface expression of class I MHC molecules.

*A major mechanism for the generation of peptides from cytosolic protein antigens is proteolysis in the **proteasome**,* a large multiprotein complex with a broad range of proteolytic activity that is found in the cytoplasm of most cells. A 700 kD form of proteasome appears as a cylinder composed of a stacked array of four inner and four outer rings, with each ring composed of seven distinct subunits. The subunits of the inner rings are the catalytic sites for proteolysis. A larger, 1500 kD proteasome is likely to be most important *in vivo* and is composed of the 700 kD structure plus several additional subunits that regulate proteolytic activity. Two catalytic subunits present in many 1500 kD proteasomes, called LMP2 and LMP7, are encoded by genes in the MHC (see Chapter 5). Both LMP2 and LMP7 expression are upregulated by IFN-γ, leading to an increase in the number of proteasomes containing these subunits. The proteasome performs a basic housekeeping function in cells by degrading many different cytoplasmic proteins. For example, NF-κB activation is dependent on proteasomal degradation of IκB (see Box 4–4, Chapter 4). Proteins are targeted for proteasomal degradation by covalent linkage of several copies of a small polypeptide called ubiquitin. This process of polyubiquitination requires adenosine triphosphate (ATP) and a variety of enzymes. Several lines of evidence suggest that the proteasome, and probably ubiquitination, are involved in antigen processing for the class I MHC pathway of antigen presentation.

1. In some experimental situations, inhibition of the enzymes required for ubiquitination also inhibits the presentation of cytoplasmic proteins to class I MHC–restricted T cells specific for a peptide epitope of that protein.

2. Modification of proteins by attachment of an N-terminal sequence which is recognized by ubiquitin-conjugating enzymes leads to enhanced ubiquitination and more rapid class I MHC–associated presentation of peptides derived from those proteins.

3. Specific inhibitors of proteasomal function, such as peptide aldehydes, block presentation of a cytoplasmic protein to class I MHC–restricted T cells specific for a peptide epitope of that protein.

4. Proteasomes typically generate peptides between five and 11 amino acids long, which includes the lengths that best fit the peptide-binding clefts of class I MHC molecules.

5. The specificity of proteolysis by LMP-2– and LMP-7–containing proteasomes from IFN-γ–treated cells favors the generation of peptides with C-terminal basic or hydrophobic amino acid residues, which are typical of many class I MHC–binding peptides.

There are many examples of protein antigens that apparently do not require ubiquitination or proteasomes in order to be presented by the class I MHC pathway. In some cases this may reflect the fact that other, less well-defined mechanisms of cytoplasmic proteolysis exist. In addition, some class I MHC–binding peptides may be generated by proteolytic enzymes resident in the ER. For example, peptides from secretory proteins with hydrophobic signal sequences are often found associated with class I MHC molecules. These proteins are targeted directly to the ER during translation and therefore may bypass cytoplasmic degradation.

Delivery of Peptides From Cytoplasm to the ER

Class I MHC molecules are assembled in the ER, and this process is dependent on peptides. Since peptides generated in the cytosol are presented by class I MHC molecules, a mechanism must exist for delivery of cytosolic peptides into the ER. The initial insights into this mechanism came from studies of cell lines that are defective in assembling and displaying peptide–class I MHC complexes on their surfaces. The mutations responsible for this defect turned out to involve two genes in the MHC, which are homologous to a fam-

ily of genes that encode proteins that mediate ATP-dependent transport of low molecular weight compounds across intracellular membranes. The two genes in the MHC that belong to this family encode proteins called **t**ransporter **a**ssociated with antigen **p**resentation-1 or **TAP-1,** and **TAP-2.** TAP-1 and TAP-2 form heterodimers, which are localized in the ER and *cis*-Golgi (Fig. 6–8). In this location they mediate the active, ATP-dependent transport of peptides from the cytosol into the ER lumen. Although the TAP heterodimer has a broad range of specificities, it optimally transports peptides ranging from eight to 12 amino acid residues long and therefore delivers to the ER peptides of the right size for binding to class I MHC molecules. Mice with targeted disruptions of the genes encoding TAP-1 or TAP-2 show defects in class I MHC expression and cannot effectively present proteins to class I MHC–restricted T cells. Rare examples of human TAP-2 gene mutations have also been identified, and predictably, the patients carrying these mutant genes also show defective class I MHC–associated antigen presentation.

Assembly and Surface Expression of Peptide–Class I MHC Complexes

The actual assembly and surface expression of stable class I MHC molecules require the presence of peptides. A variety of experimental data have indicated a particular sequence of events in assembly and expression of peptide–class I MHC complexes:

1. The class I MHC α chain and β_2 microglobulin are synthesized on the rough ER and transported into the smooth ER as separate polypeptide chains.

2. The α chain associates with molecular chaperones, which prevent degradation and promote proper folding of the protein. Two chaperones that are known to associate with the α chain in the ER are BiP, a member of the heat shock protein family, and calnexin.

3. β_2 microglobulin binds to partially or completely folded α chain and the chaperones dissociate. These newly formed α chain–β_2 microglobulin dimers are unstable and cannot be transported efficiently out of the ER.

4. The α chain–β_2 microglobulin dimers move to and become physically associated with the luminal aspects of the TAP proteins within the ER. This close association ensures that peptides transported into the ER by the TAP bind to the associated empty class I MHC molecules. It is also possible that the TAP association promotes further folding of the α chain and β_2 microglobulin.

5. Peptide binding to the class I molecule greatly enhances its stability and causes its release from the TAP protein.

6. Stable peptide–class I MHC complexes now move through the Golgi, where the MHC molecules

undergo further carbohydrate modification, and then they are transported to the cell surface by exocytic vesicles. Surface complexes can now be recognized by CD8+ T cells.

The requirement for peptides in class I MHC assembly has been clearly shown by analysis of TAP-deficient cells (either mutant cell lines or cells from TAP-1 gene knockout mice), which express significantly reduced levels of surface class I MHC (Fig. 6–9). Since TAP delivers peptides to the ER, these findings suggest that peptides in the ER are required for class I MHC assembly. Those class I MHC molecules that do get expressed in TAP-deficient cells have bound peptides that are mostly derived from signal sequences of proteins destined for secretion or membrane expression. These signal sequences are cleaved off and degraded to peptides within the ER during translation, without a requirement for TAP. There are two reasons why peptides transported into the ER preferentially bind to class I and not class II MHC molecules. First, as we have discussed, newly formed class I MHC molecules are bound to the luminal aspect of the TAP complex. Second, as mentioned previously, in the ER the class II MHC–peptide-binding cleft is blocked by the invariant chain.

The sequence of events in class I MHC mole-

cule assembly which we have discussed ensures that only properly folded, peptide-loaded class I MHC molecules are displayed for T cell surveillance. A few empty class I MHC complexes do make it out to the cell surface, but these are unstable and rapidly dissociate. It is, of course, likely that there are other steps involved in this pathway that are not yet resolved, and it is also possible that alternate pathways may exist. Nonetheless, the effects of mutations and inhibitors of this pathway, as discussed, indicate that it is critical for normal immune function. Furthermore, the importance of this pathway to anti-viral immunity is demonstrated by the evolution of viral mechanisms that interfere with it. For example, herpes simplex virus produces a protein, called ICP47, that effectively plugs up the TAP pore through which peptides are delivered to the ER and thus prevents presentation of viral antigens to T cells (see Chapter 16).

PHYSIOLOGIC SIGNIFICANCE OF MHC-ASSOCIATED ANTIGEN PRESENTATION

So far we have discussed the specificity of CD4+ and CD8+ T lymphocytes for MHC-associated foreign protein antigens and the mechanisms by which complexes of peptides and MHC molecules

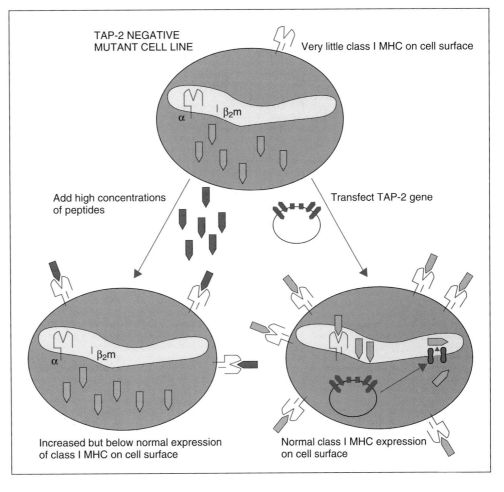

FIGURE 6–9. TAP gene products are required for assembly and cell surface expression of peptide–class I major histocompatibility (MHC) complexes. A cell line with a nonfunctional TAP-2 gene expresses very few surface class I MHC molecules. The peptides bound to these few surface class I MHC molecules are predominantly derived from the signal sequences of membrane or secreted proteins. The addition of high doses of peptides can induce some class I MHC molecule assembly and expression. In this case, it is not known whether the assembly of the peptide–class I complexes occurs at the cell surface or intracellularly. When a functional TAP-2 gene is transfected into the cell line, normal assembly and expression of peptide–class I MHC molecules are restored.

TAP-2 NEGATIVE MUTANT CELL LINE

Very little class I MHC on cell surface

β₂m

α

Add high concentrations of peptides

Transfect TAP-2 gene

β₂m

α

Increased but below normal expression of class I MHC on cell surface

Normal class I MHC expression on cell surface

are produced. There are several fundamental properties of T cell–mediated immune responses that are consequences of the fact that T cells only recognize MHC-associated antigens. In this section, we will consider the impact of MHC-associated antigen presentation on the role that T cells play in protective immunity, the nature of T cell responses to different antigens, and the limitations of what T cells will recognize in protein antigens.

T Cell Surveillance for Foreign Antigens

As we discussed throughout this chapter, *both the class I and class II MHC pathways of antigen presentation sample pools of predominantly normal self proteins for display to the T cell repertoire, which surveys these samples for the rare foreign or mutant peptide.* The recent advances in our understanding of how peptide-MHC complexes are formed confirm that MHC molecules are scaffolds for peptide display to the immune system and that antigen processing pathways have evolved to sample both extracellular and intracellular proteins in order to supply the peptides. The specialized class II MHC-expressing APCs have various characteristics, such as the phagocytic activity of macrophages, the high-affinity Ig antigen receptors on B cells, and the long cytoplasmic processes of dendritic cells, which enable them to encounter the full range of possible extracellular protein antigens. The convergence of the endocytic pathways in these cells with the exocytic pathway of class II MHC expression ensures that peptides derived from these extracellular antigens will be displayed on the cell surface for possible recognition by CD4+ T cells. The widespread expression of class I MHC in nucleated cells, and the pathway of peptide loading of class I MHC molecules which is linked to a ubiquitous mechanism for degrading cellular proteins, ensures that peptides from virtually any intracellular protein will be displayed for possible recognition by CD8+ T cells. Superimposed on this system of antigen presentation is a sensitive system of T cell surveillance of the displayed peptides, which is based on continuous recirculation of T cells to sites of APCs throughout the body, and the exquisite sensitivity of T cells, allowing them to respond to small numbers of peptide-MHC complexes. Thus, the paradox that antigen presentation mechanisms overwhelmingly display normal self peptides is actually fundamental to the ability of the immune system to find rare foreign protein antigens.

The Nature of T Cell Responses

Based on our understanding of antigen presentation to T cells, we can now explain other physiologic consequences of MHC-restricted antigen recognition that were introduced in Chapter 5.

1. Because T cells recognize only MHC-associated peptide antigens, they can respond only to antigens associated with other cells (the APCs) and are unresponsive to soluble or circulating proteins. *This unique specificity for cell-bound antigens is essential for the functions of T lymphocytes, which are largely mediated by cell-cell interactions and by cytokines that act at short distances.* For instance, helper T cells activate B cells and macrophages. Not surprisingly, B lymphocytes and macrophages are two of the principal cell types that express class II MHC genes, function as APCs for CD4+ helper T cells, and focus helper T cell effects to their immediate vicinity. Similarly, CTLs can lyse any nucleated cell producing a foreign antigen, and all nucleated cells express class I MHC molecules, which are the restricting elements for antigen recognition by CD8+ CTLs.

2. The triaging of endosomal versus cytoplasmic proteins to class II or class I MHC pathways of antigen presentation determines which subsets of T cells will respond to antigens found in those two pools of proteins (Fig. 6–10). Extracellular antigens usually end up in the endosomal pool and activate class II–restricted CD4+ T cells. These cells function as helpers to stimulate effector mechanisms such as antibodies and phagocytes that serve to eliminate extracellular antigens. Conversely, endogenously synthesized antigens are present in the cytoplasmic pool of proteins and usually activate class I–restricted CD8+ CTLs. These lymphocytes lyse cells producing intracellular antigens. *Thus, antigens from microbes that reside in different locations selectively stimulate the T cell populations that are most effective at eliminating that type of microbe.*

Immunogenicity of Protein Antigens

MHC molecules may determine the immunogenicity of protein antigens in two related ways:

1. *The immunodominant epitopes of complex proteins are often the peptides that bind most avidly to MHC molecules.* If an individual is immunized with a multideterminant protein antigen, in many instances the majority of the responding T cells are specific for one or a few linear amino acid sequences of the antigen. These are called the "immunodominant" determinants or epitopes. For instance, in H-2^k mice immunized with hen egg lysozyme (HEL), more than half the HEL-specific T cells are specific for the epitope formed by residues 46-61 of HEL in association with the I-A^k but not the I-E^k molecule. This is because HEL(46-61) binds to I-A^k better than do other HEL peptides, and does not bind to I-E^k. However, it is not yet known exactly which structural features of a peptide determine immunodominance. As mentioned earlier, for class I–restricted antigen presentation, immunodominant peptides are required to have amino acid residues whose side chains fit into pockets of the MHC molecule–peptide-binding cleft. Common features of immunodominant peptides for class II MHC–restricted antigen presenta-

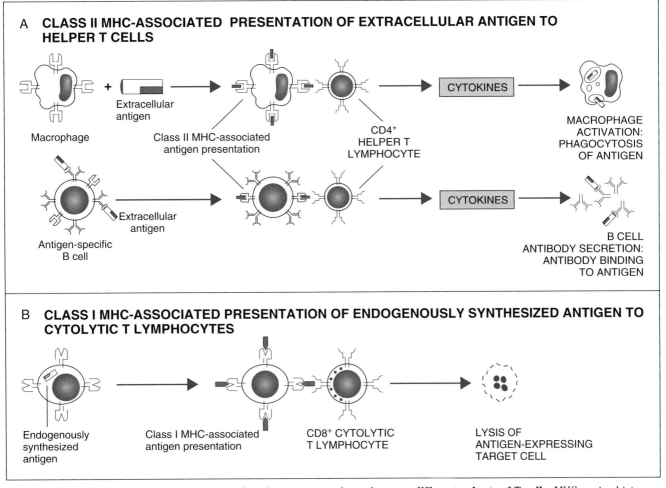

FIGURE 6–10. Presentation of exogenous and endogenous protein antigens to different subsets of T cells. MHC, major histocompatibility complex.

tion are less well defined. The question is an important one because an understanding of these features may permit the efficient manipulation of the immune system with synthetic peptides. An obvious application of such knowledge is the design of vaccines. For example, a protein encoded by a viral gene could be analyzed for the presence of amino acid sequences that would form a typical immunodominant secondary structure capable of binding to MHC molecules with high affinity. Vaccines composed of synthetic peptides mimicking this region of the protein theoretically would be effective in eliciting T cell responses against the viral peptide expressed on an infected cell, thereby establishing protective immunity against the virus.

2. *The expression of particular class II MHC alleles in an individual determines the ability of that individual to respond to particular antigens.* The phenomenon of immune response (Ir) gene-controlled immune responsiveness was mentioned in Chapter 5. We now know that Ir genes that control antibody responses are class II MHC genes. They influence immune responsiveness in part because various allelic class II MHC molecules differ in their ability to bind different antigenic peptides and,

therefore, to stimulate specific helper T cells. For instance, H-2^k mice are responders to HEL(46-61), but H-2^d mice are non-responders to this epitope. Equilibrium dialysis experiments have shown that HEL(46-61) binds to I-A^k but not to I-A^d molecules. A possible molecular basis for this difference in MHC association is suggested from the model of the class II molecule and the known amino acid sequences of I-A^k and I-A^d proteins. If the HEL(46-61) peptide is hypothetically placed in the predicted binding cleft of the I-A^k molecule, charged residues of the HEL peptide become aligned with oppositely charged residues of the MHC molecule. This would presumably stabilize the bimolecular interaction. In contrast, the I-A^d molecule has different amino acids in the binding cleft that would result in the aligning of like-charged residues with the HEL peptide. Therefore, HEL(46-61) would not bind to or be presented in association with I-A^d, and the H-2^d mouse would be a non-responder. Similar results have been obtained with numerous other peptides. MHC-linked immune responsiveness may also be important in humans. For instance, Caucasians who are homozygous for an extended HLA haplotype containing HLA-B8,DR3,

DQw2a are low responders to hepatitis B virus surface antigen. Individuals who are heterozygous at this locus are high responders, presumably because the other alleles contain one or more HLA gene that confers responsiveness to this antigen. Thus, HLA typing may prove to be valuable for predicting the success of vaccination. These findings support the **determinant selection model** of MHC-linked immune responses. This model, which was proposed many years before the demonstration of peptide-MHC binding, states that the products of MHC genes in each individual select which determinants of protein antigens will be immunogenic in that individual. We now understand the structural basis of determinant selection and Ir gene function in antigen presentation. Most Ir gene phenomena have been studied by measuring helper T cell function, but the same principles apply to CTLs. Individuals with certain MHC alleles may be incapable of generating CTLs against some viruses. In this situation, of course, the Ir genes may map to one of the class I MHC loci.

Although these concepts are based largely on studies with simple peptide antigens and inbred strains of mice, they are also relevant to the understanding of immune responses to complex multideterminant protein antigens in outbred species. It is likely that all individuals will express at least one MHC molecule capable of binding at least one determinant of a complex protein, so that all individuals will be responders to such antigens. As stated in Chapter 5, this may be the evolutionary pressure for maintaining MHC polymorphism.

This discussion of the influence of MHC gene products on the immunogenicity of protein antigens has focused on antigen presentation and has not considered the role of the T cells. We have mentioned earlier that the exquisite specificity and diversity of antigen recognition are attributable to antigen receptors on T cells. MHC-linked immune responsiveness is also dependent, in part, on the presence and absence of specific T cells. In fact, some peptides may bind to MHC molecules in a particular inbred mouse strain but do not activate T cells in that strain. It is likely that these mice lack T cells capable of recognizing the particular peptide-MHC complexes. *Thus, Ir genes may function by determining antigen presentation or by shaping the repertoire of antigen-responsive T cells.* The development of the T cell repertoire and the role of the MHC in T cell maturation are discussed in Chapter 8.

SUMMARY

T cells recognize antigens only on the surface of accessory cells in association with the products of self MHC genes. CD4+ helper T lymphocytes recognize antigens in association with class II MHC gene products (class II MHC–restricted recognition), and CD8+ CTLs recognize antigens in association with class I gene products (class I MHC–restricted recognition). Antigen processing consists of the introduction of protein antigens into APCs, the proteolytic degradation of these proteins into peptides, the binding of peptides to newly assembled MHC molecules, and the display of the peptide-MHC complexes on the APC surface for potential recognition by T cells. Antigen-processing pathways in APCs utilize basic cellular proteolytic mechanisms, which also operate independent of the immune system. Both extracellular and intracellular proteins are sampled by these antigen-processing pathways, and peptides derived from both normal self proteins and foreign proteins are displayed by MHC molecules for surveillance by T lymphocytes. Specialized APCs, including macrophages, B lymphocytes, and dendritic cells, internalize extracellular proteins into endosomes for processing by the class II MHC pathway. These proteins are proteolytically cleaved by enzymes that function at acidic pH in vesicles of the endosomal pathway. Newly synthesized class II MHC heterodimers associate with the invariant chain and are directed from the ER to the endosomal vesicles, where the invariant chain is proteolytically cleaved, and a small peptide remnant of the invariant chain is removed from the peptide binding cleft of the MHC molecule by the DM molecules. The peptides generated from extracellular proteins then bind to the class II MHC molecule, and the trimeric complex (class II MHC α and β chains and peptide) moves to the surface of the cell. Cytosolic proteins, usually synthesized in the cells, such as viral proteins, enter the class I MHC pathway of antigen presentation. The proteasome is a cytoplasmic multiprotein complex which proteolytically degrades ubiquitinated cytoplasmic proteins and probably generates a large part of the peptides destined for display by class I MHC molecules. Peptides are delivered from the cytoplasm to the ER by the TAP molecules. Newly formed class I MHC dimers in the ER associate with and bind peptides delivered by TAP. Peptide binding stabilizes class I MHC molecules and permits their movement out of the ER, through the Golgi, to the cell surface. These pathways of MHC-restricted antigen presentation ensure that most of the body's proteins are screened for the possible presence of foreign antigens. The pathways also ensure that proteins from extracellular microbes are likely to generate peptides bound to class II MHC molecules for recognition by CD4+ helper T cells, while proteins encoded by intracellular microbes generate peptides bound to class I MHC molecules for recognition by CD8+ CTLs. The immunogenicity of microbial proteins depends on the ability of antigen-processing pathways to generate peptides from these proteins which bind to self MHC molecules.

Selected Readings

Cresswell, P. Invariant chain structure and MHC class II function. Cell 84:505–507, 1996.

Germain, R. H., and D. M. Margulies. The biochemistry and cell biology of antigen processing and presentation. Annual Review of Immunology 11:403–450, 1993.

Germain, R. N. MHC-dependent antigen processing and peptide presentation: providing ligands for T lymphocyte activation. Cell 76:287–299, 1994.

Neefjes, J. J., and H. L. Ploegh. Intracellular transport of MHC class II molecules. Immunology Today 13:179–189, 1992.

Unanue, E. R. Cellular studies on antigen presentation by class II MHC molecules. Current Opinion in Immunology 4:63–69, 1992.

Unanue, E. R., and P. M. Allen. The basis for the immunoregulatory role of macrophages and other accessory cells. Science 236:551–557, 1987.

York, I. A., and K. L. Rock. Antigen processing and presentation by the class I major histocompatibility complex. Annual Review of Immunology 14:369–396, 1996.

CHAPTER SEVEN

T LYMPHOCYTE ANTIGEN

RECOGNITION AND ACTIVATION

In Chapter 6 we introduced the concept that helper T cells and cytolytic T lymphocytes (CTLs) recognize peptides that are physically associated with major histocompatibility complex (MHC) molecules on the surfaces of antigen-presenting cells (APCs) or target cells. Antigen recognition by T cells is the initiating stimulus for their activation and various functional responses, including secretion of cytokines, proliferation, and the performance of regulatory or cytolytic effector functions.

The **T cell receptor (TCR)** for antigen is composed of membrane proteins expressed only on T lymphocytes and specifically binds to peptide-MHC complexes on the surface of APCs or target cells. As would be expected, the TCR proteins differ among T cells with different antigen specificities. The TCR is expressed in association with a complex of several invariant plasma membrane proteins that transduce intracellular signals required for initiation of T cell responses. In addition, T cells express a number of other cell surface proteins, which are collectively called **accessory molecules.** Several of these accessory molecules function to strengthen the adhesion of the T cells to other cells, thereby promoting maximally effective interactions between helper T cells and APCs or between CTLs and their targets. Many accessory molecules transduce signals which enhance or modify the responses of the T cell to the antigen receptor–generated signals. Other accessory molecules regulate the migration of T lymphocytes to different anatomic sites.

In this chapter, we describe the molecules and signaling events involved in T cell responses to antigen. We discuss the structure and function of the T cell antigen receptor and accessory molecules, and the cascade of intracellular biochemical events that occur after antigen recognition.

GENERAL FEATURES OF T LYMPHOCYTE RESPONSES TO ANTIGEN

T cell antigen recognition leads to several biologic responses, which are of central importance in both cell-mediated and humoral immunity. These responses include the following (Fig. 7–1):

1. *Proliferation of T cells*, in response to antigen recognition, is mediated primarily by an autocrine growth pathway, in which the responding T cell secretes its own growth-promoting cytokines and also expresses cell surface receptors for these cytokines. The principal autocrine growth factor for most T cells is interleukin (IL)-2. The result of the proliferative response is **clonal expansion,** which generates enough antigen-specific T cells to handle foreign antigen.

2. *Differentiation* is the process that converts naive T lymphocytes to effector cells that perform various functions. This differentiation process depends on both antigen-induced activation of the T cell and the actions of other factors such as cyto-

kines and signals delivered by accessory molecules. The progeny of antigen-stimulated T lymphocytes may differentiate along several pathways leading to the generation of functionally distinct subsets of helper T cells and CTLs. The differentiation process involves changes in the transcriptional regulation of genes encoding effector molecules that are released from the T cell, such as cytokines or CTL granule proteins, as well as regulation of expression of T cell surface molecules.

3. *Effector functions* of T cells are initiated by antigen stimulation of differentiated effector cells generated by antigen-induced differentiation of naive T cells. These functions are the biologic activities that enable T cells to mount useful immune responses to foreign antigens. *The major effector function of CD4-expressing helper T cells is the secretion of cytokines*, which act on T cells and on other cell types, including B lymphocytes, macrophages, granulocytes, and vascular endothelium. These cytokines exert various effects that promote and regulate humoral and cell-mediated immune responses and inflammation. The structure and functions of T cell cytokines are discussed in detail in Chapter 12. Some of the effector functions of helper T cells are mediated by surface molecules that are expressed after antigen stimulation. These are described later in this chapter and in the context of humoral and cell-mediated immune responses in Chapters 9 and 13. *The major effector function of CTLs is to lyse antigen-bearing target cells*; in addition, CTLs secrete some cytokines. The mechanisms of action and functions of CTLs are discussed in Chapter 13.

4. Some of the progeny of antigen-responsive cells develop into antigen-specific *memory T cells.* Although functional and mitotic responses of T cells to antigenic stimulation last only for brief periods, and the responses quickly wane as the antigen is eliminated, memory T cells survive and initiate larger secondary immune responses upon subsequent exposures to the antigen (see Chapter 2).

THE $\alpha\beta$ T CELL RECEPTOR FOR MHC-ASSOCIATED PEPTIDE ANTIGENS

The activation of T lymphocytes begins with specific recognition of antigen. The molecular nature of the TCR responsible for MHC-restricted antigen recognition was elucidated in the 1980s, several years after immunoglobulin (Ig) molecules and their genes were described. An important advance in the study of T cell receptors was the development of technologies for propagating monoclonal T cell populations *in vitro*, including T-T hybridomas and antigen-specific T cell clones (Box 7–1). All the cells in a clonal T cell population are derived from a single cell, are genetically identical, and therefore express identical TCRs that are different from the receptors produced by all other clones. Antibodies specific for unique (idiotypic) determi-

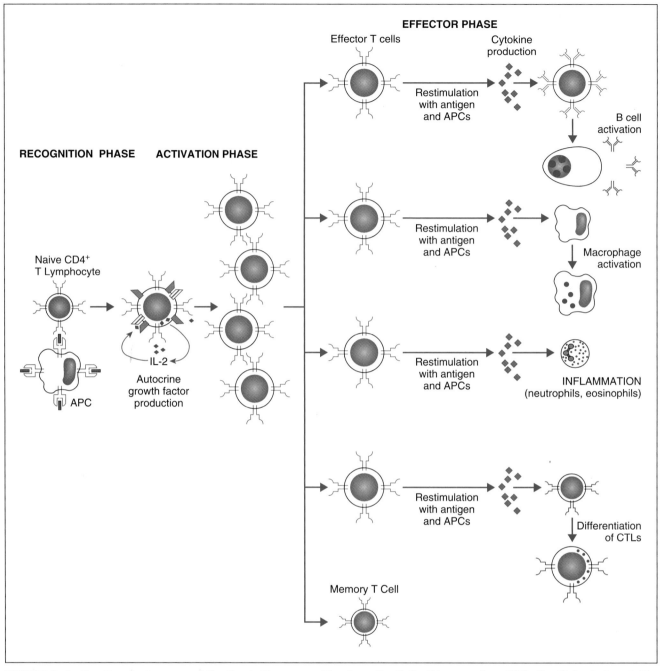

FIGURE 7–1. Functional responses of T cells. Antigen recognition by a T cell (in this example, a CD4⁺ T cell) leads to cytokine (e.g., interleukin-2 [IL-2]) production, proliferation as a result of autocrine IL-2 stimulation, and effector functions (e.g., B cell stimulation, macrophage activation, promotion of inflammation, and help for CTL differentiation). APC, antigen-presenting cell; CTL, cytolytic T lymphocyte.

nants of the antigen receptors of a clonal T cell population were used to purify TCR molecules so that they could be biochemically characterized. The receptor proteins identified in this way shared sequence homology with Ig polypeptides and contained highly variable regions that differed from one clone to another. Clonal T cell populations were also used in the first successful attempts at isolating and characterizing TCR genes, at a time when the biochemical analysis of the proteins was largely incomplete. The strategy for the identifica-

tion of TCR genes was based on the assumptions that they would be uniquely expressed in T cells, they would undergo somatic rearrangements during T cell development (like Ig genes do during B cell development), and they would be homologous to Ig genes. Some genes that were isolated on the basis of expression in T lymphocytes but not in B cells satisfied these criteria and the encoded amino acid sequences agreed with partial sequences obtained from putative TCR proteins purified with anti-idiotypic antibodies. The subsequent

BOX 7–1.   Experimental Methods for Studying T Cell Biology

Most of our current knowledge of the cellular events in T cell activation is based on *in vitro* experiments in which T cells can be stimulated in a controlled manner and their responses can be measured accurately. The T cell responses in these experiments are believed to be the same as those that occur normally *in vivo* when antigen-MHC complexes on the surface of APCs or target cells bind to the TCR. For studying the mechanisms of T cell activation *in vitro,* immunologists have employed several kinds of T cell populations and several kinds of stimuli.

Polyclonal populations of normal T cells with a wide variety of different antigen specificities can be derived from the blood and peripheral lymphoid organs of specifically immunized individuals. The immunization serves to expand the number of antigen-specific T cells, which can then be restimulated *in vitro* by adding antigen and MHC-matched APCs to the T cells. In T cell populations from unimmunized animals or humans, however, the number of cells specific for a particular peptide-MHC complex represents a very small fraction of the total cell number, and thus it is not possible to measure antigen-specific responses. Functional responses of these cells can be more easily studied by the use of polyclonal activators, which bind to many or all TCR complexes regardless of specificity and mimic antigen plus MHC-induced perturbations of the TCR complex. For example, polymeric plant proteins called lectins bind specifically to certain sugar residues on T cell surface glycoproteins, including the TCR and CD3 proteins, and thereby stimulate the T cells. Commonly used T cell–activating lectins include concanavalin-A (Con A) and phytohemagglutinin (PHA). Alternatively, polyclonal T cell populations can be stimulated by antibodies specific for invariant framework epitopes on TCR or CD3 proteins. Often, these antibodies need to be immobilized on solid surfaces or beads or cross-linked with secondary anti-antibodies in order to induce an optimal activation response.

Superantigens are another kind of polyclonal stimulus that bind to and activate all T cells that express a particular type of TCR β chain (see Box 16–1).

Monoclonal populations of T cells have also been very useful for the study of functional responses to specific antigen plus MHC. By definition, all the T cells in a monoclonal population are genetically identical to one another and therefore express identical TCRs. This provides a homogeneous population of cells for functional, biochemical, and molecular analyses. Three types of monoclonal T cell populations have been frequently utilized in experimental immunology.

(1) Antigen-specific T cell clones are derived by *in vivo* immunization of an individual with a particular antigen, isolation of T cells from either blood or lymphoid tissues, repetitive *in vitro* stimulation with the immunizing antigen plus MHC-matched APCs (which drive proliferation of T cells only with the appropriate specificities), and cloning of single antigen-MHC–responsive cells in semisolid media or in liquid media by limiting dilution. Antigen-specific responses can be easily measured in these populations, since all the cells in a cloned cell line have the same receptors and have been selected for growth in response to a known antigen-MHC complex. Both helper and CTL clones have been established from mice and humans.

(2) Antigen-specific T-T hybridomas are created in a way similar to that for B cell hybridomas (see Box 3–1, Chapter 3). Mice are immunized with the antigen of interest, lymph nodes draining the site of immunization are removed, and T cells are purified. These cells are fused with an autonomously growing T cell tumor line, and metabolic selection is applied so that only the fused cells grow. From the resultant hybridomas, cells responding to the desired antigen-APC combination are selected and cloned. Murine T-T hybridomas can be easily made, including those derived from both mature T cells and thymocytes. Human T-T hybridomas have also been made recently.

(3) Tumor lines derived from T cells have been established *in vitro* after removal of malignant T cells from animals or humans with T cell leukemias or lymphomas. Some tumor-derived lines express functional TCR complexes. Even though the antigen specificities of these lines are not known, they can be activated via their TCRs in other ways, e.g., with anti-TCR antibodies or lectins. The Jurkat line, derived from a human T cell leukemia cell, is an example of a widely used tumor line.

These monoclonal T cell populations have been extensively used to identify T cell surface molecules, to clone genes encoding such molecules (such as the T cell antigen receptor genes), and to study the biochemical and functional consequences of ligand binding to TCR. The TH1 and TH2 subsets of helper T cells were discovered by analyzing the sets of cytokines produced by mouse T cell clones (see Chapter 12).

More recently, mice expressing transgenic TCRs have provided a powerful means of obtaining large numbers of T cells with a single antigen specificity (see Box 8–1, Chapter 8). One of the unique advantages of TCR-transgenic mice is that they permit the isolation of sufficient numbers of naive T cells of defined specificity to allow functional responses to be studied. These TCR-transgenic T cell populations have been invaluable for studies of how T cells are driven to differentiate upon their first exposure to antigen.

---

analysis of many isolated TCR genes, as well as their manipulation in transgenic animals, have provided many important advances in immunology. More recently, x-ray crystallographic structures of components of the TCR have been solved, allowing a more precise understanding of how T cells recognize peptide-MHC complexes (see Color Plate V, preceding Section I). The characteristics of TCR proteins are discussed in the following sections. The genomic organization of TCR genes and the mechanisms of their rearrangements leading to TCR diversity are discussed in detail in Chapter 8

in the context of development of T cells from bone marrow–derived precursors.

Biochemical Characteristics of the $\alpha\beta$ TCR

The receptor for peptide-MHC complexes on the majority of T cells, including MHC-restricted helper T cells and CTLs, is a heterodimer consisting of two transmembrane polypeptide chains, designated α and β, covalently linked to each other by disulfide bonds (Fig. 7–2 and Table 7–1). (Another less common type of TCR, found on a small subset of T cells, is

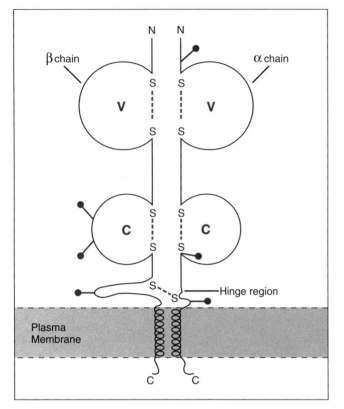

FIGURE 7–2. Schematic diagram of the T cell receptor (TCR) for peptide–major histocompatibility complex (MHC) complexes. V and C refer to Ig-like variable and constant domains, respectively, of the α and β chains; N and C refer to the amino and carboxy termini of the polypeptides, respectively. S––S indicates a disulfide bond, and ⟶• indicates approximate location of carbohydrate groups.

composed of γ and δ chains and is discussed later.) Amino acid sequence data and recent x-ray crystallographic studies indicate that there are striking similarities between the α and β chains of the TCR and Ig polypeptides. In fact, the extracellular portion of the $\alpha\beta$ heterodimer can be considered structurally similar to the Fab fragment of an Ig molecule (see Chapter 3). The extracellular portions of both α and β chains contain an N-terminal variable (V) domain and a membrane-proximal constant (C) region. The V domains have tertiary structures similar to Ig V domains, and the C domain of the β chain is similar to an Ig C domain. In Chapter 3 we introduced the concept that mem-

TABLE 7–1. Proteins in T Cell Antigen Receptor Complexes

Name	Function	Size (kD) Human	Size (kD) Mouse	Multimeric Form	Comments
TCR α	One chain of receptor for recognition of antigen-MHC complexes	40–60	44–55	$\alpha\beta$	Ig superfamily member; rearranging genes; on CD4$^+$ or CD8$^+$ T cells
TCR β	One chain of receptor for recognition of antigen-MHC complexes	40–50	40–55	$\alpha\beta$	Ig superfamily member; rearranging genes; on CD4$^+$ or CD8$^+$ T cells
TCR γ	One chain of receptor for peptide and nonpeptide antigens	45–60	45–60	$\gamma\delta$ or $\gamma\gamma$	Ig superfamily member; rearranging genes, predominantly on CD4$^-$CD8$^-$ T cells
TCR δ	One chain of receptor for peptide and nonpeptide antigens	40–60	40–60	$\gamma\delta$	Ig superfamily member; rearranging genes, predominantly on CD4$^-$CD8$^-$ T cells
CD3 γ	Signal transduction for $\alpha\beta$ and $\gamma\delta$ TCR Cell surface expression of TCR complex	25–28	21		Ig superfamily member; one ITAM in cytoplasmic tail
CD3 δ	Signal transduction for $\alpha\beta$ and $\gamma\delta$ TCR Cell surface expression of TCR complex	20	28		Ig superfamily member; one ITAM in cytoplasmic tail
CD3 ϵ	Signal transduction for $\alpha\beta$ and $\gamma\delta$ TCR Cell surface expression of TCR complex	20	25		Ig superfamily member; one ITAM in cytoplasmic tail
ζ	Signal transduction for $\alpha\beta$ and $\gamma\delta$ TCR Cell surface expression of TCR complex	16	16	$\zeta\zeta$ or $\zeta\eta$	Three ITAMs in cytoplasmic tail
η	Signal transduction for $\alpha\beta$ and $\gamma\delta$ TCR	22	22	$\zeta\eta$	Alternative splice form of ζ; one ITAM in cytoplasmic tail

Abbreviations: TCR, T cell receptor; Ig, immunoglobulin; ITAM, immunoreceptor tyrosine-based activation motif; MHC, major histocompatibility complex.

brane proteins that contain domains that are structurally homologous to Ig V and C domains are members of a family of molecules that constitute the Ig superfamily (Box 7–2). Proteins belonging to this family are thought to have evolved from one ancestral gene, and many of these proteins play important roles in cell-cell interactions. The tertiary structure of the TCR α chain C domain, however, is not typical of a classical Ig domain, since the β-strands, which would normally form the top β-sheet, are spaced too far apart to form hydrogen bonds with one another (see Box 7–2). Like Ig V domains, the V regions of the TCR α and β chains have sequences that are highly variable from one T cell clone to the next, reflecting their role in

antigen recognition. In addition, the V regions of both α and β chains contain sequences that are homologous to Ig sequences encoded by joining (J) gene segments, and the V region of the β chain also has sequences homologous to Ig sequences encoded by diversity (D) gene segments (Fig. 7–3).

Both α and β chains contain a short hinge region adjacent to the outer side of the plasma membrane. A cysteine residue in each hinge region contributes to a disulfide bond that links the two chains. A stretch of hydrophobic amino acid residues comprises the transmembrane portion of both chains. An unusual feature of these transmembrane portions is the presence of positively

---

BOX 7–2. The Ig Superfamily

Many of the cell surface and soluble molecules that mediate recognition, adhesion, or binding functions in the vertebrate immune system share partial amino acid sequence homology and tertiary structural features that were originally identified in Ig heavy and light chains. In addition, the same features are found in many molecules outside the immune system that also perform similar functions. These diverse proteins are members of the Ig superfamily. A superfamily is broadly defined as a group of proteins that share a certain degree of sequence homology, usually at least 15 per cent. The conserved sequences shared by superfamily members often contribute to the formation of compact tertiary structures referred to as domains, and most often the entire sequence of a domain characteristic of a particular superfamily is encoded by a single exon. Members of a superfamily likely derive from a common precursor gene by divergent evolution, and multidomain proteins may belong to more than one superfamily.

The criterion for inclusion of a protein in the Ig superfamily is the presence of one or more Ig domains (also called Ig homology units), which are regions of 70 to 110 amino acid residues homologous to either Ig V or C domains. The Ig domain contains conserved residues that permit the polypeptide to assume a globular tertiary structure called an antibody (Ig) fold (see Chapter 3), composed of a sandwich arrangement of two β-sheets, each made up of three to five antiparallel β-strands of five to ten amino acid residues. The sandwich-like structure is stabilized by hydrophobic amino acid residues on the β-strands pointing inward, which alternate with hydrophilic residues pointing out. Since the inward-pointing residues are essential for the stability of the tertiary structure, they are the major contributors to the regions that are conserved between Ig superfamily members. In addition, there are usually conserved cysteine residues that contribute to the formation of an intrachain disulfide-bonded loop of 55 to 75 amino acids (approximately 90 kD). Ig domains are classified as V-like or C-like on the basis of closest homology to either Ig V or C domains. V domains are formed from a longer polypeptide than C domains and, like V domains of Ig, contain an extra pair of β-strands within the β-sheet sandwich compared to C domains. A third type of Ig domain, called C2 or H, has a length similar to C domains but has sequences typical of both V and C domains.

Using several criteria of evolutionary relatedness, such as primary sequence, intron-exon structure, and ability to undergo DNA rearrangements, molecular biologists have postulated a scheme, or family tree, depicting the evolution of

members of the Ig superfamily. In this scheme, a very early event was the duplication of a gene for a primordial surface receptor followed by divergence of V and C exons. Modern members of the superfamily contain different numbers of V and/or C domains. The early divergence is reflected by the lack of significant sequence homology in Ig and TCR V and C units, although they share similar tertiary structures. A second early event in the evolution of this family was the acquisition of the ability to undergo DNA rearrangements, which has remained a unique feature of the antigen receptor gene members of the family.

Most identified members of the Ig superfamily (see accompanying figure) are integral plasma membrane proteins with Ig domains in the extracellular portions, transmembrane domains composed of hydrophobic amino acids, and widely divergent cytoplasmic tails, usually with no intrinsic enzymatic activity. There are exceptions to these generalizations. For example, the platelet-derived growth factor receptors have cytoplasmic tails with tyrosine kinase activity, and the Thy-1 molecule has no cytoplasmic tail but, rather, is anchored to the membrane by a phosphatidylinositol linkage.

One recurrent characteristic of the Ig superfamily members is that interactions between Ig domains on different polypeptide chains are essential for the functions of the molecules. These interactions can be homophilic, occurring between identical domains on opposing polypeptide chains of a multimeric protein, as in the case of $C_H:C_H$ pairing to form functional Fc regions of Ig molecules. Alternatively, they can be heterophilic, as occurs in the case of $V_H:V_L$ or $V\beta:V\alpha$ pairing to form the antigen-binding sites of Ig or TCR molecules, respectively. Heterophilic interactions can also occur between Ig domains on entirely distinct molecules expressed on the surfaces of different cells. Such interactions provide adhesive forces that stabilize immunologically significant cell-cell interactions. For example, the presentation of an antigen to a helper T cell by an APC probably involves heterophilic intercellular Ig domain interactions between at least four pairs of Ig superfamily molecules, including TCR:class II MHC, CD4:class II MHC, CD2:LFA-3, and CD28:B7. The importance of all these interactions is demonstrated by the observation that antibodies that block the binding of these molecules to one another also block antigen plus APC-induced T cell activation. Several Ig superfamily members have been identified on cells of the developing and mature nervous system, consistent with the functional importance of highly regulated cell-cell interactions in these sites.

Continued

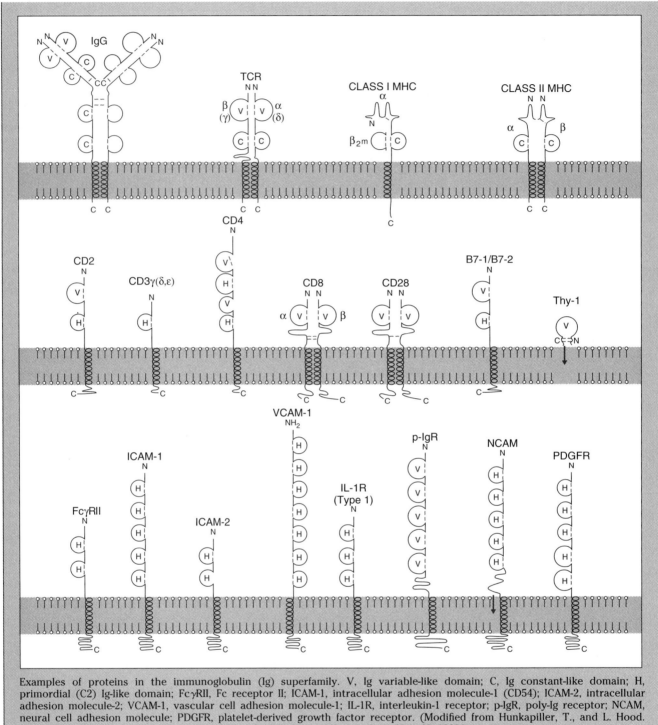

Examples of proteins in the immunoglobulin (Ig) superfamily. V, Ig variable-like domain; C, Ig constant-like domain; H, primordial (C2) Ig-like domain; FcγRII, Fc receptor II; ICAM-1, intracellular adhesion molecule-1 (CD54); ICAM-2, intracellular adhesion molecule-2; VCAM-1, vascular cell adhesion molecule-1; IL-1R, interleukin-1 receptor; p-IgR, poly-Ig receptor; NCAM, neural cell adhesion molecule; PDGFR, platelet-derived growth factor receptor. (Modified from Hunkapiller, T., and L. Hood. Diversity of the immunoglobulin gene superfamily. Advances in Immunology 44:1–63, 1989.)

charged amino acid residues, including a lysine residue (β chain) or a lysine and an arginine residue (α chain). These residues interact with negatively charged residues found in the transmembrane portions of the CD3 polypeptides (see below). Both α and β chains have carboxy terminal cytoplasmic tails 5 to 12 amino acids long. Like membrane Ig on B cells, these cytoplasmic regions

are too small to have intrinsic enzymatic activity, and in fact other molecules physically associated with the TCR are needed to provide signal-transducing functions. There are several significant differences between TCR and Ig molecules. In general, the TCR is not secreted, and therefore it does not perform effector functions on its own. Instead, upon binding peptide-MHC complexes, the TCR ini-

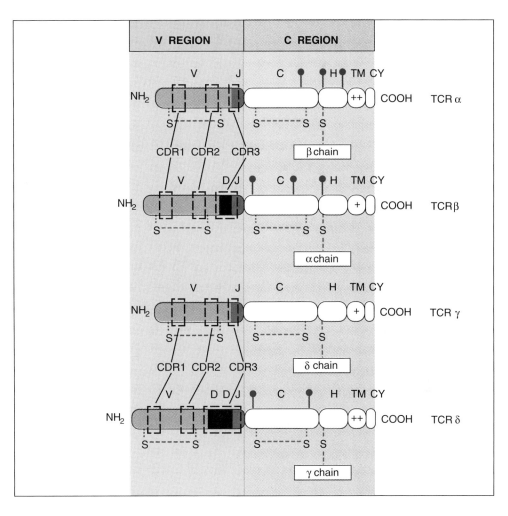

FIGURE 7–3. Relation of T cell receptor (TCR) gene segments to domains of polypeptide chains. The V region is encoded by variable (V), diversity (D), and joining (J) gene segments. Locations of disulfide binds (S--S) and carbohydrates (→) are approximate. Areas in dashed boxes are hypervariable complementarity-determining (CDR) regions. "+" refers to positively charged amino acid residues in the transmembrane regions. H, hinge region; TM, transmembrane domain; CY, cytoplasmic domain. Compare with Figure 4–5 for immunoglobulin. (Modified with permission from Davis, M., and P. J. Bjorkman. T cell antigen receptor genes and T cell recognition. Nature 334: 395–402, 1988. Copyright © 1988, Macmillan Magazines Ltd.)

tiates signals that activate T cell effector functions. Also, unlike Ig, the TCR α and β chains do not undergo changes in C region expression, i.e., "isotype switching," or affinity maturation during T cell differentiation.

Role of the $\alpha\beta$ TCR Receptor in Recognition of MHC-Associated Peptide Antigen

The $\alpha\beta$ TCR heterodimer recognizes complexes of processed peptides, generated from protein antigens, bound to self MHC molecules. Before the isolation and experimental manipulation of TCR genes, it was uncertain if T cells expressed two different types of receptors, one specific for protein antigen and the other for MHC molecules. This was shown to be unlikely by experiments indicating that self MHC restriction and foreign antigen recognition do not segregate independently. For example, when two T cells with different peptide antigen specificities and different MHC restrictions (e.g., for peptide A plus MHC allele 1 and peptide B plus MHC allele 2) are fused, the hybrid cell line recognizes each of the peptides recognized by the two parent T cells, but only with the same independent MHC restrictions as the parent T cells (i.e., there is no recognition of peptide A

plus MHC 2 or peptide B plus MHC 1). Definitive proof that the $\alpha\beta$ heterodimer is responsible for both the antigen (peptide) specificity and the MHC restriction of a T cell came from experiments with isolated TCR genes. For example, functional TCR α and β genes can be isolated from a T cell clone of defined peptide and MHC specificity. When these genes are expressed in other T cells, either by transfection or transgenic approaches, they confer both the peptide and MHC specificity of the original clone from which they were isolated (Fig. 7–4). These types of experiments have also proved that *both the α and β chains of the TCR contribute to the recognition specificity for the complex of peptide antigen and MHC molecules.*

The structural features of the $\alpha\beta$ TCR-binding site for peptide-MHC complexes have been deduced from amino acid sequence data, mutational studies, and most recently from x-ray crystallographic analyses. Perhaps the most significant difference between TCR and Ig molecules lies in the nature of the antigens which they recognize. As discussed in Chapters 5 and 6, TCR specificity is limited to peptides bound to MHC molecules, while antibodies are capable of specific binding to virtually any kind of molecular structure. Despite these differences in specificity, the variable regions of

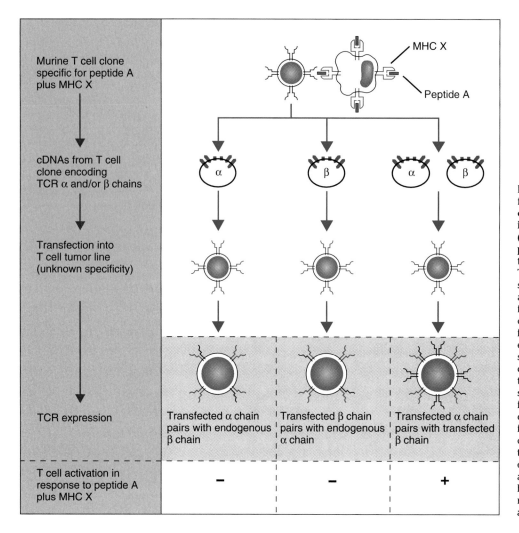

FIGURE 7-4. T cell specificity for antigen and major histocompatibility complex (MHC) is a function of T cell receptor (TCR) α and β chain gene products. In this experiment, the α and β chain genes from a T cell hybridoma of defined specificity were transfected into a T cell line of unknown specificity. Although the α and β chains encoded by the transfected genes can pair with the endogenous β and α chains, respectively, only co-transfection of both α and β chain genes led to expression of a TCR with the specificity of the original T cell from which the genes were cloned. Furthermore, specificity for antigen alone or MHC alone could not be conferred upon the T cell line by transfection of either the α or β chain genes alone, indicating that the αβ heterodimer is responsible for recognizing both the antigen and MHC molecule.

the two types of molecules are quite similar. There are at least three highly diverse regions in both the α and β chains of the TCR that correspond to the antigen-binding complementarity-determining regions (CDRs) of Ig. Thus, the antigen-binding surface of a TCR consists of three hypervariable hairpin loops contributed by the α chain and three contributed by the β chain, for a total of six CDRs. In both α and β chains, two of the hypervariable regions, CDR1 and CDR2 are contained entirely within the V domains, while CDR3 is composed of sequences encoded by V and J gene segments (in the α chain) or V, D, and J gene segments (in the β chain) (see Fig. 7-3). The CDR1 and CDR2 loops in the TCR chains are generally less variable than their counterparts in Ig chains, and therefore the variability of TCRs is even more concentrated in the CDR3 loop than it is in Ig molecules. The genetic mechanisms that account for CDR3 variability are discussed in Chapter 8. A fourth peptide hairpin loop composed of V domain amino acids has been demonstrated in the crystal structures of the TCR β chain next to the three CDR loops. This loop is called CDR4, or the fourth hypervariable region, since sequences in this region have been

shown to be hypervariable among some TCRs. The role of CDR4 in antigen recognition is uncertain.

Analysis of the crystal structure of the variable region of an αβ heterodimer which binds a peptide–class I MHC complex has confirmed the predicted similarity of this part of the TCR to the Fab region of an antibody molecule. This peptide-MHC binding site is a flat surface with a hydrophobic cavity between the two CDR3 loops. Further analysis of the crystal structure of this receptor bound to its peptide-MHC ligand has shown that the CDR3 loops are positioned over the center of the peptide, while the CDR1 and CDR2 loops contact the ends of the peptide. All the loops are positioned to make some contact with the MHC molecule (see Color Plate V, preceding Section I). The crystal structures of several different TCR-peptide-MHC complexes, including both class I and class II MHC–containing complexes, will need to be solved before generalizations can be made.

The side chains of as few as three amino acid residues of the MHC-bound peptide make contact with the TCR, a remarkable fact given the exquisite specificity of T cell antigen recognition. The actual residues that contact the TCR have been determined for many different peptides by assessing the effects of single amino acid substitutions on both T cell recognition and MHC binding (see Table 7–2).

The affinity of TCR binding to peptide-MHC complexes is believed to be significantly lower, on average, than Ig binding to antigen. This may be, in part, due to selection processes in the thymus that favor development of T cells with low affinity TCRs (see Chapter 8). Crystallographic data also suggest that the TCR is a much more rigid structure than Ig, with less flexibility at the hinge regions or between C and V domains. This is in keeping with the restricted TCR recognition of only peptide-MHC complexes and not other structures.

Whether a particular T cell is class I or class II MHC–restricted is generally not determined by the V, D, J, or C gene sequences in the α or β chain of the TCR of that cell, although there may be some predilection for the selective use of certain V genes in CD4$^+$ versus CD8$^+$ T cells. Overall, the same sets of TCR genes can be expressed in class I– and class II–restricted T cells. As we mentioned in Chapter 6 and will discuss in more detail later in this chapter, the ability of a particular T cell to respond to either class I– or class II–associated peptide is determined mainly by the expression of CD8 or CD4, respectively.

THE CD3, ζ, AND η PROTEINS ASSOCIATED WITH THE TCR COMPLEX

The $\alpha\beta$ TCR heterodimer provides T cells the ability to recognize peptide antigens bound to MHC molecules, but *both the cell surface expression of TCR molecules and their function in activating T cells are dependent on up to five other transmembrane proteins that non-covalently associate with the $\alpha\beta$ heterodimer.* Together, these proteins form the functional **TCR complex** (Fig. 7–5 and Table 7–1). Three proteins in the complex are called **CD3 molecules,** and include highly homologous Ig superfamily members designated γ, δ, and ϵ. In addition, 80 to 90 per cent of TCR complexes contain a disulfide-linked homodimer of a protein called the **ζ chain** that does not belong to the Ig gene superfamily. The remaining 10 per cent of human T cells express heterodimers consisting of the ζ chain and the highly homologous FcϵRI γ chain (see Chapter 14). In mice, 10 to 20 per cent of T cells express a heterodimer of the ζ chain and an alternative splice product of the ζ gene called the η chain. The exact stoichiometry of TCR complexes is not known but models have been proposed based on the molecular weights and molar ratios of immunoprecipitated TCR components, as well as consideration of charged residues in transmembrane segments of the TCR α and β chains and the CD3 chains. A likely stoichiometry of the most common form of TCR is $(\alpha\beta)_2\epsilon_2\gamma\delta\zeta$-$\zeta$.

Structure and Association of CD3, ζ, and η Proteins

The CD3 proteins were first identified, before the $\alpha\beta$ heterodimer, by the use of monoclonal antibodies raised against T cells, and the ζ and η chains were identified later by co-immunoprecipitation with $\alpha\beta$ and CD3 proteins. The physical association of the $\alpha\beta$ heterodimer, CD3, and ζ chains has been demonstrated in two ways.

1. Antibodies against the $\alpha\beta$ TCR heterodimer or the CD3 proteins co-precipitate both the heterodimer and the associated proteins from solubilized cell membrane preparations.
2. When intact T cells are treated with either anti-CD3 or anti-$\alpha\beta$ heterodimer antibodies, the en-

TABLE 7–2. Identification of MHC-Binding and T Cell Receptor–Binding Residues in Peptide Antigens

	HEL Peptide										Stimulation of HEL-Specific T Cells	Binding to Purified I-A^d	Competition with Native HEL for T Cell Stimulation
	Amino Acid Residue Position No.												
	52	53	54	55	56	57	58	59	60	61			
1	Asp	Tyr	Gly	Ile	Leu	Gln	Ile	Asn	Ser	Arg	+	+	NA
2	Asp	Tyr	Gly	Ile	*Ala*	Gln	Ile	Asn	Ser	Arg	−	+	+
3	Asp	*Ala*	Gly	Ile	Leu	Gln	Ile	Asn	Ser	Arg	−	+	+
4	Asp	Tyr	Gly	*Ala*	Leu	Gln	Ile	Asn	Ser	Arg	−	−	−
5	Asp	Tyr	*Ala*	Ile	Leu	Gln	Ile	Asn	Ser	Arg	+	+	NA

Synthetic peptides were produced that differed from the native hen egg lysozyme peptide HEL (52-61) (peptide 1) by substitutions for single residues, and the functional consequences of these engineered mutations were analyzed. Substitutions at positions 56 and 53 (peptides 2 and 3) result in loss of T cell stimulation, but retain I-A binding. The amino acids at these positions in the native peptide are part of the epitope recognized by the T cell receptor. Substitution of residue 55 (peptide 4) results in loss of T cell stimulation and I-A binding. This residue is in part of the peptide that binds to the class II MHC molecule. A substitution at position 54 (peptide 5) has no effect, and therefore this residue is not essential for binding of the peptide to either the MHC or T cell receptor molecules.

Abbreviations: MHC, major histocompatibility complex; HEL, hen egg lysozyme; NA, not applicable.
Adapted with permission from Unanue, E. R., and P. M. Allen. The basis for the immunoregulatory role of macrophages and other accessory cells. Science 236:551–557, 1987. Copyright 1987 American Association for the Advancement of Science.

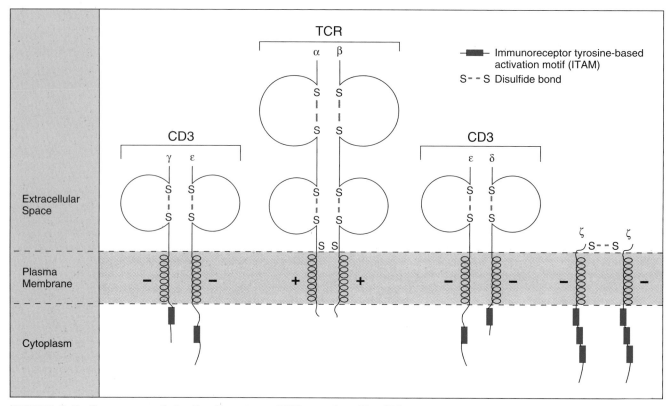

FIGURE 7–5. Components of the T cell receptor (TCR) complex. The CD3 proteins include the γ, δ, and ϵ chains, which are present as monomers non-covalently associated with the disulfide-linked $\alpha\beta$ TCR heterodimer. The presence of two copies of CD3ϵ, as shown, is suggested by experimental evidence. It is also possible that two copies of the $\alpha\beta$ TCR heterodimer are present in a single TCR complex (not shown). The ζ chain is present as a disulfide-linked homodimer. Some TCR complexes include a ζ-η heterodimer instead of the ζ-ζ homodimer (not shown). The + and − symbols refer to charged amino acid residues in the transmembrane regions, which probably mediate association of the CD3 and ζ chains with the $\alpha\beta$ heterodimer. The immunoreceptor tyrosine-based activation motifs (ITAMs) in the cytoplasmic tails of the CD3 and ζ chains are conserved sequences that include sites for tyrosine phosphorylation (see text).

tire TCR complex is endocytosed and disappears from the cell surface; i.e., the proteins are co-modulated.

The structures of these proteins associated with the $\alpha\beta$ TCR heterodimer have been elucidated by biochemical analysis and sequencing of complementary deoxyribonucleic acid (cDNA) clones that encode the molecules. The γ, δ, and ϵ chains are highly homologous to each other; their genes are located adjacent to one another on human chromosome 11, and all probably arose by duplication from a common ancestral gene.

The N-terminal extracellular regions of γ, δ, and ϵ chains each contain a single Ig-like domain, and therefore these three proteins are members of the Ig superfamily. There is no variability or polymorphism identified in the extracellular domains of the CD3 proteins or their genes, consistent with the fact that these proteins do not contribute to the specificity of antigen recognition. The transmembrane segments of all three CD3 chains contain a negatively charged aspartic acid residue. This unusual feature is important for the physical association of the CD3 proteins with the TCR α and β chains, since each of the latter polypeptides contains at least one positively charged residue in its transmembrane domain. The cytoplasmic domains of the CD3 γ, δ, and ϵ proteins range from 44 to 81 amino acid residues long and, therefore, are of sufficient size to transduce signals to the cell interior. In fact, the cytoplasmic tail of each of the CD3 proteins contains one copy of a sequence motif important for signaling functions, which is called either the **immunoreceptor tyrosine-based activation motif (ITAM)** or the antigen recognition activation motif (ARAM). ITAMs are also found in the cytoplasmic tails of several other membrane proteins involved in signal transduction, including Igα and Igβ proteins associated with membrane IgM and IgD (see Fig. 9–3, Chapter 9), components of several Ig Fc receptors (see Box 3–4, Chapter 3), and the TCR complex ζ and η chains (see below). The ITAM is composed of approximately 26 mostly unconserved amino acid residues in which the sequence tyrosine-X-X-leucine occurs twice separated by six to eight residues, where X is an unspecified amino acid. Tyrosine residues in ITAMs become phosphorylated in response to TCR antigen recognition. This permits the binding of SH2-domain containing proteins to the ITAMs and the initiation of a cascade of intracellular signaling events (discussed later in this chapter).

The ζ and η chains are encoded by alternatively spliced ribonucleic acid (RNA) transcripts of the same gene and have identical amino acids in their extracellular and transmembrane domains, but differ in their cytoplasmic tails. The extracellular domains are short (nine amino acids), the transmembrane domains contain a negatively charged aspartic acid residue (similar to the CD3 γ, δ, and ϵ chains), and the cytoplasmic domains are long (113 and 155 amino acids for ζ and η chains, respectively). The cytoplasmic tail of the ζ chain contains three ITAMs. As we will discuss in Chapter 13, the ζ chain is associated with other lymphocyte receptors such as the Fcγ receptor (FcγRIII) of natural killer (NK) cells.

Functions of CD3 and ζ Proteins in Signaling and TCR Complex Assembly

When antigen binds to the TCR, the associated CD3 and ζ chains transduce the signals to the cytoplasm of the T cell, which leads to functional activation. Several lines of evidence support this concept.

1. Antibodies against CD3 proteins can often stimulate T cell functional responses that are identical to antigen-induced responses. However, unlike antigens, which stimulate only specific T cells, anti-CD3 antibodies stimulate all T cells in a mixed population, regardless of antigen specificity. Thus, anti-CD3 antibodies are *polyclonal activators* of T cells (see Box 7–1).

2. The cytoplasmic tails of either the CD3 ϵ or ζ chain can transduce the signals necessary for T cell activation in the absence of the other compo-

nents of the TCR complex. This was shown by expressing genetically engineered chimeric molecules containing the cytoplasmic portions of the CD3 ϵ or the ζ protein fused to the extracellular and transmembrane domains of other cell surface receptors for soluble ligands, such as the IL-2 receptor. Ligand binding to these chimeric molecules expressed in T cell tumor lines resulted in activation responses identical to those induced by stimulation through the normal T cell receptor complex.

3. *The CD3 and ζ proteins are substrates of and bind tyrosine kinases.* The earliest intracellular events to follow peptide-MHC recognition are the phosphorylation of tyrosine residues within the ITAMs in the cytoplasmic tails of the CD3 and ζ chains. This permits cytosolic proteins with SH2 domains to bind to the ITAMs. These cytosolic proteins include various tyrosine kinases, which become activated upon association with the ITAMs. These signaling events are the initial steps leading to functional activation of T cells and will be discussed in more detail later in the chapter.

In the subsequent discussion we will refer to T cell antigen recognition or activation being mediated by the "TCR complex." It should be noted, however, that recognition of antigen is due to the $\alpha\beta$ TCR heterodimer only, and the signals that initiate activation are transduced not by the antigen-binding $\alpha\beta$ heterodimer but by the associated proteins in the complex. It should also be noted that although the signals transduced by the TCR complex are necessary for initiating the activation of normal T cells, they are usually not sufficient. Additional costimulators, which interact with T cell

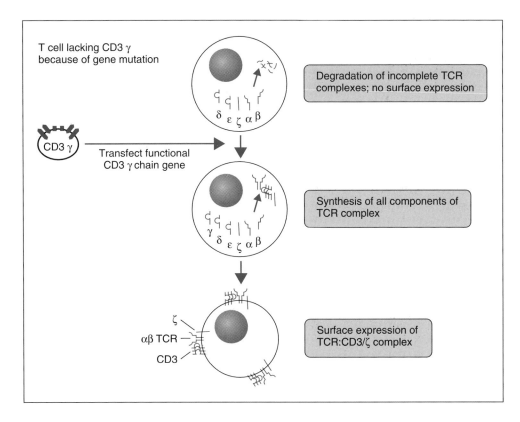

T cell lacking CD3 γ because of gene mutation

CD3 γ

Transfect functional CD3 γ chain gene

Degradation of incomplete TCR complexes; no surface expression

Synthesis of all components of TCR complex

$\alpha\beta$ TCR

ζ

CD3

Surface expression of TCR:CD3/ζ complex

FIGURE 7–6. Co-expression of T cell receptor (TCR) and CD3 molecules. If a T cell does not express one of the CD3 proteins (CD3 γ in the example shown) due to a mutant gene, the other members of the TCR complex do not assemble properly and are degraded intracellularly. No TCR complex is expressed on the surface of these cells. Transfection of a functional CD3 gene can restore expression of the entire TCR complex. The same phenomenon occurs if the TCR α or β chain is not expressed (not shown).

surface molecules other than the TCR complex, are also required, as we will discuss later.

Another function of the CD3, ζ, and η proteins is to facilitate the surface expression of the entire TCR complex. In fact, the αβ TCR heterodimer and the associated CD3 proteins are mutually dependent upon one another for cell surface expression. For example, T cells that cannot synthesize one of the CD3 proteins because of a gene mutation do not express the other CD3 proteins, the ζ chain, or the αβ TCR heterodimers on their surfaces. When a functional CD3 gene is transfected into these cells to replace the mutated gene, expression of the entire TCR complex, including the αβ TCR, is restored (Fig. 7–6).

The synthesis of the components of the TCR complex, their assembly, and their surface expression are tightly regulated and coordinated phenomena that occur during the maturation of T cells in the thymus (see Chapter 8). The genes encoding the CD3 γ, δ, and ε and ζ/η proteins are expressed by immature thymocytes, before TCR α or β chain genes are expressed. Furthermore, the protein products of the CD3 genes are post-translationally modified and form γδε core structures in the absence of TCR α or β chains. The association of the αβ TCR heterodimer with the CD3γδε complex takes place in the endoplasmic reticulum (ER), after which the complex is transported to the Golgi, where further modification of N-linked oligosaccharides takes place. Incomplete complexes do not make their way to the plasma membrane, because some unknown mechanism inhibits transport of these proteins out of the Golgi to the plasma membrane until they all are physically associated with one another. A 90 kD phosphoprotein called **calnexin** plays a role in assembly of the TCR as well as other multichain protein complexes, including MHC molecules. Calnexin may function to retain individual members of the TCR complex within the ER before they are assembled together. The ζ chain, which is synthesized in limited amounts compared with the other components of the TCR complex, also greatly facilitates surface expression of the entire complex.

THE γδ TCR

The γδ TCR is a second type of diverse, CD3- and ζ-associated disulfide-linked heterodimer expressed on a small subset of αβ-negative peripheral T cells and immature thymocytes. (It should not be confused with the γ and δ components of the CD3 complex.) Its existence was first suggested when the γ chain gene was cloned and characterized as an Ig-like gene that is somatically rearranged and expressed only in T cells, but with no sequences compatible with the N-linked glycosylation sites that were known to exist on TCR α and β chains. Subsequent cloning of the Ig-like δ chain gene, and concurrent immunochemical identification of a CD3-associated heterodimeric protein on cells not expressing α and β chains, confirmed

the existence of the γδ receptor. As is the case with the αβ receptor, much of the information available about the protein structure of the γδ receptor is predicted from the sequences of the cloned genes. Although the function of γδ-expressing T lymphocytes is not known, there are several interesting properties of these cells and the genes encoding their receptors. The protein structure, specificity, and postulated function of γδ receptors are discussed below. The genomic organization, rearrangement, and generation of diversity of γδ receptor genes are described in Chapter 8.

Biochemical Characteristics of the γδ TCR

The γδ heterodimers are expressed on αβ-negative T cells. γ and δ chains are transmembrane glycoproteins with structures similar to the α and β chains. Both γ and δ chains include extracellular Ig-like V and C regions, short connecting or hinge regions, hydrophobic transmembrane segments, and short cytoplasmic tails (see Fig. 7–3 and Table 7–1). The hinge regions usually contain cysteines involved in interchain disulfide linkages. In humans, the γδ heterodimer may be disulfide linked or non-covalently linked. The transmembrane regions of γ and δ chains, similar to α and β chains, contain positively charged amino acid residues that interact with the negatively charged residues in the transmembrane regions of the CD3 polypeptides.

Analysis of the CDR3s from γδ receptors indicates that they are structurally more similar to Ig molecules than αβ TCR receptors. This suggests that the ligands that γδ receptors recognize are different from those recognized by αβ receptors.

Specificity and Function of T Cells Expressing γδ Antigen Receptors

Our current understanding of the structure of the γδ receptor is a tribute to modern molecular biology, in that the information about the genes encoding the molecule is available before any thorough understanding of the specificity or function of the T cells that express this receptor. Two interrelated approaches to analyzing the function of γδ receptor–bearing T cells have been (1) characterization of the location and effector capabilities of cells that express the receptor, and (2) determination of what γδ receptors recognize.

The percentages of T cells expressing γδ TCRs vary widely, depending on tissue and species, but in humans less than 5 per cent of T cells express this form of antigen receptor. Studies of the rearrangements of the γ and δ genes in αβ TCR-expressing cells, as well as studies of T cell development in mice expressing γ and δ transgenes, indicate that γδ T cells are a distinct lineage from αβ T cells (see Chapter 8). Furthermore, there are subsets of γδ T cells that can be distinguished from one another on the basis of when they develop during ontogeny, which genes they use to encode their V regions, their sites of development, and the tissues they populate. Studies in mice indicate that different populations of $\gamma\delta$ T cells using particular V genes are generated at different times during fetal or postnatal life, and these populations eventually reside in different tissues. In each population, many of the cells express identical $\gamma\delta$ TCRs with the same VDJ and C sequences and virtually no junctional diversity. For example, many skin $\gamma\delta$ T cells in the mouse develop early in life and express one particular TCR, whereas many of the $\gamma\delta$ T cells in the vagina, uterus, and tongue appear later and express another TCR. These monospecific populations of $\gamma\delta$ T cells arise in the thymus. There is evidence that their lack of diversity results from both molecular constraints on rearrangements of γ and δ TCR genes and selection pressures based on their specificities (see Chapter 8). Some $\gamma\delta$ T cells develop in athymic mice, perhaps in the intestinal mucosa.

One intriguing feature of $\gamma\delta$ T cells is their predominance in various epithelial tissues of certain species. For example, greater than 50 per cent of lymphocytes within the small bowel mucosa of mice and chickens, called **intraepithelial lymphocytes,** are $\gamma\delta$-expressing T cells. In the mouse epidermis, there is a population of intraepidermal T cells, most of which express the $\gamma\delta$ receptor. Equivalent cell populations are not as abundant in humans; only 10 per cent of human intestinal T cells express the $\gamma\delta$ receptor, compared with less than 5 per cent of blood T cells.

The $\gamma\delta$ heterodimer apparently associates with the same CD3 and ζ/η proteins as do $\alpha\beta$ receptors. Furthermore, the signaling events and activation responses typical of $\alpha\beta$-expressing T cells discussed later in this chapter are also observed in $\gamma\delta$ T cells. Most of the other cell surface molecules found on $\alpha\beta$ T cells are also found on $\gamma\delta$ T cells. Two notable exceptions are CD4 and CD8, neither of which is found on the majority of $\gamma\delta$ T cells. A variety of biologic activities have been ascribed to different $\gamma\delta$ T cells that are also characteristic of $\alpha\beta$ T cells, including secretion of various cytokines and lysis of target cells.

The question of what the $\gamma\delta$ TCR recognizes remains unresolved. There are rare examples of cloned $\gamma\delta$ T cells that recognize peptides associated with classical MHC molecules, just like other T cells. However, most $\gamma\delta$ T cells apparently recognize structures unlike peptide-MHC complexes. There is evidence that some $\gamma\delta$ T cells recognize proteins directly without antigen processing; examples include nonpolymorphic MHC-like molecules described in Chapter 5, and viral glycoproteins. Many $\gamma\delta$ receptors recognize small, phosphorylated, nonpeptide molecules derived from mycobacteria, including phosphocarbohydrates and nucleotide derivatives. The limited diversity of the $\gamma\delta$ TCRs in many tissues suggests that the physiologic ligand for these receptors is not polymorphic. A working hypothesis for the function and specificity

of $\gamma\delta$ receptor–expressing T cells is that they may recognize frequently encountered antigens at epithelial boundaries between the host and external environment. Thus, they may initiate immune responses to a small number of common microbes at these sites prior to recruitment of more specific $\alpha\beta$ T cells.

GENERAL PROPERTIES OF ACCESSORY MOLECULES ON T CELLS

The proteins in the TCR complex are the key molecules involved in specific antigen recognition by and antigen-induced activation of MHC-restricted helper T lymphocytes and CTLs. In addition, T cells express several other integral membrane proteins that play significant roles in the functional responses to antigen recognition (Table 7–3). These proteins, often collectively called **accessory molecules,** were initially discovered and characterized by the use of monoclonal antibodies raised against T cells. The antibodies were first used to identify T cell surface molecules by immunofluorescence and immunoprecipitation techniques. These antibodies were also used to block or initiate functional responses of T cells and thus served as probes for studying the physiologic roles of accessory molecules. There are several common properties shared by these accessory molecules:

1. *Accessory molecules on T cells specifically bind other molecules (ligands) present on the surface of other cells, such as APCs, target cells or vascular endothelium, or molecules in the extracellular matrix, including proteins and proteoglycans.*

2. *Accessory molecules are nonpolymorphic and invariant.* Thus, unlike the TCR, accessory molecules are essentially identical on all T cells in all individuals of a species. This implies that these molecules have no capacity to specifically recognize many different, variable ligands, such as antigens.

3. *As a consequence of binding their specific ligands on the surfaces of other cells, many accessory molecules increase the strength of adhesion between a T cell and an APC or target cell.* This property helps to ensure that the T cell and APC remain attached to one another long enough to allow functional interactions to occur between the TCRs and the rare peptide-MHC molecules to which they bind on the APC. In addition, accessory cell-dependent intercellular adhesion may ensure that the APC is bound long enough to be influenced by the effector functions of the T cell.

4. *Accessory molecule binding to endothelial cell surfaces and extracellular matrix ligands contributes to T cell homing and retention in tissues (see Chapter 11).*

5. *Many accessory molecules may transduce biochemical signals to the interior of the T cell that are important in regulating functional responses.* Signal transduction occurs as a consequence of ligand

TABLE 7–3. T Cell Surface Accessory Molecules

Name	Synonyms	Biochemical Characteristics	Gene Family	Cellular Distribution	Ligand	Adhesion	Signal Transduction
							Function in T Cells
CD4	T4 (human), L3T4 (mouse)	55 kD monomer	Ig	TCR α/β positive class II MHC–restricted T cells; macrophages	Class II MHC molecules	+	+
CD8	T8 (human), Lyt-2 (mouse)	α/α homodimer with 78 kD α chain or α/β heterodimer	Ig	TCR α/β positive class I MHC–restricted T cells	Class I MHC molecules	+	+
CD49CD29	VLA-4,5,6	α/β heterodimer	Integrin	Leukocytes, other cells	Matrix molecules, VCAM-1	+	+
CD11aCD18	LFA-1	α/β heterodimer with 180 kD α chain, 95 kD β chain	Integrin	All bone marrow–derived cells	ICAM-1, ICAM-2	+	+
CD28	Tp44	80–90 kD homodimer	Ig	All CD4+ T cells; 50% CD8+ T cells	B7-1, B7-2	–	+
CD2	T11, LFA-2, Leu-5, SRBC receptor	50 kD monomer	Ig	>90% mature human T cells, >70% human thymocytes	LFA-3	+	+
CD45R	T200; leukocyte common antigen; B220 (on B cells)	180–220 kD monomers; cytoplasmic tyrosine phosphatase domain		All immature and mature leukocytes		–	+
CD40 ligand	gp39	Trimers of 39 kD chains	TNF	Activated T cells and mast cells	CD40	–	+
Fas ligand		Trimers of ~40 kD chains	TNF	Activated T cells	Fas (CD950)	–	+
CD5	T1, Leu-1 (humans), Lyt-1 (mice)	67 kD monomer		All T cells and thymocytes		–	+
Ly6		10–18 kD monomers; glycophospholipid membrane anchor		Immature and mature T and B cells; various other tissues		–	+
CD44	PgP-1	80–200 kD monomer; variably glycosylated, chondroitin sulfated	Cartilage link proteins	Thymocytes, T cells, granulocytes, macrophages, erythrocytes, fibroblasts	Collagen fibronectin, hyaluronate	+	?+
L-selectin	CD62L Mel-14 Lam-1	150 kD monomer	Selectin	B cell, T cells, monocytes, NK cells	Sialylated glycoproteins on high HEV	+	–

Abbreviations: HEV, high endothelial venule; ICAM, intercellular adhesion molecule; Ig, immunoglobulin; kD, kilodalton; LFA, lymphocyte function–associated antigen; MHC, major histocompatibility complex; NK, natural killer; TCR, T cell receptor; TNF, tumor necrosis factor; VCAM, vascular cell adhesion molecule; VLA, very late activation.

binding and may act in concert with other signals generated by the TCR complex.

6. T cell accessory molecules are useful cell surface "markers" that facilitate immunocytochemical identification of T cells in pathologic lesions, such as T cell lymphomas and leukemias. In addition, antibodies against these markers can be used to physically isolate T cells for experimental or diagnostic procedures.

Accessory molecules contribute to the regulation of immune responses in various ways. First, the expression of accessory molecules on T lymphocytes varies with the stage of differentiation of the cells and their prior encounter with antigen. The ligands for these molecules may be differentially expressed on endothelium in different tissues. Thus, the variable expression of these molecules influences the way the T cells recirculate from blood to sites of antigen in either lymphoid or peripheral tissues (see Chapter 11). Second, cells that express ligands for accessory molecules may function more efficiently as APCs or CTL targets than cells that do not express these ligands. Several different accessory molecule interactions with ligands on APCs or CTL target cells have been defined (Fig. 7–7). Therefore, variations in the expression of either the T cell accessory molecules or their ligands on APCs may influence the ability of the T cell to respond to antigen presentation. It is possible that T cell accessory molecule-dependent adhesion to APCs may precede TCR binding of peptide-MHC complexes. Once the TCR does bind its ligand, however, signals are generated within the T cell that increase the avidity of certain accessory molecules for their ligands (discussed in more detail later in the chapter). This may serve as a positive amplification step that promotes the tight adhesion of T cells to APCs only if the T cells recognize foreign antigen presented by

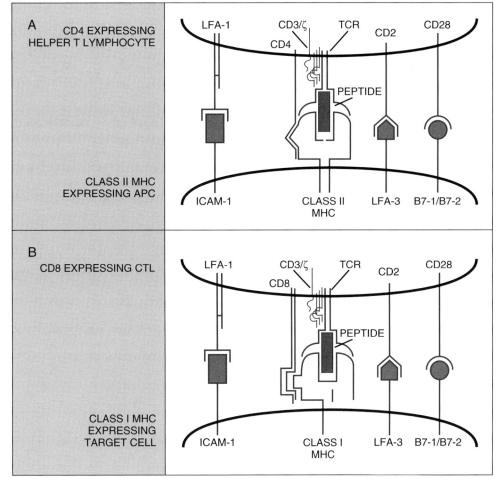

FIGURE 7-7. T lymphocyte surface molecules and their ligands involved in antigen recognition and T cell responses. Interactions between a CD4$^+$ T cell and an antigen-presenting cell (APC) (A) or a CD8$^+$ cytolytic T lymphocyte (CTL) and a target cell (B) involve multiple T cell surface proteins that recognize different ligands on the APC or target cell. Some of these interactions promote adhesion (e.g., LFA-1–ICAM-1 interactions) and some provide costimulatory signals (e.g., CD28-B7-1/B7-2 interactions). MHC, major histocompatibility complex.

the APC. The importance of accessory molecules in antigen presentation is suggested by many experiments showing that antibodies against one or more of these molecules can block T cell responses to antigens. The requirement for so many adhesion interactions during antigen presentation may reflect the generally low affinity of TCRs for their peptide-MHC ligands.

In the following sections, we will discuss T cell accessory molecules whose structures, functions, and importance in T cell activation are well understood. Several other T cell surface molecules have been identified that contribute to T cell responses, but their roles in physiologic immune responses are less well defined. Information about these other accessory molecules is included in Table 7-3 and in the Appendix.

CD4 AND CD8: CO-RECEPTORS INVOLVED IN MHC-RESTRICTED T CELL ACTIVATION

CD4 and CD8 are T cell surface glycoproteins that bind to nonpolymorphic regions of MHC molecules and are found on mutually exclusive subsets of mature T cells with distinct patterns of MHC restriction. CD4 binds directly to class II MHC molecules and is expressed on T cells whose TCRs recognize complexes of peptide and class II MHC

molecules. Most CD4$^+$ class II–restricted T cells have the functional phenotype of cytokine-producing helper cells. CD8 binds to class I MHC molecules and is expressed on T cells whose TCRs recognize complexes of peptide and class I MHC molecules. Most CD8$^+$ class I–restricted T cells are CTLs or their precursors. Both CD4 and CD8 perform a combination of adhesive and signaling functions, which greatly enhance the sensitivity of T cells to antigen. Since they are intimately associated with the TCR, and they bind MHC molecules on APCs or target cells at the time of antigen recognition, CD4 and CD8 are often called **co-receptors**. Approximately 65 per cent of peripheral $\alpha\beta$-positive T cells express CD4, and 35 per cent express CD8.

Structure of CD4 and CD8

Both CD4 and CD8 are transmembrane glycoprotein members of the Ig superfamily, but they have very distinct structures (Table 7-3). CD4 is expressed as a monomer on the surface of both peripheral T cells and thymocytes. In humans, it is also present on monocytes and macrophages. The three-dimensional structure of the CD4 molecule has been in part predicted from the sequences of cDNA clones and in part determined by x-ray crystallography. There are four extracellular Ig-like domains, including an N-terminal V-like domain and three domains that are neither C- nor V-like. In addition, there is a hydrophobic transmembrane region, and a highly basic cytoplasmic tail 38 amino acids long. Most CD8 molecules exist as disulfide-linked heterodimers composed of CD8α and CD8β. Both the α and β chains have a single extracellular Ig V-like domain, a hydrophobic transmembrane region, and a highly basic cytoplasmic tail about 25 amino acids long. Other forms of CD8 also exist, including CD8α homodimers and heterodimers of CD8α and the CD1 molecule. The biologic significance of these different forms is not known.

Functions of CD4 and CD8

The functional roles of CD4 and CD8 were initially demonstrated by the ability of antibodies specific for these molecules to block MHC-restricted antigen stimulation of T cells *in vitro* and *in vivo*. In particular, antibodies to CD4 block class II MHC–restricted helper T cell activation and antibodies to CD8 block class I MHC–restricted CTL killing of target cells. Additional proof of function came from gene transfection experiments. For example, if the TCR α and β genes are isolated from a CD8+ CTL clone and transfected into another T cell line that does not express CD8, the transfected line will not kill target cells bearing the relevant MHC-associated antigen. Cytolytic activity against such targets is restored, however, if the CD8α gene is co-transfected along with the TCR genes.

It is now clear that both CD4 and CD8 subserve two important functions in the activation of T cells.

1. *CD4 and CD8 promote the adhesion of MHC-restricted T cells to APCs or target cells by virtue of their specific affinity for MHC molecules* (Fig. 7–7). The invariant CD4 protein binds via its two N-terminal Ig-like domains to the nonpolymorphic β2 domain of the class II MHC molecule, thereby stabilizing interactions between a class II–restricted helper T cell and an APC. Similarly, the Ig domains of CD8 bind to the nonpolymorphic Ig-like α3 domains of class I MHC molecules, thereby stabilizing the interaction of a class I MHC–restricted T cell (usually a CTL) with a target (see Fig. 7–7). The ability of CD4 and CD8 interactions with MHC molecules to mediate cell-cell adhesion has been demonstrated in a variety of ways. For example, monolayers of fibroblasts that express transfected CD8 or CD4 genes bind class I or class II MHC–expressing cell lines, respectively, but untransfected fibroblasts do not. In addition, anti-CD8 antibodies block the formation of conjugates between class I MHC–restricted CTLs and class I MHC–expressing target cells. In a population of antigen-specific T cells, there is a wide spectrum of affinities of TCRs for their specific peptide-MHC molecule ligands. The adhesive roles of CD4 and CD8 are likely to

be most critical when the TCR affinity is low, thus expanding the number of T cells capable of responding to antigen.

2. *CD4 and CD8 participate in the early signal transduction events that occur upon T cell recognition of peptide-MHC complexes on APCs.* These signal-transducing functions are at least in part attributable to the fact that a T cell–specific *src* family protein tyrosine kinase called **lck** is non-covalently but tightly associated with the cytoplasmic tails of both these molecules. Studies with mutant forms of CD4 which cannot bind lck indicate that lck-CD4 interactions are required for full activation of helper T cells. It is likely that lck phosphorylates tyrosine residues on ITAMs of CD3 and ζ. In order for this to occur, the cytoplasmic tails of CD4 or CD8 would need to be brought into close proximity to the cytoplasmic tails of the TCR complex at the time of antigen recognition. In fact, there is evidence that a single CD4 or CD8 molecule forms physical associations with both the TCR molecule and the MHC molecule that the TCR engages. This permits lck to catalyze the phosphorylation of tyrosines in the nearby cytoplasmic tails of TCR complex proteins and thereby promote the T cell activation cascade. We will discuss these signaling events later in this chapter.

The combination of adhesive and signaling functions of the CD4 and CD8 coreceptors greatly enhances the efficiency of antigen stimulation of mature T cells. CD4 and CD8 also play essential roles in the development of mature class II and class I MHC–restricted T cells in the thymus, as discussed in Chapter 8. In addition to its physiologic roles, CD4 is the major receptor for the human immunodeficiency virus (see Chapter 21).

THE REQUIREMENT FOR COSTIMULATION IN T CELL ACTIVATION

A general property of lymphocytes, including both T and B cells, is the need for two distinct extracellular signals in order to induce proliferation and differentiation into effector cells. The first signal is provided by antigen binding to the antigen receptor. In the case of T cells, peptide-MHC complex binding to the TCR (and CD4 and CD8 coreceptors) provides signal 1. The second signal for T cell activation is provided by **costimulatory molecules,** which are surface molecules on APCs which bind to specific receptors on the T lymphocyte. The importance of costimulatory signals is most evident from analyses of what happens to T cells in their absence. By this approach, two important roles for costimulation have been established:

1. *Costimulatory signals are required, concurrent with antigen-induced signals, in order for T cells to achieve full activation responses.* When T cells recognize antigen in the absence of costimulation, i.e., when they receive signal 1 without signal 2, they do not become fully activated to perform effector

functions. This can be demonstrated experimentally by treating pure populations of CD4$^+$ T cells with TCR agonist ligands such as polyvalent anti-CD3 in the absence of any accessory cells. Under these conditions, the T cells produce very few cytokines and do not proliferate. If a source of costimulatory signals, such as monocytes, is added, the T cells will respond vigorously to the anti-CD3.

2. *A lack of costimulation at the time of antigen presentation may eliminate a T cell from the pool of antigen-responsive lymphocytes either by promoting its death or by inducing a state of unresponsiveness called* **anergy.** This concept is discussed further in Chapter 10.

The expression of costimulators is restricted to certain cell types and is highly regulated. Although there are various molecules that have been shown to have costimulatory properties, the most potent costimulators are expressed at high levels only on professional APCs such as mononuclear phagocytes, activated B cells, and dendritic cells. Furthermore, costimulator expression is often low on resting APCs and is upregulated by stimuli such as cytokines, which accompany inflammation. These properties of costimulator expression profoundly influence the regulation of T cell–mediated immune responses in the following ways:

1. *The regulated expression of costimulators ensures that T lymphocytes are activated at the correct time and place.* For example, at a site of infection both microbial products and cytokines elaborated by inflammatory cells will upregulate costimulator expression on local APCs and thereby promote appropriate T cell activation by microbial antigens.

2. *The expression of costimulators amplifies interactions between T cells and B cells and between T cells and macrophages.* In addition to serving as APCs, both B cells and macrophages are important recipients of T cell help, i.e., they respond to products of activated helper T cells by themselves becoming activated and performing their functions. Therefore, costimulator expression on these APCs promotes activation of both the T cells and the APCs.

3. *The absence of costimulators on unactivated or "resting" APCs in normal, uninflamed tissues contributes to the maintenance of tolerance to self antigens.* Since such APCs are capable of presenting self antigens to T cells, the lack of costimulator expression ensures that potentially self-reactive T cells are not activated and may be rendered permanently inactive (see Chapter 19).

One of the most important and best characterized costimulatory pathways in T cell activation involves the T cell surface molecule **CD28,** which binds the costimulatory molecules **B7-1** (CD80) and **B7-2** (CD86) expressed on APCs. CD28 delivers signals that enhance T cell responses to antigen. In addition, another T cell surface molecule, CTLA-4, also binds B7-1 and B7-2 but, in contrast to CD28, it transmits signals that inhibit T cell activation. CD28 and CTLA-4 are homologous disulfide-linked homodimeric glycoproteins with single extracellular Ig V-like domains in each polypeptide chain and cytoplasmic tails with some shared homologies. CD28 is constitutively expressed on 80 per cent of human CD4$^+$ T cells and 50 per cent of CD8$^+$ T cells, and the amount of CD28 expressed often increases after T cell stimulation. In contrast, CTLA-4 is not detectable on the surface of resting T cells, and its expression is induced after T cell activation, with maximal levels achieved within 48 hours. B7-1 and B7-2 are homologous single chain glycoproteins, each with two extracellular Ig-like domains, a transmembrane segment, and a cytoplasmic tail. The B7 molecules are found on professional APCs, including B cells, dendritic cells, and monocytes. They are generally absent on resting cells but are induced by various stimuli, including cross-linking of membrane Ig on B cells and by cytokines on B cells and monocytes. An exception is the high constitutive expression of B7-2 on dendritic cells. The kinetics of induction of B7-1 and B7-2 are distinct, with B7-2 appearing as early as 6 hours after stimulation and B7-1 appearing after 24 hours. Activated T lymphocytes also express B7-1 and B7-2, although the functional significance of this is not known. CD28 and CTLA-4 bind both B7 molecules; however, CTLA-4 binds with higher affinity than does CD28.

The way in which CD28 promotes T cell activation is incompletely understood and probably involves several mechanisms. There is evidence that CD28-mediated signals increase the expression of cytokine genes, including the T cell autocrine growth factor IL-2, and this may occur by a combination of enhanced transcription and stabilization of IL-2 messenger RNA (mRNA). In addition, CD28 signals apparently protect T cells from undergoing programmed cell death by increasing expression of the survival protein Bcl-x$_L$ (see Box 2–2, Chapter 2). The biochemical signals generated by CD28 and CTLA-4 are incompletely characterized, although it is known that the cytoplasmic tails of both these molecules can become phosphorylated on tyrosines, and they can recruit cytoplasmic enzymes with SH2 domains. One such enzyme known to associate with both CD28 and CTLA-4 is a lipid and protein kinase called phosphatidylinositol-3 kinase. There is some evidence that a protein tyrosine phosphatase called Syp associates with the cytoplasmic tail of CTLA-4 and may abrogate the activating influence of the tyrosine kinases involved in TCR signaling.

Although all the details are not yet known, it is clear that the CD28/CTLA-4-B7-1/B7-2 pathway superimposes on the TCR activation pathway a complex system of regulation, which ensures that immune responses are turned on when they are needed and are kept off when they are not needed. A model depicting this concept is shown in Figure 7–8.

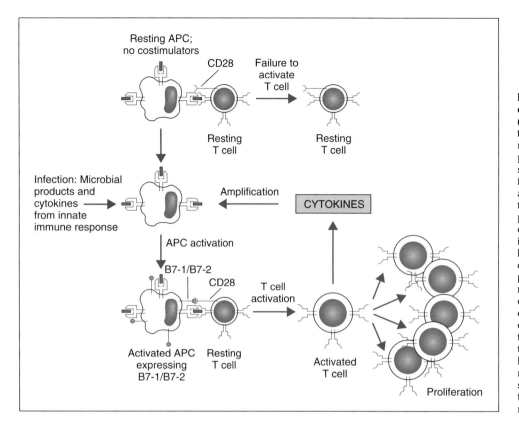

FIGURE 7–8. Costimulators on antigen-presenting cells (APCs) are required for effective T cell activation. Resting macrophage APCs do not express costimulatory molecules such as B7-1 or B7-2, and they fail to activate T cells during antigen presentation. In the setting of an infection, microbial products such as endotoxin or cytokines elaborated by innate immune responses can upregulate B7-1 and B7-2 expression on the macrophage. Antigen presentation by the activated macrophage will then lead to T cell activation characterized by cytokine production and proliferation. Cytokines secreted by the activated T cell, such as interferon-γ, can induce costimulator expression on other macrophages, enabling them to serve as effective APCs and thereby amplifying the immune response.

Other cell surface molecules expressed on APCs that may act as costimulators for T cell activation are VCAM-1, ICAM-1, and LFA-3, which bind to VLA-4, LFA-1, and CD2, respectively, on T cells. These molecules are discussed below. In addition, some soluble cytokines, including IL-12, regulate T cell growth and differentiation and may, therefore, be thought to have costimulatory properties (see Chapter 12).

CD45: A PROTEIN TYROSINE PHOSPHATASE REQUIRED FOR T CELL ACTIVATION

CD45 (T-200, leukocyte common antigen) is a cell surface integral membrane glycoprotein that plays a critical role in T cell activation. Various forms of CD45 are expressed on immature and mature leukocytes, including T and B cells, thymocytes, mononuclear phagocytes, and polymorphonuclear leukocytes. The CD45 family consists of multiple members that are all products of a single complex gene. This gene contains 34 exons, and the primary RNA transcripts of three of the exons (called A, B, and C) are alternatively spliced to generate up to eight different mRNAs and eight different protein products. The predicted amino acid sequences of the protein products include external domains of varying length, a transmembrane region, and a 705 amino acid, highly conserved cytoplasmic domain, which is one of the largest yet identified among all membrane proteins. Different glycosylation patterns of the same peptide backbone also contribute to heterogeneity of the members of this protein family. Several different monoclonal antibodies that recognize individual members have been useful for studying the distribution and functions of CD45 proteins. Isoforms of CD45 proteins that are expressed on a restricted group of cell types are designated CD45R (see Chapter 2). The expression of different forms of CD45R is regulated during the process of maturation and activation of T cells. As discussed in Chapter 2, naive human T cells express a form of CD45R called CD45RA, and memory T cells induced by prior exposure to antigen express a different isoform called CD45RO.

CD45 expression is required for optimal T cell activation. The large *cytoplasmic domain of CD45 contains an intrinsic protein tyrosine phosphatase activity* that can catalyze the removal of phosphates from tyrosines in several potentially important substrates. One such substrate is lck, the protein tyrosine kinase that is associated with the cytoplasmic tails of CD4 and CD8. Certain tyrosine residues on lck, as well as other Src family kinases, serve as negative regulators of these enzymes; phosphorylation at these sites turns off the kinase activity. One likely way CD45 may regulate T cell activation is to mediate the removal of phosphates from the negative regulatory tyrosine residues on lck and other tyrosine kinases. Lck would then be able to catalyze tyrosine phosphorylation of various substrates. This is in contrast to the role of other lymphocyte protein tyrosine phosphatases, which are thought to downregulate lymphocyte activation by counteracting the effects of protein tyrosine kinases involved in signaling cascades.

Ligands for the extracellular domain of CD45 have not been identified, and the way CD45 may be inducibly engaged into the T cell activation pathway is not known. Naive and memory T cells have different requirements for activation; in particular, naive T cells are more dependent on costimulatory signals than are memory T cells. It is interesting to speculate that the enzymatic activity of the CD45RO isoform found on memory T cells may be induced by extracellular binding to a commonly encountered ligand, while the CD45RA isotype on naive T cells does not respond to the same ligand. CD45 gene knockout mice show a block in T cell maturation and a defect in B cell and mast cell activation.

INTEGRINS INVOLVED IN T CELL ADHESION INTERACTIONS WITH OTHER CELLS

The integrin family of heterodimeric leukocyte proteins functions primarily as adhesion molecules, although they may serve signaling functions as well (Box 7–3). We will discuss two subfamilies of lymphocyte integrins, the β_2 integrins represented by LFA-1, and the β_1 integrins represented by the VLA molecules.

LFA-1 (leukocyte function-associated antigen-1, or CD11aCD18), a member of the β_2 integrin family, is expressed on more than 90 per cent of thymocytes and mature T cells, B cells, granulocytes, and monocytes. *LFA-1 is essential for a wide variety of adhesion-dependent lymphocyte functions,* including antigen and APC-induced helper T cell stimulation, CTL-mediated killing of target cells, and lymphocyte adhesion to endothelium (see Chapters 11 and 13). This has been shown by experiments in which anti-LFA-1 antibodies block these T cell functions and also block conjugate formation between T lymphocytes and other cells. *The avidity of LFA-1 for its ligands is increased shortly after stimulation of T cells through the TCR.* This phenomenon in which intracellular signals induced by one membrane receptor lead to changes in the function of the extracellular domain of another receptor is often called *inside-out signaling* and is typical of other integrins besides LFA-1. TCR-induced upregulation of LFA-1 avidity ensures that antigen recognition and integrin-mediated intercellular adhesion function coordinately to optimize T cell–APC interactions. The biochemical basis of the increased avidity likely involves phosphorylation of the integrin chain cytoplasmic tails, which leads to a conformational change in the extracellular domains. In addition, certain cytokines called chemokines (described in Chapter 12) can increase LFA-1 avidity for ligands expressed on endothelial cells, thereby promoting migration of T cells into inflammatory sites.

One specific ligand for LFA-1 is intercellular adhesion molecule-1 (ICAM-1), an integral membrane glycoprotein that contains five extracellular

---

BOX 7–3. The Integrin Family of Adhesion Proteins

The specific (nonrandom) adhesion of cells to other cells or to extracellular matrices is a basic component of cell migration and recognition and underlies many biologic processes, including embryogenesis, tissue repair, and immune and inflammatory responses. It is, therefore, not surprising that many different genes have evolved that encode proteins with specific adhesive functions. These genes display homologies indicative of a common ancestral gene. The integrin superfamily consists of about 30 structurally homologous proteins that promote cell-cell or cell-matrix interactions. The Ig superfamily, which is described in Box 7–2, is another set of homologous genes encoding proteins with adhesive and recognition functions.

All integrins are heterodimeric cell surface proteins composed of two non-covalently linked polypeptide chains, α and β. The α chain varies from 120 to 200 kD, and the β chain varies from 90 to 110 kD. The N-terminus of each chain forms a globular head that contributes to the interchain linking and to ligand binding. These globular heads contain divalent cation-binding domains. Divalent cations are essential for integrin receptor function and may interact directly with integrin ligands. Stalks extend from the globular heads to the plasma membrane, followed by transmembrane segments and cytoplasmic tails, which are usually less than 50 amino acid residues long. The extracellular domains of the two chains bind to various ligands, including extracellular matrix glycoproteins, activated complement components, and proteins on the surface of other cells. Several integrins bind to Arg-Gly-Asp (RGD) sequences in the fibronectin and vitronec-

tin molecules, but most do not. Some integrins may bind Asp-Aly-Glu-Ala (DGEA) in type I collagen, and Glu-Ile-Leu-Asp-Val (EILDV) in fibronectin. The cytoplasmic domains of the integrins interact with cytoskeletal components (including vinculin, talin, actin, α-actinin, and tropomyosin), and it is hypothesized that the integrins coordinate (i.e., "integrate") extracellular ligand binding with cytoskeleton-dependent motility, shape change, and phagocytic responses.

Three integrin subfamilies were originally defined on the basis of which of three β subunits were used to form the heterodimers. This led to a simplified organizational scheme because it was thought that each of these β chains could pair with a distinct and non-overlapping set of α chains. More recently, five additional β chains have been identified, and many examples have been found of a single α chain pairing with more than one kind of β chain. Nonetheless, the subfamily designation is still useful for consideration of integrins relevant to the immune system; the major members of these subfamilies are listed in the table.

The β_1-containing integrins are also called VLA molecules, referring to "very late activation" antigens, because $\alpha_1\beta_1$ and $\alpha_2\beta_1$ were first shown to be expressed on T cells 2 to 4 weeks after repetitive stimulation *in vitro.* In fact, other VLA integrins, including VLA-4, are constitutively expressed on some T cells or rapidly induced on others. The β_1 integrins are also called CD49a-fCD29, CD49a-f referring to different α chains (α_1-α_6) and CD29 referring to the common β_1 subunit. Most of the β_1 integrins are widely expressed on

Continued

leukocytes and non-blood cells and mediate attachment of cells to extracellular matrices. VLA-4 ($\alpha_4\beta_1$ or CD49dCD29) is expressed only on leukocytes and can mediate attachment of these cells to endothelium by interacting with VCAM-1. VLA-4 is one of the principal surface proteins that mediate homing of lymphocytes to endothelium at peripheral sites of inflammation.

The β_2 integrins, also known as the LFA-1 (leukocyte function–associated antigen-1) family, were identified by monoclonal antibodies that blocked adhesion-dependent lymphocyte functions such as killing of target cells by CTLs. LFA-1 plays an important role in the adhesion of lymphocytes with other cells, such as APCs and vascular endothelium. This family is also called CD11CD18, CD11 referring to different α chains and CD18 to the common β_2 subunit. LFA-1 itself is termed CD11aCD18. Other members of the family include CD11bCD18 (Mac-1 or CR3) and CD11cCD18 (p150,95 or CR4), both of which have the same β subunit as LFA-1. CD11bCD18 and CD11cCD18 both mediate leukocyte attachment to endothelial cells and subsequent extravasation. CD11bCD18 also functions as a fibrinogen receptor and a complement receptor on phagocytic cells, binding particles opsonized with a by-product of complement activation called the inactivated C3b (iC3b) fragment. An autosomal recessive inherited deficiency in LFA-1, Mac-1, and p150,95 proteins, called type 1 leukocyte adhesion deficiency, has been identified in a few families and is characterized by recurrent bacterial and fungal infections, lack of polymorphonuclear leukocyte accumulations at sites of infection, and profound defects in adherence-dependent lymphocyte functions. The disease is a result ·of mutations in the CD18 gene, which encodes the β chain of LFA-1 subfamily

molecules, and it demonstrates the physiologic importance of the LFA-1–related proteins.

A general characteristic of integrins is their ability to be functionally modulated by physiologic activation of the cell on which they are expressed. Most commonly, the modulation is an increase in integrin avidity for ligand. This has been called "inside-out signaling" because intracellular signals generated by other receptors on the cell lead to a change in the extracellular function of the integrin. For example, in T cells, the affinity of LFA-1 for ICAM-1 increases rapidly and transiently in response to TCR signaling. Similarly, neutrophil LFA-1 affinity for ICAM-1 increases when the cell is stimulated with cytokines (e.g., chemokines such as IL-8) or complement by-products (e.g., C5a). The increase in affinity is thought to be due to conformational changes in the extracellular domains, and it has been proposed that changes in affinity are initiated by intracellular biochemical events such as phosphorylation of cytoplasmic tails of the integrins or associated proteins. The significance of this activation-dependent increase in ligand-binding activity is that the adhesive functions of the integrins can be turned on specifically at sites where leukocyte adhesion to other cells or extracellular matrix is physiologically needed, e.g., at sites of antigen recognition or inflammation. In addition to adhesion, integrins deliver stimulatory signals to cells upon ligand binding, i.e., outside-in signaling. The mechanism of signaling involves tyrosine phosphorylation of different substrates, inositol lipid turnover, and elevated cytoplasmic calcium. The functional consequences of these integrin-mediated signals vary with cell type; in T lymphocytes, ICAM-1 binding to the β_1 integrins can provide costimulatory signals that enhance cytokine gene expression.

The Integrins				
Subunits		**Name**	**Ligands/Counter Receptors**	**Functions**
β_1	α_1	VLA-1 (CD49aCD29)	Collagens, laminin	Cell-matrix adhesion
	α_2	VLA-2 (CD49bCD29)	Collagens, laminin	Cell-matrix adhesion
	α_3	VLA-3 (CD49cCD29)	Fibronectin, collagens, laminin	Cell-matrix adhesion
	α_4	VLA-4 (CD49dCD29)	Fibronectin, VCAM-1	Cell-matrix adhesion; T cell homing; ?T cell costimulation
	α_5	VLA-5 (CD49eCD29)	Fibronectin	Cell-matrix adhesion
	α_6	VLA-6 (CD49fCD29)	Laminin	Cell-matrix adhesion
	α_7	CD49gCD29	Laminin	Cell-matrix adhesion
	α_8	CD49hCD29	?	?
	α_v		Vitronectin, fibronectin	Cell-matrix adhesion
β_2	α_L	CD11aCD18 (LFA-1)	ICAM-1, ICAM-2, ICAM-3	Leukocyte adhesion to endothelium; T cell-APC adhesion; ?T cell costimulation
	α_M	CD11bCD18 (Mac-1, CR3)	iC3b, fibrinogen, factor X, ICAM-1	Leukocyte adhesion and phagocytosis; cell-matrix adhesion
	α_X	CD11cCD18 (p150,95; CR4)	iC3b; fibrinogen	Leukocyte adhesion and phagocytosis; cell-matrix adhesion
β_3	α_{iib}		Fibrinogen, fibronectin, von Willebrand factor, vitronectin, thrombospondin	Platelet adhesion and aggregation
	α_v	Vitronectin receptor (CD51CD61)	Vitronectin, fibrinogen, von Willebrand factor, thrombospondin, fibronectin, osteopontin, collagen	Cell-matrix adhesion
β_4	α_6		Laminin (?)	
β_5	α_v		Vitronectin	Cell-matrix adhesion
β_6	α_v		Fibronectin	Cell-matrix adhesion
β_7	α_4	LPAM-1	Fibronectin, VCAM-1, Mucosal addressin	Lymphocyte homing to mucosa
	α_E		E-cadherin	Lymphocyte adhesion to mucosal epithelium
β_6	α_v		?	

Abbreviations: APC, antigen-presenting cell; VLA, "very late activation" antigen; LFA, leukocyte function–associated antigen; ICAM, intercellular adhesion molecule; iC3b, C3b inactivated; VCAM, vascular cell adhesion molecule. Adapted from Hynes, R. O. Integrins: versatility, modulation, and signaling in cell adhesion. Cell 69:11–25, 1992, © Cell Press.

Ig-like domains. ICAM-1 is expressed on a variety of hematopoietic and nonhematopoietic cells, including B and T cells, fibroblasts, keratinocytes, and endothelial cells, and the level of expression on these cells can be upregulated by various cytokines. Soluble recombinant forms of ICAM-1 can enhance TCR-mediated activation of T cells *in vitro*, suggesting that LFA-1 may deliver costimulatory signals to T cells. ICAM-1 has been shown to be a specific receptor for rhinoviruses, the etiologic agents of many cases of the common cold.

Some LFA-1–dependent cell adhesion phenomena cannot be blocked by anti-ICAM-1 antibody; in fact, another LFA-1-binding molecule called ICAM-2 has been identified. ICAM-2 is also an Ig superfamily member, with two extracellular Ig-like domains; it is constitutively expressed on vascular endothelium and not other tissues and is not regulated by cytokines. A third ligand for LFA-1, called ICAM-3, is expressed on lymphoid cells. ICAM-1, but not ICAM-2 or ICAM-3, can also bind to the β_2 integrin CD11bCD18 (Mac-1).

The VLA (very late activation) molecules, or β_1 integrins, all share the same β chain (CD29), as described in Box 7–3. Three members of this family, VLA-4, VLA-5, and VLA-6, are expressed on resting T cells. Like LFA-1, the affinities of these molecules for their ligands are increased upon T cell activation, and their levels of expression are higher on memory than naive T cells. VLA-4 mediates binding of lymphocytes to endothelium at inflammatory sites by interacting with a protein called vascular cell adhesion molecule-1 (VCAM-1) expressed on cytokine-activated endothelial cells. Thus, VLA-4–VCAM-1 interactions, like LFA-1–ICAM-1 interactions, may regulate the movement of lymphocytes out of blood vessels to inflammatory sites (see Chapters 11 and 13). In addition, T cell VLA molecules bind to extracellular matrix ligands (fibronectin for VLA-4 and VLA-5, laminin for VLA-6), and these adhesive interactions may be important for the retention of T cells in tissues. VLA-4 binding to both VCAM-1 and extracellular matrix proteins has also been demonstrated to provide costimulatory signals for T cell activation.

CD2: A MOLECULE MEDIATING ADHESIVE AND COSTIMULATORY FUNCTIONS

The CD2 protein, also called LFA-2, is an Ig superfamily glycoprotein present on more than 90 per cent of mature T cells and on 50 to 70 per cent of thymocytes. CD2 is also present on NK cells. The molecule contains two extracellular Ig domains, a hydrophobic transmembrane region and a long (116 amino acid residue) cytoplasmic tail.

CD2 functions as an intercellular adhesion molecule. The principal ligand for CD2 in humans is the structurally similar molecule called leukocyte function-associated antigen-3 (LFA-3, CD58). LFA-3 is expressed on a wide variety of hematopoietic and nonhematopoietic cells. It has a similar extracellular domain structure to CD2 but in contrast to CD2, it can be expressed either as a typical transmembrane protein or as a phosphatidylinositol-anchored surface molecule. CD2 binding to LFA-3 promotes cell-cell adhesion. This may be critical for the binding of helper T cells to APCs, CTLs to their target cells, and maturing thymocytes to thymic epithelial cells (see Fig. 7–7). Consistent with this hypothesis is the finding that anti-CD2 antibodies can block conjugate formation between T cells and other LFA-3–expressing cells, and perhaps, as a consequence of diminishing cell-cell adhesion, these antibodies can block both CTL activity and antigen-dependent helper T cell responses. In mice, the principal ligand for CD2 is CD48, which is distinct from but structurally similar to LFA-3.

In addition to its adhesive function, CD2 is also a signal-transducing molecule that mediates costimulation of human T cells *in vitro*. Certain combinations of anti-CD2 antibodies can activate T cells to secrete cytokines and to proliferate, and CD2–LFA-3 interactions greatly enhance TCR-mediated activation of human T cells. It has been difficult to assess if CD2 influences T cell activation *in vivo*. Both anti-CD2 and anti-CD48 antibodies can block certain T cell–dependent immune responses in mice, such as graft rejection, but CD2 gene knockout mice do not show detectable defects in T cell development or immune function.

CD44 AND L-SELECTIN: ADHESION MOLECULES WITH DISTINCT ROLES IN T CELL MIGRATION INTO TISSUES

CD44 (PgP-1, Ly-24) is an acidic sulfated integral membrane glycoprotein expressed in several alternatively spliced and variably glycosylated forms on a wide variety of cell types, including mature T cells, thymocytes, B cells, granulocytes, macrophages, erythrocytes, and fibroblasts. A soluble form of CD44 also circulates in the plasma. High levels of CD44 are considered a marker of memory T cells. CD44 binds hyaluronate, and this property is partially responsible for memory T cell binding to endothelium at sites of peripheral inflammation and maybe Peyer's patch endothelium (see Chapter 11). CD44 may also play a role in retention of T cells in extravascular tissues. Antibodies against CD44 can alter T cell activation responses by as yet unknown mechanisms. CD44 is also a useful marker for the study of T cell maturation and differentiation. The most immature precursors of T cells in the thymus express high levels of CD44, and expression goes down as maturation proceeds (see Chapter 8).

L-selectin is a carbohydrate-binding protein which is a member of the selectin family of proteins (see Box 11–1, Chapter 11). L-selectin is expressed at high levels on naive T lymphocytes, and expression is reduced on memory T cells. It specifically binds carbohydrate moieties found on certain glycoproteins on specialized endothelial cells in lymph nodes. Therefore, contrary to CD44, L-selectin mediates the preferential migration of naive T cells into lymph nodes. The role of L-selectin in lymphocyte recirculation is discussed further in Chapter 11.

INTRACELLULAR BIOCHEMICAL AND MOLECULAR EVENTS IN T CELL ACTIVATION

So far in this chapter, we have discussed the cell surface molecules that are required for antigen-induced activation of T cells. If TCRs bind peptide-MHC complexes on an APC, and if adhesion and costimulatory molecule interactions are also engaged, the T lymphocyte is activated to undergo a variety of functional responses (see Fig. 7–1). Antigen recognition and costimulatory signals activate naive T cells to undergo clonal expansion and differentiation into effector and memory T cells. Antigen stimulation of effector T cells leads to further clonal expansion and performance of effector functions such as cytokine secretion and cytolytic activity. All these responses are dependent on signaling pathways, which are the intracellular biochemical consequences of antigen recognition by the TCR complex. The signals generated by the TCR complex lead to increased transcription of several genes that are quiescent in unstimulated T cells. This, in turn, leads to the transient production of proteins that are essential for T cell mitosis and effector function (Fig. 7–9). Thus, T cell activation includes the following interrelated steps:

1. Early signal transduction events
2. Transcriptional activation of a variety of genes
3. Expression of new cell surface molecules

4. Secretion of effector cytokines and/or performance of cytolytic functions
5. Induction of mitotic activity
6. Downregulation of activation signals

The early events of T cell activation have largely been defined by *in vitro* models, often using monoclonal T cell populations in which the molecular consequences of ligand binding to TCR complexes are analyzed (see Box 7–1). Sophisticated biochemical and genetic approaches have uncovered a complex set of interrelated signal transduction pathways that link T cell surface molecules with nuclear transcriptional events. Although the unraveling of these pathways is incomplete, their fundamental features are known. When the TCR binds peptide-MHC, protein tyrosine kinases (PTKs) associated with the TCR complex are activated. There are many different substrates for these kinases leading to different branches of the signaling cascade. For example, PTKs activate enzymes which breakdown membrane phospholipids, leading to the generation of Ca^{2+} signals and activation of serine/threonine protein kinases. PTKs also lead to activation of guanosine triphosphate–binding proteins, which in turn initiate other kinase cascades. Eventually, the diverse signals lead to activation and new synthesis of transcription factors, which enhance the expression of genes encoding cytokines, cytokine receptors, and other proteins required for T cell effector function (Fig. 7–10). These pathways have been targeted for pharmacologic treatment of unwanted immune responses, such as graft rejection or autoimmunity, and genetic defects in components of these pathways have been identified as the causes of several immunodeficiency diseases. We will organize our discussion of these signaling pathways into the fol-

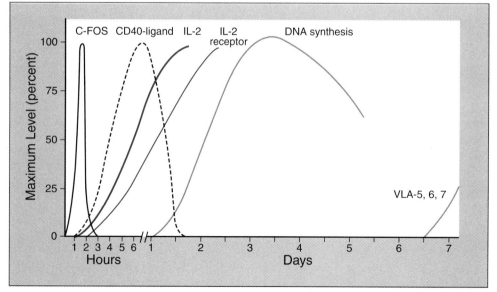

FIGURE 7–9. Time course of events in T cell activation. Representative examples of expression of new molecules are given. The time course is approximate and may vary with different populations of T cells.

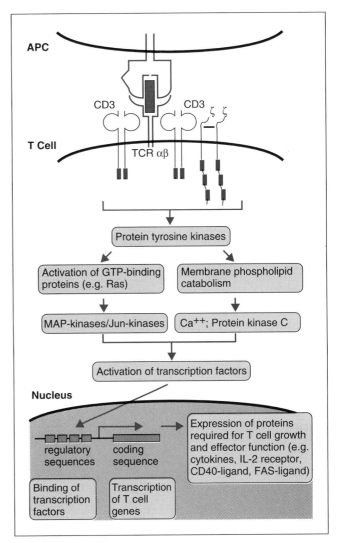

FIGURE 7–10. Overview of intracellular signaling events during T cell activation. Immediately after the T cell receptor (TCR) binds antigen on an antigen-presenting cell (APC), several protein tyrosine kinases are activated, and these enzymes phosphorylate substrates, which lead to activation of guanosine triphosphate (GTP)–binding proteins such as Ras and activation of enzymes that break down membrane phospholipids. The GTP-binding proteins activate mitogen-activated protein (MAP) kinase cascades, which in turn activate transcription factors that enhance T cell gene expression. Membrane phospholipid breakdown products lead to elevated cytoplasmic calcium ion concentrations and the activation of protein kinase C, both of which also lead to transcription factor activation, and enhanced gene expression. The details of these signaling pathways are discussed in the text and are illustrated in subsequent figures. IL-2, interleukin-2.

lowing categories: (1) recruitment and activation of PTKs; (2) inositol lipid metabolism; (3) Ras-MAP kinase and JNK pathways; and (4) activation of transcription factors and transcription of T cell genes.

Recruitment and Activation of PTKs

PTKs are enzymes that catalyze the phosphorylation of tyrosine residues in substrate proteins, and they play a pivotal role in initiating various signaling cascades in response to T cell antigen recognition. In fact, the earliest intracellular events to occur in response to TCR binding of ligands is the PTK-mediated phosphorylation of tyrosine residues within the ITAMs of the CD3 and ζ chains of the TCR complex (Fig. 7–11). As mentioned previously, one PTK that likely catalyzes the early phosphorylation of ITAMs in T cells is lck, the Src family protein that is tightly associated with the cytoplasmic tails of CD4 and CD8. TCR recognition of peptide-MHC complexes promotes the aggregation of the cytoplasmic tails of CD4 or CD8 with those of the TCR complex. This brings lck into

proximity of the ITAMs in the tails of CD3 and ζ chains and favors their phosphorylation (Fig. 7–11). **Fyn** is another another cytoplasmic PTK which is found in physical association with the TCR complex, even in resting T cells, and may play a similar role to lck. In addition to bringing coreceptor–associated lck into proximity of the cytoplasmic tails of the TCR complex, it is also possible that TCR binding to peptide-MHC induces conformational changes in the $\alpha\beta$ heterodimer. Such conformational changes may translate across the plasma membrane and increase the activity of PTKs such as fyn, which are associated with the CD3 and ζ chains.

Once the ITAMs in the ζ chain are tyrosine-phosphorylated, they become specific "docking sites," which bind cytoplasmic proteins with SH2 domains (see Box 7–4). One such protein is a 70 kD PTK called **Z**eta-**a**ssociated **p**rotein or **ZAP-70,** which is a member of a family of PTKs distinct from the Src family (see Box 7–4). Another member of the ZAP-70 family is syk, which plays an important role in membrane Ig signal transduction (see Chapter 9). As is the case for many PTKs,

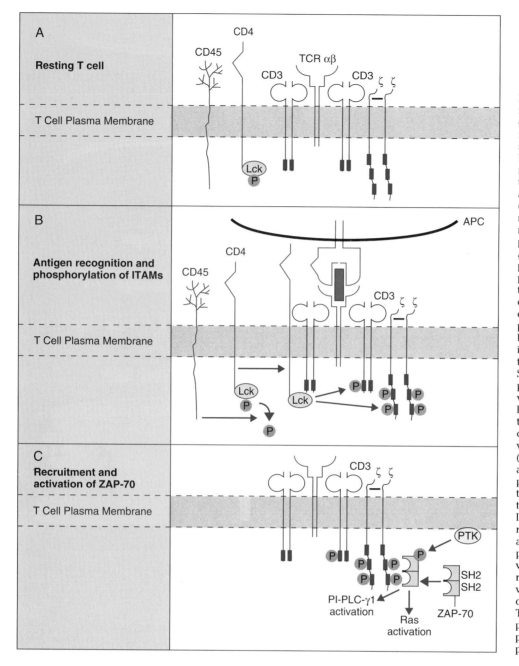

FIGURE 7–11. A model for tyrosine phosphorylation events early in T cell activation. In a resting CD4+ T cell (A), lck associated with CD4 is inactive because of phosphorylation of a negative regulatory tyrosine, the CD4 molecule is not associated with the T cell receptor (TCR) complex, and the immunoreceptor tyrosine activation motifs (ITAMs) in the cytoplasmic tails of the CD3 and ζ chains are mostly not phosphorylated. Upon antigen presentation (B), CD4 binds to the same class II MHC molecules on the antigen-presenting cell (APC) that are presenting peptides to the TCRs, thereby bringing the CD4-associated lck in proximity to the cytoplasmic tails of the CD3 and ζ chains. Simultaneously, CD45 dephosphorylates, and thereby activates, lck. Activated lck catalyzes the phosphorylation of the ITAMs in the CD3 and ζ chains. Fyn may also be involved in ITAM phosphorylation (not shown). Once the ITAMs are phosphorylated (C), cytoplasmic ZAP-70 can bind to these motifs via SH2 interactions. After docking on the ITAMs, ZAP-70 itself is phosphorylated, and thus activated, by a nearby membrane-associated protein tyrosine kinase. Activated ZAP-70 then phosphorylates tyrosine residues on various substrates in different downstream signaling pathways. The symbol ℗ refers to phosphorylated tyrosine. PI-PLC-γ1, phosphatidylinositol phospholipase Cγ1; SH2, src homology-2.

BOX 7–4. Protein Tyrosine Kinases and Src Homology Domains

Protein-protein interactions and the activities of enzymes are often regulated by the phosphorylation of tyrosine residues. It is not surprising, therefore, that the kinases that catalyze tyrosine phosphorylation are essential components of many intracellular signaling cascades in lymphocytes and other cell types. These PTKs mediate the transfer of the terminal phosphate of adenosine triphosphate (ATP) to the hydroxyl group of a tyrosine residue in a substrate protein. A particular PTK will phosphorylate only a limited set of substrates, and this specificity is determined by the amino acid sequences flanking the tyrosine as well as by tertiary structural characteristics of the substrate protein. There are several examples of receptor tyrosine kinases that are intrinsic components of the cytoplasmic tails of cell surface receptors, such as the platelet-derived growth factor

and epidermal growth factor receptors. Many PTKs of importance in the immune system, however, are cytoplasmic proteins which are non-covalently associated with the cytoplasmic tails of cell surface receptors. Sometimes this association is only transiently induced in the early stages of signaling cascades. Two families of cytoplasmic PTKs that are prominent in B and T cell antigen receptor signaling cascades, as well as Ig Fc receptor signaling, are the src and syk-ZAP-70 families. The Janus kinases are another group of nonreceptor PTKs involved in many cytokine induced signaling cascades, and are discussed in Chapter 12.

Src kinases are homologous to, and named after, the transforming gene of Rous sarcoma virus, the first animal tumor virus ever identified. Members of this family include

Continued

src, yes, fgr, fyn, lck, lyn, hck, and blk. All src family members share certain tertiary structural characteristics that are distributed among four distinct domains and are critical for the function and regulation of enzymatic activity. The N-terminal end of most src family kinases contains a consensus site for addition of a myristic acid group; this type of lipid modification serves to anchor cytoplasmic proteins to the inner side of the plasma membrane. Otherwise, the N-terminal domains of src family PTKs are highly variable among different family members, and this region confers the ability of each member to interact specifically with different proteins. For example, the N-terminal region of the lck protein is the contact site for interaction with the cytoplasmic tails of CD4 and CD8 (see text). The internal two domains are called **src-homology-2 (SH2) and SH3** domains, each of which has a distinct three-dimensional structure that permits specific non-covalent binding interactions with other proteins. SH2 domains are about 100 amino acids long and specify binding interactions with phosphotyrosines on other proteins. SH2 domains are also found on many signaling molecules outside the src family. Each SH2 domain exhibits distinct specificities for different tyrosine-phosphorylated proteins. Many src PTKs become enzymatically active only after they have bound to the cytoplasmic tail of a membrane receptor protein, via SH2-phosphotyrosine interactions. It is apparent that, in order for this to occur, the receptor protein must be the substrate of another PTK. This is the essence of what goes on early in TCR and Ig signaling cascades (see text and Chapter 9). SH3 domains are approximately 60 amino acids long and also mediate protein-protein binding, but they have a more restricted distribution and a currently less well-defined ligand specificity than SH2 domains. The function of the SH3 domain in src PTKs is not understood. SH3 domains in some proteins mediate interactions with cytoskeletal proteins. SH3 domains have also been found in several adaptor molecules which have no enzymatic activity of their own, but serve to couple receptors with downstream signaling enzymes. In fact, adaptor proteins with both SH2 and SH3 domains are found in Ras pathways (see text). The C-terminal domain in src PTKs contains a binding site for the ATP phosphate donor. Also in the C-terminal domain are at least two tyrosine residues. One of these tyrosines can be autophosphorylated by the src PTK itself, although the functional significance of this is not known. The other tyrosine serves as a regulatory switch for the PTK. When it is phosphorylated, the PTK is inactive and when the phosphate is removed by tyrosine phosphatases, the enzyme activity is turned on. This mechanism of regulation is thought to be an integral part of antigen receptor signaling (see text).

The syk-ZAP-70 family of PTKs includes only two identified members so far. Syk is abundantly expressed in B lymphocytes, but is also found in lower amounts in other hematopoietic cells, including T cells. ZAP-70 is expressed only in T lymphocytes and NK cells. Syk and ZAP-70 play similar respective roles in membrane Ig and TCR signaling cascades, respectively (see text). These PTKs have catalytic domains similar to those in the src family PTKs that contain ATP-binding sites. There are two SH2 domains in the syk-ZAP-70 PTKs, but no SH3 domain, no negative regulatory tyrosine, and no myristoylation site. Syk and ZAP-70 are inactive until they bind, via their SH2 domains, to phosphotyrosines within ITAM motifs of antigen-receptor complex proteins (see text). Furthermore, activation of syk and ZAP-70 probably requires src family PTK–mediated phosphorylation on one or more tyrosine residues.

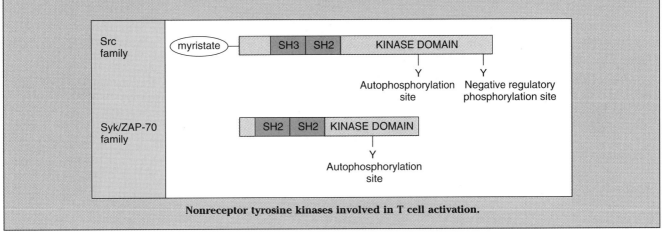

Nonreceptor tyrosine kinases involved in T cell activation.

ZAP-70 itself must be phosphorylated on an internal tyrosine residue in order to become activated, and this occurs rapidly after binding to the ζ chain. It is not yet known which PTK is responsible for the initial phosphorylation of ZAP-70, but once activated, it can autophosphorylate itself at additional tyrosine residues and then serve as a scaffolding molecule for recruitment of other SH2-containing signaling molecules. The identity of other ZAP-70 substrates in T cells is not yet known. It is clear, however, that *ZAP-70 plays a critical role in sustaining the signaling cascade that is set in motion by TCR recognition of antigen.* In fact, mutations in the ZAP-70 gene are responsible for some rare immunodeficiency diseases characterized by profound defects in T cell activation (see Chapter 21).

The activity of PTKs in T cell signaling pathways appears to be regulated by protein tyrosine phosphatases. These enzymes can remove phosphates from tyrosine residues within the PTKs, and depending on which tyrosine residue is involved, the effect may be to activate or inhibit the kinase function of the enzyme. Earlier in this chapter we described the CD45 protein, which has an intrinsic tyrosine phosphatase in its cytoplasmic tail. CD45 expression is essential for T cell activation, and its role may be to dephosphorylate inhibitory tyrosine residues in PTKs such as lck or fyn. Conversely, we have discussed how tyrosine phosphatases associated with the CTLA-4 cytoplasmic tail such as syp may function to downregulate T cell activation, perhaps by dephosphorylating stimulatory tyrosine residues on PTKs. Another tyrosine phosphatase named SHP-1 contains an SH2 domain and binds to and inhibits ZAP-70. These examples illustrate the general concept that there is a complex array of counteracting tyrosine kinase and phosphatase activities that contribute to the early regulation of T cell activation pathways.

Why there are nine ITAMs associated with each TCR complex is not understood, but there are two possible answers. First, all the ITAMs may not share the identical functions. Even though all ITAMs have two tyrosine-X-X-leucine sequences, the neighboring sequences vary considerably. In fact, there is evidence that distinct downstream signaling events are linked to each of the CD3 and ζ ITAMs, and this may reflect the specificity of each ITAM for binding different SH2 containing proteins. Second, it is possible that a certain threshold of PTK activity is required before downstream signaling events will proceed. Therefore, a critical density of phosphorylated ITAMs would be required to dock and permit activation of enough PTK molecules to achieve this threshold. There is evidence that a critical number of TCR antigen recognition events must occur before a T cell becomes functionally activated. This may be accomplished by multiple TCRs each binding one peptide-MHC complex, and by the sequential binding of individual TCRs to multiple peptide-MHC complexes.

Inositol Lipid Metabolism Leading to Protein Kinase C Activation and Calcium Signals

One of the substrates of TCR–complex associated PTKs is **phosphatidylinositol phospholipase C-γ1 (PI-PLC-γ1),** a cytoplasmic enzyme that is es-sential for generating important intermediate signals in the T cell activation cascade. Within minutes of ligand binding to the TCR, PI-PLC-γ1 is turned on and there is evidence that TCR-regulated PTKs are involved in this process in two ways. First, PTKs phosphorylate certain proteins that specifically bind PI-PLC-γ1 via its SH2 domain. The resulting complexes can then become physically associated with the plasma membrane where the appropriate PI-PLC-γ1 substrates are located. Second, PTKs directly phosphorylate PI-PLC-γ1, and this is a requirement for activation of the enzyme. Activated PI-PLC-γ1 then catalyzes the hydrolysis of a plasma membrane phospholipid called phosphatidylinositol 4,5-bisphosphate (PtdInsP2), leading to increased cytoplasmic levels of two PtdInsP2 breakdown products: **inositol 1,4,5-trisphosphate (IP$_3$)** and **diacylglycerol (DAG)** (Fig. 7–12). IP$_3$ stimulates release of membrane-sequestered intracellular calcium stores, causing a rapid rise in the cytoplasmic free calcium ion concentration. In addition, a plasma membrane calcium channel is opened in response to incompletely understood TCR-generated signals, and this leads to an influx of extracellular calcium. A sustained increase in cytoplasmic calcium is often maintained for over an hour after TCR engagement, and this is dependent on the influx from extracellular fluid. Calcium binds to a ubiquitous calcium-dependent regulatory protein called calmodulin. Calcium-calmodulin complexes can activate several enzymes, including phosphatases important for transcription factor activation, as discussed below. The DAG generated from PtdInsP$_2$ breakdown works together with Ca^{2+} to activate certain isoforms of **protein kinase C (PKC),** an enzyme that catalyzes the phosphorylation

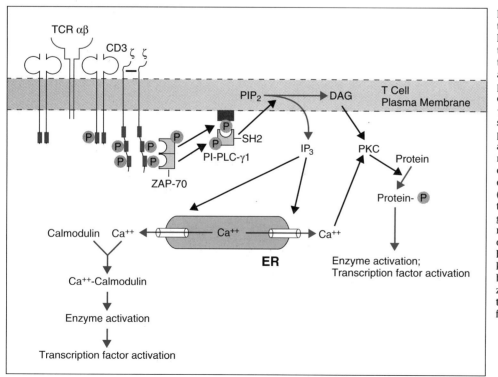

FIGURE 7–12. T cell signaling though membrane inositol lipid metabolism. T cell receptor (TCR)–associated protein tyrosine kinases activated by antigen presentation lead to the phosphorylation of phosphatidylinositol phospholipase C-γ1(PI-PLC-γ1) as well as docking sites for PI-PLC-γ1 on the plasma membrane. PI-PLC-γ1, activated by tyrosine phosphorylation, catalyzes the breakdown of membrane phosphatidylinositol 4,5-bisphosphate (PIP$_2$), generating inositol 1,4,5-trisphosphate (IP$_3$) and diacylglycerol (DAG). IP$_3$ induces the release of Ca^{2+} stored in the endoplasmic reticulum (ER), and DAG plus Ca^{2+} activate protein kinase C (PKC). Ca^{2+} and PKC both serve to activate other enzymes and eventually transcription factors. The symbol ℗ refers to phosphorylated tyrosine.

of serine/threonine residues on various substrates. The importance of PKC and calcium signals to the functional activation of T cells is supported by the fact that pharmacologic activators of PKC, such as phorbol myristate acetate (PMA), and calcium ionophores that raise cytoplasmic calcium concentrations, such as ionomycin, act together to stimulate T cell cytokine secretion and proliferation. The way calcium and PKC signals may be directly linked to gene transcription events is discussed later.

Ras-MAP Kinase and JNK Pathways in T Cell Activation

TCR-regulated tyrosine kinases not only induce inositol phospholipid breakdown, generating PKC and calcium signals, but they also activate Ras signaling pathways. **Ras** molecules comprise a family of 21 kD proteins with intrinsic GTPase activity that are involved in diverse activation responses in different cell types. Mutated *ras* genes are associated with neoplastic transformation. Ras pathways serve to link cell membrane receptors with downstream kinases, including the **mitogen activated protein (MAP) kinase** cascades. Actually, Ras itself is only one intermediate component of a series of interacting proteins that lead to the MAP kinase cascade. Ras is activated when a bound guanosine diphosphate (GDP) molecule is replaced by GTP and inactivated when the GTP is hydrolyzed to GDP. This GDP-GTP exchange is efficiently catalyzed only when Ras binds to a complex of proteins which form at the plasma membrane in response to receptor signals. In the best defined Ras pathways, there is an intermediate adaptor protein which contains both SH2 and SH3 domains. This protein binds specific tyrosine-phosphorylated proteins by its SH2 domain and a guanine-nucleotide exchange protein by its SH3 domain to form a short-lived ternary complex. The exchange protein catalyzes the formation of active GTP-Ras (Fig. 7–13). The proteins that serve these roles in T cells are not yet precisely defined. Grb-2 and crk have been implicated as intermediate adaptors containing SH2 and SH3 domains. Two substrates of TCR-regulated PTKs, shc and p36, bind to the SH2 domains of the intermediate adaptors. SOS, C3G, and vav may all bind to the SH3 domains in the adaptors and serve as the guanine-nucleotide exchange proteins in T cells (see Fig. 7–13).

MAP kinases form a large family of serine/threonine protein kinases involved in diverse cellular responses to membrane receptor signaling. Several different MAP kinases are implicated in TCR signal transduction, including **e**xtracellular signal-regulated **k**inase-1 (ERK-1) and ERK-2. These MAP kinases become fully active when they are themselves phosphorylated on neighboring threonine and tyrosine residues, and this is accomplished by specific enzymes called MAP-kinase kinases (MKKs). Several different MKKs have been identified, each with a distinct set of MAP kinase specificities. MKKs are activated in a similar fashion by MAP-kinase kinase kinases or MKKKs. One function of Ras in T cells appears to be the membrane recruitment and activation of an MKKK called Raf. Activated Ras binds Raf, and Raf then phosphoryl-

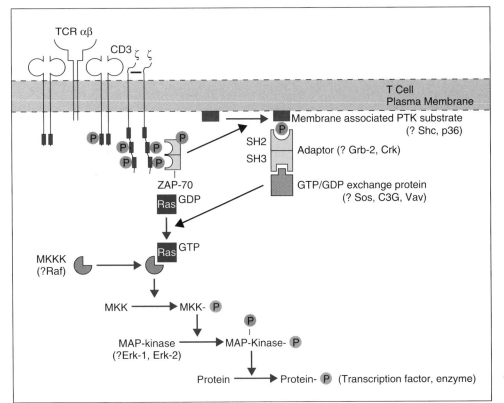

FIGURE 7–13. Ras-MAP kinase pathways in T cell activation. Activated ZAP-70 phosphorylates membrane-associated substrates, which permits the binding of SH2 containing adaptor proteins and the subsequent SH3-dependent binding of guanosine triphosphate–guanosine diphosphate (GTP/GDP) exchange proteins. The exchange proteins activate Ras, which in turn recruits the first component of the mitogen activated protein (MAP) kinase cascade. The kinases in the MAP kinase cascade include MAP-kinase kinase kinase (MKKK), MAP-kinase kinase (MKK), and MAP kinase. All these enzymes are serine/threonine kinases and MKK is also a tyrosine kinase. The symbol Ⓟ refers to phosphorylated serine or threonine, and the symbol Ⓟ refers to phosphorylated tyrosine. The exact proteins involved in these pathways in T cells are not known, but probable candidates are shown in parentheses. TCR, T cell receptor.

ates an MKK, which in turn phosphorylates one or more MAP kinases, including Erk-1 and Erk-2 (Fig. 7–13). Similarly, **Jun N-terminal kinase-1 (JNK-1)** is a serine/threonine protein kinase that is apparently downstream of another MKK/MKKK pathway that is initiated by a p21-GTPase distinct from Ras. CD28-mediated signals and PKC also appear to be linked to JNK activation. At the downstream end of these pathways, the activated MAP kinases and JNK catalyze the direct phosphorylation and activation of transcription factors or components of transcription factor complexes. Specific examples of this relevant to T cell cytokine genes are discussed below.

Activation of Transcription Factors That Regulate T Cell Gene Expression

So far we have described several signal transduction pathways initiated by ligand binding to the TCR. The ultimate importance of these pathways is that they stimulate the expression of genes encoding proteins needed for T cell effector function and clonal expansion. This is accomplished, in large part, by the activation of transcription factors that bind to regulatory regions of the relevant genes and thereby enhance promoter activity and transcription of T cell genes. Significant research effort has been directed at linking TCR signal transduction events to the regulation of cytokine gene transcription, especially IL-2 gene transcription. This is a reasonable prototype of the mechanisms by which various signal events discussed above converge to modify expression of T cell genes.

The transcriptional regulation of most cytokine genes, including IL-2 and IL-4, is largely controlled by protein binding to nucleotide sequences 5′ to where RNA-polymerase binds. The regulatory or enhancer region of the IL-2 gene, for example, includes a 5′ region of approximately 300 base pairs in which are located binding sites for several transcription factors, including **AP-1, NF-κB, Oct-1,** and **nuclear factor of activated T cells (NFAT).** These binding sites are required for the ability of TCR-induced signals to stimulate IL-2 gene transcription. (The transcription of some cytokine genes, such as interferon-γ, may also be regulated by intronic and 3′ sequences.) It has become clear from research in both T cell signal transduction and transcriptional regulation that *the initiation of cytokine gene transcription requires several signaling pathways which activate different DNA-binding proteins simultaneously.* For example, TCR-induced activation of AP-1, NFAT, and NF-κB all cooperate to enhance IL-2 gene transcription (Fig. 7–14).

TCR-generated calcium signals lead to the activation of NFAT, a TCR-activated transcription factor that binds to the enhancers, and is required for the transcription of IL-2, IL-4, tumor necrosis factor-α (TNFα), and other cytokine genes. There are actually several different NFATs each encoded by a separate gene; NFAT1 and NFAT2 are the

types found specifically in T cells. NFAT is present in an inactive, phosphorylated form in the cytoplasm of resting T lymphocytes. Activation of NFAT is dependent on increases in cytoplasmic calcium. The way calcium works to activate NFAT was discovered indirectly by studies of the mechanism of action of the immunosuppressive drugs, **cyclosporine (CsA)** and **FK506.** These drugs, which are natural products of fungi, are the main therapeutic agents used to block allograft rejection (see Chapter 17), and they function largely by blocking T cell cytokine gene transcription. CsA binds to a protein called cyclophilin in the cytoplasm of the T cell, and FK506 binds to a distinct but structurally related protein called FK506 binding protein (FKBP). Cyclophilin and FKBP are also called **immunophilins.** Their natural functions are not well understood but both CsA-cyclophilin complexes and FK506-FKBP complexes bind to and inhibit a cytoplasmic calcium-calmodulin–dependent serine/threonine phosphatase called **calcineurin.** Furthermore, both CsA and FK506 block translocation of cytoplasmic NFAT to the nucleus. These observations led to the elucidation of how TCR-generated calcium signals promote cytokine gene expression. In the simplest outline, elevated cytoplasmic calcium promotes calcium-calmodulin–mediated activation of calcineurin. Calcineurin then dephosphorylates cytoplasmic NFAT, which uncovers a nuclear-localization signal and permits NFAT translocation to the nucleus. Once in the nucleus, NFAT binds to consensus binding sequences in the enhancers of IL-2, IL-4, and other cytokine genes, usually in association with other transcription factors, such as AP-1.

AP-1 is a transcription factor found in many cell types, but it is specifically activated in T lymphocytes by TCR-mediated signals. AP-1 is actually the name for a family of DNA-binding factors that are composed of dimers of two proteins that bind to one another via a shared structural motif called a leucine zipper. The best characterized AP-1 factor is composed of Fos and Jun. TCR-induced signals lead to the appearance of transcriptionally active AP-1 in the nucleus of T cells. Unlike NFAT, activation of AP-1 does not involve nuclear translocation of pre-formed, cytoplasmically sequestered stores, but typically involves new transcription of the *fos* gene, and phosphorylation of the *jun* protein. Not surprisingly, the links between TCR-induced signals and AP-1 activation are complex and still incompletely characterized. *Fos* gene transcription can be enhanced by both PKC and MAP kinase pathways. In the latter case, the MAP kinase ERK-1 can phosphorylate a protein called elk-1, which is a component of a transcription factor complex that binds to the *fos* gene promoter and enhances *fos* transcription. Furthermore, JNK phosphorylates c-Jun in its amino terminus, and this leads to increased transcriptional enhancing activity of c-Jun-containing AP-1 complexes. Thus, AP-1 activation represents a convergence point of sev-

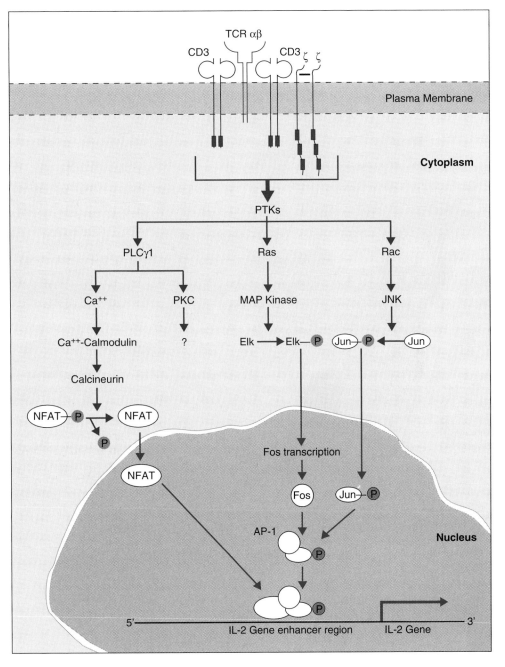

FIGURE 7–14. Activation of transcription factors in T cells. Multiple signaling pathways converge to activate transcription factors that regulate T cell genes. The formation of an active NFAT–AP-1 complex on the interleukin-2 (IL-2) gene enhancer region is shown as an example, but many other transcription factors and other genes are important in T cell activation. Calcineurin is a serine-threonine phosphatase activated by Ca²⁺ that catalyzes the removal of a phosphate from nuclear factor of activated T cells (NFAT). Dephosphorylated NFAT moves to the nucleus, where it binds to the IL-2 gene enhancer region. Elk is activated by phosphorylation by mitogen activated protein (MAP) kinase and enhances *fos* transcription. Jun is phosphorylated, and thereby activated by Jun N-terminal Kinase (JNK). Fos and Jun combine to form AP-1, which binds to the IL-2 enhancer in association with NFAT. The NFAT–AP-1 complex, along with other transcription factors, enhances IL-2 gene transcription. TCR, T cell receptor.

eral TCR-initiated signaling pathways. A further complication to this story is the fact that AP-1 physically associates with other transcription factors in the nucleus, including NFAT. Thus, signal pathways that activate AP-1 will influence NFAT-dependent transcriptional events. AP-1–NFAT cooperation has been demonstrated in the transcriptional regulation of both the IL-2 and IL-4 genes.

Nuclear factor–κB (NF-κB) is another transcription factor that is activated in response to TCR signals. A detailed discussion of NF-κB in the immune system is found in Chapter 4 (Box 4–4). TCR signals can induce the release of cytoplasmic NF-κB from the inhibitor IκB, which then promotes NF-κB nuclear translocation. The molecules that

couple TCR stimuli with NF-κB activation are not yet known but may include PKC or proteins similar to the tumor necrosis factor receptor associated factors (TRAFs).

It is clear that the signaling pathways we have discussed link antigen recognition by the TCR to the expression of genes that encode T cell effector molecules. The complexity of these pathways is in part attributable to the fact that many of the enzymes involved have several substrates, some of which may be components of different branches of the cascade. Furthermore, our knowledge of the relevant substrates *in vivo* is incomplete. The complexity of TCR signaling is likely responsible for a remarkable feature of T cell biology, namely the diversity of responses that follow antigen recognition. Although most of the research in this field has focused on signals linked to cytokine gene transcription, antigen recognition also leads to expression of genes encoding other kinds of molecules, regulation of programmed cell death pathways in developing T cells, upregulation of integrin function, and activation of cytolytic functions of CTLs. The ways in which the TCR-generated signals are linked to these diverse responses are not yet understood. Furthermore, as we shall discuss in the next section, the TCR can transduce different signals depending on variations in the ligand it binds.

PARTIAL T CELL ACTIVATION INDUCED BY ALTERED PEPTIDE LIGANDS

There is a growing body of evidence indicating that the signaling events and functional consequences of T cell antigen recognition may vary depending on the nature of the peptide that contacts the TCR. As we discussed earlier and in Chapter 5, the TCRs on an individual T cell contact only a few amino acid residues of a specific peptide, as well as amino acid residues on the MHC molecule to which the peptide is bound. Investigators have discovered that peptides in which one or two TCR contact residues have been changed induce only a subset of the functional responses seen with the unaltered peptide. For example, presentation of a wild type peptide to a T cell may result in production of IL-2, increased surface expression of IL-2 receptor, and proliferation. If, however, one of the TCR-contact amino acid residues is changed, the T cell may still express more IL-2 receptor, but it will not produce IL-2 nor proliferate. The peptides in which TCR-contact residues are changed are called **altered peptide ligands (APLs),** and the incomplete response is called **partial T cell activation.** APLs are also capable of inducing responses that are not seen at all with wild type peptide. For example, with some T cells, APLs have been shown to induce expression of a different set of cytokines from those induced with the wild type peptide. There are also examples of APLs inducing a state of anergy in T cells, such that they cannot be activated by even the wild type peptide upon secondary exposure (see Chapter 10). In addition, some APLs act as **T cell antagonists** that impart negative signals which interfere with the ability of simultaneously administered wild type peptides to activate T cells.

APLs are important for two general reasons. First, they represent a mechanism by which T cell activation can be regulated, in physiologic, pathologic, or therapeutic situations. It is possible, for example, that naturally occurring APLs are involved in the maintenance of self-tolerance or in the mechanisms by which microbes evade immune responses (see Chapter 16). APLs could theoretically be administered to patients to inhibit unwanted immune responses against defined antigens in autoimmune diseases (see Chapter 19) or in allograft rejection. Second, the analysis of intracellular biochemical events induced by APLs should provide further insights into how TCR signaling pathways work. There is already evidence that some APLs cause distinct patterns of tyrosine-phosphorylation and PTK activation events, and induce lower sustained levels of intracellular calcium than the wild type peptides from which they were derived. The biology of APL responses supports the hypothesis that the phosphorylation of different ITAMs in the TCR complex initiates distinct downstream events and therefore distinct functional responses. Thus, studies with APLs may help us better understand how the binding of peptide-MHC ligand to the extracellular domains of the $\alpha\beta$ TCR heterodimer is translated into signal transduction events in the cytoplasm.

ACTIVATION-INDUCED T CELL EFFECTOR MOLECULES: CD40 LIGAND AND FAS LIGAND

Antigen-induced activation of T cells leads to the production of many different secreted and cell surface molecules which have various effector and regulatory functions. Two activation-induced surface proteins that profoundly influence how T cells interact with other cells in the immune system are CD40 ligand and Fas ligand. Both these proteins are expressed within hours after antigen stimulation of T cells, and as their names imply, they bind to CD40 and Fas, which are surface receptors found on other cells. We will briefly discuss the general significance and basic characteristics of these molecules here, and we will return to their regulatory functions in more detail in later chapters.

Both CD40 ligand and Fas ligand are members of the TNF superfamily of molecules, most of which are integral membrane proteins that share a characteristic inverted pyramid receptor-binding structure sometimes called a TNF fold. CD40 and Fas are, not surprisingly, also members of a group of homologous transmembrane receptors called the TNF receptor superfamily. The common characteristics of the members of these superfamilies are discussed in Chapter 12.

CD40 ligand, also known as gp39, was first recognized as a membrane protein on activated CD4+ T cells that is required for helper T cell–mediated activation of B lymphocytes. In fact, in the two-signal requirement for B cell activation, signal 1 is provided by antigen binding to surface Ig, and signal 2 is provided by CD40 ligand binding to CD40. Although CD40 ligand is expressed mainly on activated T cells of the helper phenotype, it is now known that CD40 is found on many cell types, including B cells, monocytes, dendritic cells, follicular dendritic cells, epithelial cells, endothelial cells, fibroblasts, and others. In each case, binding of CD40 ligand induces activation responses in the

CD40-expressing cells. Examples of these responses include proliferation and isotype switching in B cells (see Chapter 9); enhanced monocyte cytokine secretion and microbicidal activity (see Chapter 13); and increased adhesion molecule expression on endothelial cells. Of particular relevance to T cell activation is the CD40-mediated upregulation of B7-1 and B7-2. This occurs on all three professional APC types, including dendritic cells, B cells, and monocytes. Therefore, through CD40-CD40 ligand interactions, activated T cells can further amplify T cell responses to antigen by enhancing the costimulatory function of APCs. The validity of this concept is demonstrated by the fact that there is a failure to induce effector helper T cell responses in CD40 ligand knockout mice.

Fas, which is constitutively expressed on T lymphocytes, transduces signals, leading to apoptosis of the T cell by a pathway called "activation-induced cell death" (see Box 10–1, Chapter 10). Activated T cells, which express Fas ligand, are therefore able to kill themselves or neighboring T cells. This mechanism is considered to be important for the downregulation of immune responses and for the maintenance of self-tolerance, as discussed in Chapter 10.

SUMMARY

Antigen presentation to T cells activates various functional responses including proliferation and differentiation into effector cells. When CD4$^+$ helper T cells are activated, they express new molecules, including cell surface proteins and secreted cytokines, that function to promote autocrine and paracrine T cell growth, B cell antibody production, and activation of cells of the innate immune system. Antigen presentation to CD8$^+$ CTLs induces differentiation and activation of the lytic functions of these cells. These T lymphocyte activation events begin when peptide-MHC complexes on APCs bind to TCRs. TCRs are clonally distributed, disulfide-linked heterodimeric protein receptors ($\alpha\beta$ TCRs) that are homologous to Ig molecules. These receptors specifically bind complexes of processed peptide antigen and MHC molecules. During T cell antigen recognition, the TCR makes contact with amino acid residues of the peptide as well as polymorphic residues of the presenting self MHC molecules. The $\gamma\delta$ receptor is another clonally distributed heterodimer that is expressed on a small subset of $\alpha\beta$-negative T cells. These $\gamma\delta$ T cells recognize different forms of antigen from $\alpha\beta$ T cells, including peptides in association with non-polymorphic MHC-like molecules, unprocessed proteins, and small nonpeptide molecules. The $\alpha\beta$ (and $\gamma\delta$) heterodimers are non-covalently associated with a complex of up to five distinct, invariant membrane proteins, including the CD3 and ζ proteins. The assembly of $\alpha\beta$ TCR and associated proteins is called the TCR complex. When the $\alpha\beta$ heterodimer binds peptide-MHC, the CD3 and ζ proteins transduce signals that activate T cell functional responses. In addition to the TCR complex, several accessory molecules are important in antigen-induced T cell activation. Some of these molecules bind ligands on APCs or target cells and thereby provide stabilizing adhesive forces. Accessory molecules also transduce activating or regulatory signals. CD4 and CD8 are coreceptor accessory molecules expressed on mutually exclusive subsets of mature T cells that bind nonpolymorphic determinants of class II and class I MHC molecules, respectively. CD4 is expressed on class II–restricted helper T cells, and CD8 is expressed on class I–restricted CTLs. CD28 present on T cells binds B7 molecules on APCs and delivers costimulatory signals, which are required in addition to TCR-complex generated signals for full T cell activation. The T cell integrins LFA-1 and VLA-4 bind ICAMs and VCAM-1, respectively, on the surface of other cells. These adhesive interactions are important for stable conjugation of T cells and APCs and for T cell movement from blood into tissues. CD45 is a tyrosine phosphatase that plays an important role in regulating tyrosine kinases during early T cell activation. Several interrelated signaling cascades are initiated by TCR antigen recognition. The earliest events include PTK–mediated phosphorylation of various substrates. Tyrosine residues in consensus motifs in the cytoplasmic tails of the CD3 and ζ proteins are phosphorylated and function to recruit SH2 domain–containing signaling molecules. Subsequent events include inositol phospholipid breakdown, elevated PKC activity, increases in cytoplasmic calcium, and activation of Ras-MAP kinase cascades. These signaling events lead to the activation of transcription factors that stimulate the transcription of genes required for the proliferative and effector responses of the T cells. Partial or qualitatively different signaling and functional responses can be induced by altered peptide ligands with modified TCR-contact residues. Fas ligand and CD40 ligand are T cell surface molecules whose expression is induced by T cell activation and which perform specialized effector functions.

Selected Readings

Ashwell, J. D., and R. D. Klausner. Genetic and mutational analysis of the T-cell antigen receptor. Annual Review of Immunology 8:139–167, 1990.

Bierer, B. E., B. P. Sleckman, S. E. Ratnofsky, and S. J. Burakoff. The biologic roles of CD2, CD4 and CD8 in T-cell activation. Annual Review of Immunology 7:579–600, 1989.

Cantrell D. T cell antigen receptor signal transduction pathways. Annual Review of Immunology 14:259–274, 1996.

Chan, A. C., and A. S. Shaw. Regulation of antigen receptor signal transduction by protein tyrosine kinases. Current Opinion in Immunology 8:394–401, 1996.

Chien, Y.-H., R. Jores, and M. P. Crowley. Recognition by γ/δ T cells. Annual Review of Immunology 14:511–532, 1996.

Crabtree, G. R. Contingent genetic regulatory events in T lymphocyte activation. Science 243:355–361, 1989.

Hemler, M. E. VLA proteins in the integrin family: structure, functions, and their role on leukocytes. Annual Review of Immunology 8:365–400, 1990.

Howe, L. R., and A. Weiss. Multiple kinases mediate T-cell receptor signaling. Trends in Biochemical Sciences 20:59–64, 1995.

Hunkapillar, T., and L. Hood. Diversity of the immunoglobulin gene superfamily. Advances in Immunology 44:1–63, 1989.

Hynes, R. O. Integrins: versatility, modulation, and signaling in cell adhesion. Cell 69:11–25, 1992.

Lenschow, D. J., T. L. Walunas, and J. A. Bluestone. CD28/B7 system of T cell costimulation. Annual Review of Immunology 14:233–258, 1996.

Miceli, M. C., and J. R. Parnes. The role of CD4 and CD8 in T cell activation and differentiation. Advances in Immunology 53: 59–122, 1993.

Perlmutter, R. M., S. D. Levin, M. W. Appleby, S. J. Anderson, and J. Alberola-Ila. Regulation of lymphocyte function by protein tyrosine phosphorylation. Annual Review of Immunology 11:451–500, 1993.

Raulet, D. H. The structure, function, and molecular genetics of the γ/δ T cell receptor. Annual Review of Immunology 7:175–208, 1989.

Sloan-Lancaster, J., and P. M. Allen. Altered peptide ligand-induced partial T cell activation: molecular mechanisms and role in T cell biology. Annual Review of Immunology 14:1–27, 1996.

CHAPTER EIGHT

T CELL MATURATION

IN THE THYMUS

The total number of T lymphocyte specificities for different antigens in an individual is called the **T cell repertoire.** Any individual's repertoire of mature helper and cytolytic T lymphocytes (CTLs) has two fundamental properties. First, as discussed in Chapter 6, *antigen recognition by T cells is self MHC restricted;* i.e., T cells in each individual can recognize and respond to peptide fragments of foreign antigens only in association with self major histocompatibility complex (MHC) molecules. Second, *the mature T cell repertoire is largely self-tolerant;* i.e., T cells in each individual do not respond to self antigens that are encountered during maturation in association with self MHC molecules. Failure to maintain self-tolerance leads to immune responses against one's own tissue antigens and autoimmune diseases. Therefore, understanding how the mature T cell repertoire develops is important for understanding the specificity of T cells and may help us unravel the pathogenesis of autoimmune diseases.

The thymus is the major site of maturation of both helper T cells and CTLs. This was first suspected because of immunologic deficiencies associated with the lack of a thymus. If the thymus is removed from a neonatal mouse, this animal does not develop a normal T cell repertoire and remains deficient in T cells throughout its life. The congenital absence of the thymus, as occurs in the DiGeorge syndrome in humans or in the "nude" mouse strain, is characterized by low numbers of mature T cells in the circulation and peripheral lymphoid tissues and severe deficiencies in T cell–mediated immunity (see Chapter 21). The fact that some functional T cells with a mature phenotype do exist in athymic individuals suggests that extrathymic sites of T cell maturation may exist, but the location of these sites is unknown and their contribution to the development of T cell immunity is apparently minor. Furthermore, although the thymus is clearly the principal site of T cell maturation, the organ involutes with age and is virtually undetectable in postpubertal humans. Nevertheless, some maturation of T cells continues throughout adult life. It may be that the remnant of the involuted thymus is adequate for some T cell maturation or that other tissues can assume the role of the thymus. Since memory T cells have a long life span (perhaps longer than 20 years in humans), the need for generating new T cells decreases with age.

This chapter describes the development of mature T cells in the thymus. T cell maturation consists of three closely related processes.

1. *Migration and proliferation: Precursors of mature T cells originating from the bone marrow migrate through the thymus, where some cells are stimulated to proliferate and most of the progeny die.* Immature T cells that have recently arisen from precursors in the bone marrow are committed to the T lymphocyte lineage but do not express T cell receptor (TCR) or accessory molecules, and they have no capacity to recognize antigens or perform effector functions. These precursors leave the bone marrow, circulate in the blood, and enter the thymic cortex. Within the cortex, there is a high level of proliferative activity. Selective processes result in the death of most of the newly formed cells so that only MHC-restricted, self-tolerant T cells survive. The surviving cells migrate from the thymic cortex to the medulla and are finally released as mature T cells into the periphery.

2. *Differentiation: The mature phenotype of T cells develops in the thymus.* TCR complexes are expressed early during intrathymic T cell maturation, after the formation of functional TCR genes by somatic rearrangement of different gene segments. In addition to the TCR complex, surface expression of a variety of accessory molecules occurs during thymic maturation. Some of these accessory molecules, including CD4 and CD8, play important roles in the T cell maturation process as well as in activation of mature T cells. Furthermore, programs of functional differentiation begin in the thymus so that the potential of a thymocyte to become either a helper or a cytolytic T lymphocyte is already fixed before the cell enters the circulation.

3. *Selection: The mature repertoire of foreign antigen specific, self MHC-restricted T cells is selected in the thymus from the larger set of possible specificities encoded in the germline.* All individuals contain essentially the same full sets of TCR genes in their genomes. These TCR genes code for many receptors that can recognize many different peptides in association with many MHC molecules. Therefore, in every individual, as T cells arise from bone marrow precursors, they have the potential of expressing receptors that can recognize peptides derived from virtually any protein (self or foreign) in association with any MHC molecule (also self or foreign). After different receptors are expressed on the surface of different clones of developing T cells, the repertoire is modified or shaped by two related selection processes (Fig. 8–1). **Positive selection** is the process by which the T cell repertoire becomes self MHC–restricted. This ensures that only T cells expressing useful TCRs capable of recognizing antigens on an individual's own antigen-presenting cells (APCs) will be permitted to mature. A second **negative selection** process eliminates or inactivates potentially autoreactive clones, especially those specific for antigens that are present at high concentrations in the thymus, ensuring that an individual is tolerant to these antigens. Tolerance to self antigens present in peripheral tissues is induced after T cells leave the thymus (see Chapter 19). The selective growth or death of different cells results in the self MHC–restricted, self antigen–tolerant mature T cell repertoire.

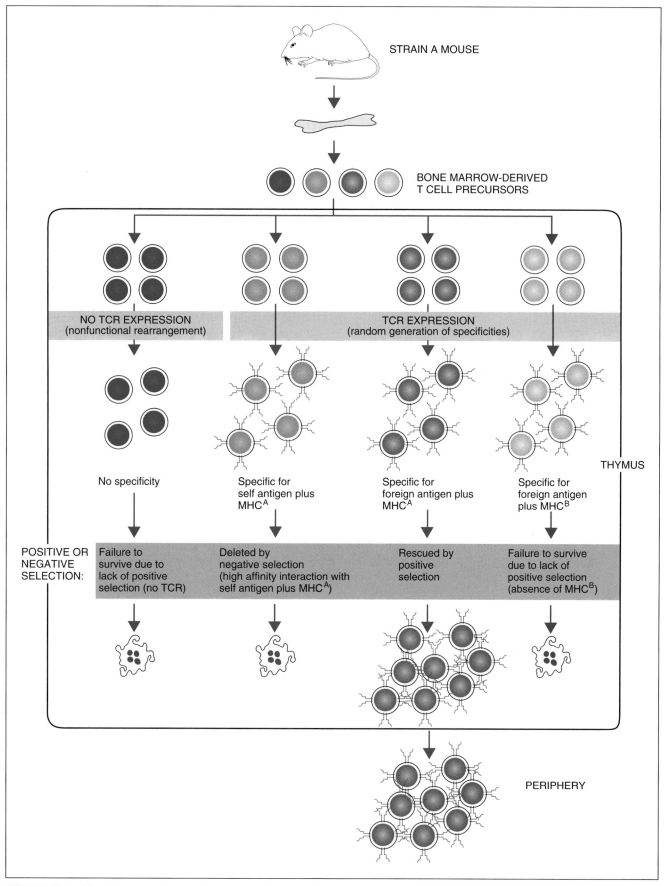

FIGURE 8–1. Thymic maturation and selection of T lymphocytes. Bone marrow–derived precursors have unrearranged T cell receptor (TCR) genes and do not express TCRs. During intrathymic maturation, TCR gene rearrangements occur and maturing thymocytes express TCRs with random specificities. Only those clones specific for foreign antigen peptides or self antigen peptides not present in the thymus bound to self major histocompatibility complex (MHC) molecules are selected to mature and leave the thymus to populate peripheral lymphoid tissues.

MIGRATION AND PROLIFERATION OF MATURING T CELLS IN THE THYMUS

T lymphocytes, like B lymphocytes, originate from hematopoietic stem cell precursors which are found in the fetal liver and adult bone marrow. At present, little is known about the marrow stem cells that give rise to T or B lymphocytes or when they become committed to mature along a particular lineage. There is some evidence that pluripotential stem cells from the bone marrow capable of forming both B and T lymphocytes, as well as other lineages, seed the thymus, but there is also evidence for the presence of cells which are pre-committed to the T lineage before entering the thymus. In mice, immature lymphocytes are first detected in the thymus on the 11th day of the normal 21-day gestation. This corresponds to about week 7 or 8 of the normal 40-week gestation in humans. Developing T cells in the thymus are called **thymocytes.** The most immature thymocytes do not express TCR, CD3, or ζ chains and are incapable of recognizing or responding to antigen or performing effector functions. Despite the absence of these markers of mature T cells, there has been extensive effort at classifying subpopulations of early T cell precursors on the basis of the expression of other surface molecules. Even though the functional significance of these markers is not always known, they are useful for following different stages of maturation in experimental systems. Three such markers are c-kit, CD44, and CD25. C-kit is the receptor for a hematopoietic stem cell growth factor called c-kit ligand, CD44 is a sulfated glycoprotein adhesion molecule (see Chapter 7), and CD25 is the α chain of the interleukin (IL)-2 receptor (see Chapter 12). The majority of the early T cell precursors in the adult mouse thymus express high levels of c-kit and CD44, but do not express CD25. C-kit and CD44 expression goes down and CD25 goes up abruptly at the time the TCR β chain gene rearranges (discussed later). Some of the early precursors may also transiently express CD4 shortly after entering the thymus, but very soon afterward, the majority of early T cell precursors do not express either CD4 or CD8, and are therefore called **double-negative** thymocytes (Fig. 8–2).

The most immature, double-negative thymocytes are found in the subcapsular sinus of the thymus. These cells are larger than more mature thymocytes, and they are actively proliferating at this stage. From the subcapsular region, the thymocytes migrate into and through the cortex where most of the subsequent maturation events occur. It is in the cortex that the thymocytes first express TCRs and mature into CD4$^+$ class II MHC–restricted or CD8$^+$ class I MHC–restricted T cells. As these thymocytes undergo the final stages of intrathymic maturation they migrate from cortex to medulla and then exit the thymus via either lymphatics or veins (Fig. 8–2).

During this migration, the thymocytes *come into close physical contact with a variety of non-lymphoid cells.* These include **thymic epithelial cells** and bone marrow–derived cells, including macrophages and dendritic cells (see Fig. 2–9, Chapter 2). The superficial cortical thymic epithelial cells also include "nurse" cells, which surround thymocytes within membrane invaginations. Deeper within the cortex, epithelial cells form a meshwork of long cytoplasmic processes, around which thymocytes must pass in order to reach the medulla. Epithelial cells are also present in the medulla. Bone marrow–derived dendritic cells are present at the corticomedullary junction and within the medulla, and macrophages are present primarily within the medulla. The migration through this anatomic arrangement allows sequential interactions between thymocytes and these other cells, and such interactions are presumably necessary for the maturation of T lymphocytes.

Two types of molecules produced by the non-lymphoid thymic cells are important for T cell maturation. The first are **MHC molecules,** which are expressed by many of the non-lymphoid cells in the thymus. Cortical macrophages, epithelial cells, and dendritic cells express high levels of class II MHC molecules; medullary epithelial and dendritic cells express both class I and class II MHC molecules; and medullary macrophages express high levels of class I MHC molecules. The interaction of maturing thymocytes with these MHC molecules within the thymus are essential for the selection of the mature T cell repertoire, as will be discussed in detail later. Second, thymic stromal cells, including epithelial cells, secrete thymic hormones and cytokines, which are postulated to promote T cell maturation. These hormones include a variety of proteins that have been well characterized biochemically but whose physiologic roles are largely unknown. They have been given various names, including thymosin, thymopoietin, thymulin, and thymic humoral factor. Thymic hormones have been shown to induce the appearance of some T lymphocyte lineage-specific surface molecules on bone marrow cells or immature thymocytes *in vitro* and to enhance T cell functional responses, such as proliferative responses to polyclonal activators. Some of these hormones have also been used in clinical trials for the treatment of immunodeficiency states in humans. Thymic stromal cells also secrete IL-7, a cytokine that stimulates the proliferation and maturation of developing T cells in the thymus. However, no combination of thymic hormones and cytokines has yet been shown to support the development of immunocompetent, TCR-expressing T cells from bone marrow precursors or from immature cortical thymocytes *in vitro* in the absence of thymic stromal cells.

There is a high rate of mitosis in the cortex, with each bone marrow–derived precursor giving rise to multiple progeny. *Nonetheless, more than 95 per cent of the cortical thymocytes die before reach-*

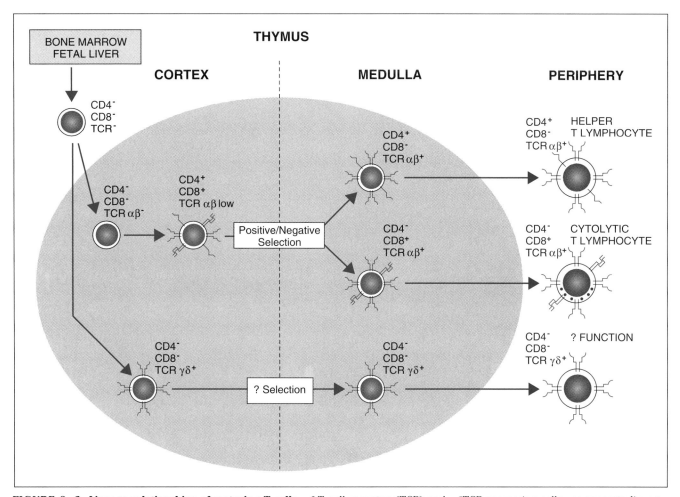

FIGURE 8-2. Lineage relationships of maturing T cells. γδ-T cell receptor (TCR) and αβTCR-expressing cells are separate lineages that develop from a common precursor. In the αβ lineage, the majority of thymocytes express both CD4 and CD8. TCR expression commences in this double-positive stage, beginning with low numbers of receptors on each cell and increasing as maturation proceeds. Single-positive (i.e., CD4 or CD8 γβTCR-expressing mature cells) are selected from this population. Some γδ cells express CD4 or CD8.

ing the medulla. Virtually all of this cell death is apoptotic (see Box 2–2, Chapter 2) with no associated inflammation. In fact, there are many "tingible body" macrophages within the normal thymus which have phagocytosed the condensed nuclear debris of apoptotic cells. The cell death is a consequence of the selection processes that preserve the minority of developing T cells that express self MHC–restricted, foreign antigen–specific TCRs, and eliminate most of the cells that express receptors of all other specificities or do not express any TCRs. Consistent with this view that thymic selection processes occur in the cortex is the fact that TCR expression, as well as CD4 and CD8 expression, is first detected on cortical thymocytes.

T CELL RECEPTOR GENES: ORGANIZATION, REARRANGEMENTS, AND GENERATION OF DIVERSITY

The selection processes that shape the mature T cell repertoire begin only after developing T cells first express randomly generated TCR mole-

cules with diverse specificities. The expression of TCRs is necessary for both positive and negative selection because both processes are dependent on specific recognition of self MHC and/or self antigens on the surfaces of thymic epithelial cells, dendritic cells, and macrophages. This portion of the chapter describes the organization of TCR genes and the mechanisms for generating the diverse repertoire of TCR specificities prior to selection.

Organization of T Cell Receptor α and β Genes in the Germline

Functional TCR α and β chain genes, which are capable of being expressed as polypeptides, are normally present only in cells of the T lymphocyte lineage. *These functional TCR genes are formed by somatic rearrangement of germline gene segments,* by a process that is very similar to immunoglobulin (Ig) gene rearrangements (see Chapter 4). The genomic organization of TCR α and β genes is fundamentally the same in all species studied (Figs. 8–3 and 8–4) and is similar to the organiza-

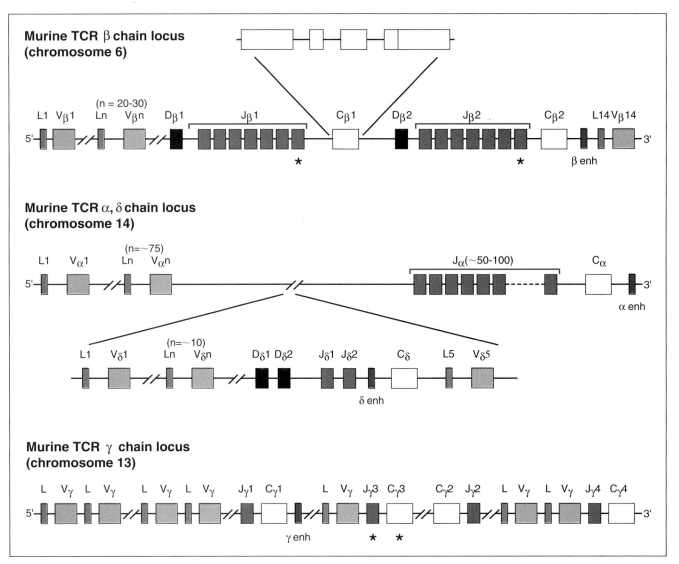

FIGURE 8-3. Organization of mouse T cell receptor (TCR) genes in the germline. Sizes of exons and intervening DNA sequences are not shown to scale. The TCR β locus has been completely sequenced; other loci are incompletely defined. Note that the δ locus is located within the α chain locus. All of the C genes are actually composed of multiple exons, as shown for the $C_\beta 1$ gene. * indicates nonfunctional pseudogenes. The numbering of $V\gamma$ genes is not yet uniformly defined. Enh, enhancer.

tion of Ig genes. Each TCR locus consists of variable (V), joining (J), and constant (C) region genes, and the β chain locus also contains diversity (D) gene segments. Complete mapping of the unrearranged β chain locus has been achieved in mice; other TCR loci in mice and humans are still incompletely described. The β chain locus is on chromosome 7 in humans and on chromosome 6 in mice. In humans, there are two nearly identical C_β genes, designated C_{β_1} and C_{β_2}, each containing four exons. Each C_β gene is associated with a 5' cluster of six or seven J segments and one D segment. In mice there are 20 to 30 V_β segments that can be grouped into 20 families, members of which are more than 75 per cent homologous in deoxyribonucleic acid (DNA) sequence. Most V_β segments are located 5' of the two clusters of C and J segments. Interestingly, some strains of mice have deletions of up to half of the V_β segments or half of the D_β and J_β segments, and yet they are immuno-

logically normal. In humans, there are approximately 50 functional V_β segments. The α chain locus is present on chromosome 14 in both humans and mice. There is a single C_α gene of four exons associated with a large 5' cluster of 60 to 70 different J segments. D segments have not been identified in the α locus. There are about 75 V_α gene segments, grouped into at least 12 families, all located 5' of the J and C regions. There is a very large region of intervening DNA between the V_α and J_α exons, which includes the entire TCR δ chain locus (discussed below).

Rearrangement and Expression of T Cell Receptor α and β Chain Genes

The TCR genes in the earliest T cell precursors are in the nonfunctional germline configuration, which is characterized by the spatial separation of V, D, J, and C gene segments on the

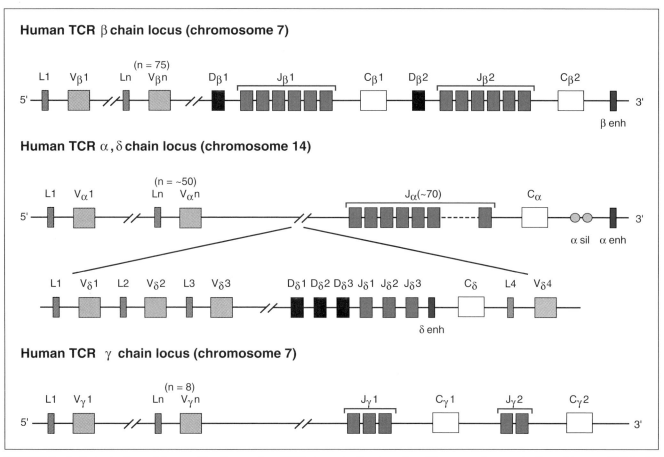

FIGURE 8–4. Organization of human T cell receptor (TCR) genes in the germline. The genomic organization of human TCR loci is similar to that of the mouse (see Fig. 8–3); the human loci have not been completely sequenced. Enh, enhancer; sil, silencer.

chromosome. During maturation of T cells in the thymus, the TCR gene segments are rearranged in a defined order, resulting in the formation of functional TCR α and β genes in which V, D, J, and C segments are in close proximity to one another (Fig. 8–5). This process of somatic rearrangement is a prerequisite to TCR gene expression and is important for the generation of TCR diversity, as we will discuss below.

Somatic rearrangements of TCR V, D, and J genes are mediated by recombinases which are thought to recognize specific sequences of nucleotides in the genome adjacent to each rearranging segment (see Chapter 4). These recognition sequences are essentially the same in Ig and TCR gene segments, and the same recombination mechanism mediates both types of receptor gene rearrangements. In fact, germline TCR genes transfected into immature B cell lines are efficiently rearranged. Furthermore, the products of the *RAG-1* and *RAG-2* genes (discussed in Chapter 4) are essential for both Ig and TCR gene rearrangements, and mice with null mutations in these genes do not develop mature B or T cells. The recognition sequences for TCR gene rearrangements include a conserved heptamer and nonamer separated by either a 12 base pair (bp) or 23 bp nonconserved spacer sequence. The locations of heptamer, nona-

mer, and spacer sequences flanking the V, D, and J gene segments in the β chain locus are such that either VDJ joining or (unlike Ig) direct VJ joining can occur. As a result, T cell receptor β transcripts do not always contain D sequences. The mechanisms that have been proposed for Ig DNA rearrangements, including deletions and inversions, may both be operational in T cell receptor DNA rearrangements. Despite the fact that Ig and TCR genes use the same recombinases, Ig genes are functionally rearranged only in B cells and TCR genes in T cells. The basis for this lineage specificity of antigen receptor gene expression is still not well understood.

The β chain locus rearranges prior to the α locus, and the product of a functionally rearranged β chain gene is essential for signaling subsequent events in T cell maturation. The process begins with the joining of one D_β segment with one J_β segment. This is followed by a second rearrangement in which the newly formed DJ segment is joined to a V_β segment, resulting in a VDJ gene. The genomic sequences between the rearranging elements, including D, J, and possibly $C_\beta 1$ genes (if $C_\beta 2$ is used), are deleted during this rearrangement process. If a nonproductive rearrangement involving the $C_\beta 1$ gene segment occurs, a subsequent rearrangement involving $C_\beta 2$ can occur, thus pro-

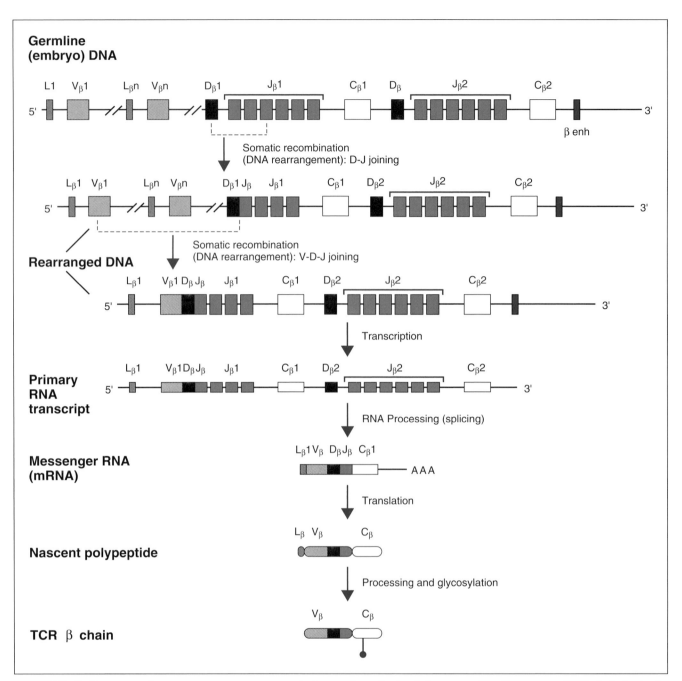

FIGURE 8–5. Sequence of gene rearrangement, transcription, and synthesis of the mouse T cell receptor (TCR) β chain. In the example shown, the V region is encoded by exons V1, D1, and the third exon in the J1 cluster; the C region is encoded by C1. Each C gene consists of multiple exons that are not shown. Unused V and J segments located between rearranged V and J genes are deleted. Note that in this example the DNA rearrangement involves DJ, followed by V joining to DJ; but direct VJ joining may also occur in the β chain locus. Enh, enhancer.

viding more chances for the cell to produce a TCR and therefore possibly survive positive selection.

The primary nuclear transcripts of the TCR β genes contain noncoding sequences (introns) between the VDJ and C genes. These are spliced out to form a mature messenger ribonucleic acid (mRNA) in which VDJ segments are juxtaposed to either one of the two C genes. The two C genes may be thought of as structurally analogous to the Ig heavy chain isotype genes; but unlike Ig genes,

the use of $C_\beta 1$ versus $C_\beta 2$ appears to be random, and there is no evidence that an individual T cell ever switches from one C gene to another. Furthermore, there is no known association of the use of either C_β gene segment with a particular function or specificity of the TCR.

β chain promoters have been identified in the 5' flanking regions of V genes. These promoters are minimally active and not T cell specific. A powerful β chain enhancer is located 3' of the $C_\beta 2$

gene, which is responsible for high-level, T cell–specific activity of the V promoters. Several nuclear-binding proteins associate with this enhancer, but most of these factors are not expressed uniquely in T lymphocytes. Nonetheless, TCR β chain transcription is T cell specific. In part, this may be due to T cell–specific rearrangements that bring β chain promoters in proximity to β chain enhancers, as proposed for the Ig locus (see Fig. 4–15, Chapter 4).

Functional rearrangement of the β chain gene results in the synthesis of β chain protein prior to the rearrangement and expression of the α chain gene, or expression of CD4 and CD8. The newly formed β chains pair with a molecule called **pre-T α chain (pTα),** and the pTαβ heterodimers are expressed on the thymocyte plasma membrane in association with CD3 proteins. The function of pTα is thought to be analogous to that of surrogate light chains, including λ₅, in B cell development (see Chapter 4). pTα is invariant, and it is not

known if the pTαβ "receptor" actually recognizes any extracellular or intracellular ligands. Nonetheless, the pTαβ heterodimer does generate intracellular signals, which lead to at least three important responses (Fig. 8–6). First, further rearrangements of the β chain locus on either chromosome are inhibited. This results in β chain allelic exclusion, i.e., each mature T cell expresses identical copies of only one β chain. Allelic exclusion of TCR genes has been demonstrated by transfection and transgene experiments, in which the introduction of an exogenous, functionally rearranged TCR β chain gene into immature T cells inhibits the rearrangement and expression of the endogenous genes. In normal development, nonproductive rearrangements of one or both alleles of β chain locus are quite common. If both alleles of the β chain locus are nonproductively rearranged, the developing T cell will die. Second, the thymocytes expressing the pTαβ receptor are stimulated to proliferate, resulting in multiple progeny cells all with a single

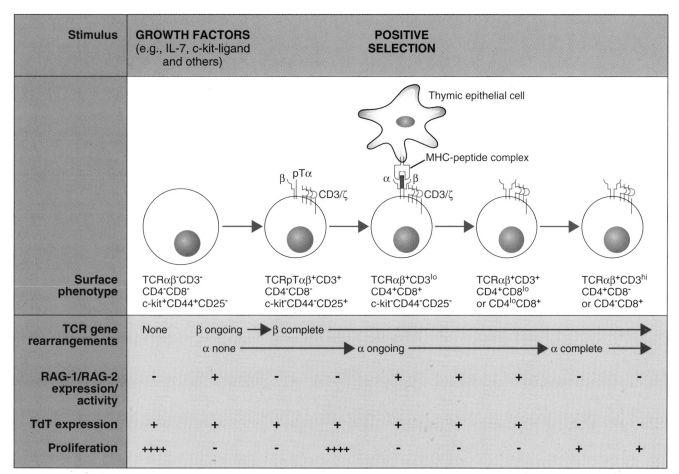

FIGURE 8–6. Sequence of events in thymocyte maturation. Bone marrow precursors enter the thymus and undergo a series of phenotypic changes as they mature into T cells. Initial proliferative activity is due to growth signals probably delivered by interleukin-7 (IL-7) and *c-kit* ligand. T cell receptor (TCR) β or α chain gene rearrangements only occur while RAG-1 and RAG-2 are active, and this occurs only when the thymocytes are not actively proliferating. Signals delivered by the preTαβ (pTαβ) receptor induce proliferation, α chain gene rearrangements, and CD4 and CD8 expression. Positive selection of major histocompatibility complex (MHC)-restricted T cells involves recognition of MHC-peptide complexes on thymic epithelial cells and promotes survival of thymocytes and differentiation from the double-positive CD4⁺CD8⁺ stage to the single-positive CD4⁺CD8⁻ or CD4⁻CD8⁺ stage. TCR α chain gene rearrangements continue for an extended time, permitting successive expression of TCRs with different specificities, until RAG-1/RAG-2 and TdT expression is extinguished.

form of β chain but with the potential to express different α chains. Third, the pT$\alpha\beta$ receptor induces CD4 and CD8 expression, a prerequisite for the selection processes that will follow later during cell maturation. The details of the signaling pathways that lead to these responses are not known, but it is clear that protein tyrosine kinase (PTK) activation is involved and that lck, the *src* family nonreceptor PTK which associates with CD4 and CD8 in mature T cells (see Chapter 7), is essential. It is therefore not surprising that in lck-deficient mice T cell maturation is arrested at an early stage, before expression of TCR α chains, CD4, or CD8.

The rearrangement of the α chain gene segments is delayed until after proliferative activity associated with pT$\alpha\beta$ expression is over. Once α chain rearrangement commences, it proceeds for 3 or 4 days (in mice) until the expression of the *RAG-1* and *RAG-2* genes is turned off. It is not known if pT$\alpha\beta$ signals contribute to the initiation of DNA rearrangements in the α chain locus. The steps in α chain gene rearrangement are similar to those in β chain gene rearrangements. Since there are no D segments in the α locus, rearrangement consists solely of the joining of V and J segments. The high number of J$_\alpha$ segments permits multiple attempts at productive VJ joining on each chromosome, thereby increasing the probability that a functional $\alpha\beta$ TCR will be produced. Once VJ joining has occurred, transcription of the α chain gene ensues. Transcriptional regulation of the α chain gene is apparently similar to that of the β chain. There are promoters 5' of each V gene that have low-level T cell–nonspecific activity but that are responsible for high-level T cell–specific transcription when brought in proximity to an α chain enhancer located 3' of the C$_\alpha$ gene. Nuclear-binding proteins and some consensus-binding sequences

have been shown to be functionally important for the α chain enhancer. Some of these factors are T cell specific but not yet well characterized, and others are expressed in other cell types as well. In addition, 5' to the α chain enhancer, there are "silencer" sequences that can inhibit α chain transcription in non-T cells as well as in cells of the $\gamma\delta$ lineage. Since there is only one C$_\alpha$ gene, RNA processing of a primary transcript gives rise to only one possible complete α chain mRNA. There is no evidence, however, that production of α chain protein suppresses further rearrangements of the α chain locus. In fact, only during the subsequent selection processes are RAG-1 and RAG-2 expression suppressed, causing a halt in further α locus rearrangement. Therefore, if productive rearrangements occur on both chromosomes, it is possible that a thymocyte will express two α chains. In fact, up to 30 percent of mature peripheral T cells do express two TCRs each with a different α chain and the same β chain. If no productive α chain rearrangements occur on either chromosome, the thymocyte will die.

Estimates of the potential size of the unselected immature T cell repertoire range from 10^{10} to 10^{15} specificities. As discussed in Chapter 7, the peptide-MHC binding site of a $\alpha\beta$ TCR heterodimer is formed by the V, D, and J regions of both chains, similar to Ig molecules. The structural diversity of the TCR heterodimers is generated by molecular mechanisms that are essentially similar to the mechanisms that generate antibody diversity (Table 8–1). These include the following:

1. *Multiple germline V, D, and J segments* can be used to create TCRs of different specificities. Although there are fewer V genes in TCR α and β loci than in Ig, the much greater number of J gene segments more than compensates for this. For in-

TABLE 8–1. Contribution of Different Mechanisms to Generation of Diversity of T Cell Receptor (TCR) and Immunoglobulin (Ig) Genes

| Mechanism | Immunoglobulin | | TCR $\alpha\beta$ | | TCR $\gamma\delta$ | |
	Heavy Chain	κ	α	β	γ	δ
Variable segments	250–1000	250	75	25	7	10
Diversity (D) segments	12	0	0	2	0	2
D segments read in all three reading frames	Rare	—	—	Often	—	Often
N-region diversification	V-D, D-J	None	V-J	V-D, D-J	V-J	V-D1, D1-D2, D1-J
Joining segments	4	4	50	12	2	2
Variable segment combinations	62,500–250,000		1875		70	
Total potential repertoire with junctional diversity	$\sim 10^{11}$		$\sim 10^{16}$		$\sim 10^{18}$	

The mechanisms are described in the text. TCR loci have fewer V gene segments than Ig loci, but there is potentially much greater junctional diversity in TCR genes. The contribution of somatic mutation, which occurs in the Ig but not the TCR genes, is not included in the table, since these mutations tend to increase affinities of Ig molecules for antigen but may or may not change specificity. The numbers are representative of mouse Ig and TCR loci.

Adapted from Davis, M. M., and P. J. Bjorkman. T-cell receptor antigen genes and T-cell recognition. Nature 334:395–402, 1988. Copyright © 1988, Macmillan Magazines Ltd.

stance, there are up to 50 J segments in mice compared with only four each in the IgH and κ chain loci. During TCR gene rearrangements, *combinatorial associations of different V, D, and J gene segments generate TCR diversity.*

2. *Junctional diversity,* involving coding sequences at VJ, VD, and DJ junctions, contributes significantly more to TCR diversity than to Ig diversity. Several different mechanisms are involved in TCR junctional diversity. First, the random addition of nucleotides that are not part of the genomic sequence, at VD, DJ, and VJ junctions, called **N-region diversification,** occurs in both α and β genes but only in Ig heavy chain and not light chain genes. N-region diversification is catalyzed by the enzyme terminal deoxyribonucleotidyl transferase (TdT). Second, imprecise recombination can lead to different sequences at V-DJ and V-J joins, as in Ig rearrangements (see Fig. 4–10, Chapter 4). For example, more than one 3' nucleotide of a V_α gene can join to J_α segments, and more than one 5' nucleotide of a J segment can join with D segments. Third, many D segments can be translated in all three possible reading frames, an uncommon feature for Ig heavy chain genes. Because of this, imprecise VD joining more often results in functional rearrangements.

3. The *pairing of α and β chains* serves to multiply the diversity generated for each chain. There are up to 75 V_α and 50 V_β genes in the human genome, contributing up to 3750 potential $V_\alpha V_\beta$ region combinations.

In summary, TCR genes fundamentally resemble Ig genes in their mechanisms of diversity generation and expression, but they also differ from Ig genes in several respects. Although there are far fewer TCR V genes than Ig V genes, the potential diversity of TCR molecules has been estimated to be greater than that for Ig, largely because of the greater number of J segments and the greater junctional diversity in TCR genes (Table 8–1). Once a TCR gene is functionally rearranged, however, there are no further genetic alterations leading to changes in function or affinity of the protein receptor. Thus, unlike Ig genes, there is no isotype switching in TCR. Furthermore, there is no evidence of functional somatic mutation in TCR genes; consequently, affinity maturation of the TCR is not observed in secondary T cell immune responses as is observed in secondary antibody responses.

T Cell Receptor γ and δ Genes

The TCR γ chain locus is located on the short arm of chromosome 7 in humans and on chromosome 13 in mice, distinct from either TCR α or β locus. The human γ locus is organized similarly to the TCR β chain locus (Fig. 8–4) with two JC clusters containing five J segments and two C segments, located 3' of multiple V gene segments (up

to 14 in total, including six nonfunctional "pseudogenes"). In mice, the arrangement is more complex, with up to seven V gene segments interspersed with four JC clusters (Fig. 8–3). (One of the murine V gene segments and one of the JC clusters are nonfunctional.) No D segments have been identified. One striking difference in the TCR γ chain locus compared with the TCR α and β chain loci is the variation among the multiple C gene segments, including differences in sequence, length, and number of the exons encoding the hinge region (connecting peptide) of the protein. This variation results in different molecular weights and different N-linked glycosylation patterns of the expressed γ chain proteins on different cells, as well as the presence of both disulfide and non-covalent linkages of the γ chain with the δ chain.

The TCR δ chain locus is unusual in that it is located entirely within the α chain locus between the V gene segments and the JC clusters (see Figs. 8–3 and 8–4). In humans, the δ locus consists of up to four V gene segments and one C gene segment associated with three J and two D segments. The arrangement of the δ chain locus in mice is fundamentally similar, with at least ten V gene segments.

Rearrangements of TCR γ and δ genes occur by the same mechanism as Ig and TCR α and β genes, utilizing the same recombinases and recognition signals as the other antigen receptor loci. Since the δ locus lies between the V and C genes, functional rearrangements of the α locus will delete the δ locus. Some of the V_α segments can join with either J_α or J_δ segments. There is no evidence for pairing of γ or δ chains with surrogate partners, like the pTα chain, during maturation of the $\gamma\delta$ lineage.

The small number of γ and δ V gene segments suggests a limited amount of combinatorial diversity among $\gamma\delta$ receptors. The limitations are even greater because of the selective use of certain V segments during different stages of life (see below). There is, however, an enormous potential diversity of $\gamma\delta$ receptors as a result of variations in junctional sequences, especially in the δ chain. This potential junctional diversity is largely due to the unique feature that either one or both D segments in tandem can be used within a single rearranged δ gene. Because imprecise joining of V, D, and J segments and N-region diversification also occur in rearranging δ and γ genes, there are theoretically more possible $\gamma\delta$ molecules than $\alpha\beta$ or Ig molecules (Table 8–1).

In the mouse, there are subsets of $\gamma\delta$ T cells that are distinguished both by their anatomic location and by the Vγ or Vδ gene segments they utilize (see Chapter 7). Two of the earliest $\gamma\delta$ T cell subsets to appear in mouse fetuses exhibit no TCR diversity, i.e., all their TCRs are identical (but of unknown specificity). For example, one of these $\gamma\delta$ T cell subsets populates mouse skin, and all the cells in this subset utilize the same rearranged V, J, and C segments, namely $V_\gamma 5J_\gamma 2C_\gamma 1$ and $V_\delta 1D_\delta 2J_\gamma 1C_\delta$. Furthermore, these $\gamma\delta$ T cells mature before TdT is expressed, and therefore no N-region diversification occurs in their TCRs. In addition to these populations with identical $\gamma\delta$ TCRs, there is a limited use of other V_γ and V_δ genes in other human and mouse $\gamma\delta$ T cells, but these populations do incorporate different J and D segments and exhibit greater junctional diversity.

Transcription of rearranged γ and δ genes is apparently controlled by locus-specific enhancers, similar to those described for α and β loci. A γ chain enhancer has been identified 3' to the $C_\gamma 1$ gene segment, and a δ chain enhancer is located between $J_\delta 3$ and C_δ of the human locus. These enhancers are active only in T cells. A γ silencer analogous to the α silencer mentioned above has been identified 3' to the $C_\gamma 1$ gene segment that is likely to be partly responsible for keeping γ gene expression turned off in $\alpha\beta$-expressing T cells. This is especially pertinent since up to one third of $\alpha\beta$ T cells have undergone potentially functional γ gene rearrangements.

THYMOCYTE DIFFERENTIATION: T CELL RECEPTOR AND ACCESSORY MOLECULE EXPRESSION

The rearrangement and expression of TCR genes during intrathymic maturation are the necessary first steps in the development of the T cell repertoire. TCR gene expression occurs coordinately with the expression of other proteins, including CD3, CD4, and CD8, which are important for the selection of MHC-restricted helper T cells and CTLs (Fig. 8–6). Subpopulations of thymocytes may be identified based on the expression of surface molecules, mainly TCR, CD4, and CD8, and these remain the most informative markers for stages of T cell maturation. The anatomic locations, sizes, and functional properties of these subpopulations, and changes induced in them by experimental manipulations, provide important insights into the mechanisms of thymic selection. We next discuss the order of events in T cell surface molecule expression and the characteristics of the best-defined thymocyte subsets.

T Cell Receptor Expression

Most studies of T cell ontogeny have relied on fetal murine thymuses, in which distinct populations of developing T cells can be identified. Recently, experimental models of human T cell maturation have been developed including thymic grafts into severe combined immunodeficiency disease mice and human thymic cultures. These models will undoubtedly be used to obtain similar kinds of information as that from the mouse models. As described above, the first T cell precursors detected in the fetal mouse thymus early in gestation do not express TCR complexes. Organ culture of embryonic thymuses, in which there is no ongoing influx of bone marrow precursors, has demonstrated that these TCR^-CD3^- cells are the precursors of all more mature forms.

The TCR genes remain in the germline configuration in T cell precursors for 2 to 3 days after they enter the thymus. In fetal thymuses, the first TCR gene rearrangements to occur involve the γ and δ loci. Surface expression of CD3-associated $\gamma\delta$ heterodimeric protein occurs 3 to 4 days after the precursor cells first arrived, reflecting concomitant expression of the newly rearranged TCR γ and δ genes and CD3 genes. *Thus, the $\gamma\delta$ receptor is the first TCR to be expressed during fetal life.*

Although nonfunctional DJ rearrangements at the β chain locus begin simultaneously with γ and δ chain rearrangements in fetal thymuses, functional β chain VDJ rearrangements and full-length 1.3 kb transcripts are first detected 2 days after γ and δ transcripts appear. The α chain genes are the last to rearrange and to generate functional mRNA a day after the β chain genes. Thus, surface expression of CD3-associated $\alpha\beta$ heterodimers is first detected about 2 days after $\gamma\delta$ expression. Expression of $\alpha\beta$ receptors rapidly overtakes expression of $\gamma\delta$ receptors, so that by birth most TCR:CD3–expressing thymocytes have $\alpha\beta$ receptors. This is consistent with the fact that more than 90 per cent of mature peripheral T cells express only the $\alpha\beta$ form of antigen receptors.

There is strong evidence that $\gamma\delta$- and $\alpha\beta$-expressing thymocytes are separate lineages with a common precursor.

1. Southern blot analysis of mature $\alpha\beta$-expressing T cells often show out-of-frame rearrangements of γ genes that are incapable of being transcribed, indicating that these cells could never have expressed $\gamma\delta$ receptors.

2. The δ chain gene segments are located between the V and J gene segments (Figs. 8–3 and 8–4), and they are deleted in a circle of DNA when the α chain gene rearranges. Analysis of these deleted DNA circles from murine thymuses shows that the δ chain gene is in the germline configuration in many cells that have rearranged their α chain genes. This finding also indicates that many $\alpha\beta$-expressing T cells have never expressed $\gamma\delta$ receptors.

3. β gene knockout mice develop normal numbers of $\gamma\delta$ T cells, and δ gene knockout mice develop normal numbers of $\alpha\beta$ T cells.

4. Interestingly, $\gamma\delta$ transgenic mice in which the silencer of the γ transgene has been deleted do not develop $\alpha\beta$ T cells, suggesting that, at some early point in T cell development, silencing of γ transcription is required in order for α and β gene expression to proceed.

This temporal pattern of early $\gamma\delta$ gene expression followed by $\alpha\beta$ expression is seen in chickens, mice, and humans. In human fetal thymuses, $\gamma\delta$

receptor expression begins at about 9 weeks of gestation, followed by $\alpha\beta$ TCR expression at 10 weeks.

A single primitive T cell precursor that enters the thymus can generate multiple cells, each with a unique TCR conferring a distinct antigen-binding specificity. This has been demonstrated by an experiment in which a single precursor cell with unrearranged $\alpha\beta$ TCR genes was introduced into a thymus previously depleted of thymocytes. The progeny of this cell showed multiple, different β chain rearrangements after 12 days. Thus the diversity of the TCR repertoire is dependent on mitotic expansion of thymocytes and independent TCR gene rearrangements in the progeny of these cell divisions. Nonetheless, recombinase activity is suppressed in cycling cells so that a cohort of thymocytes goes through sequential waves of proliferation followed by TCR gene rearrangements (Fig. 8–6). The level of $\alpha\beta$ TCR-CD3 complex expression on a thymocyte also changes during the later stages of maturation, starting out low and increasing after positive selection and migration into the medulla.

CD4 and CD8 Expression

CD4 and CD8 molecules are commonly used as markers to categorize subsets of thymocytes and to define the sequence of thymocyte maturation. Furthermore, recent experiments indicate that the expression of these molecules may be important in the selection processes that shape the T cell repertoire. Thymocytes go through three main developmental stages defined on the basis of CD4 and CD8 expression (Figs. 8–2 and 8–6). In the earliest, least mature stage, the thymocytes do not express either CD4 or CD8 and are designated CD4⁻CD8⁻, or "double-negative." Next, they become CD4⁺CD8⁺, or "double-positive." Finally, thymocytes mature into CD4⁺CD8⁻ or CD4⁻CD8⁺ "single-positive" cells. Two-color immunofluorescence staining of an adult mouse thymus for CD4 and CD8 expression will typically show four discrete populations reflecting these distinct developmental stages (Fig. 8–7). In addition to these major populations, CD4 and CD8 expression can be used to define numerically smaller groups of thymocytes which represent other steps in T cell development. For example, there is transient expression of CD4 on very immature thymocytes prior to the double-negative stage. These early CD4⁺CD8⁻ cells can be distinguished from the much more abundant mature single positive CD4⁺CD8⁻ thymocytes by the lack of expression of other molecules such as TCR and CD3 on the immature cells. In addition, subpopulations of double-positive cells can be distinguished on the basis of the relative abundance of

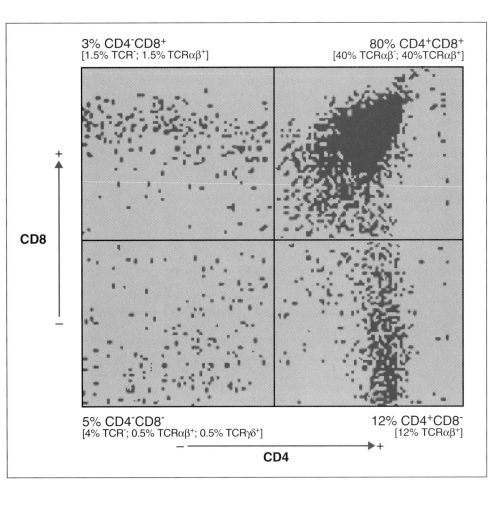

FIGURE 8–7. Subpopulations of thymocytes in the adult mouse thymus. A two-color flow cytometry analysis of thymocytes is shown using anti-CD4 and anti-CD8 antibodies, each tagged with a different fluorochrome. The percentage of all thymocytes contributed by each major population is shown in the four quadrants, and the percentages of subpopulations are indicated in the brackets. TCR, T cell receptor.

CD4 and CD8. These subpopulations represent intermediate stages of transition from double-positive to single-positive cells, and will be discussed later.

In addition to CD4 and CD8, several other surface molecules are used to further subdivide populations of thymocytes. As discussed previously, *c-kit,* CD44, and CD25 have proved to be reliable markers for early differentiation events. The earliest stem cells which enter the thymus express high levels of *c-kit* and CD44 and no CD25. Concomitant with TCR β chain gene rearrangements, *c-kit* and CD44 levels go down and CD25 is expressed transiently. Other markers used to characterize thymocyte subpopulations include CD1, CD2, CD5, thy-1, and peanut agglutinin receptor. The significance of these markers or the subsets they define is not completely known, and they will not be discussed further.

The CD4⁻CD8⁻ double-negative cells are a heterogeneous group making up about 5 per cent of the total thymocytes in an adult (Fig. 8–7). Most (80 per cent) of these cells are immature cortical thymocytes that are actively dividing or actively rearranging TCR genes but are not yet expressing TCR complexes. The remaining double-negative cells are more mature and consist mainly of γδ-positive thymocytes (most of which will never express CD4 or CD8). Little is known about the stimuli that drive the proliferation and maturation of immature double-negative thymocytes. Many of these cells express transcripts for IL-2 and IL-4 as well as the IL-2 receptor α chain (CD25). Nonetheless, T cell maturation in IL-2 and IL-4 gene knockout mice is essentially normal. Several lines of evidence suggest that the cytokine IL-7, which is secreted by thymic stromal cells, supports the growth of bone marrow–derived lymphopoietic stem cells within the thymus. For example, double-negative cells proliferate *in vitro* in response to IL-7. Furthermore, mice genetically deficient in IL-7 or either the binding or the signaling chain of the IL-7 receptor (also shared by the IL-2, IL-4, IL-9, and IL-15 receptors; see Chapter 12) have greatly reduced numbers of thymocytes and peripheral T cells, although differentiation appears to proceed normally. Both IL-1 and tumor necrosis factor (TNF) are also capable of sustaining the growth of double-negative T cells, especially the *c-kit*ˡᵒCD44ˡᵒCD25⁺ population, *in vitro.* Therefore, it is likely that cytokines produced by non-lymphoid cells in the thymus function as growth factors for the immature thymocytes.

Most of the double-negative thymocytes eventually mature into TCR αβ-expressing, MHC-restricted T cells that express either CD4 or CD8, but not both. During this maturation pathway, the double-negative cells first go through a transient stage when low levels of either CD8 or CD4 alone are expressed, and then they become double-positive cells, which express both CD4 and CD8 simultaneously (see Fig. 8–6). Double-positive cells constitute up to 80 per cent of the cells in an adult

mouse thymus (Fig. 8–7). Transgenic and knockout mice have provided valuable information about the transition from the double-negative to double-positive phenotype (Box 8–1). Studies with such genetically manipulated mice have demonstrated that rearrangement and expression of TCR β chain genes and the expression of pTαβ receptors are required for the CD4⁻CD8⁻ to CD4⁺CD8⁺ transition. For example, thymocytes in RAG-2–deficient mice cannot rearrange either α or β chain genes and fail to mature past the double-negative stage. When a functionally rearranged β chain gene is introduced as a transgene, maturation proceeds to the double-positive stage. An α chain transgene, however, does not relieve the maturation block. pTα gene knockout mice also show a drastic reduction in the number of double-positive thymocytes.

The rearrangement of the TCR α chain genes, and the expression of αβ TCR heterodimers, occurs in the double-positive population just before or during migration of the thymocytes from the cortex to medulla (Figs. 8–2 and 8–6). In the adult thymus, about half the double-positive cells are TCR αβ:CD3-positive and half are TCR αβ:CD3-negative (see Fig. 8–7). The great majority of double-positive cells die *in vivo,* because selection processes act at this stage of maturation. As we will discuss in detail later, thymocytes need to be positively selected by TCR engagement with peptide-MHC complexes in order to escape programmed death. One reason for a lack of positive selection is unproductive TCR gene rearrangements that occur frequently, leading to failure to express TCR molecules and therefore programmed cell death.

CD4⁺CD8⁺ cells give rise to mature CD4⁺ and CD8⁺ (single-positive) peripheral T cells, which also express high levels of TCR complexes (see Figs. 8–2 and 8–6). This has been established by a variety of experiments. For instance, if neonatal mice are injected with anti-CD4 antibody, the number of both CD4⁺ and CD8⁺ cells in the periphery are reduced. This is interpreted to indicate that the antibody bound to and induced the complement-dependent lysis of double-positive thymocytes and inhibited the maturation of all single-positive cells. CD4⁺CD8⁻TCR αβ⁺ thymocytes make up about 12 per cent of adult thymocytes (Fig. 8–7). These cells display functional helper activity in *in vitro* assays and migrate to the circulation and peripheral lymphoid tissues to constitute the mature, class II MHC–restricted helper T cell subset. CD4⁻CD8⁺TCRαβ⁺ cells make up about 3 per cent of thymocytes in the adult. They migrate out of the thymus to become mature, class I MHC–restricted pre-CTLs in the periphery. We do not know the stimuli that induce a particular CD4⁺CD8⁺ cell to selectively express either CD4 or CD8 and to simultaneously acquire the functional capabilities of a helper or pre-CTL, respectively. It is clear that the specificity of the TCR on a given thymocyte determines if CD4 or CD8 will ultimately be expressed. Thus, thymocytes whose TCRs are specific for class I MHC and peptide will ultimately express

BOX 8–1. Genetically Manipulated Mouse Models for the Study of T Cell Maturation

A majority of the recent experimental approaches to study T cell maturation in the thymus have utilized mice that express TCR transgenes and/or have targeted mutations in genes encoding MHC, accessory, or signaling molecules. These approaches have provided most of our current information on the processes of positive and negative selection.

Following thymic selection processes in normal mice is very difficult because only a small percentage of developing thymocytes will express TCRs that recognize a particular peptide-MHC complex. This problem is overcome by TCR transgenic mice that are created by injecting mouse embryos with functionally rearranged TCR α and β chain genes derived from a clonal T cell line with known peptide and MHC specificity. (See Box 4–1, Chapter 4, for a description of how transgenic mice are made.) The TCR transgenes are expressed appropriately only in developing and mature T cells, and virtually all the T cells in these mice express the transgenic $\alpha\beta$ TCR. Because there is little allelic exclusion of α chain gene rearrangements (see text), many T cells in TCR transgenic mice will also express receptors composed of the transgenic β chain and an endogenously encoded α chain. (These endogenous α chains can be eliminated by breeding the TCR transgene into mice incapable of rearranging their own TCR genes, such as RAG-1 or RAG-2 knockout mice.) The vast majority of TCR-expressing T cells in these transgenic mice will recognize the peptide-MHC complexes that are recognized by the T cell clone from which the TCR genes were originally isolated. Therefore, the consequences of thymocyte recognition of this peptide-MHC complex *in vivo* can easily be followed because large numbers of cells will be affected.

Homologous recombination-mediated disruption of genes encoding various molecules has also been invaluable for the study of T cell development. (See Box 4–1, Chapter 4, for a description of how these "gene knockout" mice are produced.) Two general categories of knockout mice that have been used in the study of T cell maturation are mice deficient in expression of cell surface molecules required for T cell antigen recognition and mice lacking expression of signaling molecules. In the former category are mice that lack class I or class II MHC, CD4, CD8, or TCR α, β, γ, or δ chains. In the latter category are mice deficient in lck, fyn, jak-3, Icarus, and ZAP-70. The phenotypes of some of these mice are described in the text.

Particularly powerful experimental models can be created by developing mice that lack expression of one gene and express another transgene. These mice are derived by breeding transgenic lines with gene knockout lines and selecting progeny with the desired genetic traits. There are many examples of mice expressing TCR transgenes on a gene-knockout background. For example the importance of coreceptors on the positive selection of MHC-restricted T cells has been analyzed in mice expressing a class II MHC–restricted transgenic TCR but lacking a functional CD4 gene. (In this case, very few T cells survived positive selection.) Studies with β_2 microglobulin–deficient, class I MHC–restricted TCR transgenic mice are referred to in the text.

It should be pointed out that many of these mouse models have been used for the study of peripheral T cell function as well. For example, some TCR transgenic mice are excellent sources of naive mature T cells because the antigen for which they are specific is not likely to be encountered by the animal in normal situations. Furthermore, TCR transgenic animals have been instrumental in the study of mechanisms of peripheral tolerance. MHC-deficient mice are routinely used in studies of allograft rejection. These, and many other examples, are described in other chapters of this book.

---

only CD8, whereas thymocytes whose TCRs are class II MHC–restricted will mature into cells expressing only CD4. The possible mechanisms underlying this process will be discussed later in the chapter.

This sequence of T cell maturation—from CD4$^-$CD8$^-$TCR$^-$ to CD4$^+$CD8$^+$TCR$^+$ to CD4$^+$CD8$^-$TCR$^+$ or CD4$^-$CD8$^+$TCR$^+$ cells—has been most clearly established in mice. It is difficult to examine the maturation pathway of T cells during fetal life in humans because of many obvious limitations. The information that is available suggests that the same sequence of maturation events occurs in humans as in mice. It is clear that in both mice and humans $\alpha\beta$ TCRs are first expressed on double-positive cortical thymocytes. This is when the selection processes that determine the specificities of the mature T cell repertoire begin to occur.

THYMIC SELECTION PROCESSES LEADING TO SELF MHC RESTRICTION AND SELF-TOLERANCE OF T CELLS

The next important event in intrathymic T cell maturation, after expression of the TCR, CD4, and CD8 molecules, is the selection of cells that will make up the repertoire of mature T cells in the periphery. Selection processes are responsible for the survival of T cells that express only "useful" TCRs, i.e., TCRs that recognize peptides derived from foreign proteins in association with self-allelic forms of MHC molecules. Although the cellular and biochemical mechanisms of selection are not well understood, the current model of this process has the following general features (Fig. 8–8).

1. Before selection can begin, TCRs must be expressed on the developing thymocytes. This is because thymocytes are selected to survive or to be eliminated on the basis of their specificities, i.e., the ability of their TCRs to recognize MHC molecules and peptide antigens expressed in the thymus.

2. Because of the random nature of the molecular events in TCR gene expression, all possible specificities may be represented in the TCR-expressing, preselected, immature thymocyte population. In each individual, many of these TCRs may be incapable of recognizing self MHC molecules and would therefore be of no use. Other TCRs may be able to bind complexes of self peptides and self MHC with high affinity and therefore could be harmful.

3. **Positive selection** is the process in which thymocytes whose TCRs bind with low avidity to self MHC molecules complexed with self or foreign

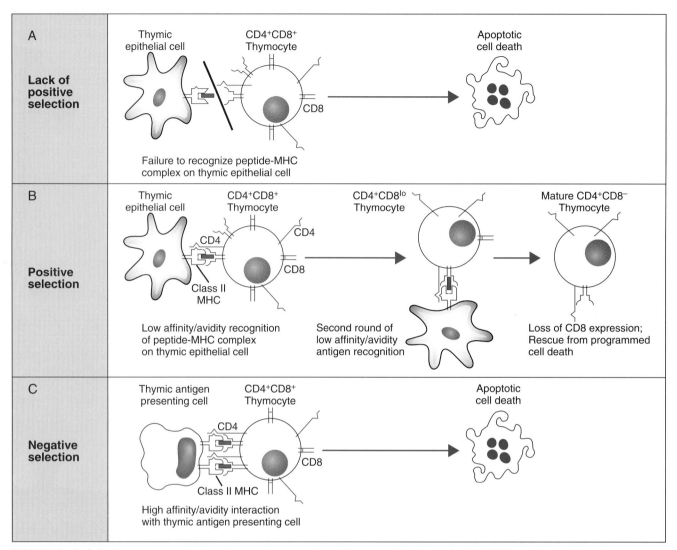

FIGURE 8–8. Selection processes during thymocyte maturation. Three possible fates of a CD4+CD8+ double-positive thymocyte are depicted. If the double-positive thymocyte fails to recognize a peptide–self major histocompatibility complex (MHC) on a thymic epithelial cell, it is not positively selected and undergoes programmed cell death (A). If the thymocyte engages in low affinity/avidity recognition of peptide–self MHC complexes on a thymic epithelial cell, it is positively selected and rescued from programmed cell death (B). If the thymocyte binds peptide–self MHC complexes on any thymic antigen-presenting cell with high affinity/avidity, it is induced to undergo apoptotic cell death (C).

peptide are permitted to survive while all those thymocytes that have no ability to bind self MHC die. This step eliminates all non-self MHC–restricted T cells which would be incapable of recognizing antigen in the periphery because the APCs obviously only express self MHC molecules. Positive selection leaves useful foreign antigen-specific, self MHC–restricted T cells as well as potentially harmful self antigen-specific, self MHC–restricted T cells.

4. During **negative selection,** clones of thymocytes whose TCRs bind with high affinity to self peptide antigens in association with self MHC molecules are eliminated (clonal deletion) or inactivated (clonal anergy). After this step, the functional thymocytes that remain are enriched for those whose TCRs bind with low affinity to self peptides plus self MHC molecules. Fortuitously, this population will include cells whose TCRs also

have high affinity for foreign peptides plus self MHC molecules. Therefore, negative selection helps ensure that the mature T cell repertoire is self MHC–restricted and tolerant to self protein antigens present in the thymus. Tolerance to self antigens present exclusively in peripheral tissues depends on other mechanisms, as discussed in Chapter 19. It is commonly assumed that negative selection follows positive selection, but it is possible that some thymocytes are negatively selected before they get a chance to be positively selected.

Positive Selection Processes in the Thymus: Development of the Self MHC–Restricted T Cell Repertoire

Positive selection is the process that results in mature peripheral T lymphocytes expressing only self MHC–restricted TCRs. Once a thymocyte expresses a T cell receptor, it cannot change its specificity. Therefore, *positive selection works by promoting the selective survival and expansion of thymocytes with self MHC–restricted TCRs and permitting thymocytes whose TCRs are not self MHC–restricted to die.* In this portion of the chapter, we will discuss the current understanding of the process of positive selection.

THE ROLES OF MHC MOLECULES, PEPTIDES, CD4, AND CD8 IN POSITIVE SELECTION

Positive selection of $CD4^+CD8^+TCR\alpha\beta^+$ thymocytes is most likely due to weak recognition of self peptide–MHC complexes presented by thymic epithelial cells. Weak recognition occurs when the avidity of the TCR interaction with peptide-MHC complexes is low. Avidity is determined by both the TCR affinity for its peptide-MHC ligand and the number of TCR-peptide-MHC ligand pairs formed at the time of antigen recognition. In contrast, most data suggest that *negative selection of thymocytes is due to strong or high avidity recognition of peptide-MHC complexes* (discussed later). Thus, the fate of a thymocyte depends on the number of TCR molecules it expresses, the number of peptide-MHC molecules on the thymic stromal cells it contacts, and the affinity of the TCR for these peptide-MHC complexes. In addition, the density of coreceptor molecules, CD4 or CD8, on the thymocyte can influence the strength of signals generated by the TCR (see Chapter 7) and therefore influence if thymocyte antigen recognition leads to positive or negative selection. Our current understanding of positive selection comes from a wide range of experimental approaches with inbred strains of animals, including, more recently, transgenic and gene knockout mice. Since the result of positive selection is to create a self MHC–restricted population of T cells, the first experiments to shed light on the process of positive selection were focused on the genetic and anatomic determinants of self MHC restriction. These are summarized below.

1. *T cells are self MHC–restricted because they encounter self MHC molecules during maturation.* Bone marrow chimeras were used to first demonstrate that MHC genes of the host animal in which T cells develop determine the MHC restriction pattern of the mature T cells. In these experiments, chimeras were created by transferring the bone marrow–derived hematopoietic stem cells from a mouse expressing two different MHC haplotypes (each inherited from a different parent) into a lethally irradiated mouse expressing only one of the two haplotypes (Table 8–2). The MHC restriction of the mature T cells in these mice was then determined, and it was found that they recognized antigen only in association with MHC molecules of the host strain.

TABLE 8–2.   Development of Major Histocompatibility Complex (MHC) Restriction in Bone Marrow Chimeric Mice

		Specific Killing of Virus-Infected Targets from	
Marrow Donor Strain	Host	*Strain A*	*Strain B*
(A × B)F1	(A × B)F1	+	+
(A × B)F1	A	+	−

Bone marrow chimeras are created as described in Chapter 2 (Fig. 2–2). Strain A and B mice have different class I MHC alleles. The MHC restriction specificity of mature T cells in these mice is tested by assaying the ability of cytolytic T lymphocytes generated in response to viral infection to kill virus-infected target cells from different mouse strains *in vitro.* These experiments demonstrate that the host MHC type, and not the bone marrow donor type, determines the restriction specificity of the mature T cells.

2. *The thymus is the critical host element for the development of the MHC restriction patterns of T cells.* This was demonstrated by creating bone marrow plus thymic chimeras. These host animals were prepared by lethally irradiating adult mice of one MHC haplotype, removing their thymuses, and transplanting into them different combinations of bone marrow and thymuses from mice with other MHC haplotypes (Table 8–3). The transplanted thymuses were either irradiated or treated with cytotoxic drugs (such as deoxyguanosine) to kill all resident macrophages, dendritic cells, and lymphoid cells, leaving only the radioresistant thymic epithelial cells. Again, mature T cells developed in these mice, and their MHC restriction was examined after immunization and *in vitro* restimulation of T cells with antigen plus APCs. Antigen recognition by these T cells was always restricted by MHC gene products expressed on the transplanted thymic epithelial cells and not necessarily by MHC gene products expressed on extrathymic cells (Table 8–3).

The T cell repertoire becomes self MHC–restricted in the thymus because only thymocytes that express TCRs which can recognize self MHC are positively selected to survive. T cell receptor transgenic mice have provided the clearest evidence that $CD4^+CD8^+TCR\alpha\beta^+$ thymocytes are positively selected by recognition of self MHC molecules in the thymus. The creation of TCR transgenic mice is described in Box 8–1. It is relatively simple to study the selection and maturation of T cells in these mice, since virtually all developing T cells express the transgenically encoded TCR. Several key features of positive selection have been demonstrated with these mice.

1. *Double-positive thymocytes must express TCRs that can recognize self MHC molecules in order to be positively selected and differentiate into single positive cells.* This was shown by expressing a

TABLE 8–3. Development of MHC Restriction Patterns in Bone Marrow Plus Thymic Chimeras

Chimera			Stimulation of Mature T Cells		
Host	*Bone Marrow Donor*	*Thymus Donor*	*Antigen*	*Strain of APC*	*Response*
A	(A × B)F1	A	+	A	+
A	(A × B)F1	A	+	B	−
A	(A × B)F1	A	−	A	−
A	(A × B)F1	B	+	B	+
A	(A × B)F1	B	+	A	−
A	(A × B)F1	B	−	B	−

In the chimeric mice, the mature T cells that develop respond to foreign antigen in association with the MHC allele expressed on the thymic epithelial cells.

Abbreviations: APC, antigen-presenting cell; MHC, major histocompatibility complex.

transgenic $\alpha\beta$ TCR with a defined MHC restriction in mice with different MHC haplotypes (Fig. 8–9). When the transgenic TCR is expressed in a mouse which also expresses the MHC allele for which the TCR is specific, T cells will mature and populate peripheral lymphoid tissue. If the mouse is of another MHC haplotype and does not express the MHC molecule that the transgenic TCR recognizes, very few transgenic TCR-expressing T cells mature. In the thymuses of these mice that cannot positively select T cells, there are normal numbers of CD4+CD8+ thymocytes but very few CD4+ or CD8+ single-positive cells. This result demonstrates that developing T cells must express TCRs that can bind self MHC molecules in order to survive. As we shall see later, the same experimental system has been used to study negative selection.

2. *Thymic epithelial cells must express class I or class II MHC molecules in order to positively select double-positive thymocytes to become CD8+ or CD4+ single-positive cells, respectively.* This has been demonstrated using strains of mice that are deficient in either class I or class II MHC using homologous recombination-targeted disruption of the appropriate genes (see Box 4–1, Chapter 4, and Box 8–1). Mice that lack class I MHC expression because of disruption of the β_2 microglobulin gene do not develop mature CD8+ T cells, but do develop CD4+CD8+ thymocytes and mature CD4+ T cells. Conversely, class II MHC–deficient mice do not develop CD4+ T cells but do develop CD8+ cells. Since CD8+ T cells are class I MHC–restricted and CD4+ T cells are class II MHC–restricted, these findings are entirely consistent with a need for TCR and coreceptor recognition of MHC in the thymus in order for MHC-restricted T cells to mature.

Peptides bound to MHC molecules on thymic epithelial cells play an essential role in positive selection. In Chapters 5 and 6, we described how cell surface class I and class II molecules almost always contain bound peptides. It is, of course, not possible that the full range of foreign antigenic peptides to which an individual can respond are present in the thymus. Therefore, foreign peptides cannot be involved in the positive selection of T

cells which ultimately may recognize these peptides. Self peptides are required for positive selection, however, and exactly what role they are playing has been the subject of intense investigation using a variety of experimental systems. Although there are few definitive conclusions that can be made at this time, the following interpretations of experimental data are widely accepted.

1. *Self peptides are required to enhance MHC stability and/or conformation, which favors low-avidity TCR recognition, but specific TCR interactions with the peptide itself are probably less important.* In order to address this issue, it was necessary to develop systems where the array of peptides presented to developing thymocytes was limited and controlled experimentally. For class I MHC–dependent positive selection, this was accomplished using TAP-1 or β_2 microglobulin–deficient mice. Thymic epithelial cells in these mice express unstable class I MHC molecules on their surfaces without peptides. These MHC molecules can be stably loaded with defined peptides by culturing thymic lobes in medium containing the peptides and exogenous β_2 microglobulin. Using these systems, it was found that addition of peptides could dramatically increase the yield of single-positive CD8+ thymocytes from the thymuses, implying effective positive selection of double-positive thymocytes expressing class I MHC–restricted TCRs. A key finding in these experiments is that one or a few peptides can result in the positive selection of a large repertoire of CD8+ T cells which can potentially recognize many different unrelated peptides. Similarly, mice genetically engineered to express only a single class II MHC allele with a single (covalently) bound peptide were able to develop a large population of CD4+ class II MHC–restricted T cells with specificities for many different foreign peptides. These data are consistent with the interpretation that peptides play a relatively nonspecific role in positive selection of T cells.

2. *Some peptides are better than others in supporting positive selection, suggesting that the structural features of some peptides are important in promoting low-avidity interactions between thymocyte TCRs and self MHC molecules.* This conclusion is

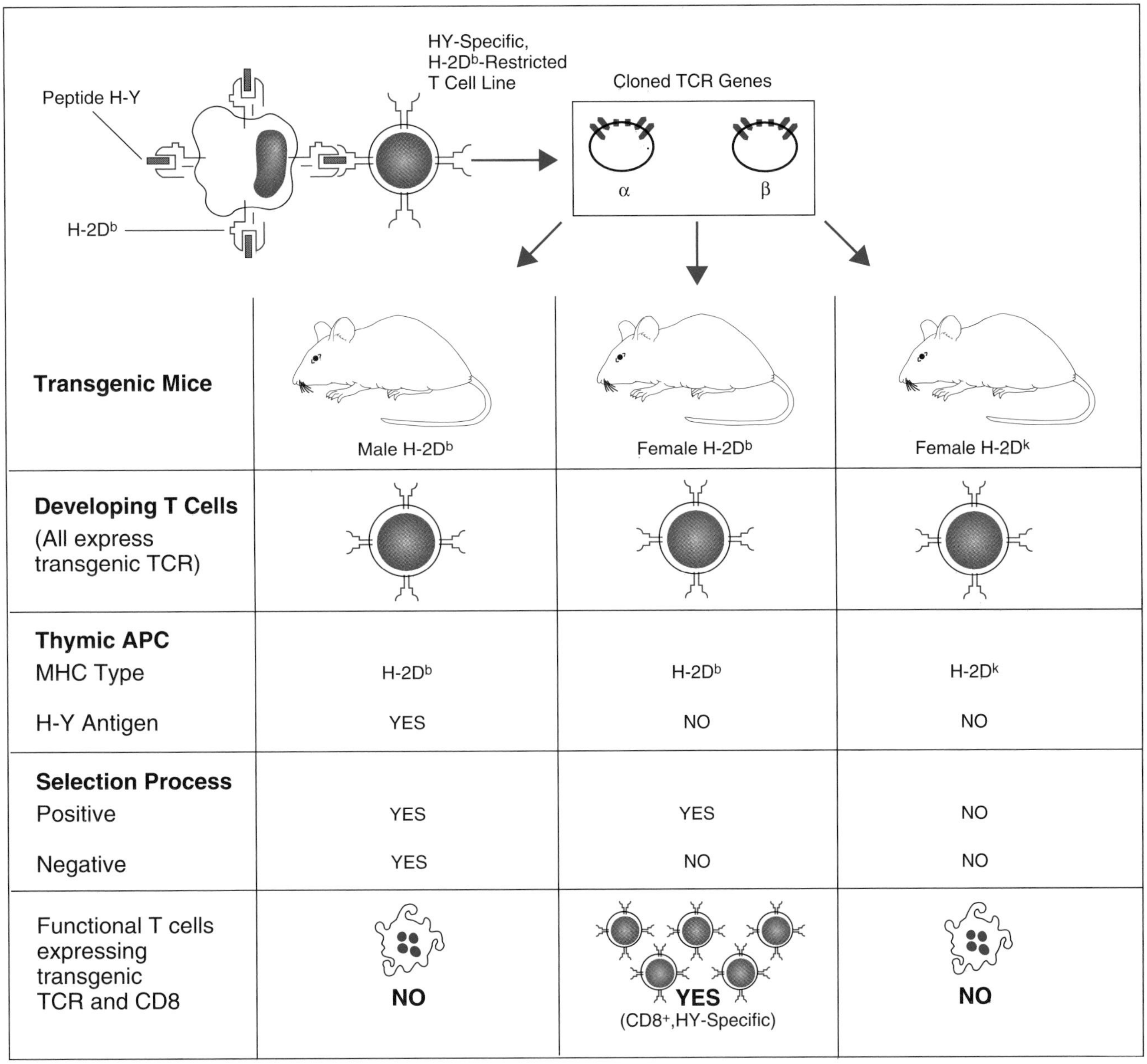

FIGURE 8–9. The use of T cell receptor (TCR) transgenic mice to study thymic selection processes. The transgenic mice express a TCR specific for an H-2D^b-associated H-Y (male-specific) antigen, which is, in effect, a foreign antigen for a female mouse. Functional T cells expressing the TCR transgenes mature in female mice expressing the D^b allele but not in H-Y male mice (because of negative selection as a result of self antigen recognition) or in female mice not expressing D^b (because of lack of positive selection of the D^b-restricted TCR-expressing cells). APC, antigen-presenting cell; MHC, major histocompatibility complex.

based on experiments on class I MHC–mediated positive selection, as described above. For example, structurally diverse mixtures of peptides are better than less diverse mixtures in supporting positive selection of CD8^+ T cells in TAP-1 or β_2 microglobulin–deficient thymic organ cultures. Similarly, some single peptides can drive positive selection while other peptides cannot. It is not known what structural features may be common to positively selecting peptides. Furthermore, most data argue against the possibility that thymic epithelial cells produce positively selecting peptides while other cell types do not. One interesting possibility is that peptides that support positive selec-

tion are partial T cell agonists similar to the altered peptide ligands discussed in Chapter 7. Such peptides might be expected to induce weak, or qualitatively different, signals in thymocytes than the peptides that do not support positive selection or that induce negative selection of thymocytes.

CD4 and CD8 molecules are required for efficient positive selection of class II– and class I–restricted thymocytes, respectively. The importance of CD4 and CD8 in positive selection has been unequivocally demonstrated by experiments with mice in which CD4–class II MHC interactions or CD8–class I MHC interactions have been eliminated. Blockade of CD8 or CD4 *in vivo* by noncytotoxic antibodies that do not deplete thymuses of double-positive cells results in a lack of development of CD8⁺ or CD4⁺ single-positive cells, respectively. Furthermore, in CD4 or CD8 gene knockout mice, there is no mature CD4⁻CD8⁻TCR $\alpha\beta^+$ population, implying that cells that normally would have matured from double-positive into either CD4 or CD8 single-positive T lymphocytes do not get positively selected in the absence of CD4 or CD8. It is likely that CD4 and CD8 function as coreceptors with TCRs during positive selection, providing both adhesive interactions with MHC molecules and signaling functions. This would be analogous to the roles of CD4 and CD8 in the activation of mature T cells (see Chapter 7).

During the transition from double-positive to single-positive thymocytes, cells with class I–restricted TCRs become CD4⁻CD8⁺, and cells with class II–restricted TCRs become CD4⁺CD8⁻. This is most convincingly demonstrated in TCR transgenic mice (see Box 8–1). For example, mice expressing a class II MHC–restricted TCR develop almost exclusively CD4⁺ mature T cells even though there are normal numbers of CD4⁺CD8⁺ thymocytes. Most experimental evidence suggests that the pairing of coreceptor and TCR expression is a stochastic/selective process in which there is a random downregulation of either CD4 or CD8 expression, and positive selection leads to the propagation of the cells on which the MHC specificity of coreceptor and TCR matches. Careful analysis of CD4 and CD8 expression during T cell maturation indicates that there is an intermediate stage between the double-positive and single-positive stages, in which one coreceptor is expressed at lower levels than the other. For example, thymocytes with class I MHC–restricted TCRs progress from CD4⁺CD8⁺ to CD4^lo CD8⁺ to CD4⁻CD8⁺, and thymocytes expressing class II MHC–restricted TCR progress from CD4⁺CD8⁺ to CD4⁺CD8^lo to CD4⁺CD8⁻. The stochastic/selective model suggests that there are at least two rounds of positive selection by thymic epithelial cells (see Fig. 8–8). The first round occurs in the double-positive stage, promoting survival and perhaps inducing the random downregulation of CD4 or CD8 gene expression. During a second round of selection, any developing T cell that has the right combination of CD4 preservation with a class II MHC–restricted TCR, or CD8 with a class I–restricted TCR, will be effectively stimulated by MHC-expressing thymic epithelium and will survive. If the wrong combination exists, the cell will not be effectively stimulated, and it will die by default, just like thymocytes with TCRs that have no affinity for any self MHC–peptide complex.

An alternative model for the appropriate pairing of CD4 and CD8 with class I– and class II–restricted thymocytes is called the instructive model. It proposes that when the TCR and CD4 or CD8 coreceptor on a double-positive thymocyte bind to a peptide-MHC complex on a thymic epi-

thelial cell, signals are generated that instruct the thymocyte and its progeny to turn off expression of the "wrong" accessory molecule. This suggests that CD4 and CD8 generate distinct biochemical signals in thymocytes, but there are no data to support this. A variety of transgenic and knockout mouse strains have been used to address which model of subset differentiation is accurate, and to date the stochastic model appears most likely to be correct.

CELLULAR MECHANISMS OF POSITIVE SELECTION

It is hypothesized that the *cellular basis of positive selection is the rescue of a thymocyte from a default pathway of death.* Once TCR expression is induced on double-positive thymocytes, they can survive about 3 days without TCR stimulation. Thymocytes whose TCRs have no affinity for self MHC will die, and the survivors will have TCRs with some affinity for self MHC. In fact, a large proportion of CD4⁺CD8⁺ thymocytes in the thymic cortex die by apoptosis, and many of these die because they were not positively selected. It is likely that the signals generated by the TCR complex that rescue a thymocyte from death during positive selection are similar to those involved in T cell activation in mature T cells, but it is not clear which signals may be critical. Although there has been great progress in characterizing cell death pathways in mature lymphocytes, little is known about the molecular mechanisms that promote or inhibit thymocyte apoptosis.

In addition to preventing cell death, *positive selection extinguishes further TCR α chain gene rearrangements in double-positive thymocytes.* This is due to the induced inhibition of RAG-1 and RAG-2 gene expression. (TCR β chain gene rearrangements are shut off earlier by pTαβ receptor signals, as discussed previously.) In addition, TdT expression is downregulated by positive selection, so that N-region diversification of rearranged VJ joins is also brought to a halt. Multiple rounds of V-J recombination can occur during the 3-day life span of a preselected double-positive thymocyte. Therefore, it is possible that a single thymocyte or its progeny may sequentially express several different α chains and go through several attempts at positive selection before either failing and dying or expressing a TCR which has some affinity for self MHC molecules expressed on thymic epithelial cells. Once the right TCR is expressed and engages a self MHC molecule, RAG and TdT gene expression ceases, and the specificity of the TCR of that thymocyte is fixed.

It is clear from chimera experiments that thymic epithelial cells are the most efficient stromal cells for inducing positive selection. At the present time, we do not know what characteristic(s) of the thymic epithelial cell make it uniquely capable of inducing the various responses discussed above. Several other cell types in the thymus express

MHC molecules and can function as APCs. It is, therefore, likely that the thymic epithelial cell alone expresses additional cell surface molecules that engage receptors on the double-positive thymocyte. These receptor-ligand interactions are hypothesized to generate signals that act in concert with TCR-generated signals to induce positive selection.

Negative Selection Processes in the Thymus: Development of Self-Tolerance

The importance of self-tolerance to the health of all individuals has been mentioned previously, and the development of autoimmunity as a result of the breakdown of self-tolerance will be described in Chapter 19. Individuals may be tolerant to self antigens because they lack B and/or T lymphocytes specific for these antigens, or because such lymphocytes are present but cannot respond to self antigens. It is now clear that *tolerance to self proteins is largely because of T cell tolerance, and one mechanism for inducing T cell tolerance is the deletion of self-reactive clones of T cells during their maturation in the thymus.* This process is called **negative selection** and is important for maintaining tolerance to self proteins which are present in the thymus, such as ubiquitous membrane and serum proteins. The inability to mount T cell responses to these self antigens as a result of thymic selection is often called **central tolerance.** Thymic negative selection cannot be important for tolerance to proteins exclusively expressed in peripheral tissues, so-called **peripheral tolerance.** The mechanisms that account for peripheral tolerance are discussed in Chapter 19. In addition to the lack of positive selection, *much of the cell death that occurs in the thymic cortex is postulated to be due to deletion of self-reactive clones.*

Negative selection in the thymus is due to high-avidity interactions between double-positive thymocytes and self peptide–MHC complexes expressed on thymic APCs. These high-avidity interactions can be a consequence of TCRs with high affinity for self peptide-MHC complexes. Alternatively, even moderate- or low-affinity receptors may engage in high-avidity interactions if the density of the specific peptide-MHC ligand on the thymic APCs is high. The latter case would be most likely to occur when the peptide is derived from a protein that is abundant in the thymus. Common serum or cellular constituents such as albumin or hemoglobin are examples of such proteins. Formal proof for clonal deletion of T cells reactive with antigens abundantly expressed in the thymus came from three experimental approaches that allowed investigators to observe the effects of self antigen recognition by a large number of developing T cells.

1. *When TCR transgenic mice are exposed to the peptide for which the TCR is specific, there is a large amount of cell death induced in the thymus and* *a block in development of mature transgenic TCR-expressing T cells.* For example, in one such study, a mouse line was derived that expressed a class I MHC–restricted transgenic TCR specific for the sex-associated H-Y molecule; H-Y is a self antigen abundantly expressed on many cell types in male mice, but not in female mice (Fig. 8–9). Female mice with this transgenic TCR had normal numbers of T cells in the periphery and normal numbers of thymocytes in the thymic medulla. In contrast, male transgenic mice had few mature peripheral T cells, and few TCR-expressing, single-positive thymocytes in the medulla. There was a block of T cell maturation after the cortical double-positive stage, presumably because all the maturing thymocytes expressed the transgenic TCR, recognized the self H-Y antigen on thymic APCs, and were deleted before they could mature into single-positive cells. Clonal deletion of thymocytes has also been demonstrated by culturing intact fetal thymuses from TCR transgenic mice *in vitro* and adding the peptide for which the TCR is specific.

2. *Mice that express an endogenously coded superantigen delete developing T cells which react with that superantigen.* Superantigens are proteins that bind to TCRs that utilize certain V_β genes irrespective of which D_β and J_β gene segments are used or which α chains are present (see Box 16–1, Chapter 16). These proteins also bind to class II MHC molecules outside the peptide-binding groove and when they are presented to T cells they stimulate responses similar to those induced by peptide-MHC complexes. Because superantigens stimulate large numbers of cells, their effects on T cell development may be analyzed to study negative selection. For example, one endogenous superantigen in mice binds exclusively to the I-E class II MHC molecule and reacts with all TCRs using a $V_\beta17a$-encoded V_β chain. In strains of mice that express I-E class II MHC molecules, there are virtually no $V_\beta17a$-expressing T cells in the thymic medulla or in peripheral lymphoid organs, but there are $V_\beta17a$-expressing thymocytes in the cortex. In contrast, strains of mice that do not express I-E molecules, because of a genetic defect in the class II MHC locus, have readily detectable $V_\beta17a$-expressing T cells in the periphery (Table 8–4). This suggests that as T cells mature in the thymus of an I-E–expressing mouse, all $V_\beta17a^+$ thymocytes encounter the self-superantigen complexed with I-E molecules and are deleted. If they do not encounter this superantigen–I-E complex, as would be the case in the I-E–deficient mice, they may mature normally.

3. The concept that peptide-specific high-avidity interactions between thymocytes and thymic epithelial cells lead to negative selection has been established by experimental systems similar to those used to demonstrate that low-avidity interactions lead to positive selection. For example, class I MHC–restricted TCR transgenes have been bred into β_2 microglobulin (class I MHC)–deficient mice,

TABLE 8–4. Clonal Deletion of V_β17a-Positive T Cells in I-E–Expressing Mice

Mouse Strain	I-E Expression	Peripheral T Cells Expressing TCRs Utilizing V_β17a (Per Cent)
A. SWR	None	14.2
SJL	None	9.4
SJA	None	8.5
B. C57BR	Yes	0.1
BALB/c	Yes	0
AKR	Yes	0
B10.TL	Yes	0

Mature T cells with V_β17a-containing TCRs are present in I-E⁻ (A) but not in I-E⁺ (B) inbred mouse strains, because these TCRs recognize an I-E–associated self antigen and are deleted in I-E⁺ mice.

Abbreviations: TCR, T cell receptor. Adapted from Kappler, J., N. Roehm, and P. Marrack. T cell tolerance by clonal elimination in the thymus. Cell 49:273, 1987. Copyright © by Cell Press.

and selection processes driven by a specific peptide have been followed using thymic organ cultures, as described previously. In this system, low doses of the peptide for which the transgenic TCR is specific would permit positive selection, but high doses would cause deletion of double-positive cells and no maturation of single-positive CD8⁺ T cells.

The clonal deletion of self-reactive T cells occurs when the TCR on a CD4⁺CD8⁺ thymocyte binds to a self peptide presented by another thymic cell. Most evidence indicates that any thymic APC can induce clonal deletion, including bone marrow–derived macrophages and dendritic cells, and thymic epithelial cells. As with positive selection, the CD4 and CD8 molecules likely play a role in negative selection because they promote effective interactions between the developing thymocytes and the "tolerizing" thymic APCs. For example, *in vivo* administration of anti-CD4 antibody to I-E–expressing mice blocks the elimination of V_β17a-expressing CD4⁺CD8⁺ thymocytes. Thus, the interaction of both CD4 and V_β17a TCR with the I-E class II MHC molecule is necessary for deletion of the V_β17a-expressing cells. Furthermore, overexpression of CD8 as a transgene in a TCR transgenic mouse can promote negative selection of thymocytes, even by low doses of peptide antigen which would not delete thymocytes with normal levels of CD8.

The cellular basis of clonal deletion in the thymus is the induction of death by apoptosis. Unlike the programmed cell death which occurs in the absence of positive selection, in negative selection active death-promoting signals are generated by the thymocyte TCR. In this way, negative selection may be similar to activation-induced cell death which plays a role in deletion of self-reactive mature T cells outside the thymus. Unlike peripheral deletion mechanisms, however, thymic negative selection does not appear to depend on Fas-Fas ligand interactions because it proceeds normally in mutant mice lacking these molecules (see Chapter 10). It is known that double-positive thymocytes are more sensitive to antigen than are peripheral mature T cells and that a concentration of peptide antigen incapable of activating a mature T cell will often induce apoptosis in a thymocyte. Furthermore, there may be fundamental differences in TCR complex signaling between thymocytes and mature T cells. For example, $\alpha\beta$ TCR heterodimers and the associated CD3 proteins appear not to be functionally coupled in immature thymocytes, since anti-$\alpha\beta$ antibodies induce increases in cytoplasmic calcium in mature T cells but not double-positive thymocytes, whereas anti-CD3 antibodies induce calcium fluxes in both thymocytes and mature T cells. This suggests that the qualitative signals generated by peptide-MHC binding to the TCR in a double-positive thymocyte may be distinctly different from those generated in a mature T cell, and these differences may be the basis for the differences in the outcome of antigen recognition in the two cell populations. There is also some evidence that thymocyte recognition of peptide-MHC complexes may induce anergy rather than deletion, although the mechanisms by which this happens or the general contribution of the phenomenon to self-tolerance are not clear.

SUMMARY

Stem cells with the potential to develop into T cells first arise in the bone marrow and migrate to the thymus during both fetal and adult life. The earliest T lineage immigrants to the thymus have unrearranged TCR genes and do not express CD4 or CD8 molecules. The developing T cells within the thymus, called thymocytes, initially populate the outer cortex, where they undergo population growth, rearrangement of TCR genes, and surface expression of CD3, TCR, CD4, and CD8 molecules. As they mature, they migrate from cortex to medulla. Different stages of maturation can be identified on the basis of sequential changes in CD4 and CD8 expression. CD4⁻CD8⁻ (double-negative) cells mature into CD4⁺CD8⁺ (double-positive) cells which develop into CD4⁺CD8⁻ or CD4⁻CD8⁺ single-positive cells. The TCR α and β polypeptides are encoded by functional genes that are created only in T cells by the somatic rearrangement of variable, diversity (β only), and joining gene segments, bringing them in the vicinity of C gene segments. Multiple combinatorial possibilities for the joining of the different gene segments, as well as several mechanisms for junctional diversity, result in the generation of a large repertoire of T cell specificities. Unlike Ig genes, there is no somatic mutation or affinity maturation in TCR genes. The functional genes encoding the TCR γ and δ polypeptides are also formed by somatic rearrangement of germline genes. The mechanisms of generation of TCR

$\gamma\delta$ diversity are similar to those described for the $\alpha\beta$ receptor, except that there are fewer V genes in the γ and δ loci and significantly more potential junctional diversity. TCR $\gamma\delta$ receptors are the first receptors to be expressed on a small subset of cortical thymocytes, followed by the more abundant expression of $\alpha\beta$ TCR receptors on a distinct lineage of developing T cells. TCR β chains are expressed before α chains, and the β chain pairs with an invariant preTα (surrogate α) chain. The pT$\alpha\beta$ receptor transduces signals that promote CD4 and CD8 expression and further proliferation of immature thymocytes. The first $\alpha\beta$ expressing cells are CD4$^+$CD8$^+$ cortical thymocytes. These cells interact with MHC-expressing cortical epithelial cells and bone marrow–derived, non-lymphoid cells and undergo selection processes that shape the T cell repertoire toward self MHC restriction and self-tolerance. Positive selection of CD4$^+$CD8$^+$TCR$\alpha\beta$ thymocytes requires low-avidity recognition of peptide-MHC complexes on thymic epithelial cells, leading to a rescue from programmed death. Central T cell tolerance is mainly due to apoptotic death of CD4$^+$CD8$^+$TCR $\alpha\beta^+$ thymocytes and involves high-avidity recognition of self peptide-MHC complexes on thymic APCs. Most of the cortical thymocytes do not survive these selection processes. As the surviving $\alpha\beta$ TCR thymocytes mature, they move into the medulla and become either CD4$^+$CD8$^-$ or CD4$^-$CD8$^+$. Medullary thymocytes acquire potential helper or cytolytic functional capabilities and finally emigrate to peripheral lymphoid tissues, where they reside as self MHC–restricted, foreign antigen–responsive helper T cells and pre-CTLs.

Selected Readings

Adkins, B., C. Mueller, C. Y. Okada, R. Reichert, I. L. Weissman, and G. J. Spangrude. Early events in T-cell maturation. Annual Review of Immunology 5:325–365, 1987.

Anderson, G., N. C. Moore, J. J. Owen, and E. J. Jenkinson. Cellular interactions in thymocyte development. Annual Review of Immunology 14:73–99, 1996.

Ashton-Rickardt, P. G., and S. Tonegawa. A differential-avidity model for T-cell selection. Immunology Today 15:362–366, 1994.

Blackman, M., J. Kappler, and P. Marrack. The role of the T cell receptor in positive and negative selection of developing T cells. Science 248:1335–1341, 1990.

Davis, M. M., and P. J. Bjorkman. T-cell antigen receptor genes and T-cell recognition. Nature 334:395–402, 1988.

Fowlkes, B. J., and D. M. Pardoll. Molecular and cellular events of T cell development. Advances in Immunology 44:207–264, 1989.

Jameson, S. C., K. A. Hogquist, and M. J. Bevan. Positive selection of thymocytes. Annual Review of Immunology 13:93–126, 1996.

Kappler, J. W., N. Roehm, and P. Marrack. T cell tolerance by clonal elimination in the thymus. Cell 49:273–280, 1987.

Leider, J. M. Transcriptional regulation of T cell receptor genes. Annual Review of Immunology 11:539–570, 1993.

Shortman, K., and L. Wu. Early T lymphocyte progenitors. Annual Review of Immunology 14:29–47, 1996.

Von Boehmer, H. Developmental biology of T cells in T cell-receptor transgenic mice. Annual Review of Immunology 8:531–556, 1990.

Von Boehmer, H., and P. Kisielow. Self-nonself discrimination by T cells. Science 248:1369–1373, 1990.

CHAPTER NINE

B CELL ACTIVATION AND

ANTIBODY PRODUCTION

Humoral immunity is mediated by antibodies, which are produced by cells of the B lymphocyte lineage. The physiologic function of antibodies is to neutralize and eliminate the antigen that induced their formation. The elimination of different antigens or microbes requires several effector mechanisms, which are dependent on distinct classes, or isotypes, of antibodies (see Chapter 3). The humoral immune system has the capacity to respond to different types of antigens by producing different classes of antibodies. For example, the antibody response to bacteria with polysaccharide-rich capsules consists largely of immunoglobulin (Ig) M, which activates the complement system, leading to opsonization and phagocytosis of the bacteria; the response to many viruses consists of high-affinity IgG antibodies, which block virus entry into host cells and also promote phagocytosis by macrophages; and the systemic response to many helminthic parasites is mainly IgE, which participates in eosinophil-mediated killing of the helminths. The humoral immune response is also different at various anatomic sites. For instance, mucosal lymphoid tissues are uniquely adapted to produce high levels of IgA in response to the same antigens that stimulate other antibody isotypes in non-mucosal lymphoid tissues.

Production of all these varied classes of antibodies is initiated by the interaction of antigens with a small number of mature IgM- and IgD-expressing B lymphocytes specific for each antigen. Mature antigen-responsive B lymphocytes develop in the bone marrow prior to antigenic stimulation. Such cells enter peripheral lymphoid tissues, which are the sites of interaction with foreign antigens. An antigen binds to the membrane IgM and IgD on specific B cells and initiates a series of responses that lead to two principal changes in that clone of B cells: **proliferation,** resulting in expansion of the clone, and **differentiation,** resulting in the progeny of the membrane Ig-expressing, antigen-responsive B cells actively secreting antibodies of different heavy chain isotypes or becoming memory cells (Fig. 9–1). The analysis of humoral immune responses is, in essence, an analysis of the growth and differentiation of B lymphocytes that occur following specific antigenic stimulation. The molecular mechanisms that regulate the expression of immunoglobulin genes have been described in Chapter 4. This chapter describes the

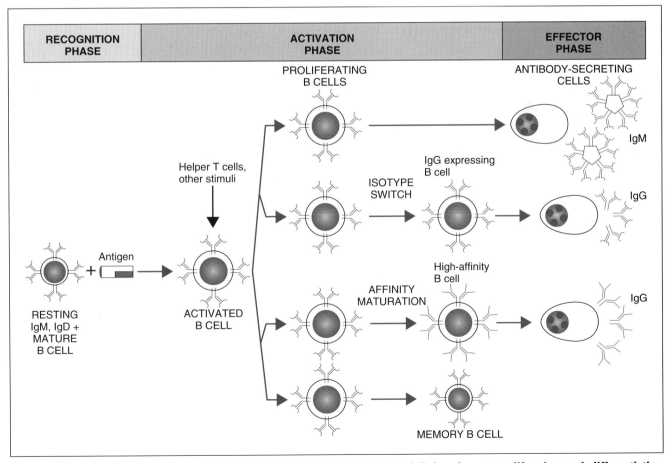

FIGURE 9–1. The humoral immune response: sequence of antigen-induced B lymphocyte proliferation and differentiation. Antigen and other stimuli, including helper T cells, stimulate the proliferation and differentiation of a specific B cell clone. Progeny of the clone may produce immunoglobulin (Ig) M or other Ig isotypes (e.g., IgG), may undergo affinity maturation, or may persist as memory cells.

cellular basis of the humoral immune response, in particular the stimuli that induce B cell growth and differentiation and the patterns of responses of B lymphocytes.

GENERAL FEATURES OF HUMORAL IMMUNE RESPONSES

The earliest studies of specific immunity were devoted to analyses of antibody responses. As a result, until the 1960s, much of our knowledge of the immune system was based on our understanding of humoral immunity. The most important general properties of antibody responses are the following:

1. Protein antigens do not induce antibody responses in the absence of T lymphocytes, for instance, in T cell–deficient individuals. For this reason, proteins are classified as "thymus-dependent" or "T-dependent" antigens. The T cells that are required for antibody production in response to protein antigens are called **helper T cells.** As we shall see later in this chapter, a great deal is now known about the mechanisms of action of helper T cells. The demonstration of their role in antibody responses is the basis of the concept that *naive B cells, which have not been previously exposed to antigen, require two distinct types of signals for their proliferation and differentiation.* One type of signal is provided by the antigen, which interacts with membrane Ig molecules on specific B cells. The second type of signal is provided by helper T lymphocytes and their secreted products. The "two-signal" requirement for lymphocyte activation applies to T lymphocytes as well, because antigen-specific helper and cytolytic T lymphocytes also need other stimuli, in addition to antigens, for their full growth and differentiation (see Chapters 7 and 13).

2. Non-protein antigens like polysaccharides and lipids induce antibody responses without a requirement for antigen-specific helper T lymphocytes. Therefore, polysaccharide and lipid antigens are called "thymus-independent" or "T-independent (TI)."

3. Antibody responses to T-independent antigens consist mainly of low-affinity IgM and some IgG antibodies, with little memory cell generation. In contrast, helper T cell–dependent responses to protein antigens are much more specialized, with different Ig isotypes, relatively high-affinity antibodies, and long-lived memory. The reason for this difference is that *heavy chain isotype switching, affinity maturation, and memory cell generation are all stimulated by helper T cells and their products,* and only protein antigens activate helper T cells.

4. *Primary and secondary antibody responses to protein antigens differ qualitatively and quantitatively* (Table 9–1). Primary responses result from the activation of previously unstimulated B cells, whereas secondary responses are due to stimulation of expanded clones of memory B cells. Therefore, the secondary response develops more rapidly than the primary response, and larger amounts of antibodies are produced in the secondary response. Heavy chain isotype switching and affinity maturation also increase with repeated exposures to protein antigens. These changes in the nature of antibody responses have many beneficial effects. The production of different Ig isotypes enables humoral immune responses to eliminate different types of microbes. Affinity maturation of antibodies makes an individual better able to combat persistent or recurrent infections and is also a reason why the optimal antigen doses required for secondary responses are lower than those for primary responses.

Humoral immune responses are initiated in peripheral lymphoid organs, such as the spleen for blood-borne antigens, draining lymph nodes for antigens entering via the skin or mucosal epithelia, and mucosal lymphoid tissues for some inhaled and ingested antigens. Our current understanding of the mechanisms of humoral immune responses is based on a variety of methods for studying B lymphocyte activation *in vitro* and *in vivo* (Box 9–1). In the following sections, we will first describe the signals generated by the binding of antigens to membrane Ig molecules on B cells and then the role of T cells in antibody production. We will return to T-independent antigens at the end of the chapter.

TABLE 9–1.   Features of Primary and Secondary Antibody Responses

	Primary Response	Secondary Response
Lag after immunization	Usually 5–10 days	Usually 1–3 days
Peak response	Smaller	Larger
Antibody isotype	Usually IgM > IgG	Relative increase in IgG and, under certain situations, in IgA or IgE
Antibody affinity	Lower average affinity, more variable	Higher average affinity ("affinity maturation")
Induced by	All immunogens	Only protein antigens
Required immunization	Relatively high doses of antigens, optimally with adjuvants (for protein antigens)	Low doses of antigens, adjuvants may not be necessary

Abbreviation: Ig, immunoglobulin.

BOX 9-1   Assays for B Lymphocyte Activation

It is technically difficult to study the effects of antigens on normal B cells because, as the clonal selection hypothesis predicted, very few lymphocytes in an individual are specific for any one antigen. In order to examine the effects of antigen binding to B cells, investigators have attempted to isolate antigen-specific B cells from complex populations of normal lymphocytes or to produce cloned B cell lines with defined antigenic specificities. The latter effort has met with little success. Recently, transgenic mice have been developed in which virtually all B cells express a transgenic Ig of known specificity, so that most of the B cells in these mice respond to the same antigen. Another approach to circumventing this problem is to use anti-Ig antibodies as analogs of antigens, with the assumption that anti-Ig will bind to constant regions of membrane Ig molecules on all B cells and have the same biologic effects as an antigen that binds to the hypervariable regions of membrane Ig molecules only on the antigen-specific B cells. To the extent that precise comparisons are feasible, this assumption appears generally correct, indicating that anti-Ig antibody is a valid model for antigens. Thus, anti-Ig antibody is frequently used as a polyclonal activator of B lymphocytes. The same concept underlies the use of antibodies against framework determinants of T cell receptors or against receptor-associated CD3 molecules as polyclonal activators of T lymphocytes (see Chapter 7).

Much of our knowledge of B cell activation is based on *in vitro* experiments, in which different stimuli are used to activate B cells, and their proliferation and differentiation can be measured accurately. The same assays may be done with B cells recovered from mice exposed to different antigens, or with homogeneous B cells expressing transgene-encoded antigen receptors. The most frequently used assays for B cell responses are the following.

ASSAYS FOR PROLIFERATION. The proliferation of B lymphocytes, like that of other cells, is measured *in vitro* by determining the amount of ³H-labeled thymidine incorporated into the replicating DNA of cultured cells. Thymidine incorporation provides a quantitative measure of the rate of DNA synthesis, which is usually directly proportional to the rate of cell division. Cellular proliferation *in vivo* can be measured by injecting the thymidine analog, bromodeoxyuridine (BrdU), into animals and staining cells with anti-BrdU antibody to identify and enumerate nuclei that have incorporated BrdU.

ASSAYS FOR ANTIBODY PRODUCTION. Antibody production is measured in two different ways: assays for cumulative Ig secretion, which measure the amount of Ig that accumulates in the supernatant of cultured lymphocytes or in the serum of an immunized individual, and single-cell assays, which determine the number of cells in an immune population that secrete Ig of a particular specificity and/or isotype. The most accurate, quantitative, and widely used techniques for measuring the total amount of Ig in a culture supernatant or serum sample are radioimmunoassay (RIA) and enzyme-linked immunosorbent assay (ELISA), described in Chapter 3. By using antigens bound to solid supports, it is possible to use RIA or ELISA to quantitate the amount of a specific antibody in a sample. In addition, the availability of anti-Ig antibodies that detect Igs of different heavy or light chain classes allows one to measure the quantities of different isotypes in a sample. Other techniques for measuring antibody levels include hemagglutination for anti-erythrocyte antibodies and complement-dependent lysis for antibodies specific for known cell types. Both assays are based on the demonstration that if the amount of antigen (i.e., cells) is constant, the concentration of antibody determines the amount of antibody bound to cells, and this is reflected in the degree of cell agglutination or subsequent binding of complement and cell lysis. Results from these assays are usually expressed as antibody titers, which are the dilution of the sample giving half-maximal effects or the dilution at which the end-point of the assay is reached.

A single-cell assay for antibody secretion that has been used in the past is the hemolytic plaque assay, in which the antigen is either an erythrocyte protein(s) or a molecule covalently coupled to an erythrocyte surface. Such erythrocytes serve as "indicator cells." They are mixed with lymphocytes, among which are the specific antibody-producing cells, and incubated in a semisolid supporting medium to allow secreted antibody to bind to the erythrocyte surface. If the antibody binds complement avidly, the subsequent addition of complement leads to lysis of the indicator cells that are coated with specific antibody. As a result, clear zones of lysis, called plaques, are formed around individual B lymphocytes or plasma cells that secrete the specific antibody. These antibody-secreting cells are also called plaque-forming cells (PFCs). This assay can also be used to detect antibodies that do not fix complement by incorporating into the medium a complement-binding anti-Ig antibody that will coat indicator cells to which the specific Ig is bound first. Another technique for measuring the number of antibody-secreting cells is the ELISPOT assay. In this method, antigen is bound to the bottom of a well, antibody-secreting cells are added in a semisolid medium, and antibodies that have been secreted and are bound to the antigen are detected by an enzyme-linked anti-Ig antibody, as in an ELISA. Each spot represents the location of an antibody-secreting cell. Single-cell assays provide a measure of the numbers of Ig-secreting cells, but they cannot accurately quantitate the amount of Ig secreted by each cell or by the total population.

SIGNAL TRANSDUCTION BY THE B LYMPHOCYTE ANTIGEN RECEPTOR COMPLEX

The activation of B lymphocytes is initiated by the binding of antigen to membrane Ig molecules on specific B cells. This antigen–membrane Ig interaction can lead to two responses (Fig. 9–2). First, the B lymphocyte antigen receptor delivers biochemical signals to the cells that initiate the process of lymphocyte activation. Second, the antigen is internalized into endosomal vesicles, and if it is a protein, it is processed and peptides are presented on the B cell surface for recognition by helper T cells. This antigen-presenting function of B cells will be discussed later.

The principles underlying signaling via B lymphocyte antigen receptors are similar to those discussed for T cell receptors in Chapter 7. Signal transduction by membrane Ig molecules requires the bringing together of two or more of these molecules on a B cell by an antigen (or experimentally by an anti-Ig antibody) that is able to cross-link Ig molecules. Membrane IgM and IgD, the antigen receptors of resting, mature B cells, have short cytoplasmic tails consisting of only three amino acids (lysine, valine, and lysine). These tails are too

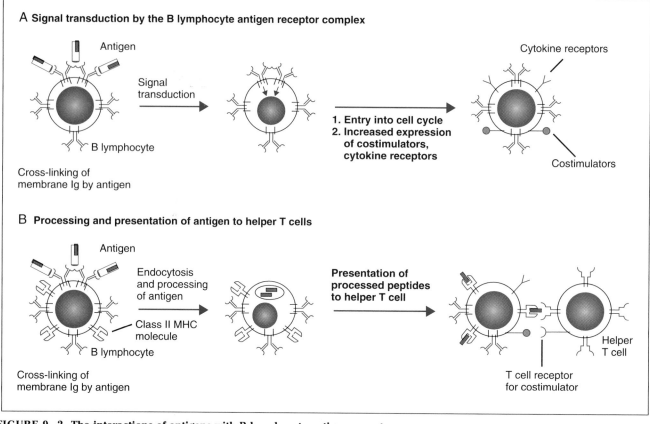

FIGURE 9–2. The interactions of antigens with B lymphocyte antigen receptors.
A. Cross-linking of membrane immunoglobulin (Ig) on B cells triggers biochemical signals that initiate the process of activation.
B. Protein antigens bound to membrane Ig are endocytosed and processed, and peptide fragments are presented in association with class II major histocompatibility complex (MHC) molecules for recognition by helper T cells.

small to transduce signals generated by the clustering of Ig. Signaling via Ig is actually transduced by two other molecules, called Igα and Igβ, which are members of the Ig superfamily and are expressed in mature B cells noncovalently associated with membrane Ig (Fig. 9–3). Igα and Igβ are also required for the surface expression of membrane Ig molecules and, together with membrane Ig, form the **B lymphocyte antigen receptor complex.** Thus, Igα and Igβ serve the same functions in B cells as the CD3 and ζ proteins do in T lymphocytes (see Chapter 7). As might be predicted from this functional similarity, Igα and Igβ contain in their cytoplasmic domains tyrosine-rich motifs (immunoreceptor tyrosine-based activation motifs, or ITAMs) that are also found in CD3 and ζ and are known, by mutational analyses, to be required for signal transduction.

The *early functional responses to antigen-mediated cross-linking of the B cell receptor complex* are the following: (1) entry of previously resting cells into the G1 stage of the cell cycle, accompanied by an increase in cell size and cytoplasmic ribonucleic acid (RNA), and increased numbers of biosynthetic organelles, such as ribosomes; (2) increased expression of class II molecules and costimulators, first B7-2 and later B7-1, because of which antigen-

stimulated B cells acquire the capacity to activate helper T lymphocytes, and (3) increased expression of receptors for several T cell–derived cytokines, thus enhancing the ability of the antigen-specific B lymphocytes to respond to T cell help. These responses are the result of several intracellular biochemical pathways that are known to be triggered by ligation of membrane Ig molecules, similar to receptor-induced signals in T lymphocytes (see Fig. 7–10, Chapter 7).

1. Within minutes after cross-linking of membrane Ig, the tyrosines in the ITAMs of Igα and Igβ are phosphorylated, probably by the action of src family kinases such as lyn, blk, and fyn. The non-src kinase syk then binds to the phosphotyrosine residues of Igα and Igβ, which provide docking sites for the SH2 domains of syk. Thus, syk is the B cell equivalent of ZAP-70 in T lymphocytes (see Box 7–4, Chapter 7). Syk needs to be phosphorylated to be active; this may be done by Ig-associated kinases or by syk phosphorylating itself. Mice in which *syk* is knocked out show a block in B cell maturation at the pre-B stage, because the enzyme also transduces signals from the antigen receptor that are required for B cell development.

2. The activated kinases may in turn activate

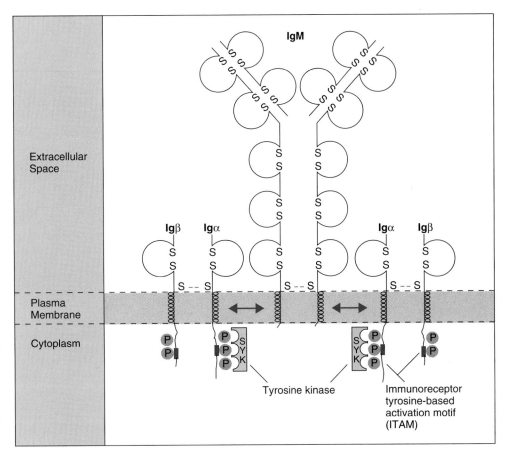

FIGURE 9-3. The B lymphocyte antigen receptor complex. Membrane immunoglobulin (Ig) M (and IgD) on the surface of mature B cells are associated with Igα and Igβ, which are phosphorylated on tyrosine residues and bind the kinase, Syk. Arrows indicate noncovalent associations of the μ heavy chain with Igα. P indicates the approximate locations of tyrosine residues that are phosphorylated upon cross-linking of the membrane Ig. Small black boxes represent ITAMs, also found in antigen receptors in T lymphocytes. Note the similarity to the T cell receptor complex (see Fig. 7–5).

numerous substrates. One of these is the $γ_1$ isoform of phosphatidylinositol phospholipase C, which breaks down membrane phosphatidylinositol bisphosphate to generate inositol trisphosphate (IP_3) and diacylglycerol (DAG). IP_3 mobilizes ionic Ca^{2+} from intracellular stores, leading to a rapid elevation of cytoplasmic Ca^{2+}, which may be augmented by an influx of Ca^{2+} from the extracellular milieu. In the presence of Ca^{2+}, DAG activates some isoforms of protein kinase C (PKC), and PKC in turn phosphorylates other proteins on serine/threonine residues.

3. The Ras protein is also activated by Ig cross-linking, and this is associated with activation of kinases of the mitogen-activated protein kinase family, as in T cells (see Fig. 7–13, Chapter 7).

4. A consequence of protein kinase activation is the activation of various transcription factors, which induce new gene expression in antigen-stimulated B lymphocytes. Some of the transcription factors that are known to be activated by antigen receptor–mediated signal transduction in B cells are c-Fos, JunB, NFκB, and c-Myc.

As in other lymphocytes, many steps in these signaling cascades are incompletely understood. The same pathways are utilized by membrane IgM and IgD on naive B cells and by other Ig heavy chains on B cells that have undergone isotype switching, because all these membrane isotypes appear to associate with Igα and Igβ. Also, other

surface molecules augment or inhibit signals transduced by the antigen receptors. These molecules include the type 2 complement receptor (CR2, or CD21) and the B cell Fc receptor, FcγRIIB. The physiologic significance and mechanisms of these regulatory interactions are discussed later in the chapter.

The importance of signaling by the B cell receptor complex for subsequent proliferation and differentiation of the cells may vary with the nature of the antigen. Many globular protein antigens may not express more than one identical epitope per molecule in their native conformation, in which case they cannot simultaneously bind to and cross-link two Ig molecules. However, even proteins may cross-link antigen receptors if they form aggregates, e.g., as a result of the binding of "natural antibodies" or complement components. Moreover, protein antigens recruit T cell help, and helper T cells and their products are capable of inducing B lymphocyte proliferation and differentiation by themselves. This is suggested by the finding that peptide antigens or monovalent anti-Ig antibodies, which cannot cross-link membrane Ig, can induce B cell responses in the presence of specific helper T cells. Therefore, protein antigens may only need to trigger minimal signals by the B cell receptor complex in order to generate humoral immune responses. In such responses, a major function of membrane Ig may be to bind and internalize the antigen for subsequent presentation to helper T cells (discussed below). In contrast, most T-independent antigens, such as polysaccharides and glycolipids, are polymeric, i.e., they contain multiple identical epitopes on each molecule. Therefore, such antigens effectively cross-link B cell antigen receptors and initiate responses even though they are not recognized by helper T lymphocytes. Signaling by the B cell antigen receptor may also contribute to enhanced survival of the B cells by blocking various pathways of apoptosis.

After antigen binding has initiated B lymphocyte activation, the subsequent proliferation and differentiation of the cells are stimulated by their interaction with helper T cells.

MECHANISMS OF HELPER T LYMPHOCYTE–DEPENDENT ANTIBODY RESPONSES

Antibody responses to protein antigens require recognition of the antigen by helper T cells and cooperation between the antigen-specific B and T lymphocytes. The helper function of T lymphocytes was discovered by experiments in the late 1960s, even before the classification of lymphocytes into T and B cell subsets was established. These experiments showed that if mouse bone marrow lymphocytes, which we now know contain mature B cells but few or no mature T cells, were adoptively transferred into irradiated syngeneic recipients, they would not produce specific antibody upon immunization with sheep red blood cells, a model protein antigen. If, however, mature thymocytes or thoracic duct lymphocytes (which contain T lymphocytes but few B cells) were transferred at the same time, antibody responses did develop following immunization (Table 9–2). This result, and later *in vitro* experiments in which purified T and B cells were mixed and stimulated with antigens, showed that B cells would proliferate and differentiate in response to protein antigens only if helper T lymphocytes were also present. Subsequent studies have established that *most helper T cells are CD4+CD8− and class II MHC–restricted* in their recognition of foreign protein antigens (see Chapters 6 and 7). Both helper T cells and B cells are stimulated by primary immunization, and both are responsible for enhanced antibody production in secondary responses.

Helper T cell–dependent antibody responses to protein antigens consist of distinct steps, which are summarized in Fig. 9–4. Each step ensures that antigen-specific B and T lymphocytes interact and function in a bidirectional manner to stimulate one another. The T–B cell interaction is called a "cognate" interaction because it is dependent on the specific recognition of antigen and several other surface molecules on the two interacting cells. The mechanisms underlying most of the phases of T-dependent B cell activation are now understood in quite precise biochemical terms.

Presentation of Protein Antigens by B Lymphocytes to Helper T Cells

Antigen-specific B lymphocytes bind the native antigen to membrane Ig molecules, internalize and process it in endosomal vesicles, and present on their surfaces peptide fragments of the antigen complexed to class II MHC molecules (see Fig. 9–2).

TABLE 9–2. Identification of Helper T Cells in Antibody Responses to Protein Antigens

A. Adoptive Transfer			
Cells Transferred into Irradiated Recipient			*Anti-SRBC Antibody-Producing Cells in Spleen*
Source of B cells	**Source of T cells**	*Antigen*	
Bone marrow cells	None	SRBC	−
None	Thoracic duct cells	SRBC	−
Bone marrow cells	Thoracic duct cells	SRBC	+
Bone marrow cells	Thoracic duct cells	—	−

B. Cell Culture		
Cells Cultured	*Antigen*	*Anti-SRBC Antibody-Producing Cells in Culture*
Unfractionated spleen cells	SRBC	+
Splenic B cells	SRBC	−
Splenic T cells	SRBC	−
Splenic B cells and T cells	SRBC	+
Splenic B cells and T cells	—	−

Mouse B lymphocytes by themselves do not produce antibody against a T cell–dependent antigen, SRBC, in vivo (A) or in vitro (B). The addition of T cells allows the B cells to respond to SRBC. A response does not occur in the absence of antigen.
Abbreviation: SRBC, sheep red blood cells.

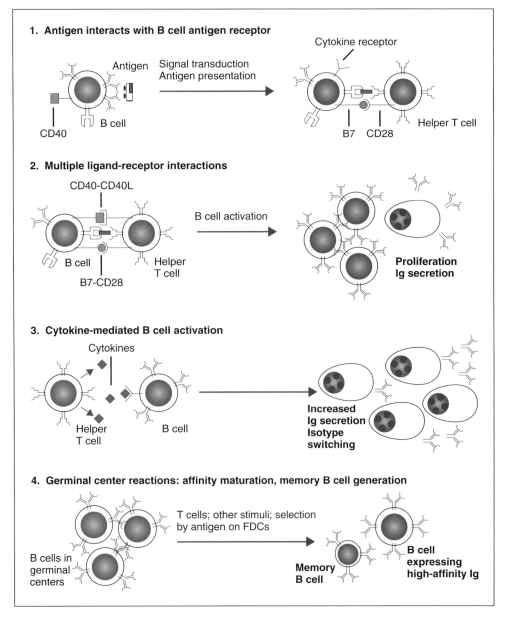

FIGURE 9–4. Phases of helper T cell–dependent antibody responses. Ig, immunoglobulin, FDCs, follicular dendritic cells.

Thus, the B cells themselves function as antigen-presenting cells (APCs). The peptide-MHC complexes can then be recognized by specific CD4$^+$ helper T lymphocytes.

Our current understanding of antigen presentation by B lymphocytes, and its role in the development of humoral immune responses, evolved from studies aimed at analyzing antibody responses to hapten-carrier conjugates. Haptens, such as dinitrophenol, are small chemicals that can be bound by B cell membrane Ig and by secreted antibodies but are not immunogenic by themselves. If, however, the haptens are coupled to proteins, which serve as carriers, the conjugates are able to induce antibody responses against the haptens. There are three important characteristics of anti-hapten antibody responses stimulated by hapten-protein conjugates. First, such responses require cooperation between hapten-specific B cells and protein (carrier)-specific helper T cells. Second, in order to stimulate a response, the hapten and carrier portions have to be physically linked and cannot be administered separately. Third, the interaction is class II MHC–restricted; i.e., the helper T cells cooperate only with B lymphocytes that express class II MHC molecules recognized as self by the T cells. The same characteristics apply not only to hapten-protein conjugates but to all protein antigens in which one intrinsic determinant is recognized by B cells (and is, therefore, analogous to the hapten) and another determinant is recognized by helper T cells (and is analogous to the carrier).

The explanations for these features of antibody responses to protein antigens and hapten-protein conjugates became clear when it was appreciated that B lymphocytes are extremely efficient APCs. Hapten-specific B cells bind the antigen via the hapten determinant and present peptides derived from the carrier protein to carrier-specific helper T lymphocytes. Thus, the two cooperating lymphocytes recognize different epitopes of the same complex antigen. The hapten is responsible for efficient carrier uptake, explaining why hapten and carrier must be physically linked. The requirement for MHC-associated antigen presentation for T cell activation accounts for the MHC restriction of T cell–B cell interactions.

At the same time that antigen is being processed and presented, the B lymphocytes have been activated by membrane Ig cross-linking. One of the consequences of activation is the enhanced expression of costimulators, including B7 molecules, on the B cells. Helper T cells can then recognize peptide-MHC complexes (signal 1 for T cells) and costimulators (signal 2) and are stimulated to perform their effector function, which is to promote B cell growth and differentiation. As we shall see later, this activation of B cells results from stimuli generated by physical contact with helper T cells and by cytokines secreted by the T cells.

In any humoral immune response, B cells specific for the antigen that initiates the response are preferentially activated as compared with bystander cells that are not specific for the antigen. There are several reasons for this. First, only B cells expressing specific membrane Ig molecules that bind the antigen receive the first signal for activation. Second, antigen-specific B cells are able to present the antigen at 10^4- to 10^6-fold lower concentrations than all other APCs. The efficiency of Ig-mediated antigen internalization and subsequent presentation is because membrane Ig molecules specifically bind the antigen at low concentrations, and antigens endocytosed with Ig are delivered to the intracellular protein processing pathway that is especially efficient at generating class II MHC–peptide complexes (see Chapter 6). As a result, whenever antigen concentrations are limiting, specific B cells are the preferential APCs. Third, as we shall discuss below, specific antigen recognition induces several surface molecules on the B cells and helper T lymphocytes that bring the two cell types into direct contact. The B cells in these T-B conjugates are exposed to signals delivered by T cell surface molecules and to the highest concentrations of T cell–derived cytokines. Therefore, *antigen-specific B lymphocytes are the preferential recipients of T cell help and are stimulated to proliferate and differentiate.* The antibodies that are subsequently secreted are specific for conformational determinants of the antigen, because membrane Ig on B cells is capable of binding conformational epitopes of native antigens. This feature determines the fine specificity of the antibody response and is independent of the fact that helper T cells recognize only linear epitopes of processed peptides. In fact, a single B lymphocyte may bind and endocytose a protein and present multiple different peptides complexed with class II MHC molecules to different helper T cells, but the resultant antibody response remains specific for the native protein.

Although it is clear that B cells need to function as APCs for MHC-restricted T cell–B cell interactions to occur, it is likely that other APCs are required for initiating the activation of helper T cells in primary antibody responses. There are several reasons for this conclusion. Naive B lymphocytes are inefficient at stimulating resting T cells and may even induce T cell tolerance, mainly because naive B cells are deficient in costimulators (see Chapters 7 and 10). In unimmunized individuals, there may be so few B cells specific for a particular antigen that they cannot effectively initiate a T cell response. Furthermore, it is known that for generating optimal primary antibody responses to protein antigens the antigens have to be administered with adjuvants, and one function of adjuvants is to enhance the antigen-presenting functions of macrophages (see Chapter 10). Therefore, when an individual is exposed to a protein antigen for the first time, APCs such as macrophages and dendritic cells may process and present the antigen to naive helper T cells. These T cells are activated, then interact with B cells that also present the antigen, and a primary response ensues. In secondary responses, on the other hand, antigen-specific memory B cells may be fully capable of functioning as APCs for the helper T cells that have previously been stimulated to become effector cells. This is supported by the observation that secondary antibody responses *in vitro* require only activated B and helper T cells and antigen.

Helper T Cell Contact-Mediated Activation of B Lymphocytes

The physical contact between B cells specific for the native antigen and helper T lymphocytes specific for processed peptides is mediated by multiple ligand-receptor pairs, some of which deliver signals to the B cells that are required for the full development of the humoral immune response. The two critical ligand-receptors pairs, in addition to antigen and antigen receptors, are B7 molecules:CD28 and CD40:CD40 ligand (Fig. 9–5). As we mentioned previously, antigen recognition by B cells enhances the expression of costimulators, first B7-2 and then B7-1. These costimulators are recognized by CD28 on helper T cells, which are, at the same time, interacting with peptide–class II MHC complexes on the B cells. The helper T cells thus receive the two sets of stimuli they need for their activation. Upon activation, helper T cells express a 34–39 kD surface molecule called **CD40 ligand** (see Chapter 7), which specifically binds to a constitutively expressed B cell receptor called **CD40.** CD40 is a member of a family of cell surface proteins that includes tumor necrosis factor (TNF) receptors and a protein called Fas (see Box 12–4, Chapter

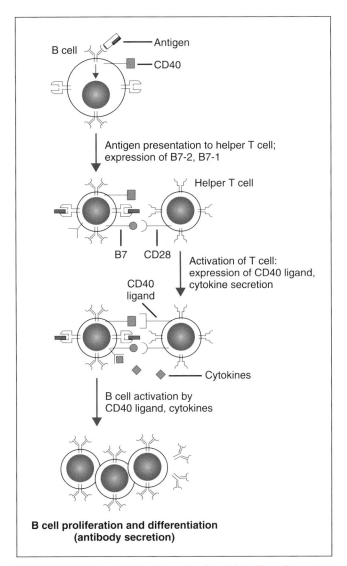

**B cell proliferation and differentiation
(antibody secretion)**

FIGURE 9–5. A model for the role of multiple ligand-receptor pairs in T cell–dependent B cell activation. In this model, antigen is presented by B cells and induces the expression of costimulators. Helper T cells recognize the antigen (in the form of peptide–MHC) and costimulators and are stimulated to express CD40 ligand. CD40 ligand binds to CD40 on the B cells and initiates B cell proliferation and differentiation. As discussed in the text, an alternative possibility is that the helper T cells have been activated by other antigen-presenting cells to express CD40 ligand, and their interaction with B lymphocytes stimulates the expression of costimulators.

12). The interaction of CD40 ligand with CD40 triggers intracellular biochemical signals in B cells that are required for the cells to continue along their process of activation (discussed in more detail below).

Some studies suggest that the B7:CD28 and CD40 ligand:CD40 pairs may interact in the reverse sequence. T cells are first activated by antigen presented by dendritic cells or macrophages, and this activation induces expression of CD40 ligand. When these helper T cells see antigen presented by specific B cells, CD40 ligand interacts with CD40

on the B cells. This not only stimulates B cell proliferation and differentiation but also leads to enhanced expression of B7 molecules, thus causing more T cell activation. In either case, the interaction between helper T cells and B cells is bidirectional. *Because the expression of both B7 and CD40 ligand is regulated, i.e., it is dependent on antigen-mediated stimulation, only lymphocytes specifically stimulated by antigen interact bidirectionally with one another, thus maintaining the specificity of the immune response.* The critical role of this interaction in antibody production is illustrated by the findings that knockout mice lacking the gene for any member of the two ligand:receptor pairs, i.e., B7 (both B7-1 and B7-2), CD28, CD40, or CD40 ligand, exhibit profound defects in isotype switching, affinity maturation, and memory B cell generation in response to protein antigens. Mutations in CD40 ligand are also responsible for an inherited immunodeficiency of humans, called the X-linked hyper IgM syndrome, in which patients are unable to produce normal amounts of most IgG, IgA, and IgE antibodies against proteins (see Chapter 21).

Role of CD40 in Humoral Immune Responses

The requirement for physical contact between helper T and B lymphocytes in the generation of antibody responses was first suspected when it was found that the helper function of T lymphocytes could not be replaced by their secreted products (cytokines). Subsequent experiments showed that if helper T cells are first activated with antigen, chemically fixed to render them incapable of producing cytokines, and then co-cultured with B cells, they are still able to induce proliferation of the B cells (Fig. 9–6). The same responses are seen if the B cells are cultured with purified plasma membranes of activated helper T lymphocytes. In the absence of cytokines, fixed T cells or membranes induce only modest B cell proliferation. The full proliferation of B cells and their differentiation into antibody-producing cells do require cytokines, as we shall discuss later.

Such results suggested that activated helper T lymphocytes express a membrane molecule(s) that binds to a receptor on B lymphocytes, and this interaction provides the stimulus that initiates B cell responses. Antibodies that blocked T-dependent B cell activation were used to identify the relevant surface molecules, which turned out to be CD40 on the B cells and its ligand on activated T cells. The function of this ligand-receptor pair has been demonstrated by numerous approaches (Fig. 9–6). Fibroblasts expressing CD40 ligand (by gene transfection) stimulate B cells much like contact with helper T lymphocytes, and antibodies to CD40 can directly activate B cells, presumably by functioning like CD40 ligand. The obligatory role of these molecules is demonstrated by gene knockouts and inherited mutations, as mentioned above.

Members of the TNF receptor/Fas/CD40 family

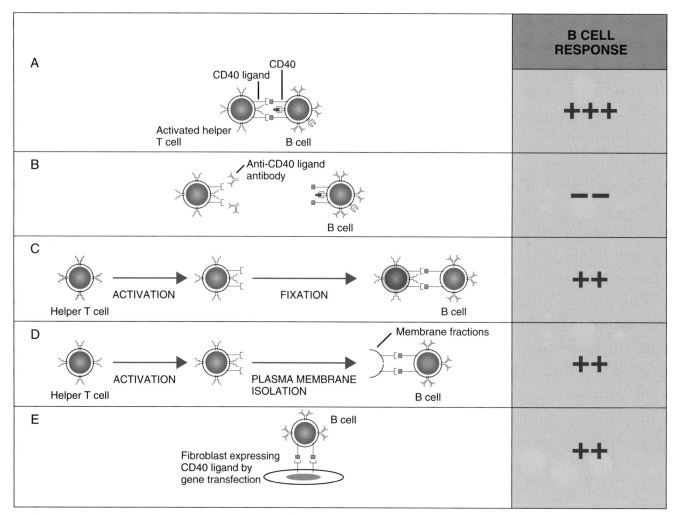

FIGURE 9–6. Role of CD40 in activation of B cells induced by contact with helper T cells. Activated helper T cells recognize antigen presented by B cells and express CD40 ligand, which binds to CD40 on B cells and initiates B cell responses (A). Blocking antibodies against CD40 ligand inhibit the T cell contact-mediated stimulation of B cells (B). Preactivated and fixed helper T cells (C), CD40 ligand-containing plasma membrane fractions of activated T cells (D), and fibroblasts expressing CD40 ligand (E) all can replace intact, viable T cells. Note that the full growth and differentiation of B cells require T cell–derived cytokines as well as T cell contact.

determine the fates of many cell populations in the immune system by regulating the choice between cell death and proliferation. Thus, these proteins serve a variety of roles in immune responses (see Box 10–1, Chapter 10, and Box 12–4, Chapter 12). CD40 ligand is a membrane protein that is structurally homologous to TNF and Fas ligand. CD40 ligand is rapidly induced on T cells upon their activation. Its binding to CD40 results in oligomerization of CD40 molecules, and this apparently induces the association of cytoplasmic proteins (called TNF receptor–associated factors, or TRAFs) to the cytoplasmic C-terminal domains of CD40. The TRAFs are in turn activated and, via kinases and biochemical intermediates that are not yet well defined, lead to the activation and nuclear translocation of transcription factors, including NFκB (see Box 12–4). The links between these transcription factors and changes in Ig gene expression, i.e., isotype switching, or induction of germinal center reactions, are not yet understood.

T cell–mediated macrophage activation may also involve the interaction of CD40 ligand on T cells with CD40 on macrophages (see Chapter 13). Thus, this pathway of contact-mediated cellular responses is a general mechanism for the activation of target effector cells by helper T lymphocytes and is not unique to antibody production.

Interestingly, a DNA virus called the Epstein-Barr virus (EBV) infects human B cells and induces their proliferation. This may lead to immortalization of the cells and the development of lymphomas (see Box 18–2, Chapter 18). The cytoplasmic domain of a transforming protein of EBV associates with the same TRAF molecules as the cytoplasmic domain of CD40, and this apparently triggers B cell proliferation. Thus, the viral protein is functionally homologous to the physiologic B cell–signaling molecule, and EBV has apparently co-opted a normal pathway of B lymphocyte activation for its own purpose, which is to promote survival of cells the virus has infected.

B Lymphocyte Growth and Differentiation Induced by Cytokines (Helper Factors)

Cytokines are soluble proteins secreted by T lymphocytes and by other cell types in response to activating stimuli. Cytokines mediate many of the effector functions of the cells that produce them and are the principal mechanisms by which various immune and inflammatory cell populations communicate with one another. Although cytokines are produced as a result of the stimulation of specific T cells by antigens, the cytokines themselves are not antigen-specific and do not bind to antigens. Their general properties, structural features, and functional effects are described in Chapter 12. The roles of these secreted proteins in humoral immunity have been most clearly established by showing that various aspects of antibody responses can be inhibited by cytokine antagonists, or are deficient in mice in which particular cytokine genes are knocked out by homologous recombination.

Cytokines serve two principal functions in antibody responses: they determine the types of antibodies produced by selectively promoting switching to different heavy chain isotypes, and they provide amplification mechanisms by augmenting B cell proliferation and differentiation. Different cytokines play distinct, but often overlapping, roles in antibody production, and their actions may be synergistic or antagonistic. Antigen recognition by B cells enhances the expression of receptors for cytokines, and B cells in direct contact with helper T lymphocytes are presumably exposed to high concentrations of these secreted proteins. As a result, antigen-specific B cells respond to cytokines more than do "bystander" B cells that are not specific for the initiating antigen but happen to be close to the antigen-stimulated lymphocytes.

Based on a variety of *in vitro* and *in vivo* studies, it is now possible to identify the cytokines that act at each stage of B cell activation (Fig. 9–7).

1. *Proliferation.* Many different cytokines have been shown to stimulate the proliferation of B cells *in vitro*. Three helper T cell–derived cytokines, interleukin (IL)-2, IL-4, and IL-5, all contribute to B cell proliferation and may act synergistically. IL-6, which is produced by macrophages, T cells, and many other cell types, is a growth factor for already differentiated, antibody-secreting B cells. Three other cytokines produced by macrophages, IL-1, IL-10, and TNF, also promote some B cell proliferation *in vitro*. The redundancy in the actions of these cytokines accounts for the observation that knockout of any one cytokine gene, or an antagonist that blocks the function of any one, has virtually no effect on B cell growth in response to protein antigens.

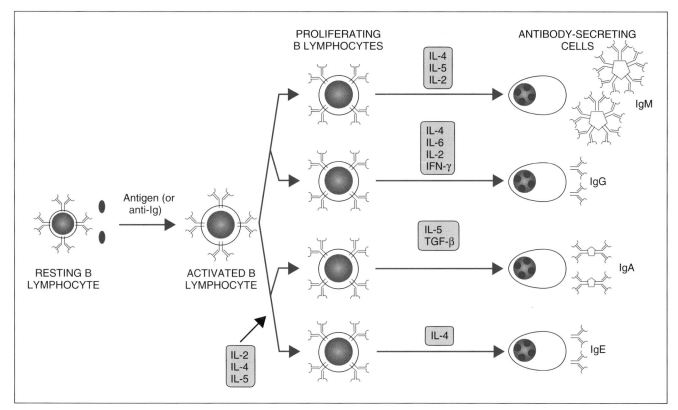

FIGURE 9–7. Functions of cytokines in B lymphocyte growth and differentiation. Various cytokines stimulate different stages of B cell proliferation and differentiation in humans and mice. The same cytokines may have less striking effects at other stages that are not shown, and there may be differences among species. Ig, immunoglobulin; IL, interleukin; TGF, transforming growth factor; IFN, interferon.

2. *Antibody secretion.* Antibody synthesis and secretion in response to protein antigens, like B cell proliferation, are also enhanced by cytokines. In the mouse, IL-4 and IL-5 are potent inducers of antibody secretion by B cells. Human B cells cultured with polyclonal activators secrete high levels of antibody if IL-2, IL-6, or IL-10 is added. It is not clear whether these differences in the cytokine responses of human and mouse B cells are due to true species variations or merely reflect differences in experimental conditions.

3. *Isotype switching. The most selective, and only obligatory, functions of different cytokines in humoral immune responses are in regulating the pattern of heavy chain isotype switching.* We have mentioned previously that heavy chain isotype switching is typically seen with protein antigens and requires helper T cells. Different T cell–derived cytokines selectively induce switching to particular Ig isotypes. For instance, IL-4 is the principal switch factor for IgE in all species examined. IgE production depends on IL-4, since IgE responses to antigen-specific or polyclonal stimulation *in vitro,* and to infection by helminthic parasites *in vivo,* are abrogated by antibodies that neutralize IL-4, and IL-4–deficient mice created by gene knockout have no serum IgE. These findings have many practical implications, since IgE antibodies are involved in host defense against parasitic infestations (see Chapter 16) and in immediate hypersensitivity (allergic) reactions (see Chapter 14). The production of IgG2a in mice is dependent on interferon-γ (IFN-γ), which is also secreted by T cells. Therefore, addition of IL-4 or IFN-γ induces switching to different isotypes in cultures of mature IgM and IgD-expressing B cells stimulated with antigens or polyclonal activators (Table 9–3). Interestingly, IFN-γ inhibits IL-4–induced B cell switching to IgE, and, conversely, IL-4 reduces IgG2a production. These are among the clearest examples of the antagonistic effects of different cytokines. Transforming growth factor-β, which is produced by many cell types, acts in concert with T cell–derived IL-5 to stimulate IgA production in mucosal lymphoid tissues (see Chapter

11). These two cytokines may, therefore, be especially important in mucosal immunity.

Cytokines induce switching to different heavy chain classes by triggering the process of switch recombination (see Fig. 4–13, Chapter 4). In naive B cells, which produce IgM, engagement of CD40 and exposure to a particular cytokine stimulates transcription through switch (S) regions located 5′ of particular heavy chain constant (C) region genes. For instance, IL-4 induces transcription through the S_ϵ-C_ϵ locus, and in mice IFN-γ through the $S_{\gamma 2a}$-$C_{\gamma 2a}$ locus. This leads first to the production of sterile or "germline" transcripts. Subsequently, the rearranged VDJ complex in that B cell recombines with the downstream C region that has become transcriptionally active, leading to the production of a new heavy chain with the same V region as the original IgM.

Germinal Center Reactions: Affinity Maturation and the Generation of Memory B Cells

The generation of high-affinity antibodies and memory B cells in response to protein antigens occurs in germinal centers. Morphologic studies of antibody production *in vivo* show that the initial B cell response to protein antigens occurs mainly at the edges of T cell–rich zones, e.g., in the periarteriolar sheaths of the spleen (Fig. 9–8). Within 4 to 7 days after antigen administration, some of the activated B cells migrate into an adjacent lymphoid follicle and begin to proliferate rapidly, forming the lightly staining central region of the follicle, called the **germinal center.** The doubling time of germinal center B cells is estimated to be 6 to 12 hours, so that within 5 days a single lymphocyte may give rise to almost 5000 progeny. The IgV genes of these rapidly dividing B cells undergo somatic hypermutation, because of which V gene mutations accumulate sequentially in the progeny of the B cells (see Fig. 4–14, Chapter 4).

As these processes are going on in lymphoid tissues, antibodies that have been secreted into the circulation bind to residual antigen molecules and may activate the complement cascade. Anti-

---

TABLE 9–3. Heavy Chain Isotype Switching Induced by Cytokines

| B Cells Cultured With | | Ig Isotype Secreted (Per cent of Total Ig Produced) | | | | |
Polyclonal Activator	*Cytokine*	*IgM*	*IgG1*	*IgG2a*	*IgE*	*IgA*
LPS	None	85	2	< 1	< 1	< 1
LPS	IL-4	70	20	< 1	5	< 1
LPS	IFN-γ	80	2	10	< 1	< 1
LPS	TGF-β + IL-5	75	2	< 1	< 1	15

Purified IgM⁺IgD⁺ mouse B cells cultured with the polyclonal activator, LPS, and various cytokines, show selective switching to different heavy chain isotypes. (The values of the isotypes shown are approximations and do not add up to 100 per cent because not all were measured.)
Abbreviations: Ig, immunoglobulin; LPS, lipopolysaccharide; IL, interleukin; TGF, transforming growth factor; IFN, interferon. Courtesy of Dr. Robert Coffman, DNAX Research Institute, Palo Alto, CA.

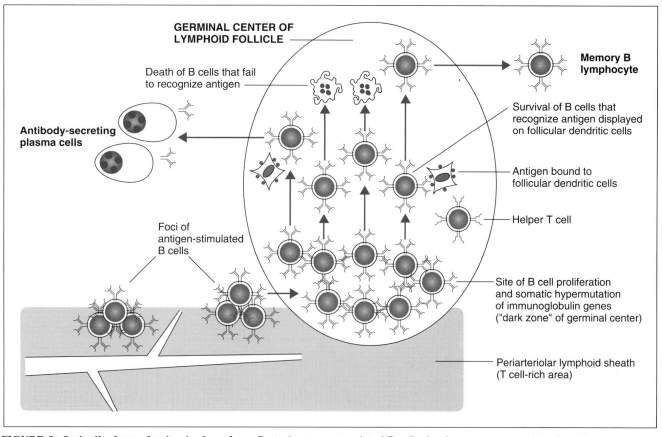

FIGURE 9–8. Antibody production in the spleen. Foci of antigen-stimulated B cells develop near periarteriolar lymphoid sheaths. In germinal centers, proliferation occurs in the basal dark zone, selection of high-affinity antigen-specific B cells occurs in the basal light zone, and antibody-secreting cells as well as memory cells exit from the apical zone.

gen-antibody complexes with attached complement proteins are avidly bound to the surfaces of **follicular dendritic cells** in the germinal centers, because these cells express high-affinity receptors for complement proteins and for the Fc regions of IgG molecules. (Follicular dendritic cells are found only in germinal centers, do not express class II MHC molecules, and are different from the dendritic cells that present peptide antigens to CD4⁺ T lymphocytes). Somatically mutated B cells whose membrane Ig molecules are able to recognize the displayed antigen are selected to survive. It is postulated that the survival signal may be delivered by antigen binding to membrane Ig and by CD40 ligand, expressed on follicular dendritic cells, binding to CD40 on the B lymphocytes. If somatic mutations have generated B cell antigen receptors that no longer recognize the antigen, these cells die by a process of apoptotic programmed cell death, because activated B lymphocytes need to recognize antigen and other stimuli to be protected from programmed death (see Box 2–2, Chapter 2). As the antibody response develops, the amount of antigen available for display on follicular dendritic cells decreases progressively. As a result, in order to be positively selected, the B cells need to express Ig receptors with progres-

sively higher affinity for the antigen. This is the basis for **affinity maturation** of the antibody response. Not surprisingly, many B cells with V region mutations lose the ability to recognize antigen and cannot be selected, so that the germinal centers are sites of tremendous apoptosis. Macrophages present in the germinal centers rapidly phagocytose and clear these dead lymphocytes. During these selection processes, B cells migrate from the basal dark zone of the germinal center, where the greatest proliferation occurs, to the basal light zone, which is the site of antigen-driven selection, and then to the apical zone, where additional isotype switching may occur. The cells then exit the germinal center and develop into high-rate antibody secretors in extra-follicular sites. Germinal centers gradually decrease in size and number and may involute by 1 to 2 weeks after immunization. At this time, affinity maturation also ceases unless the antigen is persistent or is re-administered.

Some of the B cells do not develop into antibody secretors. Instead, they acquire the ability to survive for long periods, apparently without antigenic stimulation, and to circulate freely between the blood and lymphoid tissues, including germinal centers. These are **memory cells,** capable of mounting rapid responses to subsequent introduction of antigen. It is possible that memory cells are actually continually generated and maintained by low-level stimulation provided by antigen, which may be displayed on follicular dendritic cells for months or years. We do not know why some of the progeny of an antigen-stimulated B cell clone differentiate into antibody-secreting cells with brief life spans while others become functionally quiescent, long-lived memory cells.

The biochemical mechanisms of germinal center reactions are not well understood. These reactions require helper T cells, because they are seen almost exclusively after immunization with protein antigens. A few CD4⁺ T lymphocytes are found within germinal centers and have probably migrated there after their activation in extra-follicular regions. It is not known if these T cells provide specific contact-mediated signals or cytokines that stimulate somatic hypermutation and memory B cell generation. Interactions between CD40 and CD40 ligand are required for germinal center formation, because the germinal center reactions do not develop in humans or mice with mutations or knockout of either CD40 or its ligand. As mentioned above, CD40 ligand on follicular dendritic cells may promote the survival of high-affinity B lymphocytes. The adhesion between follicular dendritic cells and activated B lymphocytes may be facilitated by the integrin VLA-4 (CD29CD49d), expressed on the lymphocytes, and its ligand, vascular cell adhesion molecule-1, on follicular dendritic cells. The same ligand-receptor pair promotes the migration of memory B cells into germinal centers, where depots of antigen are displayed.

Sequence of Events in Antibody Response to Protein Antigens *In Vivo*

Now that we have described the cellular and molecular interactions that occur in different phases of helper T cell–dependent B lymphocyte activation (see Fig. 9–4), it is useful to summarize the sequence of events that occur in humoral immune responses *in vivo*. This process is best described in the spleen but is probably similar in lymph nodes and other lymphoid tissues.

Within 1 or 2 days after administration of a protein antigen to a naive individual, the antigen is processed and presented by dendritic cells and macrophages to specific CD4⁺ T lymphocytes in the T cell–rich zones of the spleen and lymph nodes. Clusters of activated T cells that express CD40 ligand and produce various cytokines are detectable in these regions. Specific B lymphocytes also encounter the antigen, are activated, and present peptide fragments of the antigen to activated helper T lymphocytes. As a result of cell-cell contact mediated by CD40 and CD40 ligand, as well as exposure to T cell cytokines, the antigen-specific B cells begin to proliferate, differentiate into antibody-secreting cells, and undergo heavy chain class switching. Thus, by 4 to 7 days after

antigen exposure, antibody-producing B cells are found in close proximity to activated T lymphocytes. Antibody secretion and heavy chain class switching constitute the *extra-follicular pathway of B cell differentiation.* Some of the activated B cells migrate into lymphoid follicles, form germinal centers, and undergo the sequential changes, described above, that culminate in affinity maturation and memory cell generation. These comprise the *follicular pathway of B lymphocyte differentiation.* The separation of these pathways is not absolute, because some isotype switching does occur in germinal centers; however, there is no evidence for extra-follicular somatic hypermutation of Ig V genes and affinity maturation. Antibody-secreting cells are found mainly in extra-follicular sites, such as the red pulp of the spleen and the medulla of lymph nodes. These cells also migrate to the bone marrow, and at 2 to 3 weeks after immunization the marrow may be a major site of antibody production. Secreted antibodies enter the circulation, but most antibody-producing cells do not circulate actively and have finite life spans (probably a few weeks).

Circulating antibodies bind antigens to initiate the effector phase of humoral immune responses. Memory B cells that exit germinal centers circulate in the blood and have a propensity to migrate through germinal centers. Memory cells typically bear high-affinity (mutated) antigen receptors and Ig molecules of switched isotypes more commonly than do naive B lymphocytes. These processes are greatly accelerated in secondary antibody responses, because of the presence of memory cells in germinal centers and the rapid formation of immune complexes that can be concentrated by follicular dendritic cells.

Factors That Determine the Nature of Humoral Immune Responses to Protein Antigens

Different types of antigens and forms of antigen exposure stimulate antibody responses with distinct characteristics. The principal factors that determine the nature of antibody responses are the following:

1. *Types of cytokines produced.* An important general principle that has emerged from the study of cytokines is that the nature and magnitude of immune responses are influenced by the relative amounts of different cytokines produced at sites of lymphocyte stimulation. This is because, as described above, various cytokines have selective effects on B cells, especially on heavy chain class switching, and because combinations of different cytokines may be synergistic or antagonistic. It is now clear that in different immune responses, helper T lymphocytes differentiate into effectors that produce different cytokines. The two most distinct, or polarized, populations of helper T cells are called **T$_H$1 and T$_H$2 subsets** (see Chapter 12).

The principal effector cytokine of T_H1 cells is IFN-γ. IFN-γ promotes switching of B cells to isotypes, such as IgG2a in mice, that fix complement and promote phagocytosis by macrophages. IFN-γ also activates the microbicidal functions of macrophages. Therefore, T_H1 cells stimulate phagocyte-dependent host reactions and are important for the elimination of intracellular microbes (cell-mediated immunity). T_H2 cells, in contrast, produce IL-4 and IL-5. IL-4 induces switching to IgE and to other isotypes, such as IgG4 in humans or its homolog IgG1 in mice, that do not fix complement or bind to phagocyte Fcγ receptors. IL-5 activates eosinophils. Therefore, the T_H2 subset is important for phagocyte-independent, IgE and eosinophil-mediated defense, e.g., against helminths. The development and functions of these subsets will be discussed in detail in Chapter 12, after we describe the biology of cytokines.

2. *Site of antigen exposure.* Antigens that enter the blood or lymph, and subsequently the peripheral lymphoid organs, such as the spleen and lymph nodes, induce the production of antibodies of multiple isotypes. Orally administered and inhaled antigens tend to stimulate high levels of IgA production, because B cells in mucosal lymphoid tissues readily switch to IgA (see Chapter 11). This adaptation is important because IgA is the only antibody isotype that is actively secreted through epithelial cells into the lumen.

3. *Nature of the antigenic stimulus.* Large doses of antigens, especially if they are administered systemically without adjuvants, often inhibit antibody production by inducing tolerance, or unresponsiveness, in helper T or B lymphocytes. The phenomenon of immunologic tolerance to foreign antigens is described in Chapter 10.

4. *Many of the features of secondary responses to protein antigens, and the differences from primary responses (see Table 9–1), are due to the increased numbers of differentiated effector CD4$^+$ T cells that are stimulated by the first exposure to antigen.* Thus, heavy chain class switching, which is typical of secondary responses, is due to helper T cells and their cytokines. Affinity maturation, which increases with repeated antigenic stimulation, is also

a result of helper T cell–induced B cell activation. These features are usually seen in responses to protein antigens, because only proteins stimulate specific helper T cells. The more rapid and larger secondary response is because of the expansion of antigen-specific lymphocyte clones induced by prior antigenic stimulation.

ANTIBODY RESPONSES TO THYMUS-INDEPENDENT ANTIGENS

The requirement for helper T cells explains why thymus-dependent protein antigens do not induce antibody responses in T cell–deficient animals, such as congenitally athymic (nude) or neonatally thymectomized mice. In contrast, many non-protein antigens stimulate antibody production in athymic mice, and these antigens are termed **thymus-independent** (TI). Their properties and the features of the antibody responses they induce are summarized in Table 9–4.

The most important TI antigens are polysaccharides, glycolipids, and nucleic acids, all of which induce specific antibody production in T cell–deficient animals. These antigens cannot be processed and presented in association with MHC molecules, and, therefore, they cannot be recognized by helper T cells. Most TI antigens are polymeric, being composed of multiple identical antigenic epitopes; examples include dextrans and pneumococcal polysaccharide. Such multivalent antigens may induce maximal cross-linking of membrane Ig on specific B cells, leading to activation without a requirement for cognate T cell help. B cell activation by TI antigens is one situation in which membrane Ig-mediated signal transduction may be critical for subsequent responses of the cells. It is not clear if B cell responses to TI antigens are dependent on any second signals. Experiments with mice suggest that antibody responses to polysaccharides do require the presence of small numbers of macrophages and/or helper T cells. These cells may secrete cytokines that function as second signals for B cell responses to TI antigens, but neither the nature of these cytokines

TABLE 9–4.   Properties of Thymus-Dependent and Thymus-Independent Antigens

Properties	Thymus-Dependent	Thymus-Independent
Chemical Nature	Proteins	Polymeric antigens, especially polysaccharides; also glycolipids, nucleic acids
Antibody Response in		
Athymic mice	No	Yes
T cell–depleted cultures	No	May be reduced
Features of Antibody Response		
Isotype switching	Yes	No (usually)
Affinity maturation	Yes	No
Secondary response (memory B cells)	Yes	No or low
Ability to Induce Delayed-Type Hypersensitivity	Yes	No

nor the mechanism(s) leading to their production are completely understood.

Responses to most TI antigens consist largely of IgM antibodies of low affinity and do not show significant heavy chain class switching, affinity maturation, or memory. However, some TI antigens do induce Ig isotypes other than IgM, probably because these antigens stimulate cytokine production. For instance, in humans the dominant antibody class induced by pneumococcal capsular polysaccharide is IgG2. In addition, despite their inability to generate good memory, many polysaccharide vaccines, such as pneumococcal vaccine, induce quite long-lived protective immunity. A likely reason for this is that polysaccharides are not degraded efficiently, persist for long periods in lymphoid tissues, and continue to stimulate newly maturing B cells.

The practical significance of TI antigens is that many bacterial cell wall polysaccharides belong to this category, and humoral immunity is the major mechanism of host defense against such bacterial infections. For this reason, individuals with congenital or acquired deficiencies of humoral immunity are especially susceptible to life-threatening infections with encapsulated bacteria, such as pneumococcus, meningococcus, and *Haemophilus.* Other examples of responses to TI antigens are natural antibodies, which are present in the circulation of normal individuals and are apparently produced without overt antigen exposure. Most natural antibodies are low-affinity anti-carbohydrate antibodies, postulated to be induced by bacteria that colonize the gastrointestinal tract.

Antibody responses to TI antigens may occur at particular anatomic sites in lymphoid tissues. Macrophages located in the marginal zones of lymphoid follicles in the spleen are particularly efficient at trapping polysaccharides when these are injected intravenously. Such antigens may either persist for prolonged periods on the surfaces of marginal zone macrophages, where they are recognized by specific B cells, or they may be transferred from the marginal zone to the adjacent follicle. Marginal zones are present only in the spleen. This is probably why splenectomized individuals do not develop strong antibody responses to polysaccharides and are susceptible to infection by encapsulated bacteria like the pneumococcus.

Based on experiments with mice, TI antigens have been classified into two groups. Polysaccharides and the other TI antigens mentioned above are called "TI-2" antigens; rigorous depletion of T cells abolishes antibody responses to these. The prototypical "TI-1" antigen in mice is lipopolysaccharide (LPS), which stimulates B cells by itself, without a requirement for any other cells. LPS, also called endotoxin, is a major component of the cell walls of many gram-negative bacteria (see Chapter 12). At low concentrations, LPS stimulates specific antibody production in rodents. At high concentrations, it is a polyclonal B cell activator, stimulating the growth and differentiation of virtu-ally all B cells without binding to the membrane Ig. It is postulated that a component of the LPS molecule binds to B cells and directly activates the cells, i.e., the antigen provides both the first and the second signals needed for B lymphocyte activation. However, neither the nature of the B cell receptor for LPS nor the biochemical mechanisms of LPS-induced B cell stimulation is clearly established. Also, LPS is a polyclonal B cell activator in mice but not in humans and most other species. In all species, LPS is one of the most potent activators of macrophages known (see Chapter 12).

REGULATION OF HUMORAL IMMUNE RESPONSES BY COMPLEMENT, ANTIBODIES, AND ACCESSORY CELLS

The production of antibodies is regulated by signals in addition to those provided by antigen and T cells. These signals can both enhance and suppress humoral immune responses.

Role of Complement Proteins

The complement system consists of a set of proteins that are activated either by binding to antigen-complexed antibody molecules (the "classical pathway") or by binding directly to some polysaccharides and to microbial surfaces, in the absence of antibodies (the "alternative pathway"). The biochemistry and functions of complement will be described in Chapter 15. Historically, much of the interest in complement has focused on its role as an effector mechanism of humoral immunity. Recent work has shown that *complement also plays a role in the induction and amplification of antibody responses.* Mice deficient in various complement components show profound defects in antibody responses to protein antigens. Conversely, attaching activated complement proteins to protein antigens increases the immunogenicity of these antigens more than 1000-fold.

The complement system augments antibody production by two mechanisms.

1. *B lymphocytes express a receptor for a complement protein, C3d, called the type 2 complement receptor (CR2, or CD21), which transduces signals that enhance B cell proliferation and differentiation.* C3d is generated by the proteolysis of the key component of complement, C3, and becomes covalently attached to antigens that trigger complement activation via the classical or alternative pathway (see Chapter 15). Binding of C3d to the B cell complement receptor recruits another B cell surface protein, CD19, into the complex. If the antigen receptor is also engaged by simultaneous binding of antigen, the response of the B cell is greatly enhanced (Fig. 9–9A). Knockout of the C3, CD21, or CD19 gene in mice results in defects in antibody production.

2. *Complement proteins attached to antigens promote the binding of the antigens to follicular den-*

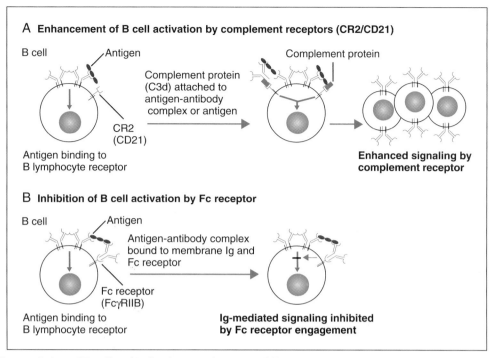

FIGURE 9–9. The regulation of B cell activation by complement and Fc receptors.
A. Binding of complement proteins (C3d) to antigens directly or via attached antibodies leads to coligation of antigen receptors (Ig) and complement receptors (CR2, CD21) on antigen-specific B cells. This enhances the immunoglobulin (Ig)-mediated signal and the subsequent B cell response.
B. Binding of IgG antibody to antigen leads to coligation of the antigen receptor and the FcγRIIB receptor on antigen-specific B cells and inhibits B cell activation.

dritic cells in germinal centers. Follicular dendritic cells express a surface receptor for another C3 breakdown product, C3b; this receptor is called the type 1 complement receptor (CR1, or CD35). Complexes of antigens, antibodies, and C3b bind to the CR1 receptor on follicular dendritic cells, providing display of the antigen in germinal centers for the selection of high-affinity B cells.

Since the complement system is usually activated after the binding of secreted antibodies to antigens, the question arises, how can complement be required for the induction of humoral immune responses? The likely answer is that in antibody responses to protein antigens, complement is essential for amplification. Thus, once a small amount of antibody is produced, it binds to the antigen, activates complement by the classical pathway, generates C3b and C3d, which bind covalently to the antigen, and the response is amplified by the two mechanisms mentioned above. Since IgM antibodies are produced first in humoral immune responses and are more potent activators of the complement system than all other Ig isotypes, they may provide the major amplification mechanism. In the case of T-independent antigens, many polysaccharides directly activate complement prior to specific antibody production, and the same sequence of events ensues. The complement system is an important component of innate immunity. Its role in specific antibody responses illustrates a concept that was introduced in Chapter 1, that the mechanisms of innate immunity may shape the magnitude and nature of specific immune responses.

Antibody Feedback: The Inhibitory Function of Fc Receptors on B cells

The simultaneous engagement of B cell Ig and Fcγ receptors by antigen-antibody complexes inhibits B cell activation. This is the explanation for a phenomenon called **antibody feedback,** which refers to the downregulation of antibody production by secreted IgG antibodies. Some antigen-complexed IgG isotypes bind to a B cell receptor for the Fc portions of these IgGs, called the Fcγ receptor II B (FcγRIIB, or CD32). The cytoplasmic domain of FcγRIIB binds to a protein tyrosine phosphatase that is found in many cell types. If the FcγRIIB is brought into close proximity to the B cell antigen receptor, the associated phosphatase may remove tyrosine residues from signaling molecules that have become attached to the B cell receptor complex and been tyrosine phosphorylated (see Fig. 9–4). This inhibits Ig-mediated signal transduction and terminates the B cell response (Fig. 9–9B).

Fc receptor–mediated antibody feedback is a physiologic control mechanism in humoral immune responses, since it is triggered by secreted antibody and blocks further antibody production. It is not clear under which circumstances secreted antibodies provide complement–mediated amplification or Fc receptor–mediated inhibition. A likely scenario is that early in humoral immune responses, IgM antibodies (which activate complement but do not bind to the Fcγ receptor) are involved in amplification, whereas increasing production of IgG leads to feedback inhibition.

Role of Accessory Cells

One of the first experiments suggesting the requirement for accessory cells in the responses of lymphocytes to antigens demonstrated that if spleen cells from unimmunized mice were depleted of adherent cells, they would not secrete antibody when stimulated by a model T cell–dependent antigen, sheep red blood cells. Responsiveness was restored by the addition of non-lymphoid cells such as macrophages, or, as shown later, by dendritic cells.

Macrophages and other accessory cells can perform various functions in the induction of humoral immunity.

1. *Macrophages and dendritic cells are necessary for presenting antigens to naive helper T lymphocytes,* thereby inducing proliferation of the T cells and their differentiation into effector cells producing various cytokines. As discussed above, this is probably most important in primary antibody responses, because in secondary responses B lymphocytes are available for antigen presentation to already differentiated effector and memory T cells. There is no evidence that B lymphocytes themselves need to recognize protein antigens presented by accessory cells. Two situations where B lymphocytes do recognize antigens displayed by other cells are in germinal center reactions, when follicular dendritic cells present antigens, and in responses to polysaccharides, which bind to marginal zone macrophages.

2. *Macrophages secrete cytokines,* which may augment the proliferation of both B and T lymphocytes. IL-1 and TNF may directly stimulate B cell proliferation, IL-6 is a growth and differentiation factor for B cells, and IL-12 stimulates the differentiation of CD4$^+$ T cells into T$_H$1 effectors.

SUMMARY

Antibody responses are initiated by the interaction of antigen with specific membrane Ig molecules on B lymphocytes. This interaction triggers intracellular signaling cascades that stimulate resting B cells to enter the cell cycle and that enhance the expression of costimulators and cytokine receptors. Thymus (or T cell)-dependent protein antigens are endocytosed and processed by the B cells, and peptides are presented in association with class II MHC molecules to antigen-specific, class II MHC–restricted CD4$^+$ helper T lymphocytes. The helper T cells are activated by recognition of antigen and costimulators on the B lymphocytes. As a result of their activation, helper T cells express CD40 ligand, which specifically recognizes CD40 on B cells, and this interaction stimulates the proliferation and differentiation of the B cells. In addition, helper T lymphocytes secrete cytokines that amplify antibody responses and regulate heavy chain class switching. Antigen-stimulated B cells undergo affinity maturation in the germinal centers of lymphoid follicles and develop into antibody-secreting cells or long-lived memory cells.

Thymus-independent antigens induce antibody responses without the participation of antigen-specific helper T cells. Most TI antigens are polymeric polysaccharides, glycolipids, and nucleic acids that efficiently cross-link membrane Ig on B cells. Additional signals may be required for antibody responses to TI antigens, but these are not well defined.

The binding of complement proteins to antigens amplifies antibody responses by increasing B cell activation and by providing an efficient mechanism for the display of the antigen in germinal centers. Antigen-antibody complexes that bind to Fcγ receptors on B cells inhibit antibody production, thus serving a negative feedback function. Accessory cells like macrophages and dendritic cells promote antibody production by various mechanisms.

Selected Readings

Banchereau, J., and F. Rousset. Human B lymphocytes: phenotype, proliferation, and differentiation. Advances in Immunology 52:125–262, 1992.

Clark, E. A., and J. A. Ledbetter. How B and T cells talk to each other. Nature 367:425–428, 1994.

Coffman, R. L., D. A. Lebman, and P. Rothman. The mechanism and regulation of immunoglobulin class switching. Advances in Immunology 54:229–270, 1993.

Fearon, D. T., and R. H. Carter. The CD19/CR2/TAPA-1 complex of B lymphocytes: linking natural to acquired immunity. Annual Review of Immunology 13:127–149, 1995.

Foy, T. M., A. Aruffo, J. Bajorath, J. E. Buhlman, and R. J. Noells. Immune regulation by CD40 and its ligand. Annual Review of Immunology 14:591–617, 1996.

Gold, M. R., and A. L. de Franco. Biochemistry of B lymphocyte activation. Advances in Immunology 55:221–295, 1994.

Kelsoe, G. *In situ* studies of the germinal center reaction. Advances in Immunology 60:267–288, 1995.

Lanzavecchia, A. Receptor-mediated antigen uptake and its effect on antigen presentation to class II MHC-restricted T lymphocytes. Annual Review of Immunology 8:773–793, 1990.

Mond, J. J., A. Lees, and C. M. Snapper. T cell-independent antigens type 2. Annual Review of Immunology 13:665–692, 1995.

Parker, D. C. T cell-dependent B cell activation. Annual Review of Immunology 11:331–360, 1993.

Rajewsky, K. Clonal selection and learning in the antibody system. Nature 381:751–758, 1996.

Reth, M. Antigen receptors on B lymphocytes. Annual Review of Immunology 10:97–121, 1992.

Stavnezer, J. Antibody class switching. Advances in Immunology 61:79–146, 1996.

van der Eertwegh, A. J. M., J. D. Laman, R. J. Noelle, W. J. Boersma, and E. Claassen. *In vivo* T cell–B cell interactions and cytokine production in the spleen. Seminars in Immunology 6:327–336, 1994.

CHAPTER TEN

REGULATION OF

IMMUNE RESPONSES

In previous chapters we have discussed how T and B lymphocytes recognize and respond to antigens, concentrating on the cellular and molecular mechanisms involved in antigen recognition and in the activation of antigen-specific clones of lymphocytes. One of the cardinal features of all immune responses is their self-limitation, which is manifested by the decline of responses with time after immunization (see Chapter 1). The principal reason for this self-limitation is that each immune response eliminates the antigen that initiated the response and thus eliminates the necessary first signal for lymphocyte activation. We have also frequently mentioned that immune responses are qualitatively and quantitatively heterogeneous, i.e., different types of microbes and other antigens tend to stimulate responses with distinct features, and some forms of antigen exposure may be inhibitory rather than stimulatory. Understanding these regulatory mechanisms is the key to developing rational strategies for manipulating immune responses. We will first discuss the mechanisms that inhibit specific immunity and then describe the factors that influence the magnitude and nature of immune responses.

MECHANISMS THAT INHIBIT IMMUNE RESPONSES

It has been known for many years that foreign antigens may be administered in ways that shut off rather than stimulate specific lymphocytes. The reason for this is that the immune system has evolved mechanisms whose principal role is to inhibit the activation and effector functions of lymphocytes. The physiologic importance of these inhibitory mechanisms is that they are used to maintain unresponsiveness to self antigens and to limit the magnitude of immune responses specific for foreign antigens in order to prevent injurious side effects of lymphocyte activation. Inhibition of specific lymphocyte responses may be exploited to treat harmful reactions, such as in autoimmune diseases and transplant rejection. The two principal inhibitory mechanisms in the immune system are immunologic tolerance and the induction of regulatory or suppressor cells.

IMMUNOLOGIC TOLERANCE TO FOREIGN ANTIGENS

Immunologic tolerance is the phenomenon of antigen-induced functional inactivation or death of specific lymphocytes, resulting in the inability of an organism to respond to that antigen. Lymphocyte activation and tolerance are two possible results of specific recognition of antigens by lymphocytes. Antigens that induce tolerance are called **tolerogens,** to be distinguished from **immunogens,** which generate immune responses. Tolerance to self antigens is a fundamental property of the immune system, and its failure leads to autoimmune diseases (see Chapter 19). Normally, all self antigens act as tolerogens. Many foreign antigens can be immunogens or tolerogens, depending on their physicochemical form, dose, and route of administration. Exposure of an individual to immunogenic antigens stimulates specific immunity, and for most immunogenic proteins subsequent exposures generate enhanced secondary responses. In contrast, exposure to a tolerogenic antigen not only fails to induce specific immunity but also inhibits lymphocyte activation by subsequent administration of immunogenic forms of the same antigen (Fig. 10–1). Thus, tolerance is not simply a passive lack of response but an actively induced unresponsiveness as a consequence of specific antigen recognition.

The phenomenon of immunologic tolerance is important for several reasons:

1. Tolerance to self antigens protects the individual from harmful autoimmune reactions.

2. Exposure of an individual to foreign antigens in particular ways may inhibit rather than stimulate specific immunity because of the induction of tolerance.

3. A balance between lymphocyte activation and tolerance may influence the magnitude of all specific immune responses.

4. Microbes may evolve ways of inducing tolerance as a means of evading host immunity.

5. Strategies for selectively inducing tolerance to defined antigens are now being tested for the treatment of autoimmune and allergic diseases and for preventing the rejection of organ transplants.

In this section we will discuss tolerance to foreign antigens. Self-tolerance is described in Chapter 19, in the context of autoimmune diseases. Despite this division, it is likely that the mechanisms of lymphocyte tolerance are essentially the same whether the tolerogen is a self antigen or a foreign antigen.

The classical studies of Peter Medawar and his colleagues in the 1950s demonstrated, for the first time, that tolerance is an immunologic phenomenon that can be analyzed experimentally. They showed that a neonatal mouse of one inbred strain could be made tolerant to the major histocompatibility complex (MHC)–encoded tissue antigens of a different strain by injection of lymphoid cells from the second strain. Once the recipient of the neonatal injection became an adult, it would accept a skin graft from the donor strain (Fig. 10–2). Moreover, lymphocytes from the recipient would not proliferate when cultured with cells from the donor strain in a mixed leukocyte reaction (see Chapter 17). This induced unresponsiveness to allogeneic MHC molecules was highly specific, since the recipient mouse would reject skin allografts and would respond to stimulator cells from all strains

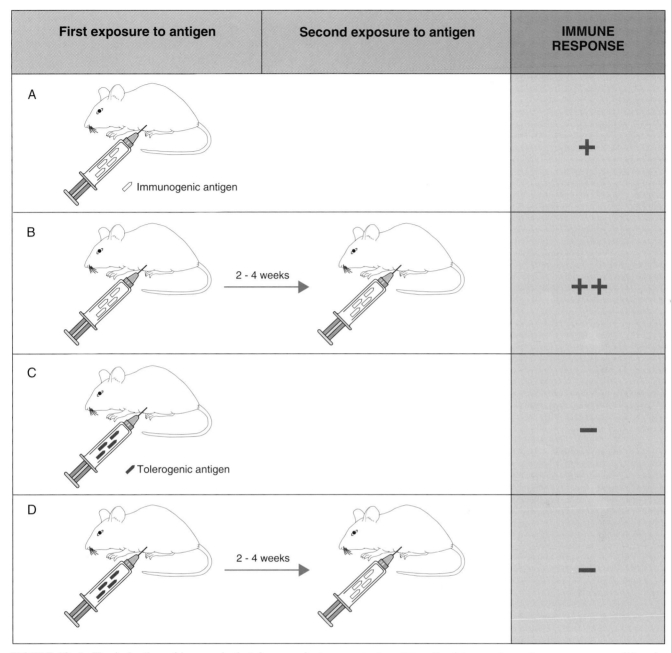

First exposure to antigen	Second exposure to antigen	IMMUNE RESPONSE
A — Immunogenic antigen		+
B — 2-4 weeks		++
C — Tolerogenic antigen		−
D — 2-4 weeks		−

FIGURE 10–1. The induction of immunologic tolerance. An immunogenic antigen stimulates a primary immune response (A) and a stronger secondary response upon subsequent immunization (B). A tolerogenic form of the same antigen does not induce an immune response (C) and prevents the response to subsequent immunization with the immunogenic antigen (D).

that differed from the donor at the MHC locus. Medawar's experiments were also the first to formally demonstrate that, although adult animals rejected grafts of foreign cells, exposure of immature (neonatal) animals to foreign antigens, in this case allogeneic MHC molecules, induced long-lived and specific unresponsiveness to these antigens. This form of tolerance, although induced by a single injection of foreign cells, is long-lived probably because some of the injected allogeneic cells survive in the recipient, which becomes a chimera. Thus, immature lymphocytes specific for the donor MHC,

which are generated throughout the life of the recipient animal, encounter the persisting donor cells, and this leads to tolerance induction.

Later studies showed that not only cell-associated MHC molecules but also soluble antigens could induce specific tolerance in immature lymphocytes, and that some forms of antigens were tolerogenic even for mature lymphocytes. Although the phenomenon of immunologic tolerance has been known for many years, the operative mechanisms are still not fully understood. Moreover, much more is known about tolerance in CD4+ T

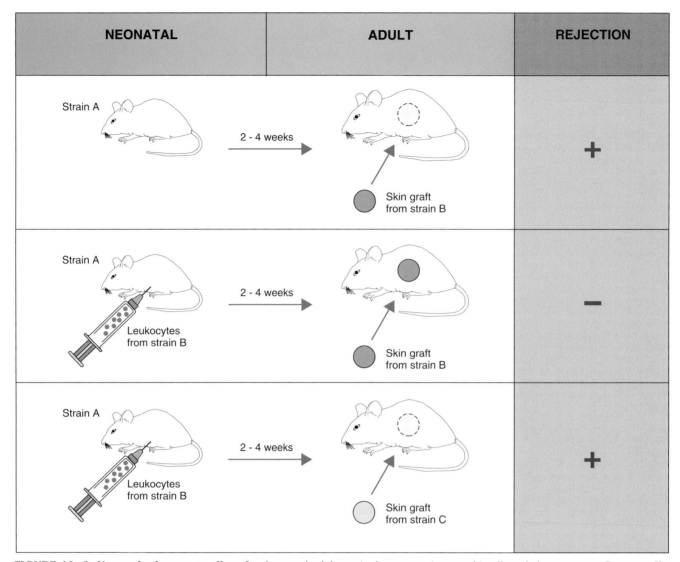

FIGURE 10–2. Neonatal tolerance to allografts. A normal adult strain A mouse rejects a skin allograft from a strain B mouse. If a neonatal strain A mouse is injected with strain B leukocytes, once the mouse becomes an adult, it fails to reject a skin graft from strain B. This form of tolerance is immunologically specific because the strain A mouse rejects a graft from a strain C donor. In this example, lymphocytes from the neonatally injected strain A mouse will not respond in vitro to strain B stimulators (in a mixed leukocyte reaction) but will respond normally to strain C stimulators.

cells than in other lymphocytes, as we shall discuss below.

General Properties of Immunologic Tolerance

Studies done in a variety of experimental systems have established the following general properties of immunologic tolerance.

1. *Tolerance is immunologically specific* and, therefore, must be due to the deletion or inactivation of antigen-specific T and/or B lymphocytes. Both lymphocyte activation and tolerance are induced by interactions of antigens with the same types of clonally distributed receptors on antigen-specific cells, i.e., membrane immunoglobulin (Ig) on B cells or the T cell receptor (TCR) on MHC-restricted T cells.

2. *Immature or developing lymphocytes are more susceptible to tolerance induction than are mature or functionally competent cells.* During their normal maturation in the generative lymphoid organs, all lymphocytes go through a stage at which antigen recognition leads to their death or inactivation. At this stage, potentially self-reactive lymphocyte clones encounter self antigens and become tolerant to these antigens (see Chapter 19). This type of tolerance, which is induced in immature lymphocytes within the generative lymphoid organs, has been called **central tolerance.** Central tolerance is important for maintaining unresponsiveness to self antigens that are present at high concentrations in the generative lymphoid organs, but it does not play a role in tolerance to foreign antigens administered in the periphery. Central tolerance will be discussed in Chapter 19.

3. *Tolerance to foreign antigens is induced even in mature lymphocytes when these cells are exposed to antigens under particular conditions.* Tolerance induced in mature lymphocytes that encounter antigens in peripheral tissues is called **peripheral tolerance.** The mechanisms of peripheral tolerance are discussed below.

Thus, whether the recognition of a foreign antigen by lymphocytes results in activation or tolerance depends on the maturational stage of the specific lymphocytes, the nature of the immunologic stimulus, and (for T cells) the nature of the antigen-presenting cells (APCs) that present the antigen.

Immunologic tolerance results from two possible consequences of the encounter of clones of antigen-specific T and B lymphocytes with antigen: **clonal deletion,** or cell death, and **clonal anergy,** or functional inactivation without cell death. It is not yet definitely established whether anergy and deletion are stages in a continuum of lymphocyte encounter with tolerogens or whether they are induced by different tolerogens under different conditions. It is also possible that an individual may develop unresponsiveness to a particular antigen despite the presence of mature, antigen-responsive lymphocytes. In these situations, the growth and differentiation of immunocompetent lymphocytes may be actively inhibited by other mechanisms, such as suppressor T cells. Thus, failure of an individual's immune system to respond to an antigen may be due to any combination of the absence (deletion) or inactivation (anergy) of T and/or B lymphocytes and the inhibitory effects of suppressor cells. We will first discuss how T and B lymphocytes are rendered unresponsive or are deleted by interacting with tolerogenic antigens in peripheral tissues; regulation by suppressor cells is discussed later in this chapter.

Mechanisms of Peripheral T Lymphocyte Tolerance

There are numerous experimental systems for inducing tolerance to protein antigens. These include the administration of large doses of antigens without adjuvants, oral antigen administration (called "oral tolerance"), repeated exposure to the antigen, and administration of mutated forms of the antigen. In all these situations, the tolerogens induce unresponsiveness by inhibiting the proliferation and differentiation of antigen-specific CD4+ T cells or by stimulating regulatory cells. Tolerogenic proteins induce unresponsiveness in CD4+ T lymphocytes by several mechanisms (Fig. 10-3).

ANERGY INDUCED BY ANTIGEN RECOGNITION WITH DEFECTIVE COSTIMULATION

If CD4+ T cells recognize processed antigens presented by APCs that are deficient in costimulators, the T cells do not proliferate and differentiate and are rendered incapable of responding to the antigen even if it is later presented by competent APCs. This type of functional unresponsiveness, or anergy, was first demonstrated by *in vitro* experiments with mouse T_H1 cell lines. If such T cell lines are exposed to peptide-MHC complexes on synthetic lipid membranes or on APCs that are treated with chemicals that destroy costimulators like B7 molecules, the T cells survive but are rendered anergic (Fig. 10-4). Anergy may be prevented by adding accessory cells that do express costimulators, or by stimulating CD28, the T cell receptor for B7 molecules, with specific agonistic antibodies. Such results laid the foundation for the concept that full activation of T cells requires the recognition of antigen ("signal 1") as well as costimulators, mainly B7-1 and B7-2 ("signal 2") (see Chapter 7). Signal 1, i.e., antigen recognition, alone may lead to no response or anergy. Anergic T cells are incapable of activating some of the transcription factors that bind to the interleukin-2 (IL-2) promoter and, therefore, fail to produce IL-2. This state of functional unresponsiveness is long-lived—*in vitro*, T cells that are rendered anergic may survive for 3 weeks or longer in this "dormant" state.

A similar form of antigen-specific functional unresponsiveness is induced *in vivo* if large doses of protein antigens are administered without adjuvants. It is postulated that, in the absence of adjuvants, APCs are not activated to express high levels of costimulators, so that T cells recognize the antigen without adequate costimulation. An experimental model that allows quantitative analyses of the responses of antigen-specific T lymphocytes to immunogenic and tolerogenic antigens is based on transferring T cells from transgenic mice expressing an antigen receptor of known specificity to normal mice, and exposing the recipients to the cognate antigen in different forms. If the protein antigen is administered in an immunogenic form, e.g., subcutaneously with adjuvants, in the draining lymph nodes antigen-specific T cells proliferate, differentiate into effectors, e.g., T_H1 cells, and migrate into lymphoid follicles, where T–B cell interactions may occur. In contrast, if the antigen is administered in a tolerogenic form (e.g., a high dose of aqueous antigen injected systemically), the antigen-specific T cells remain viable but they fail to expand, differentiate, or migrate into follicles. Thus, *anergy blocks the clonal expansion and effector functions of specific T cells.* Exposure of anergic T cells to high concentrations of the growth factor IL-2 may "break" anergy and allow the cells to respond. This may happen if adjacent T cells are stimulated by another antigen or by an infectious agent, and it is a postulated mechanism of autoimmunity (Chapter 19).

Many questions about T cell anergy remain unanswered. The fates of anergic T cells *in vivo*, i.e., how long they live, and whether or not they ever re-enter the pool of immunocompetent lymphocytes, are not known. The biochemical mecha-

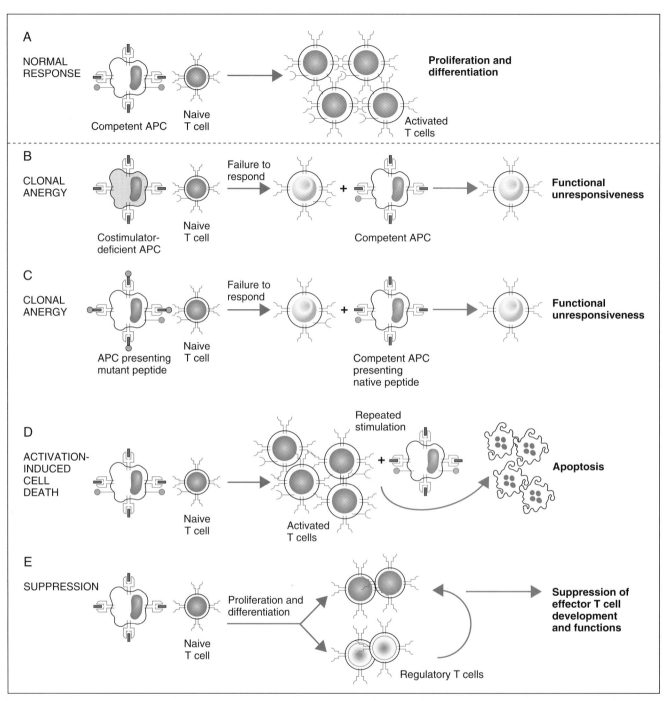

FIGURE 10–3. Mechanisms of peripheral tolerance in CD4⁺ T lymphocytes. Presentation of antigen by competent antigen-presenting cells (APCs) stimulates a normal T cell response (A). Anergy is induced if the T cells recognize the antigen presented by costimulator-deficient APCs (B) or if the antigen is mutated to alter TCR contact residues (C). Repeated stimulation of T cells results in activation-induced death (D). In some situations, the induction or functions of antigen-specific effector T cells are suppressed by regulatory cells also specific for the antigen (E).

nism of this type of *in vivo* functional unresponsiveness is also not defined. It is not even established that tolerogenic antigens are indeed presented by costimulator-deficient APCs *in vivo,* because the cells that actually present relevant peptide-MHC complexes after administration of protein antigens have not been identified in any immune response. Therefore, although T cell anergy is the postulated mechanism of peripheral tolerance in many situations of exposure to tolero-

genic antigen, its importance remains a matter of uncertainty. As we shall see in Chapter 19, the same is true of anergy as a mechanism of tolerance to self antigens.

ANERGY INDUCED BY RECOGNITION OF ALTERED PEPTIDE ANTIGENS

If CD4⁺ T lymphocytes specific for a particular peptide antigen recognize that antigen presented

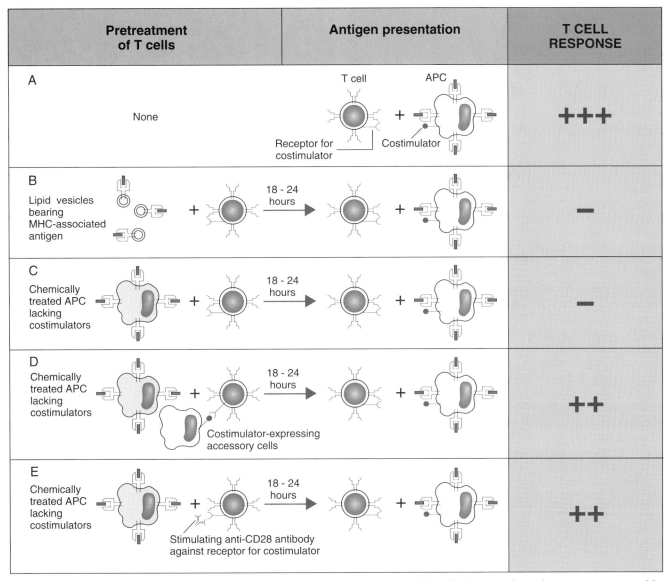

Pretreatment of T cells	Antigen presentation	T CELL RESPONSE
A None		+++
B Lipid vesicles bearing MHC-associated antigen		−
C Chemically treated APC lacking costimulators		−
D Chemically treated APC lacking costimulators		++
E Chemically treated APC lacking costimulators		++

FIGURE 10–4. Role of costimulators in T cell anergy. A peptide antigen-specific CD4$^+$ T cell responds to that antigen presented by major histocompatibility complex (MHC)–matched antigen-presenting cells (APCs) (A). Exposure of the T cell to the same peptide presented on class II MHC-bearing lipid vesicles (B) or by chemically fixed APCs (C) inhibits subsequent responses to the antigen presented by competent APCs. This form of clonal anergy is prevented by costimulator (e.g., B7)-expressing accessory cells (D) or by a stimulating antibody against a receptor for a costimulator (e.g., an anti-CD28 antibody) (E).

by competent APCs, the T cells are activated. *If, however, the same T cells first encounter a variant form of the antigen in which residues that contact the TCR are subtly altered, the cells may be rendered anergic.* Such altered antigens have been called **altered peptide ligands** (see Chapter 7). This form of anergy was also first demonstrated by *in vitro* experiments using cloned CD4$^+$ T cell lines. The APCs that present altered peptide ligands do express costimulators, so that T cells receive "signal 2," but antigen recognition (signal 1) is abnormal. It is believed that the interaction of altered peptide ligands with specific antigen receptors is a low-affinity interaction that induces partial and abortive T cell responses, with defective tyrosine phosphorylation of TCR-associated proteins and incomplete activation of downstream intermediates in the biochemical signaling cascade (see Chapter 7).

Since altered peptide ligands induce anergy in T cells specific for the native antigen, they can be used to prevent the induction of experimental autoimmune diseases in which the self antigen is known, by administering mutated versions of the self antigen. Some microbes may produce mutants of their own antigens, thus preventing specific immune responses and evading protective immunity.

DELETION OF T CELLS BY ACTIVATION-INDUCED CELL DEATH

Repeated stimulation of T lymphocytes by antigens or polyclonal activators may result in death of the activated cells by a process of apoptosis. This form of regulated cell death is called **activation-induced cell death.** Its physiologic importance is to prevent uncontrolled T cell activation. In CD4$^+$ T cells, it is usually a result of the interaction of two co-expressed molecules on activated cells, a surface receptor called **Fas** (CD95) and its ligand (Box 10–1). Fas is a member of the tumor necrosis factor (TNF) receptor family, and Fas ligand is structurally homologous to the cytokine TNF. The proteins of these receptor:ligand families regulate the choice between cellular proliferation and apoptotic death. Fas ligand is expressed primarily on activated T cells, so that this pathway of activation-induced cell death is often used by T lymphocytes. Binding of Fas ligand to Fas, or of TNF to the TNF receptor, activates a series of intracellular cysteine proteases that ultimately cause the fragmentation of nucleoproteins, apoptotic death of the cells, and the rapid removal of dead cells by phagocytes. Fas is the major mediator of activation-induced cell death in CD4$^+$ T cells, whereas the TNF receptor may be a more important mediator of death in CD8$^+$ T cells. High concentrations of the growth factor IL-2 enhance the expression of Fas ligand on antigen-stimulated T cells and the development of sensitivity to Fas-mediated apoptosis. Thus, IL-2 is both a growth factor for T cells and a feedback regulator of T cell responses.

Activation-induced cell death can be readily induced *in vitro* in T cells exposed to polyclonal activators and in antigen-specific cloned T cell lines exposed to high concentrations of their cognate antigens. *In vivo,* this type of cell death occurs if animals are given bacterial toxins, called "superantigens," that bind to and stimulate large numbers of T cells (see Box 16–1, Chapter 16). In all these situations, large numbers of T cells are activated and produce IL-2. It is not clear if activation-induced cell death can be induced by tolerogenic doses of conventional protein antigens *in vivo,* because normally very few T cells respond to any one antigen and they may not develop sensitivity to this death pathway, perhaps because they do not produce enough IL-2. Therefore, activation-induced cell death may be a mechanism of T cell tolerance and feedback regulation of immune responses only under special circumstances. Fas-mediated cell death is clearly important for the maintenance of tolerance to some self antigens, because mutations in Fas or Fas ligand are associated with autoimmune diseases (see Chapter 19). (Recall that in Chapter 2 we discussed programmed cell death as a homeostatic mechanism in immune responses, which resulted in the elimi-

---

BOX 10–1. The Fas Pathway in Activation-Induced Cell Death

Fas, also called CD95, was identified as a ~36 kD surface protein that, upon cross-linking by specific antibodies, triggered apoptosis of cells that expressed it. It belongs to a family of proteins that include receptors for the cytokine, tumor necrosis factor (TNF), and the B cell activating molecule, CD40, which was mentioned in Chapter 9. These proteins are expressed on the surfaces of many cells and share sequence homologies in their extracellular domains (see Box 12–4, Chapter 12). Members of the Fas/TNF-R/CD40 family dictate the choice between cell survival and proliferation on the one hand and apoptotic death on the other. In most cell populations and under most conditions, Fas and the type I TNF receptor (TNF-RI) are death-inducing receptors, whereas CD40 transduces anti-apoptotic and proliferative signals, and TNF-RII can promote death or growth in different cells. However, there are documented exceptions to even this simple classification.

Lymphoid and other cells express Fas, and in lymphocytes the levels of this protein increase upon stimulation by antigen. The ligand for Fas is a membrane protein that is expressed mainly on activated T lymphocytes. The activation of mature T lymphocytes results in the co-expression of Fas and Fas ligand on the cells. Reactivation, especially in the presence of IL-2, leads to engagement of Fas by Fas ligand and triggering of an apoptotic pathway. This is called **activation-induced cell death** and is a mechanism for preventing uncontrolled activation of lymphocytes, e.g., by abundant and persistent self antigens. Great interest in the function of Fas was spurred by the demonstration that in two inbred mouse strains that develop fatal autoimmune disease there are recessive mutations in the genes encoding either Fas or Fas ligand. There are rare cases of similar disorders in humans. We will return to a discussion of these autoimmune disorders in Chapter 19. In some cell populations, such as CD8$^+$ T lymphocytes, activation-induced cell death is apparently triggered not by Fas but by TNF receptors. In yet other cells, such as developing thymocytes that encounter high concentrations of self antigens, negative selection is also due to activation-induced cell death, but it does not appear to rely on either Fas or TNF receptors (see Chapter 8). Not all forms of apoptosis in lymphocytes are a consequence of activation. In fact, many lymphocytes are programmed to die unless protected by receptor-mediated stimulation or growth factors (see Box 2–2, Chapter 2). This type of programmed death is due to neglect rather than activation and does not involve the Fas/TNF receptor family.

In addition to its role in the maintenance of peripheral tolerance to self antigens, Fas:Fas ligand–mediated cell death serves a number of other functions. Cytolysis by CD8$^+$ CTLs is in part mediated by Fas ligand on the CTLs binding to Fas on target cells (see Chapter 13). Two tissues known to be sites of immune privilege, namely, the testes and the eyes, constitutively express Fas ligand. It is postulated that this kills leukocytes that enter the tissues, thus preventing local immune responses, which is the hallmark of immune privilege. (The physiologic value of immune privilege in these tissues, and in the central nervous system, is not understood.) Fas ligand induced on helper T cells may recognize Fas on B cells and function to limit antibody responses, especially if the B cells are not specifically protected by antigen recognition.

Many of the biochemical steps in Fas-mediated apoptosis have been identified (see Figure). Fas ligand is a trimer, and its binding to Fas leads to oligomerization of the receptor. The cytoplasmic region of Fas contains a conserved sequence called the "death domain" because it is required for the death signal. This domain is actually a protein-protein interaction domain. It mediates the binding of a second protein, called FADD (Fas-associated death domain), to the cytoplasmic tail of Fas, followed by a third protein whose C-terminus is homologous to a family of unusual cysteine proteases that belong to the interleukin-1 converting enzyme (ICE)/Ced-3 family of proteases (see Box 2–2, Chapter 2).

Continued

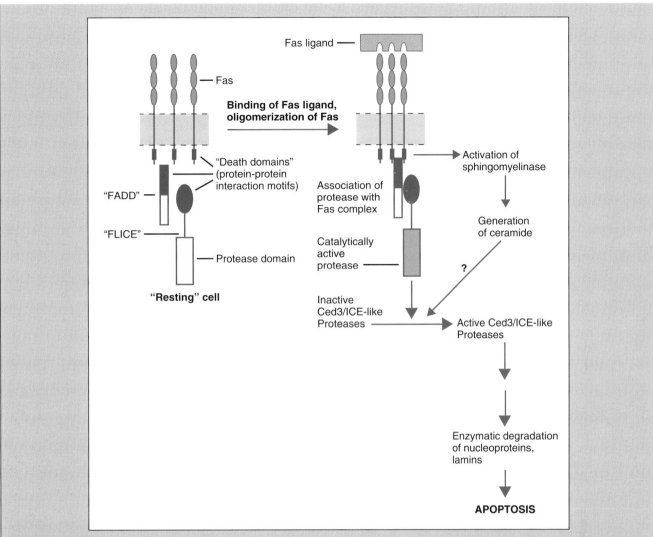

Model for Fas-mediated apoptosis. FADD (*Fas-a*ssociated *d*eath *d*omain) and FLICE (*F*ADD-*li*ke *i*nterleukin-1 *c*onverting *e*nzyme protease) are two intracellular proteins that enter into a complex with oligomerized Fas and initiate the enzymatic cascade that culminates in apoptosis. Ceramide is generated by Fas cross-linking and also induces apoptosis, but the exact location of ceramide in Fas-mediated AICD is not established.

The Fas-associated protease becomes catalytically active upon binding to the Fas complex and may activate other proteases. The substrates of these proteases are not fully defined but include nuclear proteins and other enzymes that degrade histones and matrix components, such as lamin. Fas-mediated activation of ICE/Ced-3 proteases may also occur by the sphingomyelinase-catalyzed generation of ceramide as a result of Fas cross-linking. The net result of the actions of these enzymes is the blebbing of the plasma membrane,

breakdown of nuclear deoxyribonucleic acid, and apoptosis. Apoptosis is associated with the expression on the cell surface of unidentified molecules that promote rapid phagocytosis of dead cells.

The 55 kD TNF-RI probably triggers a similar pathway of cell death, because its cytoplasmic domain binds a protein, called TRADD (TNF-receptor–associated death domain) that is homologous to FADD. This is discussed in more detail in Chapter 12.

nation of activated lymphocytes that were deprived of continuous stimulation or growth factors. This form of cell death is also due to apoptosis, but it is not mediated by Fas:Fas ligand interactions.)

Although experimental systems are providing valuable information about the mechanisms of anergy and deletion in mature CD4$^+$ T lymphocytes, the roles of these pathways in tolerance induced by different types of tolerogenic antigens,

such as antigens given without adjuvants or orally, are not firmly established. Even less is known about the mechanisms of tolerance in CD8$^+$ T cells. It is postulated that if CD8$^+$ T cells recognize class I MHC–associated peptides without costimulation or T cell help, they are inactivated. As we mentioned earlier, in some situations tolerogenic antigens may stimulate the development of lymphocytes whose main function is to suppress other lymphocytes. This, too, may be manifested as a state of immunologic unresponsiveness, or tolerance, but it is due to active suppression rather than anergy or deletion of specific lymphocytes. The properties and physiologic significance of suppressor cells are discussed later.

Mechanisms of B Lymphocyte Tolerance

Interactions of antigens with specific B lymphocytes (the first signal for B cell activation) in the absence of T cell help (the second signal) lead to tolerance in the B cells and a failure to produce antibodies. In principle, this is analogous to tolerance in T lymphocytes. Helper T cells and their secreted cytokines provide the "second signal" for B cell activation, analogous to the role of costimulators in T lymphocyte responses. There are numerous examples of B cell antigen recognition without stimulation by helper cells leading to tolerance. For instance, systemic administration of high doses of polysaccharide antigens or hapten-polysaccharide conjugates (which cannot be recognized by T cells) induces tolerance in B lymphocytes, preventing subsequent antibody responses to immunogenic doses or forms of the antigen. The binding of protein antigens to B cells in the absence of T cell help can also lead to B cell tolerance. In most of these situations the B cells become anergic owing to a block in Ig-mediated signaling and fail to migrate into lymphoid follicles, much like anergic T cells. However, it is not clear to what extent B cell tolerance contributes to deficient antibody responses after administration of tolerogenic protein antigens, because tolerogenic proteins may rapidly inactivate helper T cells, and this alone could account for inhibition of antibody responses.

The biochemical alterations that lead to B cell tolerance are poorly understood. Tolerance induction in B lymphocytes usually requires higher concentrations of antigens than in T cells, and anergic B cells recover more rapidly than do anergic T cells after the tolerogen is removed.

It is clear that much remains to be learned about immunologic tolerance to foreign antigens, especially about the mechanisms of tolerance induction *in vivo* and differences in the biochemical effects of immunogens and tolerogens on specific lymphocytes. The same problem will become apparent when we discuss self-tolerance and autoimmunity (see Chapter 19).

SUPPRESSOR T LYMPHOCYTES

Suppressor T cells were originally defined as a class of lymphocytes thought to be distinct from helper and cytolytic T lymphocytes, whose function is to inhibit the activation phase of immune responses. The existence of suppressor cells was demonstrated by experiments done in the late 1960s, using a complicated protocol that involved immunizing thymectomized and bone marrow–transplanted mice. A simpler experimental demonstration of suppressor cell effects is shown in Figure 10–5. Animals given an antigen in an immunogenic dose mount specific immune responses. If the antigen is injected under non-immunogenic conditions, e.g., high doses of aqueous antigen administered intravenously, the animals do not respond and may also be rendered unresponsive to subsequent administration of normally immunogenic antigen. This, of course, is often due to the induction of tolerance in specific lymphocytes. In other situations, however, if the lymphocytes from these unresponsive animals are adoptively transferred into syngeneic recipients, the recipients also fail to respond to a subsequent challenge with the normally immunogenic form of antigen. The inhibition is immunologically specific because the recipients of the adoptive transfer respond normally to a different antigen. Later studies showed that this form of unresponsiveness, which could be transferred from one individual to another, was mediated by T lymphocytes, which were termed **suppressor T cells.** From such studies evolved the hypothesis that the function of suppressor cells is to inhibit the activation of functionally competent antigen-specific T and/or B lymphocytes.

In the 1970s, many experimental systems were used to study suppressor cells, and their excessive or deficient function was invoked as the primary basis for a variety of immunologic abnormalities. However, for reasons that are discussed below, progress in our understanding of suppressor T cells has been slow, so that their significance and even their existence as a distinct class of T cells are doubted by many investigators.

Early studies with cultured human lymphocytes and experimental animals showed that *suppressor T cells have the following properties:*

1. Suppressor T cells are generally induced by the same immunization conditions that induce clonal anergy of lymphocytes, such as high concentrations of protein antigens or chemically reactive haptens administered without adjuvants or injected intravenously.

2. In many experimental systems, the cells that inhibit immune responses are CD8$^+$. Their growth and differentiation may be dependent on CD4$^+$ cells.

3. In mice, different populations of suppressor T cells have been shown to be specific for antigens, such as proteins or haptens, or for the idiotypic determinants of lymphocyte receptors or secreted antibodies (see below). Most studies with human suppressor T lymphocytes have examined the inhibitory effects of CD8$^+$ cells on the responses of polyclonally stimulated B or T cells, and there are few documented examples of antigen-specific or idiotype-specific human suppressor cells. It is, therefore, difficult to draw general conclusions about the immunologic specificity of suppressor cells in humans.

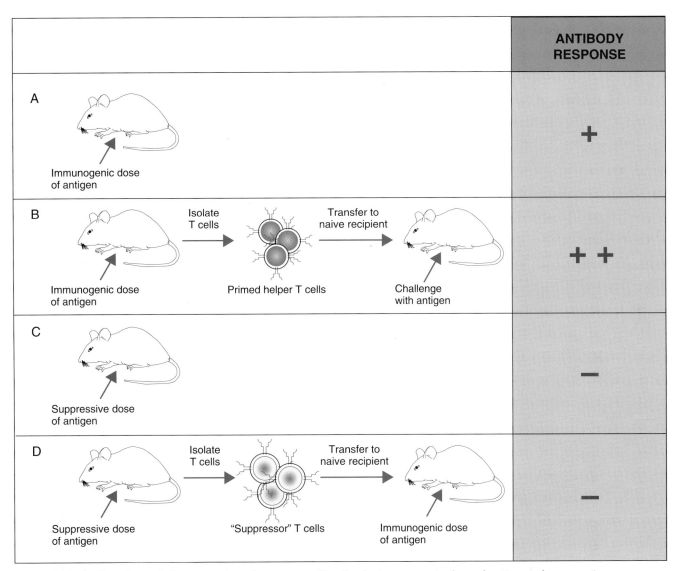

FIGURE 10-5. Experimental demonstration of suppressor T cells. An immunogenic dose of antigen induces a primary response (A), and transfer of T cells from a primed animal to a naive animal enhances the ability of the recipient to respond to challenge with immunogenic antigen (B). A suppressive dose of antigen fails to induce a response (C). Transfer of T cells from animals treated with suppressive antigen to normal recipients inhibits the ability of the recipients to respond to immunogenic antigen challenge (D). Note that the recipients would respond normally to a different immunogenic antigen.

4. The inhibitory effects of suppressor T cells are mediated by secreted proteins. Unlike the cytokines produced by other T lymphocytes, which are not antigen-specific, the suppressor factors secreted by murine suppressor T cells were found to have the same antigenic or idiotypic specificities as the T cells themselves. This led to the postulate that suppressor factors are functionally active secreted forms of T lymphocyte receptors, much as secreted and membrane Ig are two forms of the same antigen-specific B lymphocyte product. It is possible that soluble receptor molecules bind to MHC-associated antigens on APCs and competitively inhibit the activation of other T cells. However, it is not known whether this occurs *in vivo* or whether secreted T cell antigen receptors play a significant role in the physiologic downregulation of specific immune responses.

A major problem in the study of suppressor T cells has been that attempts to purify these cells in numbers sufficient for biochemical analyses of receptors and secreted products or to establish stable cloned lines or hybridomas with specific suppressive activity have been largely unsuccessful. As a result, even basic questions, such as the nature of the receptors expressed by suppressor cells, are unresolved. Isolation, biochemical characterization, and molecular cloning of suppressor factors have also not been successful despite considerable effort. It is, therefore, not possible at present to construct a model for the specificity, mode of action, or function of suppressor T cells that fits all the available data.

Despite these concerns, it is likely that some antigens do stimulate lymphocyte populations whose major effect is the downregulation of spe-

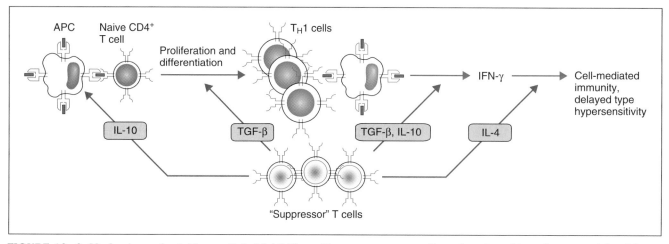

FIGURE 10-6. Mechanisms of cytokine-mediated inhibition of immune responses. Examples of cytokines that may inhibit different phases of a T_H1 response are shown. IFN, interferon; IL, interleukin; TGF, transforming growth factor; APC, antigen-presenting cell.

cific immune responses. *It may be that suppressor T cells are not a unique cell population but actually consist of lymphocytes that can inhibit immune responses in different ways.* (For this reason, and to avoid the implications of the older terminology, a better term for this cell population may be "regulatory" or "suppressive" T cells.) Like other T lymphocytes, suppressor cells may recognize antigens in a specific manner and may function by various non-specific effector mechanisms. The best documented mechanism of action of suppressor T cells is the production of an excess of cytokines with inhibitory function (Fig. 10-6). Because cytokines have both stimulatory and inhibitory effects on lymphocytes, the nature and magnitude of the overall immune response are determined by the relative concentrations of different cytokines at the site of immune activation. For instance, transforming growth factor-β (TGF-β) is a powerful inhibitor of T and B cell proliferation. Therefore, an excess of TGF-β can inhibit immune responses, and cells that secrete large amounts of this cytokine may function as suppressor cells. In other situations, different cytokines may inhibit different types of immune responses. For instance, IL-4, IL-10, and IL-13 produced by the T_H2 subset of helper T lymphocytes (see Chapter 12) inhibit macrophage activation. Interferon-γ inhibits IL-4-mediated B cell switching to IgE. Therefore, T_H2 cells may suppress cell-mediated immunity and delayed type hypersensitivity, and this regulatory role may be an important physiologic function of the T_H2 subset. The administration of high doses of protein antigens, and some altered peptide ligands, preferentially induces T_H2 responses, which may contribute to the deficient cell-mediated immunity seen in animals exposed to these tolerogenic antigens. This ability to regulate immune responses is not unique to T_H2 or TGF-β-producing T lymphocytes. IFN-γ-producing T_H1 cells may function as suppressors of IgE production, because IFN-γ inhibits IL-4-stimulated B cell switching to IgE. The important conclusion

is that *various T cell populations are capable of suppressing different immune responses,* and there may not exist one unique population of "suppressor cells."

Some suppressor cells may mediate their effects by killing other cells involved in immune responses. Antigen-specific CD8+ and some CD4+ T lymphocytes can lyse target cells bearing the stimulating antigens in association with MHC molecules (class I and class II, respectively). This cytolysis results from Fas ligand-Fas interactions and the secretion of cytokines such as TNF and lymphotoxin. It is possible that cytolytic T lymphocytes (CTLs) can specifically lyse B and helper T cells or APCs that express foreign protein antigenic determinants in association with their MHC molecules and, therefore, serve as targets for the CTLs (see Chapter 13). This could result in an inhibition of specific immune responses.

Such postulated mechanisms for the suppressive effects of certain T lymphocytes are consistent with the documented biologic activities of T cells. However, this does not explain the published reports of antigen-specific or idiotype-specific suppressor factors. It is clear that simple, quantitative experimental systems are needed to better analyze the role of suppressor T lymphocytes in the down-regulation of immune responses and the maintenance of tolerance to self and foreign antigens.

IDIOTYPIC REGULATION

The third mechanism of antigen-initiated immune regulation, in addition to tolerance and suppressor cells, was originally suggested by the observation that idiotypes, which are components of antigen receptors, are themselves potentially immunogenic, i.e., immune responses can be induced in an individual against antigen receptors expressed by the lymphocytes of that individual. Such anti-receptor responses may serve a regulatory function. The idea that cells in an individual can respond to and discriminate between receptors on other similar cells is unique to the immune system, because only the immune system is endowed with sufficient diversity to allow such reciprocal recognition. This is a theoretical idea that continues to fascinate immunologists, although there is little formal proof that regulatory mechanisms based on recognition of idiotypes are important for the physiologic control of immune responses or as primary pathogenic mechanisms in immunologic diseases.

The concept of idiotypic regulation evolved from the realization that antigen receptors on T and B lymphocytes are structurally diverse, containing variable regions that differ among different clones. These receptors are also capable of distinguishing between subtle variations in protein sequences, so that lymphocytes can recognize and respond to proteins that are only slightly different from self proteins. It is, therefore, conceivable that if one clone of lymphocytes is expanded during an immune response to a foreign antigen, other lymphocytes might specifically recognize and respond to the variable regions of the antigen receptors on the antigen-stimulated clone. Such receptor-specific lymphocytes may then interact with and alter the function of the receptor-bearing clone. The structures or determinants of antigen receptors that distinguish each clone of lymphocytes from all others are called **idiotopes,** and the collection of idiotopes on a given antigen receptor constitutes that receptor's **idiotype** (see Box 3–2, Chapter 3). Immune responses specific for idiotypes are called **anti-idiotypic.** Such responses can either augment or inhibit the activation of lymphocytes that produce the idiotypes. Thus, idiotypes and anti-idiotypes may constitute a system of self-regulation that is both stimulated by and acts on immunocompetent lymphocytes and consequently influences immune responses to foreign antigens.

The possibility that anti-idiotypic immune responses may serve a regulatory function was most clearly appreciated by Niels Jerne and enunciated in his **network hypothesis** in 1974. Jerne postulated that an antigen stimulates a specific T and/or B cell response, which in turn induces a wave of complementary anti-idiotypic responses (Fig. 10–7). Idiotype-specific antibodies can recognize the original antigen-specific responding lymphocytes and inhibit or augment their activation. Anti-idiotypic T cells may function in the same manner, perhaps by recognizing idiotypic determinants of the membrane Ig on B cells that is recycled, processed, and presented in association with MHC molecules by the idiotype-producing B cells themselves. According to Jerne's network hypothesis, a steady state in the immune system is maintained by this network of reciprocal idiotypes and anti-idiotypes. The introduction of antigen perturbs that balance and leads to detectable immune responses.

Much of the evidence in support of the regula-

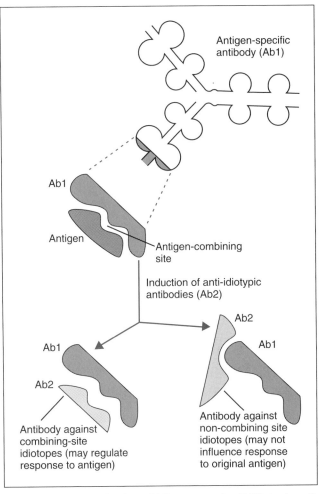

FIGURE 10–7. Production of idiotypes and anti-idiotypic antibodies. The combining site of an antibody, Ab1, specific for an antigen has a shape complementary to that of the antigen. Anti-idiotypic antibodies (Ab2) against Ab1 may be specific for the combining site of Ab1, in which case they may influence responses to the antigen, or they may be specific for idiotypic determinants of Ab1 that are not part of the combining site.

tory role of idiotypic interactions has come from experimental systems in which the immune response to an antigen is dominated by one or a few clones of responding lymphocytes. In such monoclonal or oligoclonal responses, one or a few antibodies with their unique idiotypes are dominant, so that their regulation can be manipulated and measured experimentally. The analysis of such experimental systems has shown that waves of antigen-specific antibodies and anti-idiotypic antibodies can be detected following immunization with an antigen. Moreover, administration of anti-idiotypic antibodies can inhibit or enhance immune responses to the antigen. These results support the potential regulatory function of idiotypic networks. However, it is not possible to study the importance of idiotypic regulation in conventional immune responses to most antigens, because in such responses numerous clones of lymphocytes are activated, so that a single idiotype is not dominant or even detectable. As a result, the significance of idiotypic networks in regulating most antigen-specific immune responses is uncertain, and this regulatory pathway remains largely an interesting theoretical concept.

ANTIBODY FEEDBACK

Antibodies produced in response to an antigen are capable of inhibiting further immune responses to that antigen. For instance, if antigen-specific antibodies are administered to an animal either shortly before immunization with the antigen or during an ongoing response to that antigen, subsequent antibody production is reduced. This phenomenon, called **antibody feedback,** can downregulate both humoral and cell-mediated immune responses. Feedback mediated by antibodies is important for ensuring that immune responses are self-limited and decrease in intensity with time after immunization.

This negative feedback function of antibodies is due to several mechanisms that may be operative at the same time:

1. *Antibodies eliminate and neutralize antigens and thereby remove the initiating stimulus for the immune response.* An injected antibody, or an antibody formed during an active response, complexes with the antigen. If the antigen-antibody complexes are formed by IgM or certain subclasses of IgG antibodies, they may subsequently activate the complement system (see Chapter 15). The complexes are avidly bound to and eliminated by Fcγ and/or complement receptor–bearing phagocytes and red blood cells. Clearance of antigens by enhancing their phagocytosis is one of the principal effector functions of antibodies. Antibodies also neutralize the stimulatory capacity of antigens by binding to antigenic determinants and blocking their access to specific membrane Ig on B lymphocytes. This effectively limits B cell activation.

Administration of pre-formed antibody, a form of passive immunization, is used to treat life-threatening diseases, such as tetanus infection and snake bites. Another practical application of passive immunization is in the prevention of **Rh disease** (the main cause of erythroblastosis fetalis), which remains one of the most dramatic examples of successful immunologic intervention for a serious disorder. Rh disease affects infants born of mothers who do not express Rh blood group antigens and fathers who do. The erythrocytes of such a fetus are Rh-positive because of inheritance of the paternal Rh genes. Fetal blood enters the maternal circulation in small amounts during gestation and in substantial quantities during the delivery itself. The immune system of the Rh-negative mother recognizes the fetal Rh as a foreign antigen. Therefore, the mother makes increasing amounts of anti-Rh antibodies with each successive pregnancy. During pregnancy these maternal antibodies cross the placenta, enter the fetal circulation, bind to the fetal erythrocytes, and cause hemolysis, which increases in severity with each pregnancy and can lead to fetal death. In order to prevent this disease, when an Rh incompatibility between the mother and the father is detected, the mother is injected with a large dose of an anti-Rh antibody immediately after each delivery. This antibody presumably binds to fetal Rh+ cells, inhibits the maternal immune response to fetal Rh antigens that are encountered at each delivery, and completely prevents the disease from developing.

2. *Antibodies directly inhibit B lymphocyte activation by binding to Fc receptors on B cells.* Immune complexes composed of an antigen and specific IgG antibody have been shown to inhibit the activation of B lymphocytes specific for that antigen *in vitro* and *in vivo*. Such immune complexes can form in the circulation during a humoral immune response or can be artificially produced *in vitro*. It is thought that immune complex–mediated inhibition results from simultaneous interaction of the antigenic portion of the complex with membrane Ig molecules on specific B cells and of the antibody portion of the complex with Fcγ receptors (Fcγ-RIIB) on the same cells (see Fig. 9–9, Chapter 9). This can occur only with multideterminant or aggregated antigens, in which one epitope of the antigen will participate in the formation of the immune complex and another epitope will be available for binding to membrane Ig on specific B cells. The engagement of the Fc receptor on B cells activates an intracellular phosphatase. If this Fc receptor is brought close to the antigen receptor by simultaneous binding of antigen-antibody complexes to both, the phosphatase removes phosphates from Ig-associated molecules. This blocks the increase in intracellular Ca^{2+} and the induction of intracellular second messengers that are required for B cell proliferation and differentiation (see Chapter 9). Thus, the major physiologic function of Fcγ receptors on B lymphocytes is their role in antibody-mediated feedback inhibition.

3. *Antigen-antibody complexes may regulate T cell responses and cytokine production.* It has been suggested that antigen-antibody complexes inhibit helper T cell activation and/or induce antigen-specific suppressor T lymphocytes, although there is little definitive evidence to support either mechanism. Some T lymphocytes express Fc receptors specific for IgM or IgG antibody, but the functions of these cells are not established. Antigen-antibody complexes may also influence the production of cytokines by non-lymphoid cells. For instance, immune complexes that bind to Fcγ receptors on macrophages stimulate the production of an IL-1 receptor antagonist that competitively inhibits binding of IL-1 to its receptor (see Chapter 12).

Antibody feedback is an excellent example of self-regulation because the effector molecules produced by the humoral immune response themselves serve to downregulate the response. Antibodies may also enhance immune responses. For instance, immune complexes may bind complement proteins, and these are known to stimulate further antibody production by several mechanisms (see Chapter 9).

REGULATORY EFFECTS OF CYTOKINES

Cytokines produced by T lymphocytes and accessory cells exert both stimulatory and inhibitory effects on immune responses. The **stimulatory actions** of cytokines frequently generate amplification loops that enable the small number of lymphocytes specific for any one antigen to recruit the multiple effector mechanisms required to eliminate that antigen. Cytokine cascades, in which one cytokine enhances the production of or functional responses to others, are described in Chapter 12. Among the most striking examples of cytokine-mediated amplification of immune responses are the bidirectional interactions between T lymphocytes and macrophages. For instance, CD4$^+$ T cells secrete IFN-γ, which enhances the expression of class II MHC molecules and costimulators on mononuclear phagocytes. Since these T cells recognize foreign antigens in association with class II MHC products and need costimulators for activation, increased expression of class II MHC and co-stimulatory molecules makes the macrophages better APCs and promotes T cell activation (see Fig. 6–4, Chapter 6). IL-4, secreted by CD4$^+$ T cells, increases class II MHC gene expression in B lymphocytes and may similarly enhance the avidity of antigen-specific, MHC-restricted T cell–B cell interactions. Some class I MHC–restricted CTLs secrete IFN-γ, lymphotoxin, and TNF, all of which stimulate class I MHC gene expression in target cells and enhance CTL-target interactions.

Cytokines also have profound **inhibitory effects** that might serve to regulate immune responses. The antagonistic effects of IL-4, IL-10, and IFN-γ on various immune responses, and the immunosuppressive effects of TGF-β, have been mentioned previously (see Fig. 10–6). Because of these actions of cytokines, conditions of immunization that result in dominant T$_H$1 or T$_H$2 development tend to induce responses in which some immune effector functions are active and others are suppressed (see Chapter 12). Examples of such cytokine imbalance are common in chronic infections (see Chapter 16).

FACTORS THAT DETERMINE THE NATURE AND MAGNITUDE OF IMMUNE RESPONSES

Exposure of the immune system to foreign antigens sets into motion the series of events that lead to lymphocyte activation and the generation of humoral and cell-mediated immunity. Different antigens and conditions of immunization lead to responses that vary both quantitatively and qualitatively. For instance, different antigens preferen-

tially stimulate the production of antibodies of various heavy chain isotypes, activate T cells to produce distinct sets of cytokines, or generate CTLs or other effectors of cell-mediated immunity. Such diversity of immune effector mechanisms is important because it enables the immune system to protect an individual from the many distinct types of microbes present in the environment that are most effectively eliminated by different mechanisms (see Chapter 16).

Three main factors influence the nature and magnitude of specific immune responses: (1) the type of *antigen,* as well as its dose and route of entry; (2) the numbers and types of *accessory cells* that initially interact with the antigen and induce lymphocyte activation; and (3) the nature of the *responding lymphocytes.* In the following section, we will discuss how each of these factors contributes to the development of specific immunity. Although these factors are considered individually, it should be kept in mind that antigens, accessory cells, and lymphocytes act in concert, influence one another in multiple ways, and are not separable in the induction or regulation of immune responses.

The Role of Antigen

Antigens are the obligatory first signals for lymphocyte activation. The nature of the antigen has a significant influence on the type and magnitude of the immune response that develops. These regulatory effects are of many different types:

1. *Chemically different antigens stimulate different types of immune responses.* Whereas protein antigens induce both humoral and cell-mediated immunity, polysaccharides and lipids fail to stimulate MHC-restricted T cells and to induce cell-mediated immune responses. Antibody responses to polysaccharides and lipids are typically T cell–independent and consist largely of IgM antibodies (see Chapter 9). Proteins, on the other hand, stimulate isotype switching, affinity maturation, and the generation of memory B cells. Thus, encapsulated bacteria, whose principal immunogens are capsular polysaccharides, usually stimulate low-affinity IgM and some IgG antibodies. In contrast, the protein antigens of some bacteria and most viruses induce strong humoral and cell-mediated immunity and long-lived immunologic memory. This is the reason why individuals who are naturally infected with or actively vaccinated against many viruses remain resistant for many years and often for life.

2. *The amount of antigen to which an individual is exposed influences the magnitude of the immune response generated.* Optimally immunogenic doses vary, depending on the antigen. In general, however, very large doses or repeated administration of protein antigens tend to induce specific T cell tolerance and inhibit immune responses. Large amounts of polysaccharide antigens may induce

tolerance in specific B lymphocytes and thus inhibit antibody production.

3. *The immune response to an antigen varies according to the portal of entry of that antigen.* Protein antigens that are administered subcutaneously or intradermally are usually immunogenic, because these antigens are picked up by Langerhans cells in the epidermis, which are potent APCs and efficiently transport antigens to regional lymph nodes, where immune responses are initiated (see Chapter 11). In contrast, large amounts of protein antigens administered intravenously or orally often induce specific unresponsiveness. Such unresponsiveness has been attributed to tolerance induction in lymphocytes or to the stimulation of antigen-specific suppressor T cells.

It is also known that individuals vary in their responsiveness to different foreign antigenic determinants. Unresponsiveness may be due to the absence of mature antigen-specific lymphocytes. Alternatively, individuals lacking MHC molecules capable of binding a foreign antigenic epitope cannot present this epitope to T cells, so that an immune response may not be induced against this epitope even if specific T cells are present (see Chapter 6).

The Role of Accessory Cells

As we discussed in Chapters 6 and 7, accessory cells such as dendritic cells, macrophages, and B lymphocytes are essential for the induction of T cell–dependent immune responses. Accessory cells present antigens to MHC-restricted T cells and produce membrane-associated and secreted costimulators that enhance the proliferation and differentiation of T lymphocytes. *Therefore, the presence of competent accessory cells stimulates T cell–dependent immune responses, and their absence leads to deficient responses.* Resting macrophages and naive, unstimulated B lymphocytes are deficient in costimulators. As a result, antigens presented by such APCs may fail to stimulate naive CD4+ T cells and may even induce T cell tolerance. In contrast, dendritic cells and activated macrophages are competent APCs because they express costimulators, as well as high levels of class II MHC molecules, and secrete cytokines that stimulate the proliferation and differentiation of T lymphocytes.

In order to elicit strong T cell–mediated immune responses, protein antigens need to be administered with substances called **adjuvants**. A postulated mechanism of action of adjuvants is to elicit local inflammatory reactions, which are associated with enhanced expression of costimulators on macrophages and other APCs and secretion of cytokines like IL-12 (see Fig. 7–8, Chapter 7). Because of this, the administration of protein antigens with adjuvants promotes cell-mediated immunity and T cell–dependent antibody production.

Vaccines are most effective for generating systemic immunity when administered subcutaneously or intradermally together with adjuvants. Some microorganisms contain adjuvants in their cell walls that influence the type and strength of specific immune responses that these microbes induce. For instance, the cell walls of mycobacteria contain muramyl dipeptide, which is a potent adjuvant and is at least partly responsible for the propensity of mycobacteria to stimulate strong T cell–mediated immune responses. The ability of adjuvant-induced local inflammation to promote specific immune responses illustrates an important link between innate and specific immunity, a concept that was introduced in Chapter 1. This also emphasizes the idea that *inflammation serves as the initial warning signal for the immune system to mount a specific response,* and microbes that colonize their hosts without eliciting local inflammation may evade subsequent immunologic attack (see Chapter 16).

Accessory cells may influence the development of immune responses by many other mechanisms. The role of MHC alleles expressed by APCs in the binding of antigenic peptides and the induction of T cell responses has been described in Chapter 6. It is also possible that even though an individual expresses MHC alleles capable of binding and presenting many different antigens, some of these antigens are processed and presented optimally by B cells and others by macrophages. In these cases, the magnitude of the immune response will depend on the relative numbers of B cells and macrophages that are present at the site of antigen administration or in regional lymphoid organs. Different APCs may also process the same endocytosed antigen in distinct ways, leading to the expression of different MHC-associated peptide epitopes. As a result, the type of APCs involved in initiating T cell activation may influence the fine specificity of the response to a multideterminant protein antigen.

The Types of Responding Lymphocytes

The nature of an immune response reflects the profile of antigen-specific lymphocytes that are stimulated by the immunization. *An important function of helper T lymphocytes is as regulators of the nature of specific immune responses to protein antigens.* T cells consist of populations that are activated by different types of antigens and perform different effector and regulatory functions. For instance, as discussed in Chapter 6, in viral infections viral antigens are synthesized in infected cells and presented in association with class I MHC molecules, leading to the stimulation of CD8+, class I MHC–restricted CTLs. In contrast, extracellular microbial antigens are endocytosed by APCs, processed, and presented preferentially in association with class II MHC molecules. This activates CD4+, class II MHC–restricted helper T cells, leading to antibody production and macrophage activation

TABLE 10–1. Factors That Determine the Nature and Magnitude of Immune Responses

	Factors That Favor	
	Stimulation of Immune Responses	*Inhibition or Lack of Immune Responses*
Recognition phase		
Lymphocyte repertoire	Diversity of lymphocyte receptors for foreign antigens	Deletion of self-reactive lymphocytes
Antigen presentation	Presence of MHC molecules capable of binding processed antigens	Absence of MHC molecules capable of binding certain antigenic determinants
Induction and activation phase		
Features of antigen		
Nature	Immunogenic forms	Tolerogenic forms
Amount	Optimal doses vary for different antigens	High doses favor tolerance
Portal of entry	Subcutaneous, intradermal	Intravenous, oral
Adjuvants	Recruitment and activation of accessory cells, induction of costimulators	Antigens without adjuvants are nonimmunogenic or tolerogenic
Accessory cells	Presence of costimulators (for T cells), secretion of cytokines (e.g., IL-12)	Absence of costimulators and cytokines
Antigen-specific T cells	Helper T cells	"Suppressor" T cells
Anti-idiotypic immune responses	Can be stimulating or inhibitory	Can be stimulating or inhibitory
Antibodies	Enhance antigen uptake and presentation by macrophages	Antibody feedback
Cytokines	Positive amplification loops	Antagonistic effects of different cytokines; immunosuppressive effects

Abbreviations: MHC, major histocompatibility complex; IL-12, interleukin-12.

but relatively inefficient development of CTLs. In the absence of T cell activation, e.g., with polysaccharide antigens, only relatively simple antibody responses are induced (see Chapter 9).

As we mentioned earlier, even within the population of CD4+ helper T cells there are subsets that produce distinct cytokines in response to antigenic stimulation. Because most of the effector functions of helper T cells are mediated by cytokines, cells with distinct patterns of cytokine production perform different functions. Naive CD4+ T cells produce mainly the growth factor IL-2 upon initial encounter with antigen. Antigenic stimulation may lead to the differentiation of these cells into subsets that have relatively restricted profiles of cytokine production and effector functions. In Chapter 9 we introduced the two best defined subsets of CD4+ effector T cells—T_H1 cells, which activate macrophages and stimulate the production of opsonizing and complement-fixing antibodies, and T_H2 cells, which stimulate the production of non-complement–fixing antibodies, especially IgE, activate eosinophils, and suppress macrophage responses. An important characteristic of these subsets is that they develop from the same precursors and are preferentially induced by different types of microbes or antigen exposure. Therefore, *a major determinant of the nature of immune responses to protein antigens is the ability of different microbes or immunization conditions to influence the differentiation of naive antigen-specific helper T cells into functionally distinct effector populations.* We will return to a more detailed discussion of the induction and functions of helper T cell subsets in Chapter 12.

Thus, many features of antigens, accessory cells, and lymphocytes determine the magnitude of specific immune responses (Table 10–1). In addition, different antigens and immunization conditions preferentially activate distinct subpopulations of lymphocytes, thus giving rise to responses with distinct characteristics.

SUMMARY

The magnitude and nature of immune responses to foreign antigens are finely regulated. Mechanisms that function mainly to inhibit lymphocyte activation provide important feedback control, preventing the injurious side effects of protective immunity, and are used by the immune system to avoid reactions to self antigens. Several types of antigen exposure preferentially inhibit rather than stimulate specific lymphocytes and induce tolerance to subsequent exposures to the antigen. Antigen recognition by specific T lymphocytes in the absence of costimulators, or by B lymphocytes in the absence of T cell help, induces functional lymphocyte unresponsiveness, or anergy. Recognition of mutated forms of protein antigens also induces anergy in specific T cells. Repeated stimulation of large numbers of T cells results in activation-induced apoptotic death of the cells. Some tolerogenic antigens may induce T cells, which have been called suppressor cells, whose principal function is to inhibit the activation and effector functions of antigen-specific lymphocytes. Ongoing responses may also be regulated by cells or molecules that are generated during the response itself. These include anti-idiotypic antibodies and T cells, secreted antibodies that mediate feedback inhibition, and cytokines that have inhibitory effects on lymphocyte activation and effector functions.

Factors that influence the nature of specific immunity include the type and amount of antigen, its portal of entry, the participation of accessory cells in the immune response, and the types of responding lymphocytes. These factors may determine which functionally distinct classes of lymphocytes are stimulated, and which effector responses develop as a consequence.

Many of these regulatory interactions are incompletely understood, largely because regulation often has to be studied *in vivo* and involves multiple bidirectional interactions between the cells and molecules of the immune system. Despite this complexity, elucidating the mechanisms of immune regulation is one of the major challenges facing immunologists, because of its obvious importance in normal and abnormal immune responses.

Selected Readings

Bloom, B. R., P. Salgame, and B. Diamond. Revisiting and revising suppressor T cells. Immunology Today 13:131–136, 1992.

Jerne, N. K. Towards a network theory of the immune system. Annals of Immunology (Institut Pasteur) 125C:373–389, 1974.

Matzinger, P. Tolerance, danger, and the extended family. Annual Review of Immunology 12:991–1045, 1994.

Mondino, A., A. Khoruts, and M. K. Jenkins. The anatomy of T-cell activation and tolerance. Proceedings of the National Academy of Sciences USA 93:2245–2252, 1996.

Mosmann, T. R., and R. L. Coffman. Heterogeneity of cytokine secretion patterns and functions of helper T cells. Advances in Immunology 46:111–147, 1989.

Nagata, S. Fas and Fas ligand: a death factor and its receptor. Advances in Immunology 57:129–144, 1994.

Nossal, G. J. V. Cellular and molecular mechanisms of B lymphocyte tolerance. Advances in Immunology 52:283–331, 1992.

Schwartz, R. H. Models of T cell anergy: is there a common molecular mechanism? Journal of Experimental Medicine 184:1–8, 1996.

van Parijs, L., and A. K. Abbas. Role of Fas-mediated cell death in the regulation of immune responses. Current Opinion in Immunology 8:355–361, 1996.

CHAPTER ELEVEN

FUNCTIONAL ANATOMY

OF IMMUNE RESPONSES

Much of our knowledge of specific immune responses is based on *in vitro* analyses of isolated cell populations exposed to antigens or to agents thought to be analogs of antigens, such as polyclonal activators specific for B and T lymphocytes. It is, however, obvious that in order to understand protective and pathologic immune responses, one needs to define how such responses occur *in vivo* in intact organisms. This chapter will describe our current understanding of the initiation and development of immune responses *in vivo*, with an emphasis on the interactions between the different cells of the immune system and their microenvironments.

Most infectious agents and other antigenic substances enter the body through the skin and the mucosal epithelia of the gastrointestinal and respiratory tracts. Specific T and B lymphocytes must be brought into contact with antigens and simultaneously exposed to the stimuli, such as costimulators and cytokines, that are needed to initiate immune responses. *The peripheral, or secondary, lymphoid organs are the sites where primary immune responses are initiated* (Fig. 11–1). The effector responses that function to eliminate persistent antigens must, however, be capable of locating the antigen in any tissue. In other words, *effector and memory responses are systemic and may occur in extra-lymphoid peripheral tissues.* Several anatomic features of lymphoid tissues and properties

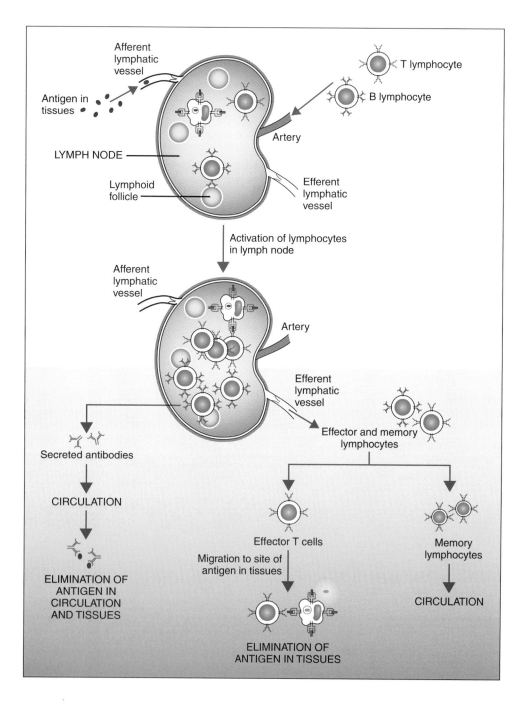

FIGURE 11–1. Sequence of events in immune responses *in vivo*. The response of antigen-specific lymphocytes to a protein antigen in a lymph node is illustrated. Effector and memory lymphocytes induced by antigen stimulation are shown exiting the node via an efferent lymphatic; it is not known whether these cells also leave through the vein. Recirculating lymphocytes shown are activated T cells because the recirculation of B cells has not been defined as well.

of lymphocytes are together responsible for determining the optimal sites for the recognition, activation, and effector phases of specific immunity.

1. *The peripheral lymphoid organs, principally lymph nodes and spleen, are the sites where antigens are concentrated and also the tissues into which naive lymphocytes preferentially migrate.* Thus, the two main participants in the initiation of specific immune responses, namely, antigen and naive lymphocytes, are brought together in the same organs. In addition, accessory cells, which are required for optimal lymphocyte activation, are also abundant in these tissues.

2. *Effector and memory lymphocytes, especially T cells, preferentially migrate to the peripheral tissues where microbes and other antigens induce local inflammatory reactions.* This property enables effector mechanisms to efficiently eliminate persistent antigens even in non-lymphoid tissues.

3. *Cooperative interactions between lymphocyte receptors for antigens and adhesion molecules control the migration and tissue retention patterns of cells that specifically recognize antigens.* Examples of such interactions will be mentioned later in the chapter.

4. *The principal portals of antigen entry, namely, the skin and the gastrointestinal and respiratory tracts, have specialized lymphoid tissues where effective immune responses can develop.* This is an adaptation that provides a local defense capability to the tissues that most commonly interact with pathogens in the external environment.

We will begin this discussion of the functional anatomy of immune responses by describing the initiation of immune responses *in vivo* and the pathways of lymphocyte recirculation. We will then discuss the sequence of events in lymphocyte activation *in vivo,* and the distinctive features of immune responses in specialized tissues, such as the cutaneous and mucosal immune systems.

INDUCTION OF IMMUNE RESPONSES *IN VIVO*

As described in Chapters 4 and 8, antigen receptors on B and T cells are generated by somatic events that lead to the expression of different and unique receptor genes in each lymphocyte clone. The immune system contains a large number of antigen-specific lymphocyte clones (estimated to be $>10^9$), and any given foreign antigen, even a complex multideterminant antigen, can be recognized only by a very small percentage of the total lymphocyte pool, on the order of 1 in 100,000 or 1 in 1,000,000 T or B cells. This poses a major logistical problem for the immune system: how can a small amount of a particular foreign antigen be efficiently recognized by the rare subpopulation of naive lymphocytes that are specific for that antigen so that a protective immune response can be initiated? Two mechanisms maximize the efficiency of induction of specific immune responses by

bringing antigens and lymphocytes together in specialized tissues, namely the peripheral (secondary) lymphoid organs. First, *antigens are collected from their portals of entry or sites of production and concentrated in the peripheral lymphoid organs.* Second, *naive lymphocytes migrate continuously from the blood* (or lymph) *through these lymphoid organs,* so that an antigen concentrated in any given organ can be seen by many more lymphocytes than those in residence in that tissue at any one time. This phenomenon of continuous, non-random migration of lymphocytes from the blood and lymphatic systems to lymphoid tissues and back to the blood stream is called **lymphocyte recirculation** and is discussed later in the chapter.

Mechanism of Antigen Collection

The function of collecting antigens from their portals of entry and delivering them to lymphoid organs is performed largely by the **lymphatic system** (Fig. 11–2). The skin, epithelia, and parenchymal organs contain numerous lymphatic capillaries that absorb and drain interstitial fluid (made of plasma filtrate) from these sites. The absorbed interstitial fluid, called **lymph,** flows through the lymphatic capillaries into convergent, ever larger lymphatic vessels, eventually culminating in one large lymphatic vessel called the thoracic duct. Lymph from the thoracic duct is emptied into the superior vena cava, thus returning the fluid to the blood stream. Liters of lymph are normally returned to the circulation each day, and disruption of the lymphatic system may lead to rapid tissue swelling.

The lymph drained from the skin, mucosa, and other sites contains a sampling of all of the soluble and particulate antigens present in these tissues. A substantial fraction of the introduced antigen may actually be picked up by antigen-presenting cells (APCs) present at the portal of entry and carried into lymphatics in this cell-associated form. Lymph nodes are interposed along lymphatic vessels and act as filters that "sample" the lymph at numerous points before it reaches the blood. Lymphatic vessels that carry lymph into a lymph node are referred to as **afferent;** those that drain the lymph from the node are called **efferent.** Since lymph nodes are connected in series along the lymphatics, an efferent lymphatic exiting from one node may also serve as the afferent vessel for another.

When lymph enters a lymph node through an afferent lymphatic vessel, it percolates through the nodal stroma, where the lymph-borne soluble antigens can be extracted from the fluid by APCs, such as macrophages and dendritic cells, that are resident in the stroma of the nodes. Macrophages, through phagocytosis, are particularly adept at extracting particulate and opsonized antigens. B cells in the node may also recognize soluble antigens. Macrophages, dendritic cells, and B cells that have taken up protein antigens can then process and

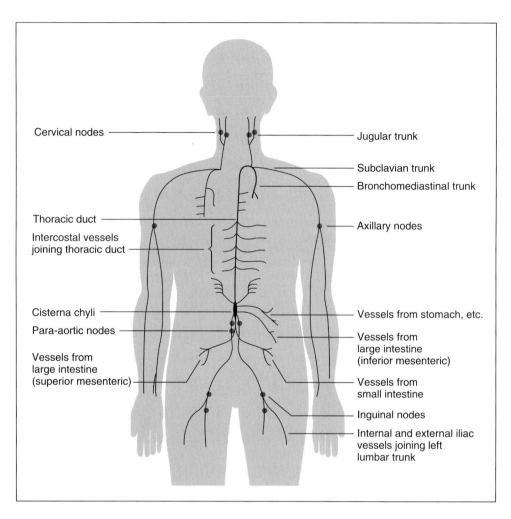

FIGURE 11–2. Schematic illustration of the human lymphatic system.

present these antigens to T cells. The net result of antigen uptake by these various cell types is to accumulate and concentrate antigen in the lymph node and display it in a form that can be recognized by specific T lymphocytes.

The collection and concentration of foreign antigens in lymph nodes are supplemented by two other anatomic adaptations that serve similar functions. First, the mucosal surfaces of the gastrointestinal and respiratory systems, in addition to being drained by lymphatic capillaries, also contain specialized collections of secondary lymphoid tissue that can directly sample the luminal contents of these organs for the presence of antigenic material. The best characterized of these mucosal lymphoid organs are the Peyer's patches of the ileum, which are described in more detail later in this chapter. Second, the blood stream itself is monitored by APCs in the spleen for any antigens that reach the circulation. Such antigens may reach the blood either directly from the tissues or by way of the lymph from the thoracic duct. Nonprotein, e.g., polysaccharide, antigens appear to be preferentially taken up by macrophages in the marginal zones of lymphoid follicles in the spleen and to induce antibody responses in this tissue (see Chapter 9).

After antigens are transported to and concentrated in lymphoid tissues, they have to be recognized by specific naive lymphocytes in order to initiate the immune response. Lymphocyte recirculation ensures that lymphocytes are taken to the sites of antigen localization.

PATHWAYS AND MECHANISMS OF LYMPHOCYTE RECIRCULATION

The pathways of migration of lymphocytes among lymphoid organs and peripheral tissues are best described for T cells (Fig. 11–3). Two features of lymphocyte recirculation are especially important in the initiation and effector phases of specific immune responses to protein antigens. First, *naive T cells and previously activated effector and memory cells exhibit distinct patterns of recirculation.* Second, *the migration of lymphocytes is mediated by homing receptors on the lymphocytes and their ligands on endothelial cells, and the expression and functions of these molecules are regulated by antigen recognition and inflammatory stimuli.* As we shall see later, this feature is critical for determining the distinct migration patterns of naive and antigen-stimulated lymphocytes to peripheral lymphoid organs and sites of inflammation, respectively.

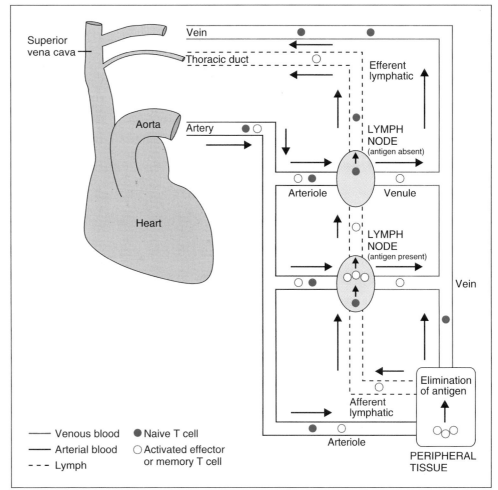

FIGURE 11–3. Pathways of T lymphocyte recirculation. Naive T cells preferentially leave the blood and enter lymph nodes across the high endothelial venules. Activated lymphocytes return to the circulation through the efferent lymphatics and the thoracic duct, which empties into the superior vena cava. Effector and memory T cells preferentially leave the blood and enter peripheral tissues through venules at sites of inflammation. Recirculation through peripheral lymphoid organs other than lymph nodes is not shown.

Recirculation of Naive T Lymphocytes Through Lymphoid Organs

The efficient homing of naive T cells to lymph nodes was initially described in the late 1950s and early 1960s by James Gowans and his colleagues. These investigators showed that lymphocyte extravasation from the blood into the stroma of a peripheral lymph node occurred selectively at modified post-capillary venules within the node. These specialized venules are lined by plump endothelial cells that, on cross-section, protrude into the vessel lumen. Because of this morphologic appearance of the endothelial cells, such vessels are called **high endothelial venules** (HEVs). HEVs are also present in mucosal lymphoid tissues, such as Peyer's patches in the gut, but not in the spleen.

Intravital videomicroscopy has shown that the *key characteristic of the HEV that leads to lymphocyte extravasation is greater adhesiveness of the high endothelium (compared to flat venular endothelium) for circulating lymphocytes* (Fig. 11–4). Normally, lymphocytes coursing through the microcirculation randomly collide with vessel walls and either immediately rebound or adhere to the lining endothelial cells for only a fraction of a second before they

are dislodged by the shear force of flowing blood. In contrast, when naive T cells collide with the endothelium lining an HEV, they remain loosely attached for several seconds. This low-affinity interaction may result in rolling along the endothelial surface as the loosely bound lymphocyte is propelled by the flowing blood. During this crucial window of time, a proportion of the T cells are able to further increase their strength of attachment, to spread out into motile forms, and to crawl between the endothelial cells into the stroma of the lymph node. If the T cells that have entered a lymph node fail to see their specific antigen within that node, they exit through an efferent lymphatic vessel and eventually return to the blood via the thoracic duct. One cycle of recirculation may take about an hour. Overall, the net flux of lymphocytes through lymph nodes is very high, and it has been estimated that approximately 25×10^9 cells pass through lymph nodes each day, i.e., each lymphocyte goes through a node once a day, on the average. As a result, at any given time only about 2 percent of the naive lymphocytes in an adult are in the blood, the remainder being mostly in peripheral lymphoid tissues. It is likely that an essentially similar sequence of events occurs dur-

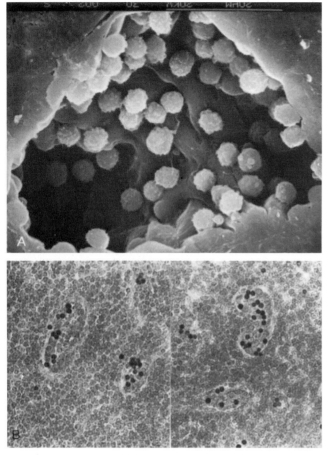

FIGURE 11–4. Lymphocyte adhesion to high endothelial venules.

A. Scanning electron micrograph of a high endothelial venule with lymphocytes attached to the lumen surface of the endothelial cells. (Courtesy of J. Emerson and T. Yednock, University of California San Francisco School of Medicine, San Francisco. From Rosen, S. D., and L. M. Stoolman. Potential role of cell surface lectin in lymphocyte recirculation. *In* K. Olden and J. Parent (eds.). Vertebrate Lectins. Van Nostrand Reinhold, New York, 1987.)

B. A frozen-section binding assay showing preferential attachment of added lymphocytes (stained darkly) to high endothelial venules, which are seen in cross-section. (Reproduced with permission from Rosen, S. D. Lymphocyte homing: progress and prospects. Current Opinion in Cell Biology 1:913–919, 1989.)

tion is a consequence of local T cell activation and cytokine production triggered by antigen. For instance, HEVs fail to develop in animals that are raised in germ-free environments and in animals that do not have functional T lymphocytes, and they revert to flat endothelial venules if delivery of antigen to a lymph node is prevented by surgically disrupting the afferent lymphatics. It is likely that HEVs are maintained by continuous, low-level stimulation of the immune system by environmental antigens and the resultant production of cytokines such as interferon-γ (IFN-γ). Analysis of neonatal mice has revealed that lymphocytes preferentially home to lymph nodes even prior to the development of HEVs, suggesting that the venular endothelium of peripheral lymphoid organs may possess an intrinsically high adhesiveness for circulating naive lymphocytes. HEVs, which develop under the influence of cytokines, may serve to increase the efficiency of extravasation at these sites.

The attachment of lymphocytes to the endothelium of peripheral lymphoid tissues is mediated by **homing receptors** on the lymphocytes and their ligands on endothelial cells. Because different ligands may be expressed by endothelial cells in different anatomic sites, they have been called **addressins,** to emphasize their role in organ-specific lymphocyte homing (Table 11–1).

The key tool in the analysis of homing receptors was the development of an *in vitro* assay for lymphocyte binding to endothelium in frozen sections of lymph nodes, under conditions in which the lymphocytes preferentially bind to the endothelium of the HEVs rather than to other elements of the node (Fig. 11–4). This frozen-section binding assay has been used to screen monoclonal antibodies raised against lymphoid cell lines for their ability to inhibit lymphocyte binding to endothelium. The specific homing receptors on T cells and the addressins on HEVs were identified and characterized using these antibodies.

*The most important homing receptor on naive T cells is **L-selectin*** (also called CD62L), a 90 kD gly-

ing lymphocyte migration through other peripheral lymphoid organs, such as the spleen, even though the spleen does not contain HEVs. Within the mucosa of the gut, the prolonged time of attachment of lymphocytes to the endothelial lining of HEVs provides such an advantage for extravasation that 50 per cent of the T cells entering the mucosa do so through the HEVs despite the fact that the surface area of flat venules exceeds that of HEVs by over 50 to one!

Because of their important place in the recirculation of lymphocytes, much study has been devoted to the characterization of HEVs. HEVs develop after birth from flat endothelial venules that are present in the secondary lymphoid organs. Several lines of evidence indicate that HEV forma-

TABLE 11–1.   Lymphocyte Homing Receptors

	T Cell Homing Receptor	**Endothelial Ligand**
A. Homing of naive T cells to peripheral lymphoid organs	L-selectin	GlyCAM-1, CD34, MadCAM-1, others
B. Homing of T cells to mucosal tissues (e.g., Peyer's patches)	$\alpha_4\beta_7$ integrin CD44	MadCAM-1 Hyaluronate
C. Homing of memory T cells to inflamed skin	CLA-1	E-selectin
D. Homing of memory and effector T cells to sites of inflammation (general)	LFA-1 VLA-4 CD44	ICAM-1, ICAM-2 VCAM-1 Hyaluronate

coprotein originally identified in mice by a monoclonal antibody called MEL-14. L-selectin is a carbohydrate-binding protein that is structurally homologous to two endothelial proteins, E-selectin and P-selectin (Box 11–1). *Selectins mediate rapid, low-affinity attachments between leukocytes and endothelium.* These attachments result in a transient tethering and, in the presence of flowing blood, rolling of the leukocytes on the endothelial surface.

The structures recognized by L-selectin are typically anionic complex carbohydrates, such as sulfated glycosaminoglycans on mammalian proteoglycans. GlyCAM-1 (for glycan-bearing cell adhesion molecule-1), a proteoglycan produced by the endothelium of lymph node HEVs, is the ligand that mediates migration of naive T cells to lymph nodes. Other L-selectin ligands are expressed on endothelium in the gastrointestinal tract and other tissues. These different addressins may contribute to the initial binding of L-selectin–expressing naive T cells to endothelium in different tissues. The subsequent firmer attachment of the lymphocytes, and their transmigration through the HEVs into the

---

BOX 11–1. Selectins

The selectins, sometimes called lectin adhesion molecules (LECAMs), are a family of three separate but closely related proteins that mediate adhesion of leukocytes to endothelial cells (see Table). One member of this family of adhesion molecules is expressed on leukocytes and two other members on endothelial cells, but all three participate in the process of leukocyte-endothelium attachment. Each of the selectin molecules is a single chain transmembrane glycoprotein with a similar modular structure. The amino terminus, expressed extracellularly, is related to the family of mammalian carbohydrate-binding proteins known as C-type lectins. Like other C-type lectins, ligand binding by selectins is calcium-dependent. The lectin domain is followed by a domain homologous to part of the epidermal growth factor, which, in turn, is followed by a number of tandemly repeated domains related to structures previously identified in complement regulatory proteins as "complement control protein repeats" (see Chapter 15). These repeats, which have six cysteine residues per domain in the selectins instead of the four found in complement regulatory proteins, are followed by a hydrophobic transmembrane region and a short cytoplasmic carboxy terminal region.

L-selectin, or CD62L, the family member that serves as a homing receptor for lymphocytes to lymph node HEVs, is also expressed on other leukocytes. On neutrophils, it serves

to bind these cells to endothelial cells that are activated by cytokines (TNF-α, IL-1, and IFN-γ) found at sites of inflammation. L-selectin recognition of its ligand is thought to happen quickly but is of low affinity; this property allows L-selectin to mediate initial attachment and subsequent rolling of neutrophils on endothelium in the face of flowing blood. L-selectin is located on the tips of microvillus projections of leukocytes, facilitating its interaction with ligands on endothelium. The rolling is followed by more stable integrin-mediated attachment of the neutrophils to endothelial cell ICAM-1, spreading, and LFA-1–mediated transmigration. At least three endothelial cell ligands can bind L-selectin: GlyCAM-1, a secreted proteoglycan found on HEVs of lymph node; MadCAM-1 (for mucosal addressin cell adhesion molecule-1), expressed on endothelial cells in the gut; and CD34, a proteoglycan on endothelial cells (and bone marrow cells). The expression of L-selectin ligands on endothelium may require extended cytokine exposure, typical of HEVs. At peripheral sites of inflammation, L-selectin on neutrophils may indirectly contribute to rolling by presenting acidic complex carbohydrates, such as sialylated Lewis-X, to E-selectin or P-selectin on the endothelial cell.

E-selectin, also known as endothelial leukocyte adhesion molecule 1 (ELAM-1) or CD62E, is exclusively expressed by cytokine-activated endothelial cells, hence the designation

The Selectin Family

Selectin	Size	Distribution	Ligand	Principal Function
L-selectin (CD62L)	90–110 kD (variation due to glycosylation)	Leukocytes (expression on T cells is high on naive cells, generally low on memory cells)	Sulfated glycosaminoglycans on GlyCAM-1, CD34, MadCAM-1, and others	Naive T cell homing
E-selectin (CD62E)	110 kD	Cytokine-activated endothelium (TNF, LT, IL-1)	Sialylated Lewis-X and related glycans (e.g., CLA-1) on L-selectin, E-selectin ligand-1, and other glycoproteins	Binding of leukocytes at sites of cytokine-induced inflammation
P-selectin (CD62P)	140 kD	1. Storage granules of endothelium 2. Storage granules of platelets	Sialylated Lewis-X and related glycans on P-selectin glycoprotein-1 and other glycoproteins	1. Binding of leukocytes early in cytokine-independent inflammation 2. Binding of activated platelets to monocytes

Abbreviations: TNF, tumor necrosis factor; LT, lymphotoxin; IL-1, interleukin-1.

Box continued on following page

"E." E-selectin recognizes complex sialylated carbohydrate groups related to the Lewis-X or Lewis-A family found on various surface proteins of granulocytes, monocytes, and certain memory T cells. The E-selection ligand on T cells has been called the cutaneous lymphocyte antigen-1 (CLA-1), and has been implicated in homing of certain T cells to the skin. Endothelial cell expression of E-selectin is a hallmark of acute cytokine-mediated inflammation, and antibodies to E-selectin can block neutrophil accumulation *in vivo*. Thus, E-selectin may be involved in the homing of memory T cells to peripheral sites of inflammation.

P-selectin (CD62P) was first identified in the secretory granules of platelets, hence the designation "P." It has since been found in secretory granules of endothelial cells, which are called Weibel-Palade bodies. When endothelial cells or platelets are stimulated to secrete stored proteins, P-selectin is translocated within minutes to the cell surface as part of the exocytic secretory process. Upon reaching the cell surface, P-selectin mediates binding of neutrophils and monocytes. The complex carbohydrate ligands recognized by P-selectin appear similar to those recognized by E-selectin.

The structural relationship of the selectins to each other has been reinforced by the finding that the three selectin genes are located in tandem on chromosome 1 in both mice and humans. The selectin structural motif clearly arose from duplication of an ancestral selectin gene. The differences among the three selectins serve to confer differences in both binding specificity and in tissue expression. However, it is thought that all three selectins mediate rapid low-affinity attachment of leukocytes to endothelium, an early and important step in leukocyte homing, although E-selectin may mediate high-affinity attachment as well.

The physiologic roles of selectins have been reinforced by studies of gene knockout mice. L-selectin–deficient mice have small, poorly formed lymph nodes and defective induction of T cell–dependent immune responses and inflammatory reactions. Mice lacking either E-selectin or P-selectin have only mild defects in leukocyte recruitment, suggesting that these two molecules are functionally redundant. Double knockout mice lacking both E- and P-selectins have significantly impaired leukocyte recruitment and increased susceptibility to infections. Humans who lack one of the enzymes needed to express the carbohydrate ligands for E- and P-selectins on neutrophils have similar problems, resulting in a syndrome called "leukocyte adhesion deficiency-2 (LAD-2)." LAD-1 is a clinically similar syndrome caused by mutations in the β chain of integrins, CD18 (see Chapter 21).

lymphoid organ, probably utilize other adhesion molecules, such as integrins. The details of this process are better understood for activated T cells than for naive lymphocytes (see Chapter 13).

There is little current information about the nature of homing receptors or addressins involved in lymphocyte homing to the spleen. This is a major gap in our knowledge since the rate of lymphocyte passage through the spleen is about half of the total lymphocyte population every 24 hours. As mentioned above, the spleen does not contain morphologically identifiable HEVs, and it is possible that homing of lymphocytes to the spleen does not show the same degree of selectivity for naive cells as do lymph nodes.

When naive T cells that have entered a lymph node encounter their specific antigen, they become activated and begin to proliferate. *During this activation process, T cells increase their expression of several cell surface proteins that mediate their adhesion to other cells and to extracellular matrix molecules.* These surface proteins include the integrins LFA-1 (CD11aCD18), VLA-4 (CD49dCD29), VLA-5 (CD49eCD29), VLA-6 (CD49fCD29), and CD44, all of which have been discussed earlier as T cell accessory molecules (see Chapter 7). Antigen recognition and activation also increase the affinities of the integrins for their ligands. As a consequence of these changes in integrin expression and affinity, activated T cells are more adherent to accessory cells and extracellular matrices and remain resident in the lymph node. The progeny of each activated T cell may differentiate into effector cells or into memory cells. As the T cells differentiate, they maintain high levels of adhesion molecules, but the affinity of the integrins for their ligands rapidly declines with time after antigenic stimulation. Consequently, differentiated effector and memory T cells are less adhesive for lymph node cells and matrix molecules than recently activated cells and return to the blood stream via the efferent lymphatics and the thoracic duct.

Migration of Effector and Memory T Lymphocytes to Sites of Inflammation

Most previously activated effector and memory T cells no longer efficiently home to lymph nodes, because they express lower levels of L-selectin than do naive cells. Instead, *effector and memory T cells preferentially home to sites of inflammation, which are often the sites of antigen entry and persistence.* The reason for this is that cytokines produced in response to inflammatory stimuli enhance the expression on endothelial cells of ligands for integrins, and integrin expression is increased on activated T cells. Following or during their migration into extra-vascular tissues, if T cells encounter their cognate antigen they are stimulated and the affinities of integrins and CD44 for extracellular matrix proteins are increased. As a result, the *antigen-stimulated T lymphocytes are preferentially retained at the site of antigen persistence,* and they mediate the effector phase of the T cell response (described in more detail in Chapter 13). The cells that migrate into peripheral tissues but do not see their antigen are picked up by lymphatics and return to the circulation. Thus, *naive T cells preferentially home to the lymphoid organs, where they recognize and respond to foreign antigens (the recognition and activation phases of primary immune responses), whereas effector and memory T cells preferentially home to inflamed peripheral tissues, where they are needed to eliminate antigens (the effector phase of immune responses).* Furthermore, the entry of memory T cells into the recirculating pool ensures that immunologic memory is systemic, i.e., that effector responses can be elicited at any site in the body, even if the memory cells were generated in the lymphoid tissues draining a single site of initial antigen entry. It should also be noted that *the recirculation and tissue-specific homing of lymphocytes are largely independent of antigen recognition,* although the patterns of homing and extravascular retention of cells are influenced by the presence of antigen and by the inflammation that often accompanies the entry of foreign antigens.

Recently it has become apparent that memory T cells are heterogeneous in their patterns of expression of adhesion molecules and in their propensity to migrate to different tissues. For instance, some memory cells express an integrin containing the same α chain as VLA-4 (called CD49d) but in a heterodimer with a unique β chain, β_7, instead of β_1 (CD29). This $\alpha_4\beta_7$ integrin, in addition to recognizing VCAM-1 and fibronectin (the ligands for VLA-4), also interacts with the Ig-like domains of the mucosal endothelial addressin, MadCAM-1, and thus mediates homing of memory T cells to mucosal lymphoid tissues. Other memory T cells may express complex acidic carbohydrates (related to sialylated Lewis-X) that mediate their adhesion to E-selectin. These T cells are especially abundant in immune reactions in the skin, and the carbohydrate-containing molecule, sometimes called CLA-1 (for cutaneous lymphocyte antigen-1), may mediate homing of the T cells to skin. Thus, *special adhesion molecules may be responsible for the migration of memory T cells to the lamina propria of the intestines or to the dermis.* Intra-epithelial lymphocytes show even more specialized differentiation. For example, T cells in the intestinal epithelium express the $\alpha_E\beta_7$ integrin, which can bind to E-cadherin molecules on epithelial cells. Naive and memory T cells differ in the expression of other surface markers. Most naive cells express the high molecular weight isoform of CD45 called CD45RA, whereas most activated and memory cells express the lower molecular weight CD45RO isoform. These molecules are not involved in lymphocyte migration, and their expression on the different cell populations may not even be fixed.

Recirculation of B Lymphocytes

In contrast to actively recirculating effector and memory T cells, both naive and activated B lymphocytes are found mainly in the peripheral lymphoid organs, such as spleen, lymph nodes, mucosal lymphoid tissues, and bone marrow. Since these cells produce antibodies that act at a distance, they need not home to sites of inflammation to mediate host defense. Memory B cells do tend to migrate to germinal centers, which are the sites where antigen is retained for long periods to elicit secondary antibody responses. One of the molecules used by memory B cells to home to germinal centers is VLA-4. Little else is known about the patterns or mechanisms of B lymphocyte recirculation.

IMMUNE RESPONSES IN SPLEEN AND LYMPH NODES

The morphology of the spleen and lymph nodes, and the anatomic compartmentalization of different cell populations in these organs, were described in Chapter 2. These organs are the principal sites for the initiation of most primary immune responses and for the activation of B lymphocytes and the production of antibodies. Recent studies using combinations of morphologic and molecular techniques, as well as analyses of homogeneous lymphocyte populations (e.g., in antigen receptor transgenic mice), are providing fascinating insights into the anatomy of immune responses in lymphoid organs.

The Fate of Antigens *In Vivo*

The localization of antigens in lymphoid tissues has been studied mainly in experimental animals. *The in vivo fate of administered antigens is determined by several factors.*

1. *The nature of the antigen.* Particulate substances, soluble proteins, and polysaccharides tend to be concentrated in different areas of lymphoid tissues, as described below.

2. *The route of antigen entry.* Intravenously administered antigens are trapped mainly in the spleen, which is the principal site of immune responses to blood-borne antigens. In contrast, antigens that enter via the skin and mucosa, or from parenchymal organs and connective tissues, are carried by the lymphatic drainage to regional lymph nodes. Some immune responses may also be initiated locally at the site of antigen entry, particularly in mucosal and cutaneous lymphoid tissues.

3. *The immunization status of the individual.* The fates of antigens vary in naive and previously immunized individuals, largely because the binding of antibodies alters the pattern of cellular uptake of antigens.

Most of our current knowledge about what happens to antigens *in vivo* is based on analyses of the spleens in animals given antigens intravenously (Fig. 11–5). Antigens enter the spleen via the central arterioles, which divide into progressively smaller branches that end in the marginal zones and from where the blood enters the vascular sinusoids of the red pulp. Intravenously injected particulate and soluble antigens are trapped largely by macrophages in the marginal zones and red pulp of the spleen. In previously unimmunized individuals, a fraction of injected soluble antigen can be found in the periarteriolar lymphoid sheaths, which are rich in interdigitating dendritic

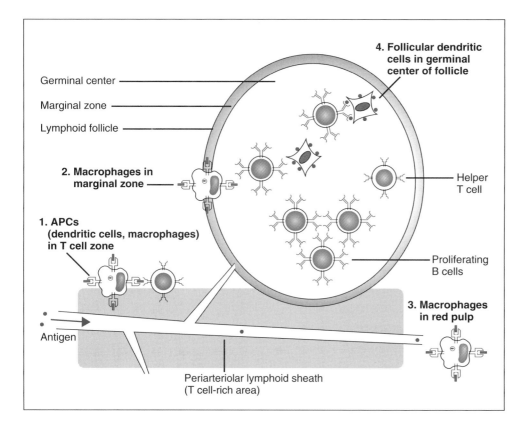

FIGURE 11–5. Fates of an antigen in the spleen, illustrating the major cell populations that bind and present antigen to lymphocytes (numbered 1–4). APCs, antigen-presenting cells.

cells and T lymphocytes. Polysaccharide antigens injected intravenously are first trapped by marginal zone macrophages and may then be transported to lymphoid follicles, where they stimulate B cell responses. In previously immunized individuals, injected antigens are rapidly complexed with antibodies, and most of the complexes are phagocytosed and destroyed. A small fraction is displayed on the surfaces of follicular dendritic cells in lymphoid follicles, and this depot of antigen may persist for many weeks or months, providing a long-lasting stimulus for memory lymphocytes.

Antigens are transported to lymph nodes either in soluble form or attached to surfaces of APCs. For instance, some cutaneously administered antigens are picked up by epidermal Langerhans cells and carried to draining lymph nodes, where they initiate T cell responses. The fate of antigens once they enter lymph nodes is not as well understood, although the general features are probably similar to those of the spleen.

Antigen Presentation in Lymphoid Tissues

In Chapter 6, we introduced the concept that several cell types, notably dendritic cells, macrophages, and B lymphocytes, have the ability to present protein antigens to CD4+ T cells. There is considerable experimental evidence supporting the importance of dendritic cells in antigen presentation to naive T cells both *in vitro* and *in vivo*. The anatomic co-localization of dendritic cells and T

cells in the periarteriolar lymphoid sheaths of the spleen and the parafollicular cortex of lymph nodes also supports the key role of this APC type in initiating T cell responses. Macrophages, which are abundant in the red pulp of the spleen and the medulla of lymph nodes, are efficient at picking up particulate antigens. These macrophages may migrate within lymphoid tissues to T cell–rich zones, where they may stimulate the development of CD4+ cells to effector cells. B cells present antigens to activated helper T cells to initiate antibody responses. As discussed in Chapters 9 and 10, B cells may be inefficient at activating naive T cells, and their role as APCs may be most important in secondary antibody responses.

Activation of T and B Lymphocytes in the Spleen and Lymph Nodes

Following intravenous administration of antigen to naive animals, T cell activation is probably initiated near the periarteriolar lymphoid sheaths of the spleen. This is the zone of greatest mitotic activity. Activated T cells may then migrate from the lymphoid sheaths into the marginal zone, and a small fraction may enter lymphoid follicles (the B cell–rich zones). As they do so, they encounter B cells migrating in the opposite direction, and this presumably maximizes the chances of cell-cell interactions. Clusters of cytokine-producing T cells, often in close proximity to antibody-producing B cells, are most numerous in the marginal zones,

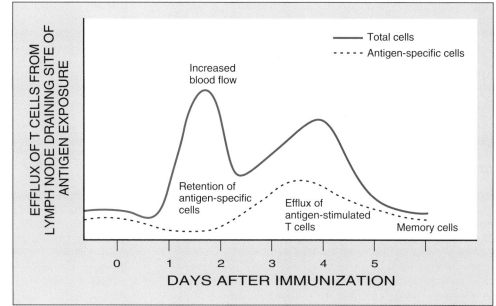

FIGURE 11–6. Trafficking of T cells through a lymph node draining a site of antigen entry. Cells are retained in the node shortly after antigen exposure. Blood flow and efflux of cells then increase, but antigen-specific cells remain in the node and appear in the efferent lymph some days later. These results are from experiments with sheep in which the efferent lymphatics of selected nodes (e.g., the popliteal node) can be cannulated.

adjacent to the terminal branches of the central arterioles. Such clusters may persist for many days after antigen exposure, suggesting that T cell activation continues in the marginal zones. The activation of B lymphocytes in the spleen was described in Chapter 9.

In lymph nodes draining sites of antigen administration, the first change is seen within a few hours after antigen exposure. Blood flow through the node increases by more than 20-fold, allowing an increased number of naive lymphocytes access to the site where antigen is concentrated (Fig. 11–6). Efflux of cells from the node may decrease at the same time. These changes are probably due to an inflammatory reaction to adjuvants associated with the antigen or due to inflammatory cytokines produced as a result of antigen entry. During the next day or so, antigen-specific cells are preferentially retained in the node, presumably because they recognize the antigen and upregulate their expression of adhesion molecules. The first wave of mitotic activity is seen within 1 or 2 days in the T cell–rich parafollicular areas of the cortex. Cytokine-producing T cells are also found in the parafollicular regions of the cortex. Shortly thereafter, there is a rapid efflux of activated cells out of the node; the mechanisms responsible for this emigration of activated cells are not fully understood. The activated cells then enter the peripheral circulation and home to different sites. For instance, activated T cells home to peripheral tissues where the antigen entered and has persisted (see Chapter 13). Some B cells may remain in the lymph nodes, where they differentiate into antibody-secreting cells and undergo affinity maturation (see Chapter 9). The output of antigen-stimulated cells decreases within about a week after antigen exposure. After this time, memory T and B cells that

have developed in the lymph node exit and enter the circulating pool. In addition, a fraction of activated lymphocytes undergo programmed cell death, although it is difficult to document the magnitude of this *in vivo* because apoptotic cells are rapidly cleared by macrophages.

Immune responses fail to occur in some tissues, such as the brain, eye, and testis, and these are called **immune privileged sites.** Several factors may contribute to immune privilege. The brain vasculature has anatomic blood-tissue barriers that may block entry of lymphocytes into the tissues. However, anatomic factors do not explain immune privilege in other tissues, which often communicate freely with the rest of the body. Some tissues, such as the eye, may contain high concentrations of immunosuppressive substances, e.g., transforming growth factor-β. The eye and testis constitutively express Fas ligand, which may kill any leukocytes that enter these tissues (see Box 10–1, Chapter 10). The physiologic importance of the immune privileged status of these organs is not clear.

These general features apply to most immune responses in peripheral lymphoid organs. The skin and mucosal epithelia are specialized, and immune responses at these sites show some distinctive characteristics.

THE CUTANEOUS IMMUNE SYSTEM

The skin is the largest organ in the body and the principal physical barrier between the organism and the external environment. In addition, the skin is an active participant in host defense, with the ability to generate and support local immune and inflammatory reactions. Many foreign antigens gain entry into the body via the skin, so that many immune responses are initiated in this tissue.

Cellular Components of the Cutaneous Immune System

The skin consists of an epidermis separated from the underlying dermis by a basement membrane. Each of these tissues contains cell populations that may play active roles in immune reactions (Fig. 11–7).

Keratinocytes. The epithelial cells of the epidermis, or keratinocytes, produce a number of cytokines, including interleukin-1 (IL-1), granulocyte-macrophage colony-stimulating factor, IL-3, tumor necrosis factor (TNF), and IL-6; and after activation by T cell–derived cytokines such as IFN-γ, keratinocytes secrete chemokines that stimulate the chemotaxis and activation of leukocytes. (The properties and actions of these cytokines are discussed in more detail in Chapter 12.) Although these findings suggest that keratinocytes may augment local inflammation and lymphocyte activation, it is not known if the cytokines produced by keratinocytes can reach the dermis, where most of the lymphocytes and other leukocytes are located.

In addition to serving as sources of cytokines, keratinocytes can be induced to express class II MHC molecules by exposure to IFN-γ. This is similar to the induction of class II expression on macrophages and other cell populations (see Chapters 5 and 6). It is, however, unclear if keratinocytes function as APCs for initiating T cell reactions.

Epidermal Langerhans Cells. Langerhans cells are a form of bone marrow–derived dendritic cells located in the suprabasilar portion of the epidermis. Although Langerhans cells constitute only about 1 per cent of the cells in the epidermis, they cover almost 25 per cent of the surface because of their long dendritic processes and their horizontal orientation. In fact, they form an almost continuous meshwork that enables them to pick up antigens that enter through the skin. Their role in antigen presentation is described below.

Intraepidermal Lymphocytes. Within the epidermis there is a small population of lymphocytes. In humans, these constitute only about 2 per cent of skin-associated lymphocytes (the rest residing in the dermis), and the majority are CD8+ T cells. Intraepidermal T cells may express a more restricted set of antigen receptors than T lymphocytes in most extracutaneous tissues. In mice (and some other species), many intraepidermal lymphocytes are T cells that express the γδ form of the antigen receptor (see Chapter 7). As we shall see later, this is also true of intraepithelial lymphocytes in the intestine, raising the possibility that γδ T cells, at least in some species, may be uniquely committed to recognizing antigens encountered at epithelial surfaces. However, neither the specificity nor the function of this T cell subpopulation is clearly defined.

Dermal Lymphocytes and Macrophages. The connective tissues of the dermis contain T lymphocytes (both CD4+ and CD8+), predominantly in a perivascular location, and scattered macrophages. This is essentially similar to connective tissues in other organs. The T cells usually express phenotypic markers typical of activated or memory cells, such as CD44, and high levels of the IL-2 receptor α chain (CD25).

Initiation of Immune Responses in the Skin

Langerhans cells provide the major pathway of presentation of cutaneously encountered antigens to naive T cells. When a protein antigen enters through the skin, Langerhans cells bind the antigen to their surfaces and may process it. If concomitant inflammatory stimuli are present, the Langerhans cells reduce their adhesiveness, leave the epidermis, and, bearing processed antigen, migrate from the epidermis into lymphatic vessels and thence into regional lymph nodes, where they come to reside in the parafollicular regions of the cortex. During their migration from the skin to the regional lymph nodes, the Langerhans cells increase their expression of class II MHC molecules

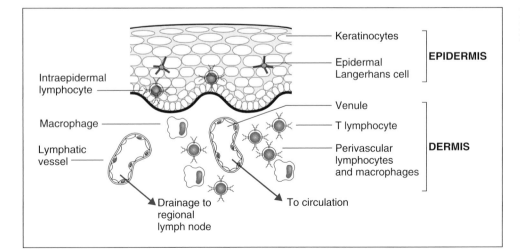

FIGURE 11–7. Cellular components of the cutaneous immune system.

and costimulators, as well as their antigen-processing capacity, thus becoming highly efficient APCs. In the lymph nodes, the Langerhans cells, which are now called interdigitating dendritic cells, reside on connective tissue fibers and present the antigen to naive CD4⁺ T cells that have entered the node from the blood through HEVs. Thus, mature lymphoid dendritic cells in peripheral lymph nodes, which are the most potent APCs in the body, are largely derived from Langerhans cells.

If protein antigens enter the dermis directly, they may be presented to T cells by dermal macrophages. Most of the T cells in the dermis are previously activated effector or memory cells. Therefore, T cell reactions to macrophage-associated antigens in the dermis may be more important for generating effector responses to antigen challenge in previously immunized individuals than for initiating primary responses to antigens encountered for the first time.

Effector Phases of Immune Responses in the Skin

A typical T cell–mediated immune response in the skin is **delayed type hypersensitivity (DTH).** This is a reaction to soluble protein antigens or to chemicals that can bind to and modify self proteins, creating new antigenic determinants. DTH is a cell-mediated immune response that results from the activation of T cells and the secretion of cytokines. It is described in more detail in Chapter 13.

Much less is known about humoral immunity in the skin. Secretory IgA is present in secretions in the skin, such as sweat, and its importance is suggested by the clinical observation that patients with IgA deficiency are susceptible to pyogenic skin infections. Since B lymphocytes are rarely encountered in cutaneous tissues, it is likely that IgA is produced by B cells activated in draining lymph nodes and is transported back to the skin via the circulation. The skin is also an important site of IgE-mediated immediate hypersensitivity reactions, which are caused by the release of chemical mediators from dermal mast cells (see Chapter 14).

THE MUCOSAL IMMUNE SYSTEM

Although lymphocytes have been recognized in the mucosa and submucosa of the gastrointestinal and respiratory tracts for many decades, the idea of a specialized mucosa-associated immune system is relatively new. Like the skin, these mucosal epithelia are barriers between the internal and external environments, and are, therefore, an important first line of defense. Moreover, *immune responses to oral antigens differ in some fundamental respects from responses to antigens encountered at other sites.* The two most striking differences are the high levels of IgA production in mucosal tissues and the tendency of oral immunization with protein antigens to induce systemic T cell tolerance rather than activation. Such observations have convincingly established both the physiologic importance and the functional uniqueness of the mucosal immune system. Much of our knowledge of mucosal immunity is based on studies of the gastrointestinal tract, and this is emphasized in the discussion that follows. In comparison, little is known about immune responses in the respiratory mucosa, even though the airways are a major portal of antigen entry. It is likely, however, that the features of immune responses are basically similar in all mucosa-associated lymphoid tissues.

Cellular Components of the Mucosal Immune System

In the mucosa of the gastrointestinal tract, lymphocytes are found in large numbers in three main regions: scattered throughout the lamina propria, in Peyer's patches, and within the epithelial layer (Fig. 11–8). Cells at each site have distinct phenotypic and functional characteristics.

Intraepithelial Lymphocytes. The majority of intraepithelial lymphocytes are T cells. In humans,

FIGURE 11–8. Cellular components of the mucosal immune system.

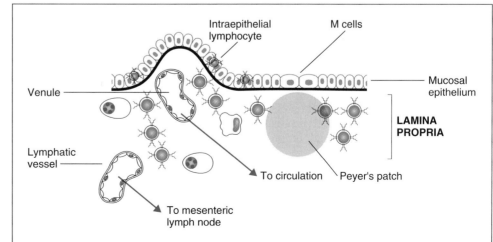

most of these are CD8$^+$ cells. Strikingly, in mice about 50 per cent of intraepithelial lymphocytes express the $\gamma\delta$ form of the T cell receptor (TCR), similar to intraepidermal lymphocytes in the skin. In most other species, including humans, only about 10 per cent of intraepithelial lymphocytes are $\gamma\delta$ cells. Although this proportion is not impressive, it is still higher than the proportions of $\gamma\delta$ cells found among T cells in other tissues. Both the $\gamma\delta$ and the $\alpha\beta$ TCR-expressing intraepithelial lymphocytes show limited diversity of antigen receptors, with a few V genes being dominant. All these findings support the idea mentioned previously that intraepithelial lymphocytes have a limited range of specificities that is distinct from that of most T cells and may have evolved to recognize commonly encountered intraluminal antigens.

Lamina Propria Lymphocytes. The intestinal lamina propria contains a mixed population of cells. These include T lymphocytes, most of which are CD4$^+$ and have the phenotype of activated cells. It is likely that T cells initially recognize and respond to antigens in regional mesenteric lymph nodes and then migrate back to the intestine to populate the lamina propria. This is similar to the postulated origin of T cells in the dermis of the skin. In addition, the lamina propria contains large numbers of activated B lymphocytes and plasma cells, as well as macrophages, eosinophils, and mast cells.

Mucosal Lymphoid Follicles. **Peyer's patches** are organized mucosal lymphoid follicles in the small intestine. Morphologically and functionally similar follicles are abundant in the appendix and are found in smaller numbers in much of the gastrointestinal and respiratory tracts. **Pharyngeal tonsils** are also mucosal lymphoid follicles analogous to Peyer's patches. Like lymphoid follicles in the spleen and lymph nodes, the central regions of the mucosal follicles are B cell–rich areas that often contain germinal centers. Peyer's patches also contain small numbers of CD4$^+$ T cells, mainly in the interfollicular regions. In adult mice, 50 to 70 per cent of Peyer's patch lymphocytes are B cells, and 10 to 30 per cent are T cells. Some of the epithelial cells overlying Peyer's patches are specialized "M (membranous) cells." M cells lack microvilli, are actively pinocytic, and transport macromolecules from the intestinal lumen into subepithelial tissues. They are thought to play an important role in delivering antigens to Peyer's patches. (Note, however, that M cells do not themselves function as APCs.)

Gut-associated lymphoid tissues are populated by cells that have emigrated from the blood. As mentioned previously, lymphocyte binding to endothelium in the intestines and cell migration into mucosal tissues involves an unusual integrin that expresses an α chain called α_4 (or the α_E chain) in association with the β_7 chain. Mice in which the gene for the β_7 chain is knocked out fail to develop gut-associated lymphoid tissues, emphasizing the essential role of this integrin in the entry of cells into mucosal tissues.

Induction of Immune Responses in Mucosal Tissues

The bulk of orally administered antigen enters from the intestinal lumen into lymphatics and is carried to draining mesenteric lymph nodes, where immune responses are initiated. Some ingested protein antigens may be transported through M cells into Peyer's patches, where they are capable of stimulating both T and B lymphocytes. Although intestinal epithelial cells can be induced to express class II MHC molecules, e.g., by exposure to IFN-γ, there is no evidence that these cells are capable of presenting antigens to and stimulating T lymphocytes. Lymphocytes that are activated in mesenteric lymph nodes may return to populate the lamina propria. Lymphocytes stimulated in Peyer's patches may also migrate to the lamina propria or into mesenteric lymph nodes and ultimately the systemic circulation. Thus, the compartments of the mucosal immune system are connected with one another and with the rest of the immune system.

Production of IgA

IgA is the major class of antibody that can be actively and efficiently secreted through epithelia. Therefore, it plays a critical role in defense against intestinal and respiratory pathogens and in the passive transfer of immunity in milk and colostrum from mothers to infants. The size of the intestinal surface accounts for the enormous quantity of IgA that is produced. It is estimated that a normal 70 kg adult secretes about 3 gm of IgA per day, accounting for 60 to 70 per cent of the total output of antibodies.

In the gastrointestinal tract, the process of IgA production is initiated by the entry of protein antigens into the Peyer's patches. Here the antigens stimulate specific T lymphocytes in the inter-follicular regions as well as B lymphocytes in the follicles. Some of the B lymphocytes differentiate into IgA-producing cells and migrate to the lamina propria. The two cytokines that are mainly responsible for switching to the IgA isotype are TGF-β (which may be produced by T cells as well as nonlymphoid stromal cells) and IL-5 (see Chapter 9). Some activated B cells migrate into the germinal centers of the Peyer's patches, where they undergo proliferation and somatic mutations of Ig genes, leading to affinity maturation of antibodies. This is probably similar to the sequence of events in lymphoid follicles in the spleen and lymph nodes. After they are generated, IgA-producing B cells may reside in the lamina propria or may migrate to other mucosal tissues or lymphoid organs.

There are several reasons why larger amounts of IgA are produced in the mucosal immune system than in other tissues. First, IgA-expressing B cells tend to home to Peyer's patches and lamina propria. This has been demonstrated by injecting antigen-stimulated, IgA-expressing B cells into the systemic circulation of normal, unimmunized animals. The homing receptors and addressins responsible for this pattern of migration are unknown. Second, IL-5–producing helper T cells are more numerous in Peyer's patches than in other lymphoid tissues. Although the mucosal immune system is the site of greatest IgA production and the source of secretory IgA, this isotype may also be produced by B cells in other lymphoid tissues and in the bone marrow, giving rise to serum IgA.

Secretory IgA forms a dimer that is held together by the coordinately synthesized and secreted J chain (see Chapter 3). (In contrast, serum IgA is usually a monomer lacking the J chain.) Once this dimer is secreted into the lamina propria, it is transported through epithelial cells into the intestinal lumen by a protein called the **secretory component** (or poly-Ig receptor). Secretory component is synthesized by mucosal epithelial cells and expressed on their basal and lateral surfaces. The membrane-associated form is a ~100 kD glycoprotein with five extracellular domains homologous to Ig domains and is thus a member of the Ig superfamily. The secreted, dimeric IgA containing the J chain binds to and forms covalent complexes with the membrane-associated secretory component on mucosal epithelial cells (Fig. 11–9). This complex is endocytosed into the epithelial cell and transported in vesicles to the luminal surface. Here the extracellular domain of the secretory component, carrying the IgA molecule, is proteolytically cleaved, leaving the transmembrane and cytoplasmic domains attached to the epithelial cell and releasing the bound IgA into the intestinal lumen. Secretory component is also responsible for the secretion of IgA into bile, milk, sputum, and saliva. Although its ability to transport IgA has been studied most extensively, secretory compo-

nent is capable of transporting IgM into intestinal secretions as well. (Note that secreted IgM is also a polymer associated with the J chain.) In secretions, IgA serves to neutralize microbes and toxins.

Oral Tolerance

The oral administration of a protein antigen often leads to a marked suppression of systemic humoral and cell-mediated immune responses to immunization with the same antigen. The inhibition is antigen-specific and fulfills the criteria for immunologic tolerance (see Chapter 10). The physiologic importance of oral tolerance may be as a means of preventing immune responses to bacteria that normally reside as commensals in the intestinal lumen and are needed for digestion and absorption. Despite extensive analysis, there is no consensus on the mechanisms responsible for oral tolerance. Some studies have shown that orally administered proteins induce T cell clonal anergy, which resembles anergy induced by antigen presentation by costimulator-deficient APCs (see Chapter 10). However, it is not known which APCs present antigens administered via the enteric route to induce T cell tolerance, or how anergy induced by oral antigen is manifested as systemic unresponsiveness.

An alternative possibility is that oral antigens may stimulate the production of cytokines, such as TGF-β, that inhibit lymphocyte proliferation, resulting in the suppression of immune responses. In fact, TGF-β may participate in both distinctive responses to oral immunization, because it is known to induce B cell switching to IgA and to inhibit lymphocyte proliferation. TGF-β–producing cells may migrate to distant sites and even inhibit immune responses in the periphery.

Anergy and suppression are not mutually exclusive mechanisms, and both may contribute to oral tolerance. It is also unclear why oral administration of some antigens, such as soluble proteins in large doses, induces systemic T cell tolerance, whereas other antigens, such as attenuated polio

FIGURE 11–9. Mechanisms of IgA transport and secretion in mucosal epithelia. IgA is produced by plasma cells in the lamina propria and binds to secretory component. The complex is actively transported through the epithelial cell, and the bound IgA is released by proteolytic cleavage.

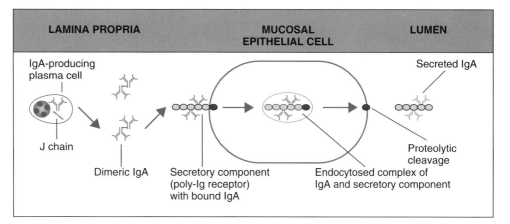

LAMINA PROPRIA

MUCOSAL EPITHELIAL CELL

LUMEN

IgA-producing plasma cell

Secreted IgA

J chain

Dimeric IgA

Secretory component (poly-Ig receptor) with bound IgA

Endocytosed complex of IgA and secretory component

Proteolytic cleavage

virus vaccines, induce excellent T cell–dependent mucosal antibody responses and long-lived memory. One possible explanation is that microbes, unlike soluble proteins, infect or activate APCs in the intestinal epithelium and trigger specific local immunity. Whatever its mechanism, the phenomenon of oral tolerance has obvious practical relevance. It is a potential treatment for autoimmune diseases in which the autoantigen is known, as well as for other clinical situations in which it may be desirable to inhibit specific immune responses to defined protein antigens.

SUMMARY

The immune system copes with the problem of how to efficiently recognize and respond to foreign antigens by collecting and concentrating antigens in the peripheral lymphoid organs and by rapidly circulating lymphocytes among these organs to sample the collected antigens. The migration of lymphocytes into the secondary lymphoid organs occurs through specialized high endothelial venules (HEVs), where tissue-specific lymphocyte homing receptors recognize endothelial ligands, called addressins. The receptor responsible for homing of lymphocytes to peripheral lymph nodes is called L-selectin. It is expressed at high levels on naive T cells, accounting for the preferential migration of these cells into lymph nodes. Most activated effector and memory T cells express lower levels of L-selectin and increased levels of adhesion molecules that promote their preferential migration through vessels at sites of antigen exposure and inflammation.

Immune responses in different tissues have distinctive characteristics. Priming of T lymphocytes occurs mainly in lymph nodes and spleen. These organs are also the principal sites of antibody production. Immune responses in the skin are initiated largely by antigen presentation by epidermal Langerhans cells. A typical response in the skin is the T lymphocyte–mediated DTH reaction. The mucosal immune system is specialized to produce large quantities of IgA, which is the only class of antibody that is efficiently secreted through epithelial cells into the lumen of the gastrointestinal and respiratory tracts.

Selected Readings

Bevilacqua, M. P. Endothelial-leukocyte adhesion molecules. Annual Review of Immunology 11:767–804, 1993.

Mondino, A., A. Khoruts, and M. K. Jenkins. The anatomy of T-cell activation and tolerance. Proceedings of the National Academy of Sciences USA 93:2245–2252, 1996.

Neutra, M. R., E. Pringault, and J.-P. Kraehenbuhl. Antigen sampling across epithelial barriers and induction of mucosal immune responses. Annual Review of Immunology 14:275–300, 1996.

Shaw, S., K. Ebnet, E. P. Kaldjian, and A. O. Anderson. Orchestrated information transfer underlying leukocyte endothelial interactions. Annual Review of Immunology 14:155–177, 1996.

Springer, T. A. Traffic signals for lymphocyte recirculation and leukocyte emigration: the multistep paradigm. Cell 76:301–314, 1994.

Van der Ertwegh, A. J. M., J. D. Laman, R. J. Noelle, W. J. Boersma, and E. Claassen. In vivo T cell–B cell interactions and cytokine production in the spleen. Seminars in Immunology 6:327–336, 1994.

Van Rooijen, N. Antigen processing and presentation in vivo: the microenvironment as a critical factor. Immunology Today 11:436–439, 1990.

Williams, I. R., and T. S. Kupper. Immunity at the surface: homeostatic mechanisms of the skin immune system. Life Sciences 58:1485–1507, 1996.

SECTION III

EFFECTOR MECHANISMS OF IMMUNE RESPONSES

The physiologic function of all specific immune responses is to eliminate foreign antigens. The mechanisms by which specific lymphocytes recognize and respond to foreign antigens, i.e., the recognition and activation phases of immune responses, were discussed in Section II. Section III describes the effector mechanisms that are recruited and stimulated by the innate immune system and by antigen-activated lymphocytes. In Chapter 12, we describe cytokines that are produced by T lymphocytes and by some non-lymphoid cells. These molecules are the soluble mediators of innate immunity and specific immunity. Chapter 13 presents the effector cells of cell-mediated immunity, including T lymphocytes, macrophages, and natural killer cells, which participate in immune-mediated inflammatory reactions and function as the primary defense mechanisms against intracellular microbes. Chapter 14 discusses immediate hypersensitivity, an inflammatory reaction caused by the activation of a specialized subset of CD4+ T cells, which leads to the production of IgE antibody, activation of mast cells and basophils, and recruitment of eosinophils. Chapter 15 deals with the complement system, one of the major effector mechanisms of the humoral immune response.

CHAPTER TWELVE

CYTOKINES

In Chapter 1, we introduced the concept that defense against foreign organisms, such as viruses or bacteria, is mediated by innate immunity and by specific (or adaptive) immunity. The effector phases of both innate and specific immunity are in large part mediated by protein hormones called **cytokines.** In innate immunity, the effector cytokines are mostly produced by mononuclear phagocytes and are therefore often called **monokines.** Monokines elicit neutrophil-rich inflammatory reactions that serve to contain and, where possible, to eradicate microbial infections. Although monokines can be elicited directly by microbes, they also can be secreted by mononuclear phagocytes in response to antigen-stimulated T cells, i.e., as part of specific immunity. Most cytokines in specific immunity are produced by activated T lymphocytes, and such molecules are commonly called **lymphokines.**

T cells produce several cytokines that serve primarily to regulate the growth and differentiation of various lymphocyte populations, and thus play important roles in the **activation phase** of T cell–dependent immune responses. Other T cell–derived cytokines function principally to activate and regulate inflammatory cells, such as mononuclear phagocytes, neutrophils, and eosinophils. These T cell–derived cytokines act in the **effector phase** of cell-mediated immunity and are also responsible for communication between the cells of the immune and inflammatory systems. Finally, both lymphocytes and mononuclear phagocytes produce cytokines, such as **colony-stimulating factors** (CSFs), which stimulate the growth and differentiation of immature leukocytes in the bone marrow, providing a source of additional leukocytes to replace the cells that are consumed during inflammatory reactions. Because many of these cytokines are made by certain populations of blood leukocytes (e.g., T cells or monocytes) and act on other leukocyte populations (e.g., monocytes, neutrophils, or eosinophils), these molecules are sometimes called **interleukins.** This term should not be construed to imply that cytokines are only synthesized by or only act upon white blood cells. However, the term "interleukin" has been useful because as new cytokines are molecularly characterized, they may be assigned a designated interleukin number (e.g., IL-1, IL-2) to assure that there is an unambiguous, shared nomenclature among investigators.

In this chapter, we discuss the structure, production, and biologic actions of cytokines. We will also discuss the increasing clinical uses of cytokines as modulators of inflammation, immunity, and hematopoiesis. In this setting, cytokines are often referred to as **biologic response modifiers.** Before describing the specific molecules, we will begin with a review of the general properties shared by cytokines that allow us to consider them as a group.

GENERAL PROPERTIES OF CYTOKINES

Although cytokines are a diverse group of proteins, a number of properties are shared by these molecules:

1. *Cytokines are produced during the activation and effector phases of innate and specific immunity and serve to mediate and regulate immune and inflammatory responses.* In innate immunity, microbial products, such as lipopolysaccharide (LPS), or viral products, such as double-stranded RNA, directly stimulate mononuclear phagocytes to secrete their cytokines. In contrast, T cell–derived cytokines are elicited primarily in response to specific recognition of foreign antigens.

2. *Cytokine secretion is a brief, self-limited event.* In general, cytokines are not stored as preformed molecules, and their synthesis is initiated by new gene transcription. Such transcriptional activation is usually transient, and the messenger ribonucleic acids (mRNAs) encoding cytokines are unstable. The combination of a short period of transcription and a short-lived mRNA transcript ensures that cytokine synthesis is transient. Some cytokines may be additionally controlled by posttranscriptional mechanisms, such as proteolytic release of an active product from an inactive precursor. Once synthesized, cytokines are usually rapidly secreted, resulting in a burst of cytokine release as needed.

3. *Many individual cytokines are produced by multiple diverse cell types.* To emphasize that the cellular source of these molecules is usually not a distinguishing characteristic, investigators have generally adopted the convention followed in this book, namely to refer to these molecules collectively as cytokines rather than as lymphokines or monokines, regardless of their cellular source in a particular setting.

4. *Cytokines act upon many different cell types.* This property is called **pleiotropism.** The early view that cytokines are primarily molecules produced by leukocytes that act particularly on leukocytes ("interleukins") is too restricted a concept.

5. *Cytokines often have multiple different effects on the same target cell.* Some effects may occur simultaneously, whereas others may occur over different time frames (e.g., minutes, hours, or days).

6. *Cytokine actions are often redundant.* Many functions originally attributed to one cytokine have proved to be shared properties of several different cytokines. This observation has been reinforced by the study of knockout mice that lack particular cytokine genes yet display only subtle abnormalities in their immune responses.

7. *Cytokines often influence the synthesis of other cytokines,* leading to cascades in which a second or third cytokine may be induced by and may mediate the biologic effects of the first cytokine. The ability of one cytokine to enhance or suppress

the production of others may provide important positive and negative regulatory mechanisms for immune and inflammatory responses.

8. *Cytokines often influence the action of other cytokines.* Two cytokines may interact to antagonize each other's action, to produce additive effects, or, in some cases, to produce greater than anticipated or even unique effects, a kind of interaction commonly referred to as synergy.

9. *Cytokines, like other polypeptide hormones, initiate their action by binding to specific receptors on the surface of target cells* (Box 12–1). The relevant target cell may be the same cell that secretes the cytokine (**autocrine** action), a nearby cell (**paracrine** action), or, like true hormones, a distant cell that is stimulated via cytokines that have been secreted into the circulation (**endocrine** action). Receptors for cytokines often show very high affinities for their ligands, with dissociation constants

(K_d) in the range of 10^{-10} to 10^{-12} M. (For comparison, recall that antibodies typically bind antigens with a K_d of 10^{-7} to 10^{-10} M, and major histocompatibility complex [MHC] molecules bind peptides with a K_d of only about 10^{-6} M.) As a consequence, only very small quantities of a cytokine need be produced to occupy receptors and elicit a biologic effect.

10. *The expression of many cytokine receptors is regulated by specific signals.* This signal may be another cytokine or even the same cytokine that binds to the receptor, permitting positive amplification or negative feedback.

11. *Most cellular responses to cytokines require new mRNA and protein synthesis.* The mechanism by which cytokine binding to cell surface receptors stimulates transcription is increasingly understood in molecular terms, as we shall discuss later in this chapter.

BOX 12–1.   Cytokine Receptor Families

The principal function of cytokine receptors is to convert an extracellular signal, namely the specific binding of a cytokine to a target cell, into an intracellular signal, such as activation of an enzyme or a transcription factor that can trigger a target cell response. All known cytokine receptors are transmembrane proteins, and the extracellular domains bind cytokine, thereby providing the means of detection of the extracellular signal. Signal transduction usually involves the intracellular portions of the cytokine receptor.

Cytokine receptors have been grouped into five large families based upon the presence of conserved folding motifs or sequence homologies (see figure, part A). The first motif to be noted in cytokine receptors was the Ig domain, and certain cytokine receptors (e.g., the type I and type II interleukin [IL]-1 receptors) contain several extracellular domains that belong to the Ig superfamily. This motif is also common in mesenchymal cell growth factor receptors, e.g., for platelet-derived growth factor or fibroblast growth factor receptors, and in receptors for certain colony-stimulating factors such as c-kit ligand and monocyte colony-stimulating factor (M-CSF). The latter two receptors also share with mesenchymal growth factor receptors a cytoplasmic protein tyrosine kinase domain that initiates the intracellular signal.

The second motif to be characterized in cytokine receptors involves a conserved extracellular sequence of five amino acid residues, tryptophan-serine-X-tryptophan-serine (written as WSXWS in the single letter amino acid code), where amino acid residue X is variable. This small motif is contained within a larger conserved domain additionally characterized by two conserved cysteine residues. All cytokine receptors bearing the two cysteine/WSXWS motif are said to belong to the "type 1 family of cytokine receptors" and bind cytokines that fold into globular domains containing four α-helical strands. The prototypic molecule for this family was growth hormone, and this structure is shared by IL-2, IL-3, IL-4, IL-5, IL-6, IL-7, IL-9, IL-11, IL-13, IL-15, granulocyte-monocyte colony-stimulating factor (GM-CSF), and granulocyte colony-stimulating factor (G-CSF). Interestingly, the IL-6 receptor contains both an Ig domain and the two cysteine/WSXWS motif.

A third group of receptors, to date defined by nucleotide sequences, includes the type I and type II interferon (IFN) receptors. The IFN receptors define the "type II family of cytokine receptors." The fourth structural motif identified in cytokine receptors is a cysteine-rich domain first identified in the two tumor necrosis factor (TNF) receptors (TNF-RI and TNF-RII), which have been called "type III cytokine receptors." These molecules are part of a larger gene superfamily that is discussed in Box 12–4. The final group of cytokine receptors is the seven transmembrane α-helical receptors, a very large family of molecules that includes the receptors for the chemokines. This motif was originally found in β-adrenergic receptors and retinal rhodopsin and is common to all receptors that are coupled to heterotrimeric guanosine triphosphate (GTP)-binding signaling proteins.

Several cytokine receptors actually consist of two or more separate transmembrane polypeptide chains that function as a complex (see figure, part B). For example, the high affinity IL-2 receptor contains three separate chains (α, β, and γ_c). Both β and γ_c, but not α, contain the two cysteine/WSXWS motif. The β chain binds IL-2 with moderate affinity and heterodimers of $\beta\gamma_c$ mediate signal transduction through a Jak/STAT mechanism (see Box 12–2). IL-2Rα, which contains structural motifs resembling complement control protein repeats (see Chapter 15), also binds IL-2 and serves to increase the affinity of the heterotrimeric $\alpha\beta\gamma_c$ form of the receptor for the cytokine, but the α chain does not transduce any signal. The γ_c subunit contributes to the signaling of several other cytokine receptors (e.g., IL-4, IL-7, IL-9, and IL-15), which is why it is now called the common γ (γ_c) chain.

Other cytokine receptors also use common signaling chains. The IL-6 receptor interacts with a separate transmembrane signal-transducing 130 kD glycoprotein that, like the IL-6 receptor itself, also contains an Ig domain and a two cysteine/WSXWS motif. At least four other protein hormones (IL-11, ciliary neurotropic factor, oncostatin M, and leukemia inhibitory factor) also have receptors that interact with the same 130 kD signal-transducing molecule. The cytokine receptors for IL-3, IL-5, and GM-CSF share a 150 kD signal-transducing subunit in humans, sometimes called the common β (β_c) chain; in mice, the IL-3 receptor has a unique but homologous signal-transducing subunit.

Continued

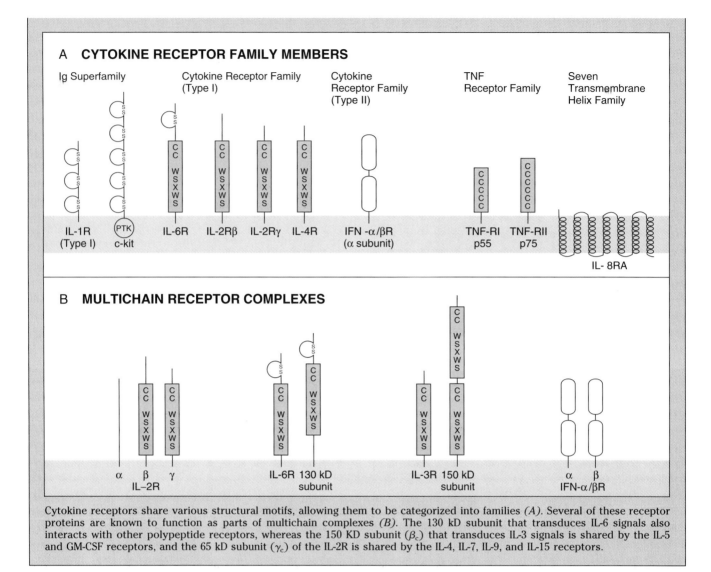

Cytokine receptors share various structural motifs, allowing them to be categorized into families *(A)*. Several of these receptor proteins are known to function as parts of multichain complexes *(B)*. The 130 kD subunit that transduces IL-6 signals also interacts with other polypeptide receptors, whereas the 150 KD subunit (β_c) that transduces IL-3 signals is shared by the IL-5 and GM-CSF receptors, and the 65 kD subunit (γ_c) of the IL-2R is shared by the IL-4, IL-7, IL-9, and IL-15 receptors.

12. *For many target cells, cytokines act as regulators of cell division, i.e., as growth factors.* Some immunologists now think that cytokines should be categorized with epithelial and mesenchymal cell growth factors into a larger functional group of polypeptide regulatory molecules. However, we will continue to distinguish those molecules whose primary actions are as mediators of host defense (i.e., cytokines) from those molecules whose primary role resides in tissue repair (i.e., the epithelial and mesenchymal cell polypeptide growth factors).

We have organized our discussion of specific cytokines into three broad categories of function: (1) *mediators and regulators of innate immunity,* which are elicited by infectious agents principally from mononuclear phagocytes and stimulate or inhibit inflammatory reactions; (2) *mediators and regulators of specific immunity,* which are elicited in response to specific antigen recognition by T lymphocytes and which serve to intensify, focus, and specialize inflammatory reactions; and (3) *stimulators of immature leukocyte growth and differentiation,* which are produced by both stimulated lymphocytes and other cells. This classification is based on what appear to be the principal biologic actions of a particular cytokine, although, as we shall see, many cytokines may function in more than one of these categories.

CYTOKINES THAT MEDIATE AND REGULATE INNATE IMMUNITY

Before discussing the characterization and function of cytokines that mediate innate immunity (Table 12–1), it is useful to begin with an overview of the events that occur in innate immune reactions. Like the specific adaptive immune reactions discussed in Section II of this book, innate immunity begins with a recognition phase that something foreign is present in the host. *There are at least four important mechanisms that are used to recognize foreign viruses and other microbes during*

TABLE 12-1. Mediators and Regulators of Innate Immunity

Cytokine	Polypeptide Size	Cell Source	Cell Target	Primary Effects on Each Target
Type I IFN	18 kD	Mononuclear phagocyte, other (α); fibroblast, other (β)	(a) All (b) NK cell	(a) Antiviral state, increased class I MHC expression (b) Activation
Interleukin-15	13 kD	Mononuclear phagocyte, other	(a) NK cell and T cell	(a) Proliferation
Interleukin-12	35 kD, 40 kD	Mononuclear phagocyte, dendritic cell	(a) NK cell and T cell	(a) IFN-γ synthesis, cytolytic function, CD4$^+$ T cell differentiation
Tumor necrosis factor	17 kD	Mononuclear phagocyte, T cell	(a) Neutrophil (b) Endothelial cell (c) Hypothalamus (d) Liver (e) Muscle, fat	(a) Activation (inflammation) (b) Activation (inflammation, coagulation) (c) Fever (d) Acute phase reactants (serum amyloid A protein) (e) Catabolism (cachexia)
Interleukin-1	17 kD	Mononuclear phagocyte, other	(a) Endothelial cell (b) Hypothalamus (c) Liver (d) Muscle, fat (e) Thymocyte	(a) Activation (inflammation, coagulation) (b) Fever (c) Acute phase reactants (serum amyloid A protein) (d) Catabolism (cachexia) (e) Costimulation (in vitro)
Interleukin-6	26 kD	Mononuclear phagocyte, endothelial cell, T cell	(a) Mature B cell (b) Liver (c) Thymocyte	(a) Growth (b) Acute phase reactants (fibrinogen) (c) Costimulation (in vitro)
Chemokines (see Table 12-2)	8-10 kD	Mononuclear phagocyte, endothelial cell; fibroblast; T cell; platelet	(a) Leukocytes	(a) Chemotaxis, chemokinesis, adhesion, activation
Interleukin-10	20 kD	Mononuclear phagocyte, T cell	(a) Mononuclear phagocyte (b) B cell	(a) Inhibition (b) Activation

Abbreviations: IFN, interferon; kD, kilodalton; MHC, major histocompatibility complex; NK, natural killer.

the innate immune response without the participation of antigen-specific T and B lymphocytes.

1. The presence of viruses can be recognized by infected cells by the presence of nucleic acids not typically associated with mammalian cell replication or protein synthesis. The best-characterized of these signals is double-stranded RNA, which is generated during replication of many virus types. The sensor by which mammalian cells recognize double-stranded RNA appears to be protein kinases that are activated by binding such atypical nucleic acids, initiating signals that lead to the transcription and secretion of type I interferons (IFN-α and IFN-β).

2. The presence of viruses can also be recognized by natural killer (NK) cells. The mechanism of this response will be discussed in more detail in Chapter 13. The sensors are NK cell receptors that deliver inhibitory signals when they engage specific allelic forms of self class I MHC molecules containing certain peptides derived from self proteins. The replacement of these self-derived peptides with virus-derived peptides results in class I MHC molecules that fail to deliver the inhibitory signal, thus leading to NK cell responses. Alternatively, certain viruses shut off class I MHC molecule synthesis or expression, again allowing NK

cells to sense the absence of the inhibitory signal. NK cells are better able to respond if they have been pre-activated by exposure to type I IFN and other cytokines of innate immunity. Activated NK cells, like certain activated T cells, synthesize and secrete interferon-γ (IFN-γ), also called type II IFN, and tumor necrosis factor (TNF). As will be described later in the chapter, these cytokines activate mononuclear phagocytes, vascular endothelial cells, and neutrophils, which act in concert to generate a local inflammatory response.

3. Bacterial lipids can be recognized because they are structurally distinct from mammalian cell lipids. Mononuclear phagocytes express a cell surface receptor that can recognize many such foreign lipids, particularly those derived from the cell walls of gram-negative bacteria, collectively called endotoxin or lipopolysaccharide (LPS) (Fig. 12-1). LPS circulates in the plasma in a complex with a specific LPS-binding protein (LBP). The mononuclear phagocyte receptor for the LPS/LBP complex is a surface protein called CD14. The engagement of CD14 by the LPS/LBP complex triggers association of occupied CD14 molecules with an as yet uncharacterized signaling receptor. This, in turn, triggers the phagocyte to synthesize a number of new proteins, including cytokines such as TNF, interleukin-1β (IL-1β), IL-6, IL-10, IL-12, and IL-15. Ac-

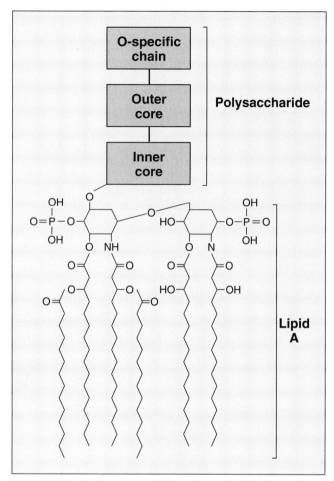

FIGURE 12–1. Structure of lipopolysaccharide. Lipopolysaccharide (LPS) is released when the cell walls of gram-negative bacteria, such as *E. coli*, are degraded. The lipid A moiety, which contains most of the biologic activity, is hydrophobic. The polysaccharide, which can contain 50 or more hexose moieties, can be divided into more conserved core regions and bacterial stain-specific ("O-specific") regions.

tivated mononuclear phagocytes also secrete chemokines that function in the recruitment of blood monocytes and neutrophils to the local site of inflammation.

4. In addition to the cellular recognition system described above, bacterial lipids can also be recognized as foreign by certain circulating plasma proteins. The best-described plasma protein system that serves this function is the complement system, which will be discussed in detail in Chapter 15. Suffice it to say here that this constituent of innate immunity, in addition to participating in the direct elimination of microbes, generates a number of mediator peptides that contribute to eliciting protective inflammatory reactions.

An important consequence of the recognition of viruses and bacteria is, as mentioned above, the secretion of cytokines. Virus-elicited type I IFNs, IL-12, and IL-15 serve to recruit and activate NK cells, the first line of defense in the antiviral response. Bacteria-elicited TNF, IL-1, and chemokines

are the principal mediators of local phagocyte recruitment and, along with IL-6, produce systemic responses that favor containment and elimination of the bacteria. IL-10 is an important feedback regulator of innate immunity.

The innate immune response not only has an important protective function but also serves to initiate and regulate subsequent specific immune responses. This concept was introduced in Chapter 1. As we shall see later in this chapter, several cytokines that are produced as components of innate immunity, notably IL-12 and IL-10, can directly or indirectly regulate the pattern of subsequent specific cell-mediated immune responses by influencing the growth and differentiation of T cells. (For this reason, these cytokines may be classified into more than one functional group.) As mentioned in Chapter 9, and discussed in more detail in Chapter 15, products of complement activation also play dual roles in innate immunity and in the regulation of subsequent humoral immune responses.

Type I Interferons

Type I interferons (IFNs) comprise two serologically distinct groups of proteins. The first group, collectively called IFN-α, is a family of about 20 structurally related polypeptides of approximately 18 kD, each encoded by a separate gene. (Some investigators now subdivide the IFN-α family into two groups, IFN-α1 and IFN-α2/IFN-ω, on the basis of the relatedness of amino acid sequences within the family.) Natural IFN-α preparations are usually a mixture of these molecules. The major cell source for production of IFN-α is the mononuclear phagocyte, and IFN-α is sometimes called leukocyte interferon. The second serological group of type I IFN consists of a single gene product, a 20 kD glycoprotein called IFN-β. The usual cell source for isolation of IFN-β is the cultured fibroblast, and IFN-β is sometimes called fibroblast interferon. However, many cells make both IFN-α and IFN-β. The most potent natural signal that elicits type I IFN synthesis is viral infection. Experimentally, production of type I IFN is commonly elicited by synthetic double-stranded RNA molecules, which may mimic a signal produced during viral replication. Both IFN-α and IFN-β are also secreted during immune responses to antigens. In this case, antigen-activated T cells stimulate mononuclear phagocytes to synthesize type I IFNs.

IFN-α and IFN-β show little structural similarity to each other. Nevertheless, all type I IFN molecules bind to the same cell surface receptor and appear to induce a similar series of cellular responses. The type I IFN receptor is a heterodimer of two structurally related polypeptides (see Box 12–1). Type I IFN acts on target cells primarily by activating new gene transcription. The intracellular signaling pathway used by type I IFN, called the **Janus kinase/signal transducer and activator of** transcription **(Jak/STAT)** pathway (Box 12–2), is now known to be used by many other cytokines as well.

There are four principal biologic actions of type I IFN:

1. *Type I IFN inhibits viral replication.* IFN causes cells to synthesize a number of enzymes, such as 2′–5′ oligoadenylate synthetase, that collectively interfere with replication of viral RNA or DNA. The antiviral action of type I IFN is primarily paracrine, in that a virally infected cell secretes IFN to protect neighboring cells not yet infected. A cell that has responded to IFN and is resistant to viral infection is said to be in an **antiviral state.**

2. *Type I IFN increases the lytic potential of NK cells.* As will be discussed in Chapter 13, a major function of NK cells is to kill virally infected cells early in the course of infection, prior to the onset of the specific immune response.

3. *Type I IFN modulates MHC molecule expression.* In general, type I IFN increases expression of class I MHC molecules and inhibits class II MHC molecule expression. Because most cytolytic T lymphocytes (CTLs) recognize foreign antigens bound to class I MHC molecules, type I IFN boosts the effector phase of cell-mediated immune responses by enhancing the efficiency of CTL-mediated killing. At the same time, type I IFN may inhibit the recognition phase of immune responses by preventing the activation of class II MHC-restricted helper T lymphocytes.

4. *Type I IFN inhibits cell proliferation.* This may be due to induction of the same enzymes that inhibit viral replication but also may involve other enzymes that prevent amino acid synthesis, especially of essential amino acids such as tryptophan. Although the mechanisms may be partly different, the antiviral effects and the antiproliferative effects of IFN cannot be uncoupled. IFN-α has been used as an antiproliferative agent for certain tumors (e.g., hairy cell leukemia and childhood hemangiomas). Paradoxically, type I IFN may act as a growth and differentiation factor for T cells in humans.

Thus, three of the principal activities of type I IFN, namely the induction of the antiviral state, the activation of NK cell lytic functions, and the increase in class I MHC molecule expression on virally infected cells, all act in concert to eradicate viral infections. IFN-α is in clinical use as an antiviral agent in certain forms of viral hepatitis. IFN-β is being used as a therapy for multiple sclerosis, but the mechanism of this action is not fully understood.

Interleukin-12

Interleukin-12 (IL-12) was originally identified as a macrophage-derived activator of NK cell cytolytic function, although it is now appreciated to be a potent inducer of IFN-γ production by T cells as well as NK cells. Active IL-12 exists as a disulfide-linked heterodimer of 35 kD (p35) and 40 kD (p40) subunits. Although these subunits contain shared regions of homology with each other, they also have unique structural features. Specifically, the p35 subunit contains motifs likely to fold into a four α-helical globular domain characteristic of cytokines that bind to type I cytokine receptors (see Box 12–1). The p40 subunit, like the receptor for IL-6, contains both an Ig-like domain and a two cysteine/WSXWS motif, the latter being characteristic of the type I cytokine receptor family. Many cells appear to synthesize p35, but p40 synthesis is largely restricted to activated mononuclear phagocytes and dendritic cells. As a result, these cell types are the principal sources of biologically active IL-12. Despite its receptor-like motifs, p40 is not the IL-12 receptor. Recently, two separate IL-12–binding polypeptides have been identified by molecular cloning, both of which contain two cysteine/WSXWS motifs; the functional receptor may be a heterodimer. IL-12 receptors are expressed primarily by NK cells, T cells, and subsets of B cells. Signaling through these receptors involves a Jak/STAT pathway; to date, IL-12 is the only cytokine known to activate STAT4.

Several biologic actions of IL-12 are seen in NK and T cell populations.

1. Interleukin-12 causes NK cells and T cells to secrete IFN-γ. Knockout mice lacking p40 (and hence functional IL-12) have reduced levels of IFN-γ production, which impairs full macrophage activation and cell-mediated immunity (see Chapter 13). However, IFN-γ synthesis is not completely abrogated in these mice.

2. IL-12 acts as a differentiation factor on CD4$^+$ T cells, promoting their specialization into IFN-γ–producing T$_H$1-like cells that help phagocyte-mediated immunity. This effect of IL-12 will be discussed again later in the chapter.

3. Interleukin-12 enhances the cytolytic functions of activated NK cells and CD8$^+$ T cells (see Chapter 13). IL-12 does not act on resting cell populations and appears to act as a differentiative signal. Knockout mice lacking p40 can still generate cytolytic cells, suggesting that this function may be redundant with other cytokines.

We have classified IL-12 as a mediator of innate immunity because it links macrophage activation by microbes to development of NK cell effector functions. At the same time, *IL-12 provides an important link between innate immunity and specific, adaptive immunity,* favoring the development of specific immune responses that can better protect the host against viruses and bacteria. Adjuvants are bacterial products that recruit and activate macrophages and thereby serve to boost specific immune responses against antigens that are co-administered. In essence, IL-12 acts as an endogenous adjuvant, and recombinant IL-12 is in clinical

BOX 12-2. Janus-family Kinases (Jaks) and Signal Transducer and Activator of Transcription (STAT) Proteins

Studies of the signal-transduction mechanisms of interferons led to the discovery of the Janus family kinase–signal transducer and activator of transcription (Jak/STAT) pathways used by many different cytokines. The initial insights into this pathway were provided by a convergence of two experimental approaches: biochemistry and genetics. From biochemical studies, it was discovered that several genes, whose transcription is directly and rapidly activated by type I interferons (IFN-α, IFN-β), shared a *cis*-acting regulatory DNA sequence element called the **interferon stimulated response element.** This element could bind a protein factor called **interferon stimulatory gene factor-3 (ISGF-3)** that is present in the nuclei of IFN-α/β–treated cells but not in the nuclei of untreated cells. ISGF-3 consists of three proteins: STAT1 (which can be synthesized either as a 91 kD isoform, called STAT1α, or as an 89 kD isoform, called STAT1β), STAT2 (a 113 kD polypeptide), and p48 (a 48 kD polypeptide). STAT1 (α or β) and STAT2 become tyrosine phosphorylated upon treatment of the cell with IFN-α or IFN-β, and each contains a src homology 2 (SH2) phosphotyrosine-binding domain (see Chapter 7, Box 7–4). The SH2 domain of STAT1 binds to the phosphotyrosine residue on STAT2 and vice versa, resulting in heterodimer formation of the activated (i.e., phosphorylated) forms of these proteins. The p48 protein binds to STAT1/STAT2 heterodimers and recognizes the ISRE motif. The heterotrimeric complex acts as a transcription factor, inducing expression of genes that contain ISRE sequences.

Biochemical studies of the response of cells to IFN-γ revealed similar but distinct results. Genes directly and rapidly activated by IFN-γ contained a different regulatory DNA sequence motif, called a **gamma activating sequence (GAS).** The nuclei of IFN-γ-treated cells, but not untreated cells, contain a protein factor that binds to this motif, called a **gamma activating factor (GAF).** Structural studies reveal that GAF is actually a homodimer of STAT1α. In cells treated with IFN-γ, STAT1α, but not STAT2, becomes phosphorylated on tyrosine residues, and in the absence of tyrosine phosphorylated STAT2, the SH2 domain of STAT1α interacts with a phosphotyrosine residue on another STAT1α protein, resulting in homodimer formation. STAT1α homodimers directly bind to GAS sequences and activate the transcription of nearby genes.

Genetic approaches were used to establish the causal relationship of these events to IFN actions. Mutant cells lacking STAT1 are unable to respond to IFN-α, IFN-β, or IFN-γ, whereas mutant cells lacking STAT2 or p48 cannot respond to IFN-α or IFN-β but still respond to IFN-γ. Analysis of other mutant cells revealed that three structurally related cytosolic protein tyrosine kinases were also involved in this response. This group of enzymes is collectively called Janus-family kinases (Jaks), named after the two-faced Roman god, because in addition to having a consensus region shared with other protein tyrosine kinases, these enzymes also contain a second consensus region shared with serine/threonine protein kinases. The latter domain is not functional, so despite the name, Jaks have only one enzymatic activity, namely phosphorylation of protein tyrosine residues. Mutant cells lacking the Jak enzyme **Tyk-2** are unresponsive to IFN-α and IFN-β, whereas mutant cells lacking **Jak-2** are unresponsive to IFN-γ, and mutant cells lacking **Jak-1** are unresponsive to both types of IFN. The Jak kinases are responsible for phosphorylation of the STAT proteins.

An obvious problem posed by these observations was

the basis of specificity: What determined which Jaks phosphorylated which STATs? The last piece in the puzzle was the discovery that the interferon receptors themselves act as critical scaffolding for putting the system together. When type I IFN receptors are clustered by binding of IFN-α or IFN-β, certain Janus-family kinases (Tyk-2 and Jak-1) activate each other and additionally phosphorylate specific tyrosine residues located within the intracellular regions of the clustered receptors. Some of these phosphotyrosine residues now stably bind the specific Janus-family kinases (i.e., Tyk-2 and Jak-1) via their SH2 domains. Other phosphotyrosine residues in the receptor bind specific STAT proteins (i.e., STAT1 and STAT2) through their SH2 domains. *The specificity for Jak kinase binding and for STAT protein binding is provided by the amino acid residues immediately adjacent to the phosphorylated tyrosine residues—i.e., the receptor sequence provides docking sites that determine which SH2-containing proteins are bound.* The bound STAT proteins become phosphorylated by the bound Jak enzymes. The SH2 domains of the STAT proteins have higher affinities for the phosphotyrosine residues of other STAT proteins than for phosphotyrosine residues of the receptor, and therefore these proteins dissociate from the receptor and dimerize once they have become phosphorylated. The dissociation from the receptor allows new STAT proteins to bind to the receptor, and these new STAT proteins, in turn, become phosphorylated, dissociate, and dimerize. STAT1/STAT2 heterodimers bind to p48 and move to the nucleus, where they bind to ISRE sequences, contributing to transcriptional activation of nearby genes. If the cells had been treated with IFN-γ instead of IFN-α or IFN-β, a similar series of events ensues involving Jak-1, Jak-2, and STAT1α, leading to STAT1α homodimer formation, migration to the nucleus, binding to GAS sequences, and transcriptional activation of a different set of genes. The ability of IFN receptors to activate Jaks and STATs appears to be further regulated by receptor-associated protein tyrosine phosphatases.

Although IFN receptors do not share extracellular structural features with the type 1 family of cytokine receptors (see Box 12–1), many of the four α-helical cytokines also use Jak/STAT signaling pathways (see Table). The common features of Jak/STAT signaling are illustrated in the figure.

Cytokine	Jak	STAT
IFN-α/β	Jak-1, Tyk-2	STAT1, STAT2
IFN-γ	Jak-1, Jak-2	STAT1α
IL-2	Jak-1, Jak-3	STAT5, (STAT3)
IL-3	Jak-1, Jak-2	STAT5
IL-4	Jak-1, Jak-3, or Jak-2	STAT6 STAT6
IL-5	Jak-1, Jak-2	STAT5
IL-6	Jak-1, Jak-2, Tyk-2	STAT1, STAT3
IL-7	Jak-1, Jak-3	STAT5
IL-10	unknown	STAT1, STAT3
IL-12	Jak-2, Tyk-2	STAT3, STAT4
IL-13	Jak-2	STAT6
IL-15	Jak-1, Jak-3	STAT5, STAT3
GM-CSF	Jak-1, Jak-2	STAT5
G-CSF	Jak-1, Jak-2	STAT3

Continued

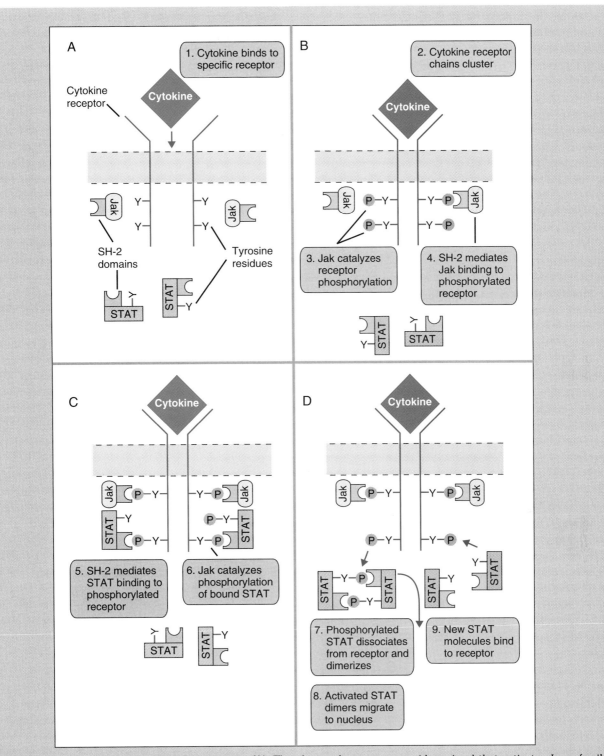

Cytokines induce the clustering of specific receptors *(A)*. The clustered receptors provide a signal that activates Janus-family kinases (Jaks), inducing autocatalytic phosphorylation of tyrosine residues. The activated Jaks further catalyze the phosphorylation of tyrosine residues in the intracellular regions of the cytokine receptor *(B)*. Some of these phosphotyrosine residues bind Jaks through their src-homology 2 (SH2) domains *(B)*, and some bind signal transducer and activator of transcription (STAT) proteins through their SH2 domains *(C)*. The bound Jaks catalyze phosphorylation of tyrosine residues on the bound STAT proteins. The phosphorylated STAT proteins dissociate from the receptor and dimerize, allowing new STAT proteins to associate with the receptor *(D)*. STAT dimers migrate to the nucleus and activate transcription of genes containing specific *cis*-regulatory STAT-binding sequences.

trials as a component of vaccines to boost protective cell-mediated immune responses.

Interleukin-15

Interleukin-15 is a 17 kD polypeptide cytokine released by mononuclear phagocytes and certain tissue cells in response to viral infection, LPS, or other signals that trigger innate immunity. Structurally, IL-15 is homologous to IL-2 and signals through the low affinity receptor complex used by IL-2, which will be discussed later in the chapter (see also Box 12–1). The binding of IL-15 to this low affinity IL-2 receptor is markedly increased by interaction with a non-signaling IL-15 binding polypeptide, called the IL-15Rα chain. The IL-15Rα chain is structurally homologous to the IL-2Rα chain, but it does not bind IL-2.

The primary function of IL-15 appears to be to promote the proliferation of NK cells. Since IL-15 is synthesized early in response to viral infections, it may mediate expansion of NK cells within the first 24 to 72 hours of viral infection. IL-15 may also act as a T cell growth factor because it binds and signals through the low affinity IL-2R, which is found on resting T cells. However, it is not known whether enough IL-15 is produced during an infection to signal through this low affinity receptor, nor is it clear whether T cells acquire IL-15Rα.

Tumor Necrosis Factor

Tumor necrosis factor (TNF) *is the principal mediator of the response to gram-negative bacteria* and may also play a role in innate immune responses to other infectious organisms. (Some investigators refer to TNF as TNF-α for historical reasons; we shall use the simpler term throughout this book). TNF was originally identified (and was so named) as a mediator of tumor necrosis that was present in the serum of animals treated with LPS. At low concentrations, LPS stimulates the functions of mononuclear phagocytes and (in mice) acts as a polyclonal activator of B cells (see Chapters 9 and 13), host responses that contribute to elimination of the invading bacteria. However, high concentrations of LPS cause tissue injury, disseminated (widespread) intravascular coagulation (DIC), and shock, often resulting in death. The Shwartzman reaction is an experimental model for studying the pathologic effects of LPS (Box 12–3). It is now clear that TNF is one of the principal mediators of these effects of LPS.

The major cellular source of TNF is the LPS-activated mononuclear phagocyte, although antigen-stimulated T cells, activated NK cells, and activated mast cells can also secrete this protein. IFN-γ, produced by T cells, augments TNF synthesis by LPS-stimulated mononuclear phagocytes. Thus, TNF is a mediator of both innate and specific immunity and an important link between specific immune responses and acute inflammation.

The mechanism of lipopolysaccharide (LPS)–mediated tissue injury was investigated by Shwartzman, who found that two intravenous injections of a sublethal quantity of LPS, administered 24 hours apart, would cause disseminated intravascular coagulation (DIC) in the rabbit. This is called the systemic Shwartzman reaction and is due to widespread intravascular thrombus formation on the surfaces of endothelial cells. If the first LPS injection is given intradermally, the second intravenous injection causes hemorrhagic necrosis of skin exclusively at the intradermal injection site. In this localized Shwartzman reaction, tissue injury is caused by activated neutrophils and by inadequate perfusion of the tissue. The inadequate tissue perfusion results from local intravascular coagulation (fibrin formation) and from cellular plugging of the microcirculation by aggregates of neutrophils and platelets. Recent studies have shown that tumor necrosis factor (TNF) can in large part substitute for LPS in eliciting both the local Shwartzman reaction and the systemic toxicity of LPS. Moreover, neutralizing antibody to TNF affords protection against both the injurious and the lethal effects of LPS. Thus, TNF is thought to be an obligatory mediator of LPS-induced tissue injury.

The Shwartzman reaction is an exaggerated form of a host response to microbes, which, under less extreme physiologic conditions, functions primarily to eliminate microbes and limit their spread. Although TNF is now known to be one of the principal cytokines involved in such host responses, TNF was first identified as a factor present in the plasma of LPS-treated animals that could cause hemorrhagic necrosis of tumors. Some of the anti-tumor action of TNF is mediated by direct tumor cell lysis, a process not well understood but believed to involve TNF binding to surface receptors on tumor cells, thereby initiating apoptotic cell death (see Chapter 18). Mostly, however, TNF induces tumor necrosis by causing a local Shwartzman-like reaction to occur in the tumor vascular bed. The basis for the selective effect on tumor blood vessels is not known, but tumor cells appear to release factors that increase the sensitivity of local endothelial cells to TNF. These tumor factors act like the first injection of LPS.

TNF is the product of a single gene located within the MHC on chromosome 6 in humans. In the mononuclear phagocyte, TNF is initially synthesized as a nonglycosylated transmembrane protein of approximately 25 kD. The orientation of membrane TNF is that of a type II membrane protein, i.e., the amino terminus is intracellular, the transmembrane segment is near the amino terminus, and the large carboxy terminus is extracellular. Membrane TNF assembles as a homotrimer; the interactions among subunits involve the extracellular carboxy terminal domains. A 17 kD fragment of each subunit, including the carboxy terminus, can be proteolytically cleaved off the plasma membrane of the mononuclear phagocyte to produce the "secreted" form, which circulates as a stable homotrimer of 51 kD. Native TNF assumes a triangular pyramidal shape such that each side of the pyramid is formed by a different monomeric subunit. The receptor binding sites are at the base of the pyramid, allowing simultaneous binding to more than one receptor.

TNF actions are initiated by binding of the soluble trimer to cell surface receptors. There are two distinct TNF receptors, of 55 and 75 kD respectively, each encoded by a separate gene. The affinity of TNF for its receptors is unusually low for a cytokine, the K_d being only approximately 5×10^{-10} M for binding to the 75 kD receptor (called TNF-RII) and 1×10^{-9} M for binding to the 55 kD receptor (called TNF-RI). However, TNF is synthesized in very large quantities and can easily saturate its receptors. Both TNF receptors are present on almost all cell types examined. Most biologic effects are mediated through TNF-RI, and TNF-RI knockout mice show much more impaired host defense than TNF-RII knockout mice. TNF-RII may preferentially mediate responses to membrane TNF. It may also serve to bind and "pass" soluble TNF to TNF-RI. Activated cells shed their TNF receptors; such soluble receptors may act as competitive inhibitors of the cell surface receptor. Structural analyses of TNF and its two receptors have revealed that these molecules are each members of large gene families containing a number of additional proteins of relevance for the immune system (Box 12–4).

Many TNF responses involve increased rates of transcription of particular target genes, often through activation of NFκB (see Chapter 4, Box 4–4) or AP-1 transcription factors. Regulation of AP-1 activity is mediated by a stress-activated protein kinase cascade, involving c-Jun N-terminal kinase, the same system that was discussed in Chapter 7 for T cell activation.

The biologic actions of TNF, like those of LPS, are best understood as a function of quantity (Fig. 12–2). *When small quantities are produced, TNF acts locally as a paracrine and autocrine regulator of leukocytes and endothelial cells.* The principal biologic actions of TNF at low concentrations are the following:

1. TNF causes vascular endothelial cells to express new surface receptors **(adhesion molecules)** that make the endothelial cell surface become adhesive for leukocytes, initially for neutrophils and subsequently for monocytes and lymphocytes. TNF also acts on neutrophils to increase their adhesiveness for endothelial cells. These actions contribute

---

BOX 12–4.   The TNF and TNF Receptor Families

Lymphotoxin (LT) was originally described in the 1960s as a T cell–derived mediator of tissue injury, and tumor necrosis factor (TNF) was described in the 1970s as the endogenous mediator of LPS-induced hemorrhagic necrosis of tumors. The purification and cloning of these two separate molecules in the early 1980s revealed that LT and TNF were structurally homologous to each other and, more surprisingly, that they competed for binding to the same cell surface receptors, now called TNF-RI and TNF-RII. The extracellular domains of these two receptors also displayed common structural motifs, namely cysteine-rich domains that were also found in the previously described low affinity receptor for nerve growth factor. In the early 1990s, it was discovered that the structural motif shared by TNF and LT, which fold as homotrimeric triangular pyramids, is present in a number of cell surface signaling proteins expressed by the cells of the immune system. These TNF-like molecules include LT-β, which is expressed as a heterotrimer with LT in a stoichiometry of $(LT)_1 (LT\beta)_2$, Fas ligand, CD40 ligand, and several others. All of these molecules, except LT, are expressed as type II membrane proteins (i.e., proteins whose carboxy termini are outside the cell and amino termini are inside the cell). Moreover, the cell surface receptors recognized by these ligands, e.g., the LT-β receptor (sometimes called TNF receptor-like protein), Fas, CD40, and others, are all type I membrane proteins (i.e., proteins that have their amino termini outside the cell and their carboxy termini inside of the cell) that share extracellular cysteine-rich structural motifs observed in TNF-RI and TNF-RII. With the exception of TNF and LT, the homology among the members of the family is sufficiently low that each ligand and its receptor do not interact with other members of the family.

The TNF receptor family was originally defined by the presence of extracellular structural motifs. More recent analyses have revealed some conserved structural motifs within the intracellular "signaling regions" as well. TNF-RI and Fas share a conserved sequence motif that has been called the "death domain," because mutation or deletion of this region prevents TNF-RI or Fas molecules from delivering apoptosis-inducing signals. The death domain is actually a protein-protein interaction domain for assembling interacting proteins with the intracellular regions of clustered TNF-RI or Fas molecules. The cell death pathway activated by death-domain containing adaptor molecules, such as Fas-associated death domain (FADD) protein, were discussed in Chapter 10 (Box 10–1). TNF-RII, LT-βR and CD40 share a different protein-protein association domain that now defines a new protein family of cytoplasmic adaptor molecules called TNF-receptor associated factors (TRAFs). To date, at least six TRAFs have been identified. TRAFs 1 and 2 interact with TNF-RII, perhaps as heterodimers. TRAFs 2 and 3 interact with CD40, and TRAF 4 interacts with LT-βR. Overexpression of TRAFs, especially TRAF 2, activates the transcription factor NFκB. TRAF-2 overexpression also activates stress-activated protein kinase pathways that regulate transcription factor AP-1. TNF-RI does not contain a TRAF domain, but it does interact with a death domain–containing protein called TRADD, which also contains a TRAF domain. TRADD can interact, through its TRAF domain, with TRAF 2, thereby coupling TNF-RI to NFκB activation. Interestingly, one of the transforming gene products of Epstein-Barr virus encodes a self-aggregating TRAF domain–containing protein that binds TRAF 1 and TRAF 2, mimicking TNF- or CD40-like signals.

It should be emphasized that these pathways are as yet not completely understood. TNF-RII, which lacks a death domain–combining motif, can sometimes deliver apoptotic signals. It is also uncertain how TRAFs lead to activation of NFκB or of stress-activated protein kinases.

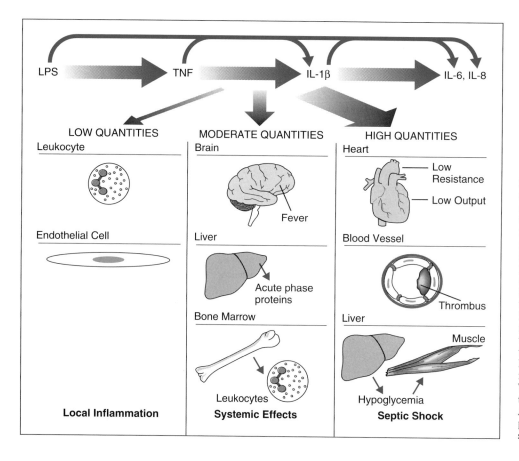

FIGURE 12-2. The LPS-induced cytokine cascade. Bacteria LPS acts on macrophages to release tumor necrosis factor (TNF). TNF induces macrophages to release interleukin-1β (IL-1β). IL-1β acts on macrophages and vascular endothelial cells to release IL-6 and IL-8. (The thinner arrows indicate that LPS directly induces IL-1β, IL-6, and IL-8 and that TNF directly induces IL-6 and IL-8, but these actions are amplified through the cascade.) When low quantities of cytokines are released, the effects are local. With moderate quantities, systemic effects can be detected. At high levels, these cytokines produce the syndrome of septic shock.

to accumulation of leukocytes at local sites of inflammation and are probably the physiologically most important local effects of TNF (see Chapter 13).

2. TNF stimulates mononuclear phagocytes and other cell types to secrete chemokines that contribute to leukocyte recruitment.

3. TNF activates inflammatory leukocytes to kill microbes. TNF is especially potent at activating neutrophils but also affects eosinophils and mononuclear phagocytes.

4. Chronic production of low concentrations of TNF leads to tissue remodeling. TNF acts both as an angiogenesis factor, inducing new blood vessel formation, and as a fibroblast growth factor, causing connective tissue deposition. If TNF production persists, such tissues may acquire the appearance of ectopic lymphoid tissue, with accumulation of organized collections of B and T cells.

These effects of TNF are critical for local inflammatory responses to microbes. If inadequate quantities of TNF are present, e.g., in animals treated with neutralizing anti-TNF antibodies or in TNF knockout mice, a consequence may be a failure to contain infections. TNF may also contribute to local inflammatory reactions that are harmful to the host, e.g., in autoimmunity (see Chapter 19). Neutralizing antibody to TNF may reduce inflammation and joint damage in patients with rheumatoid arthritis.

The original discovery of TNF was based on its

effects on tumors. The ability of TNF to kill tumors partly relates to induction of apoptotic cell death in certain malignant cell types and has led to clinical trials of TNF as an anti-tumor agent. We will discuss this aspect of TNF in Chapter 18. However, most of the anti-tumor action of TNF can be ascribed to its ability to induce inflammation in vascularized tumor implants. This action of TNF can be explained by its effects on the endothelial cells lining tumor vessels and represents a form of the Shwartzman reaction (see Box 12–3).

If the stimulus for TNF production is sufficiently strong, greater quantities of the cytokine are produced. In this setting, TNF enters the blood stream, where it can act as an endocrine hormone. *The principal systemic actions of TNF in physiologic host responses to infections are the following:*

1. TNF is an **endogenous pyrogen** that acts on cells in hypothalamic regulatory regions of the brain to induce fever. It shares this property with IL-1, and both cytokines are found in the serum of animals or people exposed to LPS, which functions as an exogenous pyrogen. Fever production in response to TNF or IL-1 is mediated by increased synthesis of prostaglandins by cytokine-stimulated hypothalamic cells. Prostaglandin synthesis inhibitors, such as aspirin, reduce fever by blocking this action of TNF or IL-1.

2. TNF acts on mononuclear phagocytes and perhaps vascular endothelial cells to stimulate secretion of IL-1 and IL-6 into the circulation (Fig. 12–

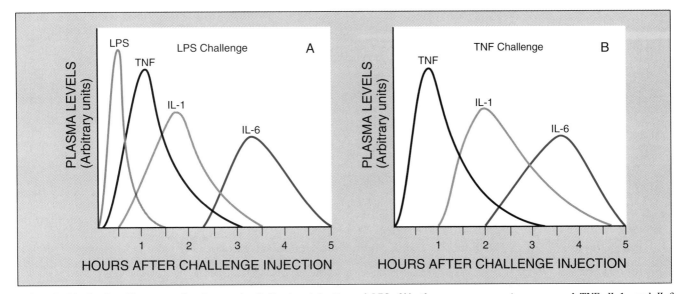

FIGURE 12–3. Cytokine cascades in sepsis. Following injection of LPS *(A)*, there are successive waves of TNF, IL-1, and IL-6 detectable in plasma. Injection of TNF *(B)* produces successive waves of IL-1 and IL-6. In the presence of antibody to TNF, LPS-induced plasma elevations of IL-1 and IL-6 are inhibited; and in the presence of antibody to IL-1, plasma elevations of IL-6 are inhibited. These data suggest that there are ordered cascades of cytokine production: LPS induces TNF, which induces IL-1, which induces IL-6 synthesis.

3). This is one example of a cascade of cytokines that share many biologic activities.

3. TNF acts on hepatocytes to increase synthesis of certain serum proteins, such as serum amyloid A protein. The spectrum of hepatocyte proteins induced by TNF is identical to that induced by IL-1 but differs from that induced by IL-6 (described below). The combination of hepatocyte-derived plasma proteins induced by TNF or IL-1 plus those induced by IL-6 constitutes the **acute phase response** to inflammatory stimuli (Box 12–5).

4. TNF activates the coagulation system, primarily by altering the balance of the procoagulant and anticoagulant activities of vascular endothelium.

5. TNF suppresses bone marrow stem cell division. Chronic administration of TNF may lead to lymphopenia and immunodeficiency.

6. Long-term systemic administration of TNF to experimental animals causes the metabolic alterations of **cachexia,** a state characterized by wasting of muscle and fat cells. The cachexia is produced largely by TNF-induced appetite suppression. TNF also suppresses synthesis of lipoprotein lipase, an enzyme needed to release fatty acids from circulating lipoproteins so that they can be utilized by the tissues. Although TNF by itself can produce cachexia in experimental animals, other cytokines, such as IL-1, may also contribute to the cachectic state accompanying certain chronic diseases such as tuberculosis and cancer. The combination of fever, elevated IL-6 levels, elevated acute phase reactants, bone marrow suppression, activation of coagulation, and appetite suppression has been noted in patients treated with intravenous TNF for cancer chemotherapy.

In the setting of gram-negative bacterial sepsis, massive quantities of TNF are produced, and serum concentrations of TNF can transiently exceed 10^{-7} M. Animals producing this much TNF die of circulatory collapse and disseminated intravascular coagulation. Neutralizing antibodies to TNF or soluble TNF receptors can prevent mortality, implicating this cytokine as a critical mediator of septic or endotoxin shock, and TNF knockout mice are resistant to lethal shock. Moreover, infusion of high levels of TNF is by itself lethal, producing a shock-like syndrome. *Several specific actions of TNF may contribute to its lethal effects at extremely high concentrations.*

1. TNF reduces tissue perfusion by depressing myocardial contractility. The mechanism of this action appears to involve induction and/or increase in activity of an enzyme in cardiac myocytes, nitric oxide synthase (NOS), that converts arginine to citrulline and NO. NO made by this enzyme inhibits myocardial contractility.

2. TNF further reduces blood pressure and tissue perfusion by relaxing vascular smooth muscle tone. TNF may act directly on smooth muscle cells and also can act indirectly by stimulating production of vasodilators, such as prostacyclin and NO, by vascular endothelial cells.

3. TNF causes intravascular thrombosis, leading to reduced tissue perfusion. This is due to a combination of endothelial and mononuclear phagocyte alterations, which promote coagulation, and activation of neutrophils to form aggregates, leading to vascular plugging by these cells. These TNF-mediated actions account for many of the effects of LPS seen in the Shwartzman reaction of rabbits

BOX 12–5. The Acute Phase Response

The acute phase response consists of a rapid adjustment of plasma protein composition in response to injurious stimuli, including infection, burns, trauma, and neoplasia. Several different plasma proteins rise in concentration, whereas others fall. Among the proteins whose levels increase are C-reactive protein, which functions as a nonspecific opsonin to augment phagocytosis of bacteria, α2 macroglobulin and other anti-proteinases, the clotting protein fibrinogen, and serum amyloid A protein, a molecule of uncertain function. Complement proteins also rise and behave as acute phase reactants. Albumin and transferrin, the iron transport protein, decline. Most of these changes in plasma concentrations can be directly attributed to alterations in the levels of synthesis of these plasma proteins by hepatocytes. Experiments using whole animals, liver slices, cultured hepatocytes, or hepatocyte-derived tumor cell lines have revealed that these changes in biosynthesis are caused by alterations in gene transcription regulated primarily by IL-6 (on fibrinogen) and IL-1/TNF (on serum amyloid protein).

The precise function of the acute phase response is largely unknown. The increases in opsonizing proteins and anti-proteinases are believed to enhance innate immunity and protect against tissue injury, respectively. Elevation in fibrinogen, caused by IL-6, is of uncertain benefit but has had major impact on clinical medicine. Specifically, elevated levels of fibrinogen can cause red blood cells to form stacks (rouleaux). When blood is collected and allowed to stand at unit gravity, rouleaux sediment more rapidly than individual red blood cells. Rouleaux in venous blood may sediment before the red blood cells are fully oxygenated, leading to a dark mass at the bottom of the container. In ancient times, this mass of dark, deoxygenated red blood cells was called "black bile," and ancient and medieval physicians would perform bleeding of patients to remove this "sickly humor." In modern times, the realization that the more rapid red blood cell sedimentation reflected the presence of illness, rather than representing its cause, allowed measure of the erythrocyte sedimentation rate to become a useful diagnostic test for the presence of the acute phase response. In the past few years, more specific measures of the acute phase response, for example, of C-reactive protein or of IL-6, have now largely supplanted this useful tool.

Although the acute phase response is characterized by rapid onset, it can persist in the setting of chronic inflammation. In some patients with chronic inflammatory disease (e.g., rheumatoid arthritis; see Chapter 20), persistent elevations of serum amyloid A protein may lead to deposition of this protein in the interstitium of tissues. Such deposited protein, in the form of fibrils rich in β-pleated sheet structure, can interfere with normal organ function (e.g., myocardial contraction, glomerular filtration). Such patients are said to have developed amyloidosis because such protein deposits stain with acidic iodine, a reaction originally developed for amylose or animal starch. Similar fibrils can develop in other settings (e.g., multiple myeloma, Alzheimer's disease, or endocrine cell tumors); however, in these cases, the protein fibrils are not of serum amyloid A protein origin and are unrelated to the acute phase response.

(see Box 12–3) and disseminated intravascular coagulation in humans.

4. TNF causes severe metabolic disturbances, such as a fall in blood glucose concentrations to levels that are incompatible with life. This is due to overutilization of glucose by muscle and failure to replace glucose by the liver.

Despite the abundant evidence that TNF is a critical mediator of septic shock in experimental animals, clinical trials with neutralizing antibodies or with soluble TNF receptors have not shown clear benefit in patients with sepsis. The cause of this therapeutic failure is not known, but it probably arises from the redundancy of cytokine actions, e.g., of IL-1 with those of TNF.

Interleukin-1

Interleukin-1 was first defined as a polypeptide derived from mononuclear phagocytes that enhanced the responses of thymocytes to polyclonal activators, i.e., as a costimulator of T cell activation. However, *it is now clear that the principal function of IL-1, similar to that of TNF, is as a mediator of the host inflammatory response in innate immunity.* Indeed, there is little evidence to support a role of IL-1 as a physiologically important costimulator of mature T cell activation.

The major cellular source of IL-1, like that of TNF, is the activated mononuclear phagocyte. IL-1 production by mononuclear phagocytes can be triggered by bacterial products such as LPS, by macrophage-derived cytokines such as TNF or IL-1 itself, and by contact with CD4+ T cells. Like TNF, IL-1 can be found in the circulation following gram-negative bacterial sepsis, where it can act as an endocrine hormone. In this case, it is produced mainly in response to TNF (see Fig. 12–3). IL-1 synthesis differs from that of TNF in two important regards. First, T cells are more effective than LPS at eliciting synthesis of IL-1 by mononuclear phagocytes. Second, IL-1 is made by many diverse cell types, such as epithelial cells (e.g., keratinocytes) and endothelial cells, providing potential local sources of IL-1 in the absence of macrophage-rich infiltrates.

There are two principal forms of IL-1, called IL-1α and IL-1β, that are products of two different genes. Both forms of IL-1 are 17 kD polypeptides but show less than 30 per cent structural homology to each other. Nevertheless, both species bind to the same cell surface receptors and their biologic activities are essentially identical. A third member of the IL-1 family, called IL-1 receptor antagonist, will be discussed below. A recently described fourth member of the family, sometimes called IL-1γ, shares with IL-12 the capacity to induce IFN-γ secretion from NK cells.

Both IL-1α and IL-1β polypeptides are synthesized as 33 kD precursors that are proteolytically cleaved to generate the mature 17 kD proteins. The 17 kD forms fold in a barrel-like structure, rich in strands of β-pleated sheet. The 33 kD IL-1α precursor is biologically active, but IL-1β must be processed to the 17 kD form before it can exert biologic effects. An IL-1–specific protease, called interleukin-1 converting enzyme (ICE), has been identified in mononuclear phagoctyes; it is responsible for most of the conversion of IL-1β to its active form. ICE was the first described mammalian member of a family of cysteine proteinases that are structurally homologous to the *C. elegans* programmed cell death gene product CED-3. Many members of this family are involved in apoptotic death in response to Fas signaling (see Chapter 10, Box 10–1). ICE itself does not appear to be involved in apoptosis. ICE knockout mice have impaired inflammatory responses, probably because these animals cannot make functional IL-1β or IL-1γ.

The amino acid sequence of both IL-1 species presents a theoretical problem: unlike conventionally secreted proteins, neither IL-1 polypeptide has a hydrophobic signal sequence to target the nascent polypeptide to the endoplasmic reticulum, and both proteins appear to be synthesized as cytoplasmic proteins. It is still unknown how these molecules are released from cells. Most of the IL-1 activity found in the circulation is IL-1β.

Two different membrane receptors for IL-1 have been characterized, both of which are members of the Ig superfamily. The type I receptor has slightly higher affinity for IL-1β than for IL-1α and is the major receptor for IL-1–mediated responses. It is expressed on almost all cell types. The type II receptor has greater affinity for IL-1α than for IL-1β, and its major function is to act as a "decoy" that competitively inhibits IL-1 binding to the type I signaling receptor. The type II receptor is expressed on B cells but may be induced on other cell types. The K_d for IL-1 binding to its receptors may be as high as 1×10^{-12} M; however, IL-1 may be active on some target cells at concentrations as low as 1×10^{-15} M, suggesting that additional IL-1 binding proteins may exist. Signals resulting from IL-1 binding to type I IL-1 receptors may involve a novel member of the TNF receptor–associated factor (TRAF) family, although this receptor does not contain a TRAF-like domain (see Box 12–4). Many IL-1–induced transcriptional effects, like those of TNF, involve NFκB and AP-1, the latter factor requiring activation by stress-activated protein kinases.

The biologic effects of IL-1, similar to those of TNF, depend on the quantity of cytokine released (see Fig. 12–2). At low IL-1 concentrations, its principal function is as a mediator of local inflammation. For example, IL-1 acts on endothelial cells to promote coagulation and to increase expression of surface molecules that mediate leukocyte adhesion. IL-1 does not directly activate inflammatory leukocytes, such as neutrophils, but it causes mononuclear phagocytes and endothelial cells to synthesize chemokines that do activate leukocytes (see below).

When secreted in larger quantities, IL-1 enters the blood stream and exerts endocrine effects. *Systemic IL-1 shares with TNF the ability to cause fever, to induce synthesis of acute phase plasma proteins (such as serum amyloid A protein) by the liver, and to initiate metabolic wasting (cachexia).*

It was initially very surprising to note the extensive similarities of IL-1 actions with those of TNF, a striking example of the redundancy of cytokine effects. However, there are several important differences between these cytokines. First, IL-1 does not produce tissue injury by itself, although it is secreted in response to LPS and can potentiate tissue injury caused by TNF. Moreover, even at very high systemic concentrations, IL-1 is not lethal. Second, although IL-1 mimics many of the inflammatory and procoagulant properties of TNF, IL-1 cannot replace TNF as a mediator of the Shwartzman reaction and does not cause hemorrhagic necrosis of tumors. Third, IL-1 does not induce apoptotic death of tumor or other cells. Fourth, IL-1 does not share with TNF an ability to increase expression of MHC molecules. Finally, IL-1 potentiates rather than suppresses the actions of CSFs on bone marrow cells.

IL-1 is the only cytokine to date for which naturally occurring inhibitors have been described. The best defined of these is produced by human mononuclear phagocytes. It is structurally homologous to IL-1 and binds to IL-1 receptors but is biologically inactive, so that it functions as a competitive inhibitor of IL-1. It is therefore commonly called **IL-1 receptor antagonist (IL-1ra).** Type I and type II IL-1 receptors are also shed by activated cells. Both IL-1ra and soluble receptors may be endogenous regulators of IL-1 action. It may also be possible to use IL-1 inhibitors as biologic response modifiers in disease states that are caused by excessive or unregulated cytokine production, such as rheumatoid arthritis. Inhibition of IL-1 action by IL-1ra or soluble receptors has not been of benefit in clinical trials of septic shock, presumably due to redundancy of IL-1 actions with those of TNF.

Interleukin-6

Interleukin-6 (IL-6) is a cytokine of approximately 26 kD that is synthesized by mononuclear phagocytes, vascular endothelial cells, fibroblasts, and other cells in response to IL-1 and, to a lesser extent, TNF. It is also made by some activated T cells. IL-6 can be detected in the circulation following gram-negative bacterial infection or TNF infusion and appears to be secreted in response to TNF or IL-1 rather than LPS itself (see Fig. 12–3). IL-6 does not cause vascular thrombosis or the tissue injury that is seen in response to LPS or TNF.

The functional form of IL-6 is probably a homodimer; each subunit forms a four α-helical globular domain observed in cytokines that bind to type I cytokine receptors. The receptor for IL-6 consists of a 60 kD cytokine binding protein and a 130 kD signal-transducing subunit. The binding subunit contains both an Ig domain and a two cysteine/WSXWS motif characteristic of type I cytokine receptors (see Box 12–1). The signal transducing subunit also contains both an Ig domain and the two cysteine/WSXWS motif, but it does not bind IL-6 and appears to serve as the signaling subunit for several other cytokines. Clustering of the 130 kD subunit induced by interactions with complexes of cytokine and specific binding protein is thought to trigger signaling via a Jak/STAT pathway (see Box 12–2). Soluble IL-6 receptors shed from cell surfaces can bind IL-6, and these soluble complexes can also signal through the 130 kD subunit.

The two best-described actions of IL-6 are on hepatocytes and B cells:

1. *Interleukin-6 causes hepatocytes to synthesize several plasma proteins, such as fibrinogen, that contribute to the acute phase response* (see Box 12–5).

2. *Interleukin-6 serves as a growth factor for activated B cells late in the sequence of B cell differentiation.* IL-6 similarly acts as a growth factor for many malignant plasma cells (plasmacytomas or myelomas), and many plasmacytoma cells that grow autonomously actually secrete IL-6 as an autocrine growth factor. Moreover, IL-6 can promote the growth of somatic cell hybrids produced by fusing normal B cells with plasmacytoma cells, i.e., the "hybridomas" that produce monoclonal antibodies (see Chapter 3, Box 3–1). Transgenic mice that over-express the IL-6 gene develop massive polyclonal proliferation of plasma cells.

In addition to these well-described *in vivo* actions, *in vitro* experiments suggest that IL-6 may serve as a costimulator of T cells and of thymocytes. IL-6 also acts as a cofactor with other cytokines for the growth of early bone marrow hematopoietic stem cells.

Chemokines

Chemokines comprise a large family of structurally homologous cytokines, approximately 8 to 10 kD in size, that share the ability to stimulate leukocyte motility **(chemokinesis)** and directed movement **(chemotaxis).** The name "chemokines" is a contraction of **chemo**tactic cyto**kines.** All of these molecules contain two internal disulfide loops. The chemokines fall into two subfamilies, based on whether the two amino terminal cysteine residues that participate in disulfide bonding are immediately adjacent (C-C) or separated by one amino acid (C-X-C). These differences correlate with organization of the two subfamilies into separate gene clusters. A distantly related member of the chemokine family, called lymphotactin, lacks one of the four cysteines and hence one of the disulfide loops. All of the chemokine receptors belong to the seven transmembrane α-helical receptor family (see Box 12–1).

The chemokines of the C-X-C subfamily (Table 12–2) are produced largely by activated mononuclear phagocytes as well as by tissue cells (endothelium, fibroblasts) and megakaryocytes (which give rise to platelets containing stored chemokine). These molecules act predominantly on neutrophils as mediators of acute inflammation. The best-characterized member of this subfamily in humans is interleukin-8. Humans express one receptor specific for IL-8 (CXC–chemokine receptor [CKR]-1) and one promiscuous receptor that binds most C-X-C chemokines (CXC–CKR-2). Knockout mice lacking CXC–CKR-2 have deficiencies in mounting acute inflammatory responses.

The C-C subfamily of chemokines (see Table 12–2) is produced largely by activated T cells. These molecules act predominantly on T cells, monocytes, eosinophils, and basophils, but not neutrophils. They may show some specificity among these cell types. For example, a chemokine called RANTES acts preferentially on memory CD4+ T cells, monocytes, and eosinophils, whereas another C-C chemokine called eotaxin acts exclusively on eosinophils. At least five separate C-C chemokine receptors (called CC-CKR1-5) exhibit overlapping specificity for C-C chemokines, and the pattern of cell expression determines which cell types respond to which chemokines. Red blood cells and certain endothelia express a promiscuous, non-signaling chemokine receptor that may serve to scavenge both C-X-C and C-C chemokines, terminating chemokine action. Chemokines of both subfamilies bind to heparan sulfate proteoglycans on the endothelial cell surface, and they may function principally to stimulate chemokinesis of leukocytes that attach to cytokine-activated endothelium through induced adhesion molecules (see Chapter 13).

It has been discovered that certain chemokine receptors, especially CC-CKR-5 and CXC–CKR-4 (also known as fusin), can act as co-receptors for the human immunodeficiency virus (see Chapter 21). Occupancy of these receptors by chemokines may interfere with viral infectivity.

Interleukin-10

Interleukin-10 (IL-10) is an 18 kD cytokine produced by activated macrophages, some lymphocytes, and some non-lymphocytic cell types (e.g., keratinocytes). IL-10 is a member of the four α-helical cytokine family and probably functions as a homodimer. The receptor belongs to the two cysteine/WSXWS family (see Box 12–1) and signals through a Jak/STAT pathway (see Box 12–2). The two major activities of IL-10 are to inhibit cytokine (i.e., TNF, IL-1, chemokine, and IL-12) production by macrophages, and to inhibit the accessory functions of macrophages in T cell activation. The latter effect is due to reduced expression of class II MHC molecules and reduced expression of costimulators, e.g., B7-1 and B7-2. The net effect of these actions is to inhibit both innate and T cell–mediated specific immune inflammation. Studies of mice in which the IL-10 gene has been disrupted by knockout technology reveal few defects in the specific immune system. However, such mice develop intestinal inflammatory lesions, presumably due to uncontrolled activation of macrophages.

TABLE 12-2. Chemokines

Family	Chemokine	Cell Target	Receptors
CXC	Interleukin 8	Neutrophil	CXC-CKR1, 2
	Gro (α, β, γ)	Neutrophil	CXC-CKR2
	Platelet basic protein (β thromboglobulin, connective tissue–activating protein 3, neutrophil activating protein 2)	Neutrophil	CXC-CKR-2
	Epithelial-derived neutrophil attractant 78	Neutrophil	CXC-CKR-2
	Platelet factor 4	?	?
	Interferon-γ–induced protein 10	Lymphocyte	CXC-CKR3
	Stromal cell–derived factor-1	Neutrophil, monocyte, lymphocyte	CXC-CKR4
CC	Monocyte chemotactic protein 1	Monocyte, T cell, eosinophil	CC-CKR 2A, 2B, 4
	Monocyte chemotactic protein 2	Monocyte, T cell, eosinophil	CC-CKR1, 2A, 2B
	Monocyte chemotactic protein 3	Monocyte, T cell, eosinophil	CC-CKR 1, 2A, 2B, 3
	RANTES	Monocyte, T cell, eosinophil	CC-CKR 1, 3, 4, 5
	Monocyte inflammatory protein 1α	Monocyte, T cell, eosinophil	CC-CKR 1, 4, 5
	Monocyte inflammatory protein 1β	Monocyte, T cell, eosinophil	CC-CKR 1, 5
	Eotaxin	Eosinophil	CC-CKR 3
C	Lymphotactin	Lymphocyte	?

Abbreviations: CXC-CKR, CXC chemokine receptor; CC-CKR, CC chemokine receptor; RANTES, regulated by activation, normal T-cell expressed and secreted.

Interestingly, the Epstein-Barr virus contains a gene homologous to IL-10, and viral IL-10 shares *in vitro* activity with the T cell–derived cytokine. This raises the intriguing possibility that acquisition of the human gene during the evolution of the virus has given the virus a selective advantage by means of inhibiting antiviral immunity.

CYTOKINES THAT MEDIATE AND REGULATE SPECIFIC IMMUNITY

As we discussed in Chapter 7, a characteristic feature of the activation phase of the specific immune response is the transcription and secretion of cytokines by T cells. These cytokines (Table 12–3) act both as regulators and as mediators of the effector phase of the response. Some of these cytokines, such as IL-2, IL-4 and TGF-β, act in an autocrine manner, regulating the response of the same T cells that synthesize these molecules. These cytokines also act in a paracrine manner on nearby T cells, B cells, and other cell types. Other cytokines, such as IFN-γ, lymphotoxin (LT), IL-5, and IL-13, act primarily on cells other than T cells, regulating effector cells such as macrophages, eosinophils, and endothelial cells. As we shall discuss at the end of this section, activated T cells may become specialized in the production of particular subsets of cytokines that favor particular types of effector cell responses. This differentiation may be influenced by the innate immune response, e.g., through production of cytokines such as IL-12.

TABLE 12-3. Mediators and Regulators of Specific Immunity

Cytokine	Polypeptide Size	Cell Source	Cell Target	Primary Effects on Each Target
Interleukin-2	14–17 kD	T cell	(a) T cell (b) NK cell (c) B cell	(a) Growth, cytokine production (b) Growth, activation (c) Growth, antibody synthesis
Interleukin-4	20 kD	CD4$^+$ T cell, mast cell	(a) B cell (b) T cell (c) Endothelial cell	(a) Isotype switching to IgE (b) Growth, differentiation (c) Activation
Transforming growth factor-β	14 kD	T cell, mononuclear phagocyte, other	(a) T cell (b) Other	(a) Inhibition (b) Growth regulation
Interferon-γ	21–24 kD	T cell, NK cell	(a) Mononuclear phagocyte (b) Endothelial cell (c) All	(a) Activation (b) Activation (c) Increased class I and class II MHC molecules
Lymphotoxin	24 kD	T cell	(a) Neutrophil (b) Endothelial cell	(a) Activation (b) Activation
Interleukin-5	20 kD	T cell	(a) Eosinophil	(a) Activation, production

Abbreviations: kD, kilodalton; NK, natural killer; Ig, immunoglobulin; MHC, major histocompatibility complex.

Interleukin-2

Interleukin-2 (IL-2), originally called T cell growth factor (TCGF), is the principal cytokine responsible for progression of T lymphocytes from the G1 to S phase of the cell cycle. IL-2 is produced by CD4$^+$ T cells and, in lesser quantities, by CD8$^+$ T cells. IL-2 acts on the same cells that produce it, i.e., it functions as an **autocrine growth factor.** IL-2 also acts on nearby T lymphocytes, including both CD4$^+$ and CD8$^+$ cells, and is also therefore a **paracrine growth factor.** During physiologic immune responses, IL-2 does not circulate in the blood to act at a distance, and thus it is not considered to be an endocrine growth factor.

Secreted IL-2 is a 14 to 17 kD glycoprotein encoded by a single gene on chromosome 4 in humans. The size heterogeneity of the mature protein is due to variable extents of glycosylation of an approximately 130 amino acid residue polypeptide. Native IL-2 is folded into a globular protein containing a paired pair of α helices. The four α-helices of IL-2 form the prototypic structure for cytokines that interact with the type 1 family of cytokine receptors characterized by the two cysteine/WSXWS motif (see Box 12–1). Normally, IL-2 is transcribed, synthesized, and secreted by T cells only upon activation by antigens. IL-2 synthesis is usually transient, with an early peak of secretion occurring about 4 hours after activation. The mechanisms of transcriptional regulation of IL-2 synthesis have been described in Chapter 7.

The action of IL-2 on T cells is mediated by binding to IL-2 receptor proteins. This system is perhaps the best understood of all cytokine receptors. Two distinct cell surface proteins on T cells bind IL-2. The first to be identified, called IL-2Rα, is a 55 kD polypeptide (p55) that appears upon T cell activation and was originally called Tac (for T activation) antigen. IL-2Rα binds IL-2 with a K$_d$ of approximately 10^{-8} M. Binding of IL-2 to cells expressing only IL-2Rα does not lead to any detectable biologic response. The second IL-2–binding protein, called IL-2Rβ, is about 70 to 75 kD (called variously p70 or p75) and is a member of the type I cytokine receptor family characterized by the two cysteine/WSXWS motif (see Box 12–1). The affinity of binding of IL-2 to this receptor is higher than to IL-2Rα, with a K$_d$ of approximately 10^{-9} M. IL-2Rβ is expressed coordinately with a 64 kD polypeptide, called the common γ (γ_c) chain because it is shared among a number of cytokine receptors. γ_c is also a member of the two cysteine/WSXWS family of receptors, and it forms a complex with IL-2Rβ, designated as IL-2R$\beta\gamma_c$. IL-2 causes growth of cells expressing IL-2R$\beta\gamma_c$, with half maximal growth stimulation occurring at the same concentration of IL-2 that produces half maximal binding (i.e., 1×10^{-9} M). IL-2R$\beta\gamma_c$ is the same low affinity signaling receptor complex that can bind and respond to IL-15, discussed earlier in this chapter. Cells that express both IL-2Rα and IL-2R$\beta\gamma_c$ can bind IL-2 much more tightly, with a K$_d$ of approximately 10^{-11} M. Growth stimulation of such cells occurs at a similarly low IL-2 concentration. Both IL-2 binding and growth stimulation can be blocked by antibodies to IL-2Rα, IL-2Rβ, or γ_c and most efficiently by a combination of antibodies to multiple receptor subunits. These observations suggest that *IL-2Rα forms a complex with IL-2R$\beta\gamma_c$, increasing the affinity of the signaling receptor for IL-2 and thereby allowing a growth signal to be delivered at significantly lower IL-2 concentrations.* It is believed that IL-2 first binds rapidly to IL-2Rα, and this facilitates association with IL-2R$\beta\gamma_c$ (Color Plate VI, preceding Section I). Resting T cells express IL-2R$\beta\gamma_c$ but not IL-2Rα and can be stimulated only by high levels of IL-2 (Fig. 12–4). Upon antigen receptor–mediated T cell activation, IL-2Rα is rapidly expressed, thereby reducing the concentration of IL-2 needed for growth stimulation. Binding of IL-2 to the IL-2R$\alpha\beta\gamma_c$ complex results in signal transduction through a Jak/STAT signaling pathway (see Box 12–2).

The principal actions of IL-2 are on lymphocytes:

1. *Interleukin-2 is the major autocrine growth factor for T lymphocytes, and the quantity of IL-2 synthesized by activated CD4$^+$ T cells is an important determinant of the magnitude of T cell–dependent immune responses.* IL-2 also stimulates synthesis of other T cell–derived cytokines such as IFN-γ and lymphotoxin. Failure to synthesize adequate quantities of IL-2 has been described as a cause of antigen-specific T cell anergy (see Chapter 10).

2. *IL-2 stimulates the growth of NK cells and enhances their cytolytic function,* producing so-called lymphokine-activated killer (LAK) cells (see Chapter 13). NK cells, like resting T cells, express IL-2R$\beta\gamma_c$ and can be stimulated by high levels of IL-2. NK cells, however, do not express IL-2Rα and therefore do not reduce their requirement of IL-2, even after activation. Thus only high concentrations of IL-2 will lead to LAK cell formation. (However, NK cells do express IL-15Rα and can respond to this cytokine at much lower concentrations.) IL-2 infusion, with or without LAK cells, has been used to treat human tumors (see Chapter 18).

3. *IL-2 acts on human B cells both as a growth factor and as a stimulus for antibody synthesis.* It does not appear to cause isotype switching. These activities of IL-2 are discussed more fully in Chapter 9.

4. IL-2 may act as a death factor for antigen-activated T cells, promoting apoptosis. It may seem paradoxical for IL-2 to serve both as a growth factor and as a death factor for T cells, but these actions occur on T cells at different stages of activation, and the overall effect of IL-2 also depends on the presence of other signals (e.g., Fas-ligand expression for activation-induced cell

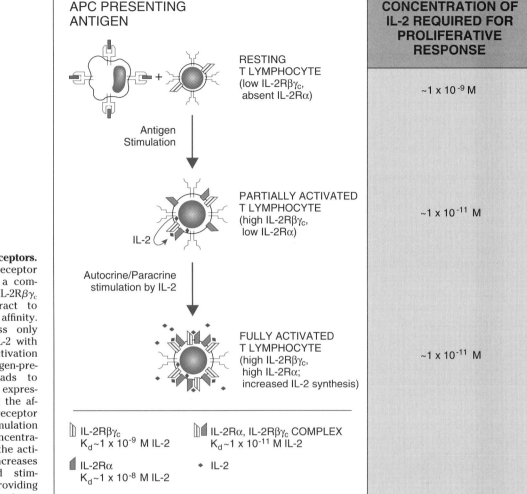

APC PRESENTING ANTIGEN

RESTING T LYMPHOCYTE
(low IL-2Rβγ$_c$,
absent IL-2Rα)

Antigen Stimulation

PARTIALLY ACTIVATED T LYMPHOCYTE
(high IL-2Rβγ$_c$,
low IL-2Rα)

IL-2

Autocrine/Paracrine stimulation by IL-2

FULLY ACTIVATED T LYMPHOCYTE
(high IL-2Rβγ$_c$,
high IL-2Rα;
increased IL-2 synthesis)

IL-2Rβγ$_c$
$K_d \sim 1 \times 10^{-9}$ M IL-2

IL-2Rα
$K_d \sim 1 \times 10^{-8}$ M IL-2

IL-2Rα, IL-2Rβγ$_c$ COMPLEX
$K_d \sim 1 \times 10^{-11}$ M IL-2

IL-2

CONCENTRATION OF IL-2 REQUIRED FOR PROLIFERATIVE RESPONSE

$\sim 1 \times 10^{-9}$ M

$\sim 1 \times 10^{-11}$ M

$\sim 1 \times 10^{-11}$ M

FIGURE 12–4. IL-2 receptors. The high-affinity IL-2 receptor (IL-2R) is composed of a complex of polypeptides (IL-2Rβγ$_c$ and IL-2Rα) that interact to bind IL-2 with high affinity. Resting T cells express only IL-2Rβγ$_c$, which binds IL-2 with lower affinity. T cell activation by antigen and an antigen-presenting cell (APC) leads to IL-2Rα synthesis and expression, thereby increasing the affinity of the IL-2Rβγ$_c$ receptor and allowing growth stimulation at physiologic IL-2 concentrations. IL-2 produced by the activated T cell further increases IL-2Rα expression and stimulates IL-2 synthesis, providing a positive amplification system.

death). The failure to eliminate cells through IL-2–dependent activation-induced cell death (see Chapter 10, Box 10–1) may explain why knockout mice lacking IL-2, IL-2Rα, or IL-2Rβ develop autoimmunity (see Chapter 19). Knockout animals lacking γ$_c$ and humans with γ$_c$ mutations instead show profound immunodeficiencies (see Chapter 21), presumably because of an inability of T cells lacking γ$_c$ to respond to other cytokines (e.g., IL-7) that utilize this polypeptide for receptor signaling.

Chronic T cell stimulation leads to shedding of IL-2Rα. Shed receptor proteins may bind free IL-2, preventing its interaction with target cells. However, the much greater affinity of IL-2Rβγ$_c$ for IL-2 compared with IL-2Rα alone suggests that serum IL-2Rα is not likely to contribute significantly to immunosuppression. Clinically, an increased level of shed IL-2Rα in the serum is a marker of strong antigenic stimulation, e.g., acute rejection of a transplanted organ. Infection of T cells by human T lymphotrophic virus–1 (HTLV-1) activates IL-2Rα synthesis and also leads to shed IL-2Rα in the serum.

Interleukin-4

Interleukin-4 (IL-4) was initially identified as a helper T cell–derived cytokine of approximately 20 kD that stimulated the proliferation of mouse B cells in the presence of anti-Ig antibody (an analog of antigen) and caused enlargement of resting B cells as well as increased expression of class II MHC molecules. It is now known that the main physiologic function of IL-4 is as a regulator of IgE- and mast cell/eosinophil-mediated immune reactions (see Chapter 14). IL-4 is a member of the four α-helical cytokine family, and its receptor is a 130 kD protein that contains the conserved two cysteine/WSXWS motif (see Box 12–1). In order to deliver signals, IL-4R must associate either with the γ$_c$ chain or with the IL-13R. These complexes signal through different Jaks but activate the same STAT protein, namely STAT6. The principal cellular sources of IL-4 are CD4$^+$ T lymphocytes, specifically of the T$_H$2 subset described later in the chapter. In fact, IL-4 production is used as the criterion for classifying CD4$^+$ T cells into this subset, with IFN-γ being the hallmark of the T$_H$1 cells. Activated mast cells and basophils, as well as some CD8+ T cells, are also capable of producing IL-4.

IL-4 has important actions on several cell types.

1. *IL-4 is required for the production of IgE and is the principal cytokine that stimulates switching of B cells to this heavy chain isotype.* The mechanisms of this effect were described in Chapter 9. IgE is the principal mediator of immediate hypersensitivity (allergic) reactions, and enhanced production of IL-4 is believed to be central to the development of allergies (see Chapter 14). IgE antibodies also play a role in defense against arthropod and helminthic infections, this being the major physiologic effector function of the T_H2 subset of helper T cells (see Chapter 16). Mice in which the IL-4 gene is disrupted fail to produce IgE. IL-4 also inhibits switching to IgG2a and IgG3 in mice, both of which are augmented by IFN-γ. This is one of several reciprocal antagonistic actions of IL-4 and IFN-γ.

2. *IL-4 is a growth and differentiation factor for T cells, in particular for cells of the T_H2 subset.* As will be discussed later in this chapter, IL-4 promotes the development of T_H2 cells from naive T cells stimulated with antigen, and it also functions as an autocrine growth factor for differentiated T_H2 cells, further promoting expansion of this subset. Knockout mice lacking the IL-4 or STAT6 gene show a deficiency in the development and maintenance of T_H2 cells, even after stimuli (such as helminthic infections) that are normally potent inducers of this subset.

3. IL-4 stimulates the expression of certain adhesion molecules, notably vascular cell adhesion molecule-1 (VCAM-1), on endothelial cells, resulting in increased binding of lymphocytes, monocytes, and especially eosinophils. IL-4–treated endothelial cells also secrete C-C family chemokines such as monocyte chemotactic protein-1 (MCP-1). As a result, high local concentrations of IL-4 induce monocyte- and eosinophil-rich inflammatory reactions.

4. IL-4 is a growth factor for mast cells and synergizes with interleukin-3 (IL-3) in stimulating mast cell proliferation.

Thus, IL-4 plays a critical role in IgE- and eosinophil-mediated inflammatory reactions.

Transforming Growth Factor-β

The original description of transforming growth factor-β was made in the field of tumor biology. It was noted that certain tumors produced activities, called transforming growth factor (TGF), that would allow normal cell types to grow in soft agar, a characteristic of malignant ("transformed") cells. Subsequently, it was found that growth stimulation was caused by one polypeptide, called TGF-α, but that survival in soft agar required a second factor, called TGF-β. TGF-α is a polypeptide growth factor for epithelial and mesenchymal cells and will not be discussed further.

TGF-β is a family of closely related molecules, encoded by distinct genes, commonly designated TGF-β1, TGF-β2, and TGF-β3. (The TGF-β family also includes other members thought to be involved in normal development rather than immunity; these will not be discussed here.) Cells of the immune system (e.g., T cells and monocytes) synthesize mainly TGF-β1, but certain anatomic sites (e.g., within the central nervous system or the anterior chamber of the eye) may contain high levels of TGF-β3, which acts to suppress local immunity. Native TGF-β1 is a homodimeric protein of approximately 28 kD. TGF-β1 is synthesized in a latent form that must be activated by extracellular proteases. Both antigen-activated T cells and LPS-activated mononuclear phagocytes secrete biologically active TGF-β1. TGF-β receptors include two high affinity polypeptide receptors (type I and type II) that may form heterodimers and that signal through an intracellular serine/threonine kinase domain in the type II receptor. TGF-β also has several low affinity (type III) receptors that may function to "present" TGF-β in a biologically potent form to the high affinity receptors.

The actions of TGF-β are highly pleiotropic. TGF-β inhibits the growth of many cell types and stimulates the growth of others. Often, TGF-β can either inhibit or stimulate growth of the same cell type, depending upon culture conditions such as degree of confluence. TGF-β causes synthesis of extracellular matrix proteins such as collagens, of matrix-modifying enzymes such as matrix metalloproteinases, and of cellular receptors for matrix proteins such as integrins. (The ability of TGF-β to induce extracellular matrix probably underlies its ability to promote cell growth in soft agar.) *In vivo,* TGF-β causes the growth of new blood vessels, a process called angiogenesis.

As a cytokine, TGF-β is potentially important because it antagonizes many responses of lymphocytes. For example, TGF-β inhibits proliferation of T cells to polyclonal mitogens or in mixed leukocyte reactions (see Chapter 17), inhibits maturation of CTLs, and inhibits the activation of macrophages. TGF-β also acts on other cells, such as polymorphonuclear leukocytes and endothelial cells, again largely to counteract the effects of pro-inflammatory cytokines. In this sense, TGF-β is an "anti-cytokine" and may be a signal for shutting off immune and inflammatory responses. Mice in which the TGF-β1 gene has been knocked out develop uncontrolled inflammatory reactions. Signals that stimulate T cells to synthesize TGF-β may cause them to behave as suppressor cells (see Chapter 10). *In vivo,* certain tumors may escape an immune response by secreting large quantities of TGF-β.

Although TGF-β is largely a negative regulator of immune responses, it may have some positive effects as well. For example, in mice, TGF-β has been shown to switch B cells to the IgA isotype, and it may therefore play a role in the generation of mucosal immune responses that are mediated by IgA (discussed in Chapter 11).

Interferon-γ

Interferon-γ (IFN-γ), also called immune or type II interferon, is a homodimeric glycoprotein containing two 21 to 24 kD subunits. The size variation of the subunit is caused by variable degrees of glycosylation, but each subunit contains an identical 18 kD polypeptide encoded by the same gene. IFN-γ is produced by activated $CD4^+$ and $CD8^+$ T cells and by NK cells. Transcription is directly initiated as a consequence of antigen activation and is enhanced by IL-2 and IL-12. IFN-γ produced by natural killer (NK) cells may function as a mediator of innate immunity and contribute to septic shock.

The receptor for IFN-γ is composed of two structurally homologous polypeptides, called IFN-γ-Rα and IFN-γ-Rβ. Both chains are related to the type I IFN receptor proteins (see Box 12–1). IFN-γ signal transduction is mediated by Jak1, Jak2, and STAT1α (see Box 12–2). STAT1 knockout mice are completely resistant to IFN-γ actions, as are IFN-γ-Rα knockout mice.

IFN-γ has several properties related to immunoregulation that separate it functionally from type I IFN:

1. *IFN-γ is a potent activator of mononuclear phagocytes.* IFN-γ directly induces synthesis of the enzymes that mediate the respiratory burst, allowing human macrophages to kill phagocytosed microbes. In murine, but not human, macrophages, IFN-γ acts in concert with TNF or LT to induce the high output isoform of nitric oxide synthase, allowing copious production of NO radicals that have effects similar to those of the reactive oxygen radicals made by human macrophages. In both species, IFN-γ up-regulates the high affinity signaling receptor for IgG, called FcγRI (see Chapter 3, Box 3–4). Cytokines that cause such functional changes in mononuclear phagocytes have been called **macrophage-activating factors** (MAFs). *IFN-γ is the principal MAF and provides the means by which T cells activate macrophages.* Other MAFs include GM-CSF, and, to a lesser extent, IL-1, TNF, and LT. Macrophage activation is described in more detail in Chapter 13. It is worth noting here that macrophage activation actually involves several different responses, and macrophages are said to be activated when they perform a particular function being assayed. For example, IFN-γ fully activates macrophages to kill phagocytosed microbes but only partly activates macrophages to kill tumor cells.

2. *IFN-γ increases class I MHC molecule expression and, in contrast to type I IFN, also causes a wide variety of cell types to express class II MHC molecules.* Thus, IFN-γ amplifies the recognition phase of the immune response by promoting the activation of class II-restricted $CD4^+$ helper T cells (see Chapter 6, Fig. 6–5). *In vivo,* IFN-γ can enhance both cellular and humoral immune responses through these actions at the recognition phase.

3. *IFN-γ acts on T lymphocytes to promote their differentiation.* As will be discussed later in this Chapter, IFN-γ promotes the differentiation of naive $CD4^+$ T cells to the T_H1 subset and inhibits the proliferation of T_H2 cells in mice. These effects may be mediated by activating mononuclear phagocytes to secrete IL-12 and T cells to express functional IL-12 receptors. IFN-γ is also one of the cytokines required for the maturation of $CD8^+$ CTLs (see Chapter 13).

4. *IFN-γ acts on B cells to promote switching to the IgG2a and IgG3 subclasses in mice and to inhibit switching to IgG1 and IgE.* The subtypes of IgG induced by IFN-γ are precisely those that bind to FcγRs on phagocytes and NK cells and are also the most potent complement-activating IgG subtypes. Thus *IFN-γ induces antibody responses that also participate in phagocyte-mediated elimination of microbes.*

5. *IFN-γ activates neutrophils,* up-regulating their respiratory burst. It is a less potent activator of neutrophils than TNF or LT.

6. *IFN-γ stimulates the cytolytic activity of NK cells,* more so than type I IFN.

7. *IFN-γ is an activator of vascular endothelial cells,* promoting $CD4^+$ T lymphocyte adhesion and morphologic alterations that facilitate lymphocyte extravasation. IFN-γ also potentiates many of the actions of TNF on endothelial cells.

The net effect of these varied activities of IFN-γ is to promote macrophage-rich inflammatory reactions, while inhibiting IgE-dependent eosinophil-rich reactions. Knockout mice in which the IFN-γ or IFN-γ receptor genes have been disrupted show several immunologic defects, including increased susceptibility to infections with intracellular microbes (which cannot be cleared because of defective macrophage activation), reduced production of nitric oxide by macrophages, reduced expression of class II MHC molecules on macrophages after infection with mycobacteria, reduced serum levels of IgG2a and IgG3 antibodies, and defective NK cell function.

Lymphotoxin

Lymphotoxin (LT) is a 21 to 24 kD glycoprotein that is approximately 30 per cent homologous to TNF and competes with TNF for binding to the same cell surface receptors. LT is sometimes called TNF-β. In humans, LT and TNF genes are located in tandem within the MHC on chromosome 6 (see Chapter 5). LT is produced exclusively by activated T lymphocytes and is often produced coordinately with IFN-γ by such cells. Human LT contains one or two N-linked oligosaccharides (accounting for the variability in molecular sizes). The functional form of LT is a homotrimer; the three-dimensional structure of LT is very similar to that of TNF, and like TNF, LT is capable of interacting with three receptors at once (see Color Plate VII, preceding Section I). In contrast to TNF and all other members of the TNF family (see Box 12–4), LT is synthesized as a true secretory protein without a membrane-spanning region.

Most studies have found little difference between the biologic effects of soluble TNF and LT, consistent with their binding to the same receptors. The most important distinctions between these cytokines are (1) that LT is synthesized exclusively by T cells, whereas TNF, although made by T cells, is predominantly derived from mononuclear phagocytes; and (2) that LT, but not TNF, can be expressed as part of a heterotrimer complexed with a type II membrane protein called LT-β. LT-β is also encoded by a gene located within the MHC, and like LT it belongs to the TNF superfamily. The heterotrimeric complex of $(LT)_1(LT\beta)_2$ binds to a receptor that is structurally related to but distinct from the TNF-RI and TNF-RII receptors utilized by LT.

In general, the quantities of LT synthesized by T cells are much less than the amounts of TNF made by LPS-stimulated mononuclear phagocytes, and LT is not readily detected in the circulation. Therefore, LT is usually a locally acting paracrine factor and not a mediator of systemic injury. Although neither TNF nor LT is toxic for normal (non-neoplastic) cells, both cytokines may induce apoptosis in some tumor cells. Like TNF, *LT is a potent activator of neutrophils and thus provides lymphocytes with a means of regulating acute inflammatory reactions.* It is more potent than IFN-γ as an activator of neutrophils, and the actions of LT are enhanced by IFN-γ. LT is also an activator of vascular endothelial cells, causing increased leukocyte adhesion, cytokine production, and morphologic changes that facilitate leukocyte extravasation. These effects, like those of TNF, are also enhanced by IFN-γ.

Several recent studies in transgenic and knockout mice have also implicated LT in the normal development of lymphoid organs. Mice overexpressing LT (or TNF) locally in the pancreas, under control of the rat insulin promoter, develop intrapancreatic ectopic secondary lymphoid tissue with discrete T and B cell zones, the latter forming germinal centers. Knockout mice lacking LT fail to form lymph nodes, Peyer's patches, and splenic white pulp. The T and B cell areas in these animals are intermixed, and few germinal centers develop in the spleen. Interestingly, TNF knockout mice only show decreased germinal center formation, emphasizing a unique role for LT. Knockout mice lacking TNF-RI form lymph nodes, but they do not form germinal centers and lack Peyer's patches. TNF-RII knockout mice appear to have normal peripheral lymphoid tissues. Knockout mice lacking LTβ fail to develop peripheral lymph nodes and

Peyer's patches, but in contrast to LT knockout mice, their spleens appear essentially normal. The emerging conclusion of these studies is that homotrimeric (LT) and heterotrimeric $(LT)_1(LT\beta)_2$ complexes appear to play different but overlapping roles in lymphoid organ development.

Interleukin-5

Interleukin-5 (IL-5) is an approximately 20 kD cytokine that functions as a homodimer. It is produced by the T_H2 subset of CD4$^+$ T cells and by activated mast cells. IL-5 belongs to the four α-helical cytokine family, with each bundle of four helices consisting of three strands from one monomer and one strand from the other. The receptor contains the two cysteine/WSXWS motif and interacts with a 150 kD signal-transducing subunit shared with IL-3 and GM-CSF, which also contains this motif (see Box 12–1). This signaling subunit is sometimes called the common β (β_c) chain. IL-5 signaling is mediated through a Jak/STAT pathway (see Box 12–2).

The major action of IL-5 is to stimulate the growth and differentiation of eosinophils and to activate mature eosinophils in such a way that they can kill helminths. In mice, neutralizing antibodies to IL-5 inhibit the eosinophilia seen in response to helminthic infection. This activity of IL-5 is complemented by the activities of IL-4 (e.g., IgE switching and eosinophil recruitment), contributing to T_H2-mediated allergic reactions and responses to helminthic and arthropod infections (see Chapter 14). IL-5 also acts as a costimulator for the growth of antigen-activated mouse B cells and was previously called either B cell growth factor 2 or T cell replacing factor. IL-5 may function synergistically with other cytokines, such as IL-2 and IL-4, to stimulate the growth and differentiation of B cells. IL-5 has also been found to act on more mature B cells to cause increased synthesis of immunoglobulin, especially of IgA. These actions are discussed in greater detail in Chapter 9. However, the importance of these actions of IL-5 on B cells is unclear, since IL-5 knockout mice show defective eosinophil responses but no abnormalities in antibody production.

Other Cytokines of Specific Immunity

Interleukin-13 is a 15 kD four α-helical cytokine made by T_H2 CD4$^+$ T cells and perhaps other cell types. The receptor for this cytokine is a heterodimer composed of a unique two cysteine/WSXWS domain–containing protein that binds IL-13 and the two cysteine/WSXWS domain–containing IL-4 receptor (see Box 12–1). This heterodimeric receptor, which can be activated by either IL-13 or IL-4, is found largely on non-lymphoid cells, especially macrophages and vascular endothelial cells. This receptor signals through Jak-2 and STAT6 (see Box 12–2). As noted earlier, lymphocytes predominantly express a different hetero-dimeric IL-4 receptor composed of the same IL-4 receptor subunit paired instead with the cytokine γ_c chain. This lymphocyte IL-4 receptor binds IL-4 but does not bind or respond to IL-13. Consequently, IL-13 mimics the effects of IL-4 on non-lymphoid cells but cannot replace IL-4 as a differentiation or growth factor for T or B cells. The endothelial cell actions of IL-13, which include induction of vascular cell adhesion molecule-1 and the production of C-C chemokines, may contribute to the recruitment of eosinophils into tissues during T_H2-mediated immune reactions.

Interleukin-16 is a 40 kD T cell–derived cytokine that acts as a specific chemoattractant of eosinophils. However, it appears to be derived from a fragment of an intracellular protein, and it is not known whether IL-16 is secreted under physiologic conditions.

Interleukin-17 is a 20 kD T cell–derived cytokine that mimics many of the pro-inflammatory actions of TNF and LT. Its *in vivo* function is unknown.

Migration inhibition factor is a T cell–derived activity that immobilizes mononuclear phagocytes, an effect that might cause them to be retained at sites of inflammation. This was the first cytokine activity to be described, but it is not clear whether any unique cytokine has such properties or whether it plays any real role in inflammation.

Functions and Development of T Cell Subsets That Produce Different Cytokines

Since cytokines are important mediators of T cell–dependent immune responses to protein antigens, the types of cytokines produced during a response determine the nature and effectiveness of that response. In Chapter 1 we introduced the concept that the specific immune response is highly specialized, and the immune system responds to different microbes in ways that are best able to eliminate these microbes. *An important basis of this functional heterogeneity is the ability of T cells responding to the protein antigens of microbes to differentiate into subsets of effector cells that produce distinct sets of cytokines and, therefore, elicit distinct effector functions.* We have referred to these subpopulations in Chapters 9 and 10 and will now describe them in more detail.

Populations of CD4$^+$ T cells that could be distinguished on the basis of their secreted cytokines were first discovered in the 1980s by studying large panels of cloned cell lines derived from mouse CD4$^+$ T cells. The two subsets that were most clearly distinguishable were called T_H1 and T_H2. It is now clear that individual T cells may express various mixtures of cytokines, but chronic immune reactions are often dominated by either T_H1 or T_H2 cytokines (see below). These subsets are distinguished solely by their cytokine profiles; to date, there are no other phenotypic markers

that provide definitive classification. Subsets of T_H1 and T_H2 effector T cells develop from the same precursor, which is a naive CD4$^+$ helper T lymphocyte. The pathway of differentiation is determined by the type of stimulation, especially by the cytokines produced at the time of antigen recognition. Moreover, once either subset develops, it produces cytokines that suppress differentiation to the other subset, resulting in increasing polarization of the immune response.

The most important effector cytokine produced by T_H1 cells is IFN-γ. IFN-γ stimulates the microbicidal activities of phagocytes, thereby promoting the intracellular destruction of phagocytosed microbes. Therefore, *the principal effector function of T_H1 cells is in phagocyte-mediated defense against infections* (Fig. 12–5). T_H1 cells also secrete IL-2, which functions as their autocrine growth factor and stimulates the proliferation and differentiation of CD8$^+$ T cells. Thus, T_H1 cells function as helpers for the development of CTLs, further promoting immunity against intracellular microbes. T_H1 cells produce LT, which promotes the recruitment and activation of neutrophils.

The characteristic cytokines produced by T_H2 cells are IL-4 and IL-5 (Fig. 12–6). Therefore, T_H2 cells are the mediators of allergic reactions and defense against helminthic and arthropod infections (Chapter 14). In addition, T_H2 cells produce cytokines (such as IL-4, IL-13, and IL-10) that antagonize the actions of IFN-γ and suppress macrophage activation. Therefore, T_H2 cells may function as physiologic regulators of immune responses by inhibiting potentially injurious T_H1 responses. Excessive or uncontrolled T_H2 development is associated with deficient cell-mediated immunity to infections with intracellular microbes, such as mycobacteria (see Chapter 16).

The differentiation of naive CD4$^+$ T cells into T_H1 or T_H2 cells is influenced by cytokines produced early in response to the microbe that triggers the immune reaction (Fig. 12–7). Many intracellular bacteria, such as *Listeria* and mycobacteria, and some parasites, such as *Leishmania,* infect macrophages, and the macrophages respond by secreting IL-12. Other microbes may trigger IL-12 secretion indirectly. For instance, viruses and some parasites stimulate NK cells to produce IFN-γ, which in turn acts on macrophages to induce IL-12 secretion. IL-12 binds to receptors on antigen-stimulated CD4$^+$ T cells and promotes their differentiation into T_H1 cells. IL-12 also directly stimulates transcription of T_H1 cytokine genes, notably IFN-γ, possibly through the activation of STAT4 protein. Thus, *the immune system responds to microbes that infect or encounter macrophages by turning on the effector pathway that promotes phagocytic elimination of these microbes.* This pathway of T_H1 development illustrates an important link between a mediator of innate immunity, IL-12, and a specific immune response. IL-12 also enhances IFN-γ production by and cytolytic activity of CTLs and natu-

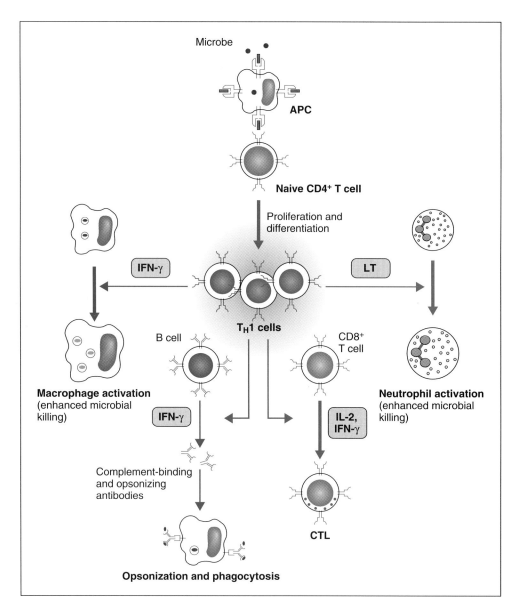

FIGURE 12–5. Functions of T$_H$1 cells. CD4$^+$ T cells that differentiate into T$_H$1 cells principally secrete interferon-γ (IFN-γ), lymphotoxin (LT), and interleukin-2 (IL-2). IFN-γ acts on B lymphocytes to produce antibodies that opsonize microbes for phagocytosis and acts on macrophages to increase phagocytosis and killing of microbes in phagolysosomes. IL-2 is the predominant autocrine growth factor made by this subset of T cells and, along with IFN-γ, can contribute to the differentiation of CD8$^+$ T cells in cytolytic T lymphocytes (CTLs). LT activates neutrophils. APC, antigen-presenting cell.

ral killer cells, further stimulating cell-mediated immunity (see Chapter 13). IFN-γ produced by T$_H$1 cells inhibits proliferation of T$_H$2 cells, thus promoting T$_H$1 dominance.

The differentiation of antigen-stimulated T cells to the T$_H$2 subset is stimulated by IL-4, which functions by activating STAT6, a transcription factor that promotes expression of IL-4 and possibly other T$_H$2 cytokine genes. The obligatory role of IL-4 in T$_H$2 differentiation raises an obvious problem. Since differentiated T$_H$2 cells are the major source of IL-4 during immune responses to protein antigens, where does IL-4 come from before T$_H$2 cells develop? The most likely explanation is that antigen-stimulated CD4$^+$ T cells secrete small amounts of IL-4 from their initial activation. If the antigen is persistent and present at high concentrations, the local concentration of IL-4 gradually increases. If the antigen also does not trigger inflammation with attendant IL-12 production, the re-

sult is increasing differentiation of T cells to the T$_H$2 subset, and a progressive accumulation of T$_H$2 effectors. Thus, in response to helminthic parasites and environmental allergens, the key determinant of T$_H$2 development may be persistent or repeated T cell stimulation with little inflammation or macrophage activation. Other explanations have been proposed for T$_H$2 cell differentiation, including production of IL-4 by other cell types and differences in antigen structure or in signals provided by antigen-presenting cells other than cytokines (e.g., costimulators). Genetic factors also influence T$_H$2 cell vs. T$_H$1 cell differentiation.

CYTOKINES THAT STIMULATE HEMATOPOIESIS

Several of the cytokines generated during both innate immunity and antigen-induced specific immune responses have potent stimulatory effects on the growth and differentiation of bone marrow pro-

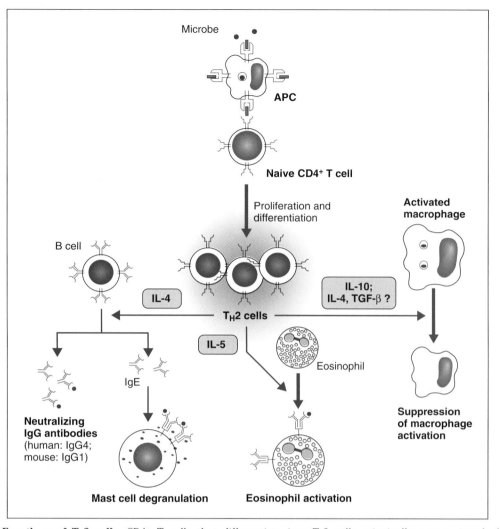

FIGURE 12–6. Functions of T$_H$2 cells. CD4$^+$ T cells that differentiate into T$_H$2 cells principally secrete interleukin-4 (IL-4) and interleukin-5 (IL-5). IL-4 acts on B cells to produce antibodies that bind to mast cells, such as IgE, and promotes mast cell development. IL-5 activates eosinophils, the principal inflammatory cell of IgE-mast cell–dependent reactions. Cytokines from T$_H$2 cells also can antagonize the effects of T$_H$1 cell–derived cytokines that activate macrophages. APC, antigen-presenting cell.

genitor cells (Table 12–4). Thus, immune and inflammatory reactions, which consume leukocytes, also elicit production of new leukocytes to replace inflammatory cells. All of the various mature leukocyte cell populations arise as a consequence of progressive expansion and irreversible differentiation of the progeny of self-renewing pluripotent stem cells. Maturation of hematopoietic cells involves commitment to a particular lineage and occurs concomitantly with loss of ability to develop into other mature cell types. This process has been depicted as a simple tree (Fig. 12–8).

The cytokines that stimulate expansion and differentiation of bone marrow progenitor cells are collectively called **colony-stimulating factors** (CSFs) because they are often assayed by their ability to stimulate the formation of cell colonies in bone marrow cultures. These colonies of cells mature during the *in vitro* assay, acquiring characteristics of specific cell lineages (e.g., granulocytes,

mononuclear phagocytes). Different CSFs act on bone marrow cells at different stages of maturation and preferentially promote development of colonies of different lineages. The names assigned to CSFs reflect the types of colonies that arise in these assays. Interestingly, many of the CSFs, including IL-3 and GM-CSF are located in a gene cluster on human chromosome 5. IL-4, IL-5, and IL-9 have been mapped to this same complex.

Some of the actions of CSFs are influenced by other cytokines. For example, TNF, LT, IFN-γ, and TGF-β may inhibit growth of bone marrow progenitor cells. In contrast, IL-1 and IL-6 enhance responses to CSFs. In general, cytokines are thought both to be necessary for normal marrow function and to provide a means of fine tuning function in response to stimulation. In this chapter, we will focus on two CSFs that function in lymphocyte development, namely c-kit ligand and IL-7, as well as IL-3, a T cell–derived multilineage CSF that is synthesized in specific immune reactions.

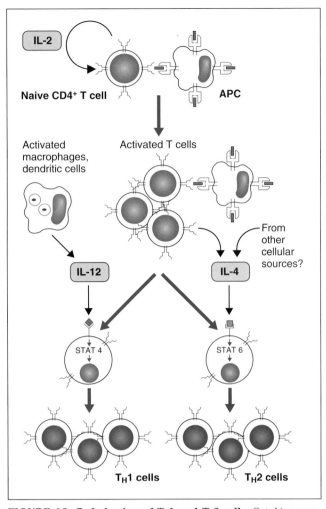

FIGURE 12–7. Induction of T_H1 and T_H2 cells. Cytokines produced in the innate immune response to microbes or early in specific immunity can influence the differentiation of naive CD4+ T cells into T_H1 or T_H2 cells. IL-12, made by activated macrophages or dendritic cells, is a strong inducer of T_H1 cell development through a STAT4–dependent pathway. IL-4, which may be made during the initial activation of naive T cells, favors induction of T_H2 cells through a STAT6–dependent pathway. APC, antigen-presenting cell.

c-Kit Ligand

The pluripotent stem cell expresses a membrane receptor with intrinsic protein tyrosine kinase activity that has been identified as the protein product of the cellular oncogene, *c-kit*. The extracellular portion of this receptor contains five Ig domains (see Box 12–1). The cytokine that interacts with the receptor has been called **c-kit ligand,** and it is also referred to as "stem cell factor." c-Kit ligand is synthesized by stromal cells of the bone marrow (including adipocytes, fibroblasts, and endothelial cells) in two forms: a transmembrane protein of about 27 kD and a secreted form of about 24 kD. These different products result from alternative splicing of the same gene. The soluble form of this ligand is absent in a mutant mouse strain called steel, and the soluble form of the c-kit ligand is thus sometimes called *steel factor.* The steel mouse has only selective gaps in its bone marrow–derived cell populations (e.g., inadequate mast cell and eosinophil production), which has led to the conclusion that the cell surface form of c-kit ligand is more important than the soluble form for stimulating stem cells to mature into various hematopoietic lineages. Elimination of both forms of c-kit ligand, by complete knockout of the gene, is lethal.

It is not yet possible to purify large numbers of stem cells for direct analysis. Many of the conclusions about c-kit ligand and other early-acting CSFs are derived from experiments in which populations enriched for stem cells are exposed to the cytokines in culture, and the types of colonies that develop are analyzed. From this kind of experiment, it is believed that c-kit ligand is needed to make stem cells responsive to other CSFs but that it does not cause colony formation by itself. Bone marrow cell cultures that contain stromal cells do not have a requirement for exogenous c-kit ligand since the stromal cells express this gene product.

In addition to its role in the bone marrow, c-kit ligand has effects on immature cells that have left the bone marrow. In particular, c-kit ligand appears to play a role in sustaining the viability and proliferative capacity of immature T cells in the thymus (see Chapter 8) and of mucosal mast cells in peripheral sites (see Chapter 14).

Interleukin-7

Interleukin-7 (IL-7) is a four α-helical cytokine secreted by marrow stromal cells that acts on hematopoietic progenitors committed to the B and T lymphocyte lineages. The IL-7 receptor consists of a unique IL-7 binding protein and utilizes the γ_c chain for signaling through a Jak/STAT pathway (see Boxes 12–1 and 12–2). Most models of hematopoiesis suggest that lymphocyte progenitors differentiate from common stem cells very early in maturation, so that IL-7 is probably acting on cells at the same level of development as c-kit ligand. IL-7 may also stimulate the growth and maturation of immature CD4+ CD8+ T cell precursors in the thymus. However, this is based on *in vitro* experiments, and the cellular source of IL-7 in the thymus is not known. Transgenic mice that overexpress IL-7 show markedly increased numbers of pre-B cells in the bone marrow and peripheral lymphoid tissues. IL-7 knockout mice are profoundly lymphopenic, with decreased numbers of T and B cells, while knockout mice lacking the IL-7 receptor not only have reduced numbers of T and B cells but completely lack γδ T cells. It is likely that the profound immunodeficiencies associated with knockout mice lacking γ_c or Jak-3 as well as humans with mutations in these genes largely result from IL-7 signal defects. *In vitro* IL-7 also acts on mature T cells, mimicking the actions of IL-2. It is not known whether IL-7 plays a physiologic role

TABLE 12–4.　Mediators of Immature Leukocyte Growth and Differentiation

Cytokine	Polypeptide Size	Cell Source	Cell Target	Primary Effects on Each Target
c-Kit ligand	24 kD	Bone marrow stromal cell	(a) Pluripotent stem cell	(a) Activation
Interleukin-7	25 kD	Fibroblast, bone marrow stromal cells	(a) Immature progenitor	(a) Growth and differentiation to lymphocytes
Interleukin-3	20–26 kD	T cell	(b) Immature progenitor	(b) Growth and differentiation to all cell lines
Granulocyte-monocyte-CSF	22 kD	T cell, mononuclear phagocyte, endothelial cell, fibroblast	(a) Immature progenitor	(a) Growth and differentiation to all cell lines
			(b) Committed progenitor	(b) Differentiation to granulocytes and mononuclear phagocytes
			(c) Mononuclear phagocyte	(c) Activation
Monocyte-CSF	40 kD	Mononuclear phagocyte, endothelial cell, fibroblast	(a) Committed progenitor	(a) Differentiation to mononuclear phagocytes
Granulocyte-CSF	19 kD	Mononuclear phagocyte, endothelial cell, fibroblast	(a) Committed progenitor	(a) Differentiation to granulocytes

Abbreviations: CSF, colony-stimulating factor; kD, kilodalton.

in the behavior of more fully differentiated T cells *in vivo.*

Interleukin-3

Interleukin-3 (IL-3), also known as **multilineage colony–stimulating factor** (multi-CSF), is a 20 to 26 kD product of CD4$^+$ T cells that acts on the most immature marrow progenitors and promotes the expansion of cells that differentiate into all known mature cell types. IL-3 is a member of the four α-helical family of cytokines. In humans, the heterodimeric receptor consists of a unique two cysteine/WSXWS domain-containing subunit and a 150 kD signal-transducing subunit shared with IL-5 and GM-CSF. In the mouse, the signal-transducing subunit of the IL-3 receptor is unique. Signal transduction in both species involves a Jak/STAT pathway. Most functional analyses of IL-3 have been performed in mice. It has been found that IL-3 also promotes the growth and development of mast cells from bone marrow–derived progenitors, an action enhanced by IL-4. IL-3 may also act as an activator of mature basophils, promoting their recruitment into late phase inflammatory reactions of allergy (see Chapter 14). IL-3 is produced by CD4$^+$ helper T cells of both the T$_H$1 and T$_H$2 subsets. Human IL-3 has been identified by the complementary DNA (cDNA) cloning of a molecule homologous to mouse IL-3. Although IL-3 is made by some human T cell clones, it has been harder to establish a role for this cytokine in experimental systems of hematopoiesis in humans. In fact, many actions attributed to murine IL-3 appear to be performed by human granulocyte-macrophage CSF. It is not known whether these experimental results reflect differences between species or in experimental conditions. If these are species differences, they may be related to the differences in IL-3 receptor structure described above. However, knockout of either the IL-3 receptor or its unique

signal-transducing subunit in mice does not cause noticeable impairment of hematopoiesis.

Other Colony-Stimulating Cytokines

Interleukin-9 (IL-9) is a 30 to 40 kD protein that supports the growth of some T cell lines and of bone marrow–derived mast cell progenitors. It may also stimulate development of other lineages from marrow-derived precursors. Like IL-7, IL-9 uses the γ_c chain signaling subunit and utilizes a Jak-3/STAT pathway. Little is known about the physiologic function of IL-9.

Interleukin-11 (IL-11) is a 20 kD cytokine produced by bone marrow stromal cells, especially after activation (which may be achieved experimentally by pharmacologic agents such as phorbol esters). IL-11 uses the same gp130 signaling subunit utilized by IL-6. IL-11 stimulates megakaryopoiesis and may prove to be of therapeutic benefit in patients with platelet deficiencies. It is also induced early in response to infection by certain viruses. Its physiologic function is unknown.

Granulocyte-monocyte colony–stimulating factor (GM-CSF), monocyte-colony–stimulating factor (M-CSF), and granulocyte-colony–stimulating factor (G-CSF) are cytokines made by activated T cells, macrophages, endothelial cells, and stromal fibroblasts that act on bone marrow to increase production of inflammatory leukocytes. In addition, G-CSF acts as an endocrine hormone, generated at the sites of inflammation, that mobilizes neutrophils from the marrow to replace those consumed in inflammatory reactions. GM-CSF is also a macrophage-activating factor and promotes the differentiation of Langerhans cells into dendritic cells. Recombinant GM-CSF and G-CSF are increasingly used to speed bone marrow recovery after cancer chemotherapy, perhaps the single most clinically successful application of any biological response modifier. These same cytokines also induce bone

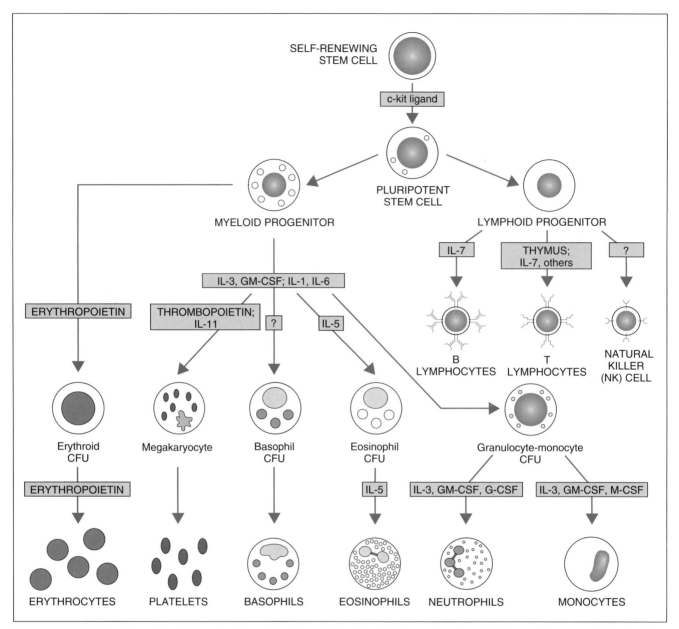

FIGURE 12–8. Functions of cytokines in the maturation of blood cells. The maturation of different lineages of blood cells is regulated by various cytokines. CFU, colony-forming unit; IL, interleukin; GM-CSF, granulocyte-monocyte colony-stimulating factor.

marrow stem cells to transiently enter the circulation, allowing stem cells to be harvested for transplantation (see Chapter 17).

Two other CSFs, of particular relevance to hematology, are erythropoietin, which stimulates red cell production, and thrombopoietin, which stimulates platelet production. As these molecules do not act directly on cells of the immune system, we will not discuss them further.

SUMMARY

Cytokines are a family of protein mediators of both innate and specific immunity. The same cyto-

kines are often made by many cell types, and individual cytokines often act on many cell types. The actions of different cytokines are often redundant and influence the actions of other cytokines. In general, cytokines are synthesized in response to inflammatory or antigenic stimuli and act locally, in an autocrine or paracrine fashion, by binding to high affinity receptors on target cells. Certain cytokines may be produced in sufficient quantity to circulate and exert endocrine actions. For many cell types, cytokines serve as growth factors. Structurally, cytokines fall into a small number of classes such as four α-helical globular domain, TNF family, and so on, that interact with receptors belonging to a limited number of structural families. Many cytokines signal through Jak/STAT signaling pathways.

We have classified cytokines into three groups, according to their principal actions:

The first group consists of those cytokines that mediate innate immunity and includes antiviral cytokines (e.g., type I interferons, interleukin-15, interleukin-12), pro-inflammatory cytokines (e.g., tumor necrosis factor, interleukin-1, interleukin-6, and chemokines) and regulatory cytokines (e.g., interleukin-10). The predominant cellular source of these molecules is mononuclear phagocytes.

The second group of cytokines is derived largely from antigen-stimulated T lymphocytes and serves to mediate and regulate specific immunity. This group includes interleukin-2, the principal T cell growth factor; interleukin-4, the major regulator of IgE synthesis; and transforming growth factor-β, which inhibits lymphocyte responses. It also includes interferon-γ, the principal activator of mononuclear phagocytes; lymphotoxin, an activator of neutrophils; and interleukin-5, an activator of eosinophils. Through the actions of these and other effector cytokines, the specific immune response is able to intensify and focus the innate immune response. CD4$^+$ T cells may differentiate into specialized effector T$_H$1 cells that secrete IFN-γ and LT, which favors phagocyte-mediated immunity, or into T$_H$2 cells that secrete IL-4 and IL-5, which favors IgE- and eosinophil-mediated immunity. The pattern of differentiation may be influenced by cytokines produced in the innate immune response or early in the specific immune response.

The third group of cytokines, collectively called colony-stimulating factors, consists of cytokines derived from marrow stromal cells and T cells, which stimulate the growth of bone marrow progenitors, thereby providing a source of additional inflammatory leukocytes. Several of these (e.g., c-kit ligand and IL-7) play a central role in lymphopoiesis.

Thus, cytokines serve many functions that are critical to host defense against pathogens and provide links between specific and innate immunity. Cytokines also regulate the magnitude and nature of immune responses by influencing the growth and differentiation of lymphocytes. Finally, cytokines provide important amplification mechanisms that enable small numbers of lymphocytes specific for any one antigen to activate a variety of effector mechanisms to eliminate the antigen. Excessive production or actions of cytokines can lead to tissue injury and even death. The administration of cytokines or their inhibitors is a potential approach for modifying biologic responses associated with disease.

Selected Readings

Abbas, A. K., K. M. Murphy, and A. Sher. Functional diversity of helper T lymphocytes. Nature 383:787–793, 1996.

Arai, K., F. Lee, A. Miyajima, S. Miyatake, N. Arai, and T. Yokota. Cytokines: coordinators of immune and inflammatory responses. Annual Review of Biochemistry 59:783–836, 1990.

Baggiolini, M., B. Dewald, and B. Moser. Human chemokines: an update. Annual Review of Immunology (in press), 1997.

Baumann, H., and J. Gauldie. The acute phase response. Immunology Today 15:74–80, 1994.

Beutler, B. TNF, immunity and inflammatory disease: lessons from the past decade. Journal of Investigative Medicine 43:227–235, 1995.

Dinarello, C. A. Biological basis for interleukin-1 in disease. Blood 87:2095–2147, 1996.

Farrar, M. A., and R. D. Schreiber. The molecular cell biology of interferon-gamma and its receptor. Annual Review of Immunology 11:571–611, 1993.

Ihle, J. N. The Janus protein tyrosine kinase family and its role in cytokine signaling. Advances in Immunology 60:1–35, 1995.

Miyajima, A., T. Kitamura, N. Harada, T. Yokata, and K. Arai. Cytokine receptors and signal transduction. Annual Review of Immunology 10:295–331, 1992.

Moore, K. W., A. O'Garra, R. de W. Malefyt, P. Vieira, and T. R. Mosmann. Interleukin-10. Annual Review of Immunology 11:165–190, 1993.

Palladino, M. A., R. E. Morris, H. F. Starnes, and A. D. Levinson. The transforming growth factor–betas. A new family of immunoregulatory molecules. Annals of The New York Academy of Sciences 593:181–187, 1990.

Schall, T. J., and K. B. Bacon. Chemokines, leukocyte trafficking, and inflammation. Current Opinion in Immunology 6:865–873, 1994.

Seder, R. A., and W. E. Paul. Acquisition of lymphokine-producing phenotype by CD4$^+$ T cells. Annual Review of Immunology 12:635–673, 1994.

Sen, G. C., and P. Lengyel. The interferon system. A bird's eye view of its biochemistry. Journal of Biological Chemistry 267:5017–5020, 1992.

Smith, K. A. Interleukin-2: inception, impact, and implications. Science 240:1169–1176, 1988.

Sugamura, K. H., H. Asao, M. Kondo, N. Tanaka, N. Ishii, K. Ohbo, M. Nakamura, and T. Takeshita. The interleukin-2 receptor γ chain: its role in the multiple cytokine receptor complexes and T cell development in XSCID. Annual Review of Immunology 14:179–205, 1996.

Thèze, J., P. M. Alzari, and J. Bertoglio. Interleukin-2 and its receptors: recent advances and new immunological functions. Immunology Today 17:481–486, 1996.

Trinchieri, G. Interleukin 12: a proinflammatory cytokine with immunoregulatory functions that bridge innate resistance and antigen-specific adaptive immunity. Annual Review of Immunology 13:251–276, 1995.

CHAPTER THIRTEEN

EFFECTOR MECHANISMS

OF T CELL–MEDIATED

IMMUNE REACTIONS

Historically, specific immunity has been divided into *humoral immunity,* which can be adoptively transferred from an immunized donor to a naive host by antibodies in the absence of cells, and *cell-mediated immunity,* which can be adoptively transferred only by viable T lymphocytes. This classification, first evident through adoptive transfer experiments, is actually quite general, because it is based on fundamental differences among various effector mechanisms in the immune system. *In humoral immunity, specific recognition of antigen in the effector phase is mediated by the binding of secreted antibody molecules to antigen.* Antibody is thus sufficient to adoptively transfer humoral immunity. *In cell-mediated immunity, in contrast, the effector phase is initiated through specific antigen recognition by memory or effector T cells.* In this chapter, we will discuss the various effector mechanisms of cell-mediated immune reactions.

In some forms of cell-mediated immunity, antigen-specific T cells directly perform the effector function, as when cytolytic T lymphocytes (CTLs) lyse specific target cells; in other forms, antigen-activated T cells secrete cytokines that recruit and activate effector cells that are not specific for the antigen, such as macrophages and natural killer (NK) cells (Fig. 13–1). When the effector cells are nonspecific, antigen specificity is conferred by

proximity to the antigen-stimulated T cells. As we discussed in Chapter 6, T cells can recognize and respond to foreign protein antigen only when a peptide fragment of the foreign protein is presented in a complex with a major histocompatibility complex (MHC) molecule on the surface of an appropriate antigen-presenting cell (APC) or target cell. *Therefore, cell-mediated immunity is directed at or near cells that bear foreign antigens on their surface, and cell-mediated immune reactions are physiologically most important for eradicating microbes or viruses that live intracellularly, i.e., within APCs.* Indeed, the original description of cell-mediated immunity was the adoptive transfer of protection against *Listeria monocytogenes,* an intracellular bacterium (see Chapter 16). Cell-mediated immune reactions may also be important for elimination of cells that express foreign MHC molecules, as in an allograft (see Chapter 17), or that express tumor-specific antigens, as in a malignant tumor (see Chapter 18).

Different types of cell-mediated immune reactions may result from T cell recognition of antigen.

1. In **delayed type hypersensitivity** (DTH), antigen-activated T cells secrete cytokines, which have several effects. Some cytokines activate venular endothelial cells to recruit monocytes and

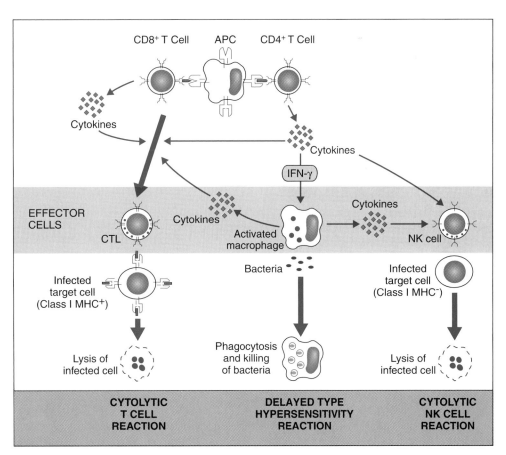

FIGURE 13–1. Types of T cell–mediated immune reactions. Antigen recognition by CD4+ and CD8+ T cells results in the release of cytokines which promote clonal expansion and which can activate a variety of effector mechanisms. CD8+ T cells may differentiate into effector cells, cytolytic T lymphocytes (CTLs), that kill infected target cells. CD4+ T cells of the T_H1 subset release cytokines that activate macrophages, increasing the capacity of these cells to kill intracellular microbes, as part of a delayed type hypersensitivity (DTH) reaction. T cell cytokines also help promote the differentiation of CD8+ T cells into CTL and increase the capacity of natural killer (NK) cells to mediate cytolysis, resulting in the destruction of virally infected tissue cells.

other leukocytes from the blood at the site of antigen challenge. Other cytokines convert the monocytes into activated macrophages that serve to eliminate the source of antigen. The T cells that mediate DTH are usually CD4$^+$ T$_H$1 cells, but cytokines produced by CD8$^+$ T cells can initiate the same reaction. DTH reactions normally involve presentation of antigens on the surfaces of macrophages that have phagocytosed bacteria or viruses. These microbe-laden macrophages become activated by T cell–derived cytokines, and the physiologic purpose of the response is to promote killing of the intracellular microbes residing in phagolysosomes of macrophages.

2. In **CTL responses** to viral infections, organ transplants, or tumor cells, antigen-activated CD8$^+$ T cells differentiate into functional CTLs, which lyse target cells expressing specific antigen-MHC complexes. This process of differentiation often requires "help" in the form of cytokines secreted by antigen-activated T cells. Synthesis of cytokines in adequate quantity to support CTL differentiation requires that antigen be presented by professional APCs, which express costimulator molecules, to either CD4$^+$ or CD8$^+$ T cells. The costimulators may also directly effect CD8$^+$ T cell differentiation into CTLs. The physiologic purpose of the CTL response is to destroy cells harboring intracellular microbes. In contrast to DTH, the CTL response is not restricted to eradicating microbes in the phagolysosomes of macrophages, and CTLs can destroy microbes residing in nonphagocytic tissue cells or in the cytoplasm of infected macrophages.

3. In the initial response to viral infections, **natural killer (NK) cells** serve to eradicate infected cells prior to the appearance of specific CTLs (see Chapter 16). NK cells may also play a role in lysis of virally infected cells or tumor cells that evade CTL-mediated lysis by reducing expression of class I MHC molecules. NK cells, stimulated by cytokines from antigen-activated CD4$^+$ T cells, differentiate into **lymphokine-activated killer (LAK) cells**, which nonspecifically lyse target cells. The physiologic role of LAK cell–mediated cytolysis is not fully understood, but it has been observed in settings of strong antigenic stimulation, e.g., in graft-versus-host disease following bone marrow transplantation (see Chapter 17).

4. In helminthic and arthopod infections and in allergic inflammation, antigen-activated CD4$^+$ T$_H$2 cells secrete cytokines that activate mast cells and recruit and activate basophils and eosinophils. This kind of immune reaction and its effector mechanisms will be discussed in Chapter 14.

This chapter begins with a discussion of the T cells that regulate cell-mediated immune reactions and then considers the various effector cell populations that are involved in each of these distinct reaction patterns.

T LYMPHOCYTES AND THE INITIATION OF CELL-MEDIATED IMMUNE REACTIONS

CD4$^+$ T cells initiate specific immunity, both cell-mediated and humoral, by recognizing portions of protein antigens (peptides) bound to self class II MHC molecules on the surface of APCs. Upon activation by specific antigen, these T cells secrete cytokines, many of which act on other cell populations involved in host defense. For example, tumor necrosis factor (TNF) and lymphotoxin (LT) activate neutrophils and vascular endothelial cells; interleukin-5 (IL-5) activates eosinophils; interferon-γ (IFN-γ) activates mononuclear phagocytes; and interleukin-2 (IL-2) activates NK cells as well as both T and B lymphocytes. *By means of cytokine secretion, T cells stimulate the function and focus the activity of nonspecific effector cells of innate immunity, thereby converting these cells into agents of specific immunity.* Indeed, the first function of T cells in the evolutionary development of specific immunity may well have been to augment and direct the effector mechanisms of innate immunity. As antigen-specific effector cells such as CTLs and B lymphocytes evolved, the general pattern of T cell function remained the same: through the secretion of cytokines, such as IL-2 and IL-4, T cells provide "help" necessary for the activation of other lymphocytes.

The multitude of cytokines produced by antigen-activated T cells, each with distinct but overlapping sets of cellular targets, raised the question of how particular antigens elicit particular types of immune reactions. As we discussed in Chapter 12, there is now considerable evidence that T cells differentiate into subsets that produce distinct patterns of cytokines and perform distinct functions. In this chapter, we will focus on reactions induced by CD4$^+$ T$_H$1 and CD8$^+$ T cells, i.e., those that produce IFN-γ, TNF, LT, and IL-2. In Chapter 14, we will discuss reactions triggered by CD4$^+$ T$_H$2 cells, i.e., those that produce IL-4 and IL-5. We do not yet fully understand how different antigens activate and expand different T cell subsets. As we discussed in Chapter 12, the likeliest explanation is that the pattern of cell-mediated immunity is determined by the effects of microbes upon the innate immune system. In particular, microbes that induce macrophages to secrete IL-12 will trigger DTH and cytolytic cell responses by favoring the differentiation of naive CD4$^+$ T cells into T$_H$1 cells and by the direct action of this cytokine upon CTLs and NK cells.

DELAYED TYPE HYPERSENSITIVITY REACTIONS

Delayed type hypersensitivity is a form of cell-mediated immune reaction in which the ultimate effector cell is the activated mononuclear phagocyte (macrophage). This type of cell-mediated immunity is part of the primary defense mechanism against intracellular bacteria, such as *Listeria monocyto-*

genes and mycobacteria. Such microbes cannot be killed by normal unactivated phagocytes and, indeed, may even preferentially survive within the phagolysosomes or cytoplasm of monocytes. Eradication of these organisms requires enhancement of the microbicidal function of phagocytes by T cell–derived cytokines. The same sequence of T cell and macrophage activation can be elicited by soluble protein antigens or chemically reactive haptens. In this situation, as in host defense, macrophage activation can cause tissue injury. If the antigen is not a microbe, DTH reactions produce tissue injury without providing a "protective" function, hence the term "hypersensitivity."

The classic animal model of DTH is the response of an immunized guinea pig to antigen applied by "skin painting" or introduced by intradermal injection. Such reactions may be induced in humans by contact sensitization with chemicals and environmental antigens, or by intradermal injection of microbial antigens in individuals immunized by prior infection (Fig. 13–2). For example, purified protein derivative (PPD), a protein prepared from *Mycobacterium tuberculosis*, will elicit a DTH response when injected into individuals who have recovered from primary tuberculosis or who have been vaccinated against tuberculosis. The characteristic response of DTH evolves over 24 to 48 hours. About 4 hours after injection of antigen, neutrophils accumulate around the post-capillary venules at the injection site. The neutrophil infiltrate rapidly subsides, and by about 12 hours the injection site becomes infiltrated by T cells and blood monocytes, also organized in a perivenular distribution (Fig. 13–3). The endothelial cells lining these venules become plump, show increased biosynthetic organelles, and become leaky to plasma macromolecules. Fibrinogen escapes from the blood vessels into the surrounding tissues, where it is converted into fibrin. The deposition of fibrin and, to a lesser extent, accumulation of T cells and monocytes within the extravascular tissue space around the injection site cause the tissue to swell and become hard ("indurated"). Induration, the hallmark of DTH, is detectable by about 18 hours after injection of antigen and is maximal by 24 to 48 hours. The lag in the onset of palpable induration is the reason for calling the response "delayed type." The development of induration 24 to 48 hours after intradermal injection of PPD is a widely used clinical indicator for evidence of tuberculosis infection.

Although experimental DTH was first described in guinea pigs, DTH-like reactions may be elicited in other experimental animals such as mice and rats. Histologically, these murine reactions differ from DTH responses in guinea pigs or humans in that the inflammatory infiltrate at 18 to 24 hours shows a paucity of T cells and activated macrophages, but rather consists largely of neutrophils. The basis of this difference is not known, but it may reside in the inflammatory functions of

vascular endothelium or species differences in the production or effects of cytokines. Nevertheless, in all species examined, DTH reactions to most protein antigens may be adoptively transferred by antigen-sensitized CD4+ T cells. CD8+ T cells are also capable of adoptively transferring DTH-like reactions. *In vivo*, it is likely that DTH reactions elicited by viruses are predominantly mediated by CD8+ T cells, whereas those that are elicited by phagocytosed bacteria are predominantly mediated by CD4+ T_H1 T cells. Presumably, this is because cytoplasmic viral proteins are largely presented to T cells in association with class I MHC molecules, whereas protein antigens in phagolysosomes are largely presented to T cells in association with class II MHC molecules (see Chapter 6). As we shall discuss later, both kinds of DTH responses may be complemented by CTL responses, predominantly involving CD8+ T cells.

In clinical practice, loss of DTH responses to universally encountered antigens (e.g., candidal antigens) is an indication of deficient T cell function, a condition known as **anergy.** (This general loss of immune responsiveness should be distinguished from "clonal anergy," a mechanism for maintaining tolerance to specific antigens, discussed in Chapter 10.) Anergic individuals are extremely susceptible to infection by microorganisms that are normally resisted by cell-mediated immunity, such as mycobacteria and fungi.

DTH reactions consist of three sequential processes.

1. The **recognition/activation phase,** in which T_H1 CD4+ and sometimes CD8+ T cells recognize foreign protein antigens presented on the surface of APCs and respond by producing cytokines. In experimental DTH, this is a memory or recall response to a previously encountered antigen. The recognition of the antigen occurs in the peripheral tissue where memory and effector T cells are recruited (see Chapter 11).

2. **Inflammation,** in which vascular endothelial cells, activated by cytokines or contact with activated T cells, recruit circulating leukocytes into the tissues at the local site of antigen challenge.

3. **Resolution,** in which macrophages activated by cytokines act to eliminate the foreign antigen. This process may be accompanied by tissue injury.

These stages will each be discussed in greater detail.

The Recognition of Antigen and Activation of T Cells in Delayed Type Hypersensitivity

Experimentally, DTH reactions are produced in two separate steps (see Fig. 13–2). In the first, or "sensitization," step, foreign antigen is presented to naive T cells, resulting in antigen-specific CD4+ T cell activation, expansion, and differentiation. *Specialized resident APCs, such as Langerhans cells in the epidermis,* carry antigen from the portal of

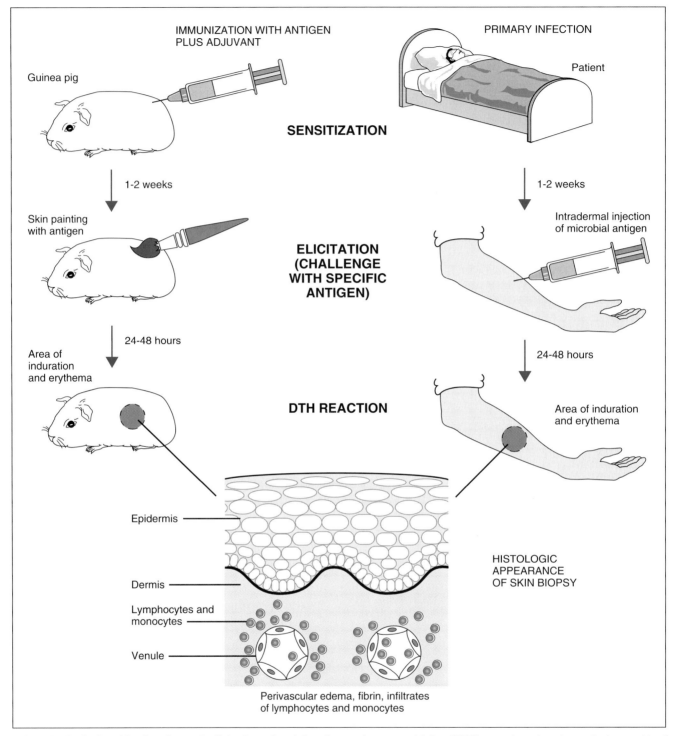

FIGURE 13-2. Sensitization for and elicitation of a delayed type hypersensitivity (DTH) reaction. A guinea pig is sensitized experimentally by injection of antigen or by skin painting with antigen (not shown); in humans, sensitization occurs through infection or by vaccination (not shown). In these two species, subsequent challenge of sensitized individuals with specific antigens then elicits histologically similar DTH reactions.

entry (e.g., the skin) to the draining lymph nodes or the spleen, where contact with antigen-specific naive T cells is more likely to occur (see Chapter 11). The role of Langerhans cells in the initial sensitization step is supported by several lines of evidence. First, agents that deplete or inhibit the egress of Langerhans cells from the epidermis reduce sensitization. Second, fluorescently labeled antigen-bearing Langerhans cells have been observed in afferent lymphatics and in draining lymph nodes within 6 to 12 hours after introduction of labeled antigen into the skin. Finally,

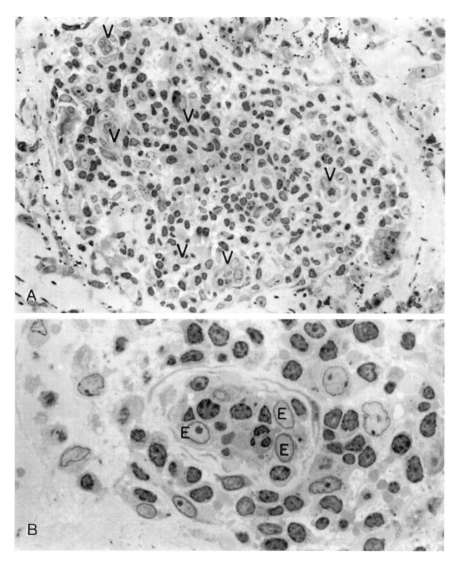

FIGURE 13-3. Morphology of a DTH reaction.

A. Low-power photomicrograph depicting mononuclear cell infiltrates surrounding venules (V) in human skin in a reaction to foreign antigen. (Reproduced with permission from Dvorak, H. F., and M. C. Mihm, Jr. Basophilic leukocytes in allergic contact dermatitis. Journal of Experimental Medicine 135:235–254, 1972. Copyright permission of the Rockefeller University Press, New York.)

B. High-power photomicrograph showing morphologically altered venular endothelium (E) in the midst of the inflammatory infiltrates. (Reproduced with permission from Dvorak, H. F., S. J. Galli, and A. M. Dvorak. Expression of cell mediated hypersensitivity *in vivo*: recent advances. International Review of Experimental Pathology 21:119–194, 1980.)

Langerhans cells cultured *in vitro* differentiate into potent APCs with high expression levels of MHC molecules and costimulator molecules, i.e., the *in vitro* counterpart of dendritic cells.

After naive T cells encounter antigen, they enter into the cell cycle, expand in number, and differentiate into effector cells. As the level of antigen subsides, some activated T cells differentiate into memory cells. Both effector and memory T cells exit from the secondary lymphoid organs and circulate in the blood. An immunized individual or experimental animal will have as many as 1 in 10,000 memory T cells in the circulation that are specific for the sensitizing antigen. In contrast, an unsensitized individual or animal will have as few as 1 in 1 million naive T cells specific for the same agent. The expansion of memory cells serves to enhance the likelihood that any specific antigen will be rapidly recognized upon its reappearance at a peripheral site.

The DTH reaction is initiated by challenging the sensitized individual or experimental animal with the specific antigen. This "elicitation" step is initiated by antigen recognition, in this case by memory CD4+ T_H1 cells (or sometimes CD8+ T cells) present in the circulation. The cells that present antigen to memory T cells and initiate the elicitation phase are not known with certainty. It is clear that a memory response to antigen can be very rapid (onset in 1 hour or less) and can occur directly in the peripheral tissue. (Secondary lymphoid organs may be involved later in the memory response, once antigen reaches these sites.) The two best candidates for presenting antigen in peripheral sites are macrophages resident in extravascular tissues and the vascular endothelial cells that line the capillaries and the post-capillary venules. Both of these cells express class I and class II MHC molecules and costimulator molecules. If macrophages are responsible for antigen presentation, then the specific memory cells must enter into the tissue site where antigen has been introduced. This could occur by chance, or the macrophages, in response to a non-specific irritant effect of the antigen, may secrete cytokines that produce inflammation and thus induce antigen-independent recruitment of T cells (see below). If antigen is presented by endothelial cells, recognition could occur in the vascular lumen, increasing the likelihood of early activation of T cells.

The Development of Inflammation in Delayed Hypersensitivity

Once memory T cells recognize their specific antigen at a peripheral site of antigen challenge, they alter the local microvascular endothelial cells in a manner that initiates recruitment of circulating leukocytes into the inflammatory reaction. Under the influence of TNF and other cytokines secreted by antigen-activated T cells or through contact-dependent signals such as engagement of endothelial cell CD40 by activated T cell CD40 ligand, *microvascular endothelial cells perform four functions that contribute to inflammation* (Fig. 13–4):

1. By production of vasodilator substances such as prostacyclin (PGI$_2$) and nitric oxide (NO), *endothelial cells cause increased local blood flow and optimize delivery of leukocytes at the site of inflammation.* TNF increases the expression and activity of enzymes in endothelial cells that synthesize prostacyclin and, in combination with IFN-γ, that synthesize NO. These changes enable endothelial cells to make greater quantities of vasodila-

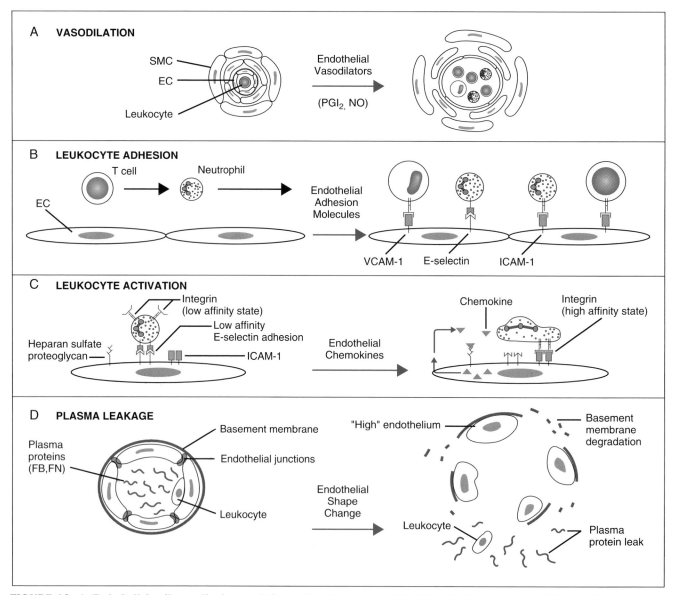

FIGURE 13–4. Endothelial cell contributions to inflammation. Vascular endothelial cells, in response to TNF and other cytokines, produce vasodilators (PGI$_2$ and NO) that increase leukocyte delivery to tissues *(A)*; express adhesion molecules (E-selectin, VCAM-1, ICAM-1) that bind leukocytes *(B)*; synthesize and express chemokines (IL-8, MCP-1, etc.) that activate leukocytes, increasing integrin affinity and cell motility *(C)*; and allow plasma proteins (fibrinogen, fibronectin) to leak into the tissues, forming a scaffolding for the migration of leukocytes *(D)*. EC, endothelial cell; FB, fibrinogen; FN, fibronectin; SMC, smooth muscle cell.

tors in response to vasoactive substances such as histamine. (The source of the histamine is likely to be perivascular mast cells [see Chapter 14], which can be activated to secrete histamine by T cell–derived chemokines or other less-well characterized "histamine-releasing factors.")

2. By expression of new or increased levels of certain surface proteins, *post-capillary venular endothelial cells become adhesive for leukocytes.* In this adhesive state, a random encounter of a circulating leukocyte with a venular endothelial cell will result in an increase in the residence time of leukocytes on the venular surface. As has been shown for lymphocyte diapedesis across high endothelial venules in organized lymphoid tissues (see Chapter 11), such increased residence time can serve to increase the likelihood of extravasation. Several leukocyte adhesion molecules on vascular endothelium are induced in peripheral tissues in response to cytokines or T cell contact-dependent signals. Three such endothelial molecules are well characterized and have been demonstrated to be important in the development of antigen-induced inflammation in DTH (Figs. 13–5 and 13–6). **E-selectin** (also called endothelial-leukocyte adhesion molecule-1 or ELAM-1) is the first molecule to appear (onset in 1 to 2 hours). E-selectin is structurally related to the peripheral lymph node homing receptor on T cells, L-selectin (see Chapter 11). On venules of peripheral tissues, E-selectin mediates

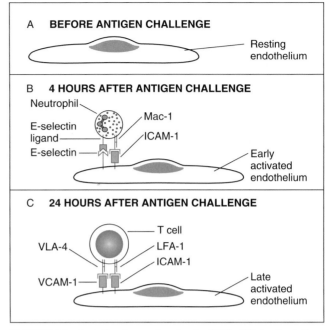

FIGURE 13–6. Adhesion molecules in DTH. *(A)* As the inflammatory reaction develops, different endothelial ligands are expressed that interact with different leukocytes. *(B)* At 4 hours, E-selectin interacts with its neutrophil ligand and ICAM-1 interacts principally with Mac-1 (CD11bCD18). Neutrophil LFA-1 may also interact with ICAM-1 or (not shown) ICAM-2. *(C)* By 24 hours, T cells interact with VCAM-1 (via VLA-4) and ICAM-1 (via LFA-1). E-selectin may play a lesser role. L-selectin, expressed on both neutrophils and T cells, may also mediate adhesion (not shown).

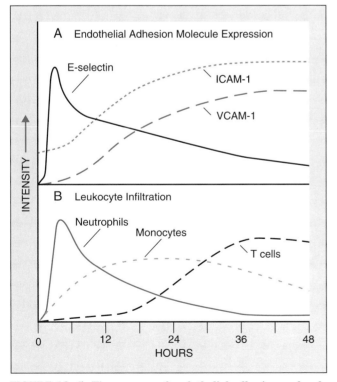

FIGURE 13–5. Time course of endothelial adhesion molecule expression and leukocyte infiltration following antigen challenge in the skin of a sensitized individual. Similar responses can be elicited with intradermal injection of TNF.

the initial attachment of neutrophils. In areas of flowing blood, E-selectin–mediated interactions with neutrophils may result in rolling of the neutrophil on the endothelial surface. E-selectin may also contribute to the initial binding of other inflammatory cell types, including CD4+ T_H1 cells, but this is less clear. Several functions of E-selectin may be shared by P-selectin. **Vascular cell adhesion molecule-1** (VCAM-1) appears at slightly later times than E-selectin (onset in 6 to 12 hours) and mediates the initial attachment of memory T cells and other leukocytes expressing the VLA-4 integrin molecule. **Intercellular adhesion molecule-1** (ICAM-1 or CD54) is induced with a time course similar to that of VCAM-1. We previously introduced ICAM-1 as an accessory molecule on APCs (see Chapter 7). On venular endothelial cells, ICAM-1 mediates firm adhesion of activated leukocytes (see below) and also plays a role in transmigration of leukocytes across the blood vessel into the tissue. T cells interact with ICAM-1 through LFA-1, whereas neutrophils may also engage ICAM-1 through Mac-1. Endothelial cells constitutively express a second ligand for LFA-1, called ICAM-2, that is not recognized by Mac-1. Functional differences between ICAM-1 and ICAM-2 in T cell recruitment are unclear. Leukocyte L-selectin may also contribute to leukocyte recruitment, but its ligand on endothelial cells of peripheral tissues is not known. *The early upregulation of E-selectin, followed at later times by VCAM-1 on the endothelial cell surface, leads to sequential adhesion, first of neutrophils (at 6 to 12 hours) and then of lymphocytes and monocytes (at 6 to 12 hours), following antigen challenge.*

3. *Antigen-activated T cells cause endothelial cells to secrete chemokines such as interleukin-8 (IL-8) and monocyte chemotactic protein-1 (MCP-1).* The secreted chemokines bind to endothelial cell surface heparan sulfate glycosaminoglycans, where they preferentially interact with leukocytes that are bound to endothelial cell adhesion molecules. The chemokines act on the leukocytes to promote extravasation.

4. *Cytokines or contact-dependent signals from activated T cells cause endothelial cells to undergo shape changes and basement membrane remodeling that favor leakage of macromolecules and extravasation of cells.* By 24 hours, endothelial cells reor-

organize their adhesion molecules to concentrate them at the intercellular junctions. Another Ig superfamily adhesion molecule, called **platelet-endothelial adhesion molecule-1** (PECAM-1, or CD31), is also localized to the junctions and plays a role in facilitating leukocyte emigration across the endothelial cell lining. Leakage of plasma macromolecules, especially fibrinogen, is the basis of induration. The deposition of fibrinogen (and its insoluble cleavage product, fibrin), as well as plasma fibronectin in the tissues, forms a scaffolding that facilitates leukocyte migration and subsequent retention in extravascular tissues.

The interplay of adhesion molecules and chemokines has led to the multistep model of leukocyte recruitment (Fig. 13–7). This model, which has been supported by videomicroscopy experiments *in vitro* and *in vivo*, involves:

1. The initial loose adhesion **(tethering)** of leukocytes to cytokine- or contact-activated endothe-

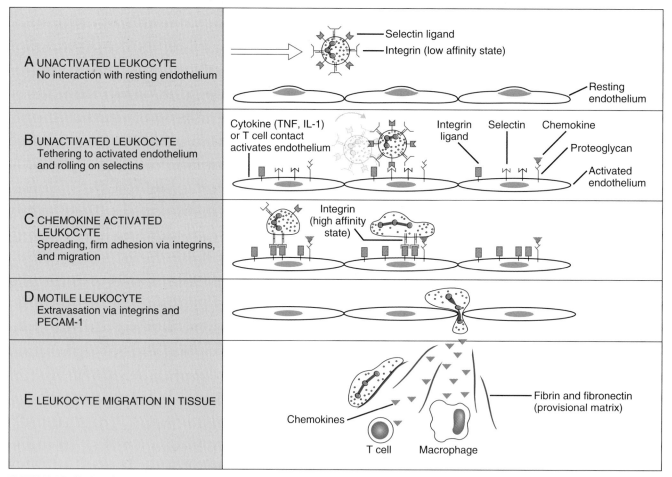

A UNACTIVATED LEUKOCYTE No interaction with resting endothelium	Selectin ligand Integrin (low affinity state) Resting endothelium
B UNACTIVATED LEUKOCYTE Tethering to activated endothelium and rolling on selectins	Cytokine (TNF, IL-1) or T cell contact activates endothelium Integrin ligand Selectin Chemokine Proteoglycan Activated endothelium
C CHEMOKINE ACTIVATED LEUKOCYTE Spreading, firm adhesion via integrins, and migration	Integrin (high affinity state)
D MOTILE LEUKOCYTE Extravasation via integrins and PECAM-1	
E LEUKOCYTE MIGRATION IN TISSUE	Fibrin and fibronectin (provisional matrix) Chemokines T cell Macrophage

FIGURE 13–7. Multistep recruitment of leukocytes. Circulating blood leukocytes do not interact with unactivated, resting endothelium *(A)*. However, when the endothelium is activated by cytokines or cell contact (e.g., via CD40), these cells express selectins, chemokines, and ligands for integrins *(B)*. Leukocytes bearing selectin ligands will tether and roll on the activated endothelium until encountering a chemokine, which induces the leukocytes to spread, become firmly adherent via their integrins, and begin to crawl along the endothelial surface *(C)*. When the leukocyte reaches the interendothelial cell junction, it extravasates through the endothelial layer in a process involving PECAM-1 *(D)*. Once in the tissue, the leukocyte may migrate, using its integrins to interact with extravasated plasma proteins such as fibrinogen and fibronectin, and possibly responding to gradients of chemokines produced by activated T cells and macrophages *(E)*.

lium via selectins (E-selectin or P-selectin on the endothelial surface or L-selectin on the leukocyte). In the microvasculature, the force of flowing blood pushes the tethered leukocytes, causing disruption of selectin-ligand interactions, which are rapidly re-formed downstream as the leukocyte begins to move. The net result of these events is **rolling** of the spherical leukocyte along the endothelial surface. The initial tethering that starts the rolling process may involve different adhesion molecules than active rolling. Moreover, the precise adhesion molecules involved in tethering and rolling will depend on which selectins and selectin ligands are expressed. In the case of T cells and eosinophils, selectin-independent tethering and rolling may occur, in this case mediated by VLA-4 engagement of VCAM-1. In order to participate in rolling, the leukocyte adhesion molecules must be concentrated on the tips of microvillous projections. This is because unactivated spherical leukocytes actually roll only on the tips of their microvilli and engage in only limited membrane contact with the endothelium.

2. As the leukocyte is rolling on an activated endothelial surface, the leukocyte becomes **activated.** The signal that causes leukocyte activation is probably a C-X-C chemokine for neutrophils or a C-C chemokine for other leukocytes. These chemokines are bound to glycosaminoglycan groups of endothelial cell proteoglycans, and thus they are displayed on the surface of the vessel lumen where they can interact with rolling leukocytes. Such chemokines may be synthesized by the activated endothelial cells, or they may be generated by leukocytes already in the tissue. The endothelial cells may also deliver contact-dependent activation signals to the rolling leukocytes, involving adhesion molecules or other membrane proteins, that may supplement or replace chemokine signals. Activated leukocytes rearrange their cytoskeletons and spread from spherical to a flattened shape, increase the affinity of their integrins for endothelial ligands, and become motile.

3. Using their high affinity integrins, especially LFA-1, Mac-1 (on neutrophils and monocytes), and VLA-4 (on leukocytes other than neutrophils), activated leukocytes **become firmly adherent** to the endothelial cell surface. Leukocytes exhibiting firm adherence do not roll and, compared with rolling leukocytes, appear fixed in place. In fact, such leukocytes are probably slowly migrating along the endothelial luminal surface until they reach an interendothelial cell junction.

4. At interendothelial cell junctions, leukocytes receive additional signals, perhaps via PECAM-1, that trigger **transmigration** through the junctions. This is the least understood aspect of the recruitment process.

5. Once in the tissue, leukocytes migrate, using their integrins to crawl along the "provisional" fibrin or fibronectin matrix that is formed from extravasated plasma proteins. Extravascular migra-tion may be influenced by gradients of chemokines that form within the tissue.

After leukocytes enter the tissues, they may die in a few days (especially characteristic of neutrophils), may become activated (especially characteristic of T cells and monocytes), or may leave, probably through lymphatic vessels. The molecules controlling these processes are not yet defined.

T cell recruitment into DTH reactions occurs in two phases. As we previously described, DTH is initiated by presentation of the eliciting antigen by macrophages or microvascular endothelial cells to circulating memory T cells. This event occurs very early (within the first hour of challenge with antigen) and results in the recruitment of a very small number of T cells. Indeed, based upon adoptive transfer experiments, the number of antigen-specific T cells that need to be activated are far too few to be detected by morphologic analyses. These antigen-activated T cells, either through cell contact or cytokine release, alter the local venular endothelium to recruit more T cells. The second phase of T cell recruitment starts at 6 to 12 hours after challenge and is antigen independent. The second phase of the response results in the recruitment of many more T cells than those specific for the eliciting antigen and is mediated by adhesion molecules and chemokines. Once in the tissues, antigen-specific memory T cells that have been recruited by antigen-independent signals may now recognize antigen presented by resident macrophages. *Antigen-activated T cells secrete several cytokines that are important for DTH reactions* (Fig. 13–8).

1. **IL-2** acts as an autocrine and paracrine growth factor and is responsible for the expansion of those T cells specific for the antigen. IL-2 also augments cytokine production by T cells, serving to amplify the response.

2. **TNF** and **LT** sustain the activated state of the venular endothelial cells, continuing to recruit new effector cells into the ongoing reaction. IFN-γ plays a secondary role in this action.

3. **IFN-γ** plays a primary role in boosting the capacities of APCs, including macrophages, endothelial cells, and possibly other cell types, to present antigens to CD4$^+$ T cells by increasing the expression of class II MHC molecules and costimulator molecules, especially B7 family molecules, on macrophages. IFN-γ also stimulates macrophages to secrete cytokines, including IL-12, which positively feed back on the T cells to increase IFN-γ production. Finally, IFN-γ plays a crucial role as an activator of macrophage effector functions, which we discuss in the next section.

T cells that are stimulated by antigen or chemokines may be retained at the site of inflammation because T cell activation increases the expression of receptors for extracellular matrix

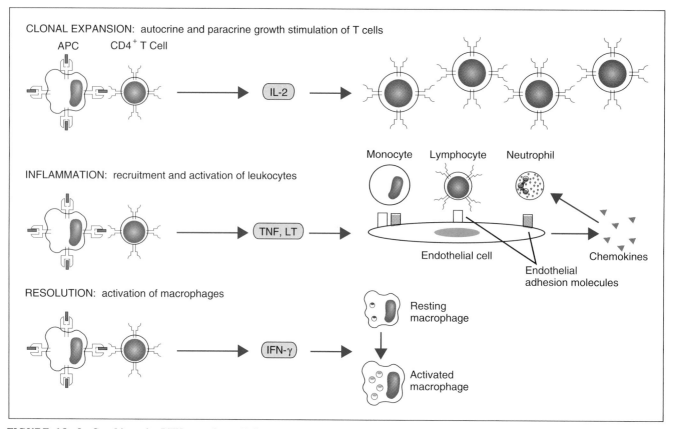

FIGURE 13-8. Cytokines in DTH reactions. Different cytokines produced by antigen-activated CD4+ T$_H$1 or (not shown) CD8+ T cells contribute to clonal T cell expansion, endothelial activation, and macrophage activation. Other cytokines that are not shown may play auxiliary roles in these processes.

molecules. The members of the β1 integrin family (see Chapter 7, Box 7-3) may be particularly important for interaction with matrix. VLA-4 and VLA-5 allow leukocytes to bind to fibronectin, and VLA-6 mediates attachment to laminin. Chronically activated T cells express additional integrin receptors (e.g., VLA-1 and VLA-2) that mediate attachment to collagen. As a result of proliferation and selective retention, the number of T cells specific for the eliciting antigen may rise from 1 in 10,000 in the blood to over 1 in 100 in the inflamed tissue. Even at this stage, however, most of the T cells in the tissue are not specific for the antigen.

Activated Macrophages and the Resolution of Delayed Type Hypersensitivity

Once blood monocytes leave the circulation and enter the extravascular tissues at sites of DTH reactions, they differentiate into macrophages, which are the ultimate effector cells of these reactions. The macrophages function to eliminate microorganisms and other sources of antigen, leading to termination **(resolution)** of the DTH reaction. The differentiation of monocytes into effector cells is called **macrophage activation.**

The activation of macrophages is not a single process. A macrophage is considered to be acti-

vated if it performs a function measured in a specific assay, e.g., killing a microbe. In other assays, e.g., initiating blood coagulation or killing a tumor cell, the same macrophage may appear to be unactivated. Because such functional assays are biologically complex, it is preferable to divide the process of activation into more simple components. The unstimulated blood monocyte is usually considered to be at rest. *Activation consists of quantitative alterations in the expression of various gene products (proteins) that endow the activated macrophage with the capacity to perform some function that cannot be performed by the resting monocyte.* In general, macrophage activation results from new or increased gene transcription. For example, activation of human macrophages may consist of increasing expression of a cytochrome enzyme that catalyzes the generation of reactive oxygen species. As a consequence of increased enzyme expression, the activated macrophage may be able to perform a function, e.g., killing of phagocytosed bacteria, that cannot be performed by the resting monocyte. The agents that cause gene transcription and thus macrophage activation are soluble cytokines, bacterial products (e.g., lipopolysaccharide [LPS]), and extracellular matrix molecules. *The most potent macrophage-activating cytokine is IFN-γ.* However, IFN-γ is not the only cytokine that can activate macrophages, and IFN-γ does not activate all possible capacities of the macrophage. Macrophages may also be activated by T cell contact through CD40, similar to the activation of B cells (see Chapter 9). Some specific examples of the functions of activated macrophages follow:

1. *Activated macrophages kill microorganisms.* Killing of bacteria by macrophages involves phagocytosis and generation of reactive oxygen species. Cytokines such as IFN-γ augment both endocytosis and phagocytosis by monocytes. Phagocytosis of specific particles can be further enhanced by opsonizing bacteria, that is, coating the bacteria with specific IgG molecules or complement components (see Chapters 3 and 15). IFN-γ causes macrophages to increase expression of high affinity receptors for the Fc portion of IgG (FcγRI, see Chapter 3, Box 3–4), promoting uptake of opsonized bacteria. Once in the cell, macrophages kill bacteria by generating reactive oxygen species, and, as noted above, IFN-γ induces transcription of the gene encoding the enzyme that generates active oxygen. *IFN-γ is thus sufficient to fully activate macrophages for killing of microorganisms.*

Mouse macrophages, but probably not human macrophages, have a second important inducible microbicidal mechanism. Upon treatment of mouse macrophages with IFN-γ in combination with LPS, TNF, or IL-1, these cells express a high-output nitric oxide synthase that catalyzes the production of nitric oxide (NO). NO made in large quantities by this enzyme can contribute to bacterial killing. NO also inhibits viral replication. The interactions of NO with reactive oxygen species are complex. Sometimes, these molecules inactivate each other but, under other circumstances, they may combine to generate highly reactive peroxynitrate radicals.

2. *Activated macrophages stimulate acute inflammation, often through secretion of short-lived inflammatory mediators.* Many of these mediators, such as platelet-activating factor (PAF), prostaglandins, and leukotrienes, are lipids. Some are synthesized by macrophages themselves, and others are generated from plasma molecules in response to enzymes and related molecules secreted by the macrophages. For example, macrophages produce a protein called tissue factor, which can initiate the extrinsic clotting cascade; thrombin, a blood protease activated during the clotting cascade, causes neutrophils and endothelial cells to synthesize PAF. Treatment with cytokines, such as IFN-γ, enhances the biosynthetic capacities of the macrophage to generate mediators such as tissue factor. The collective action of these macrophage-derived mediators is to produce local inflammation. The inflammatory reaction that results is rich in neutrophils and serves to contain and destroy infectious organisms and to get rid of injured tissue. Thus, the activated macrophage acts as an endogenous surgeon to débride the wound, leading to

elimination of antigen and resolution of the DTH reaction.

3. *Activated macrophages become more efficient APCs.* A significant part of the enhanced antigen-presenting capacity may be attributed to increased surface expression of class II MHC molecules. IFN-γ is the best known activator of the transcription of class II MHC genes, but in macrophages, granulocyte-macrophage colony-stimulating factor (GM-CSF) may also have some effect. Costimulatory functions are also enhanced in activated macrophages. Specifically, activated macrophages express B7 family molecules and increased levels of ICAM-1 and LFA-3 (see Chapter 7). Finally, activated macrophages produce cytokines, such as IL-12 or IFN-α, that stimulate lymphocyte differentiation.

4. If the inflammatory process and activated macrophages fail to eradicate the microbe, *activated macrophage products, such as cytokines and growth factors, progressively modify the local tissue environment, initially leading to destruction of tissue and later, i.e., in chronic DTH reactions, causing replacement by connective tissue.* IFN-γ facilitates synthesis of cytokines by macrophages but is not sufficient to completely activate these functions. Often, a microorganism can supply a necessary second signal for cytokine synthesis, such as LPS from a bacterial cell wall. Other T cell–derived cytokines or contact-dependent signals can substitute for LPS and work with IFN-γ to stimulate macrophages to secrete cytokines. The effects of macrophage-derived cytokines and growth factors occur in two phases. Acutely, TNF, IL-1, and macrophage-derived chemokines augment inflammatory reactions initiated by T cells. Cytokines act in concert with the inflammatory mediators described above to recruit and activate neutrophils and monocytes, causing local tissue destruction. Chronically, these same cytokines also stimulate fibroblast proliferation and collagen production. These slow actions of cytokines are augmented by the actions of macrophage-derived polypeptide growth factors. Platelet-derived growth factor, produced by activated macrophages, is a potent stimulator of fibroblast proliferation, whereas macrophage-derived transforming growth factor-β (TGF-β) augments collagen synthesis. Macrophage secretion of fibroblast growth factor causes endothelial cell migration and proliferation, leading to new blood vessel formation **(angiogenesis).** The consequence of these slow actions of cytokines and growth factors is that prolonged activation of macrophages in a tissue, e.g., in the setting of chronic antigenic stimulation, leads to replacement of differentiated tissues by fibrous tissue, a process called **fibrosis** (or scarring). *Fibrosis is the outcome of chronic DTH, when elimination of antigen and rapid resolution are unsuccessful.*

5. *Activated macrophages kill tumor cells.* Although the activated macrophage is usually thought of as an effector of host defense against infectious organisms, tumor biologists have ob-

served that activated macrophages also selectively kill malignant cells. This phenomenon is discussed in greater detail in Chapter 18. Suffice it to say here that much of the anti-tumor effect of activated macrophages may be attributed to the ability of TNF, produced by these cells, to cause tumor cell death. A second anti-tumor mechanism of murine (but probably not human) macrophages is the production of NO; this response was discussed earlier as an antimicrobial mechanism.

In chronic DTH reactions—i.e., when the source of antigen is not eradicated—activated macrophages themselves undergo changes in response to persistent cytokine signals. Such macrophages develop increased cytoplasm and cytoplasmic organelles. In standard histologic sections stained with hematoxylin and eosin, chronically activated macrophages may resemble skin epithelial cells and have therefore been called "epithelioid." Sometimes, these macrophages can fuse to form multinucleate giant cells. Individual clusters of activated macrophages, often focused about particulate sources of antigen such as *Mycobacterium*, produce palpable nodules of inflammatory tissue called **granulomas** (i.e., granular masses) (Fig. 13–9). Granulomatous inflammation is a characteristic response to some persistent microbes, such as *Mycobacterium tuberculosis*, and simply represents a form of chronic DTH. Experimentally, granulomas can be elicited by attaching soluble protein antigens to undigestible latex beads so they are particulate and persistent in tissues. Granulomatous inflammation is frequently associated with tissue fibrosis. Although fibrosis is a "healing reaction" to injury, it can also interfere with normal tissue function. Much of the respiratory difficulty associated with tuberculosis or fungal infections of the lung is caused by replacement of normal lung with scar tissue and not directly attributable to the bacteria (see Chapter 16).

The pathologic consequences of DTH raise a more general point that was alluded to earlier in the chapter. All the various effector mechanisms of the immune system have evolved as mechanisms of host defense, yet all can cause "hypersensitivity," i.e., injurious reactions. Activated macrophages contribute to T cell–mediated immunity by containing and eradicating infectious organisms, such as mycobacteria or fungi, that can survive within phagocytes. Although DTH reactions are targeted by T cells to be selective for the invading microorganisms, they are still injurious to normal host tissues. When DTH is elicited experimentally with a non-injurious agent, such as a protein or heat-killed bacteria, one sees only the destructive aspects and not the protective functions. Similarly, when the protective functions are ineffective or incompletely effective, the reaction persists and the destructive consequences can accumulate, as in granulomatous diseases.

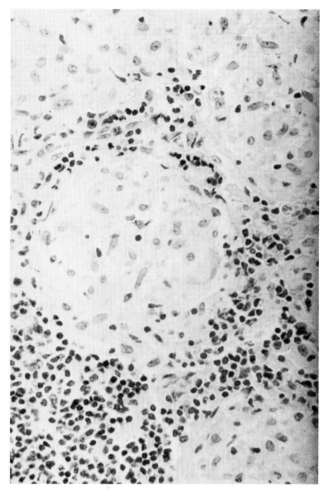

FIGURE 13–9. Granulomatous inflammation. Granulomas in human lung, showing T cells surrounding nodular collections of activated macrophages. (Courtesy of Dr. Carol Farver, Department of Pathology, Brigham and Women's Hospital, Boston.)

CYTOLYTIC T LYMPHOCYTES

Cytolytic T lymphocytes are a subset of T cells that kill target cells expressing MHC-associated peptide antigens. CTLs appear to be important effector cells in three settings: intracellular infections of non-phagocytic cells or infections that are not completely contained by phagocytosis, such as viral infections or infection by bacteria such as *Listeria monocytogenes;* acute allograft rejection; and rejection of tumors. Each of these processes is discussed in more detail in Section IV of the book. In this chapter, we focus upon the characteristics of CTLs and their mechanisms of action.

The majority of CTLs express the CD8 molecule and specifically recognize foreign peptides derived from antigens degraded in the cytosol and then expressed on the cell surface associated with self class I MHC molecules. Rare CTLs express the CD4 molecule and recognize peptides associated with class II molecules. As discussed in Chapters 7 and 8, the T cell antigen receptor genes utilized by

CD8+ CTLs are indistinguishable from those utilized by CD4+ helper T cells. The preference of CD8+ T cells for class I MHC molecules is instead related to the direct binding of the CD8 molecule to non-polymorphic regions of the class I MHC molecule. Like helper T cells, CTLs undergo maturation and selection in the thymus.

Development and Differentiation of Functional CTLs

CTLs are not fully differentiated when they exit from the thymus. Although they express functional αβ T cell receptor molecules and recognize antigen, they cannot lyse target cells. Indeed, very few, if any, functional CTLs specific for an allograft can be detected in the blood of a potential allograft recipient prior to transplantation (see Chapter 17). However, if T lymphocytes are cultured with leukocytes from the donor of an allograft, i.e., in a mixed leukocyte reaction (see Chapter 17), graft-specific CD8+ CTLs can be detected in the culture after 7 to 10 days. Similarly, if T cells from a virus-infected individual are stimulated with virus-infected syngeneic cells *in vitro*, virus-specific self MHC-restricted CD8+ CTLs are detected within 5 to 10 days. The appearance of functional CTLs depends upon a process of differentiation.

The principal features of CTL differentiation are the following:

1. *CTLs develop or differentiate from "pre-CTLs."* Pre-CTLs are T cells that are committed to the CTL lineage, have undergone thymic maturation, and are already specific for a particular foreign antigen. These cells express CD3-associated αβ T cell receptors (TCRs) and CD8, but they lack cytolytic function. The often-repeated assertion that most CD8+ T cells are CTLs is more accurately restated as most CD8+ T cells are pre-CTLs.

2. *Pre-CTLs do not require a special microenvironment for differentiation and can develop within the infected or foreign tissue.* Pre-CTLs are normally present at low frequency in the blood and peripheral lymphoid tissues and can be detected by stimulating differentiation *in vitro* before assessing cytolytic function.

3. *Differentiation of pre-CTLs to functional CTLs requires at least two separate kinds of signals: the first is specific recognition of antigen on a target cell, and the second signal may be provided either by costimulators expressed on professional APCs or by cytokines produced by helper T cells* (Fig. 13–10). There are two currently unresolved issues about the cytokine requirement for CTL differentiation. First, the precise cytokines required are not known. It is likely that IL-2 and IFN-γ are important, and IL-12, produced by APCs, also may contribute to CTL differentiation. Second, the identity of the T cells responsible for providing these cytokines is not well established. In some settings,

there may be a requirement for CD4+ T cells (presumably of the T_H1 subset) that respond to peptide antigens presented by class II+ APCs; in other models, CD8+ T cells, stimulated by peptide antigens presented by professional APCs in association with class I MHC molecules, are sufficient to provide the requisite cytokines. CD8+ CTLs that arise independent of CD4+ T cells, whether through direct interaction with costimulators or through CD8+ T cell–derived cytokines, have been referred to as "helper-independent" CTLs. Finally, it should be noted that although T_H1 cytokines promote CTL development, it is not clear that T_H2 cytokines are inhibitory, and they may even contribute to CTL differentiation.

4. *Differentiation from pre-CTLs involves the acquisition of the machinery to perform cell lysis.* Several changes occur during this process. First, CTLs develop specific membrane-bound cytoplasmic granules. These granules contain several macromolecules, including a membrane pore–forming protein called **perforin** or **cytolysin**; enzymes commonly called **granzymes** that contain reactive serines in their active site; and proteoglycans. Second, CTLs share with other activated T cells surface expression of Fas ligand, which can deliver apoptosis-inducing signals to target cells expressing Fas protein (see Chapter 10, Box 10–1). And third, CTLs share with other effector T cells the capacity to transcribe and secrete cytokines and other proteins upon activation, mostly IFN-γ, LT, TNF, and, to a lesser degree, IL-2. As discussed below, the acquisition of perforin and granzymes is most critical for CTL-mediated lysis of infected target cells, and these are the most specific features of CTL differentiation.

Mechanisms of CTL-Mediated Lysis

There are several key features of CTL-mediated lysis.

1. *CTL killing is antigen specific.* Only target cells that bear the same class I MHC-associated antigen that triggered pre-CTL differentiation can be killed by an individual CTL.

2. *CTL killing requires cell contact.* This requirement arises from the need of CTLs to be triggered by recognition of a target antigen associated with a cell surface MHC molecule. CTLs kill only those cells to which they attach, and bystander cells are not injured. This is because the lytic mechanisms of CTLs are directed toward the point of contact of the TCR molecule with antigen on the surface of the target cell.

3. *CTLs themselves are not injured during lysis of target cells.* Moreover, each individual CTL is capable of sequentially killing multiple target cells. The mechanisms that protect CTLs from lysis are not fully understood, as will be discussed below.

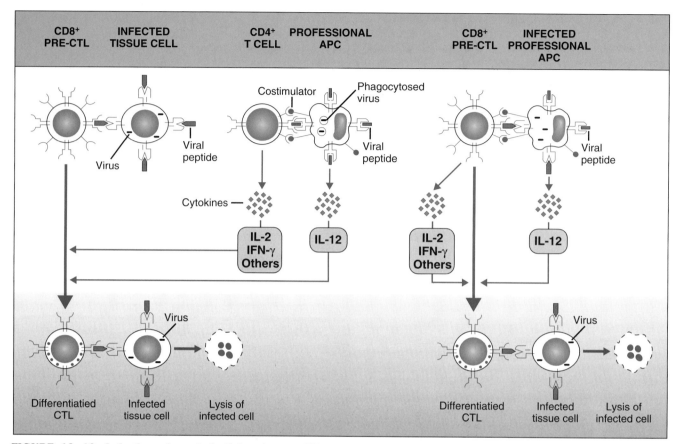

FIGURE 13-10. Induction of cytolytic T lymphocyte differentiation. CD8$^+$ T cells may encounter a foreign peptide antigen complexed with class I MHC molecules on the surface of a virally infected tissue cell that lacks costimulators. If this occurs, differentiation into a cytolytic T lymphocyte (CTL) will require additional (second) signals in the form of cytokines provided by other lymphocytes, most likely a CD4$^+$ T$_H$1 cell, recognizing a different foreign peptide antigen complexed with class II MHC molecules (presumably derived from phagocytosed viruses) on the surface of a professional antigen-presenting cell (APC) that expresses costimulators. Alternatively, CD8$^+$ T cells may encounter the foreign peptide antigen on the surface of an infected professional APC. In this case, the CD8$^+$ T cell can receive the necessary additional signals for differentiation into a CTL either directly from the costimulators expressed by the APC, or in response to cytokines whose secretion is enhanced by the costimulators expressed by the professional APC.

The process of CTL-mediated lysis consists of five steps (Fig. 13-11):

1. *Recognition of antigen and conjugate formation.* The CTL binds to the target cell, using its specific antigen receptor and other accessory molecules such as CD8, CD2, and LFA-1. Target cell recognition, therefore, involves class I MHC molecules (the ligand for CD8) complexed to specific peptide (the complex serving as the ligand for the TCR), LFA-3 (the ligand for CD2), and ICAM-1 or ICAM-2 (the ligands for LFA-1). It may be that transient conjugate formation can occur via CD2- and LFA-1–mediated adhesion in the absence of or prior to specific antigen recognition, or that antigen recognition may enhance the ability to form conjugates by augmenting the binding function of the adhesion molecules. CTL-mediated killing of target cells that do not express ICAM-1 or (in humans) LFA-3 is very inefficient.

2. *Activation of the CTL.* The CTL is activated by clustering of its antigen receptor, initiated by recognition of MHC molecule–peptide complexes on the target cell. The intracellular signals generated by the TCR may be augmented by signals delivered through the various accessory molecules. The process of CTL activation is thus entirely analogous to the activation of helper T cells described in Chapter 7.

3. *Delivery of a "lethal hit" by the activated CTL to its conjugated target* (see below). Studies of CTL-mediated killing of target cells *in vitro* had suggested that granule exocytosis-dependent (i.e., perforin/granzyme-mediated) killing and granule exocytosis-independent (i.e., Fas ligand-mediated) killing were redundant. More recent analyses of mice with mutations in the perforin/granzyme pathway or the Fas pathway have suggested that perforin and granzyme are the key mediators of CTL function in immune responses to intracellular microbes. In contrast, the Fas pathway appears to be more important for immune regulation than for CTL functions, i.e., for controlling excessive lymphocyte activation, especially against self antigens

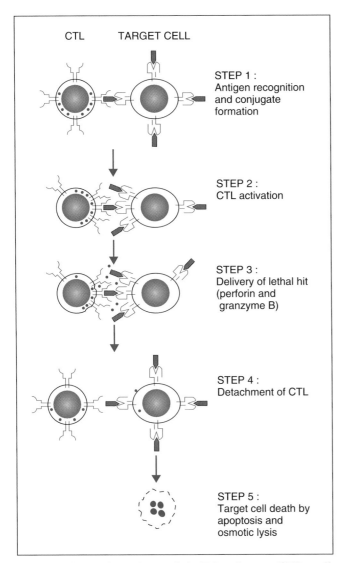

CTL   TARGET CELL

STEP 1 :
Antigen recognition
and conjugate
formation

STEP 2 :
CTL activation

STEP 3 :
Delivery of lethal hit
(perforin and
   granzyme B)

STEP 4 :
Detachment of CTL

STEP 5 :
Target cell death by
apoptosis and
osmotic lysis

FIGURE 13–11. Steps in cytolytic T lymphocyte (CTL)-mediated lysis of target cells. Note that conjugate formation (step 1) also requires interactions between CTL accessory molecules (LFA-1, CD8) on the CTLs and their specific ligands on the target cell; these are not shown (see Figure 7–8, Chapter 7).

(see Chapter 10, Box 10–1). Therefore, in this chapter, we will focus on the roles of perforin and granzyme in CTL-mediated killing.

4. *Release of the CTL.* The CTL is released from its target cell, a process that may be facilitated by decreases in the affinity of accessory molecules for their ligands.

5. *Programmed death of the target cell as a consequence of receiving a lethal hit by a combination of apoptosis and osmotic lysis.*

Delivery of the lethal hit is initiated by clustering of TCR:CD3 molecules and involves reorganization of the cytoskeleton. The microtubule organizing center of the CTL is moved to the area of the cytoplasm near the contact with the target cell, and the perforin/granzyme–containing granules become concentrated in this same region. Fusion of the granule membrane with the plasma membrane

is probably mediated by a Ras-related GTP binding protein of the Rab family, and it results in exocytosis of granule contents onto the membrane of the target cell. As a consequence of granule content exocytosis, perforin, present as a monomer in the granule, comes in contact with extracellular concentrations of calcium (typically 1 to 2 mM) and undergoes polymerization. Polymerization of perforin, leading to formation of a large water-filled channel, preferentially occurs in a lipid bilayer, such as the plasma membrane of the target cell. If a sufficient number of polymerized perforin channels are assembled, the target cell will be unable to exclude ions and water, leading to osmotic swelling and lysis. This method of cell killing is analogous to that produced by the membrane attack complex of complement, and perforin is structurally homologous to the ninth component of complement, the principal constituent of the membrane attack complex (see Chapter 15).

Although purified perforin is sufficient to cause target cell cytolysis by osmotic swelling, it has recently been appreciated that an equally important function of perforin is to provide a means for introducing granzymes into the target cell. Granzyme B is a serine protease that preferentially cleaves protein substrates at aspartic acid residues. Among the target proteins identified for granzyme B are interleukin-1 converting enzyme (ICE) and related cysteine proteases that are themselves aspartic acid–directed. ICE and ICE-like proteases become catalytically activated upon cleavage at susceptible aspartic acid residues, so that by cleaving ICE, granzyme B initiates an ICE protease cascade that is similar to the activation of the cascade induced by ligation of cell surface Fas (see Chapter 10, Box 10–1). The net result is that introduction of granzyme B into the cytoplasm of a target cell, like ligation of Fas, initiates apoptotic cell death. A late step in the apoptotic death pathway is the activation of DNA-cleaving enzymes. These enzymes are not specific for mammalian DNA. *Therefore, by activating an apoptotic death pathway, granzyme B can initiate the destruction of viral DNA genomes as well as the target cell genome, thereby eliminating potentially infectious DNA.*

As noted above, CTLs themselves are not killed during the lytic process. However, CTLs can be killed by other CTLs or by high concentrations of granule contents. This raises the question of why CTLs are not injured during target cell lysis. The answer is probably quantitative: CTLs are relatively resistant to CTL-mediated lysis. The mechanism of this relative resistance is not known; one possibility is that CTLs express high levels of membrane proteins that disassemble the perforin complex.

NATURAL KILLER CELLS

Natural killer (NK) cells are a subset of lymphocytes found in blood and lymphoid tissues, especially spleen. NK cells are derived from the bone marrow and appear as large lymphocytes with numerous cytoplasmic granules, because of which they are sometimes called **large granular lymphocytes** (LGLs). NK cell granules, like the granules of CTLs, contain perforin, granzymes, and proteoglycans. Upon directed granule exocytosis, NK cells, like CTLs, can osmotically lyse target cells and induce apoptotic cell death by perforin/granzyme B pathways. NK cells are best thought of as phylogenetically primitive CTLs that lack the specific TCR for antigen recognition. NK cells possess the ability to kill certain tumor cells, particularly of hematopoietic origin, and normal cells infected by virus.

By surface phenotype and lineage, NK cells are neither T nor B cells. NK cells do not undergo thymic maturation and may be increased in animals that lack a thymus. Moreover, NK cells do not undergo Ig or TCR gene rearrangements and do not express CD3 molecules. NK cells express a number of adhesion molecules, including CD2, LFA-1, and the neural cell adhesion molecule (NCAM or CD56). In addition, NK cells express a low-affinity receptor for the Fc portion of IgG, called FcγRIIIA (CD16). In NK cells, both CD2 and FcγRIIIA are associated with signaling subunits, namely homodimers or heterodimers involving the ζ chain of the TCR complex (Chapter 7) or the FcR γ chain (Chapter 3, Box 3–4), each of which contains immunoreceptor tyrosine activation motifs (ITAMs). Clustering of CD2 or of FcγRIIIA on the NK cell initiates secretion of cytokines, especially IFN-γ and TNF, and granule exocytosis, resulting in target cell killing. The intracellular pathways that mediate these responses are presumably similar to those activated in CTLs by clustering of the TCR complex. FcγRIIIA is clustered by interaction with the Fc regions of IgG1 or IgG3 that are bound to the surface of a target cell. NK cell–mediated lysis of antibody-coated target cells is an example of **antibody-dependent cell-mediated cytotoxicity (ADCC),** and NK cells are the major effector cells of IgG-directed ADCC.

NK cells express the signal-transducing β and γ_c subunits of the IL-2 receptor, but not the affinity-enhancing α subunit, and can be induced to proliferate only by high concentrations of IL-2. However, NK cells do express the IL-15Rα subunit and can proliferate at very low concentrations of this cytokine (see Chapter 12). In addition, NK cells express receptors for type I IFNs, IFN-γ, and IL-12, all of which act on these cells to increase their cytolytic capabilities.

Recently, the mechanism of NK cell recognition of virally infected cells has been largely elucidated (Fig. 13–12). Human NK cells express members of a family of cell surface proteins that contain two or three Ig-like domains in their extracellular regions and sequences that resemble ITAMs in the intracellular regions. Individual NK cells express one or more of these receptor proteins. Each receptor appears to recognize a spe-

cific allelic form of a class I MHC molecule (HLA-B or C) complexed to one or more abundant self protein–derived peptides. Many if not all of these receptors are inhibitory, activating a protein tyrosine phosphatase that shuts off NK cell responses. Consequently, NK cell activation occurs when a target cell is encountered that fails to engage the inhibitory receptor. Thus *NK cells recognize the absence of particular self peptides presented by self allelic forms of class I MHC molecules.* Viruses may trigger NK cell recognition either by reducing class I MHC molecule expression, a strategy used by some viruses to evade specific CTLs (see Chapter 16), or by replacing self peptides with peptides derived from viral proteins. Surprisingly, there are no homologs of these human NK cell receptors in mice. Mouse NK cells instead express members of a family of C-type lectins, collectively called Ly49, that serve a similar function, i.e., delivery of an inhibitory signal upon recognition of self allelic forms of class I MHC molecules.

Placing the basis of NK cell specificity on recognition of the absence of a signal raises a new question—namely, what is the mechanism used by NK cells to trigger granule exocytosis that is inhibited by engaging its antigen receptor? This answer is not known, but possible explanations include the existence of as yet uncharacterized stimulatory receptors or, more likely, the engagement of target cells through adhesion/accessory molecules such as CD2, LFA-1, or NCAM (CD56).

The role of NK cells in normal immunity is not fully established. It is believed that NK cells serve to lyse virally infected cells prior to the time that antigen-specific CTLs can differentiate from pre-CTLs—i.e., during the first few days of viral infection. Early in the course of a viral infection, NK cells are expanded and activated by cytokines of innate immunity, such as IL-15, type I IFN, and IL-12. In addition to lysing target cells that display reduced levels of class I molecules, activated NK cells secrete IFN-γ, which activates macrophages. In mice lacking T cells (i.e., with severe combined immunodeficiency), this NK cell–macrophage response can hold in check infection with *Listeria monocytogenes* for several weeks. In normal mice, this allows time for specific immunity to eradicate the infection through DTH and CTLs. In contrast, the T cell–deficient mice eventually succumb. A similar role for NK cells probably occurs in humans since it has been observed that rare patients who lack NK cells are more likely to develop severe viral infections.

Because NK cells can lyse certain tumor cells, it has also been proposed that NK cells serve to kill malignant clones *in vivo.* However, tumor-associated inflammatory infiltrates do not typically show significant numbers of NK cells. The one setting in which large numbers of NK cells do predominate in the lesions is in graft-versus-host disease (GVHD) in recipients of bone marrow transplants. We will discuss GVHD in greater detail

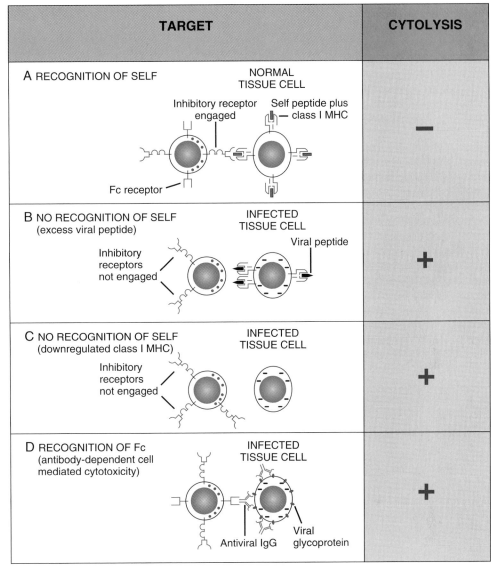

FIGURE 13–12. Natural killer (NK) cell recognition of infected cells. NK cells do not lyse uninfected cells because they receive an inhibitory signal through recognition of self peptide complexed to class I molecules on the surface of the target cell. In contrast, virally infected cells may be lysed because the inhibitory signal is absent, either because self peptides have been replaced by peptides derived from viral proteins or because the virus interferes with class I MHC molecule synthesis. The NK cell can also lyse a target cell that has bound IgG molecules, using its low affinity FcγRIIIA receptor.

in Chapter 17; suffice it to say here that NK cells infiltrate into epithelium such as skin and can be found adjacent to necrotic epithelial cells, the hallmark of GVHD. The mechanism by which NK cells lyse normal epithelial cells is not fully known. It has been observed that when NK cells are treated with sufficient concentration of IL-2 to be stimulated through IL-2R$\beta\gamma_c$, they differentiate into lymphokine-activated killer (LAK) cells. LAK cells demonstrate enhanced cytolytic capacity and a very broad target specificity, killing a wide variety of tumor cells and normal cell types, including epithelial cells. Thus, in GVHD, transplanted CD4$^+$ T cells may recognize and respond to the alloantigens of the host. These T cells produce IL-2, which may stimulate the differentiation of NK cells into LAK cells. This is another example of how, in specific cell-mediated immunity, T cells augment the functions and focus the actions of the effector cells of innate immunity. Although their physiologic role is not known, LAK cells, generated in vivo by high

concentrations of IL-2, have been used clinically to treat tumors (see Chapter 18).

SUMMARY

Cell-mediated immunity consists of immune responses that are initiated by antigen recognition by specific T lymphocytes and in which T lymphocytes participate in the effector stage as well. There are several forms of cell-mediated immune reactions, which are initiated by activation of T cells in response to specific antigen. In delayed type hypersensitivity reactions, T cells secrete tumor necrosis factor and lymphotoxin, which enable endothelial cells to recruit inflammatory leukocytes by causing local vasodilation and increased blood flow; local expression of endothelial cell adhesion molecules for leukocytes (such as E-selectin, VCAM-1, and ICAM-1); local chemokine expression on the endothelial surface; and local changes in endothelial cell junctions and basement membrane that favor extravasation of plasma macromolecules and cells. T cells also secrete interferon-γ, which activates macrophages to more effectively kill microorganisms. If the infection is not fully resolved, activated macrophages mediate tissue remodeling.

Cytolytic T lymphocytes (CTLs), which usually bear CD8, are important effector cells in settings of intracellular microbial infection. CTLs differentiate from pre-CTLs in response to two signals: (1) a target cell bearing foreign peptide antigens presented in association with self class I MHC molecules, and (2) costimulators provided by professional antigen present cells (APCs) or cytokines released by other T cells activated by professional APCs. Upon differentiation, CTLs acquire the ability to kill target cells expressing the appropriate MHC-associated peptide antigen. CTL-mediated killing involves two complementary mechanisms: (1) granule exocytosis of perforin, a membrane pore-forming protein, that polymerizes in the target cell membrane and causes osmotic lysis of target cells, and (2) entry into the target cell through the perforin channel of a serine protease (granzyme B) that activates apoptotic cell death. The DNA cleaving enzymes activated in the apoptotic death pathway also serve to destroy infectious viral genomes.

Natural killer cells are a population of large granular lymphocytes that normally serve to kill target cells that display decreased expression of self class I MHC molecule–peptide complexes (as may occur in viral infection) or cells coated with specific IgG molecules. NK cell–mediated killing uses the same perforin/granzyme pathways employed in CTL-mediated killing. NK cells are activated by cytokines of innate immunity (IL-15, type I IFN, and IL-12) and can be further activated by cytokines produced by CD4⁺ T cells. In innate immunity, NK cells form the first line of defense against intracellular microbes, prior to the differentiation of CTLs. In response to high levels of IL-2 as a consequence of specific immunity, NK cells may differentiate into lymphokine-activated killer cells that kill target cells in relatively indiscriminate fashion. By recruiting and activating NK cells, CD4⁺ T cells can cause lysis of normal cell types in response to antigen stimulation, characteristic of the reaction found in acute graft-versus-host disease.

Selected Readings

Bevilacqua, M. P. Endothelial-leukocyte adhesion molecules. Annual Review of Immunology 11:767–804, 1993.

Doherty, P. C., J. E. Allan, F. Lynch, and R. Ceredig. Dissection of an inflammatory process induced by CD8 T cells. Immunology Today 11:55–59, 1990.

Kagi, D., B. Ledermann, K. Burki, R. M. Zinkernagel, and H. Hengartner. Molecular mechanisms of lymphocyte-mediated cytotoxicity and their role in immunological protection and pathogenesis in vivo. Annual Review of Immunology 14:207–232, 1996.

Kovacs, E. J. Fibrogenic cytokines: the role of immune mediators in the development of scar tissue. Immunology Today 12:17–23, 1991.

Liu, C.-C., L. H. Y. Young, and J. D.-E. Young. Lymphocyte-mediated cytolysis and disease. New England Journal of Medicine 335:1651–1659, 1996.

Moretta, A., C. Bottino, M. Vitale, D. Pende, R. Blassoni, M. C. Mingari, and L. Moretta. Receptors for HLA class-I molecules in human natural killer cells. Annual Review of Immunology 14:619–648, 1996.

Pober, J. S., and R. S. Cotran. Immunologic interactions of T lymphocytes with vascular endothelium. Advances in Immunology 50:261–302, 1991.

Versteeg, R. NK cells and T cells: mirror images? Immunology Today 13:244–247, 1992.

CHAPTER FOURTEEN

EFFECTOR MECHANISMS OF IMMUNOGLOBULIN E–INITIATED IMMUNE REACTIONS

One of the most powerful effector mechanisms of the immune system is the reaction initiated by IgE-dependent stimulation of tissue mast cells and their circulating counterparts, the basophils. When antigen binds to immunoglobulin E (IgE) molecules preattached to the surfaces of these cells, there is a rapid release of a variety of mediators that collectively cause increased vascular permeability, vasodilation, bronchial and visceral smooth muscle contraction, and local inflammation. This reaction is called **immediate hypersensitivity** because it begins rapidly, within minutes of antigen challenge. In its most extreme systemic form, called **anaphylaxis,** mast cell–derived or basophil-derived mediators can restrict airways to the point of asphyxiation and produce cardiovascular collapse leading to death. (The term anaphylaxis was coined to indicate that antibodies, especially IgE antibodies, could confer the opposite of protection [prophylaxis] on an unfortunate individual.)

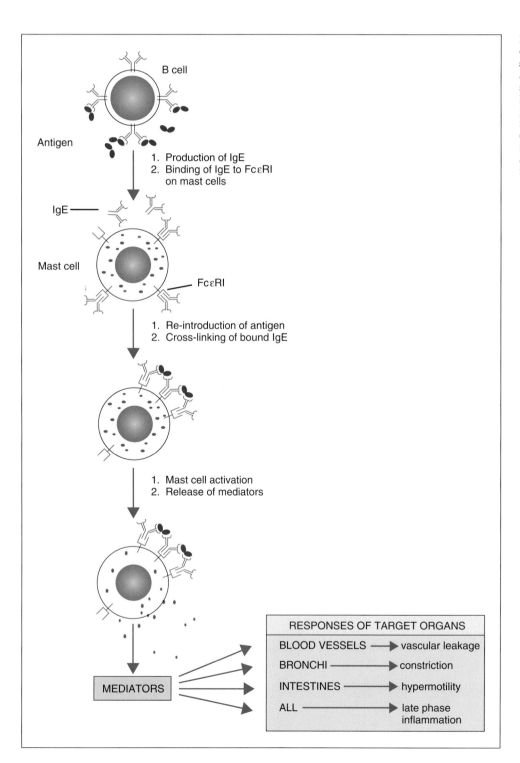

FIGURE 14–1. Sequence of events in immediate hypersensitivity. The initial contact with an antigen leads to specific IgE synthesis by B cells. Secreted IgE binds to mast cells or basophils through high-affinity Fcε receptors (FcεRI). Upon subsequent exposure to antigen, an immediate hypersensitivity reaction is triggered by cross-linking the IgE molecules.

Individuals prone to develop strong immediate hypersensitivity responses are called **atopic** and are said to suffer from **allergies.** Atopy means "unusual," but we now realize that allergy is in fact quite common. Indeed, allergy is the most common disorder of immunity, affecting 20 per cent of all individuals in the United States. In different individuals, atopy may take different forms such as hay fever, asthma, urticaria (hives), or chronic eczema (skin irritation). All of these conditions are forms of immediate hypersensitivity induced by mast cell or basophil activation.

Mast cell and basophil activation is most characteristically initiated when specific antigen binds to and cross-links preattached surface IgE molecules. Thus, *the typical sequence of events in immediate hypersensitivity* is as follows: (1) production of IgE by B cells in response to the first exposure to an antigen, (2) binding of the IgE to specific Fc receptors on the surfaces of mast cells and basophils, and (3) interaction of re-introduced antigen with the bound IgE, leading to (4) activation of the cells and release of mediators, some of which are stored in the cytoplasmic granules of the mast cells and basophils (Fig. 14–1). The clinical and pathologic manifestations of immediate hypersensitivity are due to the actions of the released mediators.

Until recently, the existence of immediate hypersensitivity posed a conundrum: For what purpose has the organism developed this potentially lethal, disease-causing arm of the immune system? The answer may be that immediate hypersensitivity is part of a larger response that leads to inflammatory infiltrates rich in eosinophils, called the **late phase reaction.** The late phase reaction is a host defense mechanism against some helminthic and arthopod infections. Although we now better appreciate that eosinophilic inflammation plays a protective function, it is also more apparent that eosinophil-mediated tissue injury is a major component of allergic diseases such as asthma. Thus, IgE-initiated protective immunity and immediate hypersensitivity are mediated by the same effector mechanisms. Recall that we made the same point when discussing cell-mediated immunity and delayed type hypersensitivity in Chapter 13.

This chapter focuses on IgE-initiated reactions. We begin by describing experimental models of immediate hypersensitivity and eosinophilic inflammation. We then turn to the biology of IgE and to the Fc receptors that bind IgE on the surface of mast cells or basophils. Next, we discuss the biology of mast cells and related cell types, including a description of how these cells are activated in response to antigen. We then describe the mediators produced by activated mast cells, basophils, and eosinophils, and the biologic effects of these mediators. We conclude with a more detailed description of different allergic diseases, integrating features of IgE-initiated reactions and the responses of various target tissues to explain specific clinical syndromes associated with immediate hypersensitivity.

FEATURES OF IMMEDIATE HYPERSENSITIVITY

The classic example of immediate hypersensitivity in humans is the "wheal and flare reaction" (Fig. 14–2). When a sensitized individual is challenged by intradermal injection of an appropriate antigen, the injection site becomes red from locally dilated blood vessels engorged with red blood cells. In the second phase, the site rapidly swells as a result of leakage of plasma from the venules. This soft swelling is called a **wheal** and can involve an area of skin as large as several centimeters in diameter. In the third phase, blood vessels at the margins of the wheal dilate and become engorged with red blood cells, producing a characteristic red rim called a **flare.** The full wheal and flare reaction can appear within 5 to 10 minutes after administration of antigen and usually subsides in less than an hour. By electron microscopy, the venules in the area of the wheal show slight separation of the endothelial cells, which accounts for the escape of macromolecules and fluid, but not cells, from the vascular lumen.

The first clue to the mechanism of this reaction also came from histologic examination. Mast cells in the area of the wheal and flare show evidence of release of pre-formed mediators, i.e., their cytoplasmic granules have been discharged. A causal association of IgE and mast cells with immediate hypersensitivity has been deduced from three kinds of experiments:

1. Immediate hypersensitivity reactions can be elicited in nonresponsive individuals if the local skin site is first injected with IgE from a responsive individual. *Thus, IgE is responsible for specific recognition of antigen and can be used to adoptively transfer immediate hypersensitivity.* Such adoptive transfer experiments were first performed with serum from immunized individuals in the 1920s. Some 40 years later, it was shown that the serum protein responsible for transferring this reaction was a class of antibody, which was named IgE. This serum factor was originally called "reagin," and for this reason IgE molecules are still sometimes called "reaginic antibodies." The antigen-initiated skin reaction that follows adoptive transfer with IgE is called **passive cutaneous anaphylaxis.**

2. *Immediate hypersensitivity reactions can be mimicked by injecting anti-IgE antibody instead of antigen.* Anti-IgE elicits a reaction both in atopic individuals who have high levels of antigen-specific IgE antibodies and in non-atopic individuals who have low but measurable levels of IgE. Anti-IgE antibodies act as an analog of antigen and directly activate mast cells and basophils that have bound IgE on their surface. This use of anti-IgE to activate mast cells is similar to the use of anti-IgM or anti-IgD antibodies as analogs of antigen to activate B cells (see Chapter 9), except that in the case of

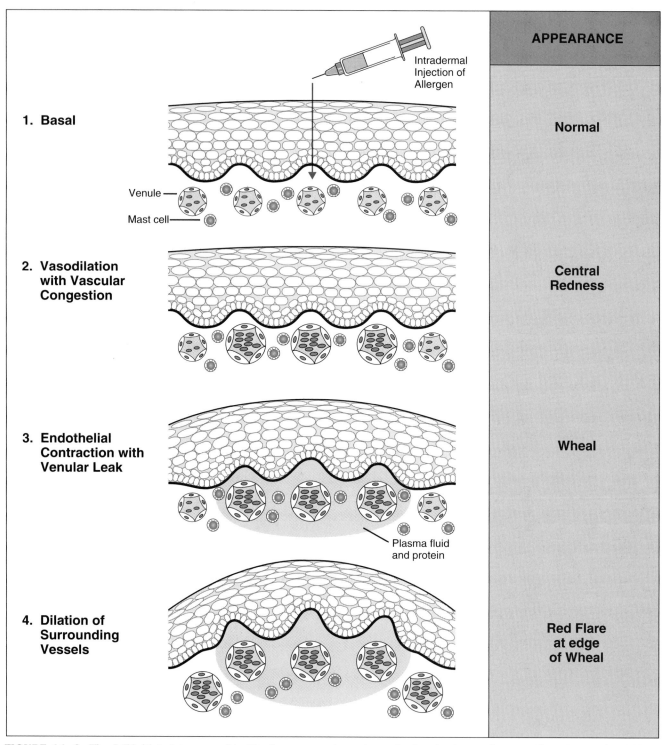

FIGURE 14–2. The IgE-initiated response in skin. In response to antigen-stimulated release of mast cell mediators, local blood vessels first dilate and then become leaky to fluid and macromolecules, producing redness and local swelling (a wheal). Subsequent dilation of vessels on the edge of the swelling produces the appearance of a red rim (the flare).

mast cells or basophils, secretory IgE, made by B cells, is bound to high-affinity Fc receptors on the cell surface rather than being synthesized as membrane IgE.

3. *Immediate hypersensitivity reactions can be mimicked by injection of other agents that directly cause mast cell activation, such as C5a or substance P, or by local trauma, which also causes degranulation of mast cells; and it can be inhibited by agents that prevent mast cell activation.*

Although these classic experiments in immediate hypersensitivity were originally performed in humans, immunologists often use animal models to permit greater ease of experimental manipulation. The guinea pig consistently mounts strong immediate hypersensitivity reactions and has proved to be a very useful animal model.

A **late phase reaction** begins between 2 and 4 hours after elicitation of many immediate hypersensitivity reactions. By this time, the wheal and flare of the immediate hypersensitivity reaction have subsided. This late phase reaction consists of accumulation of inflammatory leukocytes, including neutrophils, eosinophils, basophils, and CD4$^+$ T cells. These T cells are enriched for cells that produce interleukin-4 (IL-4) but not interferon-γ (IFN-γ), the hallmark of the T$_H$2 subset (see Chapter 12). The inflammation is maximal by about 24 hours and then gradually subsides. Late phase reactions in atopic individuals are rich in eosinophils. Atopic individuals have elevated numbers of activated eosinophils in their peripheral blood, and the composition of the infiltrate may partly reflect the blood count. However, it is also clear that IL-4 produced by mast cells or T$_H$2 T cells can selectively recruit eosinophils into the tissue. The late phase reaction is part of immediate hypersensitivity, since, like the wheal and flare reaction, it can be adoptively transferred with IgE and can be mimicked by anti-IgE antibodies or mast cell activating agents.

The principal protective function of IgE-initiated immune reactions is the eradication of parasites. Eosinophil-mediated killing of IgE-coated helminths is an effective defense against these organisms (see Chapter 16). It has also been speculated that IgE-dependent mast cell activation in the gastrointestinal tract promotes expulsion of parasites by increasing peristalsis and by an outpouring of mucus. Genetically mast cell–deficient mice show increased susceptibility to infection by tick larvae, and immunity can be provided to these mice by adoptive transfer of specific IgE and mast cells (but not by either component alone). The larvae are eradicated by the specific late phase reaction. The importance of this defense mechanism has been further underscored by studies of mice treated with anti-IL-4 antibody and of IL-4 knockout mice. As discussed in Chapter 12, such mice do not make IgE, and they appear to be less resistant than normal animals to some helminthic infections. IL-5 knockout mice, which are unable to activate eosinophils, also show increased susceptibility to some helminths.

BIOLOGY OF IgE

As we have noted, IgE antibody provides recognition of antigen for immediate hypersensitivity reactions. IgE is the isotype of immunoglobulin that contains the ϵ heavy chain (see Chapter 3). It circulates as a bivalent antibody and is normally present in plasma at a concentration of less than 1 μg/ml. In pathologic conditions, such as helminthic infections and severe atopy, this level can rise to over 1000 μg/ml. The IgE heavy chain V regions and IgE light chains are products of the same genes as other Ig molecules. The heavy chain C regions are encoded by the C$_\epsilon$ gene located in the Ig heavy chain gene cluster. Thus, IgE is produced as a result of heavy chain isotype switching (see Chapter 4).

Regulation of IgE Synthesis

There is a critical difference in IgE production between atopic and normal individuals: The former produce high levels of IgE in response to particular antigens, whereas the latter generally synthesize other Ig isotypes, such as IgM and IgG, and only small amounts of IgE. Four interacting factors contribute to regulation of IgE synthesis: (1) heredity, (2) the natural history of antigen exposure, (3) the nature of the antigen, and (4) helper T cells and their cytokines.

HEREDITY

Abnormally high levels of IgE synthesis and associated atopy often run in families. Although the full inheritance pattern is probably multigenic, family studies have shown that there is clear autosomal transmission of atopy. Two candidate loci for atopy have been mapped by family studies. One site is on chromosome 5q, near the site of the gene cluster encoding hematopoietic colony-stimulating factors, IL-4, IL-5, and IL-9. The second site is on chromosome 11p, the location of the gene encoding the β chain of the high affinity IgE receptor (see below).

Within the same family, the target organ of atopic disease is variable. In other words, hay fever, asthma, and eczema can be present to various degrees in different members of the same kindred. All of these individuals, however, will show higher than average plasma IgE levels. In addition to this general proclivity to synthesize IgE, the ability to make specific IgE antibodies to certain antigens, e.g., ragweed pollen, is also inherited and may be linked to particular class II major histocompatibility complex (MHC) alleles. This may be an example of an "immune response gene" (Ir gene) effect (see Chapter 6). In the case of a complex antigen such as ragweed pollen, multiple class II alleles may serve to present different peptides to specific T cells.

NATURAL HISTORY OF ANTIGEN EXPOSURE

The natural history of antigen exposure is an important determinant of the level of specific IgE antibodies. In general, repeated exposure to a particular antigen is necessary to develop an atopic reaction to that antigen. Individuals with allergic rhinitis or asthma often benefit from a geographic change of residence with a change in indigenous plant pollens, although local antigens in the new residence may trigger an eventual return of the symptoms. The most dramatic examples of the influence of the natural history of exposure to antigen are seen in cases of insect, e.g., bee, stings. The protein toxins in the insect venoms are usually not of concern on the first encounter because the atopic individual has no pre-existing specific IgE antibodies. However, an IgE response may occur after a single encounter with antigen, and a second sting by an insect of the same species may induce fatal anaphylaxis!

NATURE OF THE ANTIGEN

Antigens that elicit strong immediate hypersensitivity reactions are called **allergens.** Allergens may be either proteins or chemicals bound to proteins. It is not known why some antigens cause strong allergic responses whereas other antigens, which may be encountered by the same route of administration, are simply not allergenic and instead result in non-IgE humoral or cell-mediated immune responses. The property of being allergenic may reside in the antigen itself, perhaps in epitopes seen by certain T cells. Some drugs, such as penicillin, sometimes elicit strong IgE responses. It is thought that these drugs bind to self proteins, forming hapten-carrier conjugates that function as "neoantigens."

Anaphylactic responses to foods typically involve highly glycosylated small proteins. These structural features probably protect these allergens from denaturation and degradation in the gastrointestinal tract, allowing them to be absorbed intact. The intact proteins serve as a source of foreign peptides for activation of T cells and can directly interact with IgE.

HELPER T CELLS AND CYTOKINES

IgE- and eosinophil-mediated immune reactions are dependent on the activation of $CD4^+$ helper T cells of the T_H2 subset. These T cells secrete IL-4, which is required for isotype switching to IgE (see Chapter 9) and promotes eosinophil recruitment, and IL-5, which activates eosinophils. Accumulations of T_H2 cells have been demonstrated at sites of immediate hypersensitivity reactions in the skin and bronchial mucosa. Atopic individuals contain larger numbers of allergen-specific IL-4–secreting T cells in their circulation than do non-atopic persons. In addition, in atopic patients, the allergen-specific T cells produce more IL-4 per cell than in normal individuals. All these factors contribute to the increased IgE production associated with atopy. Because immediate hypersensitivity reactions are dependent on T cells, T cell–independent antigens such as polysaccharides cannot elicit such reactions unless they become attached to proteins.

As we discussed in Chapter 12, the activation of T_H1 or T_H2 cells in response to protein antigens leads to quite distinct classes of immune reactions. The cytokines produced by T_H1 cells are responsible for delayed type hypersensitivity (see Chapter 13). In contrast, T_H2 cells not only elicit IgE production and eosinophilic inflammation but also limit macrophage activation through the secretion of IL-10 and other regulatory cytokines. The effects of these helper T cell subsets on antibody production are also distinct, and they cooperate with their effects on inflammatory cells. Thus, the IFN-γ produced by the T_H1 subset promotes the secretion of antibodies, such as IgG2a in mice, that bind to Fcγ receptors on macrophages, enhancing phagocytosis of opsonized particles. At the same time, IFN-γ inhibits switching to IgE. In contrast, IL-4 produced by T_H2 cells induces IgE production, and IgE is the isotype preferentially utilized in antibody-dependent cell-mediated cytotoxicity (ADCC) mediated by eosinophils.

Fc Receptors for IgE

IgE, like all other antibody molecules, is exclusively made by B cells, yet IgE functions as an antigen receptor on the surface of mast cells and basophils. On these cells, IgE is bound by a high affinity Fc receptor specific for ϵ heavy chains, called FcϵRI, that is constitutively expressed. The dissociation constant (K_d) of FcϵRI for IgE is about 1×10^{-10} M; this binding is much stronger than that of any other Fc receptor for its ligand (see Chapter 3, Box 3–4). Consequently, the serum concentration of IgE, although quite low compared with other Ig isotypes in normal individuals (i.e., less than 1 μg/ml or $\sim 5 \times 10^{-10}$ M), is sufficiently high to allow occupancy of FcϵRI receptors. In addition to mast cells and basophils, FcϵRI has been detected on epidermal Langerhans cells and some dermal macrophages; its function on these cells is not known. It has also been observed on some blood monocytes and activated eosinophils, where it mediates IgE-directed ADCC.

Each FcϵRI molecule contains four separate polypeptides, one α, one β, and two identical γ chains (see Chapter 3, Box 3–4, and Fig. 14–3). As deduced from transfection experiments, all three subunits must be present to have cell surface expression. The α chain, which mediates binding of IgE, has a predicted size of 25 kilodaltons (kD). The 180 amino terminal residues form two extracellular 90 amino acid residue repetitive sequences that are members of the Ig superfamily. The IgE binding site, formed by these Ig domains, is highly homologous to the IgG binding sites of the FcγRII and FcγRIII receptors described in Chapter 3 (see Box 3–4). Each FcϵRI α chain has an approximately 20 amino acid residue hydrophobic sequence that is believed to cross the cell membrane once and approximately 20 carboxyl terminal amino acids that form a cytoplasmic domain. The β chain of FcϵRI is a 26 kD polypeptide whose predicted structure crosses the membrane four

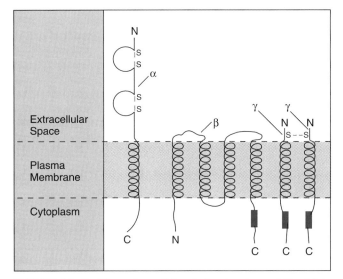

FIGURE 14–3. The polypeptide chain structure of the high-affinity IgE Fc receptor (FcεRI). IgE binds to the Ig-like domains of the α chain. The β chain and the γ chains mediate signal transduction. The boxes in the cytoplasmic region of the β and γ chains are immunoreceptor tyrosine activation motifs (ITAMs) similar to those found in the TCR complex (see Fig. 7–5).

times. Both the amino terminus and the carboxy terminus are located in the cytoplasm, and the carboxy terminus contains a single immunoreceptor tyrosine activation motif (ITAM), which is also present in the CD3 chains of the T cell receptor complex (see Chapter 7). The two identical γ chain polypeptides are only 7 kD each and are highly homologous to the ζ chain of the T cell antigen receptor complex (see Chapter 7). From its predicted structure, only five amino terminal amino acid residues of the γ chain are extracellular. Each γ chain crosses the membrane once, and the remaining residues are intracellular; the intracellular portion of the γ chain contains one ITAM. The γ chain of FcεRI is now known to serve as a common subunit for FcγRI, FcγRIIIA and FcαR and is sometimes called the FcR γ chain. Studies using chimeric receptors incorporating the ITAMs from either the β or γ chains of FcεRI suggest that the γ chain delivers activating signals, whereas the β chain may be inhibitory. We will return to the signaling functions of FcεRI when we discuss mast cell activation later in the chapter.

Two other IgE "receptors" have been described. The first of these, called FcεRII, is a 30 kD protein related to C-type mammalian lectins, a family that includes the selectins (see Chapter 11, Box 11–1). The affinity of FcεRII for IgE is much lower than that of FcεRI. Molecular cloning studies of the IgE binding chain of FcεRII suggest that two different polypeptides may be generated by use of alternative translational start sites and alternative splicing of messenger RNA (mRNA) from the same gene. One product is B cell–specific and is expressed constitutively (FcεRIIA); the other product (FcεRIIB, also called CD23) is induced on B cells,

monocytes, and eosinophils by IL-4. The functions of FcεRIIA and B are largely unknown. The second low affinity receptor for IgE is a 35 kD β-galacto-side–binding lectin called galectin 3. Galectin 3 is present in the cytoplasm of mononuclear phagocytes and other cells and may contribute to macrophage-mediated killing. Its role in IgE biology and immediate hypersensitivity is unknown.

BIOLOGY OF MAST CELLS, BASOPHILS, AND EOSINOPHILS

Properties of Mast Cells and Basophils

All mast cells are derived from progenitors present in the bone marrow. Normally, mature mast cells are not found in the circulation. Progenitors migrate to the peripheral tissues as immature cells and undergo differentiation *in situ*. Mature mast cells are found throughout the body, predominantly located near blood vessels and nerves and beneath epithelia. They are also present in lymphoid organs. By light microscopy, human mast cells may be round, oval, or even spindle-shaped. The nuclei are typically round. The cytoplasm contains membrane-bound granules and lipid bodies (Fig. 14–4). The granules contain acidic proteoglycans, which bind basic dyes. Some of these dyes assume a different color when bound by the granules than they do when staining nuclear DNA, so that the granules are sometimes called "metachromatic."

An important advance in the understanding of mast cell biology is the appreciation that in rodents, mature mast cells may assume one of two phenotypes (Table 14–1). Mast cells found in the mucosa of the gastrointestinal tract have chondroitin sulfate as their major granule proteoglycan. Such "mucosal" mast cells contain little histamine. The second phenotype has been found in the lung and in the serosa of body cavities. These "connective tissue" mast cells contain heparin as their major granule proteoglycan and produce large quantities of histamine. Mast cells may also be cultured from rodent bone marrow in the presence of IL-3.

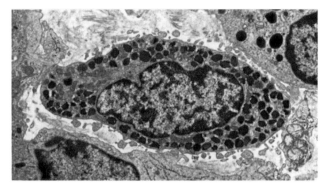

FIGURE 14–4. Electron micrograph of a human mast cell. Note the characteristic numerous cytoplasmic granules, known to contain histamine, heparin, and various enzymes. (Courtesy of Dr. Noel Weidner, Department of Pathology, Brigham and Women's Hospital, Boston.)

TABLE 14–1. Mast Cell Heterogeneity

	Connective Tissue	Mucosal
T cell–dependence	No	Yes
c-Kit ligand-dependence	Yes	No
Histamine content	High	Low
Major proteoglycan	Heparin	Chondroitin sulfate
Major arachidonate metabolite	PGD$_2$	LTC$_4$ > PGD$_2$

In rodents, mast cells isolated from connective tissue or gastrointestinal mucosa differ in several properties, the most important of which are listed here.

Abbreviations: PGD$_2$, prostaglandin D$_2$; LT, leukotrienes.

Such cultured mast cells resemble mucosal mast cells based on granule content of chondroitin sulfate and low histamine. Moreover, the presence of mucosal mast cells *in vivo* appears to depend upon T cells, the presumed source of IL-3, since they are absent in athymic mice. Bone marrow–derived mucosal mast cells can be changed to a connective tissue mast cell phenotype by co-culture with fibroblasts, or incubation with the ligand for c-kit, sometimes called stem cell factor (see Chapter 12). Repopulation experiments in mast cell–deficient mice further suggest that the mucosal and connective tissue phenotypes are not fixed and that bidirectional changes may be possible in suitable microenvironments. The key point is that *the precise nature of the mast cell and the mediators it can produce vary with its anatomic location and is probably regulated by local cytokines.*

In humans, the factors that regulate mast cell growth and development are less well defined. There appears to be a similar pattern of T cell–independent connective tissue mast cells (that depend on c-kit ligand) and T cell–dependent mucosal mast cells. Major differences between types of human mast cells reside in the composition of serine proteases found in the granules (trypsin-like or chymotrypsin-like in substrate specificity) and in the ultrastructural morphology of the granules. Nevertheless, it does appear that in humans as well as in mice the pattern of mediators produced by mast cells may vary with anatomic location.

Basophils share a number of similarities with mast cells. Like mast cells, basophils are derived from bone marrow progenitors and contain granules that bind basic dyes. Basophils are capable of synthesizing many of the same mediators as mast cells. Most significantly, both basophils and mast cells express the same high-affinity Fcε receptor (FcεRI) and can be triggered by antigen binding to IgE. Therefore, basophils, like mast cells, may mediate immediate hypersensitivity reactions to antigen. Despite these similarities, basophils appear to be a distinct cell type from mast cells. Basophils mature in the bone marrow and circulate in their differentiated form. The responsiveness of basophils to antigen may be increased by IL-3. Like other granulocytes, basophils enter tissues only when they are recruited into inflammatory sites. Basophils, like neutrophils, express a number of adhesion molecules important for homing, such as LFA-1 (CD11aCD18), Mac-1 (CD11bCD18), and CD44. Thus, basophils are best thought of as an inflammatory granulocyte, with structural and functional similarities to mast cells but derived from a different cell lineage. Because of their responsiveness to T cell–derived cytokines, basophils may serve to amplify T$_H$2 CD4$^+$ T cell–mediated reactions by supplementing the production of vasoactive and inflammatory mediators by resident mast cells.

Activation of Mast Cells and Basophils

The event that initiates immediate hypersensitivity is the binding of antigen to IgE on the mast cell or basophil surface. Mast cells and basophils are activated by cross-linking of FcεRI molecules, which is thought to occur by binding of multivalent antigens to the attached IgE molecules (Fig. 14–5). Experimentally, antigen binding can be mimicked by polyvalent anti-IgE or by anti-FcεRI antibodies. In fact, such antibodies can activate mast cells from atopic as well as non-atopic individuals, whereas allergens activate mast cells only in atopic persons. The reason for this is that in an individual allergic to a particular antigen, a significant proportion of the IgE bound to mast cells is specific for that antigen. Administration of the antigen will cross-link sufficient IgE molecules to trigger mast cell activation. In contrast, in non-atopic individuals, the mast cell–associated IgE is specific for many different antigens (all of which may have induced low levels of IgE production). Therefore, no single antigen will cross-link enough of the IgE molecules to cause mast cell activation. Anti-IgE antibodies, on the other hand, can cross-link these IgE molecules regardless of antigen specificity and lead to comparable triggering of mast cells from both atopic and non-atopic individuals.

Activation of mast cells (and basophils) results in three types of biologic responses:

1. Mast cells undergo regulated secretion in which the *pre-formed contents of their granules are released by exocytosis.*

2. Mast cells enzymatically synthesize *lipid mediators* derived from precursors stored in cell membranes and, in some cases, in the lipid bodies.

3. Mast cells initiate transcription, translation, and *secretion of cytokines.*

The mechanisms of granule exocytosis are partly understood, largely from studies of rat mast cell and basophil leukemia cell lines. The cross-linking of FcεRI results in recruitment and activation of Syk and Fyn protein tyrosine kinases in a process involving the ITAMs of the FcR γ chain. These enzymes, in turn, phosphorylate and activate other proteins, including γ isoforms of a phosphatidylinositol-specific phospholipase C (PI-PLCγ)

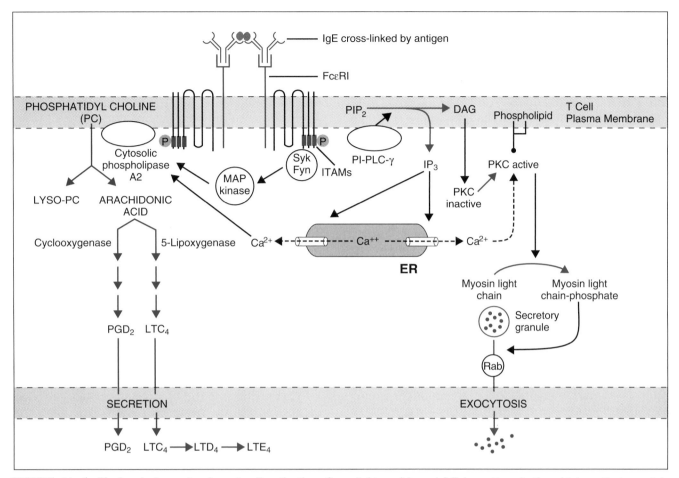

FIGURE 14–5. Biochemical events of mast cell activation. Cross-linking of bound IgE by antigen is thought to activate protein tyrosine kinases (Syk and Fyn), which in turn cause activation of a mitogen-activated protein (MAP) kinase cascade and of a phosphatidylinositol-specific phospholipase C (PI-PLCγ). PI-PLCγ catalyzes release of inositol triphosphate (IP$_3$) and diacylglycerol (DAG) from membrane phosphatidylinositol bisphosphate (PIP$_2$). IP$_3$ causes release of intracellular calcium (Ca^{2+}) from the endoplasmic reticulum. Ca^{2+} and DAG combine with membrane phospholipids to activate protein kinase C (PKC), which phosphorylates substrates such as myosin light chain protein, leading to degradation and release of preformed mediators. Ca^{2+} and MAP kinase combine to activate the enzyme cytosolic phospholipase A$_2$ (cPLA$_2$), which initiates synthesis of lipid mediators by releasing arachidonic acid from phosphatidyl choline.

that catalyze phosphatidylinositol bisphosphate breakdown to inositol 1,4,5-triphosphate (IP$_3$) and diacylglycerol (DAG). The coupling of FcϵRI to the activation of PI-PLCγ is similar to events that occur following the clustering of T and B cell antigen receptors (see Chapters 7 and 9). IP$_3$ causes elevation of cytoplasmic calcium, which, in combination with DAG, activates protein kinase C. The activated protein kinase C phosphorylates myosin light chains, leading to disassembly of actin-myosin complexes beneath the plasma membrane, thereby allowing granules to come in contact with the plasma membrane. Fusion of the granule membrane with the plasma membrane is mediated by interactions between proteins associated with the granule membrane and proteins associated with the plasma membrane; these events are regulated by the GTP-binding (activated) form of the Ras-related proteins Rab 3b or d, which are present in mast cells.

Cross-linking of FcϵRI also activates the en-

zyme adenylyl cyclase through a heterotrimeric GTP-binding protein. This, in turn, elevates cyclic adenosine monophosphate (cAMP) levels, and cAMP activates protein kinase A. Protein kinase A inhibits degranulation, suggesting that this pathway is a negative feedback loop.

The synthesis of lipid mediators (see below) is controlled by activation of the enzyme cytosolic phospholipase A$_2$ (cPLA$_2$). This enzyme is activated by two signals: elevated cytoplasmic calcium, released by IP$_3$, and phosphorylation catalyzed by a mitogen-activated protein (MAP) kinase. MAP kinases are activated as a consequence of a kinase cascade initiated through the receptor ITAMs, probably utilizing the same pathways as in T cells (see Chapter 7). Cytokine transcription in mast cells is also presumed to be similar to the events that occur in T cells. Syk appears to activate various adaptor molecules, leading to nuclear translocation of NFAT and NFκB, as well as activation of AP-1 by stress-activated protein kinases, such as cJun N-terminal kinase. Overall, this response leads to transcription of several cytokines (IL-3, IL-4, IL-5, IL-6, and tumor necrosis factor [TNF], among others), but in contrast to T cells, not IL-2. Basophils also produce T_H2-like cytokines.

Mast cells may be activated by mechanisms other than cross-linking of FcεRI. For example, certain types of mast cells or basophils may respond to mononuclear phagocyte-derived chemokines produced as part of innate immunity, to T cell–derived chemokines produced as part of specific cell-mediated immunity, and to complement-derived anaphylatoxins such as C5a produced during humoral immune responses (see Chapter 15). The T cell–derived chemokines (see Chapter 12) probably constitute the so-called "histamine releasing factors" observed in delayed type hypersensitivity reactions (see Chapter 13). Mast cells may also be activated by neutrophil granule contents or by neurotransmitters such as norepinephrine and substance P. These latter agents are potentially important as links between the nervous system and the immune system. The nervous system is known to affect the expression of immediate hypersensitivity reactions. The "flare" produced at the edge of the wheal in elicited immediate hypersensitivity reactions is in part mediated by the nervous system, as shown by the observation that the flare is markedly diminished in skin sites lacking innervation.

Eosinophils

The inflammatory infiltrates of late phase reactions are typically rich in eosinophils. Eosinophils are bone marrow–derived granulocytes whose granules contain basic proteins that bind acidic dyes such as eosin. Indeed, the two major proteins of the eosinophil granule are called major basic protein and eosinophil cationic protein. Major basic protein is toxic for helminths, and eosinophils are the principal effector cells of ADCC against helminthic infections. Major basic protein also causes damage to normal tissues.

The activation of eosinophils is under the control of the same T cells (of the T_H2 subset) that regulate IgE synthesis. IL-5 acts as an eosinophil-activating factor, converting resting eosinophils to a larger "hypodense" state that is more potent at mediating ADCC. IL-5 also augments eosinophil production, and this is supported by the ability of anti-IL-5 antibody to inhibit the eosinophilia that occurs in mice infected with helminthic parasites.

There is increasing evidence that hypodense eosinophils are selectively recruited into late phase inflammatory reactions. Eosinophils, like neutrophils, bind to endothelial cells expressing E-selectin. However, unlike neutrophils, eosinophils express VLA-4 (CD49dCD29) and also adhere to endothelial cells that express vascular cell adhesion molecule-1 (VCAM-1). IL-4 can induce endothelial cells to express VCAM-1 without inducing expression of E-selectin. IL-4 may also stimulate eosinophil transmigration across endothelial cell layers. Furthermore, although IL-5 is not directly chemotactic for eosinophils, it increases the response of eosinophils to C-C chemokines such as eotaxin (see Chapter 12). Consequently, inflammatory sites enriched for IL-4 and IL-5–producing T_H2 T cells or basophils will efficiently attract eosinophils.

Once recruited into an inflammatory site, eosinophils act as effector cells. We have already discussed the fact that these cells release toxic granule proteins that can kill helminths by ADCC or cause tissue damage in allergic reactions. ADCC may be mediated through the FcεRI or FcαR. Little is known about the mechanisms involved in eosinophil degranulation and mediator production. Once stimulated, eosinophils also produce vasoactive lipid mediators, which may augment the effects of mediators produced by mast cells and basophils, and eosinophils probably synthesize some pro-inflammatory cytokines, but the importance of eosinophil-derived mediators in immediate hypersensitivity is not completely known.

MAST CELL–DERIVED AND BASOPHIL-DERIVED MEDIATORS

This section of the chapter describes the various mediators of immediate hypersensitivity released by mast cells or basophils upon activation. It should be remembered that mast cells are heterogeneous and that not all mast cells will release the same mediators or the same combinations of mediators. Nevertheless, the same general classes of mediators appear to be made by most mast cells as well as by basophils. These may be divided into **pre-formed mediators,** which include biogenic amines and granule macromolecules, and **newly synthesized mediators,** which include lipid-derived mediators and cytokines. Some cytokines, such as TNF in human dermal mast cells, may also exist as pre-formed stores that are released during degranulation. This is an exception to the general rule that cytokines are not pre-formed.

Biogenic Amines

The granules of mast cells contain non-lipid, low molecular weight vasoactive mediators. In humans, the prototypic mediator of this class is **histamine,** but in certain rodents serotonin may be of equal or greater import. Because these substances have in common the structural features of an amine group and share common functional effects on blood vessels, they have been collectively called "biogenic" or "vasoactive" amines. Histamine acts by binding to target cell receptors, and different cell types express distinct classes of receptors (e.g., H1, H2, H3) that can be distinguished by pharmacologic inhibitors. Upon binding to cellular receptors, histamine initiates intracellular events, such as phosphatidylinositol bisphosphate breakdown to IP_3 and DAG, which cause different changes in different cell types. In vascular endothelial cells, binding of histamine leads to endothelial cell contraction and leakage of plasma into the tissues. Histamine also causes endothelial cells to synthesize vascular smooth muscle cell relaxants, such as prostacyclin (PGI_2) and nitric oxide, which cause vasodilation. These actions of histamine produce the wheal and flare response of immediate hypersensitivity. H1 histamine receptor antagonists (commonly called antihistamines) can inhibit the wheal and flare response to intradermal allergen or anti-IgE antibody.

Histamine also causes constriction of intestinal and bronchial smooth muscle. Thus, histamine may contribute to the increased peristalsis or bronchospasm associated with ingested allergens or asthma, respectively. However, in these instances, especially in asthma, antihistamines are not effective at blocking the reaction. Moreover, bronchoconstriction in asthma is more prolonged than the effects of histamine, which is rapidly removed from the extracellular milieu by amine-specific transport systems. Thus, other mast cell–derived mediators are clearly important in some forms of immediate hypersensitivity.

Granule Proteins and Proteoglycans

In addition to vasoactive amines, mast cell granules contain several enzymes, such as serine proteases and aryl sulfatase, as well as proteoglycans such as heparin or chondroitin sulfate. The enzymes may cause tissue injury when released upon mast cell degranulation. One function of the negatively charged proteoglycans may be to bind and store the positively charged biogenic amines. However, it is not clear how important these substances are as mediators of immediate hypersensitivity reactions.

Lipid Mediators

Three classes of lipid mediators are synthesized by activated mast cells (Fig. 14–6). In general, these reactions are all initiated by the actions of $cPLA_2$, which releases substrates from precursor phospholipids stored in membranes or in the lipid bodies. The substrates are then converted by enzyme cascades into the ultimate mediators.

1. The first mast cell lipid mediator to be described was **prostaglandin D_2** (PGD_2). Released PGD_2 binds to receptors on smooth muscle cells and acts as a vasodilator and as a bronchoconstrictor. Moreover, PGD_2 is released from lung mast cells during asthmatic bronchoconstriction. PGD_2 is synthesized from arachidonic acid derived from phospholipid by the sequential actions of enzymes. The synthesis of PGD_2, like that of other prostaglandins, depends upon the enzyme cyclooxygenase, and PGD_2 synthesis can be prevented by inhibitors of cyclooxygenase, such as aspirin and nonsteroidal anti-inflammatory agents. Surprisingly,

doses of these drugs that completely prevent PGD_2 synthesis paradoxically exacerbate asthmatic bronchoconstriction. Thus, PGD_2 is unlikely to be a key mediator of this form of immediate hypersensitivity.

2. The second class of mast cell arachidonic acid–derived mediators is the **leukotrienes.** Mast cells convert arachidonic acid, by the action of 5-lipoxygenase and other enzymes, into leukotriene C_4 (LTC_4), which can be subsequently degraded into LTD_4 and LTE_4. (Lesser amounts of LTB_4 may be made as well, but this leukotriene is more characteristically a product of blood phagocytes.) Mast cell–derived leukotrienes bind to specific receptors on smooth muscle cells, different from the receptors for PGD_2, and cause prolonged bronchoconstriction. When injected into skin, these leukotrienes produce a characteristic long-lived wheal and flare reaction. Collectively, LTC_4, LTD_4, and LTE_4 constitute what was once called "slow-reacting substance of anaphylaxis" (SRS-A) and are now thought to be major mediators of asthmatic bronchoconstriction. Pharmacologic inhibitors of 5-lipoxygenase also block anaphylactic reactions in experimental systems. The probable reason that aspirin exacerbates asthma is that PGD_2 and leukotriene synthesis are the two major competitive fates for arachidonic acid in mast cells, and inhibition of cyclooxygenase shunts arachidonic acid into the 5-lipoxygenase pathway, leading to increased leukotriene production.

3. The third lipid mediator produced by mast cells is called **platelet-activating factor** (PAF) for its original bioassay as an inducer of rabbit platelet aggregation. In mast cells and basophils, PAF is synthesized by acylation of lysoglyceryl ether phosphorylcholine, which is derived from a membrane phospholipid by $cPLA_2$-mediated release of a fatty acid from the *sn*2 position. (If the fatty acid is arachidonic acid, then $cPLA_2$ can generate the precursors of all three lipid mediators in one reaction.) PAF has direct bronchoconstricting actions. It also causes retraction of endothelial cells and can relax vascular smooth muscle. However, PAF is very hydrophobic and is rapidly destroyed by a plasma enzyme called PAF hydrolase, limiting its biologic actions. Pharmacologic inhibitors of PAF receptors ameliorate some aspects of immediate hypersensitivity in rabbit lung. Recently, it has been discovered that genetic deficiencies of PAF hydrolase cause asthma in humans. PAF may be of particular importance in late phase reactions, in which it can activate inflammatory leukocytes. In this situation, the major source of PAF may be basophils or the surface of vascular endothelial cells (stimulated by histamine or leukotrienes) rather than mast cells.

Cytokines

It has been appreciated that cultured mast cells and basophils are significant sources of cyto-

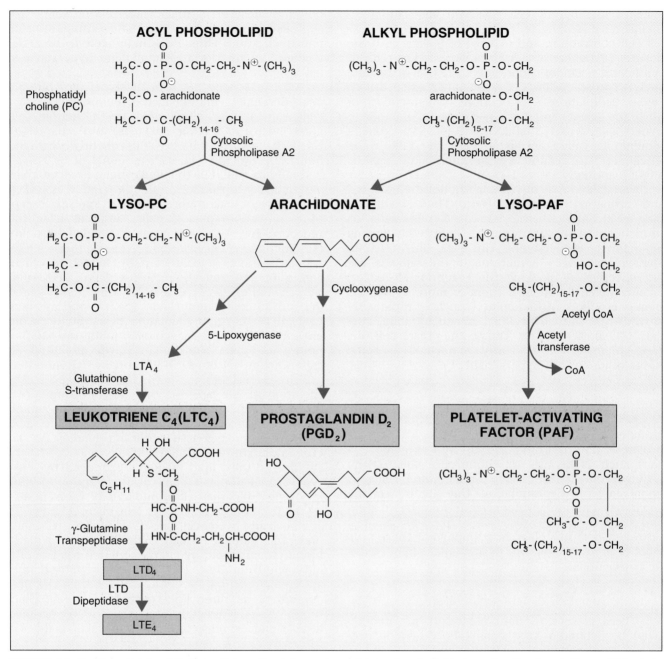

FIGURE 14-6. Biosynthesis of lipid mediators. Breakdown of membrane phospholipids by cytosolic phospholipase A2 leads to generation of leukotriene C_4, prostaglandin D_2, and platelet-activating factor.

kines, including TNF, IL-1, IL-4, IL-5, IL-6, and various colony-stimulating factors (CSFs) such as IL-3 and granulocyte-monocyte colony–stimulating factor (GM-CSF). Many of these cytokines, especially IL-3 and IL-4, were once thought to be exclusively produced by T cells, but isolated basophils may produce more IL-4 per cell than T lymphocytes. The relative contributions of mast cells or basophils versus T cells to the production of these molecules *in vivo* are not yet clear. Nevertheless, *it now appears likely that cytokines, released upon IgE-mediated mast cell or basophil activation or upon T_H2 cell recruitment, are predominantly responsible for the late phase reaction.* TNF in particular may

account for sequential polymorphonuclear and mononuclear cell infiltrates by the same mechanisms described in Chapter 13 for DTH reactions. The key differences between the inflammatory infiltrates of the late phase reaction and those of DTH are the abundance of eosinophils and T_H2 cells in the former compared with activated macrophages and T_H1 cells in DTH. As mentioned earlier, IL-4, through its selective induction of endothelial VCAM-1 and stimulation of eosinophil-selective transendothelial migration, is responsible for selective recruitment of eosinophils, and IL-5 is important for eosinophil activation. Since T_H2 cells have no surface markers that distinguish them from T_H1

cells or undifferentiated CD4⁺ memory T cells, it is not yet known how T$_H$2 cells may be selectively recruited into late phase reactions. Recently, several laboratories have developed mouse models of eosinophil inflammation and airway hyperreactivity in the lung in response to antigen. In general, antibody neutralization experiments in these models have supported important roles for both IL-4 and IL-5 in atopic disease, but protection is often incomplete, suggesting that there may be some redundancy of T$_H$2 cytokines with other mediators.

CLINICAL ALLERGY IN HUMANS

Now that we have described the cells and mediators of immediate hypersensitivity, we can return to consideration of various allergic diseases. As we noted earlier, severely atopic individuals typically have elevated levels of serum IgE. In addition, these individuals have more high-affinity Fcε receptors on each mast cell, and a larger proportion of these receptors are occupied by IgE compared with those in non-atopic individuals.

Atopic individuals characteristically present with one or more manifestations of atopic disease. The most common forms of atopic disease are allergic rhinitis (hay fever), bronchial asthma, and atopic dermatitis (eczema). The clinical and pathologic features of allergic reactions vary with the anatomic site because of several factors: (1) the point of contact with and the nature of the allergen, (2) the concentrations of mast cells in various target organs, (3) the local mast cell phenotype, and (4) the sensitivity of target organs to mast cell–derived mediators. Furthermore, only certain kinds of antigens, including plant pollens, dust mites, and some drugs and foods, typically produce immediate hypersensitivity reactions.

Immediate Hypersensitivity Reactions in the Skin

When an allergen is introduced into the skin of a sensitized, atopic individual, the reaction that ensues is mediated largely by histamine. Histamine binds to venular endothelial cells, which express numerous histamine receptors. Most immediately, the endothelial cells synthesize and release prostacyclin, nitric oxide, and PAF. These mediators cause vascular smooth muscle cell relaxation, and the injection site becomes red from local accumulation of red blood cells. The endothelial cells also retract, allowing plasma to extravasate. The initial redness (erythema) is then replaced by soft swelling, the wheal. Finally, the blood vessels on the edge of the wheal dilate in a reaction augmented by the nervous system, producing the flare. Antihistamines (H1 receptor antagonists) can block this response almost completely. Skin mast cells appear to produce little in the way of long-acting mediators, such as leukotrienes, and the wheal and flare response typically subsides after about 15 to 20 minutes. Clinically, this reaction can occur after contact of the skin with an allergen, or after an allergen enters the circulation via the intestinal tract or by injection. The cutaneous reaction to systemic allergens, called urticaria, may persist for several hours, probably because antigen persists in the plasma.

In many atopic patients, a late phase reaction begins after about 2 to 4 hours. Tumor necrosis factor, IL-4, and other cytokines, probably derived from mast cells, act on the venular endothelial cells to promote inflammation, involving the production of vasodilators, adhesion molecules, and chemokines that facilitate leukocyte diapedesis (see Chapter 13). As may be expected for a cytokine-mediated response, the late phase inflammatory reaction is not inhibited by antihistamines. It can be blocked by pretreatment with corticosteroids, which inhibit cytokine synthesis.

Chronic eczema is probably initiated as a clinical manifestation of the late phase reaction in the skin. However, as the initial stimulus subsides, the inflammatory reaction may continue. At this point, the infiltrates are characterized by both IL-4 and IFN-γ–secreting T cells. Eczema is often treated with topical corticosteroids.

Immediate Hypersensitivity Reactions in the Lung

Immediate hypersensitivity (asthmatic) reactions in the lung resemble skin reactions in some ways but differ in others. In the lung, mast cell mediators act not only on the blood vessels but also on bronchial smooth muscle. Bronchial asthma (Box 14–1) is characterized by paroxysms of bronchial constriction and increased production of thick mucus, which leads to bronchial obstruction and exacerbates respiratory difficulties. The smooth muscle cells in the airways of asthmatics are often hypertrophied and are "hyper-reactive" to constricting stimuli. Bronchitis and infections are frequent complications. Most cases of asthma are due to immediate hypersensitivity, and the bronchial mucosa contains increased numbers of eosinophils and CD4⁺ T cells of the T$_H$2 subset. However, in about 30 per cent of patients, asthma may not be associated with atopy.

Systemic Immediate Hypersensitivity

In systemic immediate hypersensitivity, such as **anaphylactic shock,** vasodilation and vascular exudation of plasma occur in vascular beds throughout the body. This usually implies the systemic presence of antigen by injection, insect sting, or absorption across an epithelial surface, such as the skin or gut mucosa. The decrease in vascular tone and leakage of plasma lead to a fall in blood pressure, or shock, that may be fatal. The cardiovascular effects are complicated by constriction of the upper and lower airways, hypersensitivity of the gut, outpouring of mucus in the gut and respiratory tree, and urticarial lesions (hives) in the skin. The most important mast cell mediators of anaphylactic shock are not fully known. The mainstay of treatment is systemic epinephrine, which can be life-saving, reversing the bronchoconstrictive and vasodilatory effects of the various mast cell mediators. Epinephrine also improves cardiac output, further aiding survival from threatened circulatory collapse. Antihistamines may also be beneficial in anaphylaxis, suggesting a role for histamine in this reaction. In some animal models, PAF receptor antagonists offer partial protection. It is also possible that TNF is an important mediator of anaphylactic shock, much as it is in septic shock (see Chapter 12).

Immunotherapy for Allergy

The mainstay of therapy for atopic disorders such as allergic rhinitis (hay fever) is to block the production or release of mediators (e.g., with corticosteroids or cromolyn) or to antagonize mediator actions on target cells (e.g., with histamine receptor antagonists or β-adrenegic agonists). In addition to therapy aimed at the consequences of immediate hypersensitivity, clinical immunologists often try to limit the onset of allergic reactions by treatments aimed at reducing the quantity of IgE present in the individual. Despite the fact that the regulation of IgE synthesis and T_H2 cell development are not fully understood, several empirical protocols have been developed to diminish specific IgE synthesis.

BOX 14–1. Bronchial Asthma

Asthma is a clinical-pathologic triad of intermittent (and reversible) airway obstruction, chronic bronchial inflammation with eosinophils, and bronchial smooth muscle cell hyper-reactivity to bronchoconstrictors. Episodes of bronchoconstriction are recurrent, and the airways of patients with asthma show chronic changes, notably increased numbers of smooth muscle cells, increased numbers of mucus-producing cells, and increased deposition of connective tissue ("basement membrane") beneath the bronchial epithelium. Strictly speaking, asthma does not cause destruction of the airways or airspaces (alveoli), but it frequently co-exists with bronchitis or emphysema, which do. Asthma affects about 10 million people in the United States. The prevalence rate is similar in other industrialized countries, but it may be lower in less-developed areas of the world. Affected individuals may suffer considerable morbidity, and asthma causes death in some patients.

Several lines of evidence have established that most asthma is a form of immediate hypersensitivity. About 70 per cent of patients show positive skin test responses (i.e., wheal and flare reactions) to injection of one or more common allergens. Furthermore, airway obstruction can be triggered in many asthmatic patients by inhalation of specific allergens. Finally, many asthmatics and their family members have other manifestations of atopy, such as rhinitis (hay fever) or eczema. In the most common form of occupational asthma exposure, to isocyanates, it is thought that these reactive chemicals act as haptens so that chemically modified (i.e., haptenated) self proteins, such as albumin, appear to function as the allergen. Even among non-atopic asthmatics, the pathophysiology of airway constriction is similar, suggesting that alternative mechanisms of mast cell degranulation (e.g., by locally produced neurotransmitters) may underlie the disease.

Most investigators now believe that asthma should be considered an inflammatory disease. In this view, the primary lesion of asthma is the accumulation of $CD4^+$ T_H2 T cells, basophils, and eosinophils in the airway mucosa, an example of a late phase reaction. Smooth muscle cell hypertrophy and hyper-reactivity are thought to result from leukocyte-derived mediators and cytokines. The pathophysiologic sequence may be initiated by mast cell activation in response to allergen binding to IgE. Mast cell–derived cytokines lead to recruitment of eosinophils, basophils, and T_H2 cells. Mast cells, basophils, and eosinophils all produce mediators that constrict airway smooth muscle. The most important of the bronchoconstricting mediators are LTC_4 and its breakdown products, LTD_4 and LTE_4, and PAF. In clinical experiments, antagonists of LTC_4 synthesis or leukotriene receptor antagonists prevent allergen-induced airway constriction. Aspirin, which blocks cyclooxygenase and shunts more arachidonic acid into leukotriene biosynthesis, can exacerbate or even trigger asthmatic attacks. Recent genetic evidence has pointed to PAF as a mediator of asthma. Specifically, individuals with genetic deficiency of PAF hydrolase, the plasma enzyme that inactivates PAF, develop asthma in early childhood. Histamine plays little role in airway constriction, and antihistamines (H1 receptor antagonists) have no role in treatment of asthma. Indeed, since many antihistamines are also anticholinergics, these drugs may worsen airway obstruction by causing thickening of mucous secretions, an action opposed by acetylcholine.

Current therapy has two major targets: prevention and reversal of inflammation, and relaxation of airway smooth muscle. In recent years, the balance of therapy has shifted toward anti-inflammatory agents as a primary mode of treatment. Two major classes of drugs are in current use: corticosteroids, which block cytokine production, and sodium cromolyn. The mechanism of action of sodium cromolyn is not clear, but it appears to antagonize IgE-induced release of mediators. Both agents can be used prophylactically as inhalants. Corticosteroids may also be given systemically, especially once an attack is under way, to reverse inflammation. Newer anti-inflammatory therapies are in pre-clinical and clinical trials, including inhibitors of leukocyte adhesion to endothelial cells (e.g., antibodies to VCAM-1, ICAM-1, or E-selectin) and PAF receptor antagonists. It remains to be seen whether these agents are effective in treating asthma.

Bronchial smooth muscle cell relaxation has principally been achieved by elevating intracellular cAMP in the smooth muscle cells, which inhibits contraction. The major drugs used are activators of adenylyl cyclase, such as epinephrine and related $\beta2$-adrenergic agents, and inhibitors of the phosphodiesterase enzymes that degrade cAMP, such as theophylline. These drugs also raise cAMP in mast cells, acting to inhibit mast cell degranulation. Inhalants of epinephrine-like drugs have been widely used as prophylactic therapy, but long-term cardiovascular side effects are limiting use of these agents. Inhibitors of leukotriene synthesis and leukotriene receptor antagonists are newer approaches to preventing bronchoconstriction.

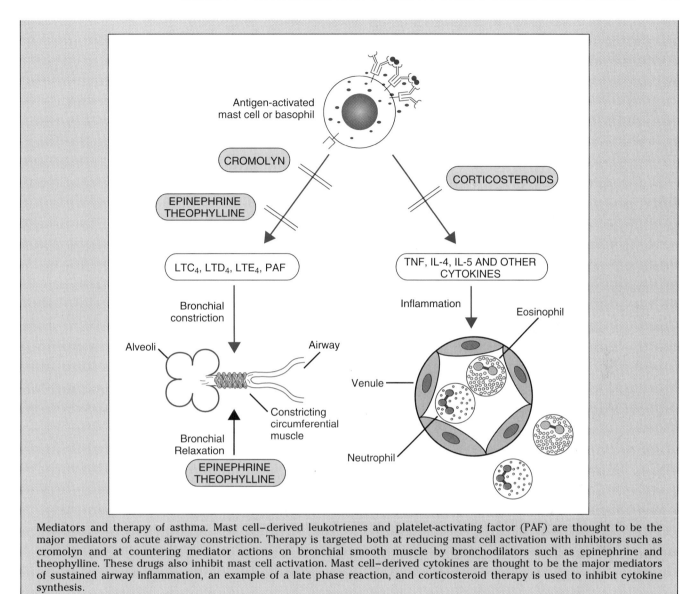

Mediators and therapy of asthma. Mast cell–derived leukotrienes and platelet-activating factor (PAF) are thought to be the major mediators of acute airway constriction. Therapy is targeted both at reducing mast cell activation with inhibitors such as cromolyn and at countering mediator actions on bronchial smooth muscle by bronchodilators such as epinephrine and theophylline. These drugs also inhibit mast cell activation. Mast cell–derived cytokines are thought to be the major mediators of sustained airway inflammation, an example of a late phase reaction, and corticosteroid therapy is used to inhibit cytokine synthesis.

In one approach, called **desensitization,** small but increasing quantities of antigen are administered subcutaneously over a period of hours or more gradually over weeks or months. As a result of this treatment, specific IgE levels decrease and IgG titers often rise, perhaps further inhibiting IgE production by neutralizing the antigen and by antibody feedback (see Chapter 10). It is also possible that desensitization may work by inducing specific T cell tolerance or by changing the predominant phenotype of antigen-specific T cells from T_H2 to T_H1; however, there are few direct data to support these hypotheses. The beneficial effects of desensitization may occur in a matter of hours, much earlier than changes in IgE levels. Although the precise mechanism is unknown, this approach has been very successful in preventing acute anaphylactic responses to protein antigens (e.g., insect venoms) or vital drugs (e.g., penicillin). It is more variable in its effectiveness for chronic atopic conditions such as hay fever.

SUMMARY

Binding of antigen to IgE attached to mast cells or basophils initiates an effector reaction that functions to protect the host against parasitic infection. The same reaction may be triggered by environmental antigens, called allergens, producing an injurious process called immediate hypersensitivity. Immediate hypersensitivity is characterized by rapid vascular leakage of plasma, vasodilation, bronchoconstriction, and, at later times, inflammation. The inflammatory infiltrates of this late phase reaction are enriched in eosinophils, basophils, and T_H2 cells. In extreme cases (anaphylaxis), death may result from asphyxiation and circulatory collapse. Individuals prone to immediate hypersensitivity reactions are called atopic and often have more IgE and more IgE receptors per mast cell than do non-atopic individuals.

IgE synthesis is regulated by heredity, exposure to antigen, and T cell cytokines. In particular, T_H2 cells, through secretion of IL-4 and IL-5, favor IgE production and eosinophil-rich inflammation. Mast cells are derived from bone marrow and mature in the tissues. In mice, many mast cells can be classified as connective tissue type, maturing under the influence of fibroblast-derived factors such as c-kit ligand, and mucosal type, responding to T cell–derived cytokines such as IL-3. Human mast cells may also consist of distinct subsets, but these are not as well defined.

Basophils are granulocytes that accumulate at inflammatory sites. They are functionally similar to mast cells but appear to be a distinct cell lineage. Eosinophils are a special class of granulocytes that are recruited into inflammatory reactions by IL-4 and are activated by IL-5. Eosinophils are effector cells of IgE-initiated reactions. They mediate IgE-directed antibody-dependent cell-mediated cytotoxicity to eradicate parasites. In allergic reactions, eosinophils contribute to tissue injury.

Upon binding of antigen to IgE on the surface of mast cells or basophils, the high-affinity Fcε receptors become cross-linked and activate intracellular second messengers. Activated mast cells and basophils produce three important classes of mediators: biogenic amines, such as histamine; lipid mediators, such as prostaglandin D_2, leukotrienes C_4, D_4, and E_4, and platelet-activating factor; and cytokines, such as TNF, IL-4, and IL-5. Biogenic amines and lipid mediators produce the rapid components of immediate hypersensitivity, such as vascular leakage, vasodilation, and bronchoconstriction. Cytokines mediate the late phase reaction.

Various organs show distinct forms of immediate hypersensitivity involving different mediators and target cell types. Drug therapy is aimed at inhibiting mast cell mediator production and at blocking or counteracting the effects of mediators on target organs. The goal of immunotherapy is to prevent or reduce T cell responses and immediate hypersensitivity reactions to specific antigens.

Selected Readings

Bochner, B. S., and L. M. Lichtenstein. Anaphylaxis. New England Journal of Medicine 324:1785–1790, 1991.

Bochner, B. S., B. J. Undem, and L. M. Lichtenstein. Immunological aspects of allergic asthma. Annual Review of Immunology 12:295–335, 1994.

Corrigan, C. J., and A. B. Kay. T cells and eosinophils in the pathogenesis of asthma. Immunology Today 13:501–507, 1992.

Drazen, J. M., J. P. Arm, and K. F. Austen. Sorting out the cytokines of asthma. Journal of Experimental Medicine 183:1–6, 1996.

Galli, S. J. New concepts about the mast cell. New England Journal of Medicine 328:257–265, 1993.

Galli, S. J., K. M. Szebo, and E. N. Geissler. The kit ligand, stem cell factor. Advances in Immunology 55:1–96, 1994.

McFadden, E. R., Jr., and I. A. Gilbert. Asthma. New England Journal of Medicine 327:1928–1937, 1992.

O'Hehir, R. E., R. D. Garman, J. L. Greenstein, and J. R. Lamb. The specificity and regulation of T-cell responsiveness to allergens. Annual Review of Immunology 9:67–95, 1991.

Paul, W. E., R. A. Seder, and M. Plaut. Lymphokine and cytokine production by FcεRI cells. Advances in Immunology 53:1–29, 1993.

Romagnani, S. Regulation and deregulation of human IgE synthesis. Immunology Today 11:316–321, 1990.

Schwartz, L. B., and K. F. Austen. Structure and function of the chemical mediators of mast cells. Progress in Allergy 34:271–321, 1984.

Valent, P., and P. Bettelheim. Cell surface structures on human basophils and mast cells: biochemical and functional characterization. Advances in Immunology 52:333–423, 1992.

Wardlaw, A. J., R. Moqbel, and A. B. Kay. Eosinophils: biology and role in disease. Advances in Immunology 60:151–266, 1995.

CHAPTER FIFTEEN

THE COMPLEMENT SYSTEM

The **complement system** comprises a group of more than 30 serum and cell surface proteins that interact with other immune system molecules and with one another in a highly regulated manner to provide many of the effector functions of humoral immunity and inflammation. The name of the complement system was derived from experiments performed by Charles Bordet shortly after the discovery of antibodies. He demonstrated that if fresh serum containing an antibacterial antibody was added to the bacteria at physiologic temperature (37° C), the bacteria were lysed. If, however, the serum was heated to 56° C or more, it lost its lytic capacity. This was not because of decay of antibody activity, because antibodies are heat stable, and even heated serum was capable of agglutinating the bacteria. Bordet concluded that the serum must contain another heat-labile component that assists or "complements" the lytic function of antibodies, and he called this component complement.

The principal biologic functions of the complement system include the following (Fig. 15–1):

1. Certain activated complement components mediate **cytolysis** by polymerizing on cell surfaces to form pores or by disrupting the integrity of the phospholipid bilayer in the membranes of these cells. In this way, if complement activation occurs on the surfaces of foreign organisms, these microbes can be killed by osmotic lysis.

2. **Opsonization** of foreign organisms or particles occurs by the binding of complement proteins to their surfaces. These complement proteins are called **opsonins.** Phagocytic leukocytes express specific receptors for these opsonins. In this way, opsonins promote phagocytosis of particles or organisms.

3. **Activation of inflammation** occurs in response to the generation of certain proteolytic fragments of complement proteins. These complement-derived peptides act on several targets. They activate mast cells, causing reactions that resemble immediate hypersensitivity (see Chapter 14); in extreme cases, this reaction can mimic anaphylaxis, and these complement fragments are sometimes called **anaphylatoxins.** Other targets of complement-derived peptides include vascular endothelium, smooth muscle, and inflammatory leukocytes.

4. The complement system promotes **solubilization and phagocytic clearance of immune complexes,** thereby minimizing the damage caused when immune complexes that are formed in the circulation deposit in tissues or vessel walls.

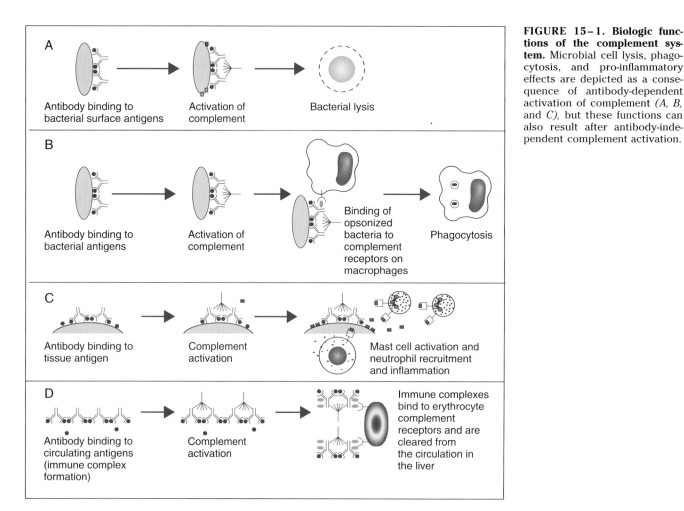

FIGURE 15–1. Biologic functions of the complement system. Microbial cell lysis, phagocytosis, and pro-inflammatory effects are depicted as a consequence of antibody-dependent activation of complement *(A, B,* and *C),* but these functions can also result after antibody-independent complement activation.

5. The complement system plays a significant role in **promoting humoral immune responses** by aiding in antigen presentation to B cells in germinal centers and by lowering the threshold of sensitivity of B cell activation by antigen. These functions are mediated by receptors for complement fragments expressed on follicular dendritic cells and on B lymphocytes.

The functions of complement proteins will be discussed in detail later in the chapter. The complement system has a number of important properties that enable it to operate efficiently in host defense against foreign invaders without injuring normal tissues:

1. *The complement system amplifies the response to microbes by means of an enzymatic cascade.* The soluble serum complement components include multiple proteolytic enzymes, which become sequentially activated by proteolysis and proteolytically cleave and activate other proteins of the complement system. Proteins that acquire proteolytic enzymatic activity by the action of other proteases are called **zymogens.** The process of sequential zymogen activation, i.e., an enzymatic cascade, is also characteristic of the coagulation and kinin systems. Proteolytic cascades allow for tremendous amplification, since each individual enzyme molecule activated at one step can generate multiple activated enzyme molecules, or active fragments, at the next step.

2. Two converging pathways of complement activation co-exist, each one initiated by a specific set of stimuli, but both pathways share homologous molecules with similar functions. The phylogenetically oldest of the two pathways is called the **alternative pathway** because it was discovered second. It is usually initiated on microbial surfaces without the requirement for specific immune responses and is an important mechanism of innate immunity against infectious organisms. The **classical pathway** is usually initiated by the binding of complement proteins to antibody-antigen complexes, and in this way serves as an effector mechanism of specific humoral immunity. There are, however, ways in which the classical pathway can be activated directly by microbes in the absence of antibodies. Both pathways converge to a common final pathway that generates an assembly of proteins with cytolytic activity called the **membrane attack complex.**

3. Although the components of the proteolytic cascades of the complement system are present as soluble serum proteins, they are inactive or have only a low level of spontaneous activation in the circulation. Two mechanisms ensure that *initiation of the complement cascades occurs only at certain localized sites where it will be most useful.* First, some complement components are only activated by binding to certain types of molecules that are present on the surfaces of infectious organisms but not on normal host cells. Second, the require-ment for immune complexes for classical pathway initiation focuses complement activation to sites where specific antibodies bind to foreign antigens.

4. *The various biologic functions of the complement system are mediated in two general ways.* First, complexes of complement proteins become directly bound to, and influence the fate of, microbes or immune complexes. Second, soluble fragments of complement proteins are generated during activation. These fragments diffuse from the sites where they were generated, bind to specific receptors on other nearby cells, and thereby activate effector functions of those cells.

5. *The complement system is tightly regulated by several soluble and cell membrane–associated proteins* that inhibit complement activation at multiple steps. These regulatory mechanisms have two main functions. First, they limit or stop complement activation after the system has appropriately performed its functions. Second, they prevent abnormal or constitutive complement activation in the absence of microbes and antibodies. Thus, these regulatory mechanisms in effect enable the complement system to distinguish self from nonself and thereby prevent damage to normal tissues.

This chapter describes the biochemistry, regulation, and functions of the complement system and the relationship of complement to other components of the immune system.

THE COMPLEMENT CASCADES

Before we describe the details of the complement cascades, it is useful to present an overview of the system (Fig. 15–2). The **C3** protein plays a central role in both pathways of complement activation. It is a disulfide-linked heterodimeric glycoprotein composed of α and β polypeptide chains. The serum concentration of C3, approximately 0.55 to 1.2 mg/ml, is higher than that of any other component of the complement system. The biologically active forms of C3 are its proteolytic cleavage products; therefore, proteolysis of C3 is a key step in complement activation. This is accomplished by enzymes called **C3 convertases,** which cleave C3 to produce C3a and C3b. (By convention, lower case letter suffixes denote proteolytic products of each complement protein.) During activation, both the classical and alternative pathways generate multiprotein complexes with C3 convertase activity. In the alternative pathway, C3b or a fluid phase equivalent called C3i is spontaneously generated at low levels in the circulation and binds a protein fragment called Bb, generated by proteolytic cleavage of a protein called Factor B. The C3bBb complex is the **alternative pathway C3 convertase,** and it functions to further break down C3, generating more C3b. In the early steps of the classical pathway, antibody molecules that are complexed with antigen also bind, via their Fc portions, a complement protein called C1. C1 then sequen-

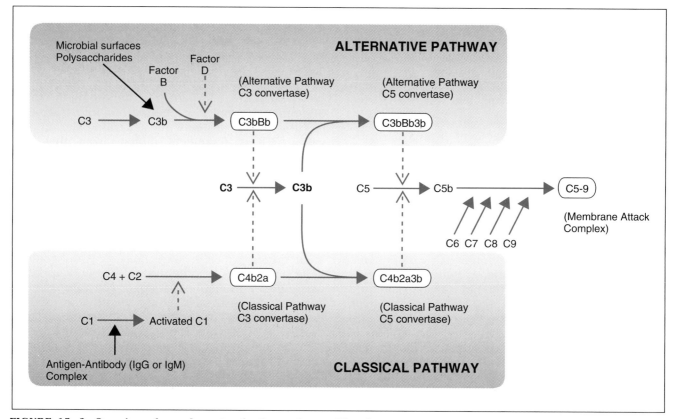

FIGURE 15–2. Overview of complement activation pathways. The alternative pathway, in a state of low-level activation in the blood, is accelerated by C3b binding to various activating surfaces, such as microbial cell walls. The classical pathway is initiated by C1 binding to antigen-antibody complexes. The C3b involved in alternative pathway initiation may be generated in several ways, including spontaneously, by the classical pathway, or by the alternative pathway itself (see text). Both pathways converge and lead to the formation of the membrane attack complex. In this and subsequent figures, dashed lines indicate proteolytic activities of various components.

tially cleaves and activates C4 and C2, leading to the formation of a C4b2a complex, which functions as the **classical pathway C3 convertase.** Since both the alternative and classical pathways are directed toward the molecular assembly of C3 convertases, it is not surprising that there are structural and functional homologies between components of each pathway. Thus, C2 and C4 of the classical pathway are equivalent to Factor B and C3, respectively, of the alternative pathway.

Both C3 and C4 contain an unusual internal thioester bond between a cysteine residue and a nearby glutamate residue (Fig. 15–3). When C3 and C4 are converted to C3b and C4b, respectively, in the alternative and classical pathways, these thioester bonds become unstable and highly reactive, and they rapidly form covalent linkages with surface molecules of microorganisms or with immune complexes. Therefore the *C3 convertases which form in each pathway are covalently bound to microbial surfaces or antigen-antibody complexes, and this ensures that the subsequent activation and amplification steps of the complement cascades will be localized to these sites.*

After the C3 convertase cleaves C3 to generate C3b, the next step in both pathways is the binding of the C3b to the C3 convertase complex, changing

it to a **C5 convertase,** which catalyzes the proteolytic cleavage of the C5 protein. Although the C5 convertases of the two pathways are molecularly distinct, they catalyze identical reactions and act on identical substrates. Once C5 is cleaved, both pathways share the same terminal steps. These terminal events do not involve proteolysis but rather the sequential binding of several soluble complement proteins, called C6, C7, C8, and C9, to the activating surface. This leads to the formation of a lipid-soluble pore structure called the **membrane attack complex** (MAC), which causes osmotic lysis of cells.

Different effector functions are mediated by different proteins produced during complement activation. Cytolysis is mediated by the MAC. Opsonization is largely due to C3b, specific receptors for which are expressed on many phagocytic leukocytes and other cells. Inflammation, consisting of the activation of mast cells and recruitment of leukocytes, is mediated by cleavage products of C3, C4, and C5, called C3a, C4a, and C5a, respectively. Immune complexes are solubilized by binding of the classical pathway proteins to the Fc regions of antibody molecules. Each of these functions will be discussed in more detail later.

In the following discussion, we describe the

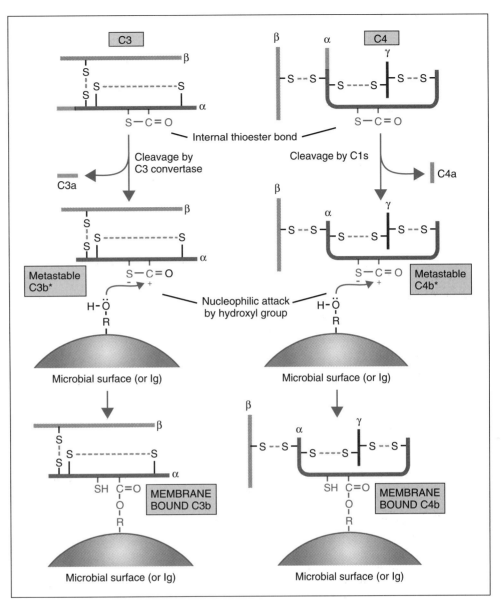

FIGURE 15–3. Internal thioester bonds of C3 and C4 molecules. A schematic view of the internal thioester bonds in the C3 and C4 molecules and their role in forming covalent bonds with other molecules is shown. Proteolytic cleavage of the α chains of C3 and C4 converts them into metastable forms (C3b* and C4b*) in which the internal thioester bonds are susceptible to nucleophilic attack by oxygen or nitrogen atoms. The result is the formation of covalent bonds with proteins or carbohydrates on cell surfaces (solid phase C3b and C4b).

various proteins that make up the complement pathways; the major biochemical and functional characteristics of these proteins are listed in Table 15–1. We will begin with the phylogenetically older alternative pathway, an understanding of which can serve as a basis for understanding the classical pathway.

The Alternative Pathway

The alternative pathway (Fig. 15–4) is activated in the absence of antibody and is therefore a mechanism of innate immunity. It generates both soluble and membrane-bound forms of C3 convertase, which catalyze the proteolysis of C3. In fact, *the alternative pathway is in a continuous state of low-level activation due to the reactivity of soluble C3 with water.* This fluid-phase activation, sometimes called "C3 tickover," goes on to promote a solid-

phase activation of the pathway on microbial surfaces, which involves basically the same steps and same proteins. The fluid-phase cascade is initiated when the internal thioester bond of C3 is hydrolyzed by water to form an unstable intermediate called C3i (or C3 [H_2O]) (Fig. 15–4). C3i binds to the alternative pathway protein **Factor B**. After it is complexed with C3i, Factor B becomes susceptible to proteolysis by another alternative pathway protein, **Factor D,** a serine protease that circulates at very low concentrations in the serum, probably in its enzymatically active form. Factor D proteolytically cleaves bound Factor B, releasing a small fragment (Ba) and leaving a larger fragment, Bb, attached to C3i. The C3iBb complex is unstable, however, and it rapidly decays unless it is stabilized by the binding of another alternative pathway member, **properdin.** Properdin may also bind to C3i before, and promote the binding of, Factor B.

TABLE 15–1. Protein Components of the Complement Cascades

Component	Molecular Size (kD)	Serum Concentration (μg/ml)	Subunit Chains; Molecular Size of Each (kD)	Activation Products	Comments on Function
Alternative Pathway					
Factor B	93	200	1	Ba	Bb is a serine protease, part of C3 and C5 convertases
Factor D	25	1–2	1		Protease that circulates in active state; cleaves Factor B bound to C3b
Properdin	220	25	4; 56		Stabilizes alternative pathway C3 convertase
C3 (also part of classical pathway)	195	550–1200	1 of α; 110 1 of β; 85	C3a C3b	C3a is an anaphylatoxin C3b covalently binds to activating surfaces, where it is part of C3 and C5 convertases and also acts as an opsonin
Classical Pathway					
C1 (C1qr$_2$s$_2$)	750				Initiates classical pathway
C1q	410	75	6 of A; 24 6 of B; 23 6 of C; 22		Binds to Fc portion of Ig
C1r	85	50	2	C1r	C1r is a serine protease; cleaves C1s
C1s	85	50	2	C1s	C1s is a serine protease; cleaves C4 and C2
C4	210	200–500	1 of α; 90 1 of β; 78 1 of γ; 33	C4a C4b	C4a is an anaphylatoxin C4b covalently binds to activating surfaces, where it is part of C3 convertase and also acts as an opsonin
C2	110	20	1	C2a C2b	C2a is a serine protease, part of C3 and C5 convertases
Terminal Lytic Components					
C5	190	70	1 of α; 115 1 of β; 75	C5a C5b	C5a is an anaphylatoxin C5b initiates MAC assembly
C6	128	60	1		Component of MAC
C7	121	60	1		Component of MAC
C8	155	60	1 of α; 64 1 of β; 64 1 of γ; 22		Component of MAC
C9	79	60	1		Component of MAC; polymerizes to form membrane pores

Abbreviations: kD, kilodalton; MAC, membrane attack complex; Ig, immunoglobulin.

The resulting stable C3iBb complex is called the initial alternative pathway C3 convertase, with the Bb fragment functioning as a serine protease capable of cleaving C3. As a result of this proteolytic step, the thioester bond in the α chain of the C3b fragment is susceptible to nucleophilic attack by amine and hydroxyl groups and thus becomes highly unstable (see Fig. 15–3). This chemically reactive form of the molecule is called metastable C3b. Most of the C3b thioester bonds rapidly react with water molecules. However, some C3b thioester bonds undergo transesterification to form covalent amide or ester bonds with microbial cell surfaces. Once these stable covalent linkages are formed, C3b goes on to behave like C3i, recruiting Factor B and Factor D, permitting proteolysis of Factor B, and binding Bb and properdin to form the stable, solid phase alternative pathway C3 convertase.

The normal function of the alternative pathway depends on two important features of the activation and regulation of its components:

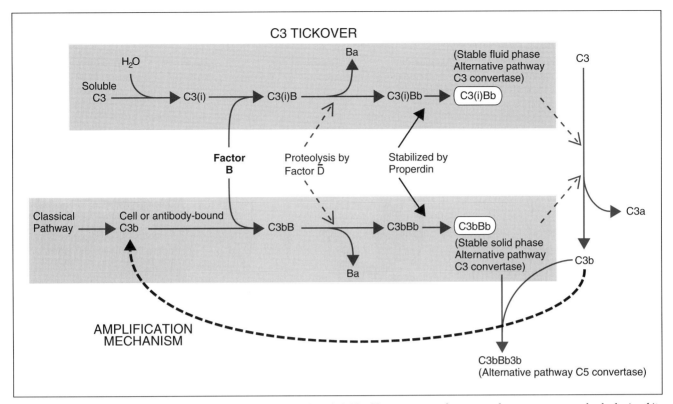

FIGURE 15–4. Alternative pathway of complement activation. Soluble C3 in serum undergoes a slow, spontaneous hydrolysis of its internal thioester bond, generating C3i, which binds Factor B and forms a fluid-phase alternative pathway C3 convertase. The resulting continuous, low-level production of C3b by fluid-phase C3 convertase is called C3 tickover. C3b can also be produced by the classical pathway. The C3b formed either way can then participate in the formation of alternative pathway C3 convertase, which generates C3b from C3. The alternative pathway has a built-in amplification mechanism because the C3b generated by the alternative pathway C3 convertase is also a component of the convertase. The C3 convertase formed by the alternative pathway proteolytically cleaves C5, just as the classical pathway C5 convertase does, to begin the formation of the membrane attack complex.

1. As discussed above, C3b is continuously generated in the circulation by fluid-phase alternative pathway C3 convertase. However, it has no deleterious effects because the C3b generated by the C3 tickover mechanism is usually hydrolyzed to an inactive form within a fraction of a second. As a result, significant complement activation does not occur in the circulation. Rare C3b molecules do form covalent bonds with amino or hydroxyl groups on nearby surfaces in a random manner. It is at this stage of the alternative pathway that the rudimentary capacity for self/non-self discrimination is apparent. *If the C3b is deposited on autologous cell surfaces, it is rapidly inactivated by the action of regulatory proteins, described later, thus stopping the cascade.* In contrast, C3b binding to the surfaces of many microbes (which lack the regulatory proteins) leads to the binding of Factor B and, as described above, to the formation of stable, enzymatically active, surface-bound alternative pathway C3 convertase, or C3bBb.

2. *The alternative pathway provides an intrinsic amplification mechanism for the complement system.* The surface-bound C3bBb enzyme complex generates many more C3b molecules, and these, in turn, are deposited on the same surface, forming more C3 convertases. Thus, C3b is both a component of

the C3 convertase enzyme complex as well as a product generated by the action of C3 convertase. This situation allows for a positive feedback amplification of the alternative pathway (Fig. 15–4). In fact, because C3b generated by the classical pathway (see below) can trigger the alternative pathway, the alternative pathway C3 convertase is also an amplification mechanism for complement activation initiated by the classical pathway. The magnitude of this amplification mechanism is demonstrated by the fact that several million C3b molecules can be generated on a cell surface within 5 minutes of initiation of the solid-phase alternative pathway.

Some of the C3b molecules generated by the alternative pathway C3 convertase bind to the C3 convertase itself to form a new complex, C3bBb3b. This is the **alternative pathway C5 convertase,** and its function is to proteolytically cleave C5. After this step, the classical and alternative pathways converge, and the same terminal sequence of events occurs, as described later.

The Classical Pathway

The classical pathway (Fig. 15–5) is one of the major effector mechanisms of humoral immunity. It

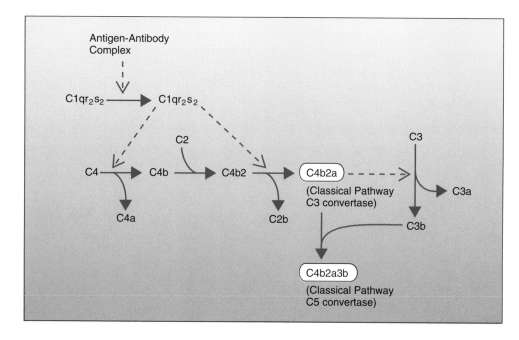

FIGURE 15–5. Classical pathway of complement activation. Antigen-antibody complexes that activate the classical pathway may be soluble or fixed on the surface of cells or trapped in interstitial spaces. The C5 convertase formed by the classical pathway proteolytically cleaves C5 to begin the formation of the membrane attack complex.

is activated principally by the binding of the first component of the classical pathway, C1, to the Fc portions of antibody molecules that have bound antigen. Stringent requirements for antibody-mediated C1 activation ensure that the classical pathway is activated only under certain conditions:

1. Activation of C1 occurs at significant levels only when it binds to the $C\gamma2$ domains of IgG or the homologous $C\mu3$ domains of IgM molecules (see Chapter 3). IgA, IgD, and IgE cannot activate the classical pathway. Furthermore, only certain subclasses of IgG are effective C1 activators. Human IgG3 and IgG1 are very effective; IgG2 is a relatively poor activator; and IgG4 does not activate complement. These differences are attributed to structural variations in the constant regions of Ig isotypes and subtypes, which influence the ability of C1 to bind and/or to be activated after binding.

2. A single C1 molecule must bind simultaneously to at least two Fc portions of Ig, each Fc portion having a single C1 binding site (Fig. 15–6). This is required in order to induce a conformational change in the C1 molecule that turns on its enzymatic activity. Because secreted IgM is a pentamer containing five Fc regions, even a single IgM molecule can bind C1 and trigger the classical pathway. In contrast, since IgG is a monomer, multiple IgG molecules must be aggregated, bringing multiple Fc regions sufficiently close together, before C1 can be activated. Such aggregation typically occurs if the IgG antibody binds to a multideterminant antigen, such as on a bacterial cell surface. Because of these structural differences in Ig isotypes, IgM is a much more efficient complement-binding (also called **complement-fixing**) antibody than is IgG.

3. Only antigen-antibody complexes and not free or soluble antibodies activate complement. For IgG, this is because of the requirement for aggregation mentioned above. Even though free IgM in the circulation is pentameric, it does not bind

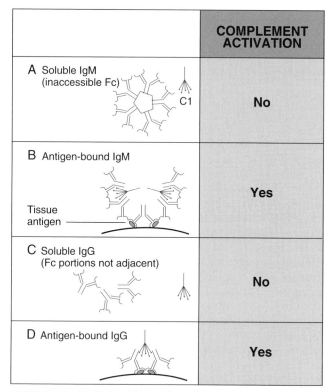

	COMPLEMENT ACTIVATION
A Soluble IgM (inaccessible Fc)	No
B Antigen-bound IgM	Yes
C Soluble IgG (Fc portions not adjacent)	No
D Antigen-bound IgG	Yes

FIGURE 15–6. C1 binding to Fc portions of IgM and IgG. C1 must bind to two or more Fc portions in order to initiate the complement cascade. The Fc portions of soluble pentameric IgM are not accessible to C1 (*A*). After IgM binds to surface-bound antigens, it undergoes a shape change that permits C1 binding and activation (*B*). Monomeric soluble Ig molecules will not activate C1 (*C*), but after binding to cell surface antigens, adjacent IgG Fc portions can bind and activate C1 (*D*).

C1, apparently because the Fc regions are inaccessible to C1 in solution. Binding of the IgM to an antigen or a cell surface induces a conformational change that exposes the Fc regions, allowing C1 to bind (Fig. 15–6).

C1 is a large, multimeric protein complex, approximately 750 kD, composed of C1q, C1r, and C1s subunits. The C1q subunit is composed of an umbrella-like radial array of six chains, each of which has a globular head connected by collagen-like arms to a central stalk (Fig. 15–7). This hexamer performs the recognition function of the molecule, binding specifically to the Fc region of Ig. The catalytic functions of C1 are mediated by a C1s-C1r-C1r-C1s (C1r$_2$-C1s$_2$) tetramer that is noncovalently associated with the C1q subunit. Both C1r and C1s are **serine esterases** that remain inactive until the C1 molecule binds to an immune complex. *Binding of two or more of the globular heads of C1q to IgM or IgG molecules induces a conformational change that leads to the enzymatic activation of the associated C1r, which then cleaves and activates C1s.* C1s, in turn, acts on the next two components of the classical pathway, C4 and C2.

A family of C1q-like molecules, called **collectins,** may participate in antibody-independent microbial defense mechanisms. The collectins are lectin (carbohydrate-binding) molecules with collagen-like stalks, which resemble C1q. Examples include mannose binding protein and conglutinin, both found in the plasma, as well as lung surfactant proteins A and D. These molecules can bind to several carbohydrates on the surfaces of infectious organisms, and this may lead to the clearance of the microbes in one of two ways. First, at least one of the collectins, mannose binding protein, is capable of triggering the classical pathway in much the same way as C1q, either by activating C1q-C1r-C1r-C1s or by associating with another serine esterase, called mannose binding protein–associated serine esterase, which cleaves C4. Thus, mannose binding protein may lead to complement-mediated killing of microbes in the absence of Ig. Second, several cell types, including leukocytes, platelets, endothelial cells, and smooth muscle cells, express molecules called **C1q receptors,** which bind both C1q and the collectins. C1q receptors transduce signals that enhance the phagocytic and metabolic activities of leukocytes and thereby may enhance clearance of microbes to which collectins have bound. In a similar fashion, Ig-C1q complexes bound to microbial surfaces may serve as opsonins for C1q receptor–bearing phagocytes.

C4 is the next serum complement protein to be activated in the classical pathway. It is homologous to C3, participates in the formation of the C3 convertase, and is critical for keeping subsequent activation steps localized to the initial site where the antibody is bound. C4 consists of three polypeptide chains, α, β, and γ, and the α chain contains an internal thioester bond similar to that found in C3. As we discussed earlier, this feature is important for localizing the functions of C4 (see Fig. 15–3). The α chain of soluble C4 is cleaved by the active C1s component of an Ig-bound C1 complex, yielding a small fragment called C4a and a large C4b molecule. (C4a diffuses from the cell surface and has biologic effects that are discussed later in the chapter.) One C1s molecule can generate multiple C4b molecules. As a result of this proteolytic step, a short-lived metastable C4b is formed in which the thioester bond in the α chain is highly susceptible to nucleophilic attack. Most of these molecules rapidly react with water and are inactivated. However, some C4b forms covalent bonds with hydroxyl or amino groups of cell surface molecules (Fig. 15–3). There are two isoforms of C4, called C4A and C4B, encoded by adjacent genes in the major histocompatibility complex, and there is evidence that C4A preferentially forms amide bonds with proteins, whereas C4B forms ester

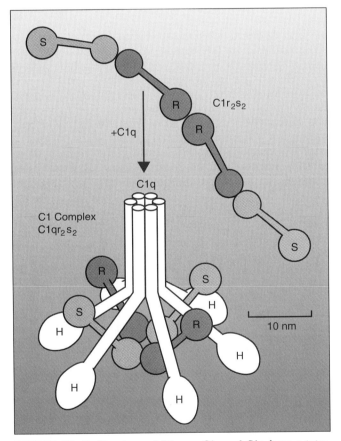

FIGURE 15–7. Structure of C1q,r,s. C1r and C1s form a tetramer composed of two C1r and two C1s molecules. The larger balls at the ends of C1r and C1s, designated by the letters R and S, are the catalytic domains of these proteins. C1q consists of six identical subunits, arranged to form a central core and symmetrically projecting radial arms. The globular heads at the end of each arm, designated H, are the contact regions for immunoglobulin. One C1r$_2$s$_2$ tetramer wraps around the radial arms of the C1q complex in a manner that juxtaposes the catalytic domains of C1r and C1s. (Modified with permission from Arlaud, G. J., M. G. Colomb, and J. Gagnon. A functional model of the human C1 complex. Immunology Today 8:107–109, 1987.)

bonds with carbohydrates. As a result, the C4b molecule becomes covalently attached to nearby cell surfaces or to Ig molecules, ensuring that complement activation occurs stably and efficiently at the site where the bound antibody is located.

C2 is the third soluble serum component of the classical pathway, and it is also involved in the formation of the C3 convertase. It is a single chain polypeptide that binds to cell surface–bound C4b molecules in the presence of Mg++. Once complexed with C4b, C2 is cleaved by a nearby C1s molecule to generate a soluble C2b fragment of unknown importance, and a larger C2a fragment that remains physically associated with C4b on the cell surface. *The resulting C4b2a complex is the* **classical pathway C3 convertase,** *having the ability to bind to and proteolytically cleave C3.* Binding to C3 is mediated by the C4b component, and proteolysis is catalyzed by the C2a component.

C3 is sequentially the fourth soluble serum component of the classical pathway. As mentioned earlier, it is also a central component of the alternative pathway. The classical pathway C3 convertase removes a small C3a fragment from the α chain of C3 and leaves C3b. (C3a has several biologic activities initiated by binding to C3a receptors on other cells; these are discussed later in the chapter.) As with C4b, newly generated metastable C3b molecules contain unstable thioester bonds that react with water, yielding inactive C3b byproducts that no longer participate in the complement cascade. However, up to 10 per cent of metastable C3b molecules that have been engaged by the C3 convertase form covalent bonds with cell surfaces or with the Ig to which the C4b2a is bound. This results in the formation of a new complex, C4b2a3b, which functions as the **classical pathway C5 convertase.** As the name implies, this complex catalyzes the enzymatic cleavage of C5, which begins the formation of the MAC, described below.

The Membrane Attack Complex

The C5 convertases, generated by either the alternative or the classical pathway, initiate the activation of the terminal components of the complement system, culminating in the formation of the cytocidal **membrane attack complex** (Fig. 15–8). The proteins involved in this terminal pathway include C5, C6, C7, C8, and C9. This process begins with the cleavage of C5 by either of the C5 convertases. This is the last enzymatic step in the complement cascade; subsequent steps involve binding and polymerization of intact proteins. **C5** is a disulfide-linked heterodimeric serum protein with homology to C3 and C4, but without an internal thioester bond. C5 binds to the C3b molecule within either alternative or classical pathway C5 convertases. C5 is then cleaved into a small C5a fragment that is released and a two-chain C5b fragment that remains bound to the cell surface. (C5a has potent biologic effects on several cells; these will be discussed later in the chapter.) The remaining components of the complement cascade, C6, C7, C8, and C9, are structurally related proteins, sharing homologous domains. C5b transiently maintains a conformation capable of binding the next protein in the cascade, **C6**, and the stable C5b,6 complex remains loosely associated with the cell surface

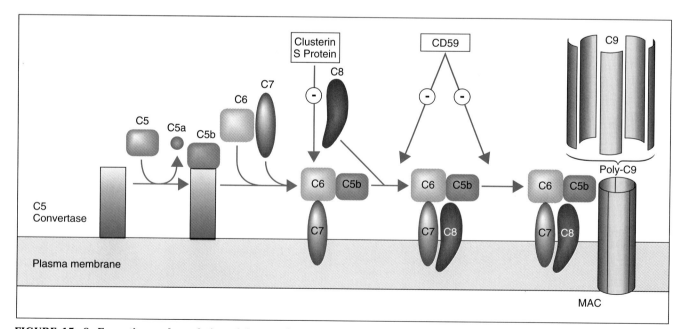

FIGURE 15–8. Formation and regulation of the membrane attack complex. A schematic view of the cell surface events leading to the formation of the MAC is shown. Cell-associated C5 convertase cleaves C5, generating C5b, which becomes bound to the convertase. C6 and C7 bind sequentially, and the C5b,6,7 complex becomes directly inserted into the lipid bilayer of the plasma membrane, followed by stable insertion of C8. Up to 15 C9 molecules may then polymerize around the complex to form lytic pores in the membrane. The sites of action of regulatory proteins, including S protein (vitronectin), clusterin, and CD59, are shown.

until it binds a single **C7** molecule. The resulting C5b,6,7 complex is highly lipophilic, and it inserts into the hydrophobic lipid bilayer of cell membranes, where it becomes a high-affinity integral membrane receptor for one **C8** molecule. The C8 protein is a trimer composed of an α chain that is disulfide-linked to a γ chain, which is non-covalently linked to a β chain. The β chain binds to the C5b,6,7 complex, and the γ chain inserts into the lipid bilayer of the membrane. This stably inserted C5b,6,7,8 complex (C5b-8) has limited ability to lyse some cells, but it does not have potent bactericidal activity.

The formation of a highly lytic and microbicidal MAC is accomplished by binding of **C9**, the final component of the complement cascades, to the C5b-8 complex. C9 is a monomeric serum protein that polymerizes at the site of the bound C5b-8. MACs with just four C9 molecules (C5b-8,9$_4$)

have full lytic capabilities for many microorganisms and eukaryotic cells. When the MAC contains between 12 and 15 C9 molecules associated with one C5b-8 complex, it is called "poly-C9" and forms pores in plasma membranes with a characteristic electron microscopic appearance. The pores have an internal diameter of about 110 Å (Fig. 15–9) and appear similar to, but smaller than, the membrane pores formed by the perforin protein, found in cytolytic T lymphocytes (CTLs) and natural killer (NK) cells (see Chapter 13).

Although it is not known whether the transmembrane pores observed in experimental situations play a role in microbial killing *in vivo*, it is clear that MACs with as few as three C9 molecules are fully bactericidal even though such structures cannot form the enclosed pores formed by MACs with 12 to 15 C9 molecules. Thus, MAC killing may be due to formation of membrane channels or by

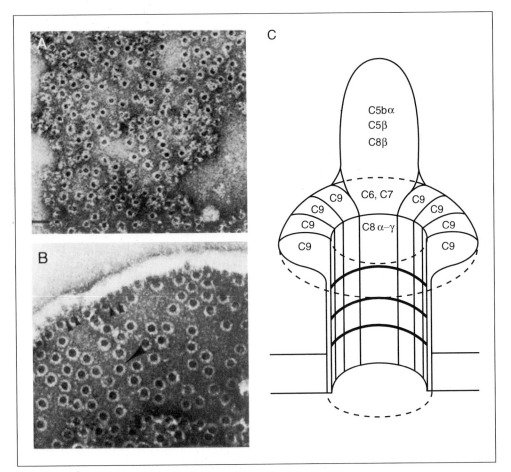

FIGURE 15–9. Structure of the membrane attack complex (MAC) in cell membranes.
A. Complement lesions in erythrocyte membranes are shown in this electron micrograph. The lesions consist of holes approximately 110 Å in diameter and are formed by poly-C9 tubular complexes.
B. For comparison, membrane lesions induced on a target cell by a cloned cytolytic T lymphocyte (CTL) line are shown in this electron micrograph. The lesions have a morphology similar to that of complement-mediated lesions, except for a larger internal diameter (160 Å). In fact, CTL and natural killer (NK) cell–induced membrane lesions are formed by tubular complexes of a polymerized protein (perforin), which is homologous to C9 (see Chapter 13).
C. A model of the subunit arrangement in the MAC is shown. The transmembrane region consists of 12 to 15 C9 molecules arranged as a tubule, in addition to single molecules of C6, C7, and C8 α and β chains. The C5ba, C5b, and C8β chains form an appendage that projects above the transmembrane pore. (From Podack, E. R. Molecular mechanisms of cytolysis by complement and cytolytic lymphocytes. Journal of Cellular Biochemistry 30:133–170, 1986. Copyright © 1986 Alan R. Liss, Inc. Reprinted by permission of Wiley-Liss, Inc., a subsidiary of John Wiley & Sons, Inc.)

other functional disturbances of membrane lipid bilayers caused by the insertion of the hydrophobic portions of the C5b-9 complex. In any case, MACs promote the passive exchange of small soluble molecules, ions, and water, thus dissipating energy-dependent gradients across the outer membranes of microorganisms or host cells that are required for cell survival. The pores formed by poly-C9 will allow ions such as K^+ and Ca^+ to enter but are too small to allow large molecules such as proteins to escape from the cytoplasm. This situation results in influx of water into the cells, leading to osmotic lysis or death due to toxic effects of high concentrations of Ca^{++}. As we will discuss later, MAC formation probably plays a minor role in defense against most microbes, and it is perhaps most important when it pathologically kills host cells.

REGULATION OF THE COMPLEMENT CASCADES

The complement cascades have evolved to permit very rapid, self-amplified activation on microbial surfaces, which is necessary for effective defense against infection. Because of the rapid kinetics and magnitude of complement activation, there is a requirement for tight regulation of the system. Without sufficient regulation, complement activation at the site of infection may lead to two deleterious consequences. First, serum complement components may be used up, or "consumed," resulting in the inability to activate complement at additional sites of infection. Second, inappropriate complement activation on normal host cells may lead to the destruction of these cells by the effects of excess inflammatory mediators and by direct MAC-mediated lysis. This does not normally happen because both the classical and alternative complement cascades are tightly regulated by several proteins that interact in specific ways with various complement components. In fact, about half the protein components of the complement system function as regulatory molecules. Several general characteristics of complement regulation are useful to consider before we discuss the details of the various mechanisms.

1. As a reflection of the fact that complement activation involves both fluid-phase and localized membrane events, the *regulatory mechanisms which have evolved include both soluble plasma protein regulators and integral membrane protein regulators.* This ensures that complement activation does not proceed in the blood or on host cells.
2. Since the formation of C3 convertases and the generation of C3b is the central feature of both alternative and classical pathways, it is not surprising that *the majority of regulatory mechanisms are directed toward controlling C3 convertase activity.*
3. *Regulation of the C3 convertases (and C5 convertases) involves two basic mechanisms. These are*

decay acceleration, in which the dissociation of the convertase complex is enhanced, and cofactor-dependent proteolysis of C3b and C4b by Factor I. Some regulatory proteins have both decay accelerator activity as well as cofactor activity. Furthermore, some regulatory proteins with these activities can inhibit both alternative and classical pathway convertases.

4. The discrete initiation step of the classical pathway by immune complex binding of C1 is regulated by a specific molecule called C1 inhibitor. In contrast, the alternative pathway, which is in a state of continuous low-level activation, is primarily regulated at the C3-convertase amplification step. MAC formation is inhibited by a different set of proteins.

Because of these regulatory mechanisms, a balance of activation and inhibition of the complement cascades is achieved that prevents damage to autologous cells and tissues but promotes the effective destruction of foreign organisms. The pathologic consequences of unregulated complement activation are apparent in various disease states in which the regulatory proteins are deficient, and these will be discussed later in the chapter. The major regulatory elements of the complement system, including individual regulatory proteins that work at various points in the cascades, are described next. The features of these regulatory proteins are listed in Table 15–2.

C1-Inhibitor Control of Initiation of the Classical Pathway

C1 inhibitor (C1INH) is a heavily glycosylated plasma protein that functions to block the proteolytic activities of C1r and C1s. In this way, it prevents the accumulation of active $C1r_2$-$C1s_2$ in the plasma and limits the extent of active $C1r_2$-$C1s_2$ on Ig bound to cell surfaces. C1INH is a member of the serpin (serine protease inhibitor) family, which includes α1-antitrypsin, antithrombin II, and α1-antichymotrypsin. As with other serpins, C1INH performs its inhibitory functions by presenting a "bait" sequence that mimics the normal substrates of C1r and C1s. Thus C1INH is cleaved by C1r or C1s and forms covalent stable ester linkages with these serine proteases. This leads to the dissociation of proteolytically inactive complexes of C1q and C1s, each with two covalently bound C1INH molecules. In this way, C1INH functions in the fluid phase to rapidly inactivate any active $C1r_2$-$C1s_2$ that may be formed on circulating immune complexes. C1INH may also prevent spontaneous activation of C1 in the plasma, which can occur at a low but significant rate in the absence of antibody. Most of the C1 in the blood is non-covalently bound to C1INH (which is present at seven times the molar concentration of C1), and this prevents the conformational changes that cause spontane-

TABLE 15-2. Soluble and Membrane Proteins That Regulate Complement Activation

Protein	Molecular Size (kD)/ Molecular Characteristics	Serum Concentration (µg/ml) or Cellular Distribution	Specifically Interacts with	Function
Soluble Serum Proteins				
C1 inhibitor (C1 INH)	105 Serpin Heavily glycosylated	200	C1r, C1s	Serine protease inhibitor covalently binds to C1r and C1s and blocks their ability to participate in classical pathway Binds to inactive C1 and prevents spontaneous activation Also inhibits kallikrein, plasmin, and factor XIIa of coagulation system
C4BP	560 7 α chains with 8 CCPRs each, 1 β chain with 3 CCPRs	250	C4b	Accelerates decay of classical pathway C3 convertase (C4b2a) Acts as cofactor for Factor I—mediated cleavage of C4b
Factor H	150 20 CCPRs	500	C3b	Accelerates decay of alternative pathway C3 convertase (C3bBb) Acts as cofactor for Factor I—mediated cleavage of C3b
Factor I	88	35	C4b, C3b	Proteolytically cleaves and inactivates C3b and C4b, using Factor H, C4BP, CR1, or MCP as cofactor
Anaphylatoxin inactivator (carboxypeptidase N)	305	35	C3a, C4a, C5a	Proteolytically removes terminal arginine residues and inactivates the anaphylatoxins
S protein	83 (vitronectin)	505	C5b-7	Binds to C5b-7 complex and prevents membrane insertion of MAC
SP-40,40	80 Heterodimer	50	C5b-9	Modulates MAC formation
Integral Membrane Proteins				
Membrane cofactor protein (MCP) (CD46)	45–70 4 CCPRs	Most blood cells (except erythrocytes), epithelial cells, endothelial cells, fibroblasts	C3b, C4b	Acts as cofactor for Factor I—mediated cleavage of C3b and C4b
Decay-accelerating factor (DAF)	70 4 CCPRs Phosphatidylinositol linkage	Most blood cells, endothelial cells, epithelial cells	C4b2b, C3bBb	Accelerates dissociation of classical and alternative pathway C3 convertases
CD59 (membrane inhibitor of reactive lysis)	18 Phosphatidylinositol linkage	Erythrocytes, lymphocytes, monocytes, neutrophils, platelets, endothelial cells, epithelial cells	C7, C8	Inhibits lysis of bystander cells (reactive lysis) Blocks C9 binding to C8, preventing MAC formation and lysis

Abbreviations: CCPR, complement control protein repeat; MAC, membrane attack complex.

ous activation. Once C1 binds to antigen-complexed antibody, via the C1q subunit, the C1INH is released; this allows the classical pathway to proceed. Subsequently, the active C1r$_2$-C1s$_2$ associated with the immune complex may covalently bind, and be inhibited by, C1INH. In addition to its effects on C1r$_2$-C1s$_2$, C1INH is also a major plasma inhibitor of other serine proteases, including kallikrein and clotting factor XIIa.

Regulation of C3 and C5 Convertases

Once fully activated, both the alternative and classical pathways have the potential to very quickly generate millions of C3b molecules by the action of C3 convertases, and this leads to formation of numerous C5 convertase complexes. These processes need to be regulated in the plasma and on host cell surfaces to avoid the pathologic consequences of rapid consumption of C3, excess generation of inflammatory mediators, and the lysis of host cells. Since the mechanisms for regulating alternative and classical pathway convertases are fundamentally similar, we will discuss them together, beginning with Factor I–mediated proteolysis of C3b and C4b, followed by decay acceleration.

Several of the proteins involved in C3/C5 convertase regulation are members of a family of homologous proteins called regulators of complement activation (RCA) (Tables 15–2, 15–3, 15–6). All the RCA members are encoded by closely linked genes on the long arm of human chromosome 1, and all have the capacity to bind to derivatives of C3 and/or C4. A shared structural feature of these proteins is the presence of multiple copies of a 60 amino acid long, cysteine-rich module called a complement control protein repeat (CCPR), or short consensus repeat. In fact, CCPRs are present in many other molecules in the immune system, including other complement proteins, the IL-2 receptor α chain, and the selectins, and they are likely involved in protein-protein interactions.

TABLE 15–3. Regulators of C3 Convertase Activity

Protein	Decay Acceleration Activity for:		Cofactors for Factor I–Mediated Proteolysis of:	
	C4b2a	*C3bBb*	*C4b*	*C3b*
Factor H	−	+	−	+
C4BP	+	−	+	−
DAF	+	+	−	−
MCP	−	−	+	+
CR1	+	+	+	+

Abbreviations: C4BP, C4 binding protein; DAF, decay-accelerating factor; MCP, membrane cofactor protein; CR1, complement receptor type 1.

The first mechanism for C3 convertase regulation that we will discuss is Factor I–mediated proteolysis. **Factor I** is a plasma serine protease that cleaves C3b, C4b, and their degradation products, thereby inhibiting the further participation of these molecules in C3 and C5 convertase formation (Fig. 15–10). In addition, some of the membrane-bound degradation products of Factor I–mediated prote-

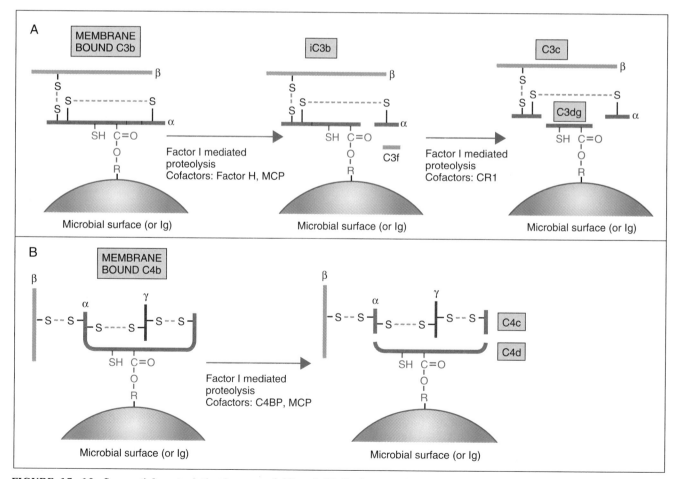

FIGURE 15–10. Sequential proteolytic cleavage of C3 and C4. Each proteolytic step generates a soluble fragment and a membrane-bound fragment. Inactive C3b (iC3b), or inactive C4b (iC4b) generated by the first Factor I–mediated cleavage, can no longer participate in the formation of C3 convertase.

olysis act as opsonins, binding to specific cell surface receptors that promote phagocytic clearance of microbes or immune complexes. Although Factor I is present in plasma in a constitutively active form, it still requires the presence of one of several cofactors in order to degrade its substrates; these cofactors are discussed below. Factor I controls convertase formation on host cell membranes by degrading C3b in two or more steps. In the presence of a cofactor, Factor I cleaves membrane-bound C3b to generate a small, soluble C3f fragment and inactive C3b **(iC3b),** which remains bound to the cell surface. iC3b is unable to contribute to C3 convertase formation, but it is further degraded by Factor I to release a soluble C3c fragment and leave membrane-bound C3dg. Both iC3b and C3dg are ligands for complement receptors and participate in the clearance of immune complexes that have fixed complement (discussed later). Factor I also cleaves membrane-bound C4b generated by the classical pathway, releasing a soluble C4c molecule and leaving a membrane-bound C4d molecule, which cannot participate in C3 convertase formation. Factor I also functions in the fluid phase to degrade soluble C3bi generated by the spontaneous hydrolysis of C3, and this, in part, is why the alternative pathway exhibits only a low level of "tickover" activity.

As we mentioned previously, Factor I activity is completely dependent on either membrane-bound or fluid-phase cofactors, and therefore these cofactors serve as key regulators of complement activation. The major membrane-bound cofactor for Factor I is called **membrane cofactor protein** (MCP, CD46). MCP is a member of the RCA family and is expressed on almost all cell types except erythrocytes. It binds to C3b and C4b monomers and acts as a cofactor for Factor I–mediated proteolysis (Fig. 15–11). There is some evidence that MCP can also bind to C3b and C4b in already formed convertases. *The presence of MCP on mammalian cell membranes and its absence on microbial surfaces results in selective inhibition of complement activation on host cells but not on microbes*, and this is one way in which the complement system distinguishes self from non-self. It is interesting that MCP has been identified as a specific receptor for measles virus and group A streptococcus.

Factor H and **C4-binding protein (C4BP)** are two plasma protein RCA family members that also serve as important cofactors for Factor I–mediated proteolysis. Factor H is a single polypeptide composed entirely of CCPRs. C4BP has a more complex structure, composed of seven identical α chains covalently bound to a single β chain to form a spider-like molecule; both α and β chains contain CCPRs. This structure suggests that C4BP may be capable of binding several C4b molecules at one time. Unlike MCP, both Factor H and C4BP are soluble plasma proteins whose principal functions are regulation of fluid phase complement activation, although they both may have activity on cell surface–bound ligands as well. In addition, both Factor H and C4BP also have decay-accelerating

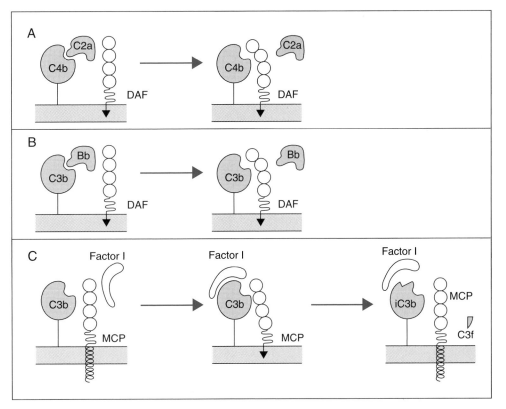

FIGURE 15–11. Membrane protein regulators of complement activation.

A. Decay accelerating factor (DAF) interferes with the formation of the classical pathway C3 convertase. CR1 and soluble C4 binding protein (C4BP) have the same activity.

B. DAF interferes with the formation of the alternative pathway C3 convertase. CR1 and soluble Factor H have the same activity.

C. MCP is a cofactor for Factor I–mediated proteolysis of C3b. CR1 and Factor H have the same activity. Similarly, MCP, CR1, and C4BP are cofactors for Factor I–mediated proteolysis of C4b.

activity, as discussed below. As its name implies, C4BP serves as a cofactor only for Factor I–mediated degradation of C4b but not C3b, and Factor H serves only as a cofactor for C3b but not C4b degradation. (Table 15–3). Factor H works by binding fluid phase C3b or C3i, greatly increasing the affinity of these molecules for Factor I. Proteolysis by Factor I generates iC3b or iC3i. Factor H does not serve as a cofactor for subsequent Factor I–mediated degradation of iC3b to C3dg. Since C3b is generated by both alternative and classical pathway C3 convertases, Factor H is important for the regulation of both pathways.

C4BP regulates fluid phase C3 convertase activity in a fashion similar to that of Factor H, acting as a cofactor for the sequential hydrolysis of C4b to an intermediate iC4b and immediately thereafter to C4d and C4c. This is important because there is a potential for C4b newly generated on a cell surface to be released into the plasma and to associate with similarly released C2a, forming C3 convertase in the fluid phase. In addition, plasma C4 undergoes spontaneous slow hydrolysis of its thioester bond to form C4i (analogous to C3i of the tickover pathway), and C4i can also potentially participate in the formation of a fluid phase classical pathway C3 convertase.

The second major mechanism of C3/C5 convertase regulation is the destabilization of the enzyme complexes, or decay acceleration. The major host cell membrane protein with decay accelerating activity is **decay-accelerating factor (DAF)**, which is another member of the RCA family. DAF is expressed on most blood cells, endothelium, and epithelial cells. It is covalently linked to the outer leaflet of the plasma membrane by a phosphatidylinositol linkage, as is the complement regulatory protein CD59, discussed later. DAF functions by binding C3 convertases of the alternative or classical pathway and causing the rapid release of Bb from C3b, or C2a from C4b (Fig. 15–11). DAF functions to block complement activation only on the same cell on which it is expressed, and it does not possess cofactor activity. *Since DAF is not expressed by microorganisms, it is also partly responsible, along with MCP, for the ability of the complement system to distinguish self from non-self.* DAF has also been identified as a specific receptor for some strains of *Escherichia coli* bacteria.

Two soluble proteins with decay-accelerating activity are Factor H and C4BP, both of which also have cofactor activity, as discussed above. Factor H displaces Bb in the alternative pathway C3 convertase C3bBb or in the initial fluid-phase alternative pathway C3 convertase C3iBb. The net result is the destruction of the convertase enzyme complex and limitation of fluid-phase, and possibly cell surface, complement activation. In addition to serving as a cofactor for Factor I–mediated proteolysis, Factor H binding to C3b or C3i inhibits their association with Bb, thereby blocking assembly of the alternative pathway C3 convertases. Factor H also interferes with C3b binding to complement re-

ceptors. In an analogous fashion to Factor H, C4BP displaces C2a from C4b2a and thereby destroys the classical pathway C3 convertase. C4BP also binds to uncomplexed C4b and inhibits its association with C2a, thereby blocking assembly of the classical pathway C3 convertase.

Complement receptor type 1 (CR1) is another membrane protein that binds both C3b and C4b, possesses both cofactor and decay-accelerating activity, and can inhibit both alternative and classical pathway C3 convertase activity. Its distribution is limited to blood cells, and it probably serves a restricted role in blocking complement activity around immune complexes that have bound to the CR1-expressing cells. CR1 will be discussed in more detail later in the chapter.

In addition to the intrinsic actions of MCP and DAF, the formation of C3 convertase on host cells may be limited by certain biochemical characteristics of mammalian plasma membranes. For example, high sialic acid content favors binding of the regulatory protein Factor H over Factor B, and, conversely, cell surfaces with reduced sialic acid content bind Factor B with greater affinity. Many bacteria contain low amounts of surface sialic acid compared with mammalian cells, and this is another reason why complement activation occurs preferentially on bacterial cell surfaces.

Regulation of Membrane Attack Complex Formation

Even after the classical or alternative pathway C3 convertases are formed, excessive complement-mediated cell lysis is prevented by a number of proteins that act at the level of the MAC (see Table 15–2). As with the regulation of the C3 and C5 convertases, the presence of MAC inhibitory molecules in host cell membranes distinguishes them from most microbes and protects cells against lysis when other regulatory mechanisms fail to inhibit the formation of C5 convertase. In addition, at the C5b-7 stage, growing MAC complexes have the ability to insert into any neighboring cell membrane besides the membrane on which they were generated. Therefore, inhibitors of MAC in the plasma and in host cell membranes are needed to ensure that innocent bystander cell lysis does not occur at the site of a microbial infection.

The major membrane inhibitor of MAC is **CD59** (also called membrane inhibitor of reactive lysis). CD59 is a single chain protein, widely distributed on many different cell types. Like DAF, it is inserted into the plasma membrane of cells by a phosphatidylinositol linkage. In addition, CD59 is found in most body fluids. CD59 can bind both C8 and C9 and works by incorporating itself into growing MAC complexes after the membrane insertion of C5b-7, thereby inhibiting C9 membrane insertion and subsequent addition of C9 molecules (Fig. 15–8). In this way, CD59 acts as an intrinsic inhibitor of MAC-mediated lysis of the cells on which it is expressed. It is also possible that soluble CD59 may insert into cells and protect them from lysis.

Another phosphatidylinositol-linked membrane protein that has been shown to have some MAC inhibitory activity is called homologous restriction factor (HRF). Full molecular characterization of HRF and evaluation of its functional significance have not been accomplished.

Soluble plasma inhibitors of MAC formation include **S protein** (also called **vitronectin**) and **clusterin** (also called SP-40-40). S protein is a glycoprotein with a complex structure, including a fibronectin-like integrin binding region and a heparan-binding domain. It functions by binding to soluble C5b-7 complexes, preventing their insertion into cell membranes near the site where the complement cascade was initiated (see Fig. 15–8). Although S protein–C5b-7 complexes can incorporate C8 and C9, they remain nonlytic. S protein can also bind C8 and C9 and inhibit C9 polymerization. In fact, nonlytic complexes of S protein and C5b-8 are often found on cells after experimental complement activation. S protein has other biologic functions not directly related to the complement system, including regulation of coagulation and promotion of cellular interaction with extracellular matrix.

Clusterin is a heterodimeric plasma glycoprotein with a thrombospondin-like amino terminal domain capable of hydrophobic interactions with C7, C8, and C9. It is present in plasma and in the cells of most human tissues. Like S protein, clusterin inhibits MAC formation by binding to soluble C5b-7 complexes and preventing membrane insertion on autologous cells. Clusterin is highly expressed in the male reproductive tract and may protect spermatozoa from MAC lysis in the female reproductive tract. Other roles for clusterin include plasma lipid transport and tissue remodeling after injury.

Our discussion of the regulators of complement activation has highlighted the fact that many proteins can inhibit the complement cascades, including several members of the RCA gene family. Furthermore, some of these proteins have redundant functions. Much of the analysis of the functions of complement regulatory proteins has relied on *in vitro* experiments, and most of these have focused on cytotoxicity assays that measure MAC-mediated lysis as an end-point. Based on these approaches, a hierarchy of importance for inhibiting complement activation is CD59 > DAF > MCP, and this may reflect the relative abundance of these proteins on cell surfaces. The function and importance of a particular regulator can also be inferred from a study of patients or animals genetically deficient in that protein. Human complement regulatory protein deficiencies are rare, but the data available indicate that CD59 is more crucial than DAF in protecting against uncontrolled autologous cell lysis. The relative importance of the different regulatory proteins in controlling the generation of complement-derived inflammatory mediators is not presently known. The details of complement deficiencies are discussed in more detail later in the chapter.

RECEPTORS FOR FRAGMENTS OF THE C3 AND C4 COMPONENTS OF COMPLEMENT

Many of the biologic activities of the complement system are mediated by the binding of complement fragments to specific integral membrane protein receptors expressed on various cell types. These complement receptors can be divided into four functional categories:

1. Receptors for fragments of C3 that are covalently bound to activating surfaces.
2. Receptors for soluble C3a, C4a, and C5a fragments (anaphylatoxins), which mediate many of the inflammatory effects of complement activation.
3. Receptors that regulate the complement cascades by binding to specific complement proteins and inhibiting their functions.
4. The C1q receptor, which mediates antibody-independent activation of the classical pathway and enhances phagocytic uptake of collectin-coated microorganisms.

Receptors for surface-bound C3 fragments are the most thoroughly characterized and will be considered in detail in this section of the chapter (Table 15–4). The molecular nature of the receptors for other complement fragments will be considered in the discussion of the biologic activities of complement fragments later in this chapter. The membrane receptors that regulate complement activation (DAF, MCP, CD59) and the C1q receptor have been described earlier in this chapter.

Complement receptor type 1 (CR1, C3b receptor, CD35) is centrally involved in several biologic functions of the complement system. It is a polymorphic, single chain, integral membrane glycoprotein encoded by a gene in the RCA cluster, and it serves as a high-affinity C3b and C4b receptor with three C4b and two C3b binding sites. In humans, CR1 is expressed mainly on blood cells, including erythrocytes, neutrophils, monocytes, eosinophils, and T and B lymphocytes; it is also found on follicular dendritic cells in lymph nodes and spleen. In mice, it is not expressed on erythrocytes, but it is present on platelets.

CR1 performs at least three important functions. First, CR1 enhances the ability of phagocytic leukocytes to ingest C3b- or C4b-coated particles and microorganisms. Not only does CR1 bind C3b and C4b, but it also transduces signals that activate the phagocytic mechanisms of leukocytes. Second, CR1 is important for the clearance of immune complexes from the circulation. This is achieved largely by erythrocyte CR1, which binds C3b and C4b on circulating immune complexes and transports them to the liver and spleen. Third, CR1

TABLE 15–4. Receptors for Complement Fragments

Receptor	Molecular Weight (kD)/Molecular Features	Ligand	Cell Distribution	Biologic Function
Complement receptor type 1 (CR1)	190–280 (several polymorphic forms) 30 CCPRs form entire extracellular region	C3b, C4b, iC3b	Erythrocyte	Clearance of circulating immune complexes Accelerates dissociation of classical and alternative pathway C3 convertases Acts as cofactor for Factor I–mediated cleavage of C3b and C4b
			Neutrophils, monocytes, macrophages	Enhances Fc receptor–mediated phagocytosis Mediates Fc receptor–independent phagocytosis
			Eosinophils	?
			B lymphocytes	?
			Glomerular epithelial cells	?Solubilization of trapped immune complexes
			Follicular dendritic cells	?Binding of immune complexes
Complement receptor type 2 (CR2)	145 16 CCPRs form entire extracellular region	iC3b, C3dg, EBV	B lymphocytes	?B cell activation Mode of EBV infection
			Nasopharyngeal epithelial cells	Mode of EBV infection
			Follicular dendritic cells	?Binding of immune complexes, memory B cell activation
Mac-1 (complement receptor type 3, CR3, CD11bCD18)	α:165 β:95	iC3b (cation-dependent)	Monocytes/macrophages, neutrophils, natural killer cells	Cellular adhesion protein required for surface adhesion in chemotaxis, phagocytosis
	Integrin, common β chain with CR4, LFA-1		Dendritic cells	?
Complement receptor type 4 (CR4, CD11cCD18, p150,95)	α:150 β:95 Integrin, common β chain with CR3, LFA-1	iC3b	Neutrophils, monocytes, platelets	Enhances Fc receptor–mediated phagocytosis Mediates Fc receptor–independent phagocytosis
C3a/C4a receptor	?	C3a, C4a	Mast cells, basophils	Degranulation releasing histamine and other mediators of inflammation
			Smooth muscle cells	Contraction (?histamine-independent)
			Lymphocytes	?
C5a receptor	40, G protein–linked, 7 transmembrane segments	C5a	Mast cells, basophils	Degranulation-releasing histamine and other mediators
			Endothelial cells	Increases vascular permeability
			Neutrophils, monocytes/macrophages	Promotes chemotaxis
			Smooth muscle cells	Promotes contraction

Abbreviations: CCPR, complement control protein repeat; LFA, leukocyte function–associated antigen; EBV, Epstein-Barr virus.

is a regulator of complement activation by virtue of its ability to inhibit C3 convertase activity, discussed previously. This regulatory activity is most likely directed toward complement activation on immune complexes that have bound to blood phagocytes. A genetically engineered, soluble form of CR1, lacking transmembrane and cytoplasmic domains, has been produced as an experimental therapeutic anti-inflammatory agent. This soluble CR1 markedly inhibits both alternative and classical pathway activation and limits tissue injury in *in vivo* models of acute inflammation.

Complement receptor type 2 (CR2, C3d receptor, CD21) is a single chain integral membrane glycoprotein present on B lymphocytes, follicular dendritic cells, and some epithelial cells. CR2 is encoded by a gene in the RCA cluster. It specifically binds iC3b and C3dg, the Factor I–generated cleavage products of surface- or immune complex–bound C3b. CR2 also serves as a receptor for Epstein-Barr virus (discussed later). CR2 promotes humoral immune responses in at least two ways. On follicular dendritic cells, CR2 serves to trap iC3b- and C3dg-coated antigen-antibody complexes in germinal centers, and displays the antigen to activated B lymphocytes. On B cells, CR2 is expressed as part of a trimolecular complex that includes two other proteins called CD19 and target of antiproliferative antibody-1 (TAPA-1). Experiments with B cell lines first suggested that this complex delivers signals to the B cell that significantly lower the threshold concentration of anti-Ig antibody required to activate B cells. A demonstration of the role of CR2 *in vivo* comes from CR2 gene knockout mice, which are markedly deficient in T cell–dependent humoral immune responses and have significantly fewer and smaller lymph node and splenic germinal centers compared with normal mice. The role of CR2 in humoral immune responses was discussed in more detail in Chapter 9.

CR2 is also the cell surface receptor for the **Epstein-Barr virus** (EBV), a human herpes virus that infects most people by adulthood and remains latent within B cells or pharyngeal epithelial cells for the duration of life. Infection may be subclinical or may cause infectious mononucleosis. In addition, EBV is linked to several human malignancies, including African Burkitt's lymphoma (a malignant B cell tumor), B cell lymphomas associated with therapeutic drug-induced immunodeficiency or human immunodeficiency virus (HIV) infection, and nasopharyngeal carcinoma (a malignant tumor of nasopharyngeal epithelium) (see Chapter 18, Box 18–2). These tumors are derived from normal cells known to express CR2. EBV is a potent polyclonal B cell activator, and *in vitro* the virus can transform normal blood B lymphocytes into immortalized lymphoblastoid cell lines.

Mac-1 (complement receptor type 3, CR3, CD11bCD18) is a specific receptor for the iC3b fragment generated by Factor I–mediated cleavage of C3b and plays an important role in the elimination of microorganisms by leukocyte phagocytosis. Mac-1 is expressed on many different bone marrow–derived cells, including neutrophils, mononuclear phagocytes, mast cells, and NK cells. It is a member of the integrin family of cell surface receptors (see Chapter 7, Box 7–3) and consists of an α chain (CD11b) non-covalently linked to a β chain (CD18), which is identical to the β chains of two closely related integrin molecules, leukocyte function–associated antigen-1 (LFA-1) and p150,95 (see below). Mac-1 mediates phagocytosis of iC3b-coated microorganisms or particles. In addition, Mac-1 on neutrophils and monocytes promotes the attachment of these cells to endothelium, even without complement activation, via binding to ICAM-1. This may be important for the accumulation of inflammatory cells at sites of tissue injury.

Complement receptor type 4 (CR4, p150,95, CD11cCD18) is another integrin, with a different α chain (CD11c) and the same β chain as Mac-1. It has a similar cellular distribution as Mac-1 and can bind iC3b as well as C3dg. The function of this receptor may be similar to that of Mac-1.

BIOLOGIC FUNCTIONS OF COMPLEMENT PROTEINS

The functions of the complement system fall into two broad categories: (1) cell lysis by the MAC, and (2) biologic effects of proteolytic fragments of complement (see Fig. 15–1). We have previously described how complement activation on cell surfaces leads to the insertion of the MAC into lipid bilayers, causing osmotic lysis of the cell. The many other effects of complement in immunity and inflammation are mediated by the proteolytic fragments generated during complement activation. These biologically active fragments may remain bound to the same cell surfaces where complement has been activated, or they may be released into the fluid phase (e.g., blood or extracellular fluid). In either case, they mediate their effects by binding to specific receptors expressed on a variety of other cell types, including phagocytic leukocytes and endothelium. In this section of the chapter, we will discuss each of these functional roles of the complement system, with reference to the particular fragments and receptors that are involved (Table 15–5).

Complement-Mediated Cytolysis

Complement-mediated lysis of foreign organisms is an important defense mechanism against certain types of microbial infection. Specific humoral responses to microbes generate antibodies that bind to the organisms; these antibodies activate complement on the surfaces of the microbes and lead to their lysis by the formation of the MAC. Some microorganisms may activate the alternative pathway or less frequently the clas-

TABLE 15–5. Biologic Functions of Complement

Function	Complement Components	Mechanisms
Lysis of cells	C5–C9	MAC forms pores in lipid membranes
Opsonization/phagocytosis	C3b, iC3b	C3b or iC3b on the surface of microorganisms binds to CR1 (and CR3, CR4) on neutrophils and macrophages, promoting phagocytosis
Inflammation		
Vascular responses	C5a > C3a ≫ C4a	C5a, C3a, and C4a stimulate mast cell histamine release
		C5a stimulates smooth muscle contraction, endothelial contraction
Polymorphonuclear leukocyte activation	C5a	C5a is a chemoattractant for neutrophils and activates neutrophil oxidative metabolism
Immune complex removal	Classical pathway, C3b	Complement activation on Ig molecules inhibits immune complex formation
		C3b on immune complexes binds to CR1 on erythrocytes, and the immune complexes are cleared from the circulation as the erythrocytes traverse the liver and spleen
B cell activation	?iC3b, C3dg	B cell responses to antigen are enhanced by CR2/CD19 signalling
		CR2-expressing follicular dendritic cells display antigen-antibody-C3dg complexes to B lymphocytes in germinal centers

Abbreviations: MAC, membrane attack complex; Ig, immunoglobulin.

sical pathway in the absence of antibody, also leading to lysis. MAC-mediated lysis may be important in defense against only some microbes, since genetic defects in MAC components increase susceptibility only to *Neisseria* bacteria. Acquired resistance to complement-mediated lysis is a mechanism by which some microbes evade host immunity (see Chapter 16). The complement system may also be involved in several pathologic conditions by causing the lysis of host cells, leading to tissue injury and disease. Some autoimmune diseases are characterized by the production of autoantibodies specific for self proteins expressed on cell surfaces. These autoantibodies may fix enough complement to overcome intrinsic regulatory mechanisms, and subsequent formation of the MAC may lead to osmotic lysis of these normal cells (see Chapter 20).

Opsonization and Promotion of Phagocytosis of Microbes

As discussed previously, activation of both alternative and classical pathways leads to the generation of C3b and iC3b covalently bound to cell surfaces. Both C3b and iC3b act as opsonins by virtue of the fact that they specifically bind to receptors on neutrophils and macrophages. C3b binds to CR1, and iC3b (but not C3b) binds to Mac-1 and CR4. In addition, activation of the classical pathway leads to generation of C4b covalently bound to cell surfaces, and C4b is also a ligand for CR1. Thus, complement activation on the surfaces of microbial cells promotes the adherence of the microbes to host leukocytes that are competent at phagocytosing and killing the microbes. *C3b- and iC3b-dependent phagocytosis of microorganisms is a major defense mechanism against systemic bacterial and fungal infections.* This is exemplified by the heightened susceptibility of C3-deficient humans and mice to lethal bacterial infections, which is attributed to a defect in phagocytic clearance.

Complement receptors and Fcγ receptors (see Chapter 3) cooperate to mediate binding and phagocytosis of opsonized particles. For example, if an unactivated neutrophil or monocyte encounters a particle coated with IgG, the particle is phagocytosed relatively slowly via Fcγ receptors. If the same particle bears iC3b, so that it simultaneously binds to CR3 on the leukocyte, Fcγ receptor–mediated phagocytosis is also greatly enhanced. Similarly, binding of opsonized particles to Fcγ receptors augments CR1- and CR3-mediated phagocytosis. In fact, Fcγ and complement receptors on leukocytes function not only to bind opsonized particles but also to transduce signals that stimulate the phagocytic capacities of the leukocytes.

Anaphylatoxins and Inflammatory Responses

C3a, C4a, and C5a are small, cationic peptides generated by the complement cascades that mediate many of the inflammatory events associated with complement activation. They are called **anaphylatoxins** because they induce the release of mediators from mast cells that cause inflammatory responses characteristic of anaphylaxis (see Chapter 14). C3a is a 9 kD peptide derived from the C3 α chain by either classical or alternative pathway C3 convertase. C4a is an 8.7 kD peptide released from the α chain of C4 in the classical pathway by

C1s-mediated cleavage. C5a is the 11 kD peptide released from the α chain of C5 by the action of either classical or alternative pathway C5 convertase.

The biologic effects of the anaphylatoxins are a function of the expression of specific anaphylatoxin receptors on various cell types. The C5a receptor is the most thoroughly characterized. It is a member of the seven α-helical transmembrane receptor family, like chemokine receptors (see Chapter 12, Box 12–1) and uses heterotrimeric G-proteins for coupling to signal transduction pathways. The C5a receptor is expressed on many cell types, including neutrophils, eosinophils, basophils, monocytes, macrophages, mast cells, endothelial cells, smooth muscle cells, epithelial cells, and astrocytes. Functional studies indicate that there are also distinct C3a and C4a receptors on blood leukocytes and mast cells, and there is evidence that they are G-protein linked. However, these receptors have not yet been fully characterized.

All three anaphylatoxins can bind to receptors on mast cells and basophils, causing granule exocytosis and release of vasoactive mediators, such as histamine (see Chapter 14). Histamine increases vascular permeability and stimulates the contraction of visceral smooth muscles. C5a is the most potent mediator of these effects, C3a is 20-fold less potent, and C4a is 2500-fold less potent. C5a also stimulates cytokine release from mast cells.

C5a has multiple other effects that promote inflammation. It acts directly on vascular endothelial cells, stimulating contraction, causing vascular leak and exocytosis, and stimulating P-selectin expression, which promotes neutrophil binding. C5a has several direct effects on neutrophils, including stimulation of movement (chemokinesis and chemotaxis), increased Mac-1 adhesiveness for ICAM-1, and, at high doses, stimulation of the respiratory burst and production of reactive oxygen intermediates. The combination of C5a actions on mast cells, endothelial cells, and neutrophils contributes to inflammation at sites of complement activation. The specificity of the response for foreign and not self tissues is imparted either by specific antibody activating the complement cascade at the appropriate location, i.e., where the foreign antigen is, or by the capacity of the alternative pathway to be activated on the surface of foreign organisms.

The actions of the anaphylatoxins are in part regulated by a plasma enzyme called carboxypeptidase-N which removes a C-terminal arginine residue common to C3a, C4a, and C5a. The resulting des-Arg forms of C3a and C4a are completely inactive, and des-Arg C5a retains only 10 per cent of the activity of intact C5a.

Solubilization and Phagocytic Clearance of Immune Complexes

Small numbers of immune complexes are constantly being formed in the circulation and may increase dramatically when an individual mounts a vigorous humoral immune response to an abundant circulating antigen. These immune complexes are potentially harmful because they may deposit in vessel walls, activate complement, and lead to inflammatory reactions that damage surrounding tissues. Formation of large, potentially harmful immune complexes may require not only the multivalent binding of Ig Fab regions to antigens but also non-covalent interactions of Fc regions of juxtaposed Ig molecules (see Chapter 3). Complement activation on Ig molecules can sterically block these Fc-Fc interactions, thereby inhibiting new formation of, or destabilizing, already formed immune complexes.

In addition to interfering with immune complex formation, the complement system can promote the clearance of immune complexes from the circulation by the mononuclear phagocyte system. In humans and other primates, this function is mediated largely by CR1 on the surface of erythrocytes, which, because of their number, account for the great majority of cell-associated CR1 molecules in the circulation. Circulating antigen-antibody complexes activate complement, and the C3b that is generated forms covalent bonds with the antibody. Such complexes can be adsorbed by CR1-expressing erythrocytes because of the high affinity of CR1 for C3b. In vivo studies have demonstrated that phagocytic cells in the liver (Kupffer cells) and spleen are responsible for removing the immune complexes from the erythrocyte membranes as the erythrocytes traverse the sinusoids of these organs. The mechanism by which this is accomplished is not clearly established but could involve Fc receptors on the mononuclear phagocytes, which would compete with the erythrocyte CR1 receptors for the binding of immune complexes. In addition, since CR1 is a cofactor for Factor I–mediated conversion of C3b to iC3b, it is also possible that the immune complexes bound to erythrocyte CR1 are tagged with iC3b, and that mononuclear phagocyte Mac-1 on liver phagocytes will compete for binding to this ligand. It is interesting that during flare-ups of autoimmune diseases such as systemic lupus erythematosus, the CR1 levels on erythrocytes are reduced. This may exacerbate the disease process by limiting the clearance of immune complexes.

Regulation of Humoral Immune Responses

Many clinical and experimental observations have suggested that complement deficiencies are associated with impaired humoral immune responses to antigens. Recent work has confirmed the functional connection of the complement system and humoral immune responses. For example, mice with targeted disruptions of the C3 or C4 genes have impaired humoral responses to T dependent antigens, and these mice do not develop normal germinal centers. This phenotype is strikingly similar to that of CR2-deficient mice, described earlier, and suggests that the interaction of C3-derived ligands (e.g., iC3b and C3dg) with their receptors (CR2) is important for promoting B cell responses to protein antigens. Current models indicate that these ligand-receptor interactions are involved in two phases of B cell activation, including: (1) germinal center follicular dendritic cell presentation of immune complex–bound antigen to B cells; and (2) the role of the CR2:CD19:TAPA complex in lowering the threshold for antigen-induced B cell activation (see Chapter 9).

COMPLEMENT GENES AND GENE EXPRESSION

The complement system proteins can be categorized as members of various gene families on the basis of sequence homologies. Members of the same gene family often have similar functional characteristics that depend on their shared protein structures. Assignment of proteins to one or another gene family (usually based on the amino acid sequences predicted from the nucleotide sequences of the cloned genes) is often useful for studying the structural basis of function. Homologies among genes encoding various complement proteins suggested that members of each gene family may have arisen by duplication of an ancestral gene, followed by the structural diversification that imparts specialized functions to the individual proteins. The various gene families represented by complement proteins are listed in Table 15–6.

The best-defined complement gene family is the RCA family mentioned earlier. All the members of this family share the ability to bind C3b and C4b, and CCPRs constitute a significant part of the structure of all these proteins. The presence of CCPRs in other complement proteins, including Factor B and C2, reflects exon duplication and fusion with otherwise unrelated genes during evolution. The five serine esterases of the complement

system, including Factor I, Factor B, C1r, C1s, and C2, all have sequence homologies with one another and with non-complement serine esterases such as trypsin and chymotrypsin. The serine protease inhibitor C1INH is a member of the serpin family. C3 and C4 are members of a small family of proteins that also includes α2-macroglobulin, characterized, in part, by the presence of internal thioester bonds. This feature allows the proteins to form covalent bonds with cell surfaces or Ig molecules. C5 is also a homologous member of this family, but it does not have the thioester bond. CR3 and CR4 are integrins sharing the same β chain as LFA-1. The membrane pore–forming C9 is homologous to the pore-forming protein (perforin) found in CTLs and NK cells.

At least three groups of complement proteins can be defined by close chromosomal linkage of their respective members. The genes encoding several members of the RCA family, including CR1, CR2, C4bp, MCP, and DAF, are all found in an 800 kb segment of DNA on the long arm of human chromosome 1. The genes encoding the MAC components C6, C7, C8, and C9 are also linked in another region on chromosome 1. The close proximity of the genes encoding C2, Factor B, and C4 within the major histocompatibility complex (MHC) in humans, mice, and other species is the most thoroughly analyzed example of complement gene linkage. In humans, these complement genes are present between the class II HLA-DR and class I HLA-B loci on chromosome 6 (Fig. 15–12; see also Chapter 5). MHC-linked complement genes are sometimes called class III genes. It is of interest that, as is characteristic of the peptide-binding class I and class II MHC molecules, the class III complement genes are polymorphic. For example, over 35 alleles are identified for the two closely linked loci that encode the C4 protein. Certain alleles of the C2, Factor B, and C4A and C4B genes are in linkage disequilibrium, so that they are often inherited *en bloc*.

Complement proteins can be categorized as acute-phase proteins, which are narrowly defined as plasma proteins synthesized by the liver whose concentrations increase in response to tissue damage and inflammation. The acute-phase response is typically due to TNF–, IL-1– or IL-6–induced increases in hepatocyte gene transcription (see Chapter 12). Like other acute-phase proteins, hepatocytes are the major source of plasma complement proteins, and several complement genes including the C3 gene are upregulated by IL-1 and/or IL-6. Two features, however, distinguish complement proteins from other acute-phase proteins. First, many cell types and tissues other than the liver are significant sources of the complement proteins, especially in response to inflammation. For example, mononuclear phagocytes can synthesize most of the complement proteins present in the serum, and it has become apparent that complement proteins can be produced by a wide vari-

TABLE 15–6. Gene Families Represented by Complement Proteins

Gene Family	Complement Protein Members/(Non-Complement Members)
Regulators of complement activity (RCA)	C4BP, Factor H, MCP, DAF, CR1, CR2
Serine esterases	Factor I, Factor B, C1r, C1s, C2
C3 family	C3, C4, C5/(α_2-macroglobulin)
Integrins	Mac-1, CR4/(LFA-1)
Membrane lytic proteins	C9/(perforin)
Serpins	C1INH/(α1-antichymotrypsin)

Abbreviations: C4BP, C4 binding protein; MCP, membrane cofactor protein; DAF, decay-accelerating factor; CR1, complement receptor type 1; CR2, complement receptor type 2; LFA-1, leukocyte function–associated antigen-1; C1INH, C1 inhibitor.

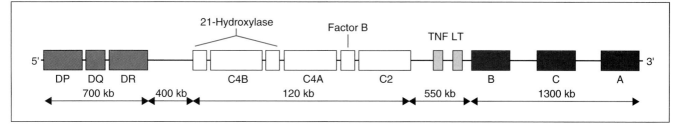

FIGURE 15–12. Complement genes in the human major histocompatibility complex (MHC). The genes encoding C2, Factor B, and C4 are located between the DR locus and the class I MHC genes 5' to genes encoding the cytokines tumor necrosis factor (TNF) and lymphotoxin (LT). Two genes encode the isoforms of C4 (C4A and C4B). Both the C2 and C4 genes are polymorphic.

ety of tissues such as brain, lung, intestines, and genitourinary tract. This may reflect a diversity of functions for these proteins that are not yet understood. Second, unlike most other acute-phase proteins of hepatic origin, interferon-γ (IFN-γ) is a major inducer of complement gene expression in both the liver and other tissues. In fact, some complement proteins, such as C4, are induced exclusively by IFN-γ and not by IL-1 or IL-6.

The regulation of synthesis of the various complement proteins is complex, and incompletely characterized. The C1 and C8 proteins are unusual in that they are complexes of polypeptide products of more than one gene. The individual chains of each protein are secreted independently, and the mature proteins assemble extracellularly.

DISEASES RELATED TO THE COMPLEMENT SYSTEM

The complement system is related to human disease in two general ways. First, deficiencies in any one of the protein components, usually due to abnormalities in gene structure, can lead to abnormal patterns of complement activation. If regulatory components are absent, too much complement activation may occur at the wrong time or wrong site. This can lead to excess inflammation and cell lysis and may also lead to consumptive lack of plasma complement proteins. The absence of an integral component of the classical, alternative, or terminal lytic pathways can result in too little complement activation and a lack of complement-mediated biologic functions. In either case, pathologic consequences may be severe. Second, an intact, normally functioning complement system may be activated in response to abnormal stimuli, such as persistent microorganisms or autoantibody responses to self antigens. In these infectious or autoimmune diseases, the inflammatory or lytic effects of complement may contribute significantly to the pathology of the disease.

Complement Deficiencies

Human deficiencies in many of the complement proteins have been described (Table 15–7), and these deficiencies are usually attributable to inherited or spontaneously mutated genes. Alternatively, acquired deficiencies in certain components can arise in individuals with normal complement genes. Deficiencies can be categorized on the basis of the functional type of protein that is lacking. Thus, there are deficiencies of the components of the classical, alternative, and terminal pathways and deficiencies in either soluble or membrane regulatory proteins. Beyond the significance to the affected individuals, an understanding of these diseases provides insights into the normal physiologic roles of the deficient proteins.

1. Genetic deficiencies in classical and alternative pathway components, including C1q, C1r, C4, C2, C3, properdin, and Factor D, all have been described; C2 deficiency is the most commonly identified human complement deficiency. The single most consistent clinical consequence of deficiencies in the early classical pathway proteins (not including C3) is the development of systemic lupus erythematosus (SLE) (see Chapter 19, Box 19–3). More than 50 per cent of patients with C2 and C4 deficiencies have SLE. The reason for this association is unknown, but it suggests that the classical pathway is normally involved in regulating the immune system. It may reflect the requirement of the classical pathway for the clearance and solubilization of circulating immune complexes. If normally generated immune complexes are not cleared from the circulation, they may deposit in blood vessel walls and tissues, where they activate the complement cascade and produce local inflammation. Such inflammatory reactions may promote the breakdown of peripheral tolerance to self antigens leading to autoimmunity. Other explanations for the association include genetic linkage of defective complement alleles in the MHC with other, as yet unidentified, disease susceptibility genes (see Chapter 19). Somewhat surprisingly, C2 and C4 deficiencies are often not associated with increased infections. This suggests that the alternative pathway may be adequate for the elimination of most bacteria. In addition, there is evidence that, with enough antibody fixing C1, either the C2 or C4 steps may be bypassed during classical pathway activation. In contrast, C3 deficiency is associated with frequent serious pyogenic bacterial infections that may be fatal. The frequency of pyogenic in-

TABLE 15–7. Complement Deficiencies and Abnormalities

	Resulting Complement Activation Abnormalities	Associated Diseases/Pathology
Classical Pathway		
C1q, C1r, C1s	Defective classical pathway activation	Systemic lupus erythematosus, pyogenic infections
C4	Defective classical pathway activation	Systemic lupus erythematosus, glomerulonephritis
C2	Defective classical pathway activation	Systemic lupus erythematosus Vasculitis Glomerulonephritis Pyogenic infections
C3	Defective classical/alternative pathway activation	Pyogenic infections Glomerulonephritis Immune complex disease
Alternative Pathway		
Properdin, Factor D	Defective alternative pathway activation	Neisserial infections Other pyogenic infections
Terminal Components		
C5, C6, C7, C8	Defective MAC formation	Disseminated neisserial infections
C9	Defective MAC formation	None

Abbreviation: MAC, membrane attack complex.

fections in these patients illustrates the importance of C3 for opsonization, enhanced phagocytosis, and destruction of these organisms. In addition, immune complex glomerulonephritis occurs frequently in patients with C3 deficiency, consistent with the role of C3 degradation products in immune complex clearance.

2. Deficiencies in members of the terminal complement components, including C5, C6, C7, C8, and C9, have also been described. These patients cannot generate the MAC and would therefore be expected not to be able to lyse foreign organisms. It is interesting that the only consistent clinical problem in these patients is a propensity for disseminated infections by the *Neisseria* bacteria, including *N. meningitidis* and *N. gonorrhoeae*, suggesting that complement-mediated bacteriolysis is particularly important for defense against these organisms. Individuals with C9 deficiencies do not show the same susceptibility to infection, suggest-

ing that poly-C9 formation is not necessary for *Neisseria* bacteriolysis.

3. Deficiencies in soluble and membrane-bound complement regulatory proteins are associated with abnormal complement activation and a variety of related clinical abnormalities (Table 15–8). The absences of classical, alternative, and terminal pathway regulatory proteins have each been described.

Hereditary angioneurotic edema (HANE) is an autosomal dominant deficiency of C1INH. Clinical manifestations of the disease include intermittent, acute accumulation of edema fluid in skin and mucosa, lasting from 24 to 72 hours. The skin of the face and extremities and the laryngeal and intestinal mucosa are most frequently involved. Serious symptoms resulting from edema in these locations include abdominal pain, nausea and vomiting, diarrhea, and potentially life-threatening airway

TABLE 15–8. Abnormalities/Deficiencies of Regulatory Complement Proteins

Protein	Resulting Complement Activation Abnormalities	Associated Diseases/Pathology	Disease/Syndrome Name
C1 inhibitor	Deregulated classical pathway activation, consumption of C3	Acute, intermittent attacks of skin and mucosal edema	Hereditary angioneurotic edema
Factor I	Deregulated classical pathway activation, consumption of C3	Pyogenic infections; immune complex disease	
Factor H	Deregulated alternative pathway activation, consumption of C3	Pyogenic infections, glomerulonephritis	
Decay-accelerating factor and CD59	Deregulated C3 convertase activity Increased susceptibility of erythrocytes to MAC-mediated lysis	Complement-mediated intravascular hemolysis	Paroxysmal nocturnal hemoglobinuria

Abbreviations: MAC, membrane attack complex.

obstruction. Attacks have been associated with antecedent emotional stress and trauma; the physiologic basis for the association is not understood. The deficiency in C1INH in HANE patients may be caused either by an absolute reduction in C1INH protein due to inheritance of a nonfunctional gene or by synthesis of a normal amount of a dysfunctional protein due to mutations in the coding sequences of the gene. Although there is usually one normal allele of the C1INH gene in these patients, the plasma levels of the C1INH protein are sufficiently reduced (20 to 30 per cent of normal) so that activation of C1 by immune complexes is not properly controlled. It is clear that during attacks there are elevated plasma levels of activated C1 and decreased levels of the substrates of activated C1, namely C2 and C4.

The mediators of edema formation in HANE patients include a proteolytic fragment of C2, called C2 kinin, and bradykinin. C1INH is a major regulator of other plasma serine proteases besides C1, including kallikrein and coagulation factor XII. Both activated kallikrein and factor XII can promote increased formation of bradykinin, which can cause vascular endothelial cell contraction leading to tissue edema. Furthermore, C1INH deficiency can lead to increased plasmin generation, which in turn can proteolytically generate a fragment from C2 with kinin activity.

Deficiencies of the soluble alternative pathway regulators Factor I and Factor H have been described. Factor I deficiency provides insights into the role of Factor I–mediated cleavage of C3b. Plasma C3 is completely consumed in these patients as a result of unregulated formation of fluid phase C3 convertase (by the tickover mechanism). The clinical consequence is increased infections with pyogenic bacteria. Similarly, rare cases of Factor H deficiency are characterized by excess alternative pathway activation, consumption of C3, and glomerulonephritis due to inadequate clearance of immune complexes, as well as direct renal deposition of complement by-products. The effects of a lack of Factor I or H are similar to the effects of an autoantibody called **C3 nephritic factor** (C3NeF), which is specific for alternative pathway C3 convertase (C3bBb). This antibody stabilizes C3bBb and protects the complex from Factor H–mediated dissociation, therefore causing unregulated consumption of C3. Patients with this antibody often have glomerulonephritis, possibly caused by inadequate clearing of circulating immune complexes.

Deficiencies of membrane protein regulators of complement activation include the absence of DAF and CD59. Both of these proteins are normally bound to the plasma membrane of erythrocytes and other cell types by phosphatidylinositol linkages. **Paroxysmal nocturnal hemoglobinuria** (PNH) is a disease in which cells lack the ability to express phosphatidylinositol-linked membrane proteins, including DAF and CD59, due to deficiency of an enzyme required to form such protein-lipid link-

ages. PNH is characterized by recurrent bouts of intravascular hemolysis, at least partially attributable to complement activation on the surface of erythrocytes. This leads to chronic hemolytic anemia, pancytopenia, and venous thrombosis. This complement-mediated lysis of erythrocytes in PNH is predictable, since DAF normally functions to inhibit C3 convertase formation on autologous cell surfaces, and CD59 normally serves to inhibit MAC formation on autologous cell membranes. Rare cases of isolated CD59 deficiency with normal DAF expression also have a PNH-like syndrome, but conversely, isolated DAF deficiency does not lead to significant hemolysis.

4. Deficiencies of complement receptors include the absence of CR3 and CR4, both resulting from rare mutations in the β chain (CD18) gene common to the CD11CD18 family of integrin molecules. The congenital disease caused by this gene defect is called **leukocyte adhesion deficiency** (see Chapter 21). This disorder is characterized by recurrent pyogenic infections, due to inadequate adherence of neutrophils to endothelium at tissue sites of infection and perhaps impaired iC3b-dependent phagocytosis of bacteria.

Pathologic Effects of a Normal Complement System

Even when it is properly regulated and appropriately activated, the complement system can cause significant tissue damage. Some of the pathology associated with bacterial infections is attributable to the biologic effects of complement activation. These pathologic effects include the bystander destruction of normal host cells when acute inflammatory responses to infectious organisms take place. For example, a bacterial infection can stimulate a humoral immune response, and antibody bound to bacterial antigens can activate complement. Direct activation of the alternative pathway on bacterial surfaces may also occur, in the absence of antibody. The generation of C3a and C5a stimulates the accumulation of neutrophils at the site of infection. The neutrophils adhere to and phagocytose the infecting organisms. In addition, neutrophils release free radicals and proteases that destroy microbes and may damage normal cells in the vicinity as well. Histamine released from mast cells in response to C5a may amplify the inflammatory response by increasing vascular permeability.

Complement activation is also associated with intravascular thrombosis, leading to ischemic injury to tissues. Several different mechanisms have been proposed for this association. Complement may be fixed by anti-endothelial antibodies in the setting of graft rejection, or by immune complexes deposited in vessel walls. In both cases, the generation of MAC can lead to damage to the endothelial surface, which favors coagulation. There is also evidence that pre-lytic MAC complexes specifically activate prothrombinases in the circulation, which will initiate thrombosis independent of MAC-mediated damage to endothelium. In addition, C5a may alter endothelial surface heparan sulfate expression in ways that favor coagulation.

An example of complement-mediated pathology occurs in immune complex diseases. Systemic vasculitis and immune complex glomerulonephritis may result from deposition of antigen-antibody complexes in the walls of vessels and kidney glomeruli (see Chapter 20). Complement activated by the Ig in these deposited immune complexes may initiate the acute inflammatory responses that destroy the vessel walls or glomeruli, leading to thrombosis, ischemic damage to tissues, and scarring. However, experiments with knockout mice lacking FcγR suggest that much of the pathology associated with immune complex disease is due to IgG Fc-mediated activation of neutrophils, and is not dependent on complement.

SUMMARY

The complement system includes serum and membrane proteins that interact in a highly regulated manner to produce biologically active protein products. The system includes two convergent proteolytic pathways, composed of several different zymogens. The alternative pathway is in a continual state of low level activation in the plasma and is amplified on the surfaces of infectious organisms. The classical pathway is usually initiated by antigen-antibody complexes. The two pathways converge, both utilizing the C3 protein to form enzymes that initiate the common terminal pathway, leading to formation of a cytolytic protein complex called the membrane attack complex. These pathways are regulated by various soluble and membrane-bound proteins, which inhibit different steps in the cascades. The biologic functions of the complement system include cytolysis, opsonization of organisms and immune complexes for phagocytic clearance, production of inflammation, enhancement of humoral immune responses, and solubilization and clearance of immune complexes. Cytolysis is mediated by MAC formation on cell surfaces. Opsonization is largely mediated by proteolytic fragments of C3, as is immune complex clearance. Inflammation is promoted by proteolytic fragments of complement proteins called anaphylatoxins (C3a, C4a, C5a). Specific membrane receptors for complement fragments are required for the opsonin and anaphylatoxin effects. The complement proteins are members of several different gene families, reflecting the common usage of duplicated exons and genes encoding functionally significant domains during the evolution of the system. Many genetic deficiencies in one or more complement components have been described, with associated infections and autoimmune diseases.

Selected Readings

Ahearn, J. M., and D. T. Fearon. Structure and function of the complement receptors CR1 (CD35) and CR2 (CD21). Advances in Immunology 46:183–219, 1989.

Asghar, S. S. Membrane regulators of complement activation and their aberrant expression in disease. Laboratory Investigation 72:254–271, 1995.

Colten, H. R., and F. S. Rosen. Complement deficiencies. Annual Review of Immunology 10:809–834, 1992.

Holers, V. M., T. Kinoshita, and H. Molina. The evolution of mouse and human complement C3-binding proteins: divergence of form but conservation of function. Immunology Today 13:231–237, 1992.

Liszewski, M. K., T. C. Farries, D. M. Lublin, I. A. Rooney, and J. P. Atkinson. Control of the complement system. Advances in Immunology 61:201–283, 1996.

Morgan, B. P., and M. J. Walport. Complement deficiency and disease. Immunology Today 12:301–306, 1991.

Volanakis, J. E. Transcriptional regulation of complement genes. Annual Review of Immunology 13:277–305, 1995.

Westel, R. A. Structure, function and cellular expression of complement anaphylatoxin receptors. Current Opinion in Immunology 7:48–53, 1995

SECTION IV

IMMUNITY IN DEFENSE

AND DISEASE

In this final section, we apply our knowledge of the fundamental mechanisms of specific immunity to understanding immunologic defenses against potential pathogens and tumors, immune reactions to transplants, and the principles of diseases that are caused by abnormalities in immune responses. We begin with a discussion of the role of specific immunity in combating microbial infections in Chapter 16. Chapter 17 is devoted to tissue transplantation, an increasingly promising therapy for a variety of diseases, in which the major limitation is immunologic rejection. Chapter 18 describes immune responses to tumors and immunological approaches for cancer therapy. Chapter 19 deals with the mechanisms that maintain tolerance to self antigens and how self-tolerance may fail, leading to autoimmunity. Chapter 20 describes the effector mechanisms of immune-mediated tissue injury and disease. In Chapter 21 we discuss the cellular and molecular bases of congenital and acquired immunodeficiencies, including the acquired immunodeficiency syndrome (AIDS), and the pathologic complications of deficient immunity.

CHAPTER SIXTEEN

IMMUNITY TO MICROBES

The principal physiologic function of the immune system is to protect the host against pathogenic microbes. Resistance to infections formed the basis for the original identification of specific immunity. As our understanding of immune responses has increased, we are better able to explain the mechanisms of antimicrobial immunity. Throughout this book we have mentioned examples of specific immune responses to particular microbes, largely to illustrate the physiologic relevance of various aspects of lymphocyte function. In this chapter, we discuss in more detail the main features of immunity to different types of pathogenic microorganisms.

The evolution of an infectious disease in an individual involves complex interactions between the microbe and the host. The key events during infection include entry of the microbe, invasion and colonization of host tissues, evasion from host immunity, and tissue injury or functional impairment. Some microbes produce disease by liberating toxins, even without extensive colonization of host tissues. Many features of microorganisms determine their virulence, and many diverse mechanisms contribute to the pathogenesis of infectious diseases. The topic of microbial pathogenesis is largely beyond the scope of this book and will not be discussed in detail. Rather, our discussion will focus on host immune responses to pathogenic microorganisms.

There are several important general features of immunity to microbes:

1. *Defense against microbes is mediated by both innate and specific immunity.* Microbial infections provide clear demonstrations of the role of specific immunity in enhancing the protective mechanisms of innate immunity and in directing these mechanisms to sites where they are needed. In the absence of specific immunity, the effector mechanisms of innate immunity are often ineffective at eliminating microbes. Furthermore, many pathogenic microbes have evolved to resist innate defense mechanisms. Protection against such microbes is critically dependent on specific immune responses.

2. *The innate immune response to microbes plays an important role in determining the nature of the specific immune response.* For instance, complement activation by some bacteria enhances the production of specific antibodies, and IL-12 secretion by macrophages is critical for the development of cell-mediated immunity.

3. *The immune system is capable of responding in distinct and specialized ways to different microbes in order to combat these infectious agents most effectively.* Because microbes differ greatly in patterns of host invasion and pathogenesis, their elimination requires diverse effector systems. It is not surprising that the magnitude and type of immune response to an infectious agent often determine the course and outcome of the infection. Our subsequent discussions will highlight the principal mechanisms of specific immunity against different bacteria, fungi, viruses, and parasites.

4. *The survival and pathogenicity of microbes in a host are critically influenced by their ability to evade or resist protective immunity.* As we shall see later in the chapter, microorganisms have developed a variety of strategies for surviving in the face of powerful immunologic defenses.

5. *Tissue injury and disease consequent to infections may be caused by the host response to the microbe and its products rather than by the microbe itself.* Immunity, like many other homeostatic mechanisms, is necessary for host survival but also has the potential of causing injury to the host.

This chapter considers five types of pathogenic microorganisms that illustrate the main features of immunity to microbes: (1) extracellular bacteria, (2) intracellular bacteria, (3) fungi, (4) viruses, and (5) protozoan and multicellular parasites. These groups of microbes illustrate the diversity of antimicrobial immunity (Fig. 16–1) and the physiologic significance of many of the responses and effector functions of lymphocytes discussed in earlier chapters.

IMMUNITY TO EXTRACELLULAR BACTERIA

Extracellular bacteria are capable of replicating outside host cells, e.g., in the circulation, in extracellular connective tissues, and in various tissue spaces such as the airways, genitourinary tract, and intestinal lumens. Examples of pathogenic extracellular bacteria include gram-positive, pus-forming, or pyogenic, cocci (*Staphylococcus*, *Streptococcus*), gram-negative cocci (meningococcus and gonococcus, two species of *Neisseria*), many gram-negative bacilli (including enteric organisms such as *Escherichia coli*), and some gram-positive bacilli (particularly anaerobes such as the *Clostridium* species).

Extracellular bacteria cause disease by two principal mechanisms. First, they induce inflammation, which results in tissue destruction at the site of infection. Pyogenic cocci are responsible for a large number of suppurative infections in humans. Second, many of these bacteria produce **toxins**, which have diverse pathologic effects. Such toxins may be **endotoxins**, which are components of bacterial cell walls, or **exotoxins**, which are actively secreted by the bacteria. The endotoxin of gram-negative bacteria, also called **lipopolysaccharide** (LPS), has been mentioned in earlier chapters as a potent stimulator of cytokine production by macrophages. Many exotoxins are primarily cytotoxic, and others cause disease by various mechanisms. For instance, diphtheria toxin inhibits protein synthesis by enzymatically modifying and thereby blocking the function of elongation factor-2. Cholera toxin stimulates cyclic adenosine monophosphate production in intestinal epithelial cells, lead-

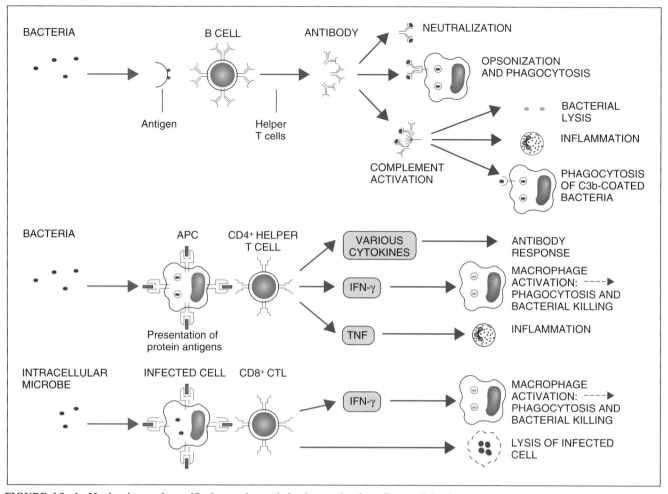

FIGURE 16–1. Mechanisms of specific immunity to infectious microbes. Extracellular bacteria and their toxins are eliminated by antibody-dependent effector mechanisms. Microbes that survive within phagocytes are eliminated by T cell–mediated stimulation of the microbicidal mechanisms of phagocytes. Microbes that infect phagocytic and other cells are also eliminated by CTL-mediated killing of infected cells.

ing to active chloride secretion, water loss, and intractable diarrhea. Tetanus toxin is a neurotoxin that binds to motor end-plates at neuromuscular junctions and causes persistent muscle contraction, which can be fatal if it affects the muscles involved in breathing. *Immune responses against extracellular bacteria are aimed at eliminating the bacteria and at neutralizing the effects of their toxins.*

Innate Immunity to Extracellular Bacteria

Because extracellular microbes are rapidly killed by the microbicidal mechanisms of phagocytes, a principal mechanism of innate immunity to these microbes is phagocytosis by neutrophils, monocytes, and tissue macrophages. The resistance of bacteria to phagocytosis and digestion within macrophages is an important determinant of virulence. Activation of the complement system, in the absence of antibody, also plays an important role in the elimination of these bacteria. Gram-positive bacteria contain a peptidoglycan in their cell walls that activates the alternative pathway of complement by promoting the formation of the alternative pathway C3 convertase (see Chapter 15). LPS in the cell walls of gram-negative bacteria was one of the first agents shown to activate the alternative complement pathway, in the absence of antibody. Bacteria that express mannose on their surface may also bind a mannose-binding protein that is homologous to C1q, leading to complement activation by the classical pathway without the participation of antibody (see Chapter 15). One result of complement activation is the generation of C3b, which opsonizes bacteria and enhances phagocytosis. In addition, the membrane attack complex (MAC) lyses bacteria, especially *Neisseria* species, and complement by-products participate in inflammatory responses by recruiting and activating leukocytes.

Endotoxins, such as LPS, stimulate the production of cytokines by macrophages and by other cells, e.g., vascular endothelium. Some of these cytokines, such as tumor necrosis factor (TNF), interleukin-1 (IL-1), IL-6, and chemokines, induce local acute inflammation. Inflammatory cells serve to eliminate the bacteria; injury to adjacent normal tissues is a pathologic side effect of these defense mechanisms. Cytokines also induce fever and stimulate the synthesis of acute phase proteins. Other cytokines, notably IL-12, stimulate the development of T_H1 cells and activate cytolytic T lymphocytes (CTLs) and natural killer (NK) cells, thus providing important links between innate and specific immunity. The functions of these cytokines have been discussed in Chapter 12.

Specific Immune Responses to Extracellular Bacteria

Humoral immunity is the principal protective specific immune response against extracellular bacteria. Some of the most immunogenic components of the cell walls and capsules of these microbes are polysaccharides, which are prototypical thymus-independent antigens. Such antigens directly stimulate B cells, giving rise to strong specific IgM responses. Some T cell–independent polysaccharides may also stimulate heavy chain isotype switching. For instance, in humans the antibody response to pneumococcal capsular polysaccharides is dominated by IgG2 antibodies. The mechanism of isotype switching induced by T-independent antigens is not known.

Antibodies perform several functions that serve to eliminate bacteria. Both IgM and IgG antibodies against bacterial surface antigens and toxins stimulate three types of **effector mechanisms** (see Fig. 16–1):

1. *IgG antibodies opsonize bacteria and enhance phagocytosis* by binding to Fcγ receptors on monocytes, macrophages, and neutrophils (see Chapter 3). Both IgM and IgG antibodies activate complement, generating C3b and iC3b, which bind to specific type 1 and type 3 complement receptors, respectively, and further promote phagocytosis. Individuals deficient in C3 are extremely susceptible to pyogenic infections.

2. *Antibodies neutralize bacterial toxins* and prevent their binding to target cells. Passive immunization against tetanus toxin by injection of antibody is a potentially life-saving treatment in acute tetanus infections. IgA antibody present in various secretions, e.g., in the gastrointestinal and respiratory tracts, is important for neutralizing the toxins of bacteria in these organs and for preventing colonization of extraluminal tissues.

3. *Both IgM and IgG antibodies activate the complement system,* leading to the production of the microbicidal MAC and the liberation of by-products that are mediators of acute inflammation. It is likely, however, that the lytic function of the MAC is important for the elimination of only some microbes. As we mentioned in Chapter 15, deficiencies of the late components of complement, C5 to C8, which are involved in the formation of the

MAC, are associated with increased susceptibility to *Neisseria* but not to other bacterial infections.

The principal T cell response to extracellular bacteria consists of CD4$^+$ T cells responding to protein antigens in association with class II major histocompatibility complex (MHC) molecules. As discussed in Chapter 6, extracellular microbes and soluble antigens are internalized by antigen-presenting cells (APCs); the antigens are processed; and fragments of the processed proteins preferentially associate with class II MHC molecules. It is not known whether macrophages, dendritic cells, or B cells are the most important APCs for such bacterial protein antigens *in vivo*. The proteins of phagocytosed bacteria may also enter the cytosolic class I MHC pathway of antigen processing and be subsequently recognized by CD8$^+$ T cells. However, this effector arm of the immune system is probably most important for microbes that live within cells, as discussed later. The effector functions of CD4$^+$ T cells are mediated by secreted cytokines, which stimulate antibody production, induce local inflammation, and enhance the phagocytic and microbicidal activities of macrophages (see Fig. 16–1). Interferon-γ (IFN-γ) is the principal cytokine responsible for macrophage activation, and TNF and lymphotoxin recruit and activate neutrophils.

The principal injurious consequences of host responses to extracellular bacteria are inflammation and septic shock, caused by cytokines produced mainly by activated macrophages. Septic shock is the most severe cytokine-induced pathologic consequence of infection by gram-negative and some gram-positive bacteria. It is a syndrome characterized by circulatory collapse, metabolic disturbances (hypoglycemia), and disseminated intravascular coagulation. As discussed in Chapter 12, TNF and IL-1 are the principal mediators of septic shock. In fact, serum levels of TNF are predictive of the outcome of severe gram-negative bacterial infections. Recently, it has been observed that some bacterial toxins stimulate large numbers of CD4$^+$ T cells. Any one of these toxins can stimulate all the T cells in an individual that express a particular set or family of V_β T cell receptor genes. Such toxins have been called **super-antigens** (Box 16–1). Their importance lies in their ability to activate many T cells, resulting in large amounts of cytokine production and clinicopathologic abnormalities that are similar to septic shock.

A late complication of the humoral immune response to bacterial infections may be the generation of disease-producing antibodies. The best-defined examples are two sequelae of streptococcal infections of the throat or skin, which are manifested weeks or even months after the infections are controlled. In **rheumatic fever,** pharyngeal infection with some serologic types of β-hemolytic streptococci leads to the production of antibodies

BOX 16–1.   Bacterial "Super-Antigens"

Staphylococcal enterotoxins (SEs) are exotoxins produced by the gram-positive bacterium *Staphylococcus aureus.* They consist of five serologically distinct groups of proteins: SEA, SEB, SEC, SED, and SEE. These toxins are the most common cause of food poisoning in humans. A related toxin, TSST, causes a disease called the toxic shock syndrome (TSS), which is often associated with wound infection and tampon use. Pyrogenic exotoxins of streptococci and exotoxins produced by mycoplasma may be structurally and functionally related to these enterotoxins.

The immune response to staphylococcal enterotoxins and related proteins has a number of features that make these bacterial products quite unique in terms of biologic and pathologic effects:

(1) *Staphylococcal enterotoxins are among the most potent naturally occurring T cell mitogens known.* They are capable of stimulating the proliferation of normal T lymphocytes at concentrations of 10^{-9} M or less. As many as one in five normal T cells in mouse lymphoid tissue or human peripheral blood may respond to a particular enterotoxin.

(2) All the T cells that respond to an enterotoxin express antigen receptors whose V_β regions are encoded by a single V_β gene or gene family. Responsiveness is apparently not related to other components of the T cell receptor (TCR), i.e., V, J, or D segments. This is because enterotoxins bind directly to the β chains of TCR molecules outside the antigen-binding regions. Different enterotoxins stimulate T cells expressing V_β genes from different families (see table). Because T cells expressing only certain TCRs recognize and respond to each enterotoxin, these proteins are called antigens and not polyclonal mitogens. However, since the frequency of enterotoxin-responsive T cells is much higher than the frequency of cells specific for conventional protein antigens, the enterotoxins have been named super-antigens.

V_β Expression in Responding T Cells

Enterotoxin	Mice	Humans
SEB	$V_\beta 7$, 8.1–8.3, 17	$V_\beta 3$, 12, 14, 15, 17, 20
SEC 2	$V_\beta 8.2$, 10	$V_\beta 12$, 13, 14, 15, 17, 20
SEE	$V_\beta 11$, 15, 17	$V_\beta 5.1$, 6.1–6.3, 8, 18
TSST-1	$V_\beta 15$, 16	$V_\beta 2$

* Abbreviations: SE, staphylococcal enterotoxin; TSST, toxic shock syndrome toxin.

(3) *Responses to staphylococcal enterotoxins require antigen-presenting cells that express class II MHC molecules.* Staphylococcal enterotoxins directly bind to class II MHC molecules on accessory cells. This complex is then recognized by CD4$^+$ T cells expressing antigen receptors with a particular V_β. Enterotoxins bind to class II MHC molecules outside the peptide-binding clefts and do not need to be processed like other protein antigens. The same enterotoxin binds to class II molecules of different alleles, indicating that the polymorphism of the MHC does not influence presentation of these antigens. Recent studies suggest that each staphylococcal enterotoxin (SEA) molecule possesses two cooperative binding sites for one class II MHC molecule. This would allow each SEA molecule to cross-link MHC molecules on APCs, and the dimers may cross-link antigen receptors on each T cell and thus initiate T cell responses.

The remarkably high frequency of staphylococcal enterotoxin-responding CD4$^+$ T cells has several functional implications. Acutely, exposure to high concentrations of enterotoxins leads to systemic reactions like fever, disseminated intravascular coagulation, and cardiovascular shock. These abnormalities are probably mediated by cytokines, such as TNF, produced directly by the T cells or by macrophages that are activated by the T cells. This is the likely pathogenesis of the toxic shock syndrome, which is characterized by shock, skin exfoliation, conjunctivitis, and severe gastrointestinal upset, and can progress to renal and pulmonary failure and death. More prolonged administration of enterotoxins to mice results in wasting, thymic atrophy, and profound immunodeficiency, also probably secondary to chronic high levels of cytokine, e.g., TNF, production.

Staphylococcal enterotoxins are proving to be useful tools for analyzing T lymphocyte development. Administration of SEB to neonatal mice leads to intrathymic deletion of all T cells expressing $V_\beta 3$ and $V_\beta 8$ TCR genes. This mimics the postulated self antigen–induced clonal deletion of self-reactive T cells during thymic maturation. Super-antigens also induce Fas- or TNF receptor–mediated activation-induced death of mature CD4$^+$ and CD8$^+$ T cells (see Chapter 10, Box 10–1). This is a model for the deletion of mature T cells that are repeatedly stimulated by self antigens.

Viral gene products may also function as super-antigens. In certain inbred strains of mice, different mouse mammary tumor virus genes have become incorporated into the genome. Viral antigens produced by the cells of one strain are capable of activating T lymphocytes from other strains that express particular V_βs in their antigen receptors. This is a form of "mixed lymphocyte reaction" that is not caused by MHC disparity. It was discovered long before viral super-antigens were identified, and the inter-strain reactions were attributed to Mls (for minor lymphocyte stimulating) loci. We now know that Mls "loci" are actually different retroviral genes that are stably inherited in different inbred strains.

---

against a bacterial cell wall protein (M protein). Some of these antibodies cross-react with myocardial sarcolemmal proteins and myosin, leading to antibody deposition in the heart and subsequent inflammation (carditis). In **post-streptococcal glomerulonephritis,** infection of the skin or throat with other serotypes of β-hemolytic streptococci leads to the formation of immune complexes of bacterial antigen and specific antibody in kidney glomeruli, and this causes nephritis. Thorough antibiotic therapy is recommended for "sore throats" caused by β-hemolytic streptococci, not because of the severity of the pharyngitis but to prevent the later development of rheumatic fever. Immune responses to bacterial infections may lead to different pathologic sequelae. The polyclonal lymphocyte activation induced by bacterial endotoxins and super-antigens may contribute to the development of autoimmunity by stimulating many lymphocytes, among which are self-reactive clones that are normally anergic to stimulation by self antigens. This concept, and the possible links between infections and autoimmune diseases, are discussed more fully in Chapter 19.

Evasion of Immune Mechanisms by Extracellular Bacteria

The virulence of extracellular bacteria has been linked to a number of mechanisms that favor tissue invasion and colonization and resist innate immunity. These include adhesive bacterial surface proteins that promote entry into cells, anti-phagocytic mechanisms, and inhibition of complement or inactivation of complement products. For instance, the capsules of many gram-positive and gram-negative bacteria contain one or more sialic acid residues that inhibit complement activation by the alternative pathway. Bacteria with polysaccharide-rich capsules also resist phagocytosis and, therefore, are much more virulent than homologous strains lacking a capsule.

One mechanism utilized by bacteria to evade specific immunity is *genetic variation of surface antigens*. Surface antigens of many bacteria, such as gonococci and *E. coli,* are contained in the pili, which are the structures primarily involved in bacterial adhesion to host cells. The major antigen of the pili is a protein of approximately 35 kilodaltons (kD), called pilin. The pilin genes of gonococci undergo extensive gene conversions, because of which one organism can produce up to 10^6 antigenically distinct pilin molecules. This mechanism helps the bacteria escape specific antibody attack, although its principal significance for the bacteria may be to select for pili that are more adherent for host cells so that the bacteria are more virulent. In other bacteria, such as *Haemophilus influenzae,* changes in the production of glycosyl synthetases lead to alterations in surface LPS and other polysaccharides, which are major antigens of these bacteria.

IMMUNITY TO INTRACELLULAR BACTERIA

A characteristic of facultative intracellular bacteria is their ability to survive, and even replicate, within phagocytes. Since these microbes are able to find a niche where they are inaccessible to circulating antibodies, their elimination requires immune mechanisms that are very different from the mechanisms of defense against extracellular bacteria. Elimination of intracellular bacteria is mediated by the mechanisms of cellular immunity. In fact, as we shall see, many of our current concepts of cell-mediated immunity are based on studies of immune responses against intracellular bacteria, such as *Listeria monocytogenes* and mycobacteria.

Innate Immunity to Intracellular Bacteria

Phagocytes ingest and attempt to destroy bacteria as part of the innate immune response. However, pathogenic intracellular bacteria are resistant to degradation within phagocytes. It is, therefore, not surprising that usually *innate immunity is ineffective in controlling colonization by and spread of these microorganisms.* Resistance to phagocytosis is also the reason why such bacteria tend to cause chronic infections that may last for years, often recur or recrudesce after apparent cures, and are difficult to eradicate.

Intracellular bacteria also activate NK cells, either directly or by stimulating macrophage production of IL-12, a powerful NK cell–activating cytokine. NK cells produce IFN-γ, which in turn activates macrophages and promotes killing of phagocytosed bacteria. Thus, NK cells provide an early defense against these microbes, prior to the development of specific immunity. In fact, T and B cell–deficient severe combined immunodeficiency (SCID) mice are able to control infection by *L. monocytogenes,* at least temporarily, by NK cell–derived IFN-γ production. As we shall discuss later in this chapter, NK cells also provide initial defense against viral infections.

Specific Immune Responses to Intracellular Bacteria

The major protective immune response against intracellular bacteria is cell-mediated immunity. Individuals with deficient cell-mediated immunity, such as AIDS patients, are extremely susceptible to infections with intracellular bacteria (and viruses). Cell-mediated immunity was first identified by George Mackaness in the 1950s as protection against the intracellular bacterium *L. monocytogenes.* This form of immunity could be adoptively transferred to naive animals with lymphoid cells but not with serum from infected or immunized animals (Fig. 16–2).

Cell-mediated immunity consists of two types of reactions—killing of phagocytosed microbes as a result of macrophage activation by T cell–derived cytokines, particularly IFN-γ, and lysis of infected cells by CD8$^+$ CTLs. Protein antigens of intracellular bacteria stimulate both CD4$^+$ and CD8$^+$ T cells. Presumably, CD4$^+$ T cells respond to released antigens that are internalized and presented by class II MHC-expressing APCs; an example of such an antigen is the purified protein derivative (PPD) of *Mycobacterium tuberculosis.* Intracellular bacteria (and viruses) are potent inducers of the differentiation of CD4$^+$ helper T cells to the T_H1 phenotype, because these microbes stimulate IL-12 production by macrophages and IFN-γ production by NK cells, and both these cytokines promote the development of T_H1 cells (see Chapter 12). T_H1 cells secrete IFN-γ, which activates macrophages to produce reactive oxygen species and enzymes

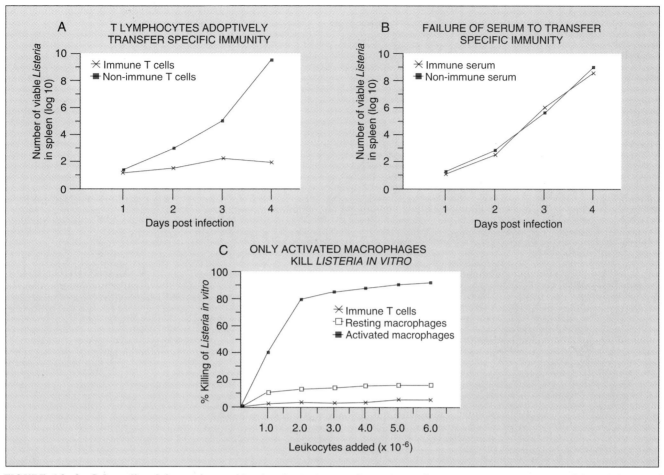

FIGURE 16-2. Cell-mediated immunity to *Listeria monocytogenes*. Immunity to *L. monocytogenes* is measured by inhibition of bacterial growth in the spleens of animals inoculated with a known dose of viable bacteria. Such immunity can be transferred to normal mice by T lymphocytes (*A*) but not by serum (*B*) from syngeneic mice previously immunized with killed *L. monocytogenes*. In this form of cell-mediated immunity, the bacteria are actually killed by activated macrophages and not by T cells, even from immune animals (*C*). In another form of cell-mediated immunity, specific T lymphocytes kill infected cells (not shown).

that kill phagocytosed bacteria. IFN-γ also stimulates the production of antibody isotypes (e.g., IgG2a in mice), which activate complement and opsonize bacteria for phagocytosis, thus aiding the effector functions of macrophages. T$_H$1 cells also produce TNF and lymphotoxin, which induce local inflammation. Thus, *cytokines produced during the innate immune response to intracellular bacteria help eliminate these organisms and stimulate protective specific immune responses* (Fig. 16–3). The importance of these cytokines in immunity to intracellular bacteria has been demonstrated in several experimental models. For instance, inhibitors of IFN-γ or TNF worsen the outcomes of such infections in mice. Furthermore, IFN-γ knockout mice and TNF receptor type I (TNF-RI) knockout mice are extremely susceptible to infections with intracellular bacteria, such as *M. tuberculosis* and *L. monocytogenes*.

If the bacteria survive within cells and release their antigens into the cytoplasm, they stimulate CD8+ CTLs. Two activities of CTLs participate in the elimination of intracellular bacteria in phago-

cytes—IFN-γ secretion with resultant macrophage activation, and lysis of infected cells. The role of different T cell subsets in defense against intracellular bacteria has been analyzed by adoptively transferring T cells from *L. monocytogenes*–infected mice to normal mice and challenging the recipients with the bacteria. By depleting particular cell types, it is possible to determine which effector population is responsible for specific immunity. Such experiments have shown that both CD4+ and CD8+ T cells are required to eliminate infection by wild-type *L. monocytogenes* (Fig. 16–4). The CD4+ cells produce large amounts of IFN-γ, which activates macrophages to kill phagocytosed microbes. However, *L. monocytogenes* produces a protein called hemolysin, which allows bacteria to escape from the phagolysosomes of macrophages into the cytoplasm. In their cytoplasmic haven, the bacteria are protected from the microbicidal mechanisms of macrophages, such as reactive oxygen species, which are produced mainly within phagolysosomes. CD8+ T cells are then activated, and they function by producing more IFN-γ and by kill-

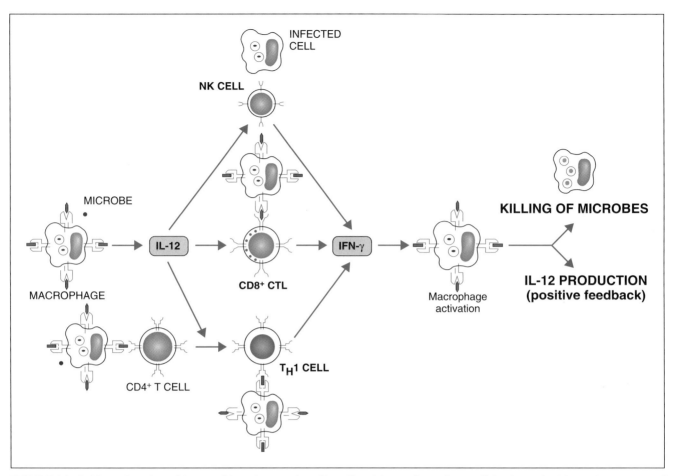

FIGURE 16–3. The roles of cytokines and lymphocytes in cell-mediated immunity to intracellular bacteria. Macrophages respond to microbes by producing IL–12, which stimulates NK cells, CTLs, and the development of T_H1 cells. NK cells may also directly respond to infected cells. All three lymphocyte populations produce IFN-γ, which activates macrophages and enhances their microbicidal activities. NK cells and CTLs also lyse infected cells; the lytic mechanism of microbial elimination is not shown.

ing any macrophages that may be harboring bacteria in their cytoplasm. A mutant of *L. monocytogenes* that lacks hemolysin remains confined to phagolysosomes. Such mutant bacteria can be completely eradicated by CD4⁺ T cell–derived IFN-γ production and macrophage activation. Thus, *the effectors of cell-mediated immunity, namely macrophages activated by CD4⁺ T cells and CD8⁺ CTLs, function cooperatively in defense against intracellular bacteria.*

The macrophage activation that occurs in response to intracellular microbes is also capable of causing tissue injury. This may be manifested as delayed type hypersensitivity (DTH) reactions to microbial protein antigens such as PPD (see Chapter 13). Since intracellular bacteria have evolved to resist killing within phagocytes, they often persist for long periods, leading to chronic antigenic stimulation and T cell and macrophage activation. This may result in the formation of **granulomas** surrounding the microbes (see Chapter 13, Fig. 13–9). The histologic hallmark of infections with some intracellular bacteria is granulomatous inflammation. This type of inflammatory reaction may serve to

localize and prevent the spread of the microbes, but it is also associated with severe functional impairment due to tissue necrosis and fibrosis. Thus, *the host immune response is the principal cause of tissue injury and disease in infections by some intracellular bacteria.* The concept that protective immunity and pathologic hypersensitivity may co-exist because they are manifestations of the same type of specific immune response is perhaps most clearly exemplified in mycobacterial infections (Box 16–2).

Differences among individuals in the patterns of immune responses to intracellular microbes are important determinants of disease progression and clinical outcome. An example of this is leprosy, caused by *Mycobacterium leprae.* There are two polar forms of the disease, although many patients fall into less clear intermediate groups. In lepromatous leprosy, patients have high specific antibody titers but weak cell-mediated responses to *M. leprae* antigens. Mycobacteria proliferate within macrophages and are detectable in large numbers. The bacterial growth and inadequate macrophage activation result in destructive lesions of skin and underlying

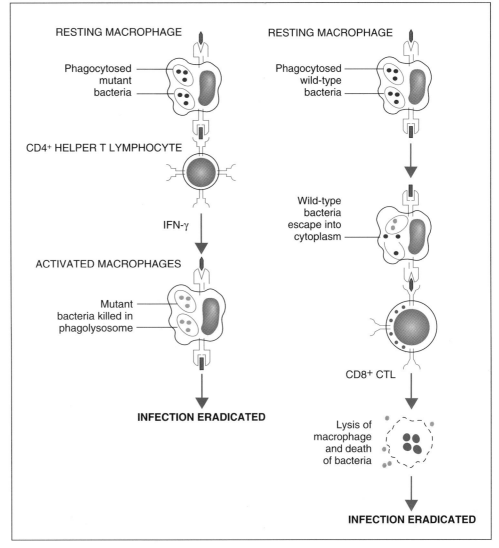

RESTING MACROPHAGE

Phagocytosed mutant bacteria

CD4+ HELPER T LYMPHOCYTE

IFN-γ

ACTIVATED MACROPHAGES

Mutant bacteria killed in phagolysosome

INFECTION ERADICATED

RESTING MACROPHAGE

Phagocytosed wild-type bacteria

Wild-type bacteria escape into cytoplasm

CD8+ CTL

Lysis of macrophage and death of bacteria

INFECTION ERADICATED

FIGURE 16–4. Cooperation of activated macrophages and CTLs in the elimination of an intracellular bacterium, *L. monocytogenes*. *L. monocytogenes*–infected macrophages may stimulate CD4+ and CD8+ T cells. Infection by mutant bacteria can be eradicated by macrophages activated by IFN-γ produced by CD4+ T cells, but wild-type bacteria are eradicated only after the colonized macrophages are lysed by CTLs.

tissues. In contrast, patients with tuberculoid leprosy have strong cell-mediated immunity but low antibody levels. This pattern of immunity is reflected in granulomas that form around nerves, giving rise to sensory peripheral nerve defects and secondary traumatic skin lesions but less tissue destruction and a paucity of bacilli in the lesions. The mechanisms responsible for the defective cell-mediated immunity in lepromatous leprosy patients are not fully known. Some studies indicate that patients with the tuberculoid form of the disease produce IFN-γ and IL-2 in lesions (suggesting T_H1 cell activation), whereas patients with lepromatous leprosy produce relatively more IL-4 and IL-10 (typical of T_H2 cells). Both the deficiency of IFN-γ and the macrophage suppressive effects of IL-10 and IL-4 may result in weak cell-mediated immunity in lepromatous leprosy. It is also possible that the lepromatous form of the disease is due to anergy in *M. leprae*–specific T cells. As one would expect, intradermal injection of IL-2 or IFN-γ has a

beneficial effect on the skin lesions of lepromatous leprosy.

Evasion of Immune Mechanisms by Intracellular Bacteria

An important mechanism for survival of intracellular bacteria is their ability to resist elimination by phagocytes. Mycobacteria do this by inhibiting phagolysosome fusion, perhaps by interfering with lysosome movement. The phenolic glycolipid of *M. leprae* functions as a scavenger of reactive oxygen species. The hemolysin produced by virulent strains of *L. monocytogenes* blocks bacterial killing in macrophages, and it may also inhibit antigen presentation by infected macrophages. *Legionella pneumophila,* the causative organism of Legionnaires' disease, inhibits phagolysosome fusion, and mutants that lose their ability to do so also lose their virulence. The outcome of infection by these organisms often depends on whether the T cell–

BOX 16–2. IMMUNITY TO *Mycobacterium Tuberculosis*

Mycobacteria are slow-growing, aerobic, facultative intracellular bacilli whose cell walls contain high concentrations of lipids. These lipids are responsible for the acid-resistant staining of the bacteria with the red dye carbol fuchsin, because of which these organisms are also called acid-fast bacilli (AFB). The two common human pathogens in this class of bacteria are *M. tuberculosis* and *M. leprae;* in addition, atypical mycobacteria such as *M. avium–intracellulare* cause opportunistic infections in immunodeficient hosts, e.g., AIDS patients. *M. bovis* infects cattle and may infect humans, and bacillus Calmette-Guerin (BCG) is an attenuated, nonvirulent strain of *M. bovis* that is used as a vaccine against tuberculosis.

Tuberculosis is an example of an infection with an intracellular bacterium in which protective immunity and pathologic hypersensitivity co-exist, and the lesions are caused mainly by the host response. *M. tuberculosis* does not produce any known exotoxin or endotoxin. Infection usually occurs by the respiratory route and is transmitted from person to person. In a primary infection, bacilli multiply slowly in the lungs and cause only mild inflammation. The infection is contained by alveolar macrophages. Over 90 per cent of infected patients remain asymptomatic, but bacteria survive in lungs and can be reactivated. By 6 to 8 weeks after infection, regional lymph nodes are involved, and CD4+ T cells are activated. These T cells produce IFN-γ, which activates macrophages and enhances their ability to kill phagocytosed bacilli. TNF produced by T cells and macrophages also plays a role in local inflammation and macrophage activation, and TNF receptor knockout mice are highly susceptible to mycobacterial infections. However, *M. tuberculosis* is capable of surviving within macrophages, because components of its cell wall inhibit the fusion of lysosomes with phagocytic vacuoles. Continuing T cell activation leads to the formation of granulomas, with central necrosis, called caseous necrosis, caused by macrophage products (lysosomal enzymes, reactive oxygen species). This is a form of delayed type hypersensitivity (DTH) reaction to the bacilli. It is postulated that necrosis serves to eliminate infected macrophages and provides an anoxic environment in which the bacilli cannot divide. Thus, even the tissue injury may serve a protective function. Caseating granulomas and the fibrosis (scarring) that accompanies granulomatous inflammation are the principal causes of tissue pathology and clinical disease in tuberculosis. Only a minority of infected individuals (probably less than 10 per cent) develop clinical symptoms. Previously infected persons show cutaneous DTH reactions to skin challenge with a bacterial antigen preparation (PPD, or tuberculin). Bacilli may survive for many years even without overt clinical manifestations, and they may be reactivated at any time. This is the reason for treating individuals who convert from PPD-negative to PPD-positive with antibiotics such as isoniazid and rifampin, even though they may have no symptoms of the disease. Tuberculosis is becoming more prevalent, in part because of the emergence of antibiotic-resistant strains. The efficacy of BCG as a prophylactic vaccine remains controversial. This might be predictable, given that the specific immune response mediates both protection against infection and tissue injury. In rare cases, *M. tuberculosis* may cause lesions in extrapulmonary sites. In chronic tuberculosis, sustained production of TNF leads to cachexia.

Different inbred strains of mice vary in their susceptibility to infection with *M. tuberculosis* (and with other intracellular microbes, such as the protozoan *Leishmania major*). Susceptibility or resistance to *M. tuberculosis* maps to a single gene, called the *bcg* or *lsh* gene, which is expressed in macrophages. This gene has been shown to be homologous to genes encoding nitrate-transport membrane proteins, raising the possibility that susceptibility is related to defective production of microbicidal nitrate derivatives such as nitric oxide. However, neither the protein product of the *bcg/lsh* gene nor its function is defined as yet.

Some studies have shown that both *in vivo* and *in vitro*, *M. tuberculosis* stimulates T cells expressing the γδ form of the antigen receptor. These T cells may be reacting to mycobacterial antigens that are homologous to heat shock proteins. Heat shock proteins are evolutionarily conserved, being present in prokaryotes and eukaryotes. They are induced upon exposure to many kinds of stress, including heat; depletion of oxygen, nutrients and essential ions; and exposure to free radicals. Their function may be as chaperones, to promote re-folding of proteins after cell injury. In infected cells, heat shock proteins may be produced by the stressed cells or by the bacteria. It is postulated that the response of γδ T cells to these proteins is a primitive defense mechanism against some microbes. Some γδ cells, as well as CD4−CD8− TCRαβ-expressing cells, have been shown to recognize mycobacterial lipid antigens presented in association with non-polymorphic CD1 molecules. These are the only known examples of T cell recognition of non-peptide antigens. However, the effector functions of these cells, their frequencies *in vivo*, and their roles in protective immunity against mycobacteria are not yet known.

Atypical mycobacteria constitute a heterogeneous group of non-tuberculous mycobacteria that are widely distributed in the animal kingdom and in nature. They frequently infect humans but rarely cause disease in immunocompetent individuals. However, they cause severe disease in immunodeficient individuals, such as AIDS patients. Unlike *M. tuberculosis* infection in patients with intact immune systems, *M. avium–intracellulare* infection in AIDS patients does not lead to well-formed granulomas, but rather pathology is due to unchecked bacterial proliferation.

M. leprae is the cause of leprosy. The nature of the T cell response, and specifically the types of cytokines produced by activated T cells, is an important determinant of the lesions and clinical course of *M. leprae* infection (see text).

---

stimulated microbicidal mechanisms of macrophages or microbial resistance to killing gain the upper hand.

IMMUNITY TO FUNGI

Fungal infections, also called mycoses, are being increasingly recognized as important causes of morbidity and mortality in humans. Some fungal infections are endemic, and these are usually caused by dimorphic fungi that are present in the environment and whose spores are inhaled by humans. Examples of these include *Histoplasma capsulatum, Blastomyces dermatiditis,* and *Coccidioides immites.* Other fungal infections are said to be opportunistic, because the causative agents cause mild or no disease in normal individuals but may infect and cause severe disease in immunodeficient persons. Examples of opportunistic fungi are *Candida, Aspergillus,* and *Cryptococcus neoformans.* There has been a recent increase in opportunistic fungal infections due to an increase in immunodeficiencies, caused mainly by AIDS and by therapy for disseminated cancers and transplant rejection.

Different fungi infect humans and may live in extracellular tissues and within phagocytes. Therefore, the features of immune responses to these microbes are often combinations of the patterns of responses to extracellular and facultative intracellular bacteria. However, much less is known about anti-fungal immunity than about immunity against bacteria and viruses. This is partly because of the paucity of animal models for mycoses, and partly because these infections frequently occur in individuals who are incapable of mounting effective immune responses.

Innate Immunity to Fungi

The principal mediator of innate immunity against fungi is the neutrophil. Patients with neutropenia are extremely susceptible to opportunistic fungal infections. Neutrophils presumably liberate fungicidal substances, such as reactive oxygen species and lysosomal enzymes, and phagocytose fungi for intracellular killing. Macrophages are also capable of combating fungal infections.

Specific Immune Responses to Fungi

Cell-mediated immunity is the major defense mechanism against fungal infections. *H. capsulatum* is a facultative intracellular parasite that lives in macrophages, and it is eliminated by the same cellular mechanisms that are effective against intracellular bacteria. CD4$^+$ and CD8$^+$ T cells cooperate to eliminate the yeast forms of *C. neoformans,* which tends to colonize the lungs and brain in immunodeficient hosts. *Candida* infections often start at mucosal surfaces, and cell-mediated immunity is believed to prevent spread of the fungi into tissues. In all these situations, T$_H$1 responses are protective and T$_H$2 responses are detrimental to the host. It is not surprising that granulomatous inflammation is an important cause of host tissue injury in some intracellular fungal infections, such as histoplasmosis.

Fungi often elicit specific antibody responses, which are useful for serological diagnosis. However, the protective efficacy of humoral immunity is not established.

Evasion of Immune Mechanisms by Fungi

Very little is known about the ability of fungi to evade host immunity. In fact, since immunocompetent individuals are not susceptible to opportunistic fungal infections, these parasites do not appear to resist immune effector mechanisms effectively. *Candida albicans* produces a proteolytic enzyme that degrades human immunoglobulins *in vitro,* but the role of this enzyme as a virulence factor is not established.

IMMUNITY TO VIRUSES

Viruses are obligatory intracellular microorganisms that replicate within cells, often using the nucleic acid and protein synthetic machineries of the host. Viruses typically infect a wide variety of cell populations by utilizing normal cell surface molecules as receptors to enter the cells. Three well-known examples of viruses that use physiologic surface proteins as receptors are (1) human immunodeficiency virus-1 (HIV-1), which binds to the CD4 molecule on human T cells and uses chemokine receptors as co-receptors; (2) Epstein-Barr virus (EBV), which binds to the type 2 complement receptor (CR2, CD21) on human B cells; and (3) rhinovirus, the agent of the common cold, which binds to intercellular adhesion molecule-1 (ICAM-1, CD54) expressed on a variety of cell types, including airway epithelium. After entering cells, viruses can cause tissue injury and disease by any of several mechanisms. Viral replication interferes with normal cellular protein synthesis and function, leading to injury to and ultimately death of the infected cell. This is one type of **cytopathic effect of viruses,** and the infection is said to be "lytic" because the infected cell is killed. Noncytopathic viruses may cause latent infections, during which they reside in host cells and produce proteins that are foreign to the host and stimulate specific immunity. Cytopathic and non-cytopathic viruses may elicit different types of specific immune responses, as we discuss below. Human immunodeficiency virus is discussed in Chapter 21.

Innate Immunity to Viruses

There are two principal mechanisms of innate immunity against viruses:

1. *Viral infection directly stimulates the production of type I IFN by infected cells.* Type I IFNs function to inhibit viral replication. The characteristics of the cytokine-induced "antiviral state" have been described in Chapter 12.

2. *Natural killer (NK) cells lyse a wide variety of virally infected cells.* NK cells (see Chapter 13) may be one of the principal mechanisms of immunity against viruses early in the course of infection, before specific immune responses have developed (Fig. 16–5). NK cells also effectively recognize and lyse infected cells in which the virus may have inhibited antigen presentation and class I MHC expression (by mechanisms discussed later in this section). This is because NK cells are preferentially activated by class I–negative targets (see Chapter 13). Type I IFN can enhance the ability of NK cells to lyse infected target cells.

Specific Immune Responses to Viruses

Immunity against viral infections is mediated by a combination of humoral and cellular immune mechanisms. *Specific antibodies are important in de-*

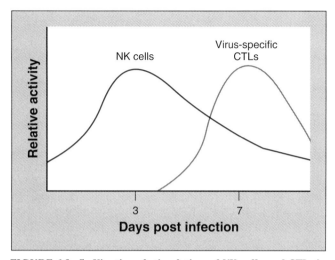

FIGURE 16–5. Kinetics of stimulation of NK cells and CTLs in a viral infection. In an acute lymphocytic choriomeningitis virus infection of mice, NK activity in the spleen is increased prior to the development of virus-specific CTLs. Type I IFN production by spleen cells parallels NK activity. (Adapted from Welsh, R. M., H. Yang, and J. F. Bukowski. The role of interferon in the regulation of virus infections by cytotoxic lymphocytes. BioEssays 8:10–13, 1988, with permission of the publisher, ICSU Press.)

fense against viruses early in the course of infection and in defense against cytopathic viruses that are liberated from lysed infected cells. Neutralizing antiviral antibodies bind to envelope or capsid proteins and prevent viral attachment and entry into host cells. Opsonizing antibodies may enhance phagocytic clearance of viral particles. Somewhat perversely, however, opsonizing antibodies may actually enhance the invasion of Fc receptor–bearing cells by viruses; this has been postulated to be a mechanism for HIV-1 infection of mononuclear phagocytes. Secretory immunoglobulins of the IgA isotype may be important for neutralizing viruses that enter via the respiratory or intestinal tract. Oral immunization against poliomyelitis works by inducing secretory immunity. Complement activation may also participate in antibody-mediated viral immunity, mainly by promoting phagocytosis and possibly by direct lysis of viruses with lipid envelopes.

The success of prophylactic vaccination with attenuated or killed viruses is largely related to the ability of these vaccines to stimulate specific antibody responses. The importance of humoral immunity is suggested by the observation that resistance to a particular virus, induced by either infection or vaccination, is often specific for the serologic type of the virus and seems to correlate with antibody specificity. An example of this is influenza virus, in which exposure to one serologic type does not confer resistance to other serotypes of the virus. However, several points about the role of humoral immunity in protection against viruses should be emphasized. First, antibodies may

be effective against viruses before the organisms enter cells, or they may block spread from cell to cell, but viruses that survive and replicate intracellularly are inaccessible to antibodies. Second, it has generally proved difficult to transfer antiviral immunity to naive animals with purified antibodies. Third, the neutralizing capacity of an antibody *in vitro* often shows little or no correlation with its protective capacity *in vivo*. Taken together, these observations suggest that although antibodies are an important component of immunity to viruses, they may not be sufficient for eliminating many viral infections.

The principal mechanism of specific immunity against established viral infections, especially with non-cytopathic viruses, is CTLs. In fact, the principal physiologic function of CTLs is surveillance against viral infections. Most virus-specific CTLs are CD8$^+$ cells that recognize endogenously synthesized cytosolic viral antigens in association with class I MHC molecules on any nucleated cell. The full differentiation of CD8$^+$ CTLs requires cytokines produced by CD4$^+$ helper cells and/or costimulators expressed on infected cells. As discussed in Chapter 13, the antiviral effects of CTLs are due to lysis of infected cells, introduction of enzymes into infected cells that degrade viral genomes, and secretion of cytokines with interferon activity.

The importance of CTLs in the outcome of viral infections has been demonstrated in several experimental systems. Mice can be protected against influenza virus by adoptive transfer of virus-specific, class I-restricted CTLs and by cloned lines of such T cells. Nevertheless, as mentioned above, actively acquired immunity to influenza virus is serotype-specific. These findings support the view that both antibodies and CTLs cooperate to protect the host against viruses—the former act to block viral binding and entry into host cells, and the latter inhibit viral replication by killing infected cells. CTLs and antibodies may be specific for different viral antigens. For instance, many influenza-specific CTLs recognize peptides derived from internal proteins (like matrix protein and nucleoprotein), whereas antibodies are specific for envelope proteins (hemagglutinin, neuraminidase) that determine the serotype of the virus strain.

In some infections with non-cytopathic viruses, CTLs may be responsible for tissue injury. The clearest example is lymphocytic choriomeningitis virus (LCMV) infection in mice, which induces inflammation of the spinal cord meninges. LCMV infects meningeal cells but does not injure them directly. It stimulates the development of specific CTLs that lyse meningeal cells during a physiologic attempt to eradicate the viral infection. Therefore, T cell–deficient mice infected with LCMV become chronic carriers of the virus, but pathologic lesions do not develop, whereas in normal mice, meningitis develops. On face value, this observation appears to contradict the usual situation, in which immunodeficient individuals are more susceptible to infectious diseases than are normal individuals. Hepatitis B virus infection in humans shows some similarities to murine LCMV, in that immunodeficient persons who become infected do not develop the disease but become carriers who can transmit the infection to otherwise healthy persons. The livers of patients with acute and chronic active hepatitis contain large numbers of CD8⁺ T cells, and hepatitis virus–specific, class I MHC-restricted CTLs can be isolated from liver biopsies and propagated *in vitro.*

Immune responses to viral infections may be involved in producing disease in two other ways. First, a consequence of persistent infection with some viruses, such as hepatitis B, is the formation of circulating immune complexes composed of viral antigens and specific antibodies. These complexes deposit in blood vessels and lead to systemic vasculitis (see Chapter 20). Second, some viruses are known to contain amino acid sequences that are also present in some self antigens. It has been postulated that because of this "molecular mimicry," antiviral immunity can lead to immune responses against self antigens. There is, however, no formal proof that viral molecular mimicry participates in the development of immune diseases.

Evasion of Immune Mechanisms by Viruses

Viruses have evolved numerous mechanisms for evading host immunity, especially surveillance by T cells:

1. *Many viruses are capable of great antigenic variation,* and large numbers of serologically distinct strains of these viruses have been identified. The influenza pandemics that occurred in 1918, 1957, and 1968 were all due to different strains of the virus, and subtler variants arise more frequently. As a result, the virus becomes resistant to immunity generated in the population by previous infections. There are so many existing serotypes of rhinovirus that specific immunization against the common cold may not be a feasible preventive strategy. HIV-1, the virus that causes AIDS, is also capable of tremendous antigenic variation (see Chapter 21). In these situations, prophylactic vaccination may have to be directed against invariant viral proteins, such as surface molecules that are required for virus entry into host cells.

2. *Some viruses inhibit class I MHC-associated presentation of cytosolic protein antigens.* Adenovirus type 12 produces a protein that suppresses the transcription of class I MHC genes. Herpes simplex virus produces a protein that binds to the peptide-binding site of the transporter in antigen processing (TAP) heterodimer, and it does not allow this transporter to bind and translocate peptides from the cytosol into the endoplasmic reticulum. Cytomegalovirus removes newly synthesized class I MHC molecules from the endoplasmic reticulum

and apparently deposits them in the cytoplasm, where they are non-functional. Inhibition of antigen processing and presentation blocks the assembly and expression of stable class I MHC molecules. As a result, cells infected by such viruses show reduced class I expression and become insusceptible to killing by CD8⁺ CTLs.

3. *Some viruses produce molecules that inhibit innate or specific immunity.* Pox viruses encode molecules that are homologous to the receptors for several cytokines, including IFNγ, TNF, and IL-1, and these are secreted by infected cells. The secreted cytokine receptor homologues may function as competitive antagonists of the cytokines. Some cytomegaloviruses produce a molecule that is homologous to class I MHC proteins, and they may compete for binding and presentation of peptide antigens. Epstein-Barr virus produces a protein that is homologous to the macrophage-suppressive cytokine IL-10 and may function to inhibit specific immunity. These examples probably represent a very small fraction of immunosuppressive viral molecules. The identification of these molecules raises the intriguing possibility that viruses have acquired inhibitors of immune responses during their passage through human hosts and have thus evolved to infect and colonize humans.

4. *Viruses may infect, and either lyse or inactivate, immunocompetent cells.* The obvious example is HIV, which survives by infecting and eliminating CD4⁺ T cells, the key regulators of immune responses to protein antigens.

IMMUNITY TO PARASITES

In infectious disease terminology, "parasitic infection" refers to infection with animal parasites, such as protozoa, helminths, and ectoparasites (e.g., arthropods, such as ticks and mites). Such parasites currently account for greater morbidity and mortality than any other class of infectious organisms, particularly in developing countries. It is estimated that about 30 per cent of the world's population suffers from parasitic infestations. Malaria alone affects almost 250 million people worldwide, with about 1 to 2 million deaths annually. The magnitude of this public health problem is the principal reason for the great interest in immunity to parasites and for the development of immunoparasitology as a distinct branch of immunology.

Most parasites go through complex life cycles, part of which is in humans (or other vertebrates) and part of which is in intermediate hosts such as flies, ticks, and snails. Humans are infected usually by bites from infected intermediate hosts or by sharing a particular habitat with an intermediate host. For instance, malaria and trypanosomiasis are transmitted by insect bites, and schistosomiasis is transmitted by exposure to water in which infected snails reside.

A fundamental feature of most parasitic infections is their chronicity. There are many reasons for this, including weak innate immunity and the ability of parasites to evade or resist elimination by specific immune responses. Furthermore, many anti-parasite antibiotics are toxic or relatively ineffective or both. Individuals living in endemic areas require repeated chemotherapy because of continued exposure, and this is often not possible due to expense and logistic problems. Because of these reasons, the development of prophylactic vaccines for parasites has long been considered an important goal for developing countries. The persistence of parasites in human hosts also leads to immunologic reactions that are chronic and may result in pathologic tissue injury as well as abnormalities in immune regulation. Therefore, some of the clinicopathologic consequences of parasitic infestations are due to the host response and not the infection itself.

Innate Immunity to Parasites

Protozoan and helminthic parasites that enter the blood stream or tissues are often able to survive and replicate because they are well adapted to resisting natural host defenses. The invertebrate stages of many parasites, which are recovered from the nonhuman intermediate hosts, activate the alternative pathway of complement and are lysed by the MAC. However, parasites recovered from the vertebrate, e.g., human, host are usually resistant to lysis by complement. This may be due to many reasons, including loss of surface molecules that bind complement or the acquisition of host regulatory proteins such as decay accelerating factor (DAF). Macrophages can phagocytose protozoa, but many pathogenic organisms are resistant to phagocytic killing and may even replicate within macrophages. The tegument of helminthic parasites makes them resistant to the cytocidal mechanisms of both neutrophils and macrophages.

Specific Immune Responses to Parasites

Different protozoa and helminths vary greatly in their structural and biochemical properties, life cycles, and pathogenic mechanisms. It is, therefore, not surprising that different parasites elicit quite distinct specific immune responses. In general, pathogenic protozoa have evolved to survive within host cells, so that protective immunity against these organisms is mediated by mechanisms similar to those that eliminate intracellular bacteria and viruses. In contrast, metazoa, such as helminths, survive in extracellular tissues, and their elimination is often dependent on special types of antibody responses.

The principal defense mechanism against protozoa that survive within macrophages is cell-mediated immunity, particularly macrophage activation by CD4+ T cell–derived cytokines. Perhaps the best-documented example of this is infection of mice with *Leishmania major,* a protozoan that survives within the endosomes of macrophages. Resistance to the infection is associated with the production of IFN-γ by the T_H1 subset of CD4+ T cells. Conversely, activation of T_H2 cells by the protozoa results in increased parasite survival and exacerbation of lesions, due to the macrophage suppressive actions of T_H2 cytokines. Inbred strains of mice that are resistant to *L. major* infection produce large amounts of IFN-γ in response to leishmanial antigens, and anti-IFN-γ antibody makes the mice susceptible to the infection. In contrast, inbred strains that are susceptible to fatal leishmaniasis produce more IL-4 in response to the infection than resistant strains, and the injection of anti-IL-4 antibody induces resistance in the susceptible strains. IFN-γ activates macrophages and enhances intracellular killing of leishmania, and high levels of IL-4 (and other T_H2-derived cytokines) inhibit the activation of macrophages by IFN-γ. Murine leishmaniasis is one of the first documented examples of dominant T_H1 or T_H2 responses determining disease resistance or susceptibility, and it remains the paradigm of this phenomenon. The gene(s) that control protective versus harmful immune responses in inbred mice, and presumably in humans as well, have not yet been identified. Attempts to alter the outcome of these infections with cytokines or cytokine antagonists are going on in many laboratories at present. A promising candidate is IL-12, which induces protective T_H1 responses even in susceptible mouse strains.

Protozoa that replicate inside cells and lyse host cells stimulate specific CTL responses, similar to cytopathic viruses. An example of such an organism is the malaria parasite. It was thought for many years that antibodies were the major protective mechanism against malaria, and early attempts at vaccinating against this infection focused on generating antibodies. It is now apparent that the CTL response is an important defense against the spread of this intracellular protozoan (Box 16–3). In fact, the poor efficacy of vaccination with malarial proteins is attributed to the inability of such immunizations to stimulate CTLs.

Defense against many helminthic infections is mediated by IgE antibodies and eosinophils. This is a special type of antibody-dependent cellular cytotoxicity (ADCC), in which IgE antibodies bind to the surface of the helminth, eosinophils then attach via Fcϵ receptors, and the eosinophils are activated to secrete granule enzymes that destroy the parasites. Production of specific IgE antibody and eosinophilia are frequently observed in infections by helminths, such as *Nippostrongylus,* filariae, *Ascaris,* and schistosomes. These responses are attributed to the propensity of helminths to stimulate the T_H2 subset of CD4+ helper T cells, which secrete IL-4 and IL-5. IL-4 stimulates the production of IgE, and IL-5 stimulates the development and activation of eosinophils. Eosinophils may be more effective at killing helminths than other leu-

BOX 16–3.   Immunity to Malaria

Malaria is a disease caused by a protozoan parasite (*Plasmodium*) that infects more than 250 million people and causes 1 to 2 million deaths annually. Malaria, especially that caused by infection with *P. falciparum,* continues to be one of the most widespread and prevalent diseases today. Infection is initiated when sporozoites are inoculated into the blood stream by the bite of an infected mosquito (*Anopheles*). The sporozoites rapidly disappear from the blood and invade the parenchymal cells of the liver. In the hepatocyte, sporozoites develop into merozoites by a multiple fission process termed schizogony. One to two weeks after infection, the infected hepatocytes burst, releasing thousands of merozoites, thereby initiating the erythrocytic stage of the life cycle. The merozoites invade red blood cells by a process that involves multiple ligand-receptor interactions, including the binding of the *P. falciparum* protein EBA-175 to glycophorin A on erythrocytes. Merozoites develop sequentially into ring forms, trophozoites, and schizonts, each of which expresses both shared and unique antigens. The erythrocytic cycle continues when schizont-infected red blood cells burst and release merozoites that invade other erythrocytes. Sexual stage gametocytes develop in some cells and are taken up by mosquitoes during a blood meal, after which they fertilize and develop into oocysts. Immature sporozoites develop in the mosquitoes within 2 weeks and travel to the salivary glands, where they mature and become infective.

The clinical features of malaria caused by the four species of *Plasmodium* that infect humans include fever spikes, anemia, and splenomegaly. Many pathologic manifestations of malaria may be due to activation of T cells and macrophages and production of TNF. The development of cerebral malaria in a murine model is prevented by depletion of CD4+ T lymphocytes or by injection of neutralizing anti-TNF antibody.

The immune response to malaria is complex and stage-specific, i.e., immunization with antigens derived from sporozoites, merozoites, or gametocytes protects only against the particular stage. Based on this observation, it is postulated that a vaccine consisting of combined immunogenic epitopes from each of these stages should stimulate more effective immunity than a vaccine that incorporates antigens from only one stage. The best-characterized malaria vaccines are directed against sporozoites. Due to the stage-specific nature of sporozoite antigens, such vaccines must provide sterilizing immunity to be effective. The protective immunity induced in humans by injection of irradiation-inactivated sporozoites is partially mediated by antibodies that inhibit sporozoite invasion of hepatoma cells *in vitro*. Such antibodies recognize the circumsporozoite (CS) protein, which mediates binding of parasites to liver cells. The CS protein contains a central region of about 40 tandem repeats of the sequence Asn-Ala-Asn-Pro (NANP)n, which makes up the immunodominant B cell epitope of the protein. Anti-(NANP)n antibodies neutralize sporozoite infectivity, but such antibodies provide only partial protection against infection. The antibody response to CS protein is dependent on helper T cells.

Most CS-specific helper T cells recognize epitopes outside the NANP region, and some of these T cell epitopes correspond to the most variable residues of the CS protein. This suggests that variation of the antigens of the surface coat may have arisen in response to selective pressures imposed by specific T cell responses.

CD8+ T cells play an important role in immunity to the hepatic stages of infection. If sporozoite-immunized mice are depleted of CD8+ T cells by injection of anti-CD8 antibodies, they are unable to resist a challenge infection. However, a role for CD8+ T cells in immunity to malaria has been demonstrated in some but not all inbred strains of mice, suggesting that host genetic factors influence the outcome of these infections. The protective effects of CD8+ T cells may be mediated by direct lysis of sporozoite-infected hepatocytes, or indirectly by the secretion of IFN-γ and activation of hepatocytes to produce nitric oxide and other agents that kill parasites. IL-12 induces resistance to sporozoite challenge in rodents and non-human primates, presumably by stimulating IFN-γ production. Conversely, resistance of *P. berghei* sporozoite-immunized mice to a challenge infection is abrogated by treatment with anti-IFN-γ antibodies.

The sexual blood stage of malaria is an important target for vaccine development. The major goal of this form of immunization is to kill infected erythrocytes or to block merozoite invasion of new red cells. Although infected individuals mount strong immune reponses against blood stage antigens during natural malaria infections, most of these responses do not appear to affect parasite survival, or they result in the selection of new antigenic variants. Nevertheless, some merozoite proteins may be good vaccine candidates. Of particular interest is the major merozoite surface antigen MSP-1, which contains a highly conserved C-terminal region. Immunization with recombinant DNA-derived C-terminal peptides has been shown to confer significant protection against malaria in rodents, and it has given promising results in primate trials with human malaria parasite strains. The resistance induced by MSP-1 vaccination appears to involve antibody-dependent mechanisms.

Transmission-blocking vaccines are being developed to act on stages of the parasite life cycle that are found in mosquitoes. Such vaccines provide no protection for the immunized individual but act to reduce the number of parasites available for development in the mosquito vector. One such vaccine is an antigen, Pfs25, that is located on the surface of zygotes and ookinetes. These parasite stages are found only in the mosquito vector and would not be under the selective pressures imposed by specific T cell responses toward antigens found in the human intermediate host. Therefore, antigens that may be useful as transmission-blocking vaccines are unlikely to show the high degree of variation that is characteristic of T cell epitopes of sporozoites.

This Box was written with the assistance of Drs. Alan Sher and Louis Miller, National Institutes of Health, Bethesda, MD.

kocytes, because the major basic protein of eosinophil granules may be more toxic for helminths than the proteolytic enzymes and reactive oxygen species produced by neutrophils and macrophages. However, an obligatory role of IgE and eosinophils in resistance *in vivo* has been formally established in very few helminthic infections, perhaps because helminths can be killed by activated macrophages, albeit less efficiently. The expulsion of some intestinal nematodes may be due to IL-4–dependent mechanisms that are not well defined but apparently do not require IgE.

Specific immune responses to parasites can also contribute to tissue injury. Some parasites and their products induce granulomatous responses with concomitant fibrosis. *Schistosoma mansoni* eggs deposited in the liver stimulate CD4+ T cells, which in turn activate macrophages and induce DTH reactions. This results in the formation of granulomas around the eggs. The granulomas serve to contain the schistosome eggs, but severe fibrosis associated with this chronic cell-mediated immune response leads to disruption of venous blood flow in the liver, portal hypertension, and cirrhosis. In lymphatic filariasis, the parasites lodge in lymphatic vessels, leading to chronic cell-mediated immune reactions and ultimately to fibrosis. This results in lymphatic obstruction and severe lymphedema. Chronic and persistent parasitic infestations are often associated with the formation of complexes of parasite antigens and specific antibodies. The complexes can deposit in blood vessels and kidney glomeruli, producing vasculitis and nephritis, respectively (see Chapter 20). Immune complex disease has been described in schistosomiasis and malaria. Malaria infections and African trypanosomiasis are also associated with the production of autoantibodies reactive with many self tissues. The myocarditis and neuropathy seen in Chagas' disease, which is caused by the protozoan *Trypanosoma cruzi,* are probably autoimmune reactions and not the result of local infection, because few or no parasites are present even in active lesions.

Evasion of Immune Mechanisms by Parasites

The ability of parasites to survive in vertebrate hosts reflects evolutionary adaptations that permit these organisms to evade or resist immune effector mechanisms. Different parasites have developed remarkably effective ways of resisting specific immunity. The most important of these fall into two categories: parasites can reduce or alter their own antigenicity, and they can actively inhibit host immune responses. Numerous *mechanisms for reducing immunogenicity* have been described with different parasites:

1. *Anatomic sequestration is commonly observed with protozoa.* Some (e.g., malaria parasites and *Toxoplasma*) survive and replicate inside cells, and others (like *Entamoeba*) develop cysts that are resistant to immune effectors. Some helminthic parasites reside in intestinal lumens and are sheltered from cell-mediated immune effector mechanisms.

2. *Antigen masking is an intriguing phenomenon in which a parasite, during its residence within a host, acquires on its surface a coat of host proteins.* The larvae of *S. mansoni* enter the skin and travel to the lungs and then into the circulation. By the time they enter the lungs, these larvae are coated with ABO blood group glycolipids and MHC molecules derived from the host. It is likely that many other host molecules attach to the surface of the schistosome larvae. It has been postulated that as a result of this coat of self proteins, parasite antigens are masked, and the organism is seen as self by the host immune system. Although this is an interesting hypothesis, the significance of antigen masking is not clear because schistosome larvae do elicit specific immunity in vertebrate hosts.

3. *Parasites become resistant to immune effector mechanisms during their residence in vertebrate hosts.* Lung stage schistosome larvae develop a tegument that is resistant to damage by antibodies and complement or by CTLs. This resistance is presumably due to a biochemical change in the surface coat. The structural complexity of the larval tegument has made it difficult to define the molecular alterations that are associated with acquired resistance. Infective forms of *T. cruzi* synthesize membrane glycoproteins, similar to decay accelerating factor, that inhibit complement activation. *L. major* promastigotes induce rapid breakdown or release of the membrane attack complex, thus reducing complement-mediated lysis. Parasites also evade macrophage killing by various mechanisms. *Toxoplasma gondii* inhibits phagolysosome fusion, and *T. cruzi* lyses the membranes of phagosomes and enters the cytoplasm before fusion with lysosomes can occur. Finally, some parasites express ectoenzymes that cleave bound antibody molecules and thus become resistant to antibody-dependent effector mechanisms.

4. *Parasites have developed effective mechanisms for varying their surface antigens during their life cycle in vertebrate hosts.* Two forms of antigenic variation are well defined.

The first is a stage-specific change in antigen expression, such that the mature tissue stages of parasites produce different antigens from the infective stages. For example, the infective sporozoite stage of malaria parasites is antigenically distinct from the merozoites that reside in the host and are responsible for chronic infection. By the time the immune system has responded to the infection, the parasite expresses new antigens and is no longer a target for immune elimination.

The most remarkable example of antigenic variation in parasites is the continuous variation of major surface antigens seen in African trypanosomes such as *Trypanosoma brucei* and *Trypanosoma rhodesiense.* Infected individuals show waves of blood parasitemia, and each wave consists of one antigenically unique parasite. The same phenomenon can be reproduced in experimental animals infected with a single clone of a trypanosome (Fig. 16–6). Thus, by the time the host produces antibodies against the parasite, an antigenically different organism has replicated. Over a hundred such recrudescent waves of parasitemia can occur in an infection. The major surface antigen of African trypanosomes is a glycoprotein dimer of approximately 50 kD, called the variable surface glycoprotein (VSG), which is attached to the surface by a phosphatidylinositol linkage. Trypanosomes contain more than 1000 different VSG genes, which

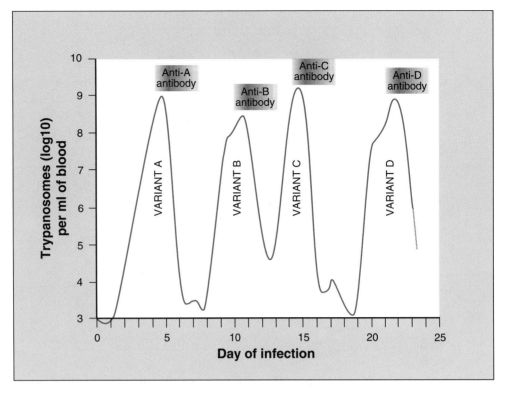

FIGURE 16–6. Parasitemia following trypanosome infections. In a mouse infected experimentally with a single clone of *Trypanosoma rhodesiense*, the blood parasite counts show cyclical waves. Each wave is due to a new antigenic variant of the parasite (labeled *A*, *B*, *C*, and *D*) that expresses a new variable surface glycoprotein, and each decline is a result of a specific antibody response to the variant. The durations of peak antibody production as shown are approximate. Similar waves of parasitemia are seen in natural infections in humans. (Courtesy of Dr. John Mansfield, University of Wisconsin, Madison.)

vary markedly in their sequences except for the most C-terminal 50 amino acids (which are responsible for the surface linkage). Any one VSG gene is expressed in a particular clone at a particular stage of infection. Expression of a new gene may involve duplication and transposition of that gene to a more telomeric chromosomal site at which active transcription ensues (Fig. 16–7). In addition, gene conversion and activation of previously silent genes may also contribute to antigenic variation. Continuous antigenic variation in trypanosomes is neither induced by nor dependent on the specific antibody response and is probably due to a programmed variation in the expression of VSG genes. The molecular mechanisms that regulate this phenomenon have been a focus of active investigation in many laboratories. One consequence of antigenic variation in parasites is that it is difficult to vaccinate individuals against these infections effectively.

5. *Parasites shed their antigenic coats, either spontaneously or after the binding of specific antibodies.* Examples of active membrane turnover and loss of surface antigens have been described with *Entamoeba histolytica,* schistosome larvae, and trypanosomes. Shedding of antigens and bound antibodies renders the parasites relatively resistant to immune effector mechanisms.

Parasites also inhibit host immune responses by multiple mechanisms. Specific anergy to parasite antigens has been described in severe schistosomiasis involving the liver and spleen and in filarial infections. The mechanisms of immunologic unre-

sponsiveness in these patients are not well understood. In lymphatic filariasis, infection of lymph nodes with subsequent architectural disruption may contribute to deficient immunity. More nonspecific and generalized immunosuppression is observed in malaria and African trypanosomiasis. This has been attributed to the production of immunosuppressive cytokines by activated macrophages and T cells, and by defects in T cell responses.

The worldwide implications of parasitic infestations for health and economic development are well appreciated. Attempts to develop effective vaccines against these infections have been actively pursued for many years. Although the progress has been slower than one would have hoped, elucidation of the fundamental mechanisms of immune responses to and immune evasion by parasites holds great promise for the future.

STRATEGIES FOR VACCINE DEVELOPMENT

The birth of immunology as a science may be dated from Edward Jenner's successful vaccination against smallpox, done in 1796. The importance of prophylactic immunization against infectious diseases is best illustrated by the fact that worldwide programs of vaccination have led to the complete or near complete eradication of many of these diseases in developed countries. Smallpox is the most impressive example. The development of effective vaccines against viruses, bacteria, and parasites remains an important goal of immunologists worldwide.

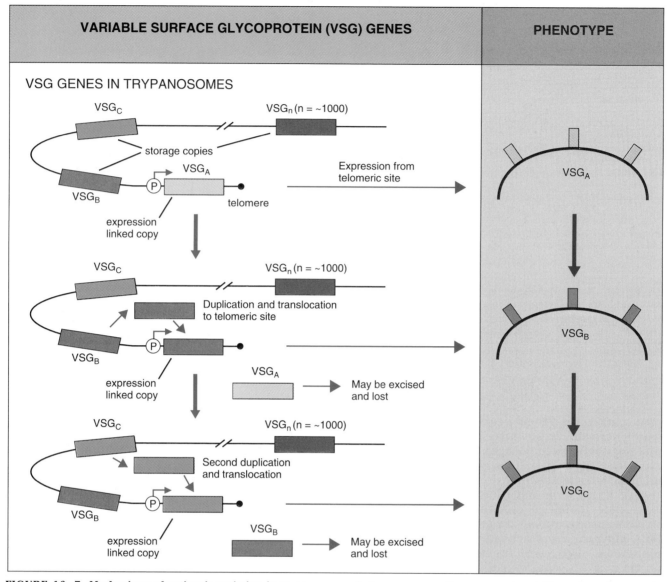

FIGURE 16–7. Mechanisms of antigenic variation in trypanosomes. In trypanosomes, the expressed variable surface glycoprotein (VSG) is encoded by a gene located close to the telomere. Another VSG gene may be duplicated and translocated to this telomeric expression site, generating a new VSG. The fate of the previously expressed VSG gene is not known, but it may be excised and lost (as shown). P, promoter.

The aim of all vaccination is to induce specific immunity that prevents microbial invasion, eliminates microbes that enter hosts, and neutralizes microbial toxins. Since effective vaccination as a public health measure requires long-lasting immunity, the ability of vaccines to stimulate memory T and B lymphocytes is an important consideration in vaccine design. The success of active immunization in eradicating infectious disease is dependent on numerous factors. For instance, infections that are limited to human hosts and are caused by poorly infectious agents whose antigens are relatively invariant are more likely to be controlled by vaccination. On the other hand, antigenic variation, the existence of animal or environmental reservoirs of infection, and high infectivity of microbes

make it less likely that vaccination alone will eradicate a particular infectious disease.

Many types of infectious agents and their products have been used as vaccines.

Attenuated and Inactivated Bacterial and Viral Vaccines. Live, attenuated bacteria were first shown by Louis Pasteur to confer specific immunity. Among the attenuated bacterial vaccines in use today are *Mycobacterium tuberculosis* bacille Calmette-Guérin (BCG), avirulent mutants of *Salmonella typhi,* inactivated *Vibrio cholerae,* and inactivated *Bordetella pertussis.* Many of these vaccines induce limited protection and are effective for relatively short periods. Live, attenuated viral vaccines are generally more effective. The most frequently used approach for producing such vaccines is to grow viruses in long-term cell culture and attenuate them by heating or chemical fixation. More recently, temperature-sensitive and deletion mutants are being generated with the same goal in mind. Polio, measles, and yellow fever are three examples of effective attenuated viral vaccines. Viral vaccines often induce long-lasting specific immunity, so that immunization of children is sufficient for life-long protection. A major advantage of attenuated live viruses as vaccines is that they infect cells, are not cytopathic, and therefore induce strong CTL responses. The principal limitation of these vaccines relates to concerns about their safety.

Purified Antigen (Subunit) Vaccines. One effective use of purified antigens as vaccines is for the prevention of diseases caused by bacterial toxins. Toxins can be rendered harmless without loss of immunogenicity, and such "toxoids" induce strong antibody responses. Diphtheria and tetanus are two infections that have been largely controlled because of immunization of children with toxoid preparations. Vaccines composed of bacterial polysaccharide antigens are used against pneumococcus and *Haemophilus influenzae*. The immunogenicity of polysaccharides may be enhanced by coupling them to protein antigens (so-called *conjugate vaccines*). Although polysaccharides are inefficient inducers of B cell memory, these vaccines often provide long-lived protective immunity, probably because the polysaccharides are not degraded easily and therefore persist in lymphoid tissues and continue to stimulate specific B lymphocytes for long periods. Subunit vaccines composed of purified polypeptides have been used for hepatitis B and influenza viruses and are in clinical trials for *Bordetella pertussis* and cholera. Purified protein vaccines stimulate helper T cells and antibody responses, but they do not generate potent CTLs.

Synthetic Antigen Vaccines. Early approaches for developing synthetic antigens as vaccines relied on the synthesis of linear and branched polymers of three to ten amino acids based on the known sequences of microbial antigens. Such peptides are weakly immunogenic by themselves and need to be coupled to large proteins to induce antibody responses. This is much like generating antibody responses to hapten-carrier conjugates (see Chapter 9). Two advances hold considerable promise for the development of synthetic peptide vaccines. First, it is now possible to deduce the protein sequences of microbial antigens from nucleotide sequence data and to prepare large quantities of proteins by recombinant DNA technology. Second, by testing overlapping peptides and by mutational analysis, it is possible to identify epitopes or even individual residues that are recognized by B or T cells or that bind to MHC molecules for presentation to MHC-restricted T lymphocytes. Empirical trial of peptides containing single or multiple amino acid substitutions has led to the construction of antigens with enhanced

binding to MHC molecules or enhanced capacity to activate T cells. To date, such studies have been done largely in inbred mice. Because T cell antigen recognition is influenced by the polymorphism of MHC molecules, it is likely that in outbred human populations it will be much more difficult to "custom design" peptides with enhanced immunogenicity. Nevertheless, there is potential in this approach for creating at will vaccines that are of high potency. Using recombinant DNA technology, synthetic peptide vaccines have been produced that correspond to immunogenic epitopes of hepatitis B virus, herpes simplex virus, and foot-and-mouth disease virus (a major pathogen for livestock). The same method is being explored in many other infectious diseases.

Live Viral Vectors. An exciting approach to vaccine development is to introduce genes encoding microbial antigens into a noncytopathic virus and infect individuals with this virus. Thus, the virus serves as a source of the antigen in the inoculated individual. One great advantage of viral vectors is that they, like other live viruses, induce strong CTL responses to the antigens they produce. This technique has been used most commonly with vaccinia virus vectors. The gene encoding the desired antigen is inserted by a process of homologous recombination into the vaccinia virus genome at the site of the nonessential viral thymidine kinase gene. Thymidine kinase–negative recombinant viruses are selected in culture medium containing bromodeoxyuridine (which kills all cells that produce thymidine kinase). At least 25,000 base pairs (bp) of foreign DNA can be inserted into the viral genome, and the upper limit of the size of the foreign gene may be much greater. With this method, recombinant vaccinia viruses producing hepatitis B surface antigen, herpes simplex virus proteins, influenza virus hemagglutinin and neuraminidase, malaria circumsporozoite protein, and many other microbial antigens have been generated. Inoculation of recombinant viruses into many species of animals induces both humoral and cell-mediated immunity against the antigen produced by the foreign gene (and, of course, against vaccinia virus antigens as well). Attempts are under way to improve the construction and efficiency of expression of viral vectors, to reduce the pathogenicity of the vaccinia virus, to enhance the immunogenicity of the expressed antigen, and to incorporate adjuvants or use delivery systems to maximize vaccine potency. The major problem with viral vectors is the development of CTL responses against virally infected cells, with resultant host cell injury. Live recombinant viruses have not been used in human trials to date because of safety concerns, but their potential is undisputed.

DNA Vaccines. The newest method for vaccination has been developed on the basis of an unexpected observation. Inoculation of a plasmid containing a cDNA encoding a protein antigen leads to strong and long-lived humoral and cell-mediated immune responses to the antigen. Presumably, some professional APCs are transfected by the plasmid and express immunogenic peptides that elicit specific responses. The unique feature of DNA vaccines is that they provide the only approach, without live microbes, for eliciting strong CTL responses. It is still not clear how long the plasmid DNA remains functional in the host, which cells need to be transfected *in vivo* to generate immune responses, or even where the responses develop (i.e., at the site of inoculation or in draining lymph nodes). Nevertheless, the ease of manipulating cDNAs to express many diverse antigens, and the ability to co-express other proteins that may enhance immune responses (such as cytokines and costimulators) make this a very promising technique.

Adjuvants and Immunomodulators. The initiation of T cell–dependent immune responses against protein antigens requires that the antigens be administered with adjuvants. Most adjuvants induce local inflammation, with increased expression of costimulators and production of cytokines, such as IL-12, that stimulate T cell growth and differentiation. Heat-killed bacteria are powerful adjuvants that are commonly used in experimental animals. However, the severe local inflammation such adjuvants trigger precludes their use in humans. Much effort is currently being devoted to developing safe and effective adjuvants for use in humans. Several are in clinical practice, including aluminum hydroxide gel (which binds proteins non-covalently and induces mild inflammation), and lipid formulations that are ingested by phagocytes. An alternative to adjuvants is to administer substances that stimulate T cell responses (and which are elicited by the adjuvants) together with antigens. For instance, IL-12 incorporated in vaccines promotes strong cell-mediated immunity and is in early clinical trials. As mentioned above, it may be possible to incorporate costimulators (B7 molecules, CD40) or cytokines into plasmid vaccines, using recombinant DNA technology.

Passive Immunization. Finally, protective immunity can also be conferred by passive immunization, e.g., by transfer of specific antibodies. In the clinical situation, this is most commonly used for diseases caused by toxins, such as tetanus. Antibodies against snake venoms can be life-saving treatments for poisonous snake bites. Passive immunity is short-lived, because the host does not respond to the immunization and protection lasts only as long as the injected antibody persists. Moreover, passive immunization does not induce specific memory, so that the immunized individual is not protected against subsequent exposures to the toxin or microbe.

Advantages and Disadvantages of Different Vaccine Approaches

The principal advantages of subunit and synthetic antigen vaccines are their safety and ease of use. However, these vaccines are often difficult and cumbersome to produce and have short shelf lives, limiting their utility, especially in developing countries. Furthermore, purified antigen vaccines stimulate antibody responses but do not generate CTL immunity, because these exogenously administered antigens do not efficiently enter the class I MHC antigen presentation pathway and are not recognized by CD8+ T cells. This greatly limits the usefulness of such vaccines for infections by many intracellular microbes.

Attenuated microbes and viral vectors are two approaches for stimulating both humoral and cell-mediated immunity, including CTLs. In fact, attenuated microbes or viral vectors should induce essentially the same immune responses as the native microbes. This is a major advantage for infections that require CTLs for effective defense, such as HIV, other viruses, and perhaps malaria. The great concern with these approaches is safety. In fact, viral vectors expressing a protein of interest can elicit potentially pathogenic responses to the vector itself. Plasmid DNA vaccines also stimulate both humoral and cell-mediated immunity and do not carry the risk of pathogenic responses to the vehicle. However, experience with DNA vaccines is limited, and numerous clinical trials have been initiated to test both safety and efficacy.

SUMMARY

The interaction of the immune system with infectious organisms is a dynamic interplay of host mechanisms aimed at eliminating infections and microbial strategies designed to permit survival in the face of powerful effector mechanisms. Different types of infectious agents stimulate distinct patterns of immune responses and have evolved unique mechanisms for evading specific immunity.

The principal protective immune response against extracellular bacteria consists of specific antibodies, which opsonize the bacteria for phagocytosis and activate the complement system. Toxins produced by such bacteria are also neutralized and eliminated by specific antibodies. Some bacterial toxins are powerful inducers of cytokine production, and cytokines account for much of the systemic pathology associated with severe, disseminated infections with these microbes.

Intracellular bacteria are capable of surviving and replicating within host cells, including phagocytes, because they have developed mechanisms for resisting lysosomal degradation. Immunity against these microbes is principally cell-mediated, and consists of CD4+ T cells activating macrophages (as in delayed type hypersensitivity) as well as CD8+ cytolytic T lymphocytes lysing infected cells. The characteristic pathologic response to infection by intracellular bacteria is granulomatous inflammation.

Protective responses to fungi consist of cell-mediated immunity as well as antibodies. Fungi are usually readily eliminated by a competent immune system, because of which these infections are often seen in immunodeficient persons.

Viruses are obligatory intracellular microbes. Natural immunity against viruses is mediated by type I interferons and NK cells. Specific antibodies protect against viruses early in the course of infection, and if the viruses are released from killed infected cells. The major defense mechanism against established infections, especially with noncytopathic viruses, consists of specific CTLs. CTLs effectively lyse infected cells and may contribute to tissue injury even when the infectious virus is not harmful by itself.

Animal parasites, such as protozoa and helminths, give rise to chronic and persistent infections, because natural immunity against them is weak and because parasites have evolved multiple mechanisms for evading and resisting specific immunity. The structural and antigenic diversity of pathogenic parasites is reflected in the heterogeneity of the specific immune responses they elicit. Protozoa that live within host cells are combated by cell-mediated immunity, whereas helminths are eliminated by IgE antibody and eosinophil-mediated killing, as well as by other leukocytes. Parasites evade the immune system by masking and shedding their surface antigens and by varying their antigens during residence in vertebrate hosts.

Vaccination is a powerful method of preventing infectious. Approaches to vaccination include the use of attenuated microbes, purified protein and polysaccharide antigens, viral vectors expressing a known antigen, and plasmid DNA encoding an antigen.

Selected Readings

Bloom, B. R., R. L. Modlin, and P. Salgame. Stigma variations: observations on suppressor T cells and leprosy. Annual Review of Immunology 10:453–488, 1992.

Chisari, F. V., and C. Ferrari. Hepatitis B virus immunopathogenesis. Annual Review of Immunology 13:29–59, 1995.

Dannenberg, A. H. Delayed-type hypersensitivity and cell-mediated immunity in the pathogenesis of tuberculosis. Immunology Today 12:228–233, 1991.

Doherty, P. C., W. Allan, P. Eichelberger, and S. R. Carding. Roles of αβ and γδ T cell subsets in viral immunity. Annual Review of Immunology 10:123–151, 1992.

Donnelly, J. J., J. B. Ulmer, J. W. Shiver, and M. A. Liu. DNA vaccines. Annual Review of Immunology 15:617–647, 1997.

Finkelman, F. D., T. Shea-Donohue, J. Goldhill, C. A. Sullivan, S. C. Morris, K. B. Madden, W. C. Gause, and J. F. Urban. Cytokine regulation of host defense against parasitic gastrointestinal nematodes. Annual Review of Immunology 15:505–534, 1997.

Herman, A., J. W. Kappler, P. Marrack, and A. M. Pullen. Superantigens: mechanisms of T-cell stimulation and role in immune responses. Annual Review of Immunology 9:745–772, 1991.

Kaufman, S. H. E. Immunity to intracellular bacteria. Annual Review of Immunology 11:129–163, 1993.

Marrack, P., and J. W. Kappler. Subversion of the immune system by pathogens. Cell 76:323–332, 1994.

Reiner, S. L., and R. M. Locksley. The regulation of immunity to *Leishmania major*. Annual Review of Immunology 13:151–177, 1995.

Romani, L., and D. Howard. Mechanisms of resistance to fungal infections. Current Opinion in Immunology 7:517–523, 1995.

Sher, A., and R. L. Coffman. Regulation of immunity to parasites by T cells and T cell–derived cytokines. Annual Review of Immunology 10:385–409, 1992.

Spriggs, M. K. One step ahead of the game: viral immunomodulatory molecules. Annual Review of Immunology 14:101–130, 1996.

Zinkernagel, R. M. Immunology taught by viruses. Science 271:173–178, 1996.

CHAPTER SEVENTEEN

IMMUNE RESPONSES TO

TISSUE TRANSPLANTS

Transplantation is the process of taking cells, tissues, or organs, called a **graft,** from one individual and placing them into a (usually) different individual. The individual who provides the graft is referred to as the **donor,** and the individual who receives the graft is referred to as either the **recipient** or the **host.** If the graft is placed into its normal anatomic location, the procedure is called **orthotopic transplantation;** if the graft is placed in a different site, the procedure is called **heterotopic transplantation. Transfusion** is transplantation of circulating blood cells and/or plasma from one individual to another.

Although attempts at transplantation date back to ancient times, the impetus behind modern transplantation was World War II and the Battle of Britain. Royal Air Force pilots were often severely burned when their planes crashed. The mortality associated with burns corresponds to the size of the area of skin that has been injured, and survival can be improved if burned skin is replaced. For this reason, British doctors turned to skin transplantation from other human donors as a mode of therapy. However, attempts to replace damaged skin with skin from unrelated donors were uniformly unsuccessful. Over a matter of several days, the transplanted skin would undergo necrosis and fall off. This problem led many investigators, including Peter Medawar, to study skin transplantation in animal models. These experiments established that the failure of skin grafting was caused by an inflammatory reaction that was called **rejection.** More importantly, several features indicated that *rejection is a form of specific immunity.* The key experimental results may be summarized as follows (Table 17–1):

1. A skin graft transplanted between genetically unrelated individuals, e.g., from a strain A mouse to a strain B mouse, is rejected by a naive host in 7 to 10 days. This process is called first set rejection and is due to a primary immune response to the graft. A subsequent skin graft transplanted from the same donor to the same recipient is rejected more rapidly, i.e., in only 2 or 3 days. This accelerated response, called second set rejection, is due to a secondary immune response. Thus, genetically disparate grafts induce immunologic memory, one of the cardinal features of specific immunity.

2. Second set rejection ensues if the first and second skin grafts are derived from the same donor or from genetically identical donors, e.g., strain A mice. However, if the second graft is derived from an individual unrelated to the donor of the first graft, e.g., strain C, there is no second set rejection; the new graft elicits only a first set rejection. Thus, the phenomenon of second set rejection shows specificity, another cardinal feature of specific immunity.

3. The ability to mount second set rejection against a graft from strain A mice can be adoptively transferred to a naive strain B recipient by immunocompetent lymphocytes taken from a strain B animal previously exposed to a graft from strain A mice. This experiment demonstrated that second set rejection is mediated by sensitized lymphocytes and provided the definitive evidence that rejection is a form of specific immunity.

Transplant immunologists have developed a vocabulary to describe the kinds of cells and tissues encountered in the transplant setting. A graft transplanted from one individual to the same individual is called an **autologous graft** (shortened to **autograft**). A graft transplanted between two genetically identical or syngeneic individuals is called a **syngeneic graft** (or **syngraft**). A graft transplanted between two genetically different individuals of the same species is called an **allogeneic graft** (or **allograft**). A graft transplanted between individuals of different species is called a **xenogeneic graft** (or **xenograft**). The molecules that are recognized as foreign on allografts are called **alloantigens,** and those on xenografts are called **xenoantigens.** The lymphocytes or antibodies that react with alloantigens or xenoantigens are described as being **alloreactive** or **xenoreactive,** respectively.

Most of this chapter focuses on allogeneic transplantation because it is far more commonly practiced and better understood than xenogeneic transplantation, which is discussed briefly near the end of the chapter. We will consider both the basic immunology and some aspects of the clinical practice of transplantation. Transplantation of organs such as kidney, heart, lung, and liver is cur-

TABLE 17–1. First Set and Second Set Allograft Rejection

Animal	Skin Graft Donor	Recipient Strain	Recipient Prior Treatment	Rejection
1	Strain A	Strain B	None	Slow (first set)
2	Strain A	Strain B	Sensitized by previous graft from strain A donor	Rapid (second set); demonstration of immunologic memory
3	Strain A	Strain B	Injected with lymphocytes from animal No. 1	Rapid (second set); demonstration of role of lymphocytes in graft rejection
4	Strain C	Strain B	Sensitized by previous graft from strain A donor	Slow (first set); demonstration of immunologic specificity

rently in widespread use, and the practice is growing. In addition, the transplantation of many other organs or cells is now being attempted. We will conclude the chapter with a discussion of allogeneic bone marrow transplantation, which raises special issues not encountered in solid organ transplants. The immunology of transplantation is important for two reasons. First, the immunologic rejection response is one of the major barriers to transplantation today. Second, although the encounter with alloantigens is unlikely in the normal life of an organism, the immune response to allogeneic molecules is very strong and has therefore been a useful model for elucidating the mechanisms of lymphocyte activation.

ALLOGENEIC TRANSPLANTATION

The immune response to alloantigens can be both cell-mediated and humoral. In general, cell-mediated immune reactions are more important for rejection of transplanted organs, but antibodies may contribute. Most studies of the immune responses to tissue transplants have focused on T cell responses to allogeneic molecules. Three major questions are addressed by these studies:

1. What antigens in grafts stimulate alloreactivity?
2. What types of lymphocytes respond to transplants?
3. Why do individuals react so strongly against tissues that they do not encounter normally?

Molecular Basis of Allorecognition

In Chapter 5, we presented evidence that recognition of transplanted cells as self or foreign is determined by inheritance of co-dominant genes. This conclusion was based on the results of experimental transplantation between inbred strains of mice.

1. Cells or organs transplanted between individuals of the same inbred strain of mice are never rejected.
2. Cells or organs transplanted between individuals of different inbred strains of mice are almost always rejected.
3. The offspring of a mating between two different inbred strains will never reject grafts from either parent. In other words, an (A × B)F1 animal will not reject grafts from an A or B strain animal.
4. A graft derived from the offspring of a mating between two different inbred strains will almost always be rejected by either parent. In other words, a graft from an (A × B)F1 animal will be rejected by either an A or a B strain animal.

These genetic experiments led to the hypothesis that certain polymorphic gene products, co-dominantly expressed on a graft, determine whether the immune system recognizes the graft

as foreign or not. Co-dominant expression means that an (A × B)F1 animal expresses both A strain and B strain alleles. This is why an (A × B)F1 animal is tolerant to both A and B strain grafts and why both A and B strain animals will recognize an (A × B)F1 graft as foreign. As described in Chapter 5, George Snell and colleagues were able to identify the polymorphic genes that served as the molecular targets of rejection in mice. Specifically, they found that polymorphic molecules encoded by genes in histocompatibility locus 2 (H-2), now known as the mouse major histocompatibility complex (MHC), were responsible for almost all strong (rapid) rejection reactions. *As many as 2 per cent of a host's T cells are capable of recognizing and responding to a single foreign MHC molecule.* This high frequency of T cells reactive with allogeneic MHC molecules is the reason why allograft rejection is a strong response *in vivo.*

At the time of these initial discoveries, immunologists were faced with two puzzles: (1) why do various cells express molecules that evoke such powerful responses from allogeneic T cells, and (2) why are so many T cells capable of recognizing alloantigens? As discussed in Chapters 5 and 6, the answer to the first question is that *MHC molecules are widely expressed because they play a critical role in the normal immune system,* namely the presentation of peptides derived from foreign protein antigens in a form that can be recognized by T cells. The role of MHC molecules as alloantigens is incidental. The answer to the second question is that *many different normal T cells, each specific for a different foreign peptide bound to a self MHC molecule, cross-react with each allogeneic MHC molecule.* The evidence that recognition of allogeneic MHC molecules is due to cross-reaction is provided by the following observations:

1. A T cell clone or hybridoma that contains one set of functionally rearranged genes encoding a TCR specific for self MHC plus a foreign peptide may also recognize one or more allogeneic MHC molecules in the absence of the specific foreign peptide.
2. Monoclonal antibodies reactive with idiotypic determinants on the TCR molecule of such a T cell clone or hybridoma may inhibit recognition of both self MHC-associated foreign peptide and allogeneic MHC molecules.
3. Transfection of rearranged α and β T cell receptor genes into a recipient T cell confers specificity both for self MHC plus foreign peptide and for allogeneic MHC molecules.

These observations posed a new question, namely, what are the structural features of allogeneic MHC molecules that make them look like self MHC plus foreign peptide? In other words, what structure is actually recognized by the cross-reactive TCR? Because in these *in vitro* systems exogenously added foreign peptides were not necessary for allorecognition, the results initially seemed to support the conclusion that bound peptide does not contribute to the determinant formed by a foreign MHC molecule. However, it is now apparent that MHC molecules expressed on cell surfaces normally contain bound peptides. Even in artificial systems, such as lipid bilayers containing purified MHC molecules, peptides remain associated with the MHC molecules through purification. Moreover, some alloreactive T cell clones have now been shown to be specific for allogeneic MHC plus a particular bound peptide. The peptides recognized in association with allogeneic MHC molecules may be self peptides because thymic education does not produce tolerance to self proteins plus allogeneic MHC.

The three-dimensional determinant recognized by a TCR includes some amino acid residues present in the α-helices of the peptide-binding cleft of the MHC molecule as well as amino acid residues in the bound peptide (see Chapter 5). As a consequence of thymic selection, the TCR repertoire is poised to detect small differences between foreign peptides bound to self MHC molecules and self peptides bound to self MHC molecules. In the normal (autologous) situation, these differences arise entirely from amino acid residues of the foreign peptides. In the allogeneic situation, differences may be contributed by either polymorphic residues of the allogeneic MHC molecule, or by a combination of the allogeneic MHC molecule plus

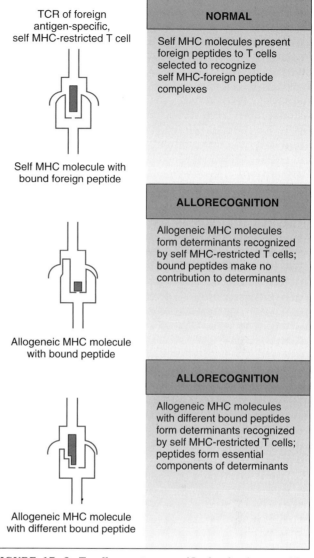

FIGURE 17–2. T cell receptors specific for foreign peptides bound to self MHC molecules cross-react with allogeneic MHC molecules or the complex of peptides associated with allogeneic MHC molecules.

different bound peptides (Figs. 17–1 and 17–2).

The conclusion that an allogeneic MHC molecule plus a peptide could mimic the determinant recognized by a self MHC molecule plus a particular foreign peptide is not surprising. The more complicated question is why each allogeneic MHC molecule is recognized by so many different TCRs, each selected for different foreign peptides. The large numbers of TCRs that cross-react with each allogeneic MHC allelic molecule have been attributed to four factors.

1. *Allogeneic MHC molecules differ from self MHC molecules at multiple amino acid residues, each of which individually or in combination may produce a determinant recognized by a different cross-reactive T cell clone.* Thus, each allogeneic MHC molecule can be recognized by multiple clones of T cells whose receptors are specific for

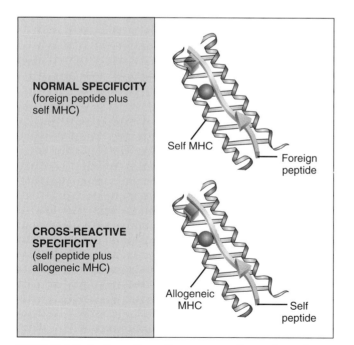

FIGURE 17–1. Formation of a cross-reactive epitope by an allogeneic MHC molecule–peptide complex. The normal specificity of a hypothetical T cell is an epitope formed from three amino acid residue side chains, depicted as ◆ and ▲ contributed by a foreign peptide, and ●, contributed by a self MHC molecule (upper figure). This T cell will cross-react with an epitope formed from the same amino acid residue side chains in which only ▲ is contributed by a self-peptide while ◆ and ● are contributed by an allogeneic MHC molecule (lower figure).

different foreign peptides in association with self MHC molecules. In this case, the cross-reactive T cell allorecognition is entirely attributable to structural polymorphic differences between self and allogeneic MHC molecules, and bound peptide serves only to ensure stable expression of the allogeneic MHC molecule. The number of TCRs that cross-react with allogeneic MHC polymorphic determinants will obviously vary with the extent of polymorphic differences that contribute to TCR contact between the self and allogeneic MHC molecules.

2. *Multiple bound peptides, in combination with one allogeneic MHC gene product, may produce determinants recognized by different cross-reactive T cells.* Any single allogeneic MHC molecule can bind only one peptide at a time, but on each allogeneic cell surface, there are many copies of each allogeneic MHC molecule, and each copy can form a complex with a different peptide. Each different complex may be recognized by a different TCR. In this case, bound peptides form essential components of the determinant recognized by the alloreactive T cell. Some of the peptides recognized in association with allogeneic MHC molecules may be foreign (and can even be derived from the allogeneic MHC molecule itself). However, *even self peptides can contribute to T cell recognition when bound to allogeneic MHC molecules because the TCRs that recognize determinants formed by allogeneic MHC plus self peptides were not eliminated during negative selection in the thymus.* Because many different self peptides form determinants with allogeneic MHC molecules that are recognized by different T cell clones, each allogeneic cell may be recognized by many different T cell clones, each having been selected to recognize a different foreign peptide. This is the principal reason for the high frequency of alloreactive T cells.

3. *Allodeterminants may be expressed on allogeneic antigen-presenting cell (APC) surfaces at higher densities than are determinants formed by specific foreign peptides bound to self MHC molecules on self APC surfaces.* This is clearly true for allogeneic determinants that are formed wholly by the allogeneic MHC molecules because all of the MHC molecules on an allogeneic APC will be allogeneic. In contrast, on self APCs, a specific foreign peptide probably never occupies more than a few per cent of the total MHC molecules expressed. The high density of these allogeneic determinants on allogeneic APCs may allow activation of T cells with low affinities for the determinant, increasing the numbers of T cells that can respond.

4. *Allogeneic MHC molecules may be recognized as conventional foreign proteins.* In the first three mechanisms described above, the allogeneic MHC molecules are presented to T cells without a requirement for antigen processing (see Chapter 6), a phenomenon called **direct presentation of alloantigen.** Allogeneic MHC molecules differ structurally from those of the host and therefore represent a source of foreign proteins. These molecules can also be presented as conventional foreign proteins, i.e., as peptides that are generated by processing within the host APCs and become associated with self MHC molecules. This is called **indirect presentation of alloantigen.** Indirect presentation usually involves allorecognition only by CD4$^+$ T cells because alloantigen is acquired by host APCs primarily through the endosomal vesicular pathway (i.e., as a consequence of phagocytosis), resulting in presentation by class II MHC molecules. In contrast, direct presentation can involve allorecognition by CD8$^+$ as well as CD4$^+$ T cells. Since MHC molecules are the most polymorphic proteins in the genome, each allogeneic MHC molecule may give rise to multiple foreign peptides, each recognized by different T cells.

Polymorphic alloantigens other than MHC molecules generally produce weak or slower (more gradual) rejection reactions and are called **minor histocompatibility antigens.** Most minor histocompatibility antigens are proteins that are processed and indirectly presented to host T cells in association with self MHC molecules on host APCs.

Cellular Basis of Allorecognition

Vigorous rejection reactions of allografts generally result from recognition of the transplanted tissues by both CD4$^+$ and CD8$^+$ T cells. The **mixed leukocyte reaction** (MLR) has been a useful model for understanding the cellular basis of alloantigen recognition by different T cell subpopulations.

THE MIXED LEUKOCYTE REACTION

As we have discussed above, MHC genes were initially identified for their role in graft rejection, which is often a T cell–mediated process. The MLR is an *in vitro* model of T cell recognition of allogeneic MHC gene products and is used as a predictive test of cell-mediated graft rejection.

The MLR is induced by culturing mononuclear leukocytes (which include T cells, B cells, natural killer [NK] cells, mononuclear phagocytes, and dendritic cells) from one individual or inbred strain with mononuclear leukocytes derived from another individual or strain. In humans, these cells are typically isolated from peripheral blood; in the mouse or rat, mononuclear leukocytes are usually purified from spleen or lymph nodes. If there are differences in the alleles of the MHC genes between the two individuals, a large proportion of the mononuclear cells will proliferate over a period of 4 to 7 days. This proliferative response, usually measured by incorporation of ^{3}H-thymidine into DNA during cell replication, is called the **allogeneic MLR** (Fig. 17–3). In the experiment described above, the cells from each donor react and proliferate against the other, resulting in a "two-way MLR." To simplify the analysis, one of the two mononuclear leukocyte populations can be ren-

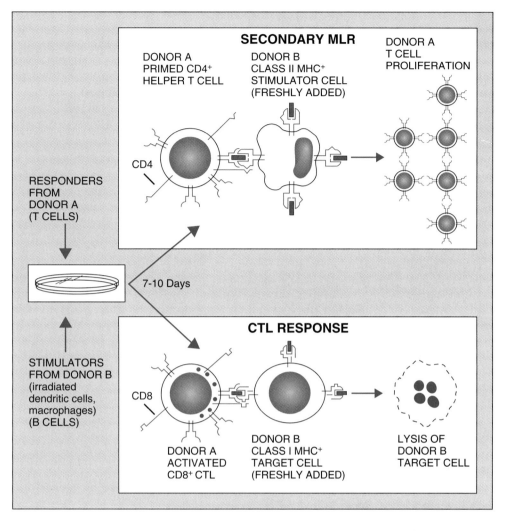

FIGURE 17–3. Responder T cells in the mixed leukocyte reaction (MLR). In a one-way primary MLR, donor B stimulator cells activate and cause the expansion of two types of donor A responder T cells: CD4⁺ helper T cells, which can be detected in a secondary MLR by rapid proliferation to antigen-presenting cells (APCs) bearing donor B class II molecules; and CD8⁺ cytolytic T lymphocytes (CTLs), which can be detected in a specific killing assay using target cells bearing donor B class I molecules.

dered incapable of proliferation, either by gamma irradiation or by treatment with the antimitotic drug mitomycin C, prior to culture. In this "one-way MLR," the treated cells serve exclusively as **stimulators** and the untreated cells, still capable of proliferation, serve as the **responders.**

Two populations of alloreactive T cells are stimulated during an allogeneic MLR, and each responding T cell subset recognizes a different MHC gene product. One type of T cell expresses the CD8 but not the CD4 molecule, usually functions as a cytolytic T lymphocyte (CTL), and is indistinguishable from self class I MHC-restricted CTLs specific for foreign protein antigens. The CTLs generated during an allogeneic MLR lyse target cells derived from the same individual or strain as the original stimulator cell population. The molecules on stimulator and target cells that are recognized by CD8⁺ CTLs are the class I MHC molecules, namely, HLA-A, -B, or -C in humans or H-2K, -D, or -L in mice. Several lines of evidence have indicated that foreign class I MHC gene products are the actual molecular targets recognized by the CD8⁺ CTLs generated in the MLR:

1. If there are no differences in class I MHC alleles between the stimulator and responder cell populations in the MLR, CD8⁺ CTLs are not generated.

2. The CTLs generated against one stimulator cell population will lyse third-party target cells only if these targets share a class I MHC allele with the original stimulators.

3. Antibodies directed against class I MHC allelic gene products on the stimulator cells protect target cells against lysis.

4. Transfection and expression of an allelic class I MHC gene can render a cell susceptible to lysis by a CTL population specific for that allele.

The full differentiation of CTLs in the MLR requires stimulation by allogeneic class I MHC molecules (signal one) as well as costimulators provided by professional APCs or by cytokines optimally provided by CD4⁺ cells present in the same culture (signal two) (see Chapter 13).

Within the CD8⁺ CTL population derived from an MLR, each individual CTL is specific for only one particular class I MHC gene product (usually involving a specific bound peptide). However, the bulk population contains CTLs directed against all class I MHC allelic differences between the original stimulator and responder populations. Furthermore, in an outbred individual, all the class I MHC alleles inherited from both parents are co-dominantly expressed on every class I–expressing cell, so that an individual target cell can be lysed by several different CTLs, each with a different class I MHC specificity.

The second type of T cell that is generated during the MLR is a cytokine-producing CD4⁺ helper T cell, indistinguishable from CD4⁺ helper T cells specific for foreign protein antigens. Such cells were initially called primed responder cells, because when such T cells are recultured with stimulator cells from the same donor individual (or strain) used in the original MLR, a secondary MLR ensues that is stronger and more rapid; i.e., peak proliferation occurs by 2 or 3 days. Alloreactive CD4⁺ helper T cells are specific for allogeneic class II MHC molecules, i.e., HLA-DR, -DP, and -DQ in humans and I-A and I-E in mice. The class II MHC molecules have been established as the molecules seen by the CD4⁺ helper cells by the same kind of evidence that established the class I MHC molecules as the targets of CD8⁺ CTLs.

1. Alloreactive CD4⁺ T cells are stimulated only if there are differences in class II MHC alleles between the original stimulator and responder cells in the primary MLR.
2. Alloreactive CD4⁺ T cells respond to third-party stimulator cells only if they share class II MHC alleles with the original stimulator population.
3. Antibodies directed against class II MHC gene products prevent development of the secondary MLR.
4. Transfection of appropriate allelic class II MHC genes can convert a cell from a non-stimulator to a stimulator of a CD4⁺ T cell population specific for that MHC allele.

Alloreactive CD4⁺ T cells, like self MHC-restricted antigen-specific helper cells, can be stimulated only by cells that express class II MHC molecules associated with specific peptides and provide costimulatory signals, i.e., professional APCs. The most efficient stimulators are dendritic cells, although B lymphocytes, mononuclear phagocytes, and, in humans, vascular endothelial cells are also able to stimulate proliferative responses of alloreactive CD4⁺ T cells. Each individual CD4⁺ helper T cell is specific for one particular class II gene product (usually with a specific bound peptide). However, the bulk population contains helper cells reactive with all class II allelic differences between the original stimulator and responder cell population (and with many different peptides bound to each allelic MHC product). Furthermore, since there is no allelic exclusion of class II genes on individual cells and all the paren-

tal alleles are co-dominantly expressed, one stimulator cell can activate several different helper T cells, each with a different class II MHC specificity.

The functional subdivision of CD4⁺ and CD8⁺ alloreactive T cells into helper cells and CTLs is not absolute. CD4⁺ class II specific CTLs can be detected in the MLR, particularly in humans. Moreover, at least some CD8⁺ T cells produce interleukin-2 (IL-2), interferon-γ (IFN-γ), tumor necrosis factor (TNF), and lymphotoxin (LT), which is similar to the cytokine profile of T_H1 CD4⁺ cells. As discussed in Chapters 6 and 7, the same exceptions have been found for self MHC-restricted T cells specific for foreign protein antigens.

A more specific analysis of the role of class I and class II molecules in allogeneic immune responses has been performed by considering the one-way MLR when only isolated class I or class II differences exist between the stimulator and responder cell populations. In the extreme case, this has been done by using cells from mouse strains that differ only by a small mutation in a single class I or class II gene product. Proliferation is strongest when stimulators and responders differ by a class II gene product and can directly stimulate CD4⁺ T cells. Fewer CTLs arise in this instance, and many of these are CD4⁺ class II–specific CTLs. When only class I differences exist between stimulator and responder cells, the proliferative response is small. Nevertheless, there is a proliferative response, and CD8⁺ CTLs specific for the class I difference do arise. In this situation, proliferation and the development of CTLs may be mediated by cytokines produced by the CD8⁺ T cells themselves responding to the foreign class I molecules. Alternatively, if class II MHC⁺ APCs are present in the responder or stimulator population, they may take up, process, and present foreign class I molecules in the form of peptides associated with self class II molecules. Such indirect presentation to CD4⁺ T cells rather than direct presentation of allogeneic class II molecules activates many fewer alloreactive CD4⁺ T cells. Consequently, less cytokine is produced. Stimulator cell populations that differ from the responders at both the class I and class II MHC loci induce many more allospecific CTLs than stimulators that differ at only class I loci. The contributions of both class I and class II MHC molecules to the allogeneic response are the reason why graft survival improves when both class I and class II alleles are matched between donor and recipient.

STIMULATION OF T CELLS BY ALLOGRAFTS *IN VIVO*

In the case of allografts that differ from hosts at both class I and class II loci, both CD8⁺ and CD4⁺ T cells are activated by recognition of alloantigens of the grafts. CD8⁺ cells recognize allogeneic class I MHC molecules, which are expressed by all the cells in the graft. However, the differentiation of these CTLs from CD8⁺ pre-CTLs is largely dependent upon professional APCs within the graft. Such APCs provide costimulators that directly contribute to CTL differentiation and increase cytokine production by alloreactive CD4$^+$ and CD8$^+$ T cells. The most important APCs stimulating an antigraft response may be dendritic cells resident in the interstitium of the graft.

The importance of professional APCs in stimulating an alloantigenic immune response *in vivo* has most clearly been demonstrated by experiments in rodents. If class II–bearing cells (which include the professional APCs) are removed from a graft prior to transplantation, such grafts are usually rejected slowly or may even be accepted despite class I MHC differences. (Experimentally, APCs may be eliminated from such grafts by several treatments, including prolonged culture; treatment with anti-class II antibody plus complement; or, in some cases, extensive perfusion of graft blood vessels to "wash out" the APCs.) When rat kidney allografts are purged of APCs by perfusion, infusion of dendritic cells derived from the organ donor concomitant with transplantation restores allorecognition and leads to rapid rejection. The professional APCs resident in the graft that are responsible for initiating graft rejection are sometimes called **passenger leukocytes.**

Although the role of passenger leukocytes is well documented in rodents, attempts to remove such cells have not been useful in human transplantation. The probable explanation is that human, but not rodent, graft endothelial cells constitutively express class I and class II MHC molecules, provide costimulator functions, activate alloreactive T cells, and are sufficient to initiate rejection, even in the absence of passenger leukocytes.

In contrast to T cell alloreactivity, much less is known about the mechanisms that lead to the production of alloantibodies against foreign MHC molecules. Presumably, B cells specific for alloantigens are stimulated by mechanisms similar to those involved in stimulation of B cells reactive with other foreign proteins.

Before we conclude this section of the chapter, we should point out that many of the issues that arise in discussions of alloreactivity and graft rejection are also relevant to maternal-fetal interactions. The fetus expresses paternal MHC molecules and is therefore semiallogeneic to the mother. Nevertheless, the fetus is not rejected by the maternal immune system. Many possible mechanisms have been proposed to account for this, and it is not yet clear which of these mechanisms are the most significant (Box 17–1).

Effector Mechanisms in Allograft Rejection

So far, we have described the molecular basis of allogeneic recognition and the cells involved in the recognition of, and responses to, allografts. We now turn to a consideration of the effector mechanisms used by the immune system to reject allo-

---

BOX 17–1.   Immunity to an Allogeneic Fetus

The mammalian fetus, except in instances in which the mother and father are syngeneic, will express paternally inherited antigens that are allogeneic to the mother. Nevertheless, fetuses are not normally rejected by the mother. An understanding of how the fetus escapes the maternal immune system may be relevant for transplantation.

Three experimental observations indicate that the anatomic location of the fetus is a critical factor in the absence of rejection:

(1) Wholly allogeneic fetal blastocysts that lack maternal genes can successfully develop in a pregnant or pseudopregnant mother. Thus, neither specific maternal nor paternal genes are necessary for survival of the fetus.

(2) Hyperimmunization of the mother with cells bearing paternal antigens does not compromise placental and fetal growth.

(3) Pregnant mothers are able to recognize and reject allografts syngeneic to the fetus placed at extrauterine sites without compromising fetal survival.

The failure to reject the fetus has focused attention on the region of physical contact between the mother and fetus. The fetal tissues of the placenta that most intimately contact the mother may be classified as vascular trophoblast, which is exposed to maternal blood for purposes of mediating nutrient exchange, and implantation site trophoblast, which diffusely infiltrates the uterine lining (decidua) for purposes of anchoring the placenta to the mother.

One simple explanation for fetal survival is that trophoblast cells fail to express paternal major histocompatibility complex (MHC) molecules. So far, class II molecules have not been detected on trophoblast. In mice, cells of implantation trophoblast, but not vascular trophoblast, do express paternal class I molecules. In humans, the situation may be more complex, in that trophoblast cells may express only a nonpolymorphic class IB molecule called HLA-G. HLA-G is not recognized by CTLs and may deliver inhibitory signals to NK cells. However, even if these cells do express classical MHC molecules, they may lack costimulator molecules and fail to act as antigen-presenting cells.

A second explanation for lack of rejection is that the uterine decidua may be an immunologically privileged site that is not accessible to functional T cells. In support of this idea is the observation that mouse decidua is highly susceptible to infection by *Listeria monocytogenes* and cannot support a delayed type hypersensitivity response. The basis of immunologic privilege is clearly not a simple anatomic barrier because maternal blood is in extensive contact with trophoblast. Rather, the barrier is likely to be functional inhibition. Cultured decidual cells directly inhibit macrophage and T cell functions, perhaps by producing inhibitory cytokines, such as transforming growth factor-β (see Chapter 12). Some of these inhibitory decidual cells may be resident anti-inflammatory T cells, although the evidence for this proposal is not convincing.

---

grafts. In different experimental models, alloreactive CD4$^+$ or CD8$^+$ T cells or specific alloantibodies are each capable of mediating allograft rejection. Furthermore, graft rejection can be inhibited by anti-CD4 or anti-CD8 antibodies. These different immune effectors cause graft rejection by different mechanisms.

1. Alloreactive CTLs, principally CD8+ T cells, directly lyse graft endothelial and parenchymal cells.

2. Alloreactive helper T cells, principally CD4+ T cells, can recruit and activate macrophages, initiating graft injury by a delayed type hypersensitivity (DTH) reaction (see Chapter 13).

3. Alloantibodies bind to endothelium, activate the complement system, and injure graft blood vessels.

For historical reasons, graft rejection is usually classified on the basis of histopathology rather than on immune effector mechanisms. Based on the experience of renal transplantation, the histopathologic pattern is called hyperacute, acute, or chronic. The names of the various forms of rejection imply a temporal sequence of events, but histology, rather than the length of time following transplantation, is the major criterion for classifying rejection reactions. However, in the current era of renal transplantation, biopsies are performed less frequently than in the past, and diagnosis is often made on the basis of clinical features and post-transplantation time without histologic confirmation.

HYPERACUTE REJECTION

Hyperacute rejection is characterized by rapid thrombotic occlusion of the graft vasculature that begins within minutes to hours after host blood vessels are anastomosed to graft vessels (Fig. 17–4). *Hyperacute rejection is mediated by pre-existing antibodies that bind to endothelium and activate complement.* Antibody and complement induce a number of changes in the graft endothelium that promote intravascular thrombosis. The endothelial cells are stimulated to secrete high molecular weight forms of von Willebrand factor that mediate platelet adhesion and aggregation. Both endothelial cells and platelets undergo membrane vesiculation, leading to shedding of lipid particles that promote coagulation. Endothelial cells lose their cell surface heparan sulfate proteoglycans that normally interact with anti-thrombin III to inhibit coagulation. Complement activation also leads to endothelial cell injury and exposure of subendothelial basement membrane proteins that activate platelets. These processes contribute to thrombosis and vascular occlusion, and the organ suffers irreversible ischemic damage within a matter of hours.

In the early days of transplantation, hyperacute rejection was often mediated by pre-existing IgM alloantibodies, which are present at high titer prior to any exposure to alloantigens. Such "natural antibodies" are believed to arise in response to carbohydrate antigens expressed by the bacteria that normally colonize the bowel. The best known examples of such alloantibodies are those directed

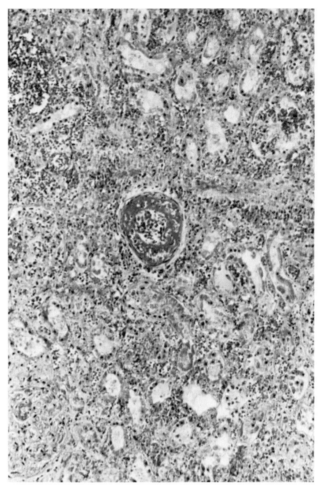

FIGURE 17–4. Hyperacute rejection in the kidney. Preformed antibodies reactive with vascular endothelium of a kidney allograft activate complement and trigger rapid intravascular thrombosis and necrosis of the vessel wall. (Courtesy of Dr. Helmut Rennke, Department of Pathology, Brigham and Women's Hospital, Boston.)

against the ABO blood group antigens expressed on red blood cells (Box 17–2). ABO antigens are also expressed on vascular endothelial cells. Today, hyperacute rejection by anti-ABO antibodies is not a clinical problem because all graft donors and recipients are selected so that they have the same ABO type. However, as we shall discuss shortly, hyperacute rejection caused by natural antibodies is the major barrier to xenotransplantation, limiting the use of animal organs for human transplantation.

In more recent clinical experience, hyperacute rejection of allografts is usually mediated by IgG antibodies directed against protein alloantigens, such as foreign MHC molecules, or against less well-described alloantigens expressed on vascular endothelial cells. Such antibodies generally arise as a result of prior exposure to alloantigens through blood transfusion, prior transplantation, or multiple pregnancies. If the titer of these alloreactive antibodies is low, hyperacute rejection may de-

BOX 17–2. ABO Blood Group Antigens

The first alloantigen system to be defined was a family of red blood cell surface antigens called ABO. Differences in the ABO system between donors and recipients limit blood transfusions by causing antibody and complement-dependent lysis of the foreign red blood cells (transfusion reactions). IgM antibodies to these red blood cell antigens pre-exist in a naive host prior to transfusion, and it is believed that they arise as responses to cross-reactive microbial antigens. The red blood cell antigen responsible for these transfusion reactions is expressed as a cell surface glycosphingolipid. All normal individuals synthesize a common core glycan, called the O antigen, that is attached to a sphingolipid. A single genetic locus encodes three common alleles of a glycosyl transferase enzyme. The O allele gene product is devoid of enzymatic activity, whereas the A allele gene product transfers a terminal N-acetylgalactosamine moiety and the B allele gene product transfers a terminal galactose moiety. Individuals who are homozygous O cannot attach terminal sugars to the O antigen and express only the O antigen. In contrast, individuals who possess an A allele (AA homozygotes, AO heterozygotes, or AB heterozygotes) form the A antigen by adding terminal N-acetylgalactosamine to some of their O antigens. Similarly, individuals who express a B allele (BB homozygotes, BO heterozygotes, or AB heterozygotes) form the B antigen by adding terminal galactose to some of their O antigens. AB heterozygotes form both A and B antigens from some of their O antigens. Because all individuals express the O antigen, all individuals are tolerant to the O antigen. Individuals with A or B glycosyltransferase alleles are also tolerant to A or B antigens, respectively. However, OO and AO individuals form anti-B IgM antibodies, whereas OO and BO individuals form anti-A IgM antibodies. If a patient receives a transfusion of red blood cells from a donor who expresses a form of the antigen not expressed on self red blood cells, a transfusion reaction may result (see below). It follows that AB individuals can tolerate transfusions from all potential donors and are therefore called universal recipients; similarly, OO individuals can tolerate transfusions only from OO donors but can provide blood to all recipients and are therefore called universal donors. The terminology has been simplified so that OO individuals are said to be blood type O; AA and AO individuals are blood type A; BB and BO individuals are blood type B; and AB individuals are blood type AB.

The same glycosphingolipid that carries the ABO determinants can be modified by other glycosyltransferases to generate minor blood group antigens that elicit milder transfusion reactions. In general, differences in minor blood groups lead to red cell lysis only after repeated transfusions produce a secondary antibody response. Almost all individuals possess a fucosyl transferase that adds a fucose moiety to a side branch of the ABO glycosphingolipid. After fucosylation, the O antigen is technically called the H antigen, and the whole antigenic system is often called ABH rather than ABO. As discussed in the text, the H fucosyl transferase is found only in humans and old world primates. All other mammals express an α-galactosyl transferase instead, which attaches to the O blood group an α-galactosyl 1,4-galactose moiety that is the major target of human natural antibodies reactive with discordant xenografts. Addition of fucose moieties at other side branch positions of the H blood group can be catalyzed by different fucosyl transferases and generates epitopes of the Lewis antigen system. Lewis antigens have received much recent attention from immunologists because these carbohydrate groups serve as ligands for E-selectin and P-selectin.

As noted above, transfusions across blood group barriers may result in harmful episodes called **transfusion reactions.** Transfusion across an ABO barrier may trigger an immediate hemolytic reaction, resulting in both intravascular lysis of red blood cells, probably mediated by the complement system, and extensive phagocytosis of antibody- and complement-coated erythrocytes by macrophages of the liver and spleen. Hemoglobin is liberated from the lysed red cells in quantities that may be toxic for kidney cells, producing acute tubular cell necrosis and renal failure. High fevers, shock, and disseminated intravascular coagulation may also develop, suggestive of massive cytokine release (e.g., of TNF or IL-1; see Chapter 12). The disseminated intravascular coagulation consumes clotting factors faster than they can be synthesized, and the patient may paradoxically die of bleeding. More delayed hemolytic reactions may result from incompatibilities of minor blood group antigens. These result in progressive loss of the transfused red cells, leading to anemia and jaundice—a consequence of overloading the liver with hemoglobin-derived pigments.

---

velop more slowly, i.e., over several days. In this case, it is sometimes referred to as "accelerated allograft rejection," because the onset is still earlier than that typical for acute rejection.

ACUTE REJECTION

Acute Vascular Rejection. Acute vascular rejection is characterized by necrosis of individual cells of the graft blood vessels. The histologic pattern is one of vasculitis (Fig. 17–5) rather than the bland thrombotic occlusion seen in hyperacute rejection. *Acute vascular rejection is often mediated by IgG antibodies against endothelial cell alloantigens (either MHC molecules or other antigens) that develop in response to the graft (i.e., post-transplantation) and involves activation of complement.* In addition, T cells may contribute to acute vascular rejection by responding to alloantigens present on

vascular endothelial cells, leading to direct lysis of these cells, or the production of cytokines that recruit and activate inflammatory cells, causing endothelial necrosis. This process usually begins after the first week of transplantation, i.e., as first set rejection.

Acute Cellular Rejection. This type of rejection, which also usually begins after the first week of transplantation, is characterized by necrosis of parenchymal cells caused by infiltrating host T lymphocytes and macrophages (Fig. 17–6). *Several different effector mechanisms may be involved in acute cellular rejection, including CTL-mediated lysis and activated macrophage-mediated lysis (as in delayed type hypersensitivity)* (see Chapter 13). *Several lines of evidence suggest that recognition and lysis of foreign cells by alloreactive CD8+ CTLs is an important mechanism of acute cellular rejection.*

1. The cellular infiltrates present in grafts undergoing acute cellular rejection are markedly enriched for CD8+ CTLs specific for graft alloantigens.

2. Cloned lines of alloreactive CD8+ CTLs can be used to adoptively transfer acute cellular graft rejection.

3. Most vascular and parenchymal cells express class I MHC molecules and are susceptible to lysis by CD8+ CTLs but are usually resistant to killing by activated macrophages.

4. Allograft rejection can occur with sparing of bystander cells within the graft, a hallmark of specific CTL killing. The best evidence of this phenomenon has come from mouse skin graft experiments using grafts that contain two distinct cell populations, one syngeneic to the host and one allogeneic to the host. (Such skins are derived from chimeric "tetraparental" animals generated by mixing the cells of fertilized blastocysts *in vitro* and reimplanting such cells in the uterus of a pregnant mouse.)

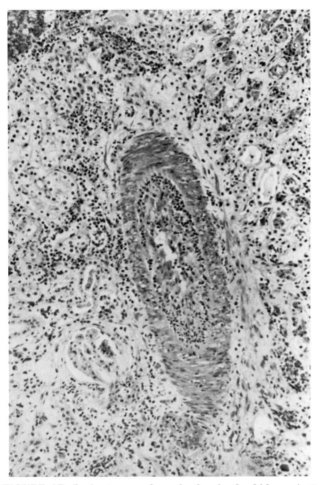

FIGURE 17-5. Acute vascular rejection in the kidney. Antibodies reactive with graft endothelial cells arise in a transplant recipient and cause a destructive inflammatory reaction in the vessel wall. T lymphocytes reactive with graft alloantigens may also participate in vascular injury. (Courtesy of Dr. Helmut Rennke, Department of Pathology, Brigham and Women's Hospital, Boston.)

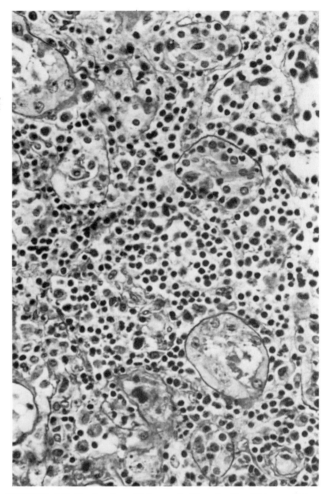

FIGURE 17-6. Acute cellular rejection in the kidney. T lymphocytes reactive with alloantigens in a kidney graft mediate necrosis of tubular epithelial cells as well as of interstitial cells and microvascular endothelial cells. (Courtesy of Dr. Helmut Rennke, Department of Pathology, Brigham and Women's Hospital, Boston.)

When these skins are transplanted, the allogeneic cells are lysed without killing of the "bystander" syngeneic cells.

The identification of CTLs as an important effector mechanism of acute graft rejection suggests that this process is similar to normal antiviral immune responses (see Chapter 16). The basis of this similarity probably arises from the fact that the foreign class I MHC molecules expressed by cells of the graft are recognized as if they were self MHC molecules associated with endogenously synthesized foreign (e.g., viral) peptides.

CHRONIC REJECTION

Chronic rejection is characterized by fibrosis with loss of normal organ structures (Fig. 17-7). As therapy for controlling acute rejection has improved, chronic rejection has emerged as the major cause of allograft loss. The pathogenesis of chronic rejection is less well understood than is

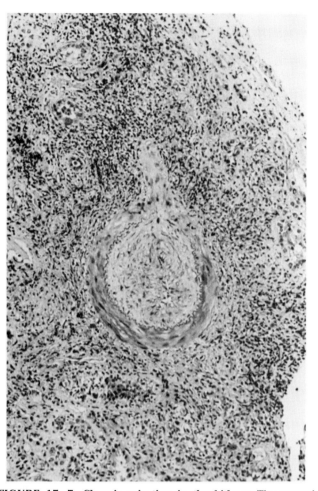

failure and may underlie chronic rejection of other organs as well. This lesion can develop in any vascularized organ transplant within 6 months to a year following transplantation. The smooth muscle cell proliferation in the vascular intima may represent a specialized form of chronic DTH, in which lymphocytes activated by alloantigens in the graft vessel wall induce macrophages to secrete smooth muscle cell growth factors. The risk of developing graft arteriosclerosis is increased in patients with cytomegalovirus infection (see below), suggesting that viral antigens may also contribute to the immune reaction. This model is similar to that proposed for chronic rejection of the organ parenchyma as a form of chronic DTH, except that in the vessel wall, smooth muscle cells rather than fibroblasts proliferate and produce collagen.

Prevention and Treatment of Allograft Rejection

If the recipient of an allograft has a fully functional immune system, transplantation almost invariably results in some form of rejection. Two

FIGURE 17–7. Chronic rejection in the kidney. The normal cells of the renal interstitium and tubules are replaced by fibrous tissue. As described in the text, this reaction may represent healing of acute rejection, chronic delayed type hypersensitivity to graft alloantigens, or chronic ischemia. (Courtesy of Dr. Helmut Rennke, Department of Pathology, Brigham and Women's Hospital, Boston.)

that of acute rejection. The fibrosis of chronic rejection may represent wound healing following the cellular necrosis of acute rejection. However, in many instances, chronic rejection develops without evidence that acute rejection ever occurred. Two other possible explanations of the fibrosis are that chronic rejection represents a form of chronic DTH in which activated macrophages secrete mesenchymal cell growth factors, such as platelet-derived growth factor or, alternatively, that chronic rejection is a response to chronic ischemia caused by injury to blood vessels. Vascular injury could result from repeated bouts of antibody-mediated acute humoral rejection or cell-mediated injury of microvascular endothelial cells.

More commonly, however, vascular occlusion is due to proliferation of intimal smooth muscle cells, called **accelerated** or **graft arteriosclerosis,** that may occur without evidence of any other vascular injury (Fig. 17–8). Graft arteriosclerosis has emerged as the major cause of cardiac transplant

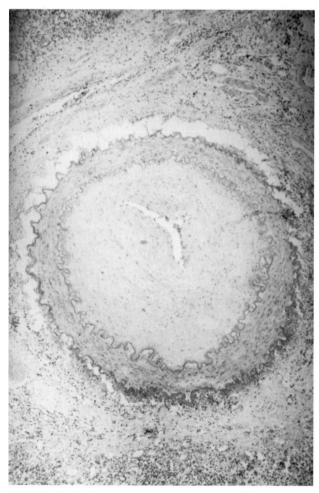

FIGURE 17–8. Graft arteriosclerosis in the kidney. In this variant of chronic rejection, the vascular lumen is replaced by accumulation of smooth muscle cells and connective tissue in the vessel intima.

general strategies have been used in clinical practice and in experimental models to avoid or delay rejection: minimizing the strength of the specific allogeneic reaction, or general immunosuppression.

Several approaches have been used to make the graft less immunogenic to the host:

1. *In human transplantation, the major strategy to reduce graft immunogenicity has been to minimize alloantigenic differences between the donor and recipient by donor selection.* For example, to avoid hyperacute rejection, the ABO blood group antigen of the graft donor is selected to be identical to that of the recipient. In addition, MHC molecule allelic differences have been considered at both class I and class II loci. For kidney transplantation, all potential donors and recipients are "tissue-typed" to determine the identity of the HLA molecules that are expressed (Box 17–3).

Analysis of the results of graft survival as a function of HLA type has led to four conclusions:

a. The larger the number of HLA-A and -B alleles that are matched between donor and recipient (e.g., three or four of four loci), the better is graft survival, especially in the first year following transplantation. (HLA-C is not routinely matched and is believed to be a less important target of T cell recognition.)

b. Matches at HLA-DR alleles are important, independent of the number of HLA-A or -B matches. Because HLA-DR and -DQ are in strong linkage disequilibrium, matching at the DR locus often also matches at the DQ locus. DP typing is not in common use, and its importance is unknown.

c. Matching is more predictive of outcome in Europe, where populations are more inbred, than in the United States, where extensive outbreeding has probably diminished linkage disequilibrium among HLA loci.

d. The recipient HLA-DR types may influence graft survival independent of the degree of matching. This effect of HLA-DR type of the recipient has been interpreted as an "immune response" gene effect, presumably because host HLA-DR molecules were involved in selecting the mature T cell repertoire. Thus, in recipients who express particular DR alleles, the T cell repertoire may not contain cells specific for some alloantigens, so that grafts bearing these antigens would fail to induce immune responses and would be accepted.

HLA matching in renal transplantation is possible because donor kidneys can be stored in organ banks prior to transplantation until a well-matched recipient can be identified and because, with dialysis, patients needing a kidney allograft can be clinically treated until a well-matched organ is available. In the case of heart and liver transplantation, organ preservation is more difficult, and potential recipients are often in critical condition. For these reasons, HLA typing is simply not considered in pairing of potential donors and recipients.

BOX 17-3. Tissue Typing

Tissue typing, also called HLA typing, is the determination of the particular MHC alleles expressed by an individual. The classical approach to tissue or HLA typing is testing whether sera collected from certain donors mediate complement-dependent lysis of an individual's lymphocytes. The sera used for this purpose are obtained from donors who have been inadvertently immunized with foreign cells bearing MHC molecules by transfusion, transplantation, or multiple pregnancies. Such sera characteristically have a low specific antibody titer and react with multiple foreign MHC molecules encoded by several loci. To determine whether an individual expresses HLA-A2, for example, lymphocytes would be tested with a panel of sera, each of which can recognize HLA-A2–bearing cells but may differ in the other specificities they recognize. Only if all of the appropriate sera react and cause lysis is the individual "typed" as HLA-A2 positive. Naturally, well-characterized human sera are in short supply. Therefore, the assays have been honed to a microscale, where 1 μl of serum plus 1 μl of complement can be tested against 50 target cells in 3 μl of solution in the bottom of a tiny well, kept from evaporating by an overlaid oil drop! In general, tissue typing is still performed this way, using standardized sera that have been tested and characterized by many different laboratories. It is hoped that conventional typing sera will be replaced by monoclonal antibodies reactive with specific HLA molecules. Unfortunately, these reagents are not yet available for most specificities.

The HLA types defined by serologic methods are not necessarily single alleles. Some common HLA types contain several different closely related alleles that may be "split" as new reagents become available that can distinguish among them. Typing with antibodies for class II alleles is especially imprecise. Alloreactive T cells often recognize some but not all of the cells that are said to share a D-related (DR) specificity. The information from secondary mixed leukocyte reactions (MLRs) can thus be useful at splitting class II types. It is interesting that some of the T cell responses that are used to split DR types are actually directed against DQ molecules present in linkage disequilibrium with a subset of the DR molecules within a type.

Recently, the polymerase chain reaction (PCR) has been used to permit more complete typing of the class II loci, replacing both serology and secondary MLRs (see Chapter 5, Box 5–3). The polymorphic residues of class II MHC molecules are largely located within exon 2 of both the α and β chains (i.e., within the $\alpha1$ and $\beta1$ polypeptide regions), and this entire region of the gene can be amplified by PCR methods using primers that bind to conserved sequences within the 5' and 3' ends of these exons. The amplified segment of DNA can then be readily sequenced. Thus the actual predicted amino acid sequence can be directly determined for the HLA-DR, -DQ, and -DP alleles of any cell, providing precise molecular tissue typing.

2. In rodents, as discussed above, *grafts may be made less immunogenic by elimination of passenger leukocytes*, an approach that has not worked for vascularized grafts in humans and other primates. Nevertheless, depletion of professional APCs may prove useful for transplanting non-vascularized grafts, such as pancreatic islets.

3. *The graft recipient may be made tolerant to*

the allograft. The original strategy for inducing tolerance was based on the observation that patients who received multiple allogeneic blood transfusions prior to renal transplantation had a better acceptance rate of their grafts than patients who had not received transfusions. Subsequently, patients were intentionally given multiple transfusions to induce "tolerance." The mechanisms of this form of tolerance are not fully known. Several new strategies to induce tolerance (see Chapter 10) are now in pre-clinical or clinical trial. For example, recipients may be treated with high doses of peptides derived from polymorphic regions of donor MHC molecules or with soluble donor MHC molecules to induce specific T cell tolerance. Alternatively, attempts are being made to induce specific T cell clonal anergy by preventing T cells from receiving costimulatory signals during their initial encounter with graft alloantigens. In particular, soluble forms of CTLA-4 have been administered to prevent interactions of graft cell B7-family molecules with host T cell CD28 (see Chapter 7). Under these circumstances, inadequate IL-2 will be produced, so that alloantigen-activated T cells should become anergic. Interference with other costimulator molecules (e.g., CD40 and CD40 ligand or CD2 and LFA-3) have also had success in experimental systems.

Immunosuppression has been the major approach for the prevention and management of transplant rejection. Several methods of immunosuppression are commonly used.

1. *T cells may be inhibited or lysed by various immunosuppressive treatments. Immunosuppressive drugs are the principal treatment regimen for graft rejection.* Commonly used immunosuppressive therapies include metabolic toxins, such as azathioprine and cyclophosphamide; specific immunosuppressive drugs, the prototype of which is cyclosporin A (also known as cyclosporine); and antibodies reactive with T cell surface molecules.

The metabolic toxins in clinical use, namely azathioprine and cyclophosphamide, kill cells that are rapidly proliferating. These agents inhibit the maturation of lymphocytes from immature precursors and also kill proliferating mature T cells that have been stimulated by alloantigens. The use of these drugs is limited by toxicity to other rapidly dividing cells, e.g., hematopoietic precursors of leukocytes in the bone marrow and enterocytes in the gut. Newer metabolic toxins, such as mycophenolate mofetil, which is in clinical trials, may be more selective for T cells.

The most important immunosuppressive agent in current clinical use is cyclosporin A. Cyclosporin A is a cyclic peptide that is a natural metabolite in a species of fungus. The major action of cyclosporin A on T cells is to inhibit transcription of certain genes, most notably the IL-2 gene. Cyclosporin A binds with high affinity to a ubiquitous cellular protein called cyclophilin. As we discussed in

Chapter 7, the complex of cyclosporin A and cyclophilin, but not either component alone, binds to and inhibits the enzymatic activity of the calcium/calmodulin–activated protein phosphatase calcineurin. Since calcineurin function is required to activate the cytoplasmic component of the transcription factor NFAT, cyclosporin A blocks NFAT activation and the transcription of IL-2 and other cytokine genes. The net result of this action is that *cyclosporin A blocks the IL-2–dependent growth and differentiation of T cells.*

The introduction of cyclosporin A into clinical practice opened the modern era of transplantation. Prior to the use of cyclosporin A, the majority of transplanted hearts and livers were rejected. Now the majority of these allografts survive for over 5 years. Nevertheless, cyclosporin A is not a panacea for transplantation. Drug levels needed for optimal immunosuppression may cause kidney damage. For this reason, there is much excitement about a recently characterized fungal metabolite called FK506. FK506 is structurally unrelated to cyclosporin A, but the complex of FK506 and its binding protein (called FKBP) share with the cyclosporin A–cyclophilin complex the ability to bind calcineurin and inhibit its action. FK506 also produces renal toxicity, but it appears to provide a wider therapeutic window between effective immunosuppressive levels and those that cause toxicity.

Another recently introduced immunosuppressive agent is the antibiotic rapamycin. Rapamycin binds to FKBP and competes with FK506 for binding to this protein. However, the rapamycin-FKBP complex does not inhibit calcineurin but instead binds to another cellular protein simply called the mammalian target of rapamycin (MTOR). The principal effect of rapamycin is to block T cell growth. The precise mechanism by which this occurs is not known. MTOR may be a regulator of a protein kinase (S6 kinase) that participates in cell cycle control, and rapamycin-treated T cells display abnormalities in the levels of certain cyclin/cyclin–dependent kinase complexes. Combinations of cyclosporin A (which blocks IL-2 synthesis) and rapamycin (which blocks IL-2-driven proliferation) are particularly potent inhibitors of T cell responses. Rapamycin is being tested in current clinical trials.

Antibodies reactive with T cell surface structures are important agents for treating acute rejection episodes. In the 1960s, commonly used agents for this purpose were polyclonal horse antisera reactive with human lymphocytes or thymocytes. Since the 1980s, mouse monoclonal antibodies to specific T cell surface markers have been more commonly used. The most widely used antibody is OKT3, the first anti-CD3 antibody to be characterized. It may seem surprising that one would use a potential polyclonal activator such as anti-CD3 to reduce T cell reactivity. *In vivo,* however, OKT3 either acts as a lytic antibody, activating the complement system to eliminate T cells, or opsonizes T cells for phagocytosis. T cells that escape probably do so by capping and endocytosing ("modulating") CD3 off their surface, and such cells may be rendered transiently nonfunctional. Newer antibodies are being tested for immunosuppressive effects without causing T cell elimination. For example, antibodies to the α subunit of the IL-2 receptor are in clinical trial because these antibodies can prevent T cell activation by blocking IL-2 binding to activated T cells. The major limitation on the use of mouse monoclonal antibodies is that human recipients rapidly develop anti-mouse Ig antibodies that eliminate the injected mouse Ig. For this reason, attempts are being made to produce human monoclonal antibodies or human-mouse chimeric ("humanized") antibodies that may be less immunogenic.

2. *B cells may be inhibited from making antibodies, and pre-formed antibodies, such as those that mediate hyperacute rejection, can be removed.* Several new immunosuppressive agents, such as rapamycin, brequinar, and 15-deoxyspergualin, effectively inhibit antibody synthesis. Since pre-formed antibodies can persist in the plasma for days to weeks, use of these agents may be preceded by plasmaphoresis to remove antibodies and other plasma proteins. In allogeneic transplantation, these modalities are reserved for patients with high levels of pre-formed antibodies against a wide variety of donor alloantigens.

3. *The effector phase of the immune response may be inhibited.* In essence, this involves reducing inflammation. The most potent anti-inflammatory agents available are corticosteroids. The proposed mechanism of action for these natural hormones and their synthetic analogs is to block the transcription and secretion of cytokines by mononuclear phagocytes. Inhibition of TNF, IL-1, and IL-6 synthesis by corticosteroids has been observed both *in vitro* and *in vivo*. Lack of TNF and IL-1 synthesis reduces graft endothelial cell activation and recruitment of inflammatory leukocytes (see Chapters 12 and 13). Corticosteroids may also block other effector mechanisms of macrophages (e.g., generation of prostaglandins, of reactive oxygen species, or, in mice, of nitric oxide [NO]). Very high doses of corticosteroids may inhibit T cell secretion of cytokines or even lyse T cells, but it is unlikely that levels of corticosteroids achieved *in vivo* can act in this way. Newer anti-inflammatory agents are in clinical trials, including soluble cytokine receptors, anti-cytokine antibodies, or antibodies that block leukocyte-endothelial adhesion (e.g., anti-LFA-1 or anti-ICAM-1).

CLINICAL ORGAN TRANSPLANTATION

We now turn our attention to some of the important clinical issues that have arisen in the practice of solid organ transplantation. Kidney transplants have been successfully performed for the longest period (since the 1950s), and the renal allograft experience has formed the basis of considering transplantation of other organs. For this reason, our discussion focuses on the kidney but refers to other organs for comparison when appropriate.

Selection of donor and recipient matches in renal transplantation is based on blood group (ABO) matching, absence of pre-formed antibodies against donor cells in the blood of the recipient (called cross-matching), and HLA typing.

In the early days of renal transplantation, immunosuppression with corticosteroids, azathioprine, and anti-T cell antibodies was sufficient to allow survival of 50 to 60 per cent of unrelated cadaveric donor grafts at 1 year and survival of 90 per cent of living related donor grafts at 1 year. Since cyclosporin A was widely introduced in the 1980s, survival of unrelated cadaveric donor grafts has approached about 80 per cent at 5 years. Heart transplantation, for which HLA matching is not possible, has about a 50 per cent 5-year success rate. Experience with other organs is more limited, but the success rates appear comparable to or better than those of heart transplantation.

Acute rejection, when present, is often managed by rapidly intensifying immunosuppressive therapy. In modern transplantation, chronic rejection has become a more common cause of allograft failure, especially in cardiac transplantation. Chronic rejection is more insidious than acute rejection, and it is much less reversible. It is likely that prevention rather than treatment will be the best approach to the problem of chronic rejection, but successful intervention will probably require a better understanding of pathogenesis.

Graft survival is dependent on sustaining adequate immunosuppression. This has introduced a clinical dilemma because transplant patients often manifest two other clinical problems caused by immunosuppressive therapy. First, they are particularly susceptible to infections, especially by viruses. Infection by cytomegalovirus, a herpes virus, is particularly common and may be fatal in the immunosuppressed patient. As noted earlier, cytomegalovirus infection may also contribute to graft arteriosclerosis and chronic rejection. For this reason, it is now common to treat organ recipients with prophylactic anti-viral therapy for cytomegalovirus infections. Second, transplant patients have an increased proclivity to the development of certain tumors (see Chapter 18). The three malignancies commonly seen in these patients are B cell lymphomas, squamous cell carcinoma of the skin, and Kaposi's sarcoma. The B cell lymphomas are thought to be sequelae of unchecked infection by Epstein-Barr virus, another herpes virus (see Chapter 18, Box 18–2). The squamous cell carcinomas of the skin are associated with human papilloma virus and probably also represent virally induced malignancies. Kaposi's sarcoma is now well known for its prevalence in patients with the acquired immunodeficiency syndrome (see Chapter 21) and may be yet another example of a herpes virus–induced or provoked malignancy.

In patients receiving immunosuppression for transplantation, clinical problems related to viral infection and virally induced or virally potentiated malignancies are not coincidental. The major thrust of transplant-related immunosuppression is to reduce CTL generation and function, the key effector mechanism of acute cellular rejection. It should thus be no surprise that defense against viruses, the physiologic function of CTLs, is preferentially undermined.

XENOGENEIC TRANSPLANTATION

A major barrier to increased use of solid organ transplantation as a clinical therapy is the lack of availability of donor organs. As many as one half of the patients who are waiting for allogeneic organ grafts never receive one because organs are not obtainable. For this reason, many transplant immunologists have become interested in the possibility of transplantation of organs from other mammals, such as pigs, into human recipients. The principal obstacle to xenogeneic transplantation is the presence of natural antibodies. As discussed earlier, many individuals develop natural IgM antibodies to non-self carbohydrate determinants of the ABO blood group system. Similarly, over 95 per cent of humans have natural antibodies that are reactive with carbohydrate determinants expressed by the cells of members of species that are evolutionarily distant, such as pig. Such species combinations that give rise to reactive natural antibodies are said to be **discordant.** Natural antibodies are rarely produced against carbohydrate determinants of closely related, **concordant** species, such as human and chimpanzee. Thus, chimpanzee or other higher primates technically can and have been used as organ donors to humans. However, both ethical and logistic concerns have limited such operations. For reasons of anatomic compatibility, pigs are the preferred xenogeneic species for organ donation to humans.

The great majority of human anti-pig natural antibodies are directed at one particular carbohydrate determinant that is formed when a fucose moiety in the H blood group antigen found on human cells (see Box 17–2) is replaced by an α-linked galactose moiety found on pig cells. Pig cells that overexpress human H antigen fucosyl transferase become much less reactive with human natural antibodies because they make human H blood group instead of the normal α-galactosyl–containing pig determinant.

The consequence of high titers of natural IgM antibodies reactive with xenogeneic organ grafts is the same as was seen in ABO-incompatible human allogeneic transplantation, namely hyperacute rejection. This reaction depends on complement proteins and involves the same basic mechanisms seen in the allogeneic reaction, namely generation of endothelial cell procoagulants and platelet-aggregating substances coupled with loss of endothelial anticoagulant mechanisms, such as heparan sulfates. However, the consequences of activating human complement on pig cells are typically more severe than activation of complement by natural antibodies on human allogeneic cells. This is because the complement regulatory proteins made by pig cells, such as decay accelerating factor or CD59, are not able to interact with human complement proteins and thus cannot limit the extent of complement-induced injury (see Chapter 15). A strategy under exploration for reducing hyperacute rejection in xenotransplantation is prevention or reversal of complement activation by infusion of the recipient with agents such as soluble complement receptor type 1 (CR1), human decay accelerating factor, or human CD59. The latter two proteins are phosphatidylinositol-linked membrane proteins and are able spontaneously to insert into the membranes of the pig endothelial cells that line the graft vessels. An alternative long-term strategy for eliminating hyperacute rejection of pig organs altogether, currently under development by several laboratories, is to construct and breed transgenic pigs expressing human proteins that inhibit the process. Such pigs could, for example, overexpress human H-group fucosyl transferase and overexpress human complement regulatory proteins such as CD59. Preliminary trials in primates using organs from such transgenic pigs have been promising, significantly delaying the onset of hyperacute rejection.

Inhibition of the complement system in experimental models of discordant xenotransplantation has revealed a new type of rejection reaction involving graft infiltration by NK cells and macrophages. This form of inflammation has been called **delayed xenograft rejection.** NK cell activation is probably initiated because the pig cells do not express self (human) class I MHC molecules that normally function to deliver inhibitory signals (see Chapter 13). Activated NK cells make cytokines, such as TNF and IFN-γ, that recruit and activate macrophages. This kind of NK cell–macrophage interaction may be similar to that which has been described following challenge with *Listeria monocytogenes* in mice lacking T cells (see Chapter 13).

T cell–mediated rejection of xenografts may occur if hyperacute and delayed xenograft rejections are avoided. Some T cell responses to xenogeneic cells appear to be weaker than responses to allogeneic cells. This has been especially striking when murine T cell responses to pig MHC molecules have been examined. Several possible reasons for the weak responses to xenoantigens have been proposed. First, T cells specific for foreign peptides associated with self MHC molecules, which are responsible for allorecognition, may fail to cross-react with xenogeneic MHC molecules because their structure is too different from self MHC molecules. In this case, xenogeneic MHC molecules may be recognized only when processed and indirectly presented in association with self MHC molecules, just like any other foreign protein antigens. As discussed earlier, the basis of the strong response to allogeneic MHC molecules depends upon direct presentation of allogeneic MHC to cross-reactive T cells, and the response is much weaker when these molecules must be indirectly presented. However, the response to indirectly presented xenoantigens may be stronger than the response to indirectly presented alloantigens because there are likely to be more foreign peptides that can be generated from a xenogeneic MHC molecule than from an allogeneic MHC molecule. A second reason for a weak xenogeneic response is that adhesion molecules and costimulators may be species-specific, so that T cells from one species are not efficiently activated by APCs from another.

Although little is known about xenoreactivity of human T cells, recent results suggest that human T cells do cross-react with pig MHC molecules, resulting in direct presentation of these molecules, and many costimulator and adhesive interactions also appear to be conserved across this species combination. It is not known whether immunosuppressive strategies developed for allogeneic transplantation will prove adequate for reducing human anti-pig xenogeneic T cell–mediated rejection.

BONE MARROW TRANSPLANTATION

Bone marrow transplantation is actually the transplantation of pluripotent hematopoietic stem cells (see Chapter 2). It is general practice to transfer stem cells as part of an inoculum of total marrow cells collected by aspiration. However, stem cells can be mobilized from the donor marrow and isolated from peripheral blood following treatment with interleukin-3, granulocyte/monocyte–colony stimulating factor (GM-CSF), or granulocyte–colony stimulating factor (G-CSF). After transplantation, stem cells repopulate the recipient's bone marrow with their differentiating progeny.

Clinically, allogeneic bone marrow transplantation may be used to remedy acquired defects in the hematopoietic system or in the immune system, since both types of cells develop from a common stem cell. It has also been proposed as a means of correcting inherited deficiencies or abnormalities of enzymes or other proteins (e.g., abnormal hemoglobin), by providing a self-renewing source of normal cells. In addition, allogeneic bone marrow transplantation may be used as part of the treatment of bone marrow malignancies, i.e., leukemias. In this case, the chemotherapeutic agents needed to destroy leukemia cells also destroy normal marrow elements, and bone marrow transplantation is used to "rescue" the patient from the side effects of chemotherapy. For other malignancies, when the marrow is not involved by tumor or when it can be purged of tumor cells, the patient's own bone marrow may be harvested and reinfused after chemotherapy. This procedure, called autologous bone marrow transplantation, lacks many of the immunologic problems associated with allogeneic bone marrow transplantation and will not be discussed further.

Several unique problems that are associated with allogeneic bone marrow transplantation lead us to consider it separately from solid organ transplantation:

1. The transplanted stem cells must "home" to establish themselves in the appropriate environment; surgeons cannot place the stem cells in a particular location in the bone marrow. Moreover, experimental and clinical experience suggests that there are only a limited number of "niches" within marrow cavities, and if these are occupied at the time of transplantation, the grafted stem cells cannot establish themselves. The recipient often must be "prepared" with radiation and chemotherapy prior to transplantation to deplete his or her own marrow cells and vacate these sites.

2. Allogeneic stem cells are readily rejected by even a minimally immunocompetent host. The mechanisms of rejection are not completely known, but in addition to specific immune mechanisms, hematopoietic stem cells may also be rejected by NK cells. The recipient's immune system must be nearly ablated to permit successful bone marrow transplantation. Again, this is accomplished by preparation of the recipient with radiation and chemotherapy.

3. Graft cells may mount a rejection response against the host. This response, called the **graft-versus-host reaction**, can injure the host and cause graft-versus-host disease (GVHD) (see below). The graft-versus-host reaction arises only after extreme injury to the host immune system, a consequence of the preparation necessary to avoid stem cell rejection.

4. Recipients of allogeneic bone marrow transplants often show prolonged and profound immunodeficiencies. In human bone marrow transplantation, this consequence is a major cause of morbidity and mortality.

Graft-Versus-Host Disease

Graft-versus-host disease is caused by the reaction of grafted mature T cells in the marrow inoculum with alloantigens of the host. It occurs when the host is immunocompromised and therefore unable to reject the allogeneic cells in the graft. In most cases, the reaction is directed against minor histocompatibility antigens of the host because bone marrow transplantation is not commonly performed when there are allogeneic differences in MHC molecules. GVHD may also develop following transplantation of solid organs that contain significant numbers of T cells, e.g., small bowel or lung. Because solid organs are transplanted across MHC barriers, GVHD in such patients may be directed against allogeneic MHC molecules.

Graft-versus-host disease is the principal limitation on the use of bone marrow transplantation. As in solid organ transplantation, GVHD may be classified on the basis of histologic patterns into acute and chronic categories.

Acute GVHD involves epithelial cell necrosis in three principal target organs: skin, liver, and the gastrointestinal tract (Fig. 17–9). In the liver, the biliary epithelial cells but not the hepatocytes are affected. Clinically, acute GVHD is characterized by skin rash, jaundice, and diarrhea. When the epithelial necrosis is extensive, the skin or lining of the gut may simply slough off. In this circumstance, acute GVHD may be fatal.

Chronic GVHD is characterized by fibrosis and atrophy of one or more of the same organs, without evidence of acute cell necrosis (Fig. 17–9). Chronic GVHD may also involve the lungs, producing obliteration of small airways. Occasionally, necrosis and fibrosis can be present at the same time, leading to a diagnosis of acute and chronic GVHD. When severe, chronic GVHD leads to complete dysfunction of the affected organ and may also be fatal.

In animal models, acute GVHD is initiated by mature T cells present in the bone marrow inoculum, and elimination of mature donor T cells from the graft can prevent development of GVHD. Efforts to eliminate T cells from human marrow inoculum have reduced the incidence of GVHD but also appear to reduce the efficiency of engraftment; mature T cells, perhaps through production of colony-stimulating factors (CSFs), significantly improve stem cell repopulation. A current approach is to combine removal of T cells with supplemental GM-CSF to promote engraftment.

Although GVHD is initiated by T cell recognition of host alloantigens, the effector cells that produce epithelial cell necrosis are less well defined. CTLs may play a role. Histologically, NK cells are often attached to the dying epithelial cells, suggesting that NK cells are the primary effector cells of acute GVHD. This hypothesis has raised the issue of how NK cells lyse normal epithelial cells, since they cannot lyse epithelial cells *in vitro*. It has been proposed that the NK cells are activated

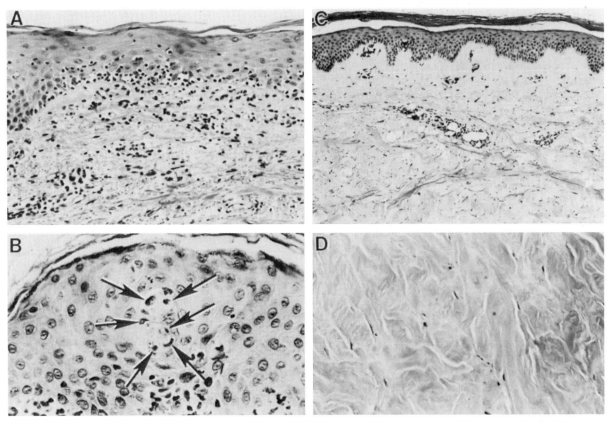

FIGURE 17–9. Acute and chronic graft-versus-host disease (GVHD) in the skin. At low magnification, acute GVHD appears as a sparse, lymphocytic infiltrate at the dermal-epidermal junction *(A);* at higher magnification *(B),* lymphocytes can be identified in the epidermis adjacent to injured epithelial cells *(arrows).* In contrast, chronic GVHD *(C)* shows fibrosis of the dermis and epidermal thinning. At higher magnification *(D),* dermal appendages can be seen to be trapped in the dense fibrosis. (Courtesy of Dr. George Murphy, Departments of Dermatology and Pathology, University of Pennsylvania, Philadelphia.)

by locally produced IL-2 to differentiate into lymphokine-activated killer (LAK) cells. As was discussed in Chapter 13, LAK cells can lyse normal cell types, including epithelium.

The relationship of chronic GVHD to acute GVHD is unknown and raises issues similar to those of relating chronic allograft rejection to acute allograft rejection. For example, chronic GVHD may represent the fibrosis of wound healing secondary to epithelial cell necrosis. However, chronic GVHD can arise without evidence of prior acute GVHD. Chronic GVHD is often associated with antibody production, and it could represent a form of autoimmunity. An alternative explanation is that chronic GVHD represents a response to ischemia caused by vascular injury.

Both acute and chronic GVHD are commonly treated with intense immunosuppression. GVHD is typically more resistant to such treatment than solid organ allograft rejection. A possible explanation is that conventional immunosuppression is targeted against T lymphocytes, especially CTLs. This works well in allogeneic rejection of solid organs but is less efficacious for NK cell–mediated or LAK cell–mediated responses. TNF and perhaps IL-1 appear to be important mediators of acute GVHD. Agents that suppress cytokine production, such as the drug thalidomide, or that antagonize cytokine action, such as soluble receptors or neutralizing antibody, have been effective at treating GVHD in pre-clinical and early clinical trials. Much effort has focused on prevention of GVHD. Cyclosporin A and the metabolic toxin methotrexate are routinely used for prophylaxis against GVHD. HLA typing is also very important for preventing GVHD. Indeed, most human bone marrow transplants are performed between siblings who are completely identical at all HLA loci, and clinical GVHD is due to differences at minor histocompatibility loci. Transplantation between parent and child may be performed with more stringent elimination of mature T cells from the marrow inoculum. Transplantation of MHC-matched unrelated donor is used only when no other option is available.

Immunodeficiency Following Bone Marrow Transplantation

As noted above, bone marrow transplantation is often accompanied by clinical immunodeficiency. Several factors may contribute to defective immune responses in recipients:

1. Bone marrow transplant recipients may be unable to regenerate a completely new T cell repertoire. The transplanted bone marrow may not contain a sufficient number and variety of self-renewing lymphoid progenitors, and the thymus gland of the recipient may have undergone irreversible changes during or after involution in early adulthood.

2. The ablation of the specific immune system in preparation for bone marrow transplantation may unmask a "natural suppression" system that prevents adequate regeneration of a specific immune system. Some immunologists have referred to specific populations of natural suppressor cells, observed after whole body irradiation. Such natural suppressor cells may be identical or related in lineage to NK cells.

3. The allogeneic host environment may overwhelm the developing immune system with alloantigenic stimuli that prevent development of a normal repertoire. An alternative statement of this explanation is that the graft-versus-host reaction preempts normal immunity. Many immunologists regard immunodeficiency as part of GVHD. However, immunodeficiency may exist in bone marrow transplant recipients who lack clinically overt or histologically detectable GVHD.

The consequence of immunodeficiency is that bone marrow transplant recipients are very susceptible to viral infections, especially cytomegalovirus, for which they receive prophylactic treatment. They are also susceptible to Epstein-Barr virus–provoked B cell lymphomas. However, the incidence of other malignancies, namely squamous cell carcinoma of the skin and Kaposi's sarcoma, has not increased as in solid organ transplant recipients. The basis for this difference is unclear. The immunodeficiencies of bone marrow transplant recipients can be more severe than those of conventionally immunosuppressed patients and may extend to bacteria and viruses. Therefore, bone marrow transplant recipients commonly receive prophylactic antibiotics and are often actively immunized against capsular bacteria such as pneumococcus.

Paradoxically, bone marrow transplant recipients also suffer from autoimmunity (see Chapter 19). Two factors may contribute to autoimmunity in the face of immunodeficiency. First, graft CD4⁺ T cells respond to alloantigens on residual host B cells, leading to inappropriate antibody (and autoantibody) production. Second, during repopulation of the immune system, the balance between regulatory and effector cells may be abnormal, leading to emergence of autoreactive lymphocytes. This has been proposed as an explanation for why recipients of autologous bone marrow transplants may occasionally show a GVHD-like syndrome. In general, systemic autoimmunity appears less frequently in human bone marrow transplant recipients than in experimental animal studies.

SUMMARY

Transplantation of tissues from one individual to a genetically nonidentical recipient leads to a specific immune response, called rejection, that can destroy the graft. The major molecular targets in transplant rejection are non-self allelic (i.e., allogeneic) forms of class I and class II MHC molecules complexed to self peptides. Many different T cell clones, specific for different foreign peptides plus self MHC molecules, cross-react with each individual allogeneic MHC molecule. This high frequency of recognition of allogeneic MHC molecules that are directly presented to T cells, i.e., without processing and presentation in association with self MHC molecules, accounts for why the allogeneic response is so much stronger than the response to conventional foreign antigens.

The cellular reaction to allogeneic class I and class II molecules can be analyzed *in vitro* in the MLR. In general, allogeneic class I molecules stimulate alloreactive CD8+ CTLs, whereas allogeneic class II molecules stimulate alloreactive CD4+ helper T lymphocytes, although the largest reactions occur when there are differences at both class I and class II loci.

In vivo rejection is mediated by T cells, including CTLs and helper T cells that cause DTH, and by antibodies. Alloreactive T cells may be stimulated by professional APCs, such as dendritic cells, in the graft.

Several patterns of rejection can occur in solid organ transplants. Pre-existing antibodies, often IgM directed against ABO antigens on endothelial cells, can cause hyperacute rejection characterized by thrombosis of graft vessels, in a matter of hours. Antibodies produced in response to the graft or alloreactive T cells cause blood vessel cell necrosis, called acute vascular rejection beginning about 1 week after transplantation. Infiltrating alloreactive CTLs cause parenchymal cell necrosis, called acute cellular rejection in the same time frame. Chronic rejection, characterized by fibrosis, may represent healing of acute rejection or may represent a chronic delayed type hypersensitivity reaction in the walls of muscular arteries, producing accelerated arteriosclerosis and ischemic injury of the graft. Chronic rejection develops months to years after transplantation.

Rejection may be prevented or treated by minimizing the immunogenicity of the graft (e.g., by limiting MHC allelic differences, or by induction of tolerance) or by general immunosuppression of the host. Most immunosuppression is directed at T cell responses, using cytotoxic drugs, specific immunosuppressive agents, or anti-T cell antibodies. The prototypic specific immunosuppressive agent is cyclosporin A, which blocks IL-2 synthesis. Immunosuppression is often combined with anti-inflammatory therapy, using agents such as corticosteroids that inhibit cytokine synthesis by macrophages.

Patients receiving solid organ transplants may experience complications related to their therapy, including viral infections, especially with cytomegalovirus, and virus-related malignancies, such as B cell lymphoma, squamous cell carcinoma of the skin, and Kaposi's sarcoma.

Xenogeneic transplantation of solid organs is a major goal of current research. It is limited by the existence of natural antibodies to surface glycans on cells of discordant species that cause hyperacute rejection. NK cells and macrophages may mediate delayed xenograft rejection. The cell-mediated immune response to xenogeneic MHC molecules is less well characterized than the response to allogeneic molecules.

Bone marrow transplant recipients are very susceptible to graft rejection and require intense preparatory immunosuppression. In addition, two unique problems not seen with solid organ transplants may develop. First, lymphocytes in the bone marrow graft may respond to alloantigens of the host, producing graft-versus-host disease (GVHD). Acute GVHD is characterized by epithelial cell necrosis in the skin, liver, and gut, causing a rash, jaundice, and diarrhea, respectively. When severe, acute GVHD may be fatal. Chronic GVHD is characterized by fibrosis and atrophy of one or more of these same target organs as well as the lungs and may also be fatal. Second, bone marrow transplant recipients often have immunodeficiencies, rendering them susceptible to infections. In some cases, patients with GVHD may develop autoantibodies.

Selected Readings

Abraham, R. T., and G. J. Wiederrecht. Immunopharmacology of rapamycin. Annual Review of Immunology 14:483–510, 1996.

Bociek, A. G., D. A. Stewart, and J. O. Armitage. Bone marrow transplantation—current concepts. Journal of Investigative Medicine 43:127–135, 1995.

Ferrara, J. L. M., and H. J. Deeg. Graft-versus-host disease. New England Journal of Medicine 324:667–674, 1991.

Krensky, A. M., A. Weiss, G. Crabtree, M. M. Davis, and P. Parham. T-lymphocyte-antigen interactions in transplant rejection. New England Journal of Medicine 322:510–517, 1990.

Lawson, J. H., and J. L. Platt. Molecular barriers to xenotransplantation. Transplantation 62:303–310, 1996.

Mason, D. W., and P. J. Morris. Effector mechanisms in allograft rejection. Annual Review of Immunology 4:119–145, 1986.

Rosenberg, A. S., and A. Singer. Cellular basis of skin allograft rejection: an *in vivo* model of immune-mediated tissue destruction. Annual Review of Immunology 10:333–358, 1992.

Sherman, L. A., and S. Chattopadhyay. The molecular basis of allorecognition. Annual Review of Immunology 11:385–402, 1993.

Sigal, N. H., and F. J. Dumont. Cyclosporin A, FK-506, and rapamycin: pharmacologic probes of lymphocyte signal transduction. Annual Review of Immunology 10:519–560, 1992.

CHAPTER EIGHTEEN

IMMUNITY TO TUMORS

Malignant tumors, or cancers, grow in an uncontrolled manner, invade normal tissues, and often metastasize and grow at sites distant from the tissue of origin. In general, cancers are derived from only one or a few normal cells that have undergone malignant transformation. The abnormal growth behavior of malignant tumors is the reflection of complex abnormalities in physiology that result from expression of mutated or viral genes and/or deregulated expression of normal genes. In addition to the abnormal expression of molecules that contribute to malignant behavior, cancer cells frequently express mutated or dysregulated genes whose products do not contribute to growth or invasive properties of the tumor. The increased frequency of mutations in such cancer cells may be related to fact that the carcinogenic agents that promote the development of cancers, such as ionizing radiation and reactive chemicals, are not selective and may damage DNA anywhere in the genome. Cancers can arise from almost any tissue in the body. Those derived from epithelial cells, called carcinomas, are the most common kinds of cancers. Sarcomas are malignant tumors of mesenchymal tissues, arising from cells such as fibroblasts, muscle cells, and fat cells. Solid malignant tumors of lymphoid tissues are called lymphomas, and marrow and blood-borne malignant tumors of lymphocytes or other hematopoietic cells are called leukemias.

A variety of clinical and pathologic evidence indicates that tumors can stimulate immune responses. A common histologic observation that suggests that tumors may be immunogenic is the presence of mononuclear infiltrates, composed of T cells, natural killer (NK) cells, and macrophages, surrounding many tumors. Such infiltrates may develop at any site of tissue injury, but they are more frequently present around certain types of tumors, including testicular seminomas, thymomas, medullary breast carcinomas, and malignant melanomas in the skin, independent of the presence of other inflammatory stimuli, such as infection or tissue necrosis. Another histopathologic indication that tumors stimulate immune responses is the frequent finding of lymphocytic proliferation (hyperplasia) in lymph nodes draining sites of tumor growth. Furthermore, there is often evidence of cytokine effects in tumors, such as increased expression of class II major histocompatibility complex (MHC) molecules and intercellular adhesion molecule-1 (ICAM-1), suggesting an active immune response at the site of the tumor. The spontaneous regression of some tumors also suggests that host immune responses to tumor cells can occur, although there are many other explanations for this phenomenon.

The first clear example that tumors can induce protective immune responses came from a now classic set of studies performed in the 1950s with

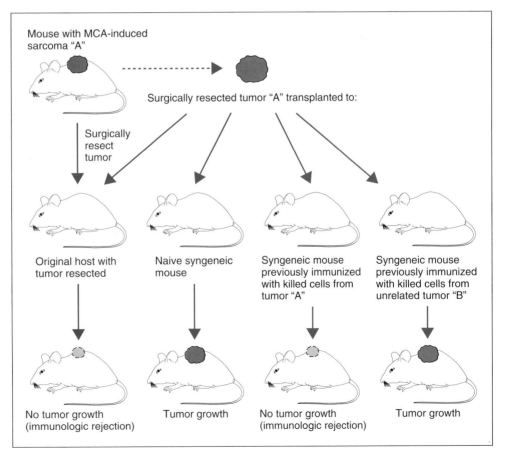

FIGURE 18–1. Experimental demonstration of tumor immunity: Specific immunologic rejection of chemically induced sarcomas. A mouse treated with the chemical carcinogen methylcholanthrene (MCA) develops a sarcoma. If this tumor is resected and transplanted into a normal syngeneic mouse, the tumor will grow. In contrast, the original tumor-bearing animal that was cured by surgical resection will reject a subsequent transplant of the same tumor. Injection of killed cells from the same tumor into a syngeneic mouse also induces protective immunity to that tumor, but injection of killed cells from an unrelated tumor does not.

Mouse with MCA-induced sarcoma "A"

Surgically resect tumor

Surgically resected tumor "A" transplanted to:

Original host with tumor resected

Naive syngeneic mouse

Syngeneic mouse previously immunized with killed cells from tumor "A"

Syngeneic mouse previously immunized with killed cells from unrelated tumor "B"

No tumor growth (immunologic rejection)

Tumor growth

No tumor growth (immunologic rejection)

Tumor growth

rodent tumors induced by chemical carcinogens or radiation. In a typical study of this kind, a sarcoma is induced in an inbred mouse by painting its skin with the chemical carcinogen methylcholanthrene (MCA). These MCA-induced sarcomas can be excised from the original host mouse and introduced into other mice, or back into the original animal. Transplantation of these tumors into other syngeneic mice is usually successful, and the tumors grow and eventually kill the new host. In contrast, reintroduction of the tumor into the original host results in a specific immunologic rejection of the tumor. Alternatively, the cells of a tumor from one mouse can be killed by irradiation and used to immunize a second syngeneic mouse. Subsequent introduction of live cells from the original tumor into the immunized mouse will result in immunologic rejection of the tumor transplant, but a mouse immunized with killed cells from an unrelated tumor will not reject a transplant of the first tumor (Fig. 18–1). Furthermore, protective immunity in one mouse, established by growth of or

immunization with a tumor, can be transferred to another tumor-free animal by CD8$^+$ cytolytic T lymphocytes (CTLs) (Fig. 18–2). These experiments demonstrated that the *rejection of the transplanted tumors displays cardinal features of specific immune responses, including dependence on lymphocytes, specificity, and memory.* Furthermore, the experiments indicate that tumor cells express antigens, called **tumor antigens,** which can be recognized by the host immune system.

If malignantly transformed cells express molecules that act as foreign antigens in the host, it is possible that a physiologic function of the immune system is to recognize and destroy these abnormal cells before they grow into tumors, or to kill tumors after they are formed. This theoretical role for the immune system is called **immunosurveillance.** The idea originated with Paul Ehrlich early in this century and was expanded in the 1950s and 1960s by Macfarlane Burnet and Lewis Thomas. If the concept of immunosurveillance is valid, then immune effector cells, such as B cells, helper T

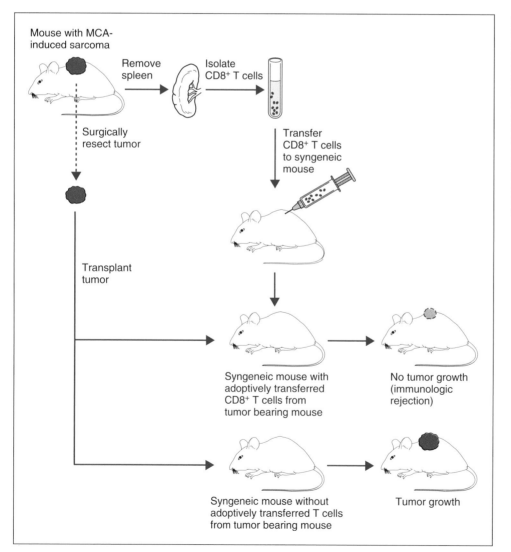

FIGURE 18–2. Immunity to transplanted tumors can be adoptively transferred by CTLs. Splenic CD8$^+$ T cells isolated from a mouse bearing a chemically induced sarcoma are transferred to a normal (tumor-free) mouse. This recipient mouse will subsequently reject tumor transplants from the original mouse, but tumor transplants will grow on another mouse which has not received any adoptively transferred CTLs. Transfer of CD4$^+$ T cells, B cells, or antibody does not transfer immunity.

cells, CTLs, and NK cells, must be able to recognize tumor antigens and mediate the killing of tumor cells. Although there still is little direct evidence that immunosurveillance protects individuals from many common tumors, certain observations support the validity of the concept. For example, immunodeficient individuals are more likely to develop certain types of tumors than normal individuals. Clinicopathologic correlations show that the presence of lymphocytic infiltrates in some tumors (e.g., medullary breast carcinomas and malignant melanomas) is associated with a better prognosis compared with histologically similar tumors without infiltrates. There is also abundant experimental evidence that tumors can stimulate specific T cell–mediated immune responses. Recently, several tumor antigens that are recognized by class I MHC–restricted CTLs *in vivo* have been identified as mutant or aberrantly expressed cellular proteins. These findings further support the idea that a function of CTLs is surveillance for and destruction of cells harboring mutated genes that could lead to, or be associated with, malignant transformation.

Immunosurveillance for tumors is often ineffective, as indicated by the fact that lethal cancers arise in immunocompetent individuals. It is therefore likely that immune responses to tumors are often weak, and the possible reasons for this are discussed later in this chapter. A major current focus of immunology and oncology research is the development of ways to augment host immune responses to tumors. This research is part of a broad field called **tumor immunology,** which encompasses the study of specific immune responses to tumors, the antigens on tumor cells that induce immune responses, immunologic effector mechanisms that kill tumor cells, and immunologic approaches to detecting, diagnosing, and treating cancers. The great progress we have made over the past 20 years in understanding the physiology of normal immune responses is already being applied to the important practical problems of prevention and treatment of tumors. In this chapter, we discuss these different aspects of tumor immunology, referring to the basic principles of recognition and effector phases of the immune response that we have already described in detail in previous chapters.

TUMOR ANTIGENS

Although tumors are derived from self tissues, they are more likely than normal cells to express molecules that are recognized as foreign by the immune system. As we have discussed, this is because of the high frequency of mutations in cancer cells, and dysregulated gene expression leading to the production of unmutated proteins not normally expressed. Furthermore, tumors caused by transforming (oncogenic) viruses may express viral proteins. The products of mutated, dysregulated, or viral genes may therefore be recognized as foreign by B and T lymphocytes because they were never expressed on self tissues prior to tumor development, or were expressed at sufficiently low levels that they did not induce self-tolerance. Such proteins can therefore stimulate specific immune responses to tumor cells. Alternatively, the products of mutated or dysregulated genes in cancer cells may contribute to the abnormal synthesis of non-protein molecules, including carbohydrates and lipids, which can be recognized by B cells as tumor antigens. In addition, surface proteins peculiar to tumors may serve as targets for effectors of innate immunity, such as NK cells.

The fact that tumor cells express antigens that can stimulate immune responses in the host has been demonstrated in both experimental animal models and in human cancer patients. It has become apparent that patterns of expression of these antigens differ among different tumors and between tumors and normal tissues. Tumor antigens are often classified into different groups based largely on these patterns of expression:

1. **Tumor-specific antigens (TSAs)** are antigens expressed on tumor cells but not normal cells. These are the antigens most likely to evoke immune responses in the host because they are perceived as foreign. **Unique tumor antigens** are TSAs that are expressed on only one or a few clonal tumors, reflecting peculiar mutations that are characteristic of those tumors alone.

2. Some tumor antigens are also expressed concurrently on normal cells in the host, and expression may or may not be restricted to the type of tissue from which the tumor originated. These are called **tumor-associated antigens (TAAs).** Although some of these TAAs may induce immune responses in the host, they often do not, because of self-tolerance. TAAs are, in fact, usually defined by antibodies generated in animals of other species by immunization with tumor cells.

This portion of the chapter describes different types of tumor antigens and their roles in the biology of tumor-host interactions. We have categorized two main groups of tumor antigens based on the types of immunologic probes used to detect them, namely those recognized by T lymphocytes and those detected by antibodies. This categorization not only reflects methodologic approaches to the study of tumor antigens, but also, in some cases, indicates the types of immune responses which such antigens evoke in tumor hosts. Within each group there are examples of tumor-specific antigens, unique tumor antigens, and tumor-associated antigens. There is also some overlap in this classification, with some types of antigens capable of eliciting, and being detected by, both T cells and antibodies.

Tumor Antigens Recognized by T Lymphocytes

Tumor antigens that are recognized by T cells are the major targets of protective anti-tumor immunity in experimental animals, and probably in humans. The identification of these antigens has been a major focus of recent tumor immunology research, and it has relied on transplantation models in animals, the development of both mouse and human cloned CTL lines as probes for the antigens, cloning of genes from tumor complementary DNA (cDNA) libraries, and elution and analysis of peptides from tumor MHC molecules. These experimental approaches are described in Box 18–1. In

Chapter 6, we discussed how MHC molecules display peptides derived from cellular proteins for surveillance by T cells. In particular, class I MHC molecules display peptides derived from cytosolic proteins in all nucleated cells. This is also true of tumor cells. Therefore, if a tumor cell produces a protein that is either not expressed by normal cells or is a mutated version of a normal protein, peptides derived from this tumor protein may be displayed by class I MHC molecules on the tumor cell surface, and recognized as foreign by host CD8⁺ T lymphocytes (Fig. 18–3). Thus, tumor cells serve as antigen-presenting cells (APCs), displaying their own antigens to T cells. These antigens could

---

BOX 18–1.   Identification of Tumor Antigens Recognized by T Lymphocytes

Tumor antigens recognized by T cells have been identified by the combination of three experimental approaches—transplantation studies of tumors in rodents, generation of cloned tumor-specific CTL lines, and the identification of peptides recognized by tumor-specific CTLs or the identification of the genes encoding these peptides.

(1) *Transplantation studies of tumors in rodents were the first experiments to indicate the types of tumor antigens recognized by T cells.* Initial studies of this type used chemically induced rodent sarcomas (see Figs. 18–1 and 18–2), and they demonstrated the existence of tumor-specific transplantation antigens (TSTAs) which elicited highly specific immune responses. Another, more recent approach to characterize antigens that stimulate tumor rejection by CTLs relies on the *in vitro* mutagenesis of an established tumorigenic mouse cell line and isolation of non-tumorigenic mutants that are immunologically rejected when transplanted into syngeneic mice. In this system, the tumorigenic cell line does not express TSTAs and therefore grows unchecked when injected into a host animal, but the mutagenized cell line does express mutant protein antigens that induce its rejection. The role of CTLs in the rejection process in this model has been established by adoptive transfer experiments, and by the propagation of CTL clones that recognize the tumor, as discussed below. The actual genes encoding the rejection antigen in a few of these tumor variants have been cloned, by methods described later, and they turn out to be cellular genes of unknown function with point mutations. This is a further demonstration that antigen-processing pathways in tumor cells can sample randomly mutated cytoplasmic proteins and thereby target the cells for lysis by CTLs.

(2) *The establishment of cloned CTL lines that recognize tumor antigens has been a key advance in the identification of tumor antigens.* This was first done in mouse models using the *in vitro* mutagenized tumor lines described earlier. Each CTL clone raised against one mutagenized tumor variant reacts against that variant but not against most others derived from the same parental tumor. This indicates that the antigens the CTLs recognize are highly diverse and unique to individual tumors, similar to the TSTAs described previously. More recently, many cloned CTL lines specific for human tumors, particularly melanomas, have been generated. Melanomas, which are malignant tumors of melanocytes, are often readily accessible, surgically resectable tumors and are relatively easy to grow in tissue culture. CTLs specific for these tumors may be propagated, and subsequently

cloned, by culturing T cells from a melanoma patient with lethally irradiated but antigenically intact cells derived from the patient's melanoma. The T cells can be isolated from peripheral blood, from lymph nodes draining the tumor, or from cells that have actually infiltrated the tumor *in vivo*. Since the T cells and the tumor are from the same individual, the MHC restriction of the T cells matches the MHC alleles expressed by the tumor. In these cocultures, CTLs that recognize peptide antigens displayed by the tumor cells are stimulated to grow, and single cell clones are propagated in interleukin-2 using limiting dilution techniques (see Figure).

(3) The identification of the peptide antigens that induce CTL responses in tumor patients, and the identification of the genes encoding the proteins from which the peptides are derived, have relied on cloned tumor-specific CTL lines. This has been accomplished in two ways. First, a direct biochemical approach has been used in which peptides bound to class I MHC molecules purified from melanoma cells are eluted by acid treatment and fractionated by reverse-phase high-performance liquid chromatography (RP-HPLC). The fractions are tested for their ability to sensitize MHC-matched nontumor target cells for lysis by a tumor-specific CTL clone. This strategy relies on having a target cell that expresses the class I MHC molecules for which the CTL clone is specific but does not normally express the tumor antigen, and on the ability to load these cell surface MHC molecules with the exogenous HPLC-purified peptides. Peptide fractions that do sensitize the target cells are then analyzed by mass spectroscopy to determine their amino acid sequences. Once the peptide sequence is known, it may be possible to compare it with databases of protein sequences to see whether it matches with any previously characterized protein, and to determine whether there are point mutations from the normal sequences. Second, a genetic approach can be used to identify genes encoding tumor antigens (see Figure). This relies on preparing cDNA libraries from a tumor cell line that contains genes encoding all the tumor proteins. The library is prepared in molecular constructs that will allow constitutive expression of the genes when they are introduced into cell lines. Pools of DNA from such a library are transfected into a cell line expressing the appropriate class I MHC allele, and the transfected cells are tested for sensitivity to lysis by an anti-tumor CTL clone. The DNA pools that sensitize the target cell line presumably contain the gene that en-

Continued

codes the protein antigen recognized by the CTL clone. Multiple additional rounds of transfections using smaller and smaller subfractions of the DNA pool can lead to exact identification of the single relevant gene. The sequence of the gene can then be determined, and comparisons can be made with known genes. Synthetic peptides corresponding to different regions of the en-

coded protein can be tested for their ability to sensitize target cells in much the same way as is done in the peptide elution approach described previously. Both the biochemical and genetic approaches have been used successfully to identify human melanoma antigens that have stimulated CTL responses in the melanoma patients.

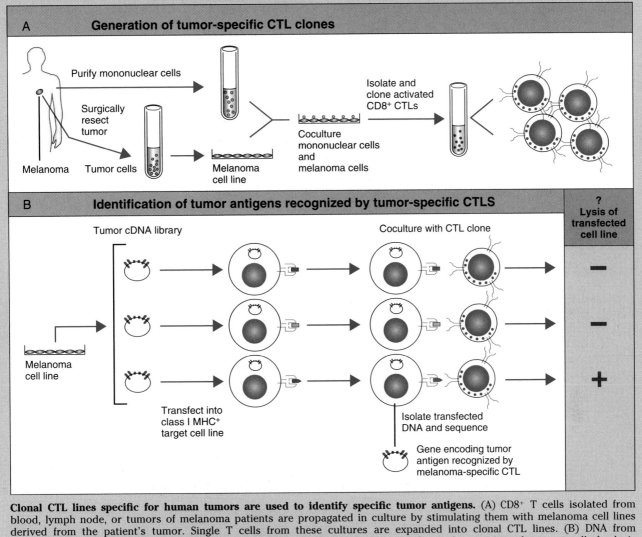

Clonal CTL lines specific for human tumors are used to identify specific tumor antigens. (A) CD8+ T cells isolated from blood, lymph node, or tumors of melanoma patients are propagated in culture by stimulating them with melanoma cell lines derived from the patient's tumor. Single T cells from these cultures are expanded into clonal CTL lines. (B) DNA from melanoma gene libraries is transfected into class I MHC–expressing target cells. Genes that sensitize the target cells for lysis by the melanoma-specific CTL clones are analyzed to identify the melanoma protein antigens recognized by the patient's CTLs.

be encoded by the tumor cell genome or by viral genomes which are carried by the tumor cells. It is also possible that proteins released into the extracellular medium by viable or dying tumor cells, or whole tumor cells themselves, may be endocytosed into the class II MHC pathway of antigen processing by professional APCs, and class II MHC–associated peptides derived from these tumor cells can be displayed for recognition by CD4+ helper T cells. It is not surprising that the tumor antigens that have been shown to be recognized by T lymphocytes include a wide variety of cellular and viral proteins.

The studies with chemically induced rodent sarcomas, such as those described in Figs. 18–1 and 18–2, established that tumors express transplantation antigens that can elicit specific immune responses. Although there was limited knowledge of antigen-presenting pathways at the time the first experiments of this sort were performed, we now understand that the tumor antigens in such studies are peptide–class I MHC complexes capable of stimulating CTLs. Experiments with multiple different rodent tumors, all induced by the same carcinogen, indicated that there was an enormous diversity of tumor transplantation antigens in that

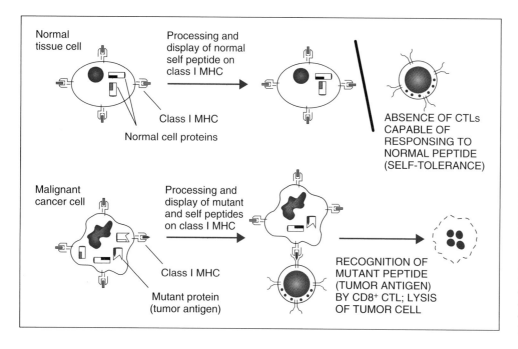

FIGURE 18–3. Immune surveillance for tumor antigens by class I MHC–restricted T cells. The class I MHC pathway of antigen presentation samples cytoplasmic proteins and displays peptides derived from those proteins on the cell surfaces. Because of self-tolerance, no T cells will respond to peptides from normal cellular proteins. CTLs will recognize peptides derived from mutant or aberrantly expressed proteins present in malignantly transformed cells, and the CTLs will kill the cancer cells expressing these proteins.

experimental tumor model, reflected by the specificity of the immune response to each individual tumor. For example, one MCA-induced sarcoma will not induce protective immunity against another MCA-induced sarcoma, even if both tumors are derived from the same mouse (Table 18–1). For this reason, such antigens are called **tumor-specific transplantation antigens (TSTAs).** In retrospect, this diversity is predictable since the carcinogens that induce these tumors may randomly mutagenize virtually any gene, and the class I MHC antigen-presenting pathway should be able to display peptides from each mutated protein in each tumor.

There are two important ways in which TSTAs

on experimental animal tumors are relevant to the study of human cancers. First, the demonstration that tumors can be killed by tumor antigen-specific CTLs raises the important possibility that CTL responses may be able to control human tumors. It has become clear from animal and some human studies that immune responses to established tumors often do occur, with the clonal expansion of tumor-specific CTLs, but these responses are frequently weak and ineffective in eradicating the tumors. When the immunogenicity of these tumors is increased by ways discussed in this chapter, specific rejection of the tumors can be observed. Second, the diversity of rodent sarcoma tumor-specific transplantation antigens, and the identification

TABLE 18–1. Transplantation Antigens on Chemically and Virally Induced Tumors

	Treatment of Mouse			
Experiment	**Immunization with Killed Tumor Cells from**	**Challenge with Live Tumor Cells from**	**Result**	**Conclusion**
1	Chemically induced sarcoma A	Chemically induced sarcoma A	No growth	Immunity to chemically induced tumors is specific for individual tumors.
	Chemically induced sarcoma A	Chemically induced sarcoma B	Growth of chemically induced sarcoma B	
2	MSV-induced sarcoma A	MSV-induced sarcoma A	No growth	Immunity to virus-induced tumors is virus-specific.
	MSV-induced sarcoma A	MSV-induced sarcoma B	No growth of sarcoma B	
	MSV-induced sarcoma A	Chemically induced sarcoma C	Growth of chemically induced sarcoma C	
	MSV-induced sarcoma A	MuLV-induced sarcoma D	Growth of MuLV-induced sarcoma D	

Abbreviations: MSV, murine sarcoma virus; MuLV, murine leukemia virus.

of tumor antigens generated by *in vitro* mutagenesis, has established the potential for CTL surveillance of many different kinds of tumors.

We will now describe the characteristics of the different types of tumor antigens recognized by T lymphocytes, and we will return to effector responses to these antigens later in the chapter. The major categories of these antigens are listed in Table 18–2.

PRODUCTS OF MUTATED NORMAL CELLULAR GENES NOT RELATED TO ONCOGENESIS

In vitro mutagenized mouse tumors were used for the first successful molecular genetic identification of tumor antigens recognized by T cells (see Box 18–1). These antigens turned out to be mutated versions of a variety of cellular proteins that

TABLE 18–2. Tumor Antigens That Stimulate T Cell Responses

Category	Examples
Tumor-specific transplantation antigens on chemically induced rodent sarcomas	No examples defined molecularly
Products of random point mutations in cellular genes not involved in tumor pathogenesis	p91A mutation in mutagenized murine mastocytoma MUM-1 melanoma antigen with a serine-to-isoleucine mutation in peptide recognized by melanoma-specific CTLs
Oncogene products	p21^{ras} proteins with point mutation at position 12 (10% of human carcinomas) p210 product of *bcr-abl* rearrangements (chronic myelogenous leukemias) HER-2/*neu*, overexpressed normal gene (various carcinomas)
Mutated tumor-suppressor gene products	p53 (~50% of human tumors)
Products of silent genes normally not expressed in most tissues	MAGE 1, -3, BAGE, GAGE (expressed by many human melanomas, many types of carcinomas, normal testis and placenta)
Viral gene products in virus-associated malignancies	SV40 T antigen (SV40-induced rat tumors) Human papillomavirus E6 and E7 gene products (human cervical carcinoma) Epstein-Barr virus EBNA-1 gene product (Burkitt's, lymphoma, and nasopharyngeal carcinoma)
Products of tissue-specific genes expressed by normal cell types from which the tumor was derived	Tyrosinase, gp100, MART-1 (expressed exclusively by melanocytes and melanomas)

Abbreviations: SV, simian virus; EBNA, Epstein-Barr nuclear antigen.
Adapted from Pardoll, D. M. Cancer vaccines. Immunology Today 14:310–316, 1993.

are expressed in many cell types but have no known function or homologies. These proteins are apparently unrelated to the malignant phenotype, and the effects of the mutations, other than imparting immunogenicity, remain unknown. A similar phenomenon has been found in some human tumors. For example, a CTL clone derived from a melanoma patient recognizes a mutated peptide sequence of a protein of unknown function, the normal form of which is expressed in many tissues. The melanoma-generated peptide recognized by the CTL is abnormal in two respects, carrying both a point mutation and intronic sequences indicative of aberrantly spliced messenger RNA (mRNA). Although these findings confirm expectations based on the biology of tumors and our understanding of how T cell immunosurveillance works, the majority of human tumor antigens identified by recognition by specific CTLs are, surprisingly, not mutated in any way. This suggests that the high frequency of mutated proteins represented in the animal tumor models may be a function of the highly mutagenic protocols used to produce the tumors, but the importance of mutations in generating antigens expressed by human tumors may not be as great. In contrast, many of the human tumor antigens are products of aberrantly expressed but unmutated genes, as discussed later.

PRODUCTS OF ONCOGENES AND MUTATED TUMOR SUPPRESSOR GENES

Virtually all tumors express genes whose products are required for malignant transformation or for maintenance of the malignant phenotype. In many instances, these genes have been identified, and often they are altered forms of normal cellular genes that control cell proliferation and differentiation. Such cellular proto-oncogenes may be altered by carcinogen-induced point mutations, deletions, or chromosomal translocations to form oncogenes whose products have transforming activity. In addition, viral integration into normal cellular proto-oncogenes can result in structurally abnormal products that have oncogenic activity. Tumor-suppressor genes also encode proteins required for normal cellular growth and differentiation, and mutations in these genes can result in nonfunctional products, leading to malignant transformation. The altered forms of proto-oncogene and tumor-suppressor gene products expressed in tumors should theoretically stimulate immune responses in the host, because these altered forms are not normally expressed as self proteins, and T cell tolerance to these abnormal proteins should not develop prior to their expression in tumors. These proteins are usually intracellular molecules that are likely to be processed and presented as peptides in association with class I MHC molecules, but phagocytosis or internalization of tumor cells or shed antigens may also result in class II MHC–associated presentation.

The mutations in some oncogenes are remarkably consistent among many different tumors, such as position 12 mutations in p21^{ras} proteins, probably because only selected mutations impart a growth advantage to the cell. Furthermore, relatively few mutations appear in a wide variety of tumors. Because their amino acid sequences are known and they are widely distributed on many different tumors, oncogene or tumor-suppressor gene products are potential targets for cancer immunotherapy. It has been established that both human CD4$^+$ and CD8$^+$ T cells can respond to the products of some of these genes, including mutated Ras, p53, and *bcr-abl* proteins. However, there is very little evidence that these antigens induce protective T cell responses in humans. One oncogene called HER-2/*neu* is not a mutated gene but transforms cells when it is overexpressed. It encodes a membrane protein that is abundant on the cell surfaces of several different types of human carcinomas, and this protein does appear to be able to stimulate both CD4$^+$ helper T cell and CD8$^+$ CTL responses to these tumors. For example, CTLs that recognize a peptide derived from the HER-2/*neu* protein have been found among T cells specific for an ovarian carcinoma, and among CD8$^+$ T cells infiltrating breast and lung tumors. Clinical trials using this HER-2/*neu*–derived peptide as an immunogen to boost anti-tumor immunity are in progress.

PRODUCTS OF NORMALLY SILENT GENES

Some genes are usually not expressed in normal tissues or are expressed only early during development, before the mechanisms of self-tolerance are operative. When these genes are dysregulated as a consequence of malignant transformation of a cell and are expressed inappropriately in tumors, they may behave as tumor antigens and evoke immune responses. These tumor antigens may be shared by many different tumors. Mouse and human genes that encode these types of tumor antigens recognized by T cells have recently been identified using tumor antigen-specific CTL clones and DNA library transfections as described in Box 18–1. The functions of the proteins encoded by these genes are unknown, but they are not required for the malignant phenotype of the cells, and their sequences are identical to the corresponding genes in normal cells, i.e., they are not mutated. One of the mouse genes identified in this way is expressed on mast cell tumors and perhaps on some immature normal mast cells but not on other cells, and its protein product stimulates CTL-mediated tumor rejection *in vivo*. The MAGE (melanoma antigen) genes, first isolated from human melanoma cells, encode cellular protein antigens recognized by many different melanoma-specific CTL clones derived from different melanoma-bearing patients. There is a family of at least 12 homologous MAGE genes, but peptides from only two MAGE proteins are known to be recognized by melanoma-specific CTLs. MAGE proteins are variably expressed on many tumors in addition to melanomas, including carcinomas of the bladder, breast, skin, lung, and prostate, and some sarcomas. MAGE expression on normal tissues is restricted to testis and placenta, both of which are immunologically privileged sites that may not support the induction of tolerance. Subsequent to the identification of the MAGE genes, two other unrelated gene families have been identified, called BAGE and GAGE, which encode melanoma antigens recognized by autologous CTL clones derived from melanoma patients. Like the MAGE genes, the BAGE and GAGE genes are normally silent in most normal tissues, except testis, but they are expressed on a variety of malignant tumors. Although there is no evidence that expression of any of these genes induces a protective tumor rejection response, it is clear that melanoma patients do have memory CTLs that are specific for several MAGE, BAGE, and GAGE peptides. Therefore, there is great interest in attempting to boost the CTL response to these antigens in melanoma patients by the use of "therapeutic" vaccines.

TUMOR ANTIGENS ENCODED BY GENOMES OF ONCOGENIC VIRUSES

Both RNA and DNA viruses are implicated in the development of tumors in both experimental animals and humans. Virally induced tumors usually contain integrated proviral genomes in their cellular genomes and often express viral genome-encoded proteins. These endogenously synthesized proteins can be processed, and complexes of processed viral peptides with class I MHC molecules may be expressed on the tumor cell surfaces. Thus, tumor cells expressing viral proteins can stimulate and become the targets of specific T cell responses. Structurally and biologically distinct antigens are produced by various DNA and RNA tumor viruses.

DNA viruses are involved in the development of various tumors in experimental animals and humans. Papovaviruses (including polyoma virus and simian virus [SV]40) and adenoviruses induce malignant tumors in neonatal or immunodeficient adult rodents. Several genes in these viruses cooperate to cause malignant transformation of infected cells. In humans, DNA viruses are associated with the development of several different tumor types. For example, the Epstein-Barr virus (EBV) is associated with B cell lymphomas, Hodgkin's lymphoma, and nasopharyngeal carcinoma (Box 18–2). Human papillomavirus (HPV) is associated with most human cervical carcinomas. The viral genes responsible for producing the malignant phenotype in these human tumors are partially defined.

In most cases, DNA virus–induced tumor cells are latently infected with virus and do not produce viral particles. Virally encoded protein antigens that are not components of infectious viral particles may be found in the nucleus, cytoplasm, or

BOX 18–2.   The Relationships Between Epstein-Barr Virus, Malignancy, and Immunodeficiency

Epstein-Barr virus (EBV) is a double-stranded DNA virus of the herpesvirus family. The virus is transmitted by saliva, infects nasopharyngeal epithelial cells and B lymphocytes, and is ubiquitous in human populations worldwide. It infects human B cells by binding specifically to the complement receptor type 2 (CR2), followed by receptor-mediated endocytosis. Two types of cellular infections can occur. In a lytic infection, viral DNA, RNA, and protein synthesis begin, followed by assembly of viral particles and lysis of the host cell. Alternatively, a latent non-lytic infection can occur, in which the viral DNA is incorporated into the host genome indefinitely. Various virally encoded antigens are detectable in infected cells. **Epstein-Barr nuclear antigens (EBNAs)** include at least six nuclear proteins that are expressed early in lytic infections and may also be expressed by some latently infected cells. **Latent membrane protein (LMP)** is expressed on the surface of latently infected cells. Other viral structural protein antigens are expressed within infected cells and on released viral particles during lytic infections, including **viral capsid antigens (VCAs).** Antibodies specific for VCAs are present in acutely infected, recovering, and remotely infected individuals.

EBV has profound effects on B lymphocyte growth characteristics *in vitro.* First, the virus is a potent, T cell–independent polyclonal activator of B cell proliferation. Second, EBV can immortalize normal human B cells so that they will proliferate in culture indefinitely. The resulting long-term B lymphoblastoid cell lines are latently infected with the virus and may express EBNA proteins, but they do not have a malignant phenotype. The molecular basis for these effects of EBV on B cells is presently unknown, but both EBNA proteins and LMP have been implicated in growth stimulation–transforming events. The binding of TNF receptor-associated factor (TRAF)-2 to the cytoplasmic tail of LMP is important for the effect of EBV on B cell proliferation (see Chapter 12, Box 12–5).

There is a wide spectrum of sequelae to infection by EBV. Most people are infected during childhood, they do not experience any symptoms, and viral replication is apparently controlled by humoral and T cell–mediated immune responses. In previously uninfected young adults, infectious mononucleosis typically develops during EBV infection. This disease is characterized by sore throat, fever, and generalized lymphadenopathy. Large, morphologically atypical T cells are abundant in the peripheral blood of infectious mononucleosis patients. These cells are activated CTLs with specificity for EBV-encoded antigens. Previously infected, healthy individuals harbor the virus for the rest of their lives in latently infected B cells and, perhaps, in nasopharyngeal epithelium. An estimated 1 of every million B cells in a previously infected individual is latently infected. EBV infection is also strongly implicated as one of the etiologic factors for the development of certain malignancies, including nasopharyngeal carcinoma in Chinese populations, Burkitt's lymphoma in equatorial Africa, and histologically variable B cell lymphomas in immunosuppressed patients.

There is compelling evidence that T cell–mediated immunity is required for control of EBV infections and, in particular, for the killing of EBV-infected B cells. First, individuals with deficiencies in T cell–mediated immunity often have uncontrolled, widely disseminated, and perhaps lethal acute EBV infections. Second, EBV-infected B cells isolated from patients with infectious mononucleosis can be propagated *in vitro* indefinitely, but only if the patient's T cells are depleted or inactivated by drugs such as cyclosporin A. In fact, immortalization of normal peripheral blood B cells by *in vitro* infection with EBV is usually successful only if the donor's T cells are removed or inactivated. Third, CTLs specific for EBV-encoded antigens, including EBNAs and LMP, are present in acutely infected and completely recovered infectious mononucleosis patients. Cloned CTL lines have been established *in vitro* that specifically lyse EBV-infected B

cells, and these CTLs most often recognize peptide fragments of EBNA and LMP proteins in association with class I MHC molecules. It is possible that EBV-specific T cells act *in vivo* to limit the polyclonal proliferation of infected B cells as well as to kill potentially immortalized clones of latently infected B cells. A loss of normal T cell–mediated immunity may allow latently infected B cells to progress toward malignant transformation. We discuss this hypothesis below.

The epidemiology and molecular genetics of Burkitt's lymphoma and other EBV-associated lymphomas have been the subject of intense investigation, and they offer fascinating insights into various aspects of viral oncogenesis and tumor immunity. Burkitt's lymphoma is a histologic type of malignant B cell tumor composed of monotonous, small, malignant B cells. The African form of the disease is endemic in regions where both EBV and malarial infection are common. In these regions, the tumor occurs frequently in young children, often beginning in the jaw. Virtually 100 per cent of African Burkitt's lymphoma patients have evidence of prior EBV infection, and their tumors almost all carry the EBV genome and express EBV-encoded antigens. Malarial infections in this population are known to cause T cell immunodeficiencies, and this may be the link between EBV infection and the development of lymphoma. Sporadic Burkitt's lymphoma occurs less frequently in other parts of the world, and although these B cell tumors are histologically similar to the endemic form, only approximately 20 per cent carry the EBV genome. Both endemic and sporadic Burkitt's lymphoma cells have reciprocal chromosomal translocations involving immunoglobulin gene loci and the cellular *myc* gene on chromosome 8 (Chapter 4, see Box 4–5).

B cell lymphomas occur at a high frequency in T cell–immunodeficient individuals, including individuals with congenital immunodeficiencies, AIDS patients, and kidney or heart allograft recipients receiving immunosuppressive drugs. Only some of these tumors can be called Burkitt's lymphomas, based on histologic appearance. Regardless of histologic appearance, many of these tumors are latently infected with EBV, like Burkitt's lymphoma. A smaller subset also contain *myc* translocations to Ig loci.

These observations can be synthesized into a hypothesis about the pathogenesis of EBV-associated B cell tumors. African children with malaria, allograft recipients, congenitally immunodeficient children, and AIDS patients all have deficiencies in normal T cell function. EBV infection proceeds unchecked in these individuals, and EBV-induced polyclonal proliferation of B cells is uncontrolled. This rapid, exuberant proliferation of B cells increases the chances of errors made during DNA replication, including translocations of oncogenes. The Ig loci are relatively accessible sites for translocations, compared with other loci in B cells. Translocation of the *myc* gene to the Ig locus leads to transcriptional deregulation and abnormal expression of *myc*, and this appears to be causally related to malignant transformation and outgrowth of a neoplastic clone of cells. Other genetic events can also lead to transformation after EBV-induced B cell proliferation, since many EBV-positive tumors do not have *myc* translocations, especially in allograft recipients and AIDS patients. In these cases, the proteins encoded by the integrated EBV genome may contribute to the malignant phenotype in EBV-positive lymphomas. This proposed scheme predicts that early in their course, EBV-associated B cell tumors may be polyclonal, since they arise from a polyclonally stimulated population of normal B cells. Later, one or a few clones may obtain selective growth advantages, perhaps because of deregulation of *myc* or other cellular or viral genes. As a result, the polyclonal proliferation evolves into a monoclonal or oligoclonal tumor. In fact, this has been shown to be the case by Southern blot analysis of Ig gene rearrangements in EBV-positive B cell tumors from immunosuppressed patients.

plasma membrane of the tumor cells. Specific immunity to DNA virus-encoded nuclear antigens protects against tumor development in animals. For example, SV40-induced tumors in mice express antigens that induce specific protective immunity against subsequent challenge with other SV40-induced tumors, but not against tumors induced by other viruses. Because these antigens are targets for tumor transplantation rejection, they are functionally defined as tumor transplantation antigens. These virally-encoded tumor antigens, however, are not unique for each tumor but are shared by all tumors induced by the same type of virus (see Table 18–1).

Although both humoral and T cell–mediated immune responses to DNA virus-encoded tumor antigens occur, only T cells specific for these antigens have been shown to mediate tumor rejection *in vivo*. One such virally encoded antigen that

stimulates tumor rejection is the T antigen, a nuclear protein in SV40-transformed cells. The T antigen is required to produce the malignant phenotype, and it is not part of infectious virus particles. Immunization of experimental animals with SV40 virus induces protective immunity against the development of SV40-induced tumors, and this immunity is mediated by T antigen–specific class I MHC–restricted CTLs. Adenovirus-induced rodent tumors express a virally encoded protein called E1A, which is found largely in the nucleus and is the principal determinant of the transformed phenotype of the infected cells. E1A is not part of infectious adenovirus particles. When class I–restricted CTLs specific for a peptide derived from the E1A protein are adoptively transferred into mice with adenovirus-induced tumors, these CTLs kill the tumors (Fig. 18–4).

A protective role of the immune system in

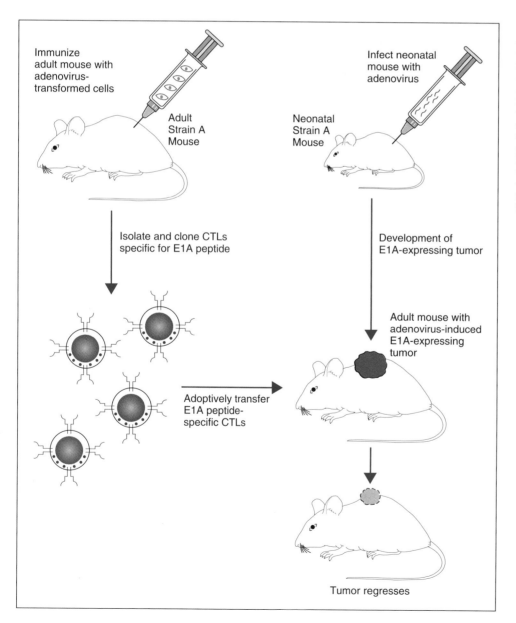

Immunize adult mouse with adenovirus-transformed cells

Adult Strain A Mouse

Isolate and clone CTLs specific for E1A peptide

Adoptively transfer E1A peptide-specific CTLs

Infect neonatal mouse with adenovirus

Neonatal Strain A Mouse

Development of E1A-expressing tumor

Adult mouse with adenovirus-induced E1A-expressing tumor

Tumor regresses

FIGURE 18–4. Viral antigen-specific cytolytic T lymphocytes (CTLs) kill virally infected tumors *in vivo*. If neonatal mice are infected with adenovirus, they develop malignant tumors as adults, and these tumors express the virally encoded E1A protein. The CTL clones isolated from a syngeneic mouse immunized with E1A-expressing cells can kill these E1A-expressing tumors when the CTLs are adoptively transferred to the tumor-bearing animal.

controlling the growth of DNA virus–induced tumors is suggested by the higher frequency of these tumors in immunodeficient individuals. In humans, EBV-associated lymphomas and HPV-associated skin cancers arise much more frequently in immunosuppressed individuals, such as allograft recipients receiving immunosuppressive therapy and acquired immunodeficiency syndrome (AIDS) patients, than in normal individuals. Adenovirus infection induces tumors much more frequently in neonatal or nude (congenitally T cell–deficient) mice, compared with normal adult mice. Thus, a competent immune system may play a role in tumor immunosurveillance because of its ability to recognize and kill virally infected cells. Although no DNA virus–encoded tumor antigen is known to induce protective immunity in human tumors, there is great interest in the products of two HPV genes, E6 and E7, which are constitutively expressed in cervical squamous cell carcinomas and are required for maintaining the malignant phenotype. Several E6- and E7-derived peptides bind to certain class I MHC alleles with high affinity, and they sensitize cells expressing these alleles for lysis by E6- or E7-specific CTL clones. Furthermore, at least one human cervical carcinoma cell line can be specifically lysed by these E6- and E7-specific CTLs. Accordingly, clinical trials of E6 and E7 peptide vaccines for treatment of HPV-positive cervical carcinoma have been started.

One of the clearest examples of viral oncogenesis is the development of tumors in animals infected with certain types of retroviruses (RNA tumor viruses). Some of these viruses carry well-defined oncogenes, induce tumors in days to weeks after infection, and are called **acute transforming retroviruses.** Examples of these acute transforming retroviruses include Rous sarcoma virus (carrying the *src* oncogene), avian myelocytomatosis virus (carrying the *myc* oncogene), and Kirsten murine sarcoma virus (carrying the K-*ras* oncogene). Other retroviruses, such as the murine leukemia viruses, cause tumors months after infection and do not carry any well-defined oncogenes. These slow-transforming retroviruses may cause tumors by inserting near, and dysregulating transcription of, cellular genes that are responsible for growth control and differentiation. The genomes of retroviruses are small, and they may express a limited number of potentially immunogenic proteins in their host tumor cells. These proteins include products of the envelope (*env*) gene, polymerase (*pol*) gene, core protein (*gag*) gene, and, in the case of acute transforming retroviruses, the oncogene. Retroviral oncogene products theoretically have the same potential antigenic properties as mutated cellular oncogenes, but they generally do not evoke strong immune responses *in vivo*. In contrast, humoral and cell-mediated immune responses to the *env* and *gag* products on tumor cells can be observed experimentally. Furthermore, *env* and *gag* products can stimulate CTL-mediated

rejection of transplanted tumors. These antigens are shared by all tumors induced by the same type of retrovirus.

The only well-established human RNA tumor virus is human T cell lymphotropic virus-1 (HTLV-1), which is the etiologic agent of adult T cell leukemia/lymphoma (ATL), an aggressive malignant tumor of CD4$^+$ T cells. Although immune responses specific for HTLV-1 encoded antigens have been demonstrated, it is not clear whether they play any role in protective immunity against development of tumors in virally infected people. Furthermore, ATL patients are often profoundly immunosuppressed, perhaps because of an effect of the virus on CD4$^+$ T cells, which the virus preferentially infects.

TISSUE-SPECIFIC DIFFERENTIATION ANTIGENS RECOGNIZED BY TUMOR-SPECIFIC T CELLS

Tissue-specific differentiation antigens are proteins expressed by normal cells and are characteristic of a particular tissue type at a particular stage of normal differentiation of that tissue. Tumors arising from one type of tissue often express the differentiation antigens of that tissue. Since these antigens are part of normal cells, they would be expected to induce self-tolerance and not to stimulate immune responses against the tumors on which they are expressed. This is, in fact, the case for many such molecules, and they have been called tumor antigens only because they stimulate antibody responses in other species, as discussed later. It is surprising that the cloning of human melanoma antigens recognized by T cells from melanoma patients has turned up several tissue-specific antigens that are expressed by normal melanocytes and melanomas. The first such antigen identified was tyrosinase, an enzyme involved in melanin biosynthesis, which is expressed only in normal melanocytes and melanomas. Tyrosinase peptides eluted from melanoma cell class I MHC molecules sensitize targets for lysis by anti-melanoma CTLs. An interesting feature of at least one of the tyrosinase peptides eluted from melanoma cell MHC molecules is the presence of an aspartate residue that represents a post-translational modification of an asparagine residue encoded by the tyrosinase gene; the gene, however, is not mutated in the melanoma cells. Such modifications may be characteristic of tumor cells but not normal cells, reflecting the aberrant expression of modifying enzymes in the tumors, and they could explain why the product of a normal tissue-specific gene induces an immune response. Both class I MHC–restricted CD8$^+$ CTL clones and class II MHC–restricted CD4$^+$ T cell clones from melanoma patients recognize peptides derived from tyrosinase, raising the possibility that tyrosinase vaccines may stimulate both helper and cytolytic T cell responses to melanomas. MART-1 and gp100 or Pmel17 are two other proteins that are expressed exclusively in melanocytes and melanomas and that are recognized by melanoma-specific CTLs. Both are transmembrane proteins of unknown function, and peptides derived from both are recognized by many independently derived melanoma-specific CTL lines. Adoptive transfer of gp100-specific T cells into melanoma patients is effective in reducing the tumor burden in some patients.

Tumor Antigens Recognized by Antibodies

Some molecules on tumor cell surfaces can elicit autologous antibody responses. In addition, some tumor molecules can be recognized by xenogeneic antibodies produced by immunizing an animal of one species with tumor cells from another species. Such molecules do not necessarily stimulate immune responses in the tumor host, but the antibodies that bind to them are potentially valuable in the diagnosis and therapy of tumors. Many cell surface molecules on tumors that have been identified by xenogeneic antibodies are called tumor antigens, but almost all of them are shared by different tumors arising from the same types of cells, and most, if not all, may also be found on some normal cells or benign tumor cells. Therefore, these molecules are tumor-associated antigens and not tumor-specific antigens. Most of these molecules do not stimulate immune responses in tumor hosts because they are normal self proteins, and they should be fully capable of inducing tolerance. It is surprising that there are cases in which antibodies and T cells specific for these tumor-associated antigens are found in tumor hosts, but there is no evidence that the antibody responses are protective. Several classes of these antigens exist, and many different ones may be expressed on the same tumor. We will consider some of the tumor antigens recognized by antibodies, dividing our discussion on the basis of biochemical characteristics or patterns of tissue distribution of these antigens.

ONCOFETAL ANTIGENS

Oncofetal antigens are proteins normally expressed on developing (fetal) but not adult tissues. They are expressed on tumor cells as a result of the derepression of genes by unknown mechanisms. The importance of oncofetal antigens is that they provide markers that aid in tumor diagnosis. As techniques for detecting these antigens have improved, it has become clear that their expression in adults is not strictly limited to tumors. The proteins can be found in tissues in various inflammatory conditions, and even in small quantities in normal tissues. Furthermore, oncofetal antigens are not antigenic in the host, since they are expressed as self proteins during development. Nonetheless, the study of oncofetal antigens has proved useful for diagnostic purposes and has provided some insights into tumor biology. The two most thoroughly characterized oncofetal antigens are **carcinoembryonic antigen** (**CEA**, CD66) and **alpha-fetoprotein (AFP).**

CEA is a highly glycosylated 180 kD integral membrane protein that is a member of the Ig superfamily. CEA is also released into the extracellular fluid. Normally, high CEA expression is restricted to the gut, pancreas, and liver during the first two trimesters of gestation, and reduced expression is found in normal adult colonic mucosa and lactating breasts. CEA expression is greatly increased in colonic, pancreatic, gastric, and breast carcinomas, resulting in a rise in serum levels. Furthermore, post-translational processing of CEA may be altered in tumor cells. Serum CEA, detected by xenogeneic antibodies, is accordingly used to monitor the occurrence or recurrence of metastatic carcinoma after primary treatment. There is very little evidence for any significant anti-CEA humoral response in human cancer patients. Recent studies have demonstrated that CEA functions as an intercellular adhesion molecule, promoting the binding of tumor cells to one another. Thus, CEA may play a role in the way tumor cells interact with one another and with the tissues in which they are growing. The usefulness of CEA as a diagnostic marker for cancer is somewhat limited by the fact that serum CEA can also be elevated in the setting of non-neoplastic diseases such as chronic inflammatory conditions of the bowel or liver.

AFP is a 70 kD α-globulin glycoprotein normally synthesized and secreted in fetal life by the yolk sac and liver. Fetal serum concentrations can be as high as 2 to 3 mg/ml, but in adult life the protein is replaced by albumin, and only low levels are present in the serum. Serum levels of AFP can be significantly elevated in patients with hepatocellular carcinoma, germ cell tumors, and, occasionally, gastric and pancreatic cancers. An elevated serum AFP level is a useful indicator of advanced liver or germ cell tumors, or of recurrence of these tumors after treatment. Furthermore, the detection of AFP in tissue sections by immunohistochemical techniques can help in the pathologic identification of tumor cells. The diagnostic value of AFP as a tumor marker is limited by the fact that elevated serum levels are also found in non-neoplastic liver diseases, such as cirrhosis. As with CEA, there is little evidence that AFP-specific anti-tumor immune responses occur in cancer patients.

ALTERED GLYCOLIPID AND GLYCOPROTEIN ANTIGENS

Most human and experimental tumors express higher levels of and/or abnormal forms of surface glycoproteins or glycolipids, including gangliosides, blood group antigens, and mucins. The abnormal forms are a result of alterations in the sequential addition of carbohydrate moieties to core protein or lipid molecules. Some aspects of the malignant phenotype of tumors, including tissue invasion and metastatic behavior, may in part be a function of altered cell surface properties that result from abnormal glycolipid and glycoprotein synthesis. Many antibodies have been raised in animals that recognize carbohydrate groups or abnormally exposed peptide cores of these molecules. Although most of the epitopes recognized by these antibodies are not specifically expressed on tumors, there is a relative abundance on cancer cells. This class of TAAs continues to be a preferred target for antibody-based approaches to cancer therapy.

The gangliosides are neuraminic acid–containing glycosphingolipids, certain forms of which are expressed at particularly high levels on melanomas and some brain tumors compared with the normal cells from which these tumors are derived. For example, the gangliosides GM$_2$ and GD$_2$ are present at high density on the surface of melanomas and therefore are considered potential targets for immunotherapy. The tumor gangliosides have been studied with xenogeneic antibodies, but in addition, normal individuals and melanoma patients appear not to be tolerant to these molecules, since they make their own anti-ganglioside antibodies. In fact, there have been successful attempts at inducing high IgM antibody responses in melanoma patients by immunizing with GM$_2$ plus adjuvant vaccine, and this has correlated with prolonged clinical remission of the tumors.

Blood group antigens are carbohydrate epitopes on glycosphingolipid or glycoprotein molecules expressed on the surfaces of blood cells and epithelial cells (see Box 17–2, Chapter 17). Different epitopes are created depending on the expression of glycosyltransferases, and some of these antigens are aberrantly expressed in carcinomas, particularly the T and sialylated Tn antigens. Soluble forms of these antigens are being used as experimental cancer vaccines. These and other carbohydrate antigens are not likely to stimulate T cells required for isotype switching and memory B cell generation, and therefore their long-term usefulness as immunogens for cancer immunotherapy may be limited.

Mucins are high-molecular-weight proteoglycans containing numerous O-linked carbohydrate side chains on a core polypeptide. Because of their complicated branching structure, there is potential for great antigenic variability. Dysregulated expression in tumors of the numerous enzymes that synthesize these carbohydrate side chains can lead to the appearance of relatively tumor-specific epitopes, either on the carbohydrate side chains themselves or on the abnormally exposed polypeptide core. Several mucins have been the focus of diagnostic and therapeutic studies, including CA-125 and CA-19-9, expressed on ovarian carcinomas, and MUC-1, expressed on breast carcinomas. Unlike many mucins, MUC-1 is an integral membrane protein that is normally expressed only on the apical surface of breast ductal epithelium, a site that is relatively sequestered from the immune system. In ductal carcinomas of the breast, however, the

TABLE 18–3. Examples of Tissue-Specific Tumor Antigens Used in Clinicopathologic Analysis of Tumors

Tissue of Origin	Tumor	Antigens
B lymphocytes	B cell leukemias and lymphomas	CD10 (CALLA) Immunoglobulin
T lymphocytes	T cell leukemias and lymphomas	Interleukin-2 receptor (α chain), T cell receptor, CD45R, CD4$^+$/CD8$^+$
Prostate	Prostatic carcinoma	Prostate-specific antigen Prostatic acid–phosphatase
Neural crest–derived	Melanomas	S-100
Epithelial cells	Carcinomas	Cytokeratins

Abbreviations: CALLA, common acute lymphoblastic leukemia antigen.

molecule is expressed in an unpolarized fashion and contains new, highly tumor-specific carbohydrate and peptide epitopes detectable by mouse monoclonal antibodies. The peptide epitopes induce both antibody and T cell responses in cancer patients and are therefore being considered as candidates for anti-tumor vaccines.

TISSUE-SPECIFIC DIFFERENTIATION ANTIGENS

We have discussed tissue-specific antigens recognized by T cells earlier in the chapter. The clinical significance of differentiation antigens on tumors relates to their use as targets for immunotherapy, and also as diagnostic markers of the tissue of origin of tumors. The histologic appearance of a tumor may not be characteristic enough to permit a diagnosis of the type of normal tissue from which the tumor arose. Therefore, antibody probes for the expression of tissue-specific antigens may be required. For example, lymphomas arising from the malignant transformation of developing B cells often may be diagnosed as B cell lineage tumors by the detection of a surface marker characteristic of normal pre-B cells, called CD10 (previously called common acute lymphoblastic leukemia antigen, or CALLA). Tumors arising from more mature B cells can be characterized by the presence of surface immunoglobulin. Examples of tissue-specific antigens expressed on tumors are listed in Table 18–3.

EFFECTOR MECHANISMS IN ANTI-TUMOR IMMUNITY

Both humoral and cell-mediated immune responses to tumor antigens have been demonstrated *in vivo,* and many immunologic effector mechanisms have been shown to kill tumor cells *in vitro.* The challenge for tumor immunologists is to determine which, if any, of these effector mechanisms contributes to protective immune responses against human tumors and to enhance these effector mechanisms in ways that are relatively tumor specific. In this section, we briefly review the evidence for tumor killing by these various effector mechanisms and discuss which are the most likely to be relevant to human tumors.

T Lymphocytes

CTLs provide effective anti-tumor immunity *in vivo,* as demonstrated by the experimental tumor transplantation studies discussed earlier. In fact, CTL-mediated rejection of transplanted tumors is the only established example of completely effective specific anti-tumor immunity *in vivo.* In these cases, the effector cells are predominantly CD8+ CTLs, which are phenotypically and functionally identical to the CTLs responsible for killing virus-infected or allogeneic cells described in Chapters 13 and 17. As discussed previously, CTLs may perform a surveillance function by recognizing and killing potentially malignant cells that express peptides which are derived from mutant cellular or oncogenic viral proteins and which are presented in association with class I MHC molecules.

The importance of this form of immunosurveillance for common, non-virally induced tumors is not established since such tumors do not arise more frequently in T cell–deficient animals or people, or in patients with drug- or virus-induced suppression of T cell immunity. On the other hand, as discussed earlier, tumor-specific CTLs can be isolated from animals and humans with already established tumors. For example, peripheral blood lymphocytes from patients with advanced carcinomas and melanomas contain CTLs that can lyse explanted tumors from the same patients. Furthermore, mononuclear cells derived from the inflammatory infiltrate in human solid tumors, called **tumor-infiltrating lymphocytes (TILs),** also include CTLs with the capacity to lyse the tumor from which they were derived. Although these CTL responses may not be effective in eradicating most tumors on their own, enhancement of CTL responses is a promising approach for anti-tumor therapy in the near future. CTL-mediated surveillance against cells infected with oncogenic viruses probably does occur naturally, as suggested by the fact that tumors associated with viral infections occur more frequently in immunosuppressed patients.

Although CD4+ helper T cells are not generally cytotoxic to tumors, they may play a role in anti-tumor responses by providing cytokines for effective CTL development (see Chapter 13). In addition, helper T cells that are activated by tumor antigens may secrete tumor necrosis factor (TNF) and interferon-γ (IFN-γ), which can increase tumor cell class I MHC expression and sensitivity to lysis by CTLs. It is also possible that tumor-specific helper T cells may promote delayed type hypersensitivity (DTH) responses against tumors. A minority of tumors that express class II MHC molecules may directly activate tumor-specific CD4+ helper T cells. More commonly, class II–expressing professional antigen-presenting cells (APCs) process and present internalized proteins derived from dying or phagocytosed tumor cells. CD4+ helper T cells from some tumor-bearing individuals are specific for tumor antigens, including tyrosinase peptides and mutated Ras protein, but a thorough analysis of other antigens that tumor-specific helper T cells may recognize has not been accomplished.

Natural Killer Cells

NK cells may be effector cells of both innate and specific immune responses to tumors. NK cells can be activated by direct recognition of tumors, or as a consequence of cytokines produced by tumor-specific T lymphocytes. They use the same lytic mechanisms as CTLs to kill cells, but they do not express T cell antigen receptors, and they have a broad range of specificities (see Chapter 13). NK cells can lyse both virally infected cells and certain tumor cell lines, especially hematopoietic tumors, *in vitro.* In fact, lysis of such cell lines serves as the major bioassay for NK activity. There appears to be a degree of specificity to NK killing, since many virally infected cells or tumor cells and most normal cells are not susceptible to NK lysis *in vitro.* NK cell recognition of MHC molecules on a potential target inhibits NK cell lysis of that target (see Chapter 13). Thus, the down-regulation of MHC expression on many tumor cells, which may allow them to escape CTL lysis, makes them particularly good targets for NK cells. In addition, NK cells can be targeted to antibody-coated cells because they express low-affinity Fc receptors (FcγRIII or CD16) for IgG molecules. The tumoricidal capacity of NK cells is increased by cytokines, including interferons, TNF, interleukin-2 (IL-2), and interleukin-12. Therefore, their role in anti-tumor immunity may depend on the concurrent stimulation of T cells and macrophages that produce these cytokines. There is great interest in the role of IL-2-activated NK cells in tumor killing. These cells, called **lymphokine-activated killer (LAK) cells,** are derived *in vitro* by culture of peripheral blood cells or tumor-infiltrating lymphocytes from tumor patients with high doses of IL-2 (see Chapter 13). LAK cells exhibit a markedly enhanced and nonspecific capacity to lyse other cells, including tumor cells. The use of LAK cells in adoptive immunotherapy of tumors will be discussed later.

A role for NK cells in tumor immunity *in vivo* is suggested by a variety of indirect evidence. For example, the incidence of tumors in different strains of inbred mice, or in mice of different ages, correlates inversely with the functional capacity of NK cells in these mice. It is interesting that T cell–deficient nude mice have normal or elevated numbers of NK cells, and they do not have a high incidence of spontaneous tumors. NK cells may play a role in immunosurveillance against developing tumors, especially those expressing viral antigens. However, a high level of NK activity is not present in the cellular infiltrates associated with solid human tumors, before *in vitro* expansion with IL-2.

Macrophages

Macrophages are potentially important cellular mediators of anti-tumor immunity. Their role is largely inferred from the demonstration that activated macrophages can preferentially lyse tumor cells, and not normal cells, *in vitro*. Like NK cells, macrophages express $Fc\gamma$ receptors (high affinity $Fc\gamma RI$ and $Fc\gamma RIII$), and they can be targeted to tumor cells coated with antibody. There are probably several mechanisms of macrophage killing of tumor target cells that are essentially the same as the mechanisms of macrophage killing of infectious organisms. These mechanisms include the release of lysosomal enzymes, reactive oxygen metabolites, and, in mice, nitric oxide.

Activated macrophages also produce the cytokine **tumor necrosis factor (TNF)**, which, as its name implies, was first characterized as an agent that can kill tumors but not normal cells. The various actions of TNF were discussed in Chapter 12. There is some evidence that TNF plays a role in macrophage-mediated tumor killing. For example, tumor cells selected *in vitro* for resistance to killing by TNF are often also resistant to killing by macrophages. Killing by both macrophages and TNF is slow (24 to 48 hours), can be augmented by protein or RNA synthesis inhibitors, and is due to apoptosis.

TNF kills tumors by direct toxic effects and indirectly by effects on tumor vasculature. Direct toxicity depends on binding of TNF to high-affinity cell surface receptors onto tumor cells. The toxicity is in part due to activation of a cell death pathway similar to that induced by Fas ligand binding to Fas (see Chapter 10). The toxicity may also be a result of the production of free radicals. Normal cells respond to TNF by synthesizing superoxide dismutase, an enzyme that participates in the inactivation of free radicals. In contrast, many tumor cells fail to make superoxide dismutase in response to TNF. Direct toxic effects of TNF may also involve disruption of cytoskeletal proteins, or interference with gap junction formation. TNF can indirectly cause tumor necrosis by inducing thrombosis in tumor vessels *in vivo*. This is suggested by the observations that even tumor cells lacking TNF receptors can be eradicated in mice by treatment with TNF, and that TNF selectively eradicates vascularized tumors and is much less effective in killing avascular tumors. Histologically, the response to TNF, described as hemorrhagic necrosis, looks very much like the localized Shwartzman re-

action described in Chapter 12 (see Box 12–3). This resemblance has led to the suggestion that TNF acts selectively on tumor vessels to produce a Shwartzman-like reaction, causing thrombosis of the vessels and ischemic necrosis of tumors. Tumor vessels may be already "primed" to trigger the Shwartzman response once they encounter TNF. Some tumor-derived angiogenic factors, such as vascular endothelial growth factor, potentiate endothelial cell responses to TNF. These tumoricidal effects of TNF have been exploited in clinical practice, as we will discuss later.

Antibodies

Antibodies are probably less important than T cells in mediating effective anti-tumor immune responses, but, as we have discussed, tumor-bearing hosts do produce antibodies against various tumor antigens. In some instances, these antibody responses are specific for viral antigens. For example, patients with EBV-associated lymphomas have serum antibodies against EBV-encoded antigens expressed on the surface of their tumor cells. In other cases, human cancer patients produce antibodies against their own tumors that can be used for *in vitro* "autologous typing" to identify tumor antigens. In these cases, the antigens recognized are almost always present on normal tissues as well. No evidence exists for a protective role of such humoral responses against tumor development or growth. Hybridomas have been prepared from the B cells of tumor patients that produce monoclonal antibodies reactive with antigens on the patients' tumors. Again, these antibodies are not specific for antigens expressed exclusively on tumor cells. The potential for antibody-mediated destruction of tumor cells has largely been demonstrated *in vitro* and is attributable to complement activation, or to antibody-dependent cell-mediated cytotoxicity in which Fc receptor–bearing macrophages or NK cells mediate the killing. Whether or not these Ig-dependent mechanisms of tumor killing play a role *in vivo* remains unknown.

MECHANISMS OF EVASION OF THE IMMUNE SYSTEM BY TUMORS

Although malignant tumors may express protein antigens that are recognized as foreign by the tumor host, and although immunosurveillance may limit the outgrowth of some tumors, it is clear that the immune system often does not prevent the occurrence of lethal human cancers. The simplest explanation for this may be that the rapid growth and spread of a tumor overwhelms the effector mechanisms of immune responses. In addition, many tumors acquire the ability to evade immune responses, much like microbes. A major focus of tumor immunology is to understand the ways in which tumor cells may evade immune destruction, with the hope that interventions can be designed to increase the immunogenicity of tumors or the responses of the host. The process of evasion, often called **tumor escape,** may be a result of several mechanisms.

1. Class I MHC expression can be downregulated on tumor cells so that they cannot form complexes of processed tumor antigen peptides and MHC molecules required for CTL recognition. There are clear demonstrations that increasing class I MHC expression on tumor cells results in increased susceptibility of these cells to CTL lysis *in vitro* and decreased tumorigenicity *in vivo* (Fig. 18–5). Furthermore, transfecting class I MHC genes into murine tumor cells often decreases their ability to form tumors when they are transplanted into healthy animals. Viruses have evolved ways to decrease class I MHC gene expression and assembly with peptides, thereby blocking the presentation of viral antigens to CTLs. Exam-

ples include the effects of adenovirus E1A on class I transcription, and herpes simplex virus inhibition of TAP function (see Chapters 6 and 16). These same mechanisms that inhibit surface expression of peptide–class I MHC complexes in infected normal cells may be operative in virally induced tumors. However, when the level of MHC expression on a broad range of experimental or human tumor cells is compared with the *in vivo* growth of those cells, no clear correlation exists. For example, metastatic tumors, which presumably have evaded immune attack, do not on the average express any fewer MHC proteins than non-metastatic tumors.

2. Even tumors that express peptide–class I MHC complexes that are recognized by host CTLs may fail to activate the CTLs for two reasons. First, because most human tumor cells do not express class II MHC molecules, they cannot directly activate tumor-specific CD4+ helper T cells. Anti-tumor CTL activity is likely to be partly dependent on

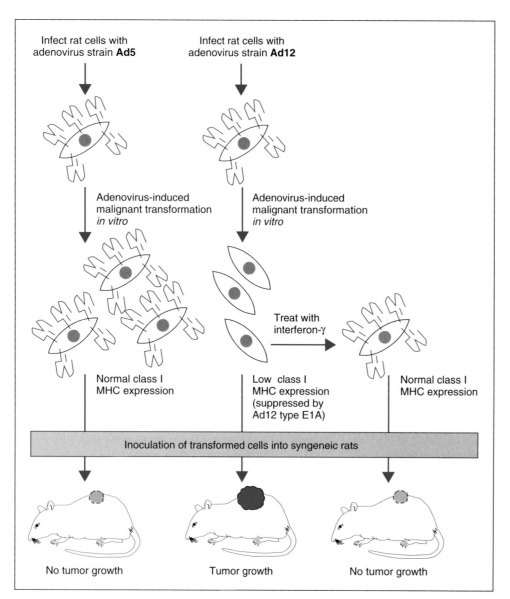

FIGURE 18–5. Relationship between class I MHC expression and tumorigenicity in adenovirus-induced tumors. Rat cells that are malignantly transformed *in vitro* by infection with the Ad5 strain of adenovirus express normal levels of class I MHC molecules and are not tumorigenic in syngeneic rats. In contrast, rat cells that are malignantly transformed *in vitro* by infection with the Ad12 strain of adenovirus express low levels of class I MHC molecules and are tumorigenic. Ad12-infected tumors can be induced to express higher levels of class I MHC molecules by interferon-γ, and this treatment renders them non-tumorigenic. The interpretation of this experiment is that class I MHC expression on a virally induced tumor permits the host animal to mount a protective immune response, presumably against a virally encoded antigen presented by the tumor cell in association with class I MHC molecules.

Within figure:

Infect rat cells with adenovirus strain **Ad5**

Infect rat cells with adenovirus strain **Ad12**

Adenovirus-induced malignant transformation *in vitro*

Adenovirus-induced malignant transformation *in vitro*

Treat with interferon-γ

Normal class I MHC expression

Low class I MHC expression (suppressed by Ad12 type E1A)

Normal class I MHC expression

Inoculation of transformed cells into syngeneic rats

No tumor growth

Tumor growth

No tumor growth

signals provided by helper T cells (see Chapter 13). If professional APCs do not adequately infiltrate these tumors, take up and present tumor antigens, and activate helper T cells, then maximal anti-tumor CTL differentiation will not occur. Second, a lack of costimulators on tumor cells may impair T cell activation. Most tumors are derived from tissues that do not express the costimulators that provide second signals for helper T cell activation (see Chapter 7). Furthermore, CTL activation also requires costimulation by cell surface molecules, such as B7-1 or B7-2, that are lacking on tumor cells. Tumor cell antigen presentation to T cells in the relative absence of costimulators may induce peripheral tolerance (clonal anergy) in tumor-specific T lymphocytes (see Chapter 10).

3. Tumor products may suppress anti-tumor immune responses. An example of an immunosuppressive tumor product is transforming growth factor-β (TGF-β), which is secreted in large quantities by many tumors and which inhibits a wide variety of lymphocyte and macrophage functions (see Chapter 12). Some tumors secrete IL-10, which also downregulates macrophage function.

4. A host may be tolerant to some tumor antigens. This may be because of neonatal exposure to such antigens or because the tumor cell may present its antigens to the immune system in a tolerogenic form. Neonatally induced tolerance has been demonstrated for tumors caused by the murine mammary tumor virus. This virus causes breast tumors in adult mice that have acquired the viral infection during neonatal life by nursing. Although these tumors are not seen as foreign in these mice and do not stimulate an immune response (because of neonatal tolerance), they are highly immunogenic when transplanted to syngeneic, virus-free adult mice. Another example of the relationship between neonatally induced tolerance and the growth of virally induced tumors is seen in mice that express the SV40 viral genome as a transgene. Transgenic mice that express the SV40 genes during early development have a high incidence of tumors, and this is correlated with tolerance to SV40 T antigen. In contrast, other SV40-transgenic mice in which expression of the transgene is delayed until later in life are not tolerant to the SV40 T antigen and have a low incidence of tumors.

5. Anti-tumor immunity can result in selection of mutant tumor cells that no longer express immunogenic peptide-MHC complexes. This could occur as a result of mutations or deletions in the genes encoding the tumor antigens, especially if the protein products of such genes are not critical for the malignant phenotype of the tumor. Alternatively, immunoselection may favor the growth of tumor cells with mutations or deletions in MHC genes needed for presentation of antigenic peptides. Given the generally high mitotic rate of tumor cells and their relative genetic instability, such mutations or deletions are theoretically likely.

Analysis of tumors that are serially transplanted from one animal to another has shown that the loss of antigens recognized by tumor-specific CTL clones correlates with increased growth and metastatic potential.

6. The loss of surface expression of tumor antigens as a result of antibody binding, called antigenic modulation, leads to acquired resistance to immune effector mechanisms. Antigenic modulation is due to endocytosis or shedding of the antigen-antibody complexes. If antigenic modulation is caused by an anti-tumor antibody that does not fix complement, it may protect tumor cells from other complement-activating antibodies. Antigenic modulation is perhaps most relevant as a problem that complicates attempted passive immunotherapy with anti-tumor antibodies.

7. The kinetics of tumor growth may allow for the establishment of immunologically resistant tumors before an effective immune response develops. This phenomenon, called "sneaking through," can be experimentally modeled by transplantation studies. Transplantation of small numbers of tumor cells leads to establishment of lethal tumors (i.e., lack of rejection), whereas larger transplants of the same tumor are rejected. One presumed reason for this apparent contradiction is that small doses of tumor antigens are not sufficiently stimulatory to the immune system, and by the time many tumor cells grow in the transplant recipient, mutations in tumor antigen genes may have occurred that reduce the chance of immune recognition.

8. Tumor cell surface antigens can be hidden from the immune system by glycocalyx molecules, including sialic acid–containing mucopolysaccharides. This process is called antigen masking, and it may be a consequence of the fact that tumor cells often express more of these glycocalyx molecules than do normal cells. Similarly, some tumors may shield themselves from the immune system by activating the coagulation system, thereby investing themselves in a "fibrin cocoon."

IMMUNOTHERAPY OF TUMORS

The potential for treating cancer patients by immunologic approaches has held great promise for immunologists and cancer biologists over much of this century. Advances in our understanding of the immune system and in defining antigens on tumor cells that are targets of T cell immunity have encouraged many new strategies. Several of these approaches are aimed at augmenting weak host immune responses to tumor antigens. In this section, we describe some of the modes of tumor immunotherapy that have been tried in the past or are currently being investigated. The discussion is divided into sections on: induction and augmentation of active immune responses against an individual's tumor; and passive immunotherapy in which tumor-specific cells or antibodies raised *ex vivo* are administered to tumor patients.

Stimulation of Active Host Immune Responses to Tumors

NONSPECIFIC STIMULATION OF THE IMMUNE SYSTEM

Nonspecific immune stimulation of tumor patients with adjuvants, such as the bacille Calmette-Guérin (BCG) mycobacterium, injected at the sites of tumor growth has been tried for many years. This treatment serves mainly to activate macrophages and thereby promotes macrophage-mediated tumor cell killing. Oncologists are still assessing the potential of local BCG administration in bladder carcinomas and melanomas. An experimental approach to nonspecific immune stimulation is the administration of low doses of anti-CD3 antibodies to mice with transplanted fibrosarcomas. This treatment results in polyclonal activation of T cells and, consequently, prevention of tumor growth. Cytokine therapies, discussed below, represent another method of enhancing immune responses in a nonspecific manner.

VACCINATION WITH KILLED TUMOR CELLS

Induction of protective immunity to tumors can theoretically be accomplished by active immunization procedures with purified tumor antigens or cells expressing these antigens. One method is to inject killed or irradiated tumor cells together with nonspecific adjuvants. The rationale for this approach is that antigen-bearing tumor cells may be able to induce a strong immune response if they are delivered to the immune system under conditions that favor lymphocyte activation. Memory T cells expanded by such immunizations, it is hoped, would limit the growth of already established tumors. A wide variety of protocols has been tried, but the efforts have been largely unsuccessful, probably because the tumor cell vaccines do not effectively activate specific CTL responses. A variation on this approach, which we will discuss later, is immunization with live tumor cells that have been transfected with genes that render the cells more immunogenic.

VACCINATION WITH TUMOR ANTIGENS OR PEPTIDES

The identification of peptides recognized by human tumor-specific CTLs, and the cloning of genes that encode tumor-specific antigens recognized by CTLs, described earlier, has provided the basis for another form of active anti-tumor immunization, namely the introduction of specific tumor antigens into tumor-bearing patients using immunogenic vaccine preparations. The rationale for this "therapeutic" vaccine approach is the likelihood that the tumor-specific memory T cells in tumor patients could be actively and specifically stimulated by such vaccines, and further development of memory T cells could be enhanced. Immunization with these antigens can be theoretically accomplished by direct administration of purified peptide or protein, by immunization with cells expressing recombinant genes which encode these antigens, or by injection of expression vectors encoding tumor antigens (DNA vaccines). One drawback of peptide vaccines is that the administered tumor peptides are unlikely to efficiently replace other non-immunogenic peptides already bound to MHC molecules on tumor cell surfaces. Intact proteins, on the other hand, are most likely to enter the class II MHC pathway of antigen presentation, and to activate CD4+ helper T cells, but not CD8+ CTLs. DNA vaccines may be the best way to induce CTL responses because the DNA can be taken up by host cells and the encoded antigens will be translated in the cytoplasm and enter the class I MHC pathway of antigen presentation. For unique tumor antigens, such as may occur with random point mutations in cellular genes, such immunization protocols may be impractical because they would require initial identification of these antigens from individual tumors by use of T cell probes. On the other hand, tumor antigens shared by many tumors, such as the MAGE, tyrosinase, and gp100 antigens on melanomas or position 12-mutated Ras proteins, are potentially useful immunogens for many different cancer patients. It would be possible to determine whether a patient's tumor expresses a particular common tumor antigen by use of polymerase chain reaction (PCR)-based detection of the relevant mutated gene, and human leukocyte antigen (HLA) typing could be used to determine whether the patient expresses MHC molecules that bind the immunodominant peptide from that antigen. As we have mentioned earlier in the chapter, vaccine trials are already under way for a variety of tumors.

Immunization with cells expressing high levels of tumor antigen genes has the advantage that such antigens will have ready access to class I MHC pathways of antigen presentation and may be more efficiently presented to CTLs than will antigens in cell-free vaccines. This process could be accomplished by linking the genes encoding the antigens to active promoters, transfecting the constructs into host tumor cells or other cell types *ex vivo,* and reintroducing the transfected cells into the patient. Alternatively, recombinant vaccinia virus vectors with tumor antigen gene inserts can be used to achieve expression of tumor antigens in the patient. Infection of animals with recombinant vaccinia virus–containing tumor antigen genes has already been shown to establish protective immunity to subsequent challenges with tumors expressing those antigens, but this method is not yet approved for humans.

The development of virally induced tumors can be blocked by vaccination with viral antigens or attenuated live virus. This approach has been successful in reducing the incidence of feline leukemia virus–induced hematologic malignancies in cats and in preventing the herpesvirus-induced lymphoma called Marek's disease in chickens. In humans, the ongoing vaccination program against the hepatitis B virus (HBV) may reduce the incidence of hepatocellular carcinoma, a cancer associated with HBV infection of the liver.

AUGMENTATION OF HOST IMMUNITY TO TUMORS WITH COSTIMULATORS AND CYTOKINES

As we have discussed previously in this chapter, tumor cells may induce weak immune responses because they lack costimulators, and usually they do not express class II MHC molecules, so they do not activate helper T cells. Two potential approaches for boosting host responses to tumors are to provide costimulation for tumor-specific T cells artificially, and to provide exogenous cytokines that can enhance T cell growth and activation, thus replacing helper T cell functions. Many cytokines also have the potential to enhance nonspecific inflammatory responses which by themselves may have anti-tumor activity.

The efficacy of enhancing T cell costimulation for anti-tumor immunotherapy has been demonstrated by animal experiments in which tumor cells were transfected with genes that encode costimulatory molecules (see Chapter 7). For example, tumor cells transfected with the gene encoding the B7-1 costimulatory molecule are potent stimulators of anti-tumor immune responses, compared with unmodified tumor cells (Fig. 18–6). When injected into naive syngeneic hosts, these B7-1–expressing tumor cells are rejected whereas unmodified tumor cells are not. In addition, injection of B7-1–expressing tumors at one site induces protective immunity against unmodified tumor cells injected later at a distant site. The reason for this probably is that B7-1 molecules enhance specific CTL responses, but the cytolytic activity of fully differentiated (effector) CTLs is not dependent on B7-1. These successes with experimental tumor models may lead to therapeutic trials in which a sample of a patient's tumor is propagated *in vitro*,

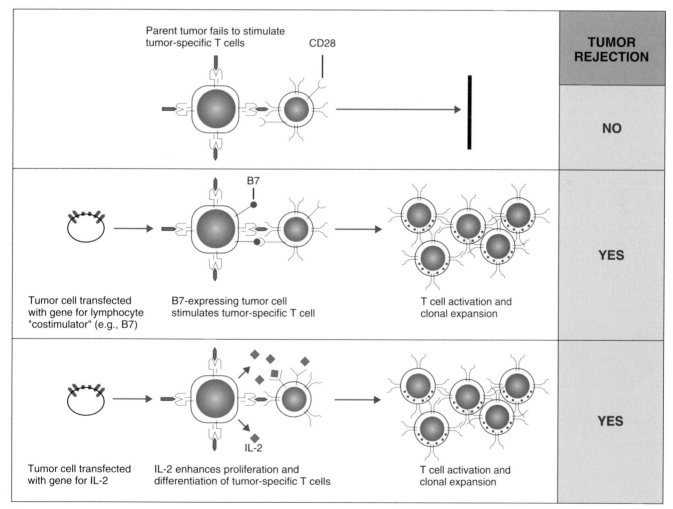

FIGURE 18–6. Enhancement of tumor cell immunogenicity by transfection of costimulator or cytokine genes. Tumor cells that do not adequately stimulate T cells when transplanted into an animal will not be rejected and therefore will grow into tumors. Transfection of these tumor cells with constitutively active genes encoding costimulators or cytokines can lead to enhanced immunogenicity of the tumors, T cell–mediated rejection, and therefore no tumor growth. In addition, after rejection of these modified tumor cells, the animals retain specific immunity to subsequent challenges by the parent (untransfected) tumor cells.

transfected with costimulator genes, irradiated, and reintroduced into the patient. Such approaches may succeed even if the rejection antigens expressed on tumors are not defined.

The potential of various cytokines to enhance both specific and innate immune responses against tumors has also been demonstrated in experimental models and has been realized in clinical practice. In several experimental protocols, tumor cells are transfected with cytokine genes in order to localize the cytokine effects to where they are needed. For instance, when rodent tumors transfected with IL-2, IL-4, IFN-γ, or granulocyte-macrophage colony–stimulating factor (GM-CSF) genes are injected into animals, the tumors are rejected or regress. In some cases, intense inflammatory infiltrates accumulate around the cytokine-secreting tumors, and the nature of the infiltrate varies with the cytokine (Table 18–4). Eosinophils and macrophages accumulate around IL-4–producing tumors, macrophages dominate infiltrates around IFN-γ–secreting tumors, and IL-2–producing tumors are surrounded by massive lymphocytic infiltrates. In essence, the cytokines elaborated by the transfected tumor cells are acting as adjuvants. The importance of these findings is that the type of inflammatory cells recruited by different cytokines may provide different effector functions as well as accessory cell functions required for optimal activation of T cells. An important fact is that, in several of these studies, the injection of cytokine-secreting tumors induced specific, T cell–dependent immunity to subsequent challenges by unmodified tumor cells. This is most evident with tumors expressing transfected GM-CSF genes, and it may reflect the ability of GM-CSF to induce the differentiation of immature dendritic cells into mature APCs. Thus, the local production of cytokines may augment specific T cell responses to tumor antigens, and cytokine-expressing tumors may act

as true tumor vaccines. Several clinical trials with cytokine gene–transfected autologous or allogeneic tumors are under way in advanced cancer patients. Another approach to achieve cytokine production around tumor cells is the use of biodegradable polymer microspheres that release encapsulated cytokines gradually over extended time periods. The microspheres are mixed with irradiated tumor cells from a patient *ex vivo,* and the mixture is then reinjected. This method avoids the technically cumbersome requirement of transfecting tumor cells.

Cytokines are also administered systemically for the treatment of various human tumors. This type of experimental therapy became feasible when highly purified or recombinant cytokines were made available in sufficient quantities. Many of the protocols for cytokine therapy are continually being modified to take into account toxic systemic effects, and the application of this approach to different tumors is continually being explored.

1. IL-2, administered in high doses, is used alone or in conjunction with adoptive cellular immunotherapy (discussed later). After administration of IL-2, there is an increased number of blood lymphocytes and NK cells, an increase in NK and LAK cell activity, and increases in serum TNF, IL-1, and IFN-γ. Presumably, the IL-2 works by stimulating the anti-tumor activity of NK cells and/or CTLs, i.e., by inducing LAK cell differentiation *in vivo.* The treatment can be highly toxic, causing fever, pulmonary edema, and often shock. These effects occur because the IL-2 stimulates production of other cytokines by T cells, which have deleterious effects at high doses. IL-2 has been effective in inducing measurable tumor regression responses in about 10 to 15 per cent of patients with advanced melanoma and renal cell carcinoma and is currently an approved treatment for these cancers.

TABLE 18–4. Modification of Mouse Tumors by Transfected Cytokine Genes

Cytokine	Enhanced Rejection of Transfected Tumor	Inflammatory Infiltrate	Distant Immunity Against Parental Tumor
Interleukin-2	Yes (dependent on CD8+ and not CD4+ T cells)	Lymphocytes	Sometimes
Interleukin-4	Yes	Macrophages and eosinophils	Sometimes (dependent on both CD8+ and CD4+ T cells)
Interferon-γ	Sometimes (varies from tumor to tumor)	Variable	Sometimes
Tumor necrosis factor	Sometimes	Mixed neutrophils and lymphocytes	No
GM-CSF	Yes	Immature mononuclear cells	Yes (long-lived immunity; dependent on both CD8+ and CD4+ T cells)
Monocyte chemotactic protein-1	No	Macrophages	No
Interleukin 3	Sometimes	Immature mononuclear cells	Sometimes (long-lived immunity; dependent on both CD8+ and CD4+ T cells)

Abbreviations: GM-CSF, granulocyte-macrophage colony–stimulating factor.
Adapted from Pardoll, D. M. Cancer vaccines. Immunology Today 14:310–316, 1993.

2. TNF clearly has potent anti-tumor effects *in vitro,* and clinical trials of TNF in advanced cancer patients have been performed. Unfortunately, high doses of TNF produce many undesirable pathologic effects (see Chapter 12), and TNF can be highly toxic at the doses required for tumor killing *in vivo*. Local injection of TNF is used to treat limb sarcomas, and in these cases, venous return from the extremity is reduced during the treatment to minimize systemic effects.

3. Alpha-interferon (IFN-α) is a type I interferon, produced largely by leukocytes (see Chapter 12). It has antiproliferative effects on cells *in vitro,* increases the lytic potential of NK cells, and increases class I MHC expression on various cell types. Clinical trials of this cytokine indicate that it can induce regression of renal carcinomas, melanomas, Kaposi sarcomas, various lymphomas, and hairy cell leukemias (a B cell lineage tumor). In fact, IFN-α treatment of hairy cell leukemia was used routinely in many medical centers until a new and successful chemotherapeutic drug was recently introduced. Currently, IFN-α is used in conjunction with chemotherapy for the treatment of metastatic melanoma.

4. IFN-γ treatment of various hematopoietic and solid tumors has been intermittently successful, and intraperitoneal administration of IFN-γ for the treatment of ovarian carcinomas has also been evaluated, but this cytokine is not part of any established therapeutic regimen. The rationale for using IFN-γ is that the macrophage- and NK-activating properties of this cytokine, as well as its ability to upregulate MHC molecule expression, would help enhance anti-tumor immunity.

5. Hematopoietic growth factors, including GM-CSF and granulocyte colony–stimulating factor (G-CSF), are used in cancer treatment protocols, although not strictly to enhance immune responses against tumors. Rather, they shorten periods of neutropenia after chemotherapy or autologous bone marrow transplantation by stimulating maturation of granulocyte precursors.

6. There is a lot of interest in the potential of IL-12 to enhance anti-tumor NK- and T cell–mediated immune responses, and toxicity trials in advanced cancer patients are now under way.

Passive Tumor Immunotherapy

Passive immunotherapy involves the transfer of immune effectors into a patient from an external source. Antisera specific for bacterial or snake toxins are examples of passive therapy that are widely used in clinical practice. The same approaches have been used in tumor immunotherapy, with variable success, and have included passive transfer of lymphocytes and antibodies.

ADOPTIVE CELLULAR IMMUNOTHERAPY

Adoptive cellular immunotherapy refers to the transfer of cultured immune cells that have anti-tumor reactivity into a tumor-bearing host. Two variations to this approach have been tested in clinical trials.

1. **Lymphokine-activated killer (LAK) cell therapy** involves the *in vitro* generation of LAK cells by removing peripheral blood leukocytes from tumor patients and culturing the cells in high concentrations of IL-2. The LAK cells are then injected back into the cancer patients. As discussed previously, LAK cells are predominantly derived from NK cells. Adoptive therapy with autologous LAK cells, in conjunction with *in vivo* administration of IL-2 or chemotherapeutic drugs, has yielded impressive results in mice, with regression of solid tumors. Human LAK cell therapy trials have so far been largely restricted to advanced cases of metastatic tumors, and the efficacy of this approach appears to be highly variable from patient to patient.

2. **Tumor-infiltrating lymphocyte (TIL) therapy** involves the generation of LAK cells from mononuclear cells originally isolated from the inflammatory infiltrate present in and around solid tumors, obtained from surgical resection specimens. The rationale for this approach is that TILs may be enriched for tumor-specific CTLs and NK cells. In fact, TILs do include activated NK cells and CTLs, but only some of the cells in these mixed populations of cells are specific for the tumors from which they were isolated. As we have discussed previously, many melanoma-specific CTL clones have been derived from melanoma TILs. Human trials with TIL therapy are ongoing. One approach for local delivery of cytokines to tumors is transfection of TILs with cytokine genes; this has been attempted with TNF in a small number of patients with limited success.

THERAPY WITH ANTI-TUMOR ANTIBODIES

Many variations on the use of passively administered antibodies in cancer therapy have been tried. One approach is to use antibodies that bind tumor antigens as vehicles to bring toxic agents to, and selectively kill, tumor cells. The potential of using antibodies as "magic bullets" has been alluring to investigators for many years and is still a very active area of research. In addition, *in vivo* administration of antibodies specific for T cells may be employed to augment cellular responses nonselectively or to target immune effector cells to tumors. Several types of these antibody treatments are described below.

1. Anti-tumor antibodies coupled to toxic molecules, radioisotopes, and drugs have all been used in immunotherapy trials in cancer patients or in experimental animals. Toxins such as ricin or diphtheria toxin are highly potent inhibitors of protein synthesis and can be useful at extremely low doses if they are bound to antibodies to form **immunotoxins.** This approach requires the co-

valent attachment of the toxin (lacking its cell-binding component) to an anti-tumor antibody molecule without loss of toxicity or antibody specificity. The systemically injected immunotoxin must be endocytosed by tumor cells and delivered to the appropriate intracellular site of action. Another approach is to covalently attach antineoplastic drugs or cytocidal radioisotopes to anti-tumor antibodies.

Several practical difficulties must be overcome for this technique to be successful. The specificity of the antibody must be such that it does not significantly bind to non-tumor cells. As we have discussed, there are few truly tumor-specific antigens to select when an antibody-based immunotherapy approach is designed. Many TSAs are peptides derived from cytoplasmic proteins that are presented bound to class I MHC molecules, and antibodies that will recognize these bound peptides are difficult to make. Most anti-tumor antibodies are directed at cell surface TAAs that are more highly expressed on tumor cells than on normal tissues. It is also difficult to ensure that a sufficient amount of antibody reaches the appropriate tumor target before it is cleared from the blood by Fc receptor–bearing phagocytic cells. Such clearance may not only reduce anti-tumor effectiveness but may also damage phagocytic cells. The toxins, drugs, or radioisotopes attached to the antibody may have systemic effects as the result of circulation through normal tissues. For example, hepatotoxicity and vascular leak syndromes are common problems with immunotoxin reagents. Since the anti-human tumor antibodies used in clinical trials are usually made in other species, as are conjugated plant or bacterial toxins, there is frequently an immune response resulting in anti-antibodies or anti-toxins that may cause increased clearance rates or block binding of the therapeutic reagent to its target. One way to diminish this problem is to use recombinant, "humanized" antibodies comprising the variable regions of a mouse monoclonal antibody specific for the tumor antigen combined with human Fc portions (see Chapter 3, Box 3–1). Another problem with the use of anti-tumor antibodies is the outgrowth of mutant tumor cells that no longer express the antigens that the antibody recognizes. This is particularly likely to happen if the target antigens are not required for the malignant phenotype. One way to avoid this problem is to use cocktails of antibodies with specificities for different TAAs expressed on the same tumor.

The results of clinical trials with anti-tumor antibody conjugates (immunotoxins) have been variable. Toxin- and radionuclide-conjugated antibodies with specificities for various TAAs on melanomas and carcinomas have been tried. In addition, antibody conjugates specific for CD19, CD22, and CD30 have been used to treat B cell lymphomas. Several clinical trials have used antibodies specific for the human interleukin-2 receptor α

(IL-2Rα) chain for treatment of adult T cell leukemias, which usually express high levels of IL-2Rα. Mouse and humanized anti-IL-2Rα antibodies have been used in unconjugated forms. Such antibodies could cause complement-mediated lysis of IL-2R–expressing tumor cells. Anti-IL-2 receptor antibodies have also been conjugated to various agents, including diphtheria toxin and the α-particle-emitting radionuclide yttrium 90. In a related strategy, a chimeric protein in which IL-2 itself is linked to the effector chain of *Pseudomonas* toxin has been used to treat T cell lymphomas. Anti-IL-2R therapy is not tumor-specific and may be immunosuppressive because normal, activated T cells would be rendered nonfunctional. In general, the efficacy of these various agents is limited, and only a small percentage of patients show significant reductions in tumor burden. Nonetheless, intermittent successes have encouraged further refinement of the reagents. Furthermore, trials of these reagents in immunodeficient mice with xenografted human tumors continue to generate new candidates for human trials. For example, a monoclonal antibody specific for a Lewis Y–related antigen conjugated to the chemotherapeutic agent doxorubicin cures widely metastatic human carcinomas transplanted into athymic mice.

2. Anti-idiotypic antibodies have been used in the treatment of B cell lymphomas that express surface Ig with particular idiotypes. The idiotype is a highly specific tumor antigen since it is expressed only on the neoplastic clone of B cells, and there was once great hope that anti-idiotypic antibodies would be effective therapeutic reagents with absolute tumor specificity. (Anti-idiotypic antibodies are raised by immunizing rabbits with a patient's B cell tumor and depleting the serum of reactivity against all other human immunoglobulins.) This strategy relies on complement fixation or antibody-dependent cell-mediated cytotoxicity (ADCC) to kill the lymphoma cells. The approach has not proved generally successful, and there are many reasons why it may not work. Because surface Ig expression is not functionally related to the malignant phenotype of the cell, selective outgrowth of non–Ig-expressing tumor cells can occur. Moreover, the high degree of somatic mutation known to occur in Ig genes results in the selective outgrowth of tumor cells with altered idiotypes no longer reactive with the anti-idiotypic antibody. Attempts to circumvent these problems with cocktails of several different antibodies have also not proved successful. More recently, immunization of animals with DNA encoding Ig idiotypes has been shown to induce anti-idiotypic antibody responses, and this strategy is being considered for the treatment of lymphomas.

3. Heteroconjugate antibodies may allow targeting of cytotoxic effector cells onto tumor cells. In this approach, an antibody specific for a tumor antigen is covalently coupled to an antibody directed against a surface protein on cytotoxic effec-

tor cells, such as NK cells or CTLs. Such hetero-conjugates can promote binding of the effector cells to tumor cells. A heteroconjugate consisting of an anti-CD3 antibody coupled to an antibody against a tumor cell surface protein has been used to enhance CTL-mediated lysis of the tumor cell. In this case, the anti-CD3 antibody not only served to bring the CTL into contact with the target cell, but it also activated the CTL. A related approach is to use conjugates of antibodies specific for effector cells with hormones whose receptors are expressed on tumor cells. For example, anti-CD3 antibodies coupled to melanocyte-stimulating hormone enhance *in vitro* destruction of human melanoma cells by CTLs. These types of antibody therapies have so far been tried only in experimental animal studies.

Another way anti-tumor antibodies have been used for cancer treatment is for the *in vitro* depletion of bone marrow tumor cells by antibody plus complement-mediated lysis. This is useful for autologous bone marrow transplants in B cell lymphoma patients. In this protocol, some of the patient's bone marrow is removed, and the patient is given lethal doses of radiation and chemotherapy, which destroy tumor cells and the remaining normal marrow cells in the patient. The bone marrow removed from the patient is then treated with antibodies directed against B lymphocyte-specific antigens, which are known to be expressed on the B cell–derived lymphoma cells. Complement is then added to promote lysis of the lymphoma cells that have bound antibody. The treated marrow, having been purged of lymphoma cells, is transplanted back into the patient and can reconstitute the hematopoietic system destroyed by irradiation and chemotherapy.

SUMMARY

Malignant tumors express antigens that may stimulate and serve as targets for anti-tumor immunity. Protective anti-tumor immune responses occur in experimental animal models. However, it has been more difficult to demonstrate that innate or specific immune responses can protect humans against tumor growth. The development of virally induced tumors, which express virally encoded antigens, can be inhibited by specific immune responses. Antigens unique to individual tumors, which stimulate specific rejection of transplanted tumors, do occur in experimental animal tumors. Other tumor antigens that can stimulate immune responses are shared by different tumors, and these include viral antigens, products of mutated or rearranged oncogenes or tumor suppressor genes, and products of derepressed genes. Human CTL clones specific for tumor antigens have been

isolated from tumor patients, and the peptide antigens or genes encoding these antigens have been identified. Tumors may also express tissue differentiation antigens or embryonic antigens to which the host is tolerant; these molecules are useful diagnostic markers. Many immunologic effector mechanisms can destroy tumor cells *in vitro*. One or more of these mechanisms may work on tumor cells *in vivo*, and different mechanisms may be effective on different tumors. CTLs are probably the most important effectors of anti-tumor immunity *in vivo*, although NK cells and macrophages may also be involved. Various mechanisms may explain how antigen-expressing tumors escape destruction by the immune system. These mechanisms include poor immunogenicity of tumors due to lack of co-stimulators and/or inability to stimulate class II MHC–restricted helper T cells, downregulation of MHC molecules, induction of tolerance to tumor antigens, loss of expression of immunogenic proteins due to mutations, modulation of tumor antigens by anti-tumor antibodies, and immunosuppression of the host. Treatment of tumors by immunologic approaches has not yet succeeded on a large scale, but new techniques are being tested. Strategies to enhance T cell–mediated anti-tumor immunity, adoptive cellular immunotherapy, and cytokine treatment all continue to be investigated.

Selected Readings

Boon, T., J. Cerottini, B. Van den Eynde, P. Van der Bruggen, and A. Van Pel. Tumor antigens recognized by T lymphocytes. Annual Review of Immunology 12:337–365, 1994.

Burnet, F. M. The concept of immunological surveillance. Progress in Experimental Tumor Research 13:1–27, 1970.

Cheever, M. A., M. L. Disis, H. Bernhard, J. R. Gralow, S. L. Hand, E. S. Huseby, H. L. Quin, M. Takahashi, and W. Chen. Immunity to oncogenic proteins. Immunological Reviews 145:33–60, 1995.

Hellstrom, K. E., I. Hellstrom, and I. Chen. Can co-stimulated tumor immunity be therapeutically efficacious? Immunological Reviews 145:167–178, 1995.

Henderson, R. A., and O. J. Finn. Human tumor antigens are ready to fly. Advances in Immunology 62:217–256, 1996.

Kumer, U., and U. Staerz. Concepts of antibody-mediated cancer therapy. Cancer Investigation 11:174–184, 1993.

Lanzavecchia, A. Identifying strategies for immune intervention. Science 260:937–944, 1993.

Pardoll, D. M. Paracrine cytokine adjuvants in cancer immunotherapy. Annual Review of Immunology 13:399–415, 1995.

Rosenberg, S. A., and M. T. Lotze. Cancer immunotherapy using interleukin-2 and interleukin-2 activated lymphocytes. Annual Review of Immunology 4:681–709, 1986.

Urban, J. L., and H. Schreiber. Tumor antigens. Annual Review of Immunology 10:617–644, 1992.

Van Pel, A., P. van der Bruggen, P. G. Coulie, V. G. Brichard, B. Lethe, B. van der Eynde, C. Uyttenhove, J.-C. Renauld, and T. Boon. Genes coding for tumor antigens recognized by cytolytic T lymphocytes. Immunological Reviews 145:229–250, 1995.

Vitetta, E. S., P. E. Thorpe, and J. Uhr. Immunotoxins: magic bullets or misguided missiles? Immunology Today 14:253–259, 1993.

CHAPTER NINETEEN

SELF-TOLERANCE AND

AUTOIMMUNITY

One of the cardinal properties of the immune system is its ability to recognize and respond to foreign antigens but not to self antigens. This is called "self/nonself discrimination." The unresponsiveness of the immune system to antigenic stimulation is called **immunologic tolerance.** The phenomenon of tolerance was introduced in Chapter 10, when the mechanisms by which foreign antigens may inhibit the development of specific immune responses were discussed. The necessity for maintaining tolerance to self antigens, which is referred to as **self-tolerance,** was appreciated from the early days of modern immunology. Self-tolerance exists partly because of the absence of self-recognizing lymphocytes from the repertoire of mature lymphocytes, and partly because of the way that mature lymphocytes recognize and respond to self antigens. Loss of self-tolerance results in immune reactions against one's own, or autologous, antigens. Such reactions are called **autoimmunity,** and the diseases they cause are called **autoimmune diseases.**

In this chapter we first review the mechanisms of self-tolerance in T and B lymphocytes. We then discuss the factors that contribute to the development of autoimmunity. These topics have been important areas of investigation for several decades. Recent advances in techniques for studying the immune system are providing many new insights into self-tolerance and autoimmunity. The mechanisms of cell and tissue injury in diseases caused by autoimmunity and other abnormal immunologic reactions are described in Chapter 20.

MECHANISMS OF SELF-TOLERANCE

Potentially immunogenic antigens are present on the cells and in the circulation and connective tissues of every individual. Such antigens are freely accessible to each individual's lymphocytes, yet normally one's lymphocytes do not react against one's own antigens. *Unresponsiveness to self antigens, or self-tolerance, is maintained by mechanisms that actively prevent the maturation or stimulation of potentially self-reactive lymphocytes.* Several fundamental concepts are relevant to our understanding of self-tolerance.

1. *Tolerance to self antigens is an actively acquired process (rather than an inherited property), in which self-reactive lymphocytes either are prevented from becoming functionally responsive to self antigens or are inactivated after encountering these antigens.* All individuals inherit roughly the same antigen receptor genes, and these genes recombine and are expressed in lymphocytes as the lymphocytes arise from stem cells. The molecular mechanisms that generate functional antigen receptors are not influenced by what is foreign or self for each individual. As a result, immature lymphocytes may express receptors capable of recognizing various foreign and self antigens. Tolerance to self is induced after antigen receptors are expressed and

is actually a consequence of the recognition of self antigens by specific lymphocytes under special conditions.

2. *Self-tolerance may be induced at various stages of lymphocyte development and activation* (Fig. 19–1). **Central tolerance** is the induction of tolerance in generative lymphoid organs as a consequence of immature self-reactive lymphocytes recognizing self antigens. **Peripheral tolerance** is the induction of unresponsiveness in peripheral sites as a result of mature self-reactive lymphocytes encountering self antigens under particular conditions.

3. *Central tolerance: During their maturation, all lymphocytes go through a stage in which encounter with antigen leads to tolerance rather than activation.* The concept that immature lymphocytes are more sensitive than mature cells to tolerance induction was introduced in Chapter 10. The most tolerance-sensitive stages of lymphocyte maturation are largely anatomically confined to the generative lymphoid organs—the thymus for T cells, and the bone marrow for B lymphocytes. The antigens normally present at high concentrations in these organs are self antigens, because foreign antigens that enter from the external environment are captured and transported to peripheral lymphoid organs, such as the lymph nodes and spleen (see Chapter 11). Therefore, immature lymphocytes normally encounter only self antigens at high concentrations, and clones of lymphocytes whose receptors may be specific for these self antigens are killed (deleted). Thus, central tolerance is due to selection of the lymphocyte repertoire based on recognition of self antigens. It ensures that the lymphocytes that attain functional maturity, leave the generative lymphoid organs, and populate peripheral tissues have been depleted of clones that can recognize ubiquitous self antigens that are present in many tissues, including the generative lymphoid organs (Fig. 19–1). Central tolerance was thought to be the major determinant of self/nonself discrimination, but it cannot explain the ability of the immune system to distinguish between foreign and self antigens that are present only in peripheral tissues.

4. *Peripheral tolerance: Mature self-reactive lymphocytes may be rendered tolerant in peripheral tissues if they encounter self antigens under conditions that favor tolerance rather than activation.* This implies that the mature lymphocyte repertoire contains cells capable of recognizing self antigens, and the conditions of antigen exposure and the regulation of lymphocyte responses are critical determinants of self-tolerance versus self-reactivity. Peripheral tolerance is most important for maintaining unresponsiveness to self antigens that are expressed in peripheral tissues and not in the generative lymphoid organs. As we shall discuss later, recent discoveries are emphasizing the importance of peripheral mechanisms in the maintenance of self-tolerance.

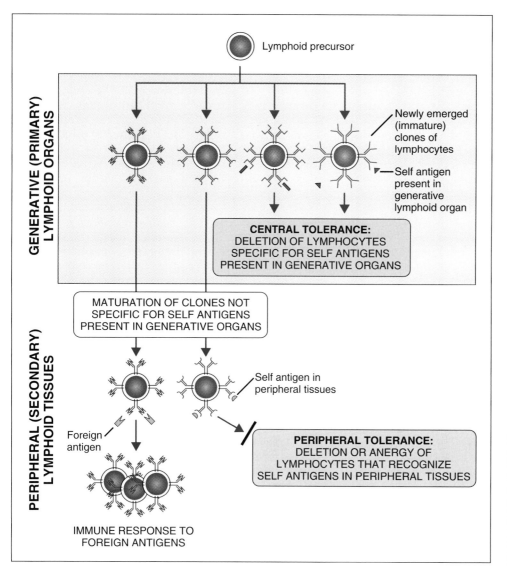

FIGURE 19–1. Central and peripheral tolerance to self antigens. Immature lymphocytes specific for self antigens may encounter such antigens in the generative lymphoid organs and are deleted or inactivated. Mature self-reactive lymphocytes may be inactivated or deleted by encounter with self antigens in peripheral tissues.

5. *The principal mechanisms of tolerance in self antigen–specific clones of lymphocytes are* **clonal deletion,** *by a process of activation-induced cell death, and* **clonal anergy,** *or functional inactivation without cell death.* Both deletion and anergy are induced by the binding of antigens to specific receptors. Central tolerance is mainly due to deletion, whereas both mechanisms contribute to peripheral tolerance. Functional unresponsiveness may also be due to the induction of **regulatory T cells** that suppress the activation and effector functions of mature self-reactive lymphocytes.

6. *Clonal ignorance: Some self-reactive lymphocytes encounter self antigens but are neither activated nor rendered tolerant.* This form of unresponsiveness has been called **clonal ignorance.** The reason why some self-reactive lymphocytes co-exist with self antigens without reacting in any detectable way is not understood. The importance of clonal ignorance as a mechanism for preventing autoimmunity is also unclear. In other words, we do not know how many or which self antigens induce deletion, anergy, or suppression of specific

lymphocytes, or are simply ignored by the immune system.

7. *Tolerance to self proteins may be due to tolerance of self antigen–reactive T or B lymphocytes.* Since antibody responses to protein antigens are helper T cell–dependent, T cell tolerance may be sufficient to prevent autoantibody production. T cell–independent self antigens, such as polysaccharides and glycolipids, must directly induce B cell tolerance.

8. *Several features of self antigens make them tolerogenic and determine whether they induce central or peripheral tolerance.* These include (1) the concentration of the self antigen in generative lymphoid organs, (2) recognition of the antigens without inflammation and costimulators, (3) the persistence of self antigens, and (4) the nature and strength of the signals that self antigens trigger in specific lymphocytes.

Historically, studying self-tolerance has been difficult because lymphocytes specific for self antigens are either not present in a particular individual or experimental animal or are functionally silent. In either case, one cannot identify these cells by examining their responses to a self antigen. Technical developments, especially the creation of transgenic mice, have provided valuable experimental models for analyzing self-tolerance, and many of our current concepts are based on studies with such models.

Mechanisms of T Lymphocyte Tolerance to Self Antigens

CENTRAL TOLERANCE

The induction of self-tolerance in immature T lymphocytes (a form of central tolerance) was described in Chapter 8, when the maturation of T cells in the thymus was discussed. To reiterate the key points, self proteins are processed and presented in association with major histocompatibility complex (MHC) molecules on thymic antigen-presenting cells (APCs). Among the T cells that develop in the thymus are some whose receptors specifically recognize self peptide–MHC complexes. If clones of immature T cells that are specific for self antigens encounter these antigens in the thymus, the result is deletion or anergy of the clones. This process is called **negative selection**

and is responsible for the fact that the repertoire of mature T cells that leave the thymus and populate peripheral lymphoid tissues is tolerant to self antigens that are present in many tissues, including the thymus.

How central T cell tolerance is maintained as individuals age is not known, because the thymus involutes after puberty and is virtually undetectable in adults. T lymphocytes continue to arise from bone marrow stem cells in adults, and they must continue to be selected. In postpubertal humans, negative selection of T cells may occur in extrathymic tissues, or the thymic remnants that persist in adults may be sufficient to serve this function.

PERIPHERAL TOLERANCE

Peripheral tolerance is the mechanism of T cell tolerance to tissue-specific antigens that are not present in the thymus. It may also provide "back-up" or "fail-safe" mechanisms for preventing the activation of lymphocytes that escape central tolerance, for any reason. The mechanisms of peripheral tolerance to self proteins are probably similar to those of tolerance to foreign antigens (see Chapter 10, Fig. 10–3). These include the following (Fig. 19–2):

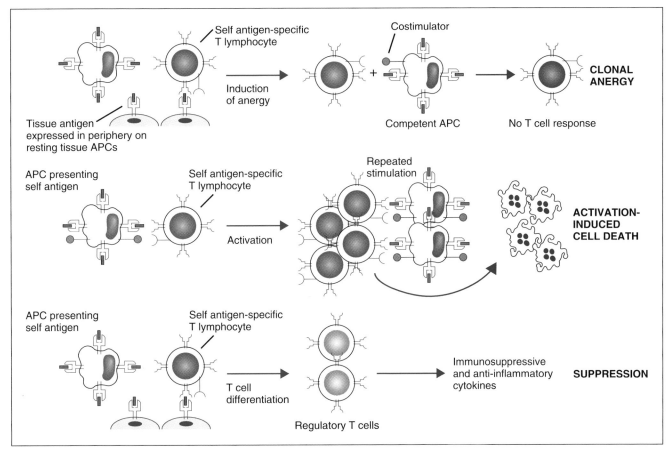

FIGURE 19–2. Mechanisms of peripheral T cell tolerance to self antigens expressed in peripheral tissues.

1. *Clonal anergy due to self antigen recognition without costimulation:* Normally, APCs in tissues are in a "resting" state and express low levels of costimulators and cytokines that stimulate T cell responses. If such APCs present self antigens to specific T cells, the T cells may become anergic. As we shall discuss later, tissue injury and inflammation may activate resident APCs, leading to increased expression of costimulators and cytokines, loss of self-tolerance, and autoimmune reactions against the tissue antigens. According to this concept, *the activation status of tissue APCs is an important determinant of whether self tolerance or autoimmunity develops.* The biochemical mechanisms of T cell anergy are not clearly defined.

2. *Activation-induced cell death due to stimulation by self antigens:* Antigens that are present at high concentrations may repeatedly stimulate specific T cells and eliminate them by Fas-mediated apoptosis (see Chapter 10, Box 10–1). The importance of this pathway for self-tolerance is indicated by the observation that inherited defects in Fas or Fas ligand result in systemic autoimmunity (discussed later). It is also possible that activation-induced death of some T cells is due to production of tumor necrosis factor (TNF) and its binding to TNF receptors, which are members of the same protein family as Fas. However, mice in which the genes for TNF receptors are knocked out do not develop autoimmunity.

3. *Suppression of self-reactive T lymphocytes by regulatory T cells:* Some self antigen–reactive T lymphocytes may be neither rendered anergic nor deleted, but their activities may be inhibited by immunosuppressive cytokines, such as IL-10 or transforming growth factor-β (TGF-β), produced by other self-reactive T cells. However, to date there is little formal evidence that self antigens normally elicit protective or regulatory T cell responses.

4. *Clonal ignorance:* Strictly speaking, clonal ignorance is not a mechanism of tolerance, because it is not actively induced immunologic unresponsiveness but rather a lack of response. Nevertheless, it may be the mechanism that prevents autoimmune reactions to many self antigens. Clonal ignorance has been demonstrated in experimental models in which a protein antigen, e.g., a viral glycoprotein, is expressed in pancreatic islet β cells of mice as a transgene, and these mice are bred with other transgenic mice expressing a T cell antigen receptor specific for the viral protein. In some cases, the T cells do not react in any detectable way, i.e., they are neither killed nor activated. The T cells also do not become anergic, because they are able to respond to the antigen *ex vivo,* and they can even be activated *in vivo* by infection with the glycoprotein-producing virus.

As we mentioned previously, these pathways of peripheral self-tolerance are postulated largely on the basis of studies with transgenic mouse models or experiments with tolerogenic forms of foreign antigens. Little is known about how different, normally produced self antigens actually induce peripheral tolerance. It is also not known whether the same self antigen can trigger multiple mechanisms of self-tolerance, or whether different antigens use different mechanisms to prevent autoimmune reactions. The main reason for these gaps in our knowledge is that it is difficult to identify lymphocytes specific for self antigens in normal individuals, and there are no methods for detecting cells that are anergic or suppressed.

Mechanisms of B Lymphocyte Tolerance to Self Antigens

Much of our knowledge of self-tolerance in B lymphocytes is based on studies with transgenic mice. In one such model, a set of mice are created in which an Ig receptor specific for a model protein antigen, hen egg lysozyme (HEL), is expressed as a transgene. Most of the B cells in these transgenic mice are HEL-specific and can be stimulated to produce anti-HEL antibody. In another set of transgenic mice the antigen HEL is expressed; because this protein is present throughout development, in effect it is a "self" protein. Moreover, in the HEL antigen–expressing transgenic mice, HEL-specific helper T cells are tolerant (deleted or anergic). If the two sets of transgenic mice are mated, in the F1 offspring immature B cells specific for HEL may be exposed to "self" HEL, and this may lead to central tolerance (Fig. 19–3). Alternatively, mature B cells expressing the HEL-specific Ig may be transferred into mice expressing the antigen, to examine peripheral tolerance. This experimental system allows one to follow the development and functional responses of specific B cells that are exposed to a model "self" antigen in the absence of T cell help *in vivo.* By expressing the HEL in membrane-bound or soluble forms, or at different concentrations, or in different tissues, one can mimic the interactions of self-reactive B cells with different self antigens. Such models have led to the concept that *there are multiple "checkpoints" during B cell development and activation at which encounter with self antigen in the absence of helper T cells may abort these processes.*

CENTRAL TOLERANCE

The fate of immature B cells that encounter self antigens in the generative lymphoid tissue, i.e, the bone marrow, is largely determined by the valence and concentration of the antigens (Fig. 19–3). In general, multivalent antigens (e.g., membrane-associated proteins) or antigens present at high concentrations induce B cell death. Some immature B cells that encounter self antigens in the bone marrow may proceed to maturity, but they become permanently unable to enter lymphoid follicles in peripheral lymphoid organs and thus to respond to antigen in the periphery. Lower con-

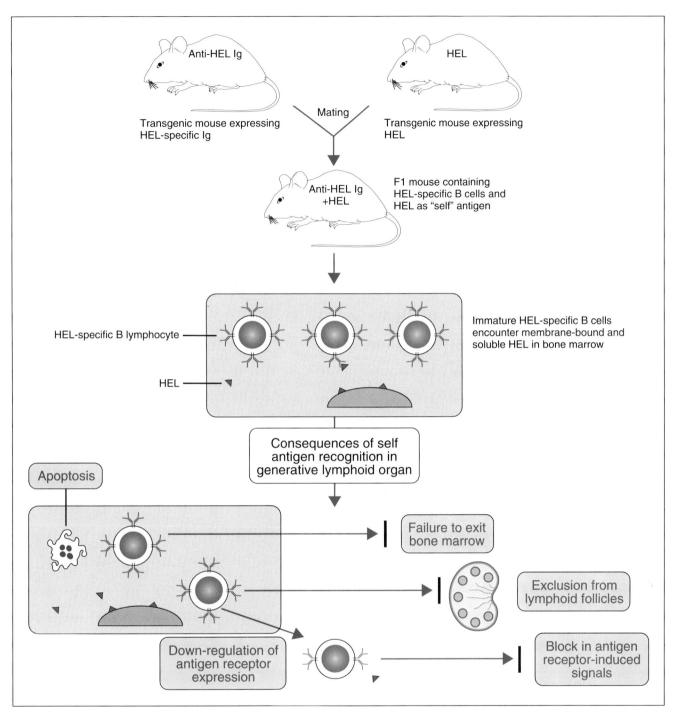

FIGURE 19–3. B cell tolerance to a "self" antigen in transgenic mice. A transgenic mouse expressing anti-HEL Ig is bred with another transgenic mouse that produces HEL. In the "double transgenic" F1 hybrid mouse, most B cells express HEL-specific Ig, and HEL is a "self" antigen. The HEL-specific B cells encounter HEL in the bone marrow and are killed, or rendered incapable of responding to HEL, by several mechanisms.

centrations of soluble self antigens induce functional anergy, which may be due to (1) decreased expression of membrane Ig receptors on B cells, or (2) inability of the antigen receptor to transduce activating signals as a result of antigen binding ("desensitization"). These different fates of B cells suggest that the strength of the antigen-induced signal may determine the consequence of antigen

recognition. Strong signals (e.g., high concentration of multivalent antigen) lead to cell death, and weaker signals cause functional unresponsiveness. However, the biochemical mechanisms of B cell death, follicular exclusion, and anergy are not known. Rarely, B cells in the bone marrow that express receptors specific for self antigens may respond to these antigens by re-activating their

RAG-1 and *RAG-2* genes and expressing a new Ig molecule, thus acquiring a new specificity. This process is called "receptor editing" and is a potential mechanism for self-reactive B cells to lose this reactivity and survive.

PERIPHERAL TOLERANCE

Mature B lymphocytes that recognize self antigens in peripheral tissues in the absence of specific helper T cells may be rendered unresponsive by many of the same mechanisms that account for functional anergy in immature cells that see antigen in the bone marrow. Some B cells become incapable of activating tyrosine kinases, such as Syk, or increasing intracellular Ca^{++}, in response to stimulation by the antigen *ex vivo*. Other B cells down-regulate their expression of antigen receptors as a result of antigen encounter. Thus, self-reactive B cells cannot proliferate or increase their expression of costimulators in response to the self antigen and cannot stimulate or respond to T cell help. Other mature B cells that have encountered self antigens become incapable of terminal differentiation into antibody secretors, due to an undefined block(s) in activation pathways, or are excluded from germinal centers. Finally, self-reactive B lymphocytes may remain functionally competent but may be incapable of producing autoantibodies because self antigen–specific helper T lymphocytes are deleted or anergic.

Experiments with transgenic mice indicate that self-tolerance in B cells develops at 10 to 100 times higher concentration of self antigen than that required for inducing tolerance in T cells. This emphasizes the statement, made earlier, that T cell tolerance may be adequate for preventing the production of high-affinity autoantibodies specific for many self proteins. B cells, however, must themselves be tolerant to autologous T-independent antigens. For instance, individuals with a particular ABO blood group type do not produce antibodies against their own ABO antigens, which are glycolipids and behave as T cell–independent antigens. There are no good models for studying the mechanisms of B cell tolerance to self polysaccharides and lipids. It is, however, likely that the general principles and mechanisms are much like those described above for protein antigens.

The mechanisms of self-tolerance in T and B lymphocytes are similar in many respects, but there are also important differences (Table 19–1). In the past few years, much has been learned from animal models, especially transgenic mice. Application of this knowledge to understanding tolerance to normally expressed self antigens, and to defining why tolerance fails, giving rise to autoimmune diseases, are areas of active investigation.

MECHANISMS OF AUTOIMMUNITY: GENERAL PRINCIPLES

The possibility that an individual's immune system may react against autologous antigens and cause tissue injury was appreciated by immunologists from the time that the specificity of the immune system for foreign antigens was recognized. In the early 1900s, Paul Ehrlich coined the rather melodramatic phrase "horror autotoxicus" for harmful ("toxic") immune reactions against self. When Macfarlane Burnet proposed the clonal selection hypothesis about 50 years later, he added the corollary that clones of autoreactive lymphocytes were deleted during development to prevent autoimmune reactions. We now know that this postulate is partly correct, although it has been modified and expanded in many ways.

Autoimmunity is an important cause of disease in humans, estimated to affect at least 1 to 2 per cent of the United States population. The clinical designation of autoimmunity is often erroneous or unsubstantiated. Many diseases in which immune reactions accompany tissue injury have been

TABLE 19–1. Self-Tolerance in T and B Lymphocytes

	T Lymphocytes	B Lymphocytes
Principal sites of tolerance induction	Thymus (cortex); periphery	Bone marrow; periphery
Tolerance-sensitive stage of maturation	CD4+CD8+ (double-positive) thymocyte	Membrane IgM+IgD− B lymphocyte
"Stimuli" for tolerance induction	Central (thymic): high-avidity antigen recognition, perhaps on bone marrow–derived thymic APCs Peripheral: antigen presentation by APCs lacking costimulators; repeated stimulation by self antigen	High-avidity antigen recognition, especially of multivalent antigen, without T cell help
Mechanism of tolerance*	Clonal deletion: cell death (apoptosis) Clonal anergy: block in proliferation and differentiation; suppression	Clonal deletion: cell death (apoptosis) Clonal anergy: reduced membrane Ig; block in signal transduction; failure to enter follicles

* Intracellular biochemical alterations induced by tolerogenic antigens are not fully known.
Abbreviations: APC, antigen-presenting cell; Ig, immunoglobulin.

called "autoimmune," but this appellation may not be correct. For example, tissue injury that occurs during normal immune responses against foreign antigens is not a type of autoimmune disease. Furthermore, the presence of antibodies or T cells reactive with self antigens may be the consequence and not the cause of tissue injury. For instance, some patients with myocardial infarction develop antibodies against their own myocardial antigens that were previously concealed from the immune system. Obviously, in this case, the autoantibodies are not the cause of the infarction. Nevertheless, true autoimmune diseases are clearly serious health concerns worldwide.

Autoimmunity results from a failure or breakdown of the mechanisms normally responsible for maintaining self-tolerance. The potential for autoimmunity exists in all individuals because all individuals inherit genes that code for lymphocyte receptors that may recognize self antigens, and many self antigens are readily accessible to the immune system. As discussed earlier, autoimmunity is normally prevented by selection processes that prevent the maturation of some self antigen–specific lymphocytes, and by mechanisms that inactivate self-reactive lymphocytes, which do mature. *Loss of self-tolerance may result from abnormal selection or regulation of self-reactive lymphocytes, and from abnormalities in the way self antigens are presented to the immune system.* Several important general concepts have emerged from analyses of autoimmunity during the past 20 years.

1. *Multiple interacting factors contribute to the development of autoimmunity.* These factors include immunologic abnormalities affecting APCs or lymphocytes, genetic backgrounds that predispose to autoimmunity, gender, tissue injury, and microbial infections. Because combinations of these factors may be operative in different disorders, it is not surprising that autoimmune diseases comprise a heterogeneous group of clinical and pathologic abnormalities.

2. *Autoimmune diseases may be either systemic or organ-specific, and these may be caused by different types of antigens and different immunologic abnormalities.* For instance, the formation of circulating immune complexes typically produces systemic diseases (see Chapter 20). In contrast, autoantibody or T cell responses against antigens with restricted tissue distribution lead to organ-specific injury.

3. *Various effector mechanisms are responsible for tissue injury in different autoimmune diseases.* These mechanisms include circulating autoantibodies, immune complexes, and autoreactive T lymphocytes, and they are discussed in Chapter 20.

A major difficulty in defining the mechanisms of many human autoimmune diseases has been the inability to identify the antigens that initiate autoimmune responses and the lymphocytes that medi-

ate these reactions. As a result, the specific etiologies of most autoimmune diseases are not known. In the remainder of this chapter, we describe the general principles of the pathogenesis of autoimmune diseases, with an emphasis on the immunologic, genetic, and other factors that contribute to the development of autoimmunity.

LYMPHOCYTE ABNORMALITIES CAUSING AUTOIMMUNITY

Autoimmunity may result from primary abnormalities of B cells, T cells, or both. Much recent attention has focused on the role of T cells in autoimmunity for two main reasons: (1) helper T cells are critical regulators of all immune responses to proteins, and (2) several autoimmune diseases are genetically linked to the MHC (the human leukocyte antigen [HLA] complex in humans), and the function of MHC molecules is the presentation of peptide antigens to T cells. Failure of self-tolerance in T lymphocytes may result in autoimmune diseases in which the lesions are caused by cell-mediated immune reactions. Helper T cell abnormalities may also lead to autoantibody production because helper T cells are necessary for the production of high-affinity antibodies against protein antigens.

Abnormalities in lymphocytes that may result in autoimmunity could affect any of the mechanisms that normally maintain self-tolerance. Different aberrations may give rise to systemic or organ-specific autoimmunity. In the following discussion, we will consider immunologic abnormalities that have the potential for causing autoimmunity, using examples of animal and human diseases to illustrate key points.

Failure of Central Tolerance

It is often hypothesized that autoimmunity results from a failure of the selection processes that normally delete immature self antigen–specific lymphocytes. However, little formal evidence supports this hypothesis in any human or experimental autoimmune disease. We do not know if this is because even if central tolerance fails, peripheral mechanisms are adequate for maintaining unresponsiveness to many self antigens.

Failure of Peripheral Tolerance

Numerous experimental models and genetic abnormalities, either spontaneously arising or created by targeted gene disruptions, support the idea that autoimmunity may result from a failure of peripheral tolerance, i.e., anergy or deletion of mature, self antigen–specific T lymphocytes. Much less is known about how, or whether, loss of peripheral tolerance in B cells may contribute to autoimmunity.

Breakdown of T Cell Anergy. Earlier in this chapter, we discussed the concept that resting, costimulator-deficient tissue APCs may present autologous tissue antigens and induce anergy in self-reactive T cells. It follows that *conditions that activate tissue APCs may break T cell anergy, by enhancing the expression of costimulators and the production of cytokines.* This may stimulate T cell proliferation and the differentiation of T cells into tissue-injurious pro-inflammatory effectors, resulting in autoimmune reactions against the tissue. The activation of APCs may be due to infections, tissue necrosis, and local inflammation. The role of inflammation in the induction of autoimmunity is suggested by the observation that experimental T cell–mediated autoimmune diseases, such as **encephalomyelitis** (Box 19–1) and thyroiditis, develop only if the self antigens (myelin proteins and thyroglobulin, respectively) are administered with strong adjuvants. Such adjuvants may activate macrophages to express B7-1 and B7-2, resulting in the breakdown of anergy and the development of effector T cells reactive with the self antigen. (Administration of self antigens with adjuvants may also overcome clonal ignorance, if the self antigens are ones that are normally ignored by self-reactive lymphocytes.)

More formal demonstration of autoimmune tissue injury resulting from the breakdown of T cell anergy has come from transgenic mouse models of the human disease **insulin-dependent diabetes mellitus** (Box 19–2). Various genes can be expressed selectively in pancreatic islet β cells as transgenes under the control of insulin promoters. Expression of the costimulator, B7-1, by itself does not lead to autoimmune insulitis, but it strongly increases susceptibility to insulitis in combination with other local abnormalities. In one experimental

---

BOX 19–1.   Experimental Allergic Encephalomyelitis

Experimental allergic encephalomyelitis (EAE) is probably the best-characterized experimental model of an organ-specific autoimmune disease mediated exclusively by T lymphocytes. The disease is induced by immunizing mice, rats, or guinea pigs with myelin basic protein (MBP) or proteolipid protein (PLP) with an adjuvant containing pertussis bacteria. About 1 to 2 weeks after immunization, the animals develop encephalomyelitis, characterized by perivascular infiltrates composed of lymphocytes and macrophages associated with demyelination in the brain and spinal cord. The neurologic lesions can be mild and self-limited or chronic and relapsing. The chronic form of the experimental disease bears some, albeit superficial, resemblance to multiple sclerosis in humans. This similarity suggests that multiple sclerosis is an autoimmune disease. However, the antigens that initiate multiple sclerosis are not definitively established.

In mice, EAE is caused by CD4$^+$ T$_H$1 cells specific for MBP or PLP. This has been established by many lines of experimental evidence. Mice immunized with either myelin antigen contain CD4$^+$ T cells that produce IL-2 and IFN-γ and proliferate in response to that antigen *in vitro*. The disease can be transferred to unimmunized animals by CD4$^+$ T cells from MBP- or PLP-immunized syngeneic animals, or with MBP- or PLP-specific cloned CD4$^+$ T cell lines. Development of the disease can be prevented by injecting antibodies specific for CD4 or for class II MHC molecules into immunized mice. Most disease-producing clones belong to the IL-2 and IFN-γ–producing T$_H$1 subset, and the lesions of EAE usually have the characteristics of delayed-type hypersensitivity (DTH) reactions. Encephalitogenic clones also express high levels of the β1 integrin VLA-4. The infiltration of T cells into the central nervous system is thought to depend on the binding of VLA-4 to its ligand, VCAM-1, on microvascular endothelium. This is one of the mechanisms responsible for the homing of activated and memory T cells to peripheral tissues (see Chapter 11).

Why immunization with autologous or heterologous myelin antigens induces specific autoimmunity is still not clear. It has been postulated that T cells specific for autologous myelin antigens are not deleted during thymic maturation, and tolerance to self myelin is maintained by T cell anergy or clonal ignorance. Immunization with a myelin antigen together with an adjuvant leads to breakdown of anergy or ignorance, and a T cell response develops against epitopes of autologous myelin proteins. As the T cells respond to these proteins adjacent to blood vessels in the central nervous system, cytokines are released that recruit and activate macrophages. This leads to destruction of the myelin.

Much interest in this disease has focused on the analysis of the fine specificity of MBP-reactive encephalitogenic T cells in different inbred mouse strains. For instance, in mice of the H-2^u strain, most encephalitogenic T cells recognize an MBP peptide consisting of the 9 amino terminal residues in association with the I-A^u molecule. Mutational analysis of various MBP peptides has shown that some amino acid residues are critical for binding to MHC molecules and others for recognition by T cells. Altered peptide ligands (see Chapter 7) may be produced by introducing conservative substitutions in the T cell receptor (TCR) contact residues of MBP or PLP. Administration of such mutant peptides blocks the induction of EAE. It is thought that this is due to the induction of anergy in T$_H$1 cells specific for native MBP or PLP, and/or the development of anti-inflammatory T cells specific for these myelin antigens. Such findings raise the exciting possibility that if one can identify self antigens that cause autoimmune diseases, administration of altered forms of these antigens may be a rational and specific immunotherapy for the diseases. However, this has not been achieved yet in any human disease, and its feasibility in humans is uncertain since many peptide antigens may be pathogenic in different patients even with the same disorder, and identifying the MHC- and TCR-binding residues of these peptides is a daunting task. It is also possible to administer native myelin antigens in ways that might induce peripheral T cell tolerance and ameliorate disease. Oral administration of MBP has already been shown to inhibit T cell responses and reduce the severity of EAE. Clinical trials of oral tolerance with MBP in patients with multiple sclerosis are now underway.

EAE has also been induced spontaneously, i.e., without immunization, by expressing an MBP-specific T cell receptor as a transgene in H-2^u mice. This proves that MBP-specific T cells can escape clonal deletion, mature, and enter peripheral tissues. In this model transgenic mice exposed to infectious microbes develop more severe disease than do pathogen-free animals. This result emphasizes a link between infections and autoimmunity that will be discussed in more detail in the text.

BOX 19-2.   Insulin-Dependent Diabetes Mellitus

Diabetes mellitus is a metabolic disease caused by a deficiency of insulin or its inadequate function, leading to abnormalities in glucose metabolism that result in ketoacidosis, thirst, and increased urine production. The late stage of the disease is characterized by progressive atherosclerotic vascular lesions, which can lead to gangrene of extremities due to arterial obstruction, renal failure due to glomerular and arterial injury, and blindness due to arterial aneurysms and increased fragility of proliferating vessels in the retina. Insulin-dependent diabetes mellitus (IDDM, also called juvenile or type I diabetes) affects about 0.2 per cent of the United States population, with a peak age of onset of 11 to 12 years. These patients have a deficiency of insulin resulting from destruction of the insulin-producing β cells of the islets of Langerhans in the pancreas, and continuous hormone replacement therapy is needed.

Several mechanisms may contribute to β cell destruction, including cytolytic T lymphocyte (CTL)-mediated lysis of islet cells, local cytokine (IFN-γ, TNF, and IL-1) production, and autoantibodies against islet cells. In the rare cases in which the pancreatic lesions have been examined at the early active stages of the disease, the islets show cellular necrosis and lymphocytic infiltrations. This lesion is also called insulitis. The infiltrates consist of both CD4+ and CD8+ T cells. Surviving islet cells often express class II MHC molecules. This aberrant expression of MHC molecules is probably an effect of local production of IFN-γ by the T cells. It has been suggested that abnormally high expression of MHC molecules may amplify T cell responses and worsen islet cell injury, but there is no evidence proving that islet cells can function as competent APCs. Autoantibodies against islet cells and insulin are also detected in the blood of these patients. These antibodies may participate in causing the disease or may be a result of T cell-mediated injury and release of normally sequestered antigens. In susceptible children who have not developed diabetes (such as relatives of patients), the presence of antibodies against islet cells is predictive of the development of IDDM. This suggests that the anti-islet cell antibodies contribute to injury to the islets.

Multiple genes are involved in IDDM. Recently, a great deal of attention has been devoted to the role of HLA genes. Ninety to 95 per cent of Caucasians with IDDM have HLA-DR3, or DR4, or both, in contrast to about 40 per cent of normal subjects, and 40 to 50 per cent of patients are DR3/DR4 heterozygotes, in contrast to 5 per cent of normal subjects. An interesting finding is that susceptibility to IDDM is actually associated with a linked DQ allele called DQ3.2 that is often in linkage disequilibrium with DR4. Sequencing of DQ molecules showed initially that all DQβ chains that are more common in IDDM patients than in control subjects have one of three amino acids (alanine, valine, or serine) at position 57 near the peptide-binding cleft, whereas the DQβ chains present at lower frequencies in IDDM patients than in control subjects have aspartic acid (Asp) at this position. The diabetes-prone NOD mouse strain has serine at position 57 in its I-Aβ chain (the murine homolog of DQβ). In contrast, most other (normal) mouse strains, including the closely related non-obese normal, have Asp at this position. These findings have led to the intriguing hypothesis that Asp57 in the DQβ chain protects against IDDM, and its absence increases susceptibility. However, several exceptions to this association of Asp57 with IDDM have emerged. For instance, Japanese patients with the disease frequently have Asp57 in their DQβ chains. Moreover, if I-A molecules containing either Ser or Asp at position 57 in the β chain are expressed as transgenes in NOD mice, the incidence of disease is reduced. It is likely, therefore, that the residue at position 57 is only one determinant of the function of the MHC molecule,

since it is one of the residues forming the peptide-binding cleft. Development of IDDM may be influenced by the structure of the entire cleft, with residue 57 playing a significant but not exclusive role. The HLA linkage of IDDM is also unusual in that it is inherited as a recessive trait—DR3/4 heterozygotes have a higher relative risk of the disease than individuals with either allele. The likely reason for this is that increased susceptibility is associated with the absence of certain residues in the class II MHC molecules, and the presence of these residues in either chain is protective. Finally, despite the high relative risk of IDDM in individuals with particular class II alleles, most persons who inherit these alleles do not develop the disease. For instance, the frequency of DQ3.2 in the general population is approximately 27 per cent, but only a very small fraction of these individuals develops IDDM. Non-HLA genes also contribute to the disease, but these are undefined. Furthermore, viral infections, particularly with coxsackievirus B-4, may precede the onset of IDDM, perhaps by initiating islet cell injury, altering self antigens, and triggering an autoimmune response. However, the nature of the antigens that initiate islet-specific immune responses is not known.

Animal models of spontaneous IDDM have been described. The inbred BB rat strain develops a T cell-mediated insulitis that is linked to both MHC and non-MHC genes. The non-obese diabetic (NOD) mouse strain also develops a spontaneous T cell-mediated insulitis. Disease can be transferred to young NOD mice with T cells from older, affected animals. As mentioned earlier, the linkage of this disease with the H-2 complex is remarkably similar to the HLA linkage of IDDM in humans. The NOD mouse expresses I-A but not I-E class II MHC molecules, and breeding with I-E-positive strains or the expression of various transgenic I-A or I-E molecules in NOD mice reduces the incidence of the disease. A postulated mechanism for this protection is that the transgenic class II MHC molecules may bind the diabetogenic self peptide antigen, thus competitively inhibiting its binding to endogenous class II and presentation to self-reactive T cells. In NOD mice, disease is induced by diabetogenic T cells that may recognize various islet antigens, including insulin and an islet cell enzyme called glutamic acid decarboxylase. Induction of T cell tolerance to these antigens retards the onset of diabetes in NOD mice.

Several experimental models have been created by expressing transgenes in pancreatic islet cells by introducing these genes into mice under the control of insulin promoters. Allogeneic class I and class II MHC molecules can be expressed in islet cells to test the hypothesis that this will create an endogenous "allograft" that should be attacked by the immune system. Insulitis does not develop in these mice, because T cells specific for the allogeneic MHC molecules become tolerant. Expression of the costimulator, B7-1, in islets increases susceptibility to insulitis (see text). Mice expressing IFN-γ in the islets develop insulitis and IDDM. In some models, if both allogeneic MHC molecules and IL-2 are expressed in islets, insulitis does develop, probably because local IL-2 production breaks T cell anergy and initiates an allogeneic reaction against the islets. Finally, if a T cell receptor specific for an islet cell antigen is expressed as a transgene in mice, the mice develop insulitis and diabetes. It is interesting that the transgene-expressing T cells are not deleted in the thymus, indicating that islet antigens do not cause central tolerance. In addition, insulitis develops many weeks before overt diabetes, but the factors responsible for disease progression are not known. These models are clearly valuable for studying T cell responses to tissue antigens, but their relevance to human IDDM is unclear.

model a foreign antigen, such as a viral protein, is expressed as a transgene in islet β cells. By itself this "self" antigen does not elicit an autoimmune reaction. However, co-expression of the viral antigen and B7-1 in islet cells breaks peripheral tolerance in viral antigen-specific T cells, triggers a response to the antigen, and results in insulitis (Fig. 19–4). Thus, aberrant expression of costimulators may predispose to autoimmune tissue injury by breaking anergy in self-reactive T cells. (In some similar models, the viral antigen expressed in islet cells does not induce anergy but rather seems to be ignored by the immune system. It is not known why particular self antigens may induce anergy or elicit no response.) To date, however, it has not been possible to prove that abnormal production of costimulators or inflammatory cytokines is an important initiating event in any spontaneous human or animal autoimmune disease. The main reason for this is that by the time clinical disease is apparent, the tissue is severely damaged by the autoimmune reactions, and the initiating stimuli cannot be identified.

T cell anergy may also fail because of abnormalities in the T cells themselves. For instance, in mice, knockout of the gene encoding CTLA-4, the inhibitory T cell receptor for the costimulators

B7-1 and B7-2, results in fatal autoimmunity with T cell infiltrates and tissue destruction involving the heart, pancreas, and other organs. The explanation for this finding may be that in order to be activated, T cells use the CD28 receptor to recognize B7 molecules on APCs, whereas recognition of B7-1 or B7-2 by the alternate receptor, CTLA-4, inhibits T cell responses and induces anergy. In the absence of CTLA-4, the T cells are resistant to anergy, and autoimmunity develops. To date there are no examples of spontaneous CTLA-4 defects associated with autoimmunity in experimental animals or humans.

Failure of Activation-Induced Cell Death. The first clear evidence demonstrating that failure of activation-induced cell death results in autoimmunity came from studies of two homozygous inbred mouse strains called *lpr/lpr* (for *l*ympho*pr*oliferation) and *gld/gld* (for *g*eneralized *l*ymphoproliferative *d*isease). These mice die by the age of 6 months from severe systemic autoimmune disease with multiple autoantibodies and nephritis. Phenotypically, the mouse disease resembles a human autoimmune disease called **systemic lupus erythematosus** (Box 19–3). (The mice also develop lymphadenopathy and splenomegaly, which is apparently unrelated to the autoimmune disease.)

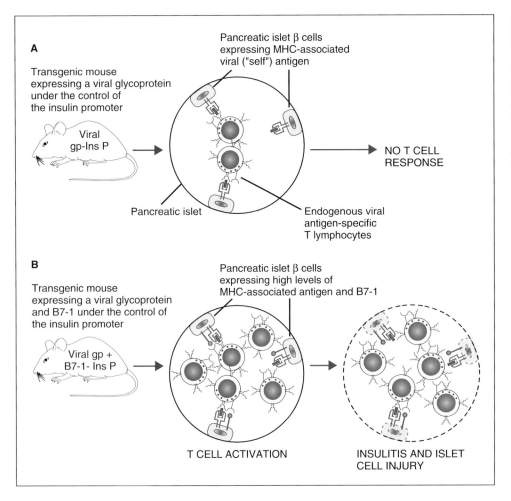

A

Transgenic mouse expressing a viral glycoprotein under the control of the insulin promoter

Viral gp-Ins P

Pancreatic islet β cells expressing MHC-associated viral ("self") antigen

Pancreatic islet

Endogenous viral antigen-specific T lymphocytes

NO T CELL RESPONSE

B

Transgenic mouse expressing a viral glycoprotein and B7-1 under the control of the insulin promoter

Viral gp + B7-1- Ins P

Pancreatic islet β cells expressing high levels of MHC-associated antigen and B7-1

T CELL ACTIVATION

INSULITIS AND ISLET CELL INJURY

FIGURE 19–4. Role of costimulators in T cell–mediated tissue-specific autoimmunity. The presence of an MHC-associated "self" antigen (e.g., a viral protein expressed throughout life) on pancreatic islet β cells does not induce an autoimmune response (*A*). Co-expression of the costimulator B7-1 with the antigen may break T cell anergy and result in autoimmune disease (*B*). Ins P, insulin promoter.

BOX 19-3. Systemic Lupus Erythematosus

Systemic lupus erythematosus (SLE) is a chronic, remitting and relapsing, multisystem autoimmune disease that affects predominantly women, with an incidence of 1 in 700 among women between the ages of 20 and 60 years (about 1 in 250 among black women) and a female:male ratio of 10:1. The principal clinical manifestations are skin rashes, arthritis, and glomerulonephritis, but hemolytic anemia, thrombocytopenia, and central nervous system involvement are also common. Many different autoantibodies are found in patients with SLE. The most frequent are antinuclear, particularly anti-DNA, antibodies; others include antibodies against ribonucleoproteins, histones, and nucleolar antigens. Immune complexes formed of these autoantibodies and their specific antigens are thought to be responsible for glomerulonephritis, arthritis, and vasculitis involving small arteries throughout the body. Hemolytic anemia and thrombocytopenia are due to autoantibodies against erythrocytes and platelets, respectively. The principal diagnostic test for the disease is the presence of antinuclear antibodies; antibodies against double-stranded native DNA are quite specific for SLE. The production of high-affinity antinuclear antibodies is dependent on helper T cells. The major population of pathogenic helper T cells seems to be reactive with peptides derived from nucleosomal proteins. Presumably, B cells specific for self DNA bind nucleosomal protein-DNA complexes, process the proteins, and present peptide epitopes to helper T cells, resulting in stimulation of the B cells and the production of anti-DNA autoantibodies. It is not known whether the primary pathogenic defect is in the B lymphocytes (failure of central or peripheral tolerance) or in helper T cells or both. Genetic factors also contribute to the disease. The relative risk for individuals with HLA-DR2 or -DR3 is 2 to 3, and if both haplotypes are present, the relative risk is about 5. Deficiencies of classical pathway complement proteins, especially C2 or C4, are seen in about 10 per cent of SLE patients but in only 1 per cent of the normal population.

Animal models of lupus provide valuable experimental systems for analyzing the pathogenesis of this disease. Several inbred mouse strains have been discovered spontaneously to develop autoimmune diseases that resemble human SLE to varying degrees.

The first to be described, and the one most like SLE, is the NZB strain and the (NZB × NZW)F1. Female mice develop kidney lesions and hemolytic anemia and produce anti-DNA autoantibodies spontaneously. Extensive breeding studies have shown that MHC genes from the NZW parent and non-MHC genes inherited from both parental strains contribute to the evolution of the disease. The B cells of (NZB ×

NZW)F1 mice are hyperresponsive to foreign antigens as well as to polyclonal activators and cytokines, and autoantibody production is dependent on pathogenic helper T cells reactive with nucleosomal peptides.

Two other mouse models with some resemblance to SLE are mice homozygous for the *lpr* (for *lymphoproliferation*) and *gld* (for *generalized lymphoproliferative disease*) genes. Both strains develop multiple autoantibodies, especially anti-DNA antibodies, and fatal immune complex nephritis. These inbred strains also develop severe lymphadenopathy because of the accumulation of an unusual population of functionally inert, CD3$^+$CD4$^-$CD8$^-$ T cells that also express a CD45 isoform called B220 that is normally found on B cells. The *lpr* gene is a defective *fas* gene, caused by a retrotransposon insertion into an intron, leading to abnormal transcription and greatly reduced Fas expression. The *gld* gene is a point mutation in the *fas ligand* gene, resulting in the inability of the expressed Fas ligand to deliver an apoptotic signal. In both *lpr* and *gld* homozygous mice, activation-induced cell death of mature CD4$^+$ T lymphocytes is defective, and this apparently results in the accumulation of self-reactive helper T cells. In addition, anergic B cells cannot be eliminated by Fas-mediated killing. Both functional abnormalities contribute to autoantibody production. The lymphadenopathy can be segregated from, and appears to be unrelated to, the autoimmune disease. The disease of *lpr* and *gld* mice is also influenced by background genes, being more severe in an inbred strain called MRL than in any other. Some children with a phenotypically similar disease are known to have mutations in *fas*, or inherit dominant negative mutants of *fas*. There is no evidence for Fas or Fas ligand abnormalities in typical SLE, and by this criterion these inbred mouse strains are not true animal models of human SLE.

A third inbred strain that develops a lupus-like disease is a recombinant called BXSB, in which disease susceptibility is linked to the Y chromosome, and only the males are affected. These mice produce anti-DNA antibodies and develop severe nephritis and vasculitis.

Another mutant mouse strain, called the viable motheaten mouse (because of skin lesions), also produces high levels of autoantibodies. The mutation in this strain affects a tyrosine phosphatase that is associated with several receptors thought to be important in the functions of immune cells (such as the type I IFN receptor). The signal transduction abnormalities caused by the phosphatase mutation, and the mechanism of autoantibody production in moth-eaten mice, are not defined.

The *lpr* defect is due to an abnormality in the gene encoding the cell death–inducing molecule, Fas, that reduces expression of this protein, and the *gld* defect is due to a point mutation in Fas ligand that abolishes the signaling capacity of this molecule. Defects in Fas or Fas ligand result in an inability to delete mature CD4$^+$ T cells by activation-induced cell death. This apparently results in the survival and persistence of helper T cells specific for self antigens that normally induce tolerance by repeatedly activating and thus deleting specific T cells. (The self antigens these T cells recognize are not known.) Thus, these mouse diseases are due to failure of the deletion mechanism of peripheral tolerance; central tolerance mechanisms do not ap-

pear to involve Fas and are apparently normal in these mice. Functional B cell abnormalities may also contribute to autoimmunity in Fas– or Fas ligand–defective mice. At least some anergic B cells are normally eliminated by Fas-dependent death resulting from interactions with T cells in peripheral lymphoid tissues, and this pathway of B cell deletion is defective in *lpr* and *gld* homozygous mice. Therefore, abnormalities in both helper T lymphocytes and B lymphocytes contribute to autoantibody production. A few dozen children with a phenotypically similar disease have been identified and shown to carry mutations in *fas* or in genes in the Fas-mediated death pathway, which result in a failure of activation-induced cell death.

No abnormalities in Fas or Fas ligand have been identified in typical systemic lupus erythematosus in humans.

Knockout mice lacking IL-2, or the α or β chain of the IL-2 receptor, develop severe splenomegaly and lymphadenopathy, as well as an autoimmune syndrome characterized by autoimmune hemolytic anemia and, occasionally, anti-DNA autoantibodies; some of these knockout mice also develop inflammatory bowel disease. (Mutations in the IL-2 receptor γ chain lead to immunodeficiency, mainly because of the loss of the lymphopoietic activity of IL-7, a cytokine whose receptor uses the same γ chain as IL-2.) The pathogenesis of this autoimmune disease is not known. It may result from failure of Fas-dependent cell death, since IL-2 potentiates Fas-mediated apoptotic signals. Such a hypothesis suggests that IL-2 may function as a "death-promoting factor" in some settings. (Several other knockout mouse strains develop inflammatory bowel disease due to various abnormalities; these are mentioned in Chapter 20.)

Failure of T Cell–Mediated Suppression. If self antigens normally induce regulatory T cells that produce immunosuppressive cytokines that function to maintain self-tolerance, then decline of these regulatory T cells may result in autoimmunity. There is, however, no human or animal autoimmune disease in which the primary abnormality is known to be a loss of regulatory T cells.

Polyclonal Lymphocyte Activation

Autoimmunity may result from antigen-independent stimulation of self-reactive lymphocytes that are not deleted during development. Polyclonal activators stimulate many T or B lymphocyte clones, irrespective of antigenic specificity and, in some cases, by interacting with surface molecules other than antigen receptors. A good example is bacterial lipopolysaccharide (LPS), which functions as a polyclonal B cell activator in mice (but not humans). Exposure to LPS may stimulate many clones of B lymphocytes, including self-reactive B cells that are anergic. Such anergic B cells are incapable of responding to the specific self antigens, but they may have retained their ability to proliferate and differentiate in response to antigen receptor-independent stimuli, such as LPS (Fig. 19–5). In fact, mice injected with LPS produce multiple autoantibodies (although in the absence of T cell help, most of these are low-affinity antibodies that do not cause disease). Polyclonal B cell activation may be induced by microbial products that act like LPS, and this may be one link between infections and autoimmunity. Polyclonal T cell activation by bacterial "super-antigens" (see Chapter 16, Box 16–1) is also a postulated mechanism for autoimmunity, but there is no good evidence causally linking super-antigens and autoimmunity.

Multiple autoimmune phenomena are associated with chronic graft-versus-host disease, especially in animal models. This disease develops after the transplantation of bone marrow containing allogeneic T cells into immunodeficient recipients (see Chapter 17). In the recipients of such transplants, grafted helper T cells may recognize host B lymphocytes as foreign (a form of alloreactivity), leading to cognate T-B cell interactions, polyclonal B cell activation, and autoantibody production in the absence of specific antigenic stimulation (Fig. 19–5).

Immunologic Cross-Reactions of Self and Foreign Antigens

Some autoimmune diseases are initiated by quite normal immune responses to foreign anti-

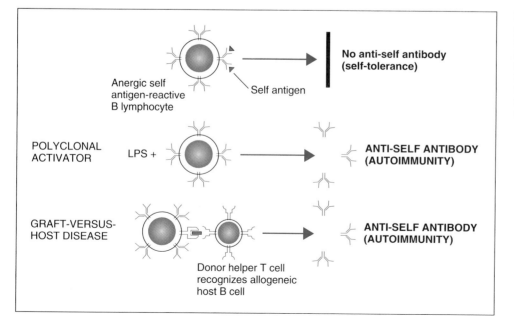

FIGURE 19–5. Polyclonal lymphocyte activation and autoantibody production. Self antigen–reactive B cells fail to respond to self antigens but may be stimulated by polyclonal activators such as LPS or by alloreactive T cells, as in graft-versus-host disease.

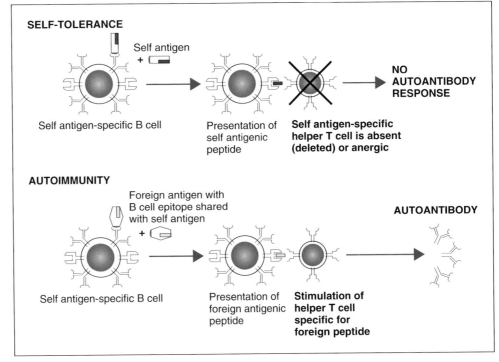

FIGURE 19–6. The induction of autoantibodies by antigenic cross-reactions. B cells specific for a self antigen may not be stimulated if T cells specific for the self antigen are absent or anergic (self-tolerance). However, these B cells may produce autoantibody if stimulated with a partially cross-reactive antigen containing foreign epitopes that are recognized by specific helper T cells (autoimmunity).

gens, such as microbes, but the antibodies or T cells that are stimulated happen to recognize a similar (cross-reactive) self protein. One example is rheumatic fever, which develops after streptococcal infections and is caused by anti-streptococcal antibodies that cross-react with human myocardial proteins, resulting in myocarditis. Strictly speaking, such diseases are not truly autoimmune but are sequelae of immune responses against foreign antigens. In other situations, autoreactive B cells may be present but may fail to produce autoantibodies because of the absence of helper T cells. Exposure to a multideterminant antigen, in which one epitope is self and binds to the B cells and another epitope is foreign and stimulates helper T cells, may lead to B cell activation and autoantibody production (Fig. 19–6). Molecular sequencing techniques have revealed numerous short stretches of homology between various microbial and self proteins. This homology is called **molecular mimicry,** and it is postulated to be one reason why immune responses against foreign antigens can lead to reactivity against self. However, the significance of such limited sequence homologies is unknown, and there is no clear evidence to support a role of molecular mimicry in human autoimmune diseases.

GENETIC FACTORS IN AUTOIMMUNITY

From the earliest studies of autoimmune diseases in patients and experimental animals, it has been appreciated that these diseases have a strong genetic component. For instance, insulin-dependent

diabetes mellitus, an autoimmune disease of pancreatic islets (see Box 19–2), shows a concordance of 35 per cent to 50 per cent in monozygotic twins and 5 per cent to 6 per cent in dizygotic twins. Family studies and the analysis of animal models by breeding and genotyping have shown that multiple susceptibility genes may contribute to an autoimmune disease. Often these genes show complex interactions, low and incomplete penetrance, and non-mendelian inheritance patterns. As a result, there is considerable genetic heterogeneity in autoimmune diseases. An important conclusion that has emerged is that *most susceptibility genes may increase the probability of getting a particular disease, but they alone do not determine whether an individual will or will not develop an autoimmune disorder.*

Among all the genes believed to be associated with autoimmunity, the strongest associations are with MHC genes, especially class II MHC genes, and we will discuss these next. Examples of non-MHC genes that may contribute to autoimmunity are discussed later in the chapter.

Role of MHC Genes in Autoimmunity

HLA typing of large groups of patients with various autoimmune diseases has shown that some HLA alleles occur at higher frequency in these patients than in the general population. From such studies, the relative risk of developing a disease in individuals who inherit various HLA alleles can be estimated (Table 19–2). The strongest such association is between ankylosing spondylitis, an inflam-

TABLE 19–2. Examples of Human Leukocyte Antigen (HLA)-Linked Immunologic Diseases

Disease	HLA Allele	Relative Risk*
Rheumatoid arthritis	DR4	6
Insulin-dependent dia-	DR3	5
betes mellitus	DR4	5–6
	DR3/DR4	20
Chronic active hepatitis	DR3	14
Sjögren's syndrome	DR3	10
Celiac disease	DR3	10
Dermatitis herpetiformis	DR3	50
Ankylosing spondylitis	B27	90–100

* Relative risk is defined as the probability of individuals with a particular HLA allele(s) to develop a disease compared with individuals lacking that HLA allele(s).

matory, presumed autoimmune, disease of vertebral joints, and the class I HLA allele B27. Individuals who are HLA-B27-positive have a 90 to 100 times greater chance of developing ankylosing spondylitis than individuals lacking B27. Neither the mechanism of this disease nor the basis of its association with HLA-B27 is known. Recently, much interest has focused on the polymorphic class II loci HLA-DR and HLA-DQ in autoimmune diseases, because these are most strongly associated with autoimmunity. Moreover, we now know that class II MHC molecules are involved in the selection and activation of CD4+ T cells, and CD4+ T cells regulate both humoral and cell-mediated immune responses to protein antigens.

Several issues about HLA associations with autoimmune diseases are worth emphasizing:

1. An HLA-disease association may be identified by serologic typing of one HLA locus, but the actual association may be with other alleles that are linked to the typed allele and inherited together. For instance, individuals with a particular HLA-DR allele (hypothetically, DR1) may show a higher probability of inheriting a particular HLA-DQ allele, hypothetically DQ2, than the probability of inheriting these alleles separately and randomly (i.e., at equilibrium) in the population. This is an example of "linkage disequilibrium." Thus, a disease may be found to be DR1-associated by HLA typing, but the causal association may actually be with the co-inherited DQ2. This realization has emphasized the concept of "extended HLA haplotypes," which refers to sets of linked genes, both classical HLA and adjacent non-HLA genes, that tend to be inherited together as a single unit (see Chapter 5).

2. The HLA molecules that are actually disease associated may be a subset of an HLA allele identified by serology. The reason for this is that a single serologically defined allele may actually consist of a family of related HLA alleles that differ slightly from one another in their polymorphic residues. Such differences can be identified only by more

detailed molecular studies, such as nucleotide sequencing. Sequencing of HLA genes in patients with autoimmunity has shown that some HLA-disease associations are much stronger than they appeared to be when calculations were based on less precise serologic typing. Such analyses also indicate that in many autoimmune diseases, the HLA molecules that show increased frequencies differ only in the peptide-binding clefts from HLA molecules that are not disease-associated. In some ways, this finding is not surprising, because polymorphic residues of MHC molecules are located within and adjacent to the clefts (see Chapter 5). Nevertheless, because the structure of the cleft is the key determinant of both functions of MHC molecules, namely peptide antigen display and recognition by T cells (see Chapter 6), these results support the general concept that *MHC molecules influence the development of autoimmunity by controlling T cell selection and activation.*

3. The inheritance of some HLA genes may predispose to particular autoimmune diseases, whereas others may be protective, i.e., their absence may be associated with increased incidence of disease. Examples of both are described below.

4. Although some diseases show strong associations with certain HLA alleles (see Table 19–2), there are also many reports of weak associations, with relative risks ranging from 1.5 to 2. In fact, studying HLA-disease associations has become a popular exercise in clinical and research laboratories. The significance of these weak associations is uncertain, at best.

The example of **insulin-dependent diabetes mellitus** (see Box 19–2) illustrates many of these features of HLA-disease associations. A particularly interesting finding is that in Caucasians, the expression of HLA-DQ alleles that encode an aspartic acid residue at a particular position (position 57 of the β chain) in the peptide-binding cleft protects against the disease. A mouse model of IDDM shows the same protective effect of aspartic acid in the same location of I-A, the murine homolog of DQ.

Several mechanisms have been postulated to explain the association of autoimmune diseases with the inheritance of particular MHC sequences.

1. The structures of MHC molecules may determine which clones of T lymphocytes are negatively selected during their maturation. For instance, if the MHC molecules in the thymus of an individual cannot bind a self protein with high affinity, immature T cells reactive with this self antigen may escape negative selection and mature to functional competence. Such a mechanism may explain the HLA-DQβ–associated resistance to IDDM—failure to inherit "protective" HLA alleles may result in the failure to select autoreactive T cells negatively. This hypothesis, however, raises an obvious problem. If the relevant MHC molecule is absent, how can autoreactive T cells that mature and enter peripheral tissues ever recognize the self

antigen that must be presented in association with this MHC molecule? At present, there is no answer to this question.

2. Class II MHC molecules may influence the activation of regulatory T cells whose normal function is to prevent autoimmunity. For instance, in IDDM, disease-producing T cells may be specific for a self peptide presented in association with HLA-DQ molecules lacking aspartic acid at position 57 of the β chain, whereas T cells that prevent tissue injury may be specific for a complex of the self peptide and HLA-DQ with aspartic acid at position 57. This would explain why inheritance of a DQ allele with aspartic acid in a particular position might protect against IDDM. However, no evidence exists to support a difference in the MHC restriction or fine specificity of functionally distinct self-reactive T cell subpopulations.

3. The disease-associated gene may not be an HLA allele itself but another gene located in the HLA complex. The search for such disease susceptibility genes in the MHC locus has not provided clear answers yet.

4. Similarities between microbial antigens and self MHC molecules may result in autoimmune reactions following infections. Although examples of such molecular mimicry have been described, their pathogenic significance is uncertain.

Thus, the available data do not permit firm conclusions about the mechanisms by which inherited MHC genes contribute to autoimmunity. Disease-associated HLA sequences are found in healthy individuals. In fact, as stated previously, if all individuals bearing a particular disease-associated HLA allele are followed prospectively, the vast majority will never develop the disease. Therefore, *the expression of a particular HLA gene is not by itself the cause of any autoimmune disease, but may be one of several factors that contribute to autoimmunity.*

Association of Non-MHC Genes with Autoimmunity

Rapid progress in gene mapping techniques is leading to better localization of susceptibility genes for autoimmune diseases. It is clear that multiple non-MHC genes contribute to autoimmunity. For instance, in human IDDM and its mouse counterpart, the inbred non-obese diabetic strain, about 20 genes appear to be associated with disease susceptibility. The MHC is the only one common to humans and mice, and its contribution is almost as much as that of all the other loci combined. Nevertheless, there is great interest in identifying the non-MHC genes associated with various autoimmune diseases. To date, very few are actually known. The role of *fas* and *fas ligand* in systemic autoimmunity was mentioned earlier. Genetic deficiencies of several complement proteins, including C2 and C4 (see Chapter 15), are associated

with lupus-like autoimmune diseases. There has also been considerable interest in identifying T cell receptor V genes, e.g., Vβ genes, that may be associated with autoimmunity. These studies have failed to show an association of autoimmunity with a particular TCR (or Ig) V gene polymorphism. In fact, it is apparent that autoreactive lymphocytes express the same antigen receptor V, (D), and J gene segments as are present in lymphocytes specific for foreign antigens. In many organ-specific autoimmune diseases, T cells in lesions tend to express limited numbers of V genes. This is likely due to the fact that each disease is caused by reactions against relatively few self antigens, so that relatively few lymphocyte specificities are expanded in the lesions.

At the present time, many of the genetic associations with autoimmune diseases are simply chromosomal locations, based on polymorphic markers. Nevertheless, the locations of some of these genes raise intriguing possibilities. For instance, one susceptibility gene for IDDM is located in a cluster of genes encoding several cytokines, including IL-4 and the p40 chain of IL-12. It is possible that this susceptibility gene regulates the balance between pro-inflammatory and anti-inflammatory cytokines. Another IDDM susceptibility gene is close to CTLA-4, and the possible role of this molecule in T cell anergy has been mentioned previously. Although these are interesting speculations, defining the roles of various susceptibility genes in autoimmunity will require more precise identification of the genes. This should also help elucidate the pathogenetic mechanisms of autoimmune diseases.

INFECTIONS, ANATOMIC ALTERATIONS, AND OTHER FACTORS IN AUTOIMMUNITY

The development of autoimmunity is related to several factors in addition to primary immunologic abnormalities and susceptibility genes.

1. *Viral and bacterial infections* are associated with autoimmunity, and infectious prodromes often precede the clinical manifestations of autoimmune diseases. In most of these cases, the infectious microorganism is not present in lesions and is not even detectable in the individual when autoimmunity develops. Therefore, the lesions of autoimmunity are not due to the infectious agent itself but result from host immune responses that may be triggered or dysregulated by the microbe. The many possible effects of infections include polyclonal lymphocyte activation, local tissue inflammation leading to enhanced expression of costimulators, alterations of self antigens to create partially cross-reactive neo-antigens, and tissue injury leading to release of anatomically sequestered antigens.

2. *Anatomic alterations in tissues,* such as inflammation (possibly secondary to infections), is-

chemic injury, or trauma, may lead to the exposure of self antigens that are normally concealed from the immune system. Such sequestered antigens may not have induced self-tolerance. Therefore, if previously sequestered self antigens are released, they can interact with immunocompetent lymphocytes and induce specific immune responses. Examples of anatomically sequestered antigens include intraocular proteins and sperm. Post-traumatic uveitis and orchitis, and orchitis following vasectomy, are thought to be due to autoimmune responses to self antigens that are released from their normal location.

Tissue inflammation may also cause structural alterations in self antigens and the formation of new determinants capable of inducing autoimmune reactions. Inflammation may result in macrophage activation by locally produced cytokines, and if these cytokines stimulate the expression of costimulators, the result may be loss of peripheral tolerance.

3. *Hormonal influences* play a role in human and experimental autoimmune diseases. For instance, SLE affects females about ten times as frequently as males. The SLE-like disease of (NZB × NZW)F1 mice develops only in females and is retarded by androgen treatment. Many other autoimmune diseases have a higher incidence in females. Whether this predominance results from the influence of sex hormones or other factors is not known.

Autoimmune diseases are among the most challenging clinical and scientific problems in immunology. The current knowledge of operative mechanisms remains incomplete, so that theories and hypotheses continue to outnumber facts. The application of new technical advances and the rapidly improving understanding of self-tolerance will lead, it is hoped, to clearer and more definitive answers to the enigmas of autoimmunity.

SUMMARY

The immune system responds to foreign antigens but is unresponsive (tolerant) to each individual's self antigens. Self-tolerance is maintained by selection processes that kill or block the maturation of potentially self-reactive lymphocytes and by mechanisms that inactivate self-reactive lymphocytes in peripheral tissues. Central tolerance is due to death or inactivation of immature lymphocytes that specifically recognize self antigens in the generative lymphoid organs. Peripheral T cell tolerance is mainly due to inactivation (anergy) of mature T cells that recognize self antigens presented by costimulator-deficient, resting APCs, or death of T cells that are repeatedly stimulated by persistent self antigens. The mechanisms of peripheral B cell tolerance are less well understood.

Autoimmunity develops as a result of multiple interacting factors, which collectively lead to a failure or breakdown of self-tolerance. The immunologic mechanisms that may contribute to autoimmunity include abnormalities in lymphocyte selection, mechanisms that overcome peripheral tolerance, polyclonal lymphocyte stimulation, and cross-reactions between foreign and self antigens. The strongest genetic association of autoimmunity is with MHC genes, and multiple mechanisms have been proposed to account for such associations. Infections, injury to tissues, and hormonal factors may also contribute to the development of autoimmune diseases. Recent advances in the understanding of self-tolerance, and in techniques for analyzing disease susceptibility genes, hold great promise for elucidating the mechanisms of autoimmunity, and for developing rational therapeutic strategies for this group of diseases.

Selected Readings

Goodnow, C. C., J. G. Cyster, S. B. Hartley, S. E. Bell, M. P. Cooke, J. I. Healy, S. Akkaraju, J. C. Rathmell, S. L. Pogue, and K. P. Shokat. Self-tolerance checkpoints in B lymphocyte development. Advances in Immunology 59:279–368, 1995.

Kotzin, B. L. Systemic lupus erythematosus. Cell 85:303–306, 1996.

Kroemer, G., J. L. Andrew, J. A. Gonzalo, J. C. Gutierrez-Ramos, and C. Martinez. Interleukin-2, autotolerance, and autoimmunity. Advances in Immunology 50:147–235, 1991.

Kruisbeek, A. M., and D. Amsen. Mechanisms underlying T-cell tolerance. Current Opinion in Immunology 8:233–244, 1996.

Liblau, R. S., S. M. Singer, and H. O. McDevitt. T_H1 and T_H2 cells in the pathogenesis of organ-specific autoimmune diseases. Immunology Today 16:34–38, 1995.

Miller, J. F. A. P., and R. A. Flavell. T cell tolerance and autoimmunity in transgenic models of central and peripheral tolerance. Current Opinion in Immunology 6:892–899, 1994.

Nagata, S., and T. Suda. Fas and Fas ligand: *lpr* and *gld* mutations. Immunology Today 16:39–43, 1995.

Nepom, G. T., and H. Erlich. MHC class II molecules and autoimmunity. Annual Review of Immunology 9:493–525, 1991.

Nossal, G. J. V. Negative selection of lymphocytes. Cell 76:229–239, 1994.

Rose, N. R., and C. Bona. Defining criteria for autoimmune diseases (Witebsky's postulates revisited). Immunology Today 14:426–430, 1993.

Tisch, R., and H. O. McDevitt. Insulin-dependent diabetes mellitus. Cell 85:291–297, 1996.

Vyse, T. J., and J. A. Todd. Genetic analysis of autoimmune disease. Cell 85:311–318, 1996.

CHAPTER TWENTY

IMMUNE-MEDIATED TISSUE

INJURY AND DISEASE

Specific immunity is a powerful homeostatic mechanism for eliminating pathogenic microbes and other foreign antigenic substances. The effector mechanisms of specific immunity, such as complement, phagocytes, inflammatory cells, and cytokines, are not themselves specific for foreign antigens. Therefore, immune responses and attendant inflammation are often accompanied by local and systemic injury to self tissues. Usually, however, such pathologic side effects are controlled and self-limited, and they abate as the foreign antigen is eliminated. Furthermore, normal individuals are tolerant of their own antigens and do not develop immune responses against autologous tissues. *Failure to control physiologic immune responses against foreign antigens or to maintain self-tolerance leads to diseases in which the primary pathogenic mechanism is immunologic.* Disorders that result from aberrant, excessive, or uncontrolled immune reactions are also called **hypersensitivity diseases.** This term arises from the clinical definition of immunity as "sensitivity," which is based on the observation that an individual who has been exposed to an antigen exhibits a detectable reaction to, or is "sensitive to," subsequent encounters with that antigen. As applied to the historical definition of immunity to microbes, such a "sensitive" individual, of course, would usually be resistant to infection by that microbe. Protective immunity and hypersensitivity may co-exist because both are manifestations of specific immune responses. Immunologic diseases that are thought to be due to immune responses against self antigens are called **autoimmune diseases** (see Chapter 19).

In this chapter we discuss the mechanisms by which humoral and cell-mediated immune responses lead to diseases, using examples of clinical and experimental disorders to illustrate the current understanding of their etiology and pathogenesis. We also touch on the principles of diagnosis and treatment of such diseases.

TYPES OF IMMUNOLOGIC DISEASES

Immunologic diseases comprise a clinically heterogeneous group of disorders. The two principal factors that determine the clinical and pathologic manifestations of such diseases are (1) the type of immune response that leads to tissue injury, and (2) the nature and location of the antigen that initiates or is the target of this response.

The most frequently used *classification of immunologic diseases is based on the principal pathogenic mechanism responsible for cell and tissue injury* (Table 20–1). Immediate hypersensitivity caused by IgE antibodies and mast cells, which is also called type I hypersensitivity, has been described in Chapter 14. Antibodies other than IgE can cause tissue injury by recruiting and activating inflammatory cells and the complement system (see Chapter 15). These antibodies may be specifi-

TABLE 20–1. Classification of Immunologic Diseases

Type of Hypersensitivity	Pathologic Immune Mechanisms	Mechanisms of Tissue Injury and Disease
Type I: immediate hypersensitivity	IgE antibody	Mast cells and their mediators (vasoactive amines, lipid mediators, cytokines)
Type II: Antibody-mediated	IgM, IgG antibodies against tissue or cell surface antigen	1. Complement activation 2. Recruitment and activation of leukocytes (neutrophils, macrophages) 3. Abnormalities in receptor functions
Type III: immune complex–mediated	Immune complexes of circulating antigens and IgM or IgG antibodies	1. Complement activation 2. Recruitment and activation of leukocytes
Type IV: T cell–mediated	1. CD4+ T cells (delayed-type hypersensitivity) 2. CD8+ CTLs (T cell–mediated cytolysis)	1. Activated macrophages, cytokines 2. Direct target cell lysis, cytokines

Abbreviations: Ig, immunoglobulin; CTL, cytolytic T lymphocyte.

cally reactive with one's own antigens or with foreign antigens that are deposited in, or are antigenically cross-reactive with, self antigens. Such disease-producing antibodies may be detectable in two forms. Some can be found bound to their target antigens or in the circulation in a free form, and the diseases they cause are called type II hypersensitivity. Other antibodies may form immune complexes in the circulation, and the complexes subsequently deposit in tissues, particularly in blood vessels, and cause injury. Diseases caused by immune complexes are classified under type III hypersensitivity. Finally, tissue injury may be due to activated T lymphocytes and the effector cells of delayed-type hypersensitivity (DTH), mainly activated macrophages; these are called type IV hypersensitivity disorders. In some T cell–mediated diseases, cell lysis by cytolytic T lymphocytes (CTLs) is the principal cause of tissue injury or it co-exists with DTH.

In our discussion, we use descriptions that identify the pathogenic mechanisms rather than the less informative numeric designations. This classification is useful because distinct types of pathogenic immune responses show quite different patterns of tissue injury and may vary in their tissue specificity. As a result, they produce disorders with distinct clinical and pathologic features. However, immunologic diseases in the clinical situation are often complex and are due to various combinations of humoral and cell-mediated immune responses and multiple effector mechanisms. This is not surprising, given that a single antigen may normally stimulate both humoral and cell-mediated immune responses.

Immunologic diseases may be due to immune responses to self (autologous) or foreign antigens. Immune responses against self antigens are usually abnormal. The mechanisms of autoimmunity have been described in Chapter 19. *Immune responses to foreign antigens may be pathogenic in several situations.* First, some microbes persist for prolonged periods because they resist elimination by immune and inflammatory mechanisms. This leads to persistent antigenic stimulation, resulting in a response of increasing magnitude associated with severe tissue injury. Second, some foreign antigens may share antigenic determinants with self tissues, and elicit immune responses that cross-react with self antigens. Third, the foreign antigen may be deposited, or "planted," in a particular tissue because of a physicochemical affinity with normal tissue components, so that an immune response directed against the foreign antigen becomes targeted to the tissue in which this antigen is fixed. Fourth, normal immune responses may become defective in their self-regulation, so that they continue unabated even after the initiating foreign antigen is eliminated. Examples of diseases caused by these different kinds of immune responses to foreign antigens are mentioned later in this chapter.

DISEASES CAUSED BY ANTIBODIES

The first immunologic diseases in which the pathogenic mechanisms were identified were diseases caused by the deposition of antibodies in tissues. This discovery occurred largely because techniques for detecting abnormal circulating autoantibodies and immunoglobulins (Ig) deposited in tissues were developed well before methods for identifying and isolating T cells from lesions or from the blood of patients. Moreover, in experimental models of immunologic diseases, it was possible to cause tissue injury by transferring purified antibodies against tissue antigens before pure or clonal populations of tissue-reactive T cells became available. For historical reasons, therefore, many of the general principles of immunologic diseases are based on antibody-mediated disorders.

Antibody-mediated diseases are of two types, which differ in their clinicopathologic manifestations and are due to the deposition of antibodies in distinct forms (Fig. 20–1):

1. *Immunologic diseases may be produced by immune complexes composed of a soluble antigen and specific antibody; such complexes are formed in the circulation and may deposit in vessel walls virtually anywhere in the body.* This leads to local activation of leukocytes and the complement system, with resultant tissue injury. The antigens that induce the pathogenic humoral immune response can be foreign or self antigens, and the antibodies in the complexes are usually IgM or IgG because these isotypes are most efficient at activating complement and/or inflammatory cells. The pathologic features of such diseases reflect the site(s) of immune complex deposition and are not determined by the cellular source of the antigen. Therefore, immune complex–mediated diseases tend to be systemic, with little or no specificity for a particular tissue or organ.

2. *Antibodies against circulating cells or fixed tissue antigens cause diseases that are specific for that cell or tissue.* The lesions are due to the binding of specific antibodies and not to the deposition of immune complexes formed in the circulation. In most cases, such antibodies are autoantibodies, although occasionally they may be produced against a foreign antigen that is immunologically cross-reactive with a component of self tissues. Such antibodies are usually of the IgM or IgG class, and they cause disease by activating the same effector mechanisms as immune complexes. Some immunologic diseases are due to antibodies specific for cellular structures, such as hormone receptors, that are important for normal function. In these situations, diseases may occur because of interference with the normal functions of these structures and not because of antibody- or complement-mediated inflammation leading to actual tissue injury.

To prove that a particular disease is caused by antibodies, one would need to demonstrate that the lesions can be induced in a normal animal by the adoptive transfer of Ig purified from the blood or affected tissues of individuals with the disease. An experiment of nature is occasionally seen in children of mothers suffering from antibody-mediated diseases. These infants may be born with transient expression of the diseases because of transplacental passage of antibodies. However, in the usual clinical situations the diagnosis of antibody-mediated disease is based on the following criteria: (1) the demonstration of antibodies or immune complexes deposited in tissues, (2) the presence of anti-tissue antibodies or immune complexes in the circulation, and (3) clinicopathologic similarities with experimental diseases that are proved to be antibody-mediated by adoptive transfer.

Mechanisms of Antibody-Mediated Tissue Injury and Functional Abnormalities

In normal immune responses, the protective functions of antibodies are mediated by neutralization of the antigen, activation of the complement system, and recruitment of host inflammatory cells. The same effector mechanisms are responsible for the pathologic consequences of antibody or immune complex deposition. Which effector systems are involved in mediating the protective functions or pathologic effects of different antibodies is

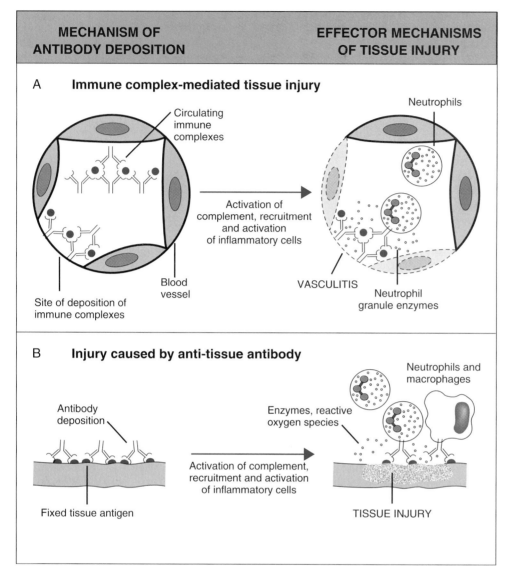

FIGURE 20-1. Types of antibody-mediated diseases. Antibodies may be deposited as immune complexes that are formed in the circulation (*A*) or by binding specifically to tissue antigens (*B*). In both cases, similar effector mechanisms lead to tissue injury at the sites of antibody deposition.

determined largely by the isotype of the Ig and the nature of the target antigen:

1. *Complement-mediated lysis of cells* occurs after IgM and some classes of IgG antibodies bind to their specific antigens (see Chapter 15). Complement activation leads to the generation of the membrane attack complex, which causes osmotic lysis of cells (Fig. 20-2*A*).

2. *Phagocytosis of antibody-coated cells* (Fig. 20-2*B*) may lead to selective depletion of the cells. For instance, in autoimmune hemolytic anemia, autoantibodies are produced against self erythrocytes. The opsonized erythrocytes bind to Fcγ receptors on macrophages, mainly in the liver and spleen, and are phagocytosed. This leads to depletion of the erythrocytes and hence gives rise to anemia.

3. *Recruitment and activation of inflammatory cells,* mostly neutrophils and monocytes, occur at sites of antibody deposition. There are two mecha-

nisms of antibody-dependent leukocyte recruitment and activation. First, antigen-antibody complexes activate the complement cascade, leading to the local generation of complement by-products, particularly C5a, which is chemotactic for leukocytes (Fig. 20-2*C*). Second, neutrophils and macrophages express surface receptors specific for the Fc portions of γ heavy chains and can therefore bind to and be activated by antigen-complexed IgG antibodies even in the absence of complement activation (Fig. 20-2*D*). In fact, knockout mice lacking the signaling (γ) chain of leukocyte Fcγ receptors (see Chapter 3, Box 3-4) do not develop immune complex–mediated lesions, whereas mice lacking complement proteins (C3 or C4) do. This suggests that, at least in mice, Fc receptor–mediated leukocyte recruitment and activation is the major pathway of antibody-induced tissue injury and inflammation. Cytokines, such as tumor necrosis factor (TNF), and interleukin-1 (IL-1), which are produced by activated mast cells, macrophages, and neutro-

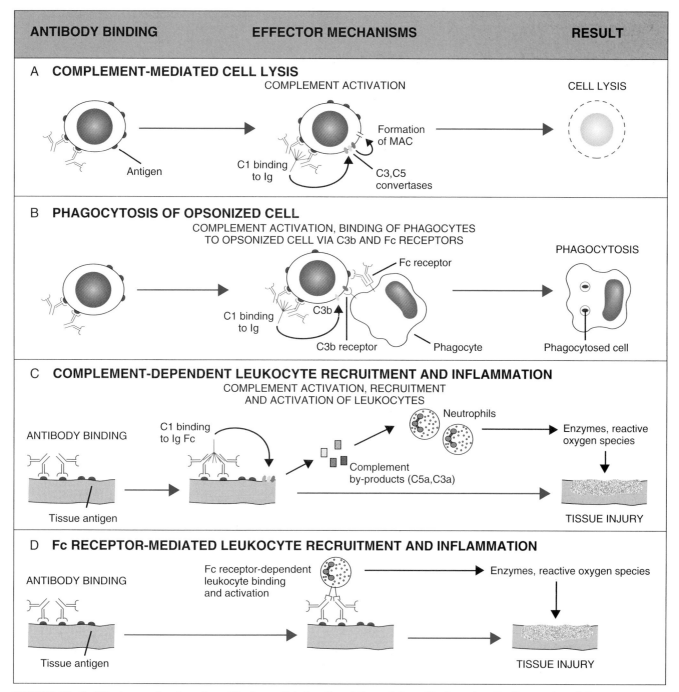

FIGURE 20-2. Effector mechanisms in antibody-mediated cell and tissue injury. Binding of antibodies, such as IgG, to antigens on a cell or tissue may cause injury by different effector mechanisms (*A–D*).
A. Activation of complement and formation of the cytocidal membrane attack complex (MAC).
B. Phagocytosis of opsonized cells by macrophages or neutrophils.
C. Recruitment and activation of neutrophils by complement by-products, followed by neutrophil degranulation and release of cytocidal substances.
D. Activation of neutrophils by binding of Fcγ receptors to antibodies.

phils, may also play a role in the recruitment of leukocytes to sites of antibody deposition. Antagonists to these cytokines inhibit antibody and immune complex–mediated tissue injury in some experimental models. Activated neutrophils and macrophages produce hydrolytic enzymes, reactive oxygen species, lipid mediators, and nitric oxide (in mice), which can all contribute to cell and tissue injury.

4. *Antibodies can cause pathologic effects by binding to functionally important molecules and altering cellular functions without causing tissue injury.* Examples of such diseases are described later in the chapter.

Immune Complex–Mediated Diseases

The occurrence of diseases due to immune complexes was suspected as early as 1911 by an astute physician named Clemens von Pirquet. At that time, diphtheria infections were being treated with serum from horses immunized with the diphtheria toxin. This is an example of passive immunization against diphtheria toxin by the transfer of serum containing anti-toxin antibodies. Von Pirquet noted that patients injected with the anti-toxin–containing horse serum developed joint inflammation (arthritis), skin rash, and fever. Two clinical features of this reaction suggested that it was not due to the infection or a toxic component of the serum itself. First, these symptoms appeared even after the injection of horse serum not containing the anti-toxin, so that the lesions could not be attributed to the anti-diphtheria antibody. Second, the symptoms appeared at least a week after the first injection of the horse serum and more rapidly with each repeated injection. Von Pirquet concluded that this disease was due to a host response to some component of the serum. He suggested that the host made antibodies to horse serum proteins, these antibodies formed complexes with the injected proteins, and the disease was due to the antibodies or immune complexes. We now know that his conclusions were entirely accurate. He called this disease "serum disease"; it is now more commonly known as **serum sickness** and is the prototype for systemic immune complex–mediated disorders.

EXPERIMENTAL MODELS OF SERUM SICKNESS

Much of our current knowledge of immune complex diseases is based on analyses of experimental models of serum sickness, performed in detail by Frank Dixon and his associates in the 1960s using techniques for accurately measuring the levels of antigens and antibodies in the blood and tissues. These investigators showed that if a rabbit is injected intravenously with a single dose (greater than 50 mg/kg of body weight) of a foreign protein antigen, bovine serum albumin (BSA), within a few days the rabbit begins to produce specific anti-BSA antibodies (Fig. 20–3). These antibodies complex with circulating BSA, leading to enhanced phagocytosis and clearance of the antigen by macrophages in the liver and spleen. Immune complexes are initially detected in the circulation and then deposit in tissues, where they activate complement, with a concomitant fall in serum complement levels. Complement activation leads to recruitment and activation of inflammatory cells, predominantly neutrophils, at the sites of immune complex deposition, and the neutrophils cause tissue injury. Since the complexes deposit mainly in arteries, renal glomeruli, and the synovia of joints, the clinical and pathologic manifestations are vasculitis, nephritis, and arthritis. The clinical symptoms are usually short-lived, and the lesions heal unless the antigen is injected again. This type of disease is an example of **acute serum sickness.** It is produced by the administration of a single large dose of a foreign antigen and

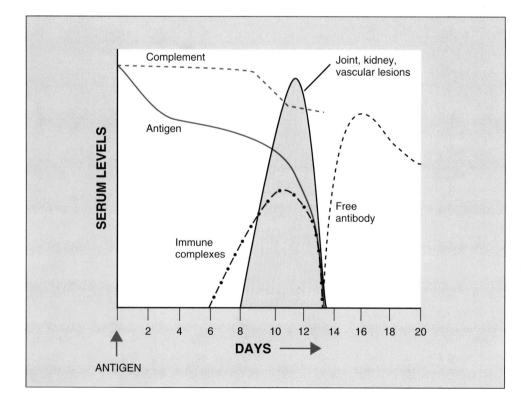

FIGURE 20–3. Sequence of immunologic responses in experimental acute serum sickness. Injection of bovine serum albumin into a rabbit leads to the production of specific antibody and the formation of immune complexes. These complexes deposit in tissues, activate complement (leading to a fall in serum complement levels), and cause lesions, which resolve as the complexes as well as the remaining antigen are removed. (Adapted with permission from Cochrane, C. G. Immune complex–mediated tissue injury. In S. Cohen, P. A. Ward, and R. T. McCluskey (eds.). Mechanisms of Immunopathology. Werbel & Peck, New York, 1979, pp. 29–48.)

is characterized by the deposition of large immune complexes. A more chronic disease, called **chronic serum sickness,** is produced by multiple injections of antigen, which lead to the formation of smaller complexes that deposit most often in kidneys, arteries, and lungs.

A localized form of experimental immune complex–mediated vasculitis is called the **Arthus reaction.** It is induced by injecting an antigen subcutaneously into a previously immunized animal. In some models, immunization is passive, i.e., the animal is first given intravenous antibody specific for the antigen. Circulating antibodies rapidly bind to the injected antigen, forming immune complexes that deposit in the walls of small arteries at the injection site. This gives rise to a local cutaneous vasculitis with necrosis. A similar reaction can be elicited by injecting antibody into the tissue and antigen intravenously; this reaction is called a "reverse passive Arthus reaction." As we shall discuss later, various diseases in humans are believed to be the clinical counterparts of acute and chronic serum sickness and the Arthus reaction.

FACTORS THAT INFLUENCE IMMUNE COMPLEX DEPOSITION

From analyses of these experimental models of immune complex–mediated diseases, it is now known that several factors determine the extent of immune complex deposition.

1. The *size of circulating immune complexes* is a major factor, because very small complexes are not deposited, and large ones are phagocytosed by mononuclear phagocytes and cleared. Usually, small and intermediate-sized immune complexes are prone to tissue deposition, but this may vary with different combinations of antigens and antibodies.

2. The extent of immune complex deposition in tissues is inversely proportional to the *ability of the host to clear immune complexes from the circulation.* Removal of circulating immune complexes is determined by the functional integrity of the mononuclear phagocyte system and the binding of complement proteins, which enhance the clearance of the complexes. Defective phagocytosis may promote the persistence and subsequent tissue deposition of immune complexes. There is a high incidence of immune complex diseases in patients with genetic deficiencies of proteins of the classical complement pathway, such as C2 and C4 (see Chapter 15), mainly because defective production of C3b by antigen-antibody reactions and the absence of complement receptor-mediated phagocytosis lead to persistence of immune complexes in the blood. In this situation, immune complexes that deposit in tissues presumably recruit inflammatory cells by Fcγ receptor-mediated, complement-independent mechanisms.

3. The *physicochemical properties of antigens and antibodies,* including charge, valence, avidity of interaction, and Ig isotype, may influence immune complex formation and deposition. For instance, complexes containing cationic antigens bind avidly to negatively charged components of the basement membranes of kidney glomeruli. Such complexes typically produce severe and long-lasting tissue injury.

4. *Anatomic and hemodynamic factors* are important determinants of the sites of immune complex deposition. Capillaries in renal glomeruli and synovia are vessels in which plasma is ultrafiltered (to form urine and synovial fluid, respectively) by passing through the capillary wall at high hydrostatic pressure, and these are among the most common sites of immune complex deposition.

5. Finally, immune complexes are thought to bind to inflammatory cells and stimulate *local secretion of cytokines and vasoactive mediators,* which cause increased adhesion of leukocytes to the endothelium, increased vascular permeability, and enhanced deposition of immune complexes in vessel walls by enlarging interendothelial spaces. This may lead to amplification of tissue injury and disease.

ROLE OF IMMUNE COMPLEXES IN TISSUE INJURY AND DISEASE

Antigen-antibody complexes are produced during many immune responses but are of pathologic significance only if the quantity, structure, or clearance of the complexes are such that abnormally large amounts are deposited in tissues. The *morphologic hallmarks of immune complex–mediated tissue injury are (1) necrosis,* which often contains fibrin because of leakage of plasma proteins and is also called *fibrinoid necrosis,* and *(2) cellular infiltrates composed predominantly of neutrophils.* Irregularly shaped (granular) deposits of antibody and complement components can be detected in these tissues by immunofluorescence, and if the antigen is known, it is possible also to identify antigen molecules in the deposits.

Compelling evidence supports a primary *pathogenic role of immune complexes in many human systemic immunologic diseases* (Table 20–2). **Systemic lupus erythematosus** (SLE) (Chapter 19, Box 19–3) is an autoimmune disease in which numerous autoantibodies are produced. Its many clinical manifestations include glomerulonephritis and arthritis, which are attributed to the deposition of immune complexes composed of self DNA or nucleoprotein antigens and specific antibodies (Fig. 20–4). The glomerular lesions of SLE often resemble the lesions seen in chronic serum sickness. Some cases of a form of systemic vasculitis called **polyarteritis nodosa** occur as a late sequel of hepatitis B virus infection and are due to arterial deposition of immune complexes composed of hepatitis virus surface antigen and specific antibodies. **Post-streptococcal glomerulonephritis** is a kidney disease that develops 1 to 3 weeks after streptococcal skin and throat infections. It is thought to

TABLE 20–2. Examples of Human Immune Complex Diseases

Disease	Antigen	Antibody	Clinicopathologic Manifestations
Post-streptococcal glomerulonephritis	Streptococcal cell wall antigen(s)	Anti-streptococcal antibody	Nephritis with glomerular lesions
Systemic lupus erythematosus	DNA, nucleoproteins, others	Autoantibodies (various)	Nephritis, arthritis, vasculitis (disseminated)
Polyarteritis nodosa	Hepatitis virus surface antigen (e.g., HBsAg)	Anti-HBs antibody	Arteritis (disseminated)

be due to glomerular deposits of immune complexes composed of streptococcal antigen and anti-streptococcal antibodies. However, post-streptococcal glomerulonephritis differs from typical immune complex–mediated diseases because it involves only the kidneys and there are no systemic manifestations. Therefore, this disease may be due to initial binding of streptococcal antigen to glomeruli, followed by deposition of the antibody, so that the pathogenic mechanism is not deposition

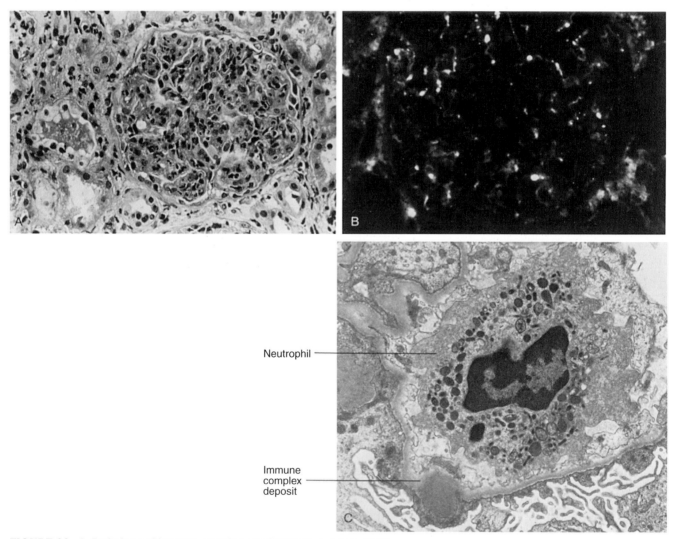

Neutrophil

Immune complex deposit

FIGURE 20–4. Pathology of immune complex–mediated glomerulonephritis.
A. Light micrograph of a kidney glomerulus, showing hypercellularity caused by infiltration of leukocytes.
B. Immunofluorescent stain for IgG showing granular deposits in a glomerulus. (Complement proteins, including C3b, would co-localize with the antibody.)
C. Electron micrograph of a glomerular capillary, showing an immune complex deposited in the wall and a neutrophil in the lumen.
(Courtesy of Dr. Helmut Rennke, Department of Pathology, Brigham and Women's Hospital, Boston; reproduced with permission from Brenner, B. M., F. L. Coe, and F. C. Rector. Clinical Nephrology. W. B. Saunders Co., Philadelphia, 1987.)

of circulating complexes but free antibody against a "planted" bacterial antigen. Granular deposits of antibody and complement have been demonstrated in injured tissues in many other forms of cutaneous necrotizing vasculitis, arthritis, and glomerulonephritis. Some skin diseases associated with vasculitis are morphologically similar to experimental Arthus reactions. Such diseases are postulated to be due to immune complexes, but the nature of the antigens is unknown.

Diseases Mediated by Antibodies Against Fixed Cell and Tissue Antigens

As we mentioned earlier, antibodies produced against fixed cellular or tissue antigens are usually autoantibodies. Less frequently, the antibodies may be produced against extrinsic antigens but may bind to immunologically similar or cross-reactive antigens present in autologous cells or tissues.

Many tissue- or organ-specific immunologic diseases are associated with the production of, and are thought to be caused by, autoantibodies (Table 20–3). In most of these diseases, specific circulating antibodies can be found in the blood, but the mechanisms responsible for autoantibody production are not known. Autoimmune hemolytic anemia and immune thrombocytopenia are due to autoan-

tibodies against erythrocytes and platelets, respectively. The antibodies cause complement-dependent lysis of the circulating cells and opsonize the cells, leading to enhanced phagocytosis by mononuclear phagocytes. Autoimmune hemolytic anemia and thrombocytopenia are usually idiopathic and may sometimes be associated with other immunologic abnormalities, e.g., in SLE. Similar diseases occur during idiosyncratic reactions to some drugs and may be due to binding of the drugs to cell surfaces, leading to the creation of neo-antigens that elicit specific antibody responses. **Goodpasture's syndrome** is a disease characterized by lung hemorrhages and severe glomerulonephritis. It is caused by an autoantibody that binds to a non-collagenous domain of type IV collagen found in the basement membranes of pulmonary alveoli and glomerular capillaries. Binding of this antibody leads to local activation of complement and neutrophils. On microscopic examination, necrosis, leukocytic infiltrates, and linear deposits of antibody and complement along basement membranes can be seen (Fig. 20–5). A number of skin diseases are due to antibodies against adhesion molecules of epidermal cells or basement membrane antigens. Binding of these antibodies disrupts the attachment of epithelial cells to one another or to the underlying basement membrane, and leads to

TABLE 20–3. Examples of Human Diseases Caused by Autoantibodies

Disease	Principal Clinical Features	Effector Mechanisms	Autoantibody Detected: Specificity	Method of Detection
Glomerulonephritis (Goodpasture's syndrome)	Nephritis with proteinuria, renal failure; lung hemorrhages	Complement, neutrophils	Type IV collagen in basement membranes of kidney glomeruli and lung alveoli	Immunofluorescence
Autoimmune hemolytic anemia	Hemolysis, anemia	Complement- and FcR-dependent phagocytosis, lysis	Erythrocyte membrane proteins	Hemagglutination
Autoimmune thrombocytopenic purpura	Platelet deficiency (thrombocytopenia), bleeding disorders	Complement- and FcR-dependent phagocytosis	Platelet membrane proteins (e.g., gp IIb/IIIa)	Immunofluorescence
Pemphigus vulgaris	Decreased adhesions between keratinocytes; skin vesicles (bullae)	Antibodies stimulate epithelial proteases, leading to disruption of intercellular adhesions	Intercellular junctions of epidermal cells	Immunofluorescence
Bullous pemphigoid	Detachment of epidermal cells; skin vesicles	Disruption of dermal-epidermal junction	Epidermal basement membrane proteins	Immunofluorescence
Myasthenia gravis	Muscle weakness	Blocking and down-modulation of acetylcholine receptor; local inflammation?	Acetylcholine receptor	Immunoprecipitation
Graves' disease (hyperthyroidism)	Hyperthyroidism due to increased production of thyroid hormones	Stimulation of TSH receptor	Thyroid-stimulating hormone receptor on thyroid follicular epithelial cells	Bioassay
Insulin-resistant diabetes mellitus	Diabetes unresponsive to insulin therapy	Inhibition of insulin binding to receptor; down-modulation of receptor?	Insulin receptor	Inhibition of insulin binding to cultured cells
Pernicious anemia	Abnormal erythropoiesis due to vitamin B_{12} deficiency	Neutralization of intrinsic factor	Intrinsic factor; gastric parietal cells	Bioassay; immunofluorescence

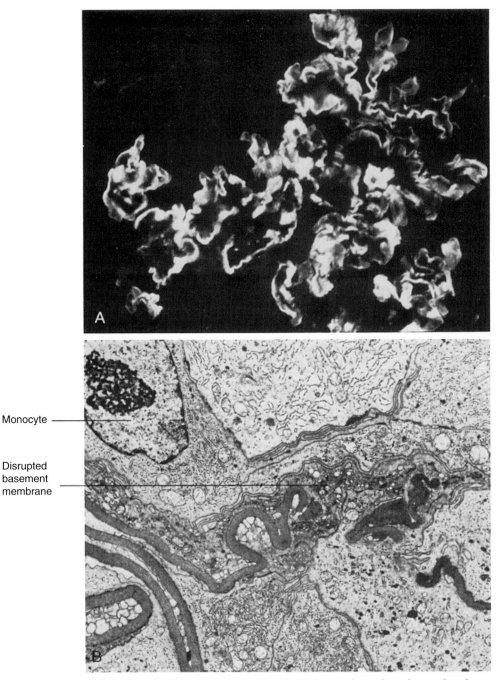

FIGURE 20-5. Pathology of glomerulonephritis induced by an antibody against the glomerular basement membrane (Goodpasture's syndrome).
A. Immunofluorescent stain for IgG, showing linear deposition of antibody along the capillary basement membrane of a glomerulus. (This pattern is very different from that of immune complex–mediated diseases; see Fig. 20–4B.)
B. Electron micrograph of a glomerular capillary, showing destruction of the basement membrane without evidence of immune complex deposition.
(Courtesy of Dr. Helmut Rennke, Department of Pathology, Brigham and Women's Hospital, Boston; reproduced with permission from Brenner, B. M., F. L. Coe, and F. C. Rector. Clinical Nephrology. W. B. Saunders Co., Philadelphia, 1987.)

blister formation. Many forms of **vasculitis** are associated with autoantibodies reactive with proteinases in the granules of neutrophils. Such **anti-neutrophil cytoplasmic antigen (ANCA) antibodies** may react with neutrophils that have been partially activated, causing complete neutrophil degranulation and injury to surrounding blood vessels. A disorder called **anti-phospholipid syndrome,** characterized by venous thromboses and recurrent abortions, is associated with, and believed to be caused by, autoantibodies against various phospholipids. The mechanisms of thrombosis or abortions are not known.

Autoantibodies against cell surface receptors

may lead to functional abnormalities without the involvement of any other effector mechanisms. For instance, some antibodies against cell surface hormone receptors bind to these receptors and lead to aberrations in cellular physiology without inflammation or tissue injury. These functional abnormalities may result from receptor-mediated stimulation of target cells or inhibition due to interference with receptor function (Fig. 20–6).

One example of stimulation by an antibody mimicking a physiologic molecule is **Graves' disease,** an autoimmune disease of the thyroid gland characterized by hyperthyroidism. The clinical syndrome results from excessive production of thyroid hormones such as thyroxine. This disease is usually caused by an autoantibody specific for the receptor for thyroid-stimulating hormone (TSH) on thyroid epithelial cells. TSH is a pituitary hormone whose normal function is to stimulate the production of thyroid hormones by thyroid epithelial cells. Binding of antibody to the TSH receptor has the same effect as TSH itself, leading to unregulated stimulation of thyroid epithelial cells and excess thyroid hormone production, even in the absence of TSH.

An example of anti-receptor antibody-mediated functional inhibition is **myasthenia gravis,** a disease of progressive muscle weakness caused by autoantibodies reactive with acetylcholine receptors in the motor end-plates of neuromuscular junctions. Binding of the antibodies interferes with acetylcholine-mediated neuromuscular transmission and may lead to a reduction in receptor numbers as a consequence of endocytosis and intracellular degradation ("down-modulation") of the receptors. The result is a failure of muscle to respond to normal neural impulses, leading to progressive muscle weakness. Experimentally, a disease resembling myasthenia gravis can be produced in rats and mice by immunizing them with purified acetylcholine receptors. The experimental disease can be adoptively transferred to normal animals by antibodies against the acetylcholine receptor. It is interesting that in some of these experimental models, the disease can also be transferred to normal animals with acetylcholine receptor–specific CD4$^+$ T cells. This illustrates a concept introduced in Chapter 19, that helper T cells may play a key role even in diseases caused by autoantibodies, and conversely, that tolerance in helper T cells may prevent the production of autoantibodies. Some patients with diabetes mellitus who are unresponsive to insulin have autoantibodies against insulin receptors that block the binding and the physiologic effects of the hormone.

Autoantibodies against physiologically important circulating molecules, such as hormones, may also lead to functional abnormalities and disease in the absence of cell or tissue destruction. Some cases of **pernicious anemia** are associated with autoantibodies against intrinsic factor, which is a cofactor for the intestinal absorption of vitamin B$_{12}$. The antibodies are thought to bind to and inhibit the function of intrinsic factor, resulting in vitamin B$_{12}$ deficiency. This causes abnormal hematopoiesis and megaloblastic anemia. In other cases, autoantibodies against physiologic molecules may trigger inflammation. For instance, some cases of nephritis are associated with an autoantibody, called C3 nephritic factor, that binds to and stabilizes C3 convertase (see Chapter 15). This leads to uncontrolled complement activation and tissue injury.

Some human diseases are caused by antibodies produced against foreign antigens that cross-react with self proteins. Perhaps the best example is **acute rheumatic fever,** which, like post-streptococcal glomerulonephritis, is a late sequela of throat infection caused by streptococci. The bacterial

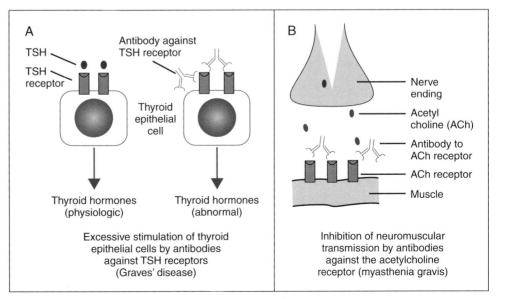

FIGURE 20–6. Effector mechanisms in antibody-mediated diseases: functional abnormalities induced by antibodies against hormone receptors.
A. Antibodies against thyroid-stimulating hormone (TSH) receptors stimulate thyroid epithelial cells by binding to the receptor and by mimicking the effects of TSH, the physiologic ligand.
B. Antibodies against the acetylcholine (ACh) receptor inhibit neuromuscular transmission by binding to the ACh receptor, leading to down-modulation of receptors and competitive inhibition of ACh binding.

strains associated with rheumatic fever are usually different from those that lead to glomerulonephritis. Rheumatic fever is characterized by arthritis, endocarditis resulting in lesions of heart valves, myocarditis, and neurologic abnormalities, but no kidney abnormalities. The myocardial injury is thought to be due to an antibody against a streptococcal cell wall protein that binds to a cross-reactive antigen in cardiac muscle cells.

Despite the numerous examples of circulating autoantibodies associated with immunologic diseases, it is important to reiterate that it is often not clear whether a particular antibody is the cause of the disease or is produced as a result of cell or tissue injury. Furthermore, autoantibodies may be present but may not be responsible for pathologic abnormalities. For instance, patients with **rheumatoid arthritis** (Box 20–1) frequently have a circulating IgM antibody that is specific for their own IgG molecules, usually the Fc portions of these molecules. Such autoantibodies are called rheumatoid factors. Although their presence is a useful diagnostic test for rheumatoid arthritis, there is no evidence that rheumatoid factors are involved in the formation of injurious immune complexes or contribute to the joint lesions in this disease.

DISEASES CAUSED BY T CELLS

The potential importance of T lymphocytes as mediators of human immunologic diseases was increasingly recognized in the 1980s. This insight was largely because of two technological advances—the production of monoclonal antibodies that identify phenotypically and functionally distinct subsets of T cells, and methods for isolating and propagating T cells from lymphoid tissues and lesions. As discussed in Chapter 19, the demonstration that T lymphocytes are critical for maintaining self-tolerance to many protein antigens has led to increasing interest in their role in autoimmune disorders.

Mechanisms of T Cell–Mediated Tissue Injury

The T cells that cause tissue injury may be autoreactive, or they may be specific for foreign protein antigens that are present in, or bound to, one's own cells or tissues. T lymphocyte–mediated tissue injury may also accompany strong protective immune responses against persistent microbes, especially intracellular microbes that resist eradication by phagocytes and antibodies. The pathologic lesions vary, depending on the types of T cells that produce these lesions. T cells injure tissues by the same two mechanisms that are responsible for cell-mediated immunity against microbes (see Chapter 13):

1. Both CD4$^+$ T cells of the T$_H$1 subset and CD8$^+$ cells secrete cytokines that activate macrophages, giving rise to **DTH reactions.** Acute tissue

---

BOX 20–1. Rheumatoid Arthritis

Rheumatoid arthritis is a destructive disease involving primarily the joints of the extremities, particularly of the fingers. As the disease progresses, more of the large joints are affected. Rheumatoid arthritis is characterized by destruction of the joint cartilage and inflammation of the synovium, with a morphologic picture suggestive of a local immune response. Both cell-mediated and humoral immune responses may contribute to development of lesions. CD4$^+$ T cells, activated B lymphocytes, and plasma cells are found in the inflamed synovium, and in severe cases, well-formed lymphoid follicles with germinal centers may be present. Numerous cytokines, including interleukin-1 (IL-1), IL-8, TNF, and IFN-γ, have been detected in the synovial (joint) fluid. Cytokines are believed to activate resident synovial cells to produce hydrolytic enzymes, such as collagenase and metalloproteinases, that mediate destruction of the cartilage, ligaments, and tendons of the joints. Many of the cytokines thought to play a role in initiating joint destruction are probably produced as a result of local T cell and macrophage activation. Clinical trials of anti-TNF antibody therapy are giving encouraging results. The specificity of the T cells that may cause arthritis, and the nature of the initiating antigen(s), is not known. Significant numbers of T cells expressing the $\gamma\delta$ antigen receptor have also been detected in the synovial fluid of patients with rheumatoid arthritis. However, the pathogenic role of this subset of T cells, like their physiologic function, is obscure.

Systemic complications of rheumatoid arthritis include vasculitis, presumably caused by immune complexes, and lung injury. The nature of the antigen or the antibodies in these complexes is not known. Patients with the adult form of rheumatoid arthritis frequently have circulating antibodies, which may be IgM or IgG, reactive with the Fc (and rarely Fab) portions of their own IgG molecules. These autoantibodies are called **rheumatoid factors,** and their presence is used as a diagnostic test for rheumatoid arthritis. Rheumatoid factors seem to play no role in the joint pathology or in the formation of injurious immune complexes. Although activated B cells and plasma cells are often present in the synovia of affected joints, the specificities of the antibodies produced by these cells or their roles in causing joint lesions are not known. Susceptibility to rheumatoid arthritis is linked to the HLA-DR4 haplotype and less so with DR1 and DRW1D. In all these alleles, the amino acid sequences from positions 65 to 75 of the β chain are nearly identical. These residues are located in or close to the peptide-binding clefts of the HLA molecules, suggesting that they influence antigen presentation or T cell recognition. However, their precise role in rheumatoid arthritis is not yet known (see Chapter 19).

There are several experimental models of arthritis. MRL/*lpr* mice develop spontaneous arthritis and have high serum levels of rheumatoid factors. The immune mechanisms of joint disease in MRL/*lpr* are not known. T cell–mediated arthritis can be induced in susceptible strains of mice and rats by immunization with type II collagen (the type found in cartilage), and the disease can be adoptively transferred to unimmunized animals with collagen-specific T cells. However, in the human disease there is no convincing evidence for collagen-specific autoimmunity. Experimental arthritis can also be produced by immunization with various bacterial antigens, including mycobacterial and streptococcal cell wall proteins. However, such diseases bear only a superficial resemblance to human rheumatoid arthritis.

injury results from the products of activated macrophages, such as hydrolytic enzymes, reactive oxygen species, and pro-inflammatory cytokines. Chronic DTH reactions often produce fibrosis as a result of the secretion of cytokines and growth factors by the macrophages (see Chapter 13).

2. CD8+ **cytolytic T lymphocytes** directly lyse target cells bearing class I major histocompatibility complex (MHC)-associated foreign antigens, without the participation of macrophages or any other effector mechanisms.

A role for T cells in causing a particular immunologic disease is suspected largely because of the demonstration of T cells in lesions and the isolation of T cells specific for self antigens from the tissues or blood of patients. Furthermore, cytokines secreted by activated T cells induce alterations in adjacent tissues that are used as indicators of local T cell stimulation. One such cytokine is interferon-γ (IFN-γ), which induces the expression of class II MHC molecules on cells that do not express these molecules constitutively, such as epithelial and mesenchymal cells. Abnormal expression of class II MHC molecules in a tissue suggests that T cells have been activated in the immediate environment. Aberrant expression of class II MHC molecules may also lead to excessive T cell activation because many more cells may acquire the ability to present antigen. However, the importance of aberrant MHC expression in exacerbating immunologic tissue injury is unclear because epithelial and mesenchymal cells that are induced to express class II molecules may not produce the costimulators necessary for T cell activation and, therefore, may not be efficient at stimulating T cells (see Chapter 7).

The presence of activated T cells in the blood or tissues of patients is not always associated with disorders of cell-mediated immunity. As mentioned previously, CD4+ helper T cells may be abundant in lesions that are mediated by antibodies and not by the T cells themselves. Moreover, as for antibodies, the identification and even isolation of T cells are not, by themselves, proof of their pathogenic role. The most definitive proof is the ability to transfer the disease adoptively to normal recipients, and this is not possible in the clinical situation. However, in experimental models of several immunologic diseases, the lesions have been transferred to normal syngeneic animals by purified T cells or by antigen-specific cloned lines of T cells (Table 20–4). The similarities between these experimentally induced lesions and clinical diseases further support a primary pathogenic role of T cells in the latter. Finally, alterations in the ratio of circulating CD4+ and CD8+ cells (normal being about 2:1) have been used as diagnostic indices for T cell–mediated immunologic diseases. Such assays, however, are of limited usefulness because they are not specific for particular immunologic abnormalities.

Diseases Caused by Delayed-Type Hypersensitivity

A variety of cutaneous diseases that result from topical exposure to foreign antigens or are sequelae of skin infections are due to T cell–mediated DTH reactions. These include skin rashes as a result of *contact sensitivity to chemicals*, such as drugs, cosmetics, and environmental antigens. The rashes usually appear hours or even days after exposure to the contact sensitizing agent. The lesions may be due to T cell responses to neo-antigens created by binding of the chemicals to normal cell surface proteins on epidermal keratinocytes or Langerhans cells. Skin biopsy

TABLE 20–4. Identification of Antigen-Specific T Cells in Immunologic Diseases

| Disease | Specificity of T Cell Clone/Line | Specific T Cells Isolated from Lesions or Blood of | | Ability to Transfer Disease in Animal Models |
		Patients	*Animal Models*	
Insulin-dependent (type I) diabetes mellitus	Islet cell antigens (including glutamic acid decarboxylase)	No	Yes	Yes
Experimental allergic encephalomyelitis	Myelin basic protein, proteolipid protein		Yes	Yes
Experimental allergic neuritis	P2 protein of peripheral nerve myelin		Yes	Yes
Experimental autoimmune myocarditis	Myosin		Yes	Yes
Myasthenia gravis*	Acetylcholine receptor	Yes	Yes	Yes
Some cases of Graves' disease,* autoimmune thyroiditis	Thyroid follicular epithelial cells	Yes	Yes	Yes

* In these cases, the T cells may be helper cells that stimulate local production of autoantibodies, which are responsible for inducing lesions.

specimens show dermal perivascular infiltrates of lymphocytes and macrophages, and edema and fibrin deposition resulting from leakage of plasma from dermal capillaries and venules (see Chapter 13). Vascular endothelial cells in the lesions may express enhanced levels of cytokine-regulated surface proteins, such as adhesion molecules and class II MHC molecules. These DTH reactions are quite different mechanistically and morphologically from two other types of immunologic skin lesions, IgE-mediated immediate hypersensitivity and immune complex–mediated Arthus reactions (Table 20–5).

Many organ-specific autoimmune diseases are caused by autoreactive T cells. In **insulin-dependent diabetes mellitus** (IDDM) (see Chapter 19, Box 19–2), there are infiltrates of lymphocytes and macrophages around islets of Langerhans in the pancreas, with destruction of insulin-producing β cells in the islets and a resultant deficiency in insulin production. Residual islet cells in these lesions express class II MHC molecules, again suggesting local cytokine production. Similar findings have been observed in spontaneous diabetes in rats and mice, in which the lesions have been adoptively transferred to normal animals with CD4+ T cells from diseased animals. CD8+ CTLs may also contribute to insulitis and islet cell destruction in human and experimental diabetes mellitus. The specificity of the T cells that cause insulitis and destroy islet cells, and the nature of the initiating antigen, are largely unknown. In the nonobese diabetic (NOD) animal model, the initial T cell response may be directed against a number of islet proteins, including insulin itself and an islet cell surface enzyme called glutamic acid decarboxylase. **Experimental allergic encephalomyelitis** (EAE) (see Chapter 19, Box 19–1) is a neurologic disease that can be induced in experimental animals by immunization with protein antigens of central nervous myelin in adjuvant. Such immunization leads to an autoimmune T cell response against myelin, culminating in activation of macrophages around nerves in the brain and spinal cord, destruction of the myelin (Fig. 20–7), abnormalities in nerve conduction, and neurologic deficits. EAE can be transferred to naive animals with myelin antigen-specific CD4+ T$_H$1 cells, and the experimental disease can be blocked by antibodies specific for class II MHC or for CD4 molecules, indicating that CD4+ class II MHC-restricted T cells play an obligatory role in this disorder. The induction of myelin antigen-specific T$_H$2 responses protects against EAE, emphasizing the anti-inflammatory function of this subset. It is interesting that once EAE is induced with one myelin antigen, immune responses to other myelin proteins often develop. This phenomenon is called "epitope spreading," and it may be a mechanism for progressively worsening immunologic tissue injury in T cell–mediated diseases. It has been postulated that EAE is an experimental counterpart of the progressive neurologic disease multiple sclerosis.

Cell-mediated immune responses to microbes and other foreign antigens may also lead to considerable injury of the tissues at the sites of infection or antigen exposure. Intracellular bacteria such as *My-*

TABLE 20–5. Lesions and Mechanisms of Different Forms of Immunologic Reactions in the Skin

	Immediate Hypersensitivity	Immune Complex–Mediated Injury	Delayed-Type Hypersensitivity
Induced by	Antigens that evoke IgE response (genetic predisposition?)	Antigens that induce IgM, IgG antibodies	Protein antigens; chemicals that bind to self proteins
Form of cutaneous reaction	Urticaria, wheal	Arthus reaction	Contact sensitivity, tuberculin reaction
Onset after antigen challenge	Minutes*	Usually 2–6 hr	Usually 24–48 hr
Pathologic lesion	Edema, vascular dilatation, local smooth muscle contraction	Necrotizing vasculitis	Perivascular cellular infiltrates and edema
Transferred to normal animals by	Serum	Serum	Lymphocytes
Antibody involved	IgE	IgG (usually complement-fixing subclasses), IgM	None
Effector cells	Mast cells with IgE bound to Fc receptors	Neutrophils, monocytes (recruited by complement-dependent and Fcγ receptor-dependent mechanisms)	CD4+ T cells, macrophages (activated by cytokines)
Secreted mediators, effector molecules	Mast cell–derived mediators: vasoactive amines, lipid mediators	Products of complement activation: membrane attack complex, C3a, C5a	Cytokines, particularly IFN-γ and TNF

* Note that the late phase reaction of immediate hypersensitivity is a cytokine-mediated inflammatory reaction that develops in 6 to 24 hours.
Abbreviations: Ig, immunoglobulin; IFN, interferon; TNF, tumor necrosis factor.

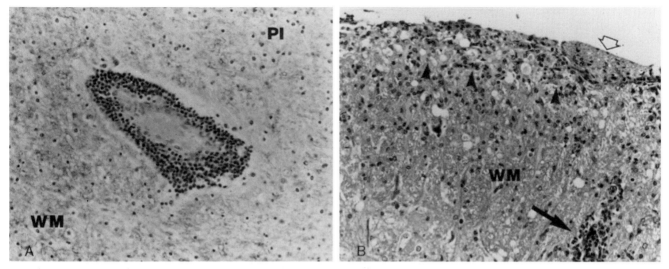

FIGURE 20–7. Pathology of multiple sclerosis and experimental allergic encephalomyelitis (EAE).
A. Multiple sclerosis: perivascular mononuclear cell infiltrate at the edge of a demyelinated plaque (Pl) in the brain. Myelin is intact in the white matter (WM) but is lost in the plaque (pale staining).
B. EAE: Perivascular mononuclear cell infiltrate *(black arrow)* associated with demyelination *(arrowheads)* in a mouse immunized with a peptide of myelin proteolipid protein with adjuvant. The *white arrow* indicates the surface of the tissue. (Courtesy of Dr. Raymond A. Sobel, Department of Pathology, Stanford University School of Medicine, Palo Alto, CA.)

cobacterium tuberculosis induce strong T cell and macrophage responses, resulting in the formation of granulomas and fibrosis caused by the production of cytokines that stimulate fibroblast proliferation and collagen synthesis (described in Chapter 13). Therefore, mycobacterial infections often result in extensive tissue destruction and scarring that can cause severe functional impairment, for instance, in the lungs. Tuberculosis is a good example of a disease in which protective cell-mediated immunity and tissue injury due to DTH may co-exist (see Chapter 16). Sarcoidosis is a disease of unknown etiology in which granulomas develop in the lungs, lymphoid tissues, liver, and spleen. This disease is probably due to a T cell–mediated immune response to a foreign antigen that has eluded identification.

Inflammatory bowel disease (IBD) is a heterogeneous group of disorders, including Crohn's disease, in which an immunologic etiology has been suspected for many years. Several knockout mouse models of IBD illustrate the different pathogenetic mechanisms that can result in phenotypically similar disorders. Mice in which the genes for the cytokines IL-10, IL-2, or TGF-β1 are knocked out, and mice lacking the T cell receptor (TCR) α chain, all develop severe IBD. It is thought that in IL-10 and TGF-β1 knockouts, the mechanism of the lesions is unregulated macrophage and neutrophil responses to enteric bacteria, because the disease does not develop in germ-free knockout mice. IL-2 knockouts may develop autoimmune lesions that are triggered or exacerbated by enteric microbes, and the pathogenesis of the disease in TCRα knockouts is unknown. These findings have spurred consider-

able interest in defining the roles of T cells and cytokines in human IBD, and in trying cytokine (e.g., IL-10) therapy for the human disease.

Diseases Caused by Cytolytic T Lymphocytes

The principal physiologic function of CTLs is to eliminate intracellular microbes, primarily viruses. It follows, therefore, that infected cells are lysed during CTL-mediated protective immune responses. Some viruses directly injure infected cells or are cytopathic, whereas others are not. Since CTLs cannot *a priori* distinguish between cytopathic and non-cytopathic viruses, they lyse virally infected cells whether or not the infection itself is harmful to the host. *Therefore, CTL responses to viral infections can lead to tissue injury even if the virus itself has no pathologic effects.* Examples of viral infections in which the lesions are due to the host CTL response and not the virus itself include lymphocytic choriomeningitis in mice and certain forms of viral hepatitis in humans (see Chapter 16).

There are few documented examples of autoimmune diseases mediated by CTLs. In mice infected with the coxsackievirus B, myocarditis develops, with infiltration of the heart by CD8+ T cells. These animals contain virus-specific, class I MHC-restricted CTLs as well as CTLs that lyse uninfected myocardial cells. It is postulated that the heart lesions are initiated by the virus infection and virus-specific CTLs, but myocardial injury leads to the exposure or alteration of self antigens and the development of autoreactive CTLs. As mentioned above, CTLs may also contribute to tissue injury in many of the disorders that are caused primarily by CD4+ T cells, such as insulitis in IDDM.

PRINCIPLES OF THERAPY FOR IMMUNE-MEDIATED DISEASES

Therapy for immune-mediated diseases is modeled after approaches that are used to prevent graft rejection, another form of injurious immune response (see Chapter 17). The mainstay of treatment for diseases caused by immune responses is anti-inflammatory drugs, particularly corticosteroids. Such drugs are targeted at reducing tissue injury, i.e., the effector phases of the pathologic immune responses. Antagonists against pro-inflammatory cytokines, such as IL-1 and TNF, and agents that block leukocyte emigration into tissues are also being tested for their anti-inflammatory effects. Recent results of clinical trials of anti-TNF antibody administration in patients with rheumatoid arthritis are encouraging. In severe cases, immunosuppressive drugs like cyclosporin A are used to block T cell activation. Plasmapheresis has been used during exacerbations of antibody-mediated diseases to reduce circulating levels of antibodies or immune complexes.

Many other therapies are being attempted in cases that are resistant to conventional regimens. T cells can be depleted by injecting antibodies against CD3 or the T cell receptor. Immunoconjugates of IL-2 and toxins (e.g., ricin) may bind to activated T cells that express high-affinity IL-2 receptors and kill these cells. More specific treatment aimed at the disease-producing T cell clones or the antigen that initiates the disease requires accurate identification of the antigen. Approaches include attempts to induce tolerance, e.g., by oral administration of antigens that cause autoimmunity, or administration of altered peptide ligands, as in EAE (see Chapter 19, Box 19–1). Although the value of such therapies has been demonstrated in various experimental (animal) models, their application to clinical disease has not been established.

SUMMARY

Diseases in which tissue injury and pathophysiologic abnormalities are due to immunologic mechanisms may be initiated by immune responses to foreign or self (autologous) antigens. Pathogenic mechanisms include antigen-antibody complexes formed during humoral immune responses, autoantibodies against fixed tissue or cell surface antigens, and T lymphocytes. The effector mechanisms by which antibodies and immune complexes induce tissue injury include activation of the complement system and various host inflammatory cells. Antibodies against physiologic agents such as hormones or against cell surface receptors for hormones induce functional abnormalities without the involvement of any other effector systems. T lymphocytes recruit and activate macrophages as the principal effectors of DTH and tissue injury, and CD8+ cytolytic T lymphocytes themselves lyse antigen-bearing target cells. The therapy of immune-mediated diseases is aimed at reducing immune responses and the attendant inflammation.

Selected Readings

Castano, L., and G. S. Eisenbarth. Type I diabetes: a chronic autoimmune disease of human, mouse and rat. Annual Review of Immunology 8:647–679, 1990.

Charreire, J. Immune mechanisms in autoimmune thyroiditis. Advances in Immunology 46:263–334, 1989.

Feldmann, M., F. M. Brennan, and R. N. Maini. Rheumatoid arthritis. Cell 85:307–310, 1996.

Naparstek, Y., and P. H. Poltz. The role of autoantibodies in autoimmune diseases. Annual Review of Immunology 11:79–104, 1993.

Zamvil, S. S., and L. Steinman. The T lymphocyte in experimental allergic encephalomyelitis. Annual Review of Immunology 8:579–621, 1990.

CHAPTER TWENTY-ONE

CONGENITAL AND ACQUIRED

IMMUNODEFICIENCIES

The integrity of the immune system is essential for defense against infectious organisms and their toxic products and, therefore, for the survival of all individuals. Defects in one or more components of the immune system can lead to serious and often fatal disorders, which are collectively called immunodeficiency diseases. These diseases are broadly classified into two groups. The **primary** or **congenital immunodeficiencies** are genetic defects that result in an increased susceptibility to infections that is frequently manifested early in infancy and childhood but is sometimes clinically detected later in life. It is estimated that in the United States approximately 1 in 500 individuals is born with a defect in some component(s) of the immune system, although only a small proportion are affected severely enough to develop life-threatening complications. **Secondary** or **acquired immunodeficiencies** develop as a consequence of malnutrition, disseminated cancers, treatment with immunosuppressive drugs, or infections of cells of the immune system, most notably with the human immunodeficiency virus (HIV), the etiologic agent of the acquired immunodeficiency syndrome (AIDS). This chapter describes the major types of congenital and acquired immunodeficiencies, with an emphasis on their pathogenesis and on the components of the immune system that are involved in each.

Before beginning our discussion, it is important to emphasize some general features of immunodeficiencies:

1. *The principal consequence of immunodeficiency is an increased susceptibility to infections.* The nature of the infections in a particular patient depends largely on the component of the immune system that is defective. For instance, deficient humoral immunity usually results in increased susceptibility to infections by pyogenic bacteria, whereas defects in cell-mediated immunity lead to infections by viruses and other intracellular microbes. Combined deficiencies in both humoral and cell-mediated immunity result in infections by all classes of microorganisms. Specific examples of these will be mentioned later in the chapter. There are, however, exceptions to these generalizations, which likely reflect our incomplete understanding of normal immune responses to some species of microbes.

2. *Patients with immunodeficiencies are also prone to certain types of cancers,* many of which are caused by oncogenic viruses. This is generally seen in T cell immunodeficiencies because, as we discussed in Chapter 18, T cells play an important role in surveillance against tumors. In addition, paradoxically, certain immunodeficiencies are associated with an increased incidence of autoimmunity; the mechanism underlying this association is not known.

3. *Immunodeficiency diseases are clinically and pathologically heterogeneous.* In large part, this is because different diseases involve different components of the immune system. However, such heterogeneity is also seen in various diseases involving the same cells or molecules and even in different patients suffering from the same disorder. The reason for this variability is not understood.

Deficient immune responses may result from abnormalities in specific or innate immunity. Defective specific immunity is due to abnormal development, activation, or function of specific T or B lymphocytes, or both (Fig. 21–1). Among the examples of impaired innate immunity are defects in phagocytes and the complement system. In this chapter, we will first describe congenital immunodeficiencies, including defects in the humoral and cell-mediated arms of the specific immune response, and defects in phagocytes of the innate immune system. We conclude the chapter with a discussion of AIDS and other acquired immunodeficiencies.

PRIMARY DEFECTS IN B LYMPHOCYTES AND ANTIBODY PRODUCTION

Congenital abnormalities in B lymphocyte development and function result in deficient antibody production. These diseases have been recognized for many years, because assays for measuring serum antibodies have been in routine clinical use since the 1950s. A large number of congenital deficiencies that selectively affect humoral immune responses are now known (Table 21–1). Clinically, these disorders are characterized by recurrent infections with encapsulated pyogenic bacteria, such as pneumococci, streptococci, meningococci, and *Haemophilus influenzae.* These infections often cause otitis, sinusitis, pneumonia, meningitis, osteomyelitis, septic arthritis, and generalized sepsis. In addition, patients are susceptible to certain viral diseases such as vaccine-acquired poliomyelitis, echovirus and enterovirus infections of the central nervous system, and severe hepatitis B infection. Infections by certain intestinal parasites, such as *Giardia,* are also frequent in antibody deficiencies. It is not surprising that immunity to these microbes is normally mediated principally by antibodies (see Chapter 16).

In different antibody deficiencies, the primary abnormality may be at different stages of B lymphocyte maturation or in the responses of mature B cells to antigenic stimulation (see Fig. 21–1). Abnormal helper T cell function may also result in deficient antibody production. One of the impressive recent achievements of molecular immunology is the identification of the genetic basis of several immunodeficiency diseases; these are mentioned below. In the following section we describe some examples of antibody immunodeficiencies, emphasizing the mechanisms of B cell defects.

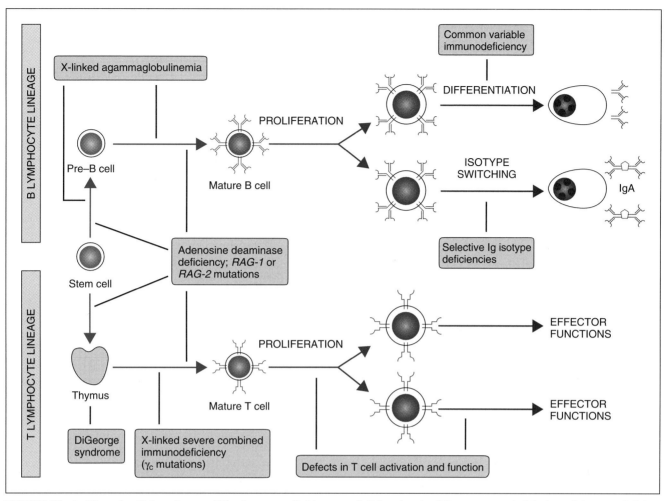

FIGURE 21–1. Sites of cellular abnormalities in congenital immunodeficiencies. In different congenital (primary) immunodeficiencies, the maturation or activation of B or T lymphocytes may be blocked at different stages.

X-Linked Agammaglobulinemia

This disease, also called Bruton's agammaglobulinemia, is characterized by the absence of γ globulins in the blood, as the name implies. It is one of the most common congenital immunodeficiencies and the prototype of selective B cell defects. It was also the first humoral immunodeficiency disease to be clinically well characterized, in 1952, and studies of affected patients were use-

TABLE 21–1. Examples of Congenital B Cell Immunodeficiencies

Disease	Functional Deficiencies	Presumed Mechanism of Defect
X-linked agammaglobulinemia	All Ig isotypes decreased; reduced B cells	Mutation in B cell tyrosine kinase, block in early steps of B cell maturation
Selective IgA deficiency	Decreased serum IgA1 and IgA2; normal B cells	Failure of terminal differentiation of IgA B cells
Ig deficiency with increased IgM (hyper-IgM syndrome)	Increased IgM; normal or increased IgD; other Ig isotypes decreased	Mutation in CD40 ligand Defect in heavy chain isotype switching
Selective IgG subclass deficiencies	Decrease in one or more IgG subclasses	Defect in isotype switching or terminal B cell differentiation
Ig heavy chain deletions	IgG1, IgG2, or IgG4 absent; sometimes associated with absent IgA or IgE	Chromosomal deletion at 14q32 (Ig heavy chain locus)
Transient hypogammaglobulinemia of infancy	IgG and IgA decreased; detectable levels of antibacterial antibodies; normal B cells	Unknown; ? delayed maturation of helper T cells in some patients
Common variable immunodeficiency	Variable reductions in multiple Ig isotypes; normal or decreased B cells	Defect in B cell maturation, usually due to intrinsic B cell abnormality

ful in proving that plasma cells produced antibodies (both of which were absent in the patients). It is an X chromosome–linked disease, so that females who carry the defective gene on one X chromosome are phenotypically normal because the other X chromosome has a normal gene, but males who inherit the abnormal X chromosome manifest the disease. (In addition to X-linked agammaglobulinemia, four other immunodeficiency diseases are linked to the X chromosome; see Tables 21–1 and 21–2.) All somatic cells in females randomly inactivate one or the other X chromosome. B cells of female carriers of X-linked agammaglobulinemia mature only if they inactivate the X chromosome bearing the mutant allele, whereas T cells and other somatic cells show a normal random pattern of X chromosome inactivation. Thus, by assessing X chromosome utilization in B cells, and comparing this with other cells, it is possible to identify carriers of the disease. X-linked agammaglobulinemia is transmitted from phenotypically normal female carriers to their male offspring. Affected boys are susceptible to bacterial and some viral infections, especially echovirus and enterovirus. Infections by most intracellular microbes and fungi are handled normally. These children suffer from recurrent pyogenic bacterial infections of the conjunctiva, throat, skin, middle ear, bronchi, and lungs. Newborn infants are often normal because maternally derived antibodies provide adequate protection, and the disease is usually recognized between 6 months and 2 years of life. If untreated, the disease is usually fatal.

Patients with X-linked agammaglobulinemia usually have low or undetectable serum immunoglobulin (Ig), reduced or absent B cells in peripheral blood and lymphoid tissues, no germinal centers in lymph nodes, and no plasma cells in tissues. Underlying these abnormalities is a profound arrest in the development of B cells in the bone marrow. Some patients do have mature peripheral B cells and even have elevated levels of serum IgG and IgA. However, in these patients, B cell numbers may be 100-fold lower than normal, and antibody responses to immunization are seriously deficient, suggesting that only a very limited repertoire of B cells is present. The maturation, numbers, and functions of T cells are generally normal. Some studies have revealed reduced numbers of activated T cells in patients, which may be a consequence of reduced antigen presentation due to the lack of B cells. Almost 20 per cent of patients develop autoimmune disorders, for unknown reasons.

TABLE 21–2. Examples of Congenital T Cell and Combined Immunodeficiencies

Disease	Functional Deficiencies	Presumed Mechanism of Defect
DiGeorge syndrome	Decreased T cells; normal B cells; normal or decreased serum Ig	Anomalous development of 3rd and 4th branchial pouches, leading to thymic hypoplasia
T cell receptor complex expression or signalling defects	Decreased T cells or abnormal ratios of subsets; decreased cell-mediated immunity	Rare cases due to mutations or deletions in genes encoding CD3 proteins, ZAP-70
Class II MHC deficiency	Normal lymphocyte numbers; normal or decreased serum Ig; deficient cell-mediated immunity	Deficiency of X-box binding factors causing defective transcription of class II MHC genes
Class I MHC deficiency	Normal lymphocyte numbers; deficient CD8$^+$ development and function	Rare cases due to TAP-2 gene mutations
SCID, X-linked γ_c defects	Markedly decreased T cells; normal or increased B cells; reduced serum Ig	Cytokine receptor common γ chain gene mutations, defective T cell maturation
SCID, Autosomal recessive Deficiencies in molecules required for early lymphocyte development	Decreased T and B cells; reduced serum Ig	Defective maturation of T and B cells; rare cases due to Jak-3 kinase gene mutations
ADA deficiency	Progressive decrease in T and B cells (mostly T); reduced serum Ig	ADA deficiency leading to accumulation of toxic metabolites
PNP deficiency	Decreased T cells; normal B cells and serum Ig	PNP deficiency leading to accumulation of toxic metabolites in developing T cells
Reticular dysgenesis	Markedly decreased T and B cells and other blood cells; reduced serum Ig	Defective maturation of hematopoietic stem cells
Wiskott-Aldrich syndrome	Progressive decrease in T cells; normal B cells; decreased IgM; deficient antibody responses to polysaccharide antigens	Defective glycosylation of membrane proteins, defective maturation of hematopoietic stem cells
Ataxia-telangiectasia	Decreased T cells; normal B cells; variable reduction in IgA, IgE, and IgG subclasses	?Defect in DNA repair

Abbreviations: SCID, severe combined immunodeficiency disease; ADA, adenosine deaminase; PNP, purine nucleoside phosphorylase; Ig, immunoglobulin; MHC, major histocompatibility complex.

The gene that is defective in X-linked agammaglobulinemia is on the long arm of the X chromosome and encodes a cytoplasmic protein tyrosine kinase called B cell tyrosine kinase (Btk). The normal *btk* gene is expressed at all stages of B cell differentiation, as well as in other hematopoietic cells. Btk is structurally related to other cytoplasmic tyrosine kinases involved in lymphocyte activation and contains SH2 and SH3 domains, as well as other domains, which are likely involved in interactions with other proteins. Patients with X-linked agammaglobulinemia have point mutations or deletions in various parts of the *btk* gene and therefore do not produce a functional form of this protein kinase. This results in abnormalities in several stages of B cell maturation in the bone marrow, including pro-B to pre-B and pre-B to mature B cell transitions, suggesting a fundamental role for Btk in signal transduction pathways required at multiple steps of B cell differentiation. An inbred mouse strain called **CBA/N** has an X-linked defect in B cell development that is also a result of a point mutation in the *btk* gene. The block in B cell maturation is much less severe in CBA/N mice than in humans with X-linked agammaglobulinemia, and this is also true when the mouse *btk* gene is inactivated in gene knockout mice. The reason for this species difference in severity of disease associated with *btk* mutations is not understood.

The infectious complications of X-linked agammaglobulinemia are greatly reduced by periodic (e.g., monthly) intravenous or intramuscular injections of pooled gamma globulin preparations. Such preparations contain pre-formed antibodies against common pathogens and provide effective passive immunity. This therapy has prolonged life for patients beyond the third or fourth decade. Death usually results from chronic, often respiratory tract, infections.

Selective Immunoglobulin Isotype Deficiencies

Many immunodeficiencies that selectively involve one or a few Ig isotypes have been described. The most common is **selective IgA deficiency,** which affects about 1 in 700 individuals of Caucasian descent and is thus the most common primary immunodeficiency. IgA deficiency usually occurs sporadically, but there are many familial cases, with either autosomal dominant or recessive patterns of inheritance. Patients with IgA deficiency are more likely than normal individuals to express certain major histocompatibility complex (MHC) alleles, especially human leukocyte antigen (HLA)–B8 or HLA-DW3, as well as deletions of the C4A complement gene. The clinical features are extremely variable. Many patients are entirely normal, others have occasional respiratory infections and diarrhea, and, rarely, patients have severe, recurrent infections leading to permanent intestinal and airway damage, with associated autoimmune disorders.

IgA deficiency is characterized by abnormally low serum IgA, usually less than 50 μg/ml (normal, 2–4 mg/ml), with normal or elevated levels of IgM and IgG. The defect in these patients is a block in the differentiation of surface IgA-expressing B cells to antibody-secreting plasma cells. The α heavy chain genes and the expression of membrane-associated IgA are normal. It is not known whether the block of B cell differentiation is due to an intrinsic B cell defect or to an abnormality in T cell help, such as the production of cytokines that enhance IgA secretion (e.g., transforming growth factor-β [TGF-β] and interleukin-5 [IL-5]), or in B cell responses to these cytokines. No gross abnormalities in the numbers, phenotypes, or functional responses of T cells have been noted in these patients.

Selective IgG subclass deficiencies have been described in which total serum IgG levels are normal but concentrations of one or more subclasses are below normal. Deficiency of IgG$_3$ is the most common subclass deficiency in adults, and IgG$_2$ deficiency, associated with IgA deficiency, is the most common in children. Some individuals with these deficiencies have recurrent pyogenic bacterial infections, but many do not have any clinical problems. Selective IgG subclass deficiencies are usually due to abnormal B cell differentiation and, rarely, are due to homozygous deletions of various constant region (Cγ) genes. Individuals with single Cγ gene deletions are usually normal, which attests to the capacity of the immune system to compensate for selective antibody deficiencies.

IgG and IgA deficiency with increased IgM (hyper-IgM syndrome) is usually inherited as an X-linked disorder. Affected male children produce only IgM antibodies and are therefore susceptible to severe bacterial infections. *Pneumocystis* pneumonia is also common in this disease. Furthermore, many of the IgM antibodies are autoantibodies reactive with the patient's own red blood cells, leukocytes, and platelets. This leads to secondary deficiencies of these blood cells, further reducing resistance to infections. The Cγ and Cα genes are structurally normal, as are the switch regions located 5' of these genes. However, heavy chain class switching to IgG and IgA does not occur, so that the patients lack B cells with surface IgG or IgA and do not produce these isotypes. The X-linked form of this disease is due to mutations in the gene encoding the CD40 ligand. The mutant forms of CD40 ligand produced in these patients do not bind to or transduce signals via CD40, and therefore do not stimulate B cells to differentiate and undergo switch recombinations (see Chapter 9). The susceptibility of these patients for *Pneumocystis* infection suggests a defect in T cell–mediated immunity, as well. This may reflect the role of CD40 ligand in activating effector functions of macrophages. CD40 ligand and CD40 knockout mice have a phenotype very similar to the human disease phenotype. Non–X-linked forms of hyper-IgM syndrome are attributed to defects in the CD40 signaling pathway in B cells, but these defects have not yet been precisely characterized.

Common Variable Immunodeficiency

This group of heterogeneous disorders is defined by the presence of hypogammaglobulinemia, impaired antibody responses to infection or vaccines, and increased incidence of infections. The diagnosis is usually one of exclusion when other primary immunodeficiency diseases are ruled out. The presentation and pathogenesis are, as the name suggests, highly variable. Although immunoglobulin deficiency and associated pyogenic infections are major components of these disorders, autoimmune diseases, including pernicious anemia, hemolytic anemia, and rheumatoid arthritis, may be just as significant. There is also a high incidence of malignancies associated with common variable immunodeficiency. These disorders may be diagnosed early in childhood or late in life. Both sporadic and familial cases occur, the latter with both autosomal dominant and recessive inheritance patterns. As with selective IgA deficiency, familial cases are associated with inheritance of certain MHC alleles and C4A gene deletions.

The hypogammaglobulinemia in common variable immunodeficiency has been attributed to multiple abnormalities, including intrinsic B cell defects, deficient T cell help, and excessive "suppressor cell" activity. It is likely that in the majority of patients the primary abnormality responsible for low antibody production is a defect in the terminal differentiation of B lymphocytes to antibody-secreting cells. In lymphoid tissues, the B cell areas (i.e., lymphoid follicles) are often hyperplastic but plasma cells are absent. These findings suggest that B cells proliferate in response to stimulation by antigens and helper T cells but fail to differentiate normally. It is not known whether this is due to defective responses to helper T cell–derived stimuli or to more distal block(s) in the program of B cell activation. In some patients, IgM production can be induced *in vitro* by transformation with Epstein-Barr virus, which functions as a T cell–independent stimulus. However, even in these individuals, switching to other isotypes, such as IgG and IgA, usually does not occur. There are some indications that common variable immunodeficiencies and selective IgA deficiency are caused by the same genetic defects since they often occur in the same families and both are associated with mutations or rare alleles in the major histocompatibility complex.

PRIMARY DEFECTS IN T LYMPHOCYTES

Severe defects in T cell maturation, such as the DiGeorge syndrome (discussed below), have been recognized for many years. As our understanding of T cell biology has improved, we can attribute more clinical immunodeficiency syndromes to primary defects in the activation and function of T lymphocytes (see Fig. 21–1; Table 21–2). Most of these defects lead to impaired cell-mediated immune reactions manifested by increased susceptibility to infections with viruses, fungi, intracellular bacteria, and protozoa. Such microorganisms are often capable of surviving and even replicating inside cells, including phagocytes (which is why their eradication is dependent on T cell immunity, as discussed in Chapter 16). As a result, these infections are usually severe and difficult to control and may be fatal. Patients with T cell deficiencies may also be susceptible to malignancies. These aspects are discussed more fully later in this chapter when we discuss the clinical features of AIDS, the prototypical acquired T cell immunodeficiency. T cell immunodeficiencies are diagnosed by reduced numbers of peripheral blood T cells, abnormally low proliferative responses to polyclonal T cell activators, e.g., phytohemagglutinin (PHA), and deficient cutaneous delayed-type hypersensitivity (DTH) reactions to ubiquitous microbial antigens, such as *Candida* antigens. In addition, as we mentioned in the previous section of the chapter, some antibody deficiencies may, in fact, be due to abnormal T cell help. In fact, many abnormalities in T cell function lead to impairment of both cell-mediated and humoral immunity.

The DiGeorge Syndrome (Thymic Hypoplasia)

This selective T cell deficiency is due to a congenital malformation that results in defective development of the third and fourth branchial pouches. These structures give rise to the thymus and the parathyroid glands at weeks 6 to 8 of gestation and to the aortic arch and portions of the lips and ears at 12 weeks of fetal life. Developmental anomalies induced at this stage of gestation lead to partial or complete DiGeorge syndrome, manifested by hypoplasia or agenesis of the thymus (leading to deficient cell-mediated immunity), absent parathyroid glands (causing abnormal calcium homeostasis and muscle twitching, or tetany), abnormal development of the great vessels, and facial deformities. Different patients may show varying degrees of these abnormalities. The nature of the developmental insult is usually not known. Some cases are associated with maternal alcohol consumption, and rare cases show autosomal dominant patterns of inheritance or are associated with translocations involving chromosome 22.

The hypoplasia of the thymus leads to defective maturation of all T lymphocytes, because of which peripheral blood T lymphocytes are absent or greatly reduced in number. Sometimes the total peripheral blood lymphocyte count is near normal, but most of the cells are B lymphocytes (which make up only 10 to 20 per cent of the blood lymphocytes in normal individuals). Peripheral blood lymphocytes do not respond to polyclonal T cell activators or to alloantigens in mixed leukocyte reactions (MLRs). Antibody levels are usually normal but may be reduced in severely affected patients. In the peripheral lymphoid tissues, the B cell areas appear normal. As in other severe T cell deficiencies, patients are susceptible to mycobacterial, viral, and fungal infections.

The disease can be corrected by fetal thymic transplantation or by HLA-identical bone marrow transplantation using marrow which has not been depleted of mature T cells. This is usually not necessary, however, because T cell function tends to improve with age and is often normal by 5 years. This is probably because of the presence of some thymic tissue or because extrathymic sites assume the function of T cell maturation. The existence of extrathymic sites of T cell development has been suspected, but no such tissue has been defined anatomically. It is also possible that as these patients grow older, typical thymus tissue develops at ectopic sites (i.e., other than the normal location). Similarly, ectopic parathyroids develop with age, with consequent improvement of tetany.

An example of T cell immunodeficiency due to abnormal development of the thymus in animals is the **nude (athymic) mouse.** These mice have an inherited defect of epithelial cells in the skin, leading to hairlessness, and in the lining of the third and fourth branchial pouches, causing thymic hypoplasia. The disorder is due to mutations in a gene on chromosome 1 that encodes a transcription factor that is apparently required for epithelial cell development. It is inherited in a recessive manner, and the affected homozygote mice (called nu/nu) have rudimentary thymuses in which T cell maturation cannot occur normally. As a result, there are few or no mature T cells in peripheral lymphoid tissues and a failure of all cell-mediated immune reactions, including allograft rejection, DTH, and antibody responses to T cell–dependent protein antigens. As the mice age to about 1 year, some mature T cells do develop, but the site of T cell maturation is not defined. Nu/nu mice are susceptible to many infections, but somewhat surprisingly they are able to eradicate some intracellular bacteria. This is because of normal or even increased numbers of natural killer (NK) cells, which produce interferon-γ (IFN-γ) that activates macrophages and serves to eliminate the microbes. NK cells may also account for the lack of susceptibility of nu/nu mice to spontaneous tumors (see Chapter 18). A similar inherited abnormality has been observed in rats, but nu/nu rats have not been analyzed in as much detail. Several mouse strains have been developed with targeted knockout of genes required for T cell development (see Chapter 8), and these mice are useful models for T cell immunodeficiency diseases.

Defects in Molecules Required for T Cell Activation and Function

There are many examples of rare immunodeficiency diseases that are caused by defects in the expression of molecules made by T cells which are required for T cell activation and function. Often these defects are found in only a few isolated cases or in one or a few families, and the clinical features and severity vary widely. Modern biochemical and molecular analyses of the affected individuals have revealed genetic defects in various T cell proteins. Examples include (1) defective T cell receptor (TCR) complex expression due to mutations in the CD3 γ or ϵ genes; (2) defective TCR complex–mediated signaling due to mutations in the ZAP-70 gene; (3) defective synthesis of cytokines such as IL-2 and IFN-γ (in some cases due to defects in transcription factors); and (4) defective expression of IL-2 receptor. The patients with these abnormalities may have deficiencies predominantly in T cell function or mixed T cell and B cell immunodeficiencies, despite normal or even elevated numbers of blood lymphocytes. Some of these defects are associated with abnormal ratios of the CD4$^+$ and CD8$^+$ T cell subsets, which may reflect impaired development of one subset in the thymus or impaired expansion of one subset in the peripheral immune system. For example, patients with ZAP-70 deficiency have more CD8$^+$ than CD4$^+$ T cells, although the reason for this is not clear.

Another potential way T cell activation may be impaired is if there is a defect in antigen presentation. This could theoretically occur if antigen-presenting cells fail to express peptide-MHC complexes, costimulators, or other surface molecules involved in effectively stimulating T lymphocytes. There are rare examples of human immunodeficiency due to defective class II or class I MHC molecule expression. Class II MHC deficiency, also called **bare lymphocyte syndrome,** is a heterogeneous group of autosomal recessive diseases in which patients express little or no HLA-DP, DQ, or DR on B lymphocytes, macrophages, and dendritic cells and fail to express class II MHC molecules in response to IFN-γ. They express normal or only slightly reduced levels of class I MHC molecules and β_2 microglobulin and do synthesize the invariant chain upon stimulation with IFN-γ. Class II MHC deficiency has been attributed to mutations in the genes encoding transcription factors essential for class II MHC expression. One of these genes encodes RFX5, a subunit of a transcription factor that binds to the "X box" regulatory element located 5' of the class II locus (see Chapter 5). Another gene encodes a transcription factor called class II transactivator (CIITA), which does not bind directly to DNA but transactivates class II MHC transcription through interaction with other protein factors. In many cases of bare lymphocyte syndrome, the production of either RFX5 or CIITA is reduced, or the proteins are structurally abnormal and, therefore, inactive. This results in reduced transcription of class II MHC genes. The consequence of deficient class II expression is a failure of antigen presentation to CD4$^+$ T cells. As a result, affected individuals are deficient in DTH responses and in antibody responses to T-dependent protein antigens, and they are susceptible to viral, bacterial, fungal, and protozoal infections. In some patients, the numbers of mature CD4$^+$ T cells in peripheral blood and tissues are reduced. This may be because, in the absence of class II MHC expression in the thymus, positive selection of CD4$^+$ T cells cannot occur (see Chapter 8). The disease presents within the first year of life and is usually fatal unless treated by bone marrow transplantation.

Autosomal recessive class I MHC deficiencies have also been described and are characterized by decreased CD8$^+$ T cell numbers and function. In some clinical cases, the failure to express class I MHC molecules is due to mutations in the gene encoding the transporter associated with antigen presentation–2 (TAP-2) subunit of the TAP complex which normally pumps peptides into the endoplasmic reticulum where they are required for class I MHC assembly (see Chapter 6). These TAP-deficient patients express very few cell surface class I MHC molecules, a phenotype similar to TAP-gene knockout mice. Surprisingly, these patients suffer mainly from respiratory tract bacterial infections rather than viral infections.

COMBINED IMMUNODEFICIENCIES (MIXED B AND T CELL DEFECTS)

In this section we will consider a clinically and mechanistically heterogeneous group of combined immunodeficiencies affecting both the B and T cell compartments (see Table 21–2). These diseases may arise as a result of primary abnormalities in both B and T cell development, or in T cell development with secondary effects on B cell function. We have separated these disorders of lymphocyte development from those disorders of T cell activation and function, discussed earlier, but the latter may also lead to clinically similar diseases with combined deficiencies in cell-mediated and T cell dependent humoral immunity. In addition, some combined immunodeficiencies are found in association with other congenital diseases.

Severe Combined Immunodeficiencies

The term "severe combined immunodeficiency disease (SCID)" is given to a heterogeneous group of disorders characterized by defective development or function of B and T lymphocytes, profound lymphopenia, and deficient humoral and cell-mediated immunity. Some types of SCID are characterized by a complete lack of both B and T lymphocytes, whereas others have few or no T lymphocytes but relatively normal numbers of B lymphocytes. These disorders manifest within the

first year of life with oral candidiasis, chronic diarrhea, failure to thrive, and eventually serious infections with bacteria, viruses, fungi, protozoa, and other parasites. SCID patients may also succumb to graft-versus-host disease subsequent to blood transfusions or intrauterine maternal-fetal transfer of lymphocytes. Unless treated with bone marrow transplantation, patients usually die from infections during the first year of life. The first example of the disease was discovered in Switzerland in the 1950s (because of which this disorder was originally called Swiss type agammaglobulinemia).

The transmission of SCID may be autosomal recessive or X-linked recessive. About 40 per cent of SCID cases are inherited in an autosomal recessive manner. In about half of these cases, the particular genetic defect is not known. It is suspected that many of these cases may be due to defects in the rearrangement and expression of antigen receptor genes. The discovery of the *RAG-1* and *RAG-2* genes that mediate V-D-J recombination of the Ig and TCR loci (see Chapters 4 and 8) led to the speculation that some SCID cases may be due to mutations in the *RAG* genes. This idea was bolstered by the finding that *RAG-1* or *RAG-2* gene knockout mice have a complete block in lymphocyte development. More recently, mutations in *RAG-1* and *RAG-2* genes have been identified in several patients with autosomal recessive SCID in which there is an absence of both B and T lymphocytes. These mutations lead to impaired V-D-J recombination activity. An instructive experimental model is the **SCID mouse,** which arose as a spontaneous mutant of an inbred strain called CB-17. In the SCID mutant strain, both B and T cells are absent because of an early block in maturation from bone marrow precursors. The defect in SCID mice is an abnormality in the DNA repair mechanisms that, in addition to *RAG-1* and *RAG-2* activity, are required for rearrangement of TCR and immunoglobulin (Ig) genes. In cell lines derived from immature B and T cells of SCID mice, aberrant recombination events interfere with functional V-D-J rearrangements, and, as a result, antigen receptors are not expressed and lymphocytes do not mature. The *scid* mutation has been localized to a gene on mouse chromosome 16 that encodes a subunit of an enzyme called DNA-dependent protein kinase. This enzyme is involved in repair of DNA double-stranded breaks and in V(D)J recombination, both of which are defective in SCID mice. It is not known whether any human SCID patients have defects in the homologous gene. About 15 per cent of inbred SCID mice are "leaky," in that they produce reduced but readily detectable numbers of mature B and T cells. The lymphocytes in these mice express limited repertoires of antigen receptor genes, suggesting that normal Ig or TCR rearrangements can occur in some developing clones.

Most of the remaining 50 per cent of the autosomal recessive forms of SCID are due to deficiency of an enzyme called **adenosine deaminase (ADA).** ADA functions in the salvage pathway of purine metabolism and catalyzes the irreversible deamination of adenosine and 2′-deoxyadenosine to inosine and 2′-deoxyinosine, respectively (Fig. 21–2). The ADA gene is located on chromosome 2, and deficiency of the enzyme can be due to deletions or mutations in the gene. The enzyme is expressed in many cell types, but the harmful effects of its absence are largely restricted to the developing immune system. ADA deficiency leads to the accumulation of deoxyadenosine, S-adenosyl homocysteine, and deoxy-adenosine triphosphate (deoxy-ATP). Various toxic effects have been attributed to these by-products. In particular, deoxy-ATP is an inhibitor of deoxynucleotide synthesis and therefore DNA synthesis. Compared with most somatic cells, developing lymphocytes are relatively inefficient at degrading deoxy-ATP into 2′-deoxyadenosine, and therefore lymphocyte maturation is particularly sensitive to ADA deficiency. Thus, ADA deficiency leads to reduced numbers of B and T cells and a resultant immunodeficiency. Some patients may have near normal numbers of T cells but the cells do not respond to antigenic stimulation. ADA deficiency is a prime candidate for treatment by specific gene transfer. The ADA gene has been cloned and constitutively expressed in transfected cells. The disease is manifested in bone marrow–derived cells and may thus be treatable by transfection of a functional gene into autologous self-renewing marrow cells and transplantation of these cells back into the patient.

A much rarer autosomal recessive form of SCID is due to the deficiency of another enzyme,

called **purine nucleoside phosphorylase (PNP),** that is involved in purine catabolism. PNP catalyzes the conversion of inosine to hypoxanthine and guanosine to guanine (see Fig. 21–2). Deficiency of PNP leads to accumulation of deoxyguanosine and deoxy-guanosine triphosphate (deoxy-GTP), with toxic effects more on T cells than on B cells. The gene for PNP is located on chromosome 9. Enzyme deficiency due to deletions or mutations in this gene usually results in deficient T cell immunity of variable severity, with normal B cell function. Patients present with variable susceptibility to infections.

Approximately 60 per cent of SCID cases are X-linked, and most of these are due to mutations in the gene encoding the common γ chain (γ_c) shared by the receptors for interleukins-2, -4, -7, -9, and -15 (see Chapter 12). X-linked SCID is characterized by impaired T cell differentiation, but there are usually normal or increased numbers of B cells. The humoral immunodeficiency in this disease is attributed to a lack of T cell help for antibody production. It is interesting that some cases of autosomal recessive SCID in which there are also few T cells but normal numbers of B cells have been attributed to mutations in the gene encoding Jak-3 kinase, an enzyme involved in signaling by γ_c-containing cytokine receptors (see Chapter 12). It is possible that mutations in genes encoding other components of the signaling cascades used by these receptors, such as STAT genes, are the cause of certain cases of autosomal recessive SCID. The phenotype of human γ_c and Jak-3 mutations suggests that one or more of the cytokines whose receptors utilize these signaling

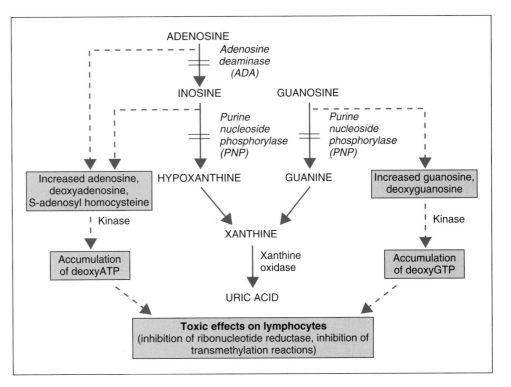

FIGURE 21–2. Congenital abnormalities in purine metabolism. The major pathways for metabolism of purines (adenosine and guanosine) are shown in solid lines. Deficiencies of the enzymes adenosine deaminase (ADA) and purine nucleoside phosphorylase (PNP) block these metabolic pathways at different steps. This leads to shunting into secondary metabolic pathways, indicated by dashed lines, and the accumulation of toxic metabolites (in boxes).

components are required for T cell but not B cell development. IL-7 is a lymphopoietic cytokine that stimulates the growth of immature thymocytes, and the phenotype of γ_c and Jak-3 deficiencies may, in large part, be due to the failure of IL-7 to signal via its receptor. It is curious that mice carrying null mutations of the γ_c, Jak-3, or IL-7 genes show defects in both B and T cell development, suggesting that IL-7 is required for both T and B lymphocyte development in mice but only T cell maturation in humans.

The most severe form of SCID is a disease called **reticular dysgenesis.** This disorder is characterized by the absence of T and B lymphocytes and myeloid cells, such as granulocytes, and is presumably due to a defect at the level of the hematopoietic stem cell. This is a very rare disease, and neither its molecular basis, nor its mode of inheritance, is clear.

Immunodeficiency Associated with Other Inherited Diseases

Variable degrees of B and T cell immunodeficiency occur in certain congenital diseases in which there is a wide spectrum of abnormalities involving multiple organ systems. One such disorder is called the **Wiskott-Aldrich syndrome,** an X-linked disease characterized by eczema, thrombocytopenia (reduced blood platelets), and susceptibility to bacterial infections. In the initial stages of the disease, lymphocyte numbers are normal, and the principal defect is an inability to produce antibodies in response to polysaccharide antigens, which are typical thymus-independent antigens (see Chapter 9). These patients are especially susceptible to infections with encapsulated pyogenic bacteria. The lymphocytes (and platelets) are smaller than normal. With increasing age, the patients show reduced numbers of lymphocytes and more severe immunodeficiency. The defective gene responsible for the Wiskott-Aldrich syndrome is in the short arm of the X chromosome and encodes a protein with little identifiable homology to other proteins. Recent studies indicate that the Wiskott-Aldrich protein is involved in cytoskeletal reorganization and may thus be important for lymphocyte activation responses and migration. Also associated with this syndrome is reduced expression of many cell surface glycoproteins, including the sialic acid–rich glycoprotein called CD43 (or sialophorin), which is normally expressed on lymphocytes (both B and T), macrophages, neutrophils, and platelets. The contribution of decreased CD43 expression to the abnormal lymphocyte function associated with this disease is not clear.

Another disease associated with immunodeficiency is **ataxia-telangiectasia,** an autosomal recessive disorder characterized by abnormal gait (ataxia), vascular malformations (telangiectasias), various neurologic deficits, increased incidence of tumors, and immunodeficiency. The immunologic defects are of variable severity and may affect both B and T cells. The most common humoral immune defect is IgA and IgG2 deficiency. The T cell defects, which are usually less pronounced, are associated with thymic hypoplasia. Patients experience sinus and pulmonary bacterial infections, multiple autoimmune phenomena, and increasingly frequent cancers with advancing age. Most cases of ataxia-telangiectasia appear to be caused by mutations in a gene called ATM, which normally encodes a 370 kDa nuclear phosphoprotein of unknown function. It is likely that this protein is involved in DNA repair mechanisms. There is an increased susceptibility of all cell types in these patients to the mutational effects of ionizing radiation. In some patients, many cell types, including lymphocytes, contain numerous chromosomal deletions and translocations that may involve Ig or TCR loci. Such defects may contribute to abnormal lymphocyte development as well as abnormal proliferative responses to antigenic stimulation.

CONGENITAL DISORDERS OF PHAGOCYTES AND OTHER CELLS OF INNATE IMMUNITY

Innate immunity is mediated principally by phagocytes and complement, and it constitutes the first line of defense against infectious organisms. In addition, phagocytes and complement participate in the effector phases of specific immunity. Therefore, congenital disorders of phagocytes and the complement system result in recurrent infections of varying severity. Complement deficiencies have been described in Chapter 15. In this section of the chapter, we discuss some examples of congenital phagocyte disorders.

Chronic Granulomatous Disease

Chronic granulomatous disease (CGD) is rare, estimated to affect about 1 in 1 million individuals in the United States. About two thirds of the cases show an X-linked recessive pattern of inheritance, and the remainder are autosomal recessive. The disease is characterized by recurrent bacterial and fungal infections, usually from early childhood. The infections are usually not controlled by neutrophilic inflammation and may result in the formation of granulomas composed of activated macrophages. The disease is often fatal, even with aggressive antibiotic therapy.

CGD is due to a defect in the production of superoxide anion, a reactive oxygen intermediate that constitutes a major microbicidal mechanism of phagocytes. In response to an encounter with bacteria, neutrophils and macrophages rapidly consume oxygen and liberate superoxide, which is a precursor of the reactive oxygen species that serve to kill bacteria. Superoxide generation is mediated by a multicomponent enzyme complex called nicotinamide adenine dinucleotide phosphate (NADPH)-oxidase, which upon activation catalyzes the 1-electron reduction of oxygen (O_2) to superoxide (O_2^-). Mutations or deletions in each of four different components of this complex can lead to CGD. Most cases are due to defects in the gene encoding the neutrophil-specific membrane protein, cytochrome b_{558}, which forms part of the active NADPH-oxidase. The gene, which is located in band p21 of the X chromosome, may be absent, truncated, or mutated so that either it is not transcribed or the RNA is unstable. Most cases of the rarer autosomal recessive form of CGD have been attributed to defects in the genes encoding one of three cytosolic proteins that associate with cytochrome b_{558} in active NADPH-oxidase complexes. It is interesting that CGD patients do not suffer from infections with catalase-negative bacteria because these organisms release enough hydrogen peroxide into phagocytic vacuoles to allow killing by the CGD phagocytes.

It has been found that IFN-γ stimulates the production of superoxide by normal neutrophils as well as CGD neutrophils, especially in cases where the cytochrome b genes are present, but their transcription is reduced. IFN-γ enhances transcription of the cytochrome b genes and also stimulates other components of the enzyme system that catalyze superoxide generation. Once neutrophil superoxide production is enhanced to within 10 to 12 per cent of normal, there is greatly improved resistance to infection. IFN-γ therapy is now commonly used for treatment of X-linked CGD.

Leukocyte Adhesion Deficiencies

Leukocyte adhesion deficiency-1 (LAD-1) is a rare autosomal recessive disorder characterized by recurrent bacterial and fungal infections and impaired wound healing. In these patients, most adhesion-dependent functions of leukocytes are abnormal. These functions include adherence to endothelium, neutrophil aggregation and chemotaxis, phagocytosis, and cytotoxicity mediated by neutrophils, NK cells, and T lymphocytes. The molecular basis of the defect is absent or deficient expression of the β_2 integrins, or the CD11CD18 family of glycoproteins, which includes leukocyte function–associated antigen-1 (LFA-1 or CD11aCD18), Mac-1 (CD11bCD18), and p150,95 (CD11cCD18). These proteins participate in the adhesion of leukocytes to other cells (see Chapter 7, Box 7–3) and in the phagocytosis of complement-coated particles (see Chapter 15). In all patients studied so far, the defect has been mapped to the 95 kD β chain (CD18). The gene encoding this chain may be mutated, producing an aberrant transcript, or its transcription may be reduced.

Leukocyte adhesion deficiency-2 (LAD-2) is another disorder described in a very small number of patients; it is clinically very similar to LAD-1 but is not due to integrin defects. In contrast, LAD-2 results from an absence of sialyl-Lewis X, the carbohydrate ligand on neutrophils that is required for binding to E-selectin and perhaps P-selectin on cytokine-activated endothelium (see Chapter 11, Box 11–1). This is attributed to mutations in a gene encoding a fucosyl transferase enzyme involved in making the carbohydrate moiety of the E-selectin ligand.

To date, fewer than 100 patients with these disorders have been described. Like other genetic defects affecting leukocytes, leukocyte adhesion deficiency patients are candidates for bone marrow transplantation and ultimately specific gene therapy.

Chediak-Higashi Syndrome

This rare autosomal recessive disorder is characterized by recurrent infections by pyogenic bacteria, partial oculocutaneous albinism, and infiltration of various organs by non-neoplastic lymphocytes. Early studies showed that the neutrophils, monocytes, and lymphocytes of these patients contained giant cytoplasmic granules. It is now thought that this disease is due to a more generalized cellular abnormality leading to increased fusion of cytoplasmic granules. Defective vesicular transport to and from lysosomes and endosomes, as well as missorting of proteins among these subcellular compartments, has been found in cells from Chediak-Higashi patients. This affects the lysosomes of neutrophils and monocytes (causing reduced resistance to infections), melanocytes (causing albinism), cells of the nervous system (causing nerve defects), and platelets (leading to bleeding disorders). Chediak-Higashi syndrome is caused by mutations in a gene called CHS, found on chromosome 1, which encodes a protein that may regulate intracellular protein trafficking.

The giant lysosomes found in neutrophils form during the maturation of these cells from myeloid precursors. Some of these neutrophil precursors die prematurely, resulting in moderate leukopenia. Surviving neutrophils may contain reduced levels of the lysosomal enzymes that normally function in microbial killing. These cells are also defective in chemotaxis and phagocytosis, further contributing to their deficient microbicidal activity. NK cell function in these patients is impaired, probably because of an abnormality in the cytoplasmic granules that store proteins that mediate cytolysis (see Chapter 13). An interesting note is that cytolytic T lymphocyte (CTL)-mediated killing is normal. A mutant mouse strain called the **beige mouse** is an animal model for the Chediak-Higashi syndrome and is caused by a mutation in the mouse homolog of the CHS gene. This strain is characterized by deficient NK cell function and giant lysosomes in leukocytes as well as by platelet abnormalities.

THERAPEUTIC APPROACHES FOR CONGENITAL IMMUNODEFICIENCIES

In theory, the therapy of choice for congenital disorders of lymphocytes is to replace the defective gene in self-renewing precursor cells. This remains a distant goal for most human immunodeficiencies at present, despite considerable effort. The main obstacles to this type of gene therapy are (1) our inability to identify and purify self-renewing stem cells, which are the ideal target for introduction of the replacement gene, and (2) a method of introduction of the genes, such as retroviral infection, that does not alter the normal properties of the stem cell. Thus, current treatment for immunodeficiencies has two aims: to minimize and control infections, and to replace the defective or absent components of the immune system by adoptive transfer and/or transplantation. Among the agents that have proved useful as replacement therapy are the following:

1. Pooled gamma globulins are enormously valuable for agammaglobulinemic patients and have been lifesaving for many boys with X-linked agammaglobulinemia.

2. Bone marrow transplantation is currently the treatment of choice for various immunodeficiency diseases and has been successful for the treatment of SCID with ADA deficiency, Wiskott-Aldrich syndrome, bare lymphocyte syndrome, and leukocyte-adhesion deficiency. It is most successful with careful T cell depletion from the marrow and HLA matching to prevent graft-versus-host disease (GVHD) (see Chapter 17). If a child with SCID is given a transplant of semi-syngeneic marrow cells from a parent (also called a "haploidentical" transplant because it is identical to the recipient at one of the two HLA haplotypes), the T cells that arise from the marrow must now develop in the partly foreign host. These T cells become restricted to recognizing foreign antigens in association with HLA molecules of the host, which are the HLA molecules that the donor T cells encounter during their maturation in the thymus. This, of course, is analogous to mouse bone marrow chimeras, which provided the initial evidence for the role of MHC molecules in the thymus in determining the selection of the T cell repertoire (see Chapter 8).

3. Enzyme replacement therapy for ADA and PNP deficiencies has been attempted using red blood cell transfusions as a source of the enzymes. This approach has produced temporary clinical improvements in several SCID patients. Injection of bovine ADA, conjugated to polyethylene glycol to prolong serum half-life, has proved successful in some cases. Ultimately, enzyme replacement therapies will have to be based on stable expression of a transfected gene encoding ADA or PNP in self-renewing bone marrow stem cells.

HUMAN IMMUNODEFICIENCY VIRUS AND THE ACQUIRED IMMUNODEFICIENCY SYNDROME

Acquired immunodeficiency syndrome is a disease first described in the early 1980s, characterized by profound immunosuppression with diverse clinical features, including opportunistic infections, malignancies, wasting, and central nervous system (CNS) degeneration. AIDS is one of a group of clinical syndromes caused by a retrovirus called **human immunodeficiency virus (HIV).** *HIV primarily infects CD4-expressing cells, including helper T cells and macrophages.* The degree of morbidity and mortality caused by HIV and the global impact of HIV infection on health care resources and economics are already enormous and continue to grow. More than 3 million people have already developed AIDS. The number of people worldwide who are infected with HIV is estimated to be about 20 million, and it is predicted to reach 40 to 60 million by the year 2000. More than half of the world's infected population live in Africa, but the number of infected Asians is rising rapidly. Currently, there is no prophylactic immunization or cure for AIDS, although promising new therapies are being developed. In this section of the chapter we describe the molecular and biologic properties of HIV, the nature and possible causes of HIV-induced immunosuppression, and the clinical and epidemiologic features of HIV-related diseases.

Molecular and Biologic Features of HIV

HIV is a member of the lentivirus family of animal retroviruses. Lentiviruses, including visna virus of sheep, and the bovine, feline, and simian immunodeficiency viruses (SIV), are capable of long-term latent infection of cells and short-term cytopathic effects, and they all produce slowly progressive, fatal diseases, which include wasting syndromes and central nervous system degeneration. Two closely related types of HIV, designated HIV-1 and HIV-2, have been identified. HIV-1 and HIV-2 differ in genomic structure and antigenicity, sharing only 40 per cent nucleic acid sequence homology. HIV-2 is more closely related to SIV than to HIV-1. Nonetheless, both forms of HIV cause similar clinical syndromes. HIV-1 is a far more common cause of AIDS than HIV-2 in the United States, and HIV-2 is more common in West Africa.

HIV STRUCTURE AND GENES

An infectious HIV particle consists of two identical strands of RNA, each approximately 9.2 kilobases (kb) long, packaged within a core of viral proteins, surrounded by a phospholipid bilayer envelope derived from the host cell membrane but including virally encoded membrane proteins (see Color Plate VIII, preceding Section I). The HIV genome shares the basic arrangement of nucleic acid sequences characteristic of all known retroviruses (Fig. 21–3), including the following elements: (i) long terminal repeats (LTRs) at each end of the genome, which regulate viral integration into the host genome, viral gene expression, and viral replication; (ii) *gag* sequences, which encode core structural proteins; (iii) *env* sequences encoding the envelope glycoproteins gp120 and gp41, required for infection of cells; and (iv) *pol* sequences encoding reverse transcriptase, endonuclease, and viral protease enzymes required for viral replication. In addition to these typical retrovirus genes, HIV-1 also includes at least six other regulatory genes, including *vpr, vif, tat, rev, nef,* and *vpu* genes, whose products regulate viral reproduction in various ways. HIV-2 contains these genes with the exception of *vpu,* and it has another gene, *vpx,* not present in HIV-1. Of these various regulatory genes, the functions of the *tat* and *rev* products are best defined and are discussed below.

VIRAL LIFE CYCLE

HIV infection occurs when viral particles in blood, semen, or other body fluids from one individual bind to cells of another individual. The HIV envelope glycoproteins gp120 and gp41 are critical for HIV infection. (Conventional notation of viral and cellular proteins includes a "p" for protein, or "gp" for glycoprotein, followed by a number designating the molecular size in kilodaltons.) The first step in HIV infection is the high-affinity binding of gp120 to CD4 molecules on the surface of a T cell or mononuclear phagocyte. HIV does not bind to CD4 molecules in non-primate species. Free HIV particles released from one infected cell can bind to an uninfected cell. Alternatively, gp120, which is expressed on the plasma membrane of infected cells before virus is released, can bind to CD4 on another cell, initiating a membrane fusion event, and HIV genomes can be passed between the fused

cells directly. Detailed molecular mapping studies, including x-ray crystallographic analysis, have defined which regions of the CD4 and gp120 molecules are involved in binding to one another. This information is potentially useful in the design of therapeutic agents that block virus interaction with CD4-expressing cells. Three well-conserved, noncontiguous regions in the carboxy terminus of gp120 of HIV-1 and HIV-2 are required for CD4 binding. It is interesting that these regions are separated by sequences that are extremely variable from one HIV isolate to another; this is significant to the way HIV may evade the host immune system, which we will discuss later. Such studies have already determined that various residues in the amino terminal IgV-like domain of human CD4 are critical for binding. Genetically engineered soluble CD4 molecules can block HIV infection of cells, but soluble CD4 has not proved to be effective in the treatment of AIDS patients.

After binding to CD4 molecules, HIV particles enter cells by direct fusion of the virus membrane with the host cell membrane. This process is facilitated by gp41 molecules on the viral membrane and is dependent on coreceptors expressed on the target cell membrane. The best defined coreceptors, as shown by *in vitro* assays of cellular infection, are members of the chemokine receptor family (see Chapter 13), and include CC-CKR5 and CXC-CKR4 (Fusin). These molecules are expressed on T cells, monocytes, and tissue macrophages, as well as other cell types. A role for the chemokine receptors in the HIV life cycle was first suspected because of the observation that chemokines can inhibit HIV infection of cells, and this role has been subsequently proven by several experimental techniques including transfection of chemokine receptor genes into cells. The importance of CC-CKR5 in HIV infection *in vivo* is suggested by the finding that individuals who do not express this receptor because of genetic mutations are highly

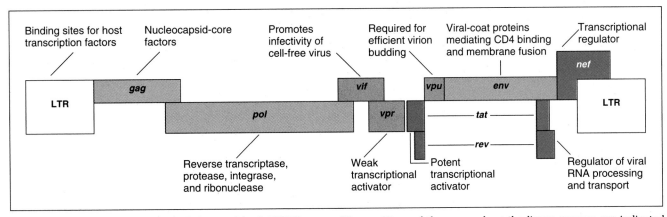

FIGURE 21–3. Human immunodeficiency virus-1 (HIV-1) genes. The positions of the genes along the linear genome are indicated as differently shaded blocks. Some genes use some of the same sequences as other genes, as shown by overlapping blocks, but are read differently by host cell RNA polymerase. Similarly shaded blocks separated by lines indicate genes whose coding sequences are separated in the genome and require RNA splicing to produce functional messenger RNA (mRNA). LTR, long terminal repeat. (Adapted from AIDS and the immune system, Greene, W. C. Copyright © 1993 by Scientific American, Inc. All rights reserved.)

resistant to HIV infection. The mechanisms by which chemokine receptors mediate virus membrane fusion with the target cell, and the way gp41 contributes to this process, are not fully understood.

Once an HIV virion enters a cell, the enzymes within the nucleoprotein complex become active and begin the viral reproductive cycle (Fig. 21–4). The nucleoprotein core of the virus becomes disrupted, the RNA genome of HIV is transcribed into a double-stranded DNA form by viral reverse tran-

scriptase, and the viral DNA enters the nucleus. The viral integrase also enters the nucleus and catalyzes the integration of the viral DNA into the host cell genome. There is evidence that the integration event is enhanced by concomitant T cell activation by antigens or super-antigens. The integrated DNA form of HIV is called the **provirus.** The provirus may remain transcriptionally inactive for months or years, with little or no production of new viral proteins or virions, and in this way HIV infection of an individual cell can be latent.

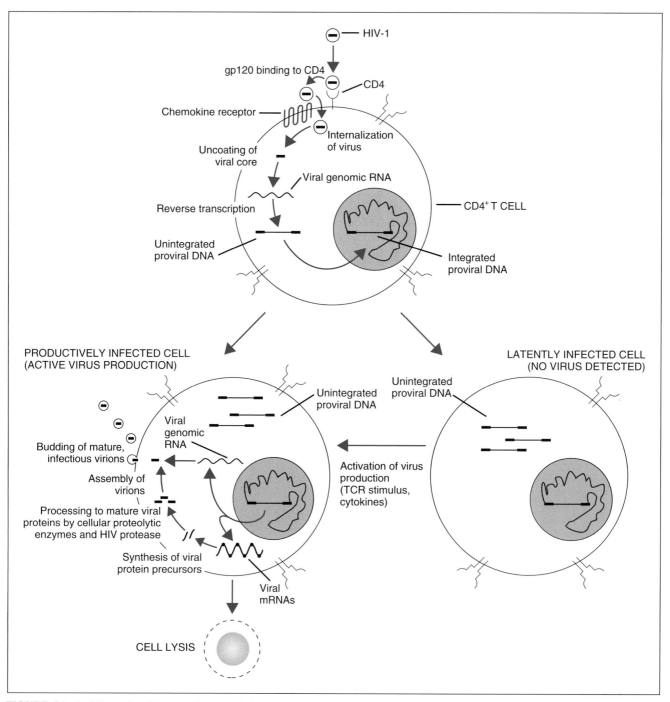

FIGURE 21–4. Life cycle of human immunodeficiency virus-1 (HIV-1). HIV-1 infection of a CD4-expressing T cell is schematically depicted, including the relationship between latent infection and productive lytic infection.

Transcription of the genes of the integrated DNA provirus is regulated by the LTR upstream of the viral structural genes. The LTRs contain polyadenylation signal sequences, TATA box promoter sequence, and binding sites for two host-cell transcription factors, NFκB and SP1. *Initiation of HIV gene transcription in T cells is linked to physiologic activation of the T cell by antigen or cytokines.* For example, TCR-binding lectins, tumor necrosis factor (TNF), and lymphotoxin all stimulate HIV gene expression in T cells, and IL-1, IL-3, IL-6, TNF, lymphotoxin, IFN-γ, and granulocyte-macrophage colony–stimulating factor stimulate HIV gene expression and viral replication in monocytes and macrophages. TCR and cytokine stimulation of HIV gene transcription probably involves the induction of nuclear factors that bind to the NFκB-binding sequences in the LTR. This phenomenon may be significant to the pathogenesis of AIDS in two ways. First, *physiologic activation of a latently infected T cell may be the way in which latency is ended and virus production begins.* Second, *the multiple infections that AIDS patients acquire lead to elevated TNF production; this, in turn, may stimulate HIV production and infection of additional cells.* It is interesting that when the regulatory sequences of the HIV LTR are linked to other genes and transfected into various cell types, they efficiently enhance the transcription of those genes. Therefore, the tissue specificity of productive HIV infection is not a function of these regulators but, rather, reflects the specificity of virus binding to and internalization by CD4+, chemokine receptor–expressing cell types.

HIV gene expression may be divided into an early stage, during which regulatory genes are expressed, and a late stage, during which structural genes are expressed and full length viral genomes are packaged (Fig. 21–5). Expression of the early regulatory genes, including *tat* and *rev,* requires sequential RNA splicing events in the nucleus generating short mRNAs, which are then transported to the cytoplasm and subsequently translated into proteins. The *tat* gene encodes a protein that binds to sequences present in the LTR called the transactivating response element (TAR), resulting in enhanced viral gene expression. Binding of the Tat protein to the viral LTR causes a 1000-fold increase in cellular RNA polymerase II–catalyzed transcription of the provirus. Synthesis of structural components of HIV occurs in the late stage when longer viral transcripts, which have only undergone a single splicing event, are transported to the cytoplasm and translated into proteins. The Rev protein is critical for the transition from the early to late stage because it binds to viral transcripts and facilitates their transport to the cytoplasm. Thus, when Rev protein levels are low early after viral infection, there is inefficient transport of transcripts out of the nucleus and more complete splicing and processing of short mRNAs that encode the regulatory proteins. When Rev levels accumulate, longer transcripts that encode the structural genes, as well as full-length viral genomes, are efficiently transported out of the nucleus.

Synthesis of mature, infectious viral particles begins after the various viral genes are expressed as proteins, and full-length genomic viral RNA transcripts are produced. The *pol* gene product is a precursor protein that is sequentially cleaved to form reverse transcriptase, protease, ribonuclease, and integrase enzymes. The reverse transcriptase and integrase proteins are required for producing a DNA copy of the viral RNA genome and integrating it as a provirus into the host genome. The *gag* gene encodes a 55 kD protein that is proteolytically cleaved into p24, p17, and p15 polypeptides by the action of the viral protease encoded by the *pol* gene. These polypeptides are the mature core proteins that are required for assembly of infectious viral particles. The primary product of the *env* gene is a 160 kD glycoprotein (gp160) that is cleaved by cellular proteases within the endoplasmic reticulum into the CD4-binding protein gp120, expressed on the external surface of the envelope, and the gp41 transmembrane glycoprotein. Gp120 does not contain a transmembrane domain, but it remains bound to the cell surface by non-covalent interactions with gp41. As mentioned above, both these molecules are crucial for viral infectivity. In addition, a soluble (virus-free) form of gp120 may be responsible for some of the immunopathology caused by HIV (see below).

After transcription of these various viral genes, viral proteins are synthesized in the cytoplasm. Assembly of infectious viral particles then begins by packaging full-length RNA transcripts of the proviral genome within a nucleoprotein complex that includes the *gag* core proteins and the *pol*-encoded enzymes required for the next cycle of integration. This nucleoprotein complex is then enclosed within a membrane envelope and released from the cell by a process of budding from the plasma membrane. *Production of mature virus is associated with lysis of the cell and is the basis of the cytopathic effect of HIV.*

Immunology of HIV Infection

The clinical course of HIV infection reflects a complex interplay between the effects of the virus on the function of immunocompetent cells and the host's immune response to the virus. We next describe what is known or hypothesized to be the basis for HIV-induced immunosuppression, and discuss the role of the host's immune response in both aggravating and limiting the pathologic effects of the virus (Table 21–3).

NATURE AND MECHANISMS OF IMMUNOSUPPRESSION

HIV infection ultimately results in impaired function of both the specific and innate immune systems. The most prominent defects are in cell-

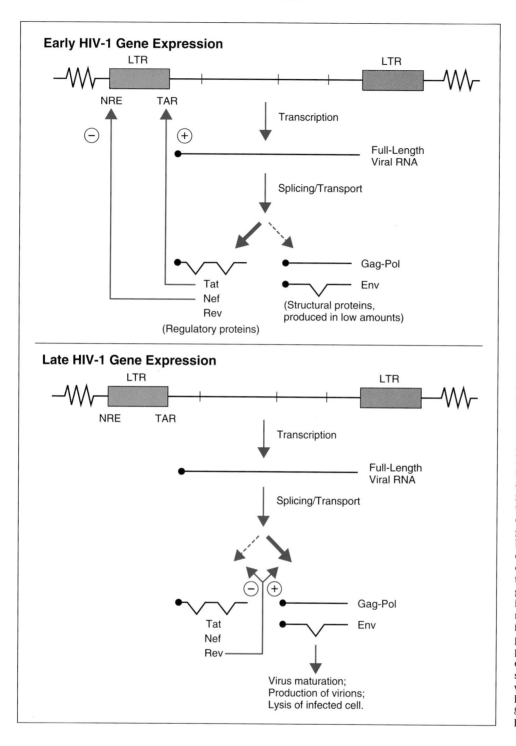

FIGURE 21-5. Early and late phases of human immunodeficiency virus-1 (HIV-1) gene expression. During early HIV-1 gene expression, the predominant types of transcripts transported out of the nucleus are multiply spliced messenger RNA (mRNA) species encoding regulatory proteins Tat, Nef, and Rev. Tat and Nef act on genomic sequences to regulate further transcription. Rev causes a change to late phase gene expression by altering the balance of mRNA transport out of the nucleus. Specifically, Rev causes a relative increase in the transport and translation of singly or unspliced mRNAs encoding viral structural and enzymatic proteins that are required for production of mature viral particles. LTR, long terminal repeat; NRE, negative regulatory element; TAR, trans-acting responsive sequences. (Modified, with permission, from Annual Review of Immunology, Volume 8, © 1990, by Annual Reviews Inc.)

mediated immunity and they can be attributed to two major causes:

1. *Most manifestations of immunodeficiency, including infections and tumors, are due to a lack of CD4+ T cells.* The hallmark of the progression of HIV-induced disease is the diminishing number of CD4+ T cells in the peripheral blood, from a normal of about 1000/mm³ to less than 100/mm³ in fully developed AIDS. This may occur over several months or take longer than 10 years in different individuals. Since CD4+ helper T cells are essential for both cell-mediated and humoral immune responses to various microbes, the loss of these lymphocytes is a major reason why AIDS patients become susceptible to so many infections. In fact, most of the opportunistic infections and neoplasias associated with HIV infection do not occur until after the blood CD4+ T cell count drops below 200 cells/mm³. The mechanisms by which HIV infection leads to CD4+ T cell depletion are discussed below.

TABLE 21–3. Mechanisms of Human Immunodeficiency Virus (HIV)-Induced Immunosuppression

Pathologic Effect	Mechanism
Depletion of CD4$^+$ T cells: direct effects of virus on infected cells	Lysis of CD4$^+$ T cells caused by viral budding and/or Env glycoprotein insertion
	Cytopathic effect of intracellular binding of gp120 to newly synthesized or recycled CD4
Depletion of CD4$^+$ T cells: indirect effects	Lysis of virally infected cells due to immune response to HIV (CTL and antibody-dependent cellular cytoxicity)
	Inhibition of CD4$^+$ T cell maturation in thymus
	Apoptosis due to soluble gp120 cross-linking of CD4
Functional impairment of immune system	Soluble gp120 blocks interaction of CD4$^+$ T cells with class II MHC on APCs
	Impaired macrophage and NK cell function, unknown mechanism
	Destruction of follicular dendritic cell network and architecture of lymph nodes and spleen
	Selective impairment of memory T cells and bias toward T$_H$2 differentiation, unknown mechanism

Abbreviations: MHC, major histocompatibility complex; APC, antigen-presenting cell; CTL, cytolytic T lymphocyte.

2. Immunodeficiency in HIV-infected individuals is, in part, a result of functional disturbances of the immune system independent of depletion of lymphocytes. In other words, there are some abnormalities in immune system function in HIV infection with normal CD4$^+$ T cell counts.

The loss of CD4$^+$ T cells in HIV-infected people is mainly due to direct toxic effects of infection of these cells by HIV. However, there is much speculation and some evidence that one or more indirect mechanisms also contribute to the depletion of CD4$^+$ T cells in HIV-infected individuals.

Several direct cytopathic effects of HIV on infected CD4$^+$ cells have been described:

1. The process of virus production, with expression of gp41 in the plasma membrane and budding of viral particles, may lead to increased plasma membrane permeability and influx of lethal amounts of calcium or osmotic lysis of the cell.
2. The plasma membranes of HIV-infected T cells fuse with uninfected CD4$^+$ T cells by virtue of gp120-CD4 interactions, leading to the formation of multinucleated giant cells or syncytia. The process of HIV-induced syncytial formation can be lethal to the HIV-infected T cells as well as the uninfected CD4$^+$ T cells that fuse to the infected cells. The

phenomenon has largely been observed *in vitro,* and syncytia are rarely seen in the tissues of AIDS patients.
3. Unintegrated viral DNA in the cytoplasm of infected cells, or large amounts of nonfunctional viral RNA, may be toxic to the infected cells.
4. Viral production can interfere with cellular protein synthesis and expression, leading to cell death.
5. Gp120 binding to newly synthesized intracellular CD4 may have toxic effects.

Mechanisms other than direct lysis of CD4$^+$ T cells by virus have been proposed for the depletion of these cells in HIV-infected individuals. Many of these mechanisms have been invoked to explain progressive loss of CD4$^+$ T cells even though there are very few infected T cells in the blood at any one time during most of the course of HIV disease. It is now known, however, that there are abundant infected T cells in lymphoid organs even when virus cannot be detected in most circulating T cells, and that direct cytotoxic effects of the virus account for much of the CD4$^+$ T cell depletion. Nonetheless, it is likely that indirect mechanisms do account for some of the loss of CD4$^+$ T cells. For example, HIV-infected cells elicit normal immune responses that lead to their destruction. HIV-specific cytolytic T lymphocytes (CTLs) are present in many AIDS patients, and these cells can kill infected CD4$^+$ T cells. In addition, antibodies against HIV envelope proteins may bind to HIV-infected CD4$^+$ T cells and target the cells for antibody-dependent cellular cytotoxicity. Other proposed mechanisms include apoptotic effects of viral proteins, including gp120 and HIV-encoded super-antigens. The role of any of these indirect mechanisms in CD4$^+$ T cell depletion in HIV-infected patients is uncertain and controversial.

Defects in the immune system of HIV-infected individuals are detectable prior to significant depletion of CD4$^+$ T cells, although these are more subtle than the immunodeficiency of full-blown AIDS. These defects include a decrease in memory T cell responses to antigen, poor CTL responses to viral infections, and weak humoral immune responses to defined antigens even though total serum Ig levels may be elevated. The mechanisms of these defects are not well understood, and there are many theories based on often controversial data. The defects may be a result of direct effects of systemic HIV infection on CD4$^+$ T cells, including the effects of soluble gp120 released from infected cells, which binds to uninfected cells. For example, CD4 that has bound gp120 may not be available to interact with class II MHC molecules on antigen-presenting cells (APCs), and thus T cell responses to soluble antigens would be inhibited. Alternatively, gp120 binding to CD4 may deliver signals that downregulate helper T cell function. There is some evidence that the proportion of IL-2– and IFN-γ–secreting (T$_H$1-like) T cells decreases gradually in HIV-infected patients, and the proportion of IL-4 and IL-10 secreting (T_H2-like) T cells increases. This may partially explain the susceptibility of HIV-infected individuals to infections by intracellular microbes, since IFN-γ activates, and IL-4 and IL-10 inhibit, macrophage-mediated killing of such microbes.

Two other cell types, besides CD4$^+$ T cells, that play an important role in HIV infection and the progression of immunosuppression are mononuclear phagocytes and follicular dendritic cells. Macrophages express much lower levels of CD4 than helper T lymphocytes, but they do express chemokine receptors, including CC-CKR5, and are susceptible to HIV infection. Nonetheless, macrophages are relatively resistant to the cytopathic effects of HIV. This may be because high CD4 expression is required for virus-induced cytotoxicity. Macrophages may also be infected by a gp120/gp41-independent route, such as phagocytosis of other infected cells or by Fc receptor–mediated endocytosis of antibody-coated HIV virions. Since macrophages can be infected but are generally not killed by the virus, they may become a major reservoir for the virus. In fact, the quantity of macrophage-associated HIV far exceeds T cell–associated virus in most tissues from AIDS patients, including brain and lung. HIV-infected macrophages may be impaired in antigen presentation functions and cytokine secretion. Follicular dendritic cells (FDCs) in the germinal centers of lymph nodes and spleen trap large amounts of HIV on their extended surfaces, in part by Fc receptor–mediated binding of antibody-coated virus. Although the FDCs are not efficiently infected, they contribute to the pathogenesis of HIV-associated immunodeficiency in at least two significant ways. First, the FDC surface is a reservoir for HIV that can infect macrophages and CD4$^+$ T cells in the lymph node. Second, the normal functions of FDCs in immune responses are impaired, and they may eventually be destroyed by the virus. Although the mechanisms of HIV-induced death of FDCs are not understood, the net result of the loss of the FDC network in lymph nodes and spleen is a dissolution of the functional architecture of the peripheral lymphoid system.

Because of the presence of abundant CD4$^+$ T cells, macrophages, and FDCs, lymph nodes and spleen are sites of continuous HIV replication and progression of HIV-induced disease even during the long, clinically silent period after acute infection. More than 90 per cent of the body's T cells are normally found in lymphoid tissues, such as lymph nodes and spleen, and infection of cells in lymph nodes and spleen occurs rapidly after HIV infection. The lymphoid tissues are sites of continual viral production, T cell infection, and T cell destruction, even when HIV is not detectable in peripheral blood T cells and patients are asymptomatic. This accounts for the steady decline in peripheral blood CD4$^+$ T cell numbers during the clinical latent period, since blood T cells and lymph node T cells are part of the same recirculating pool of lymphocytes. Eventually, in advanced disease, lymph node T cells are also depleted, and few HIV-positive cells are present.

IMMUNE RESPONSES TO HIV

Both humoral and cell-mediated immune responses specific for viral gene products have been observed in HIV-infected patients. The early response to HIV infection is, in fact, similar in many ways to immune responses to other viruses, and it serves effectively to clear most of the virus present in blood and blood T cells. Nonetheless, it is clear that these immune responses fail to eradicate all the virus, and in most individuals the infection eventually overwhelms the immune system. This is, of course, partly due to the fact that the CD4$^+$ T cells required to initiate protective immune responses are killed or inactivated by the virus, and therefore the immune responses may be too compromised to eliminate the virus. In addition, the HIV genome displays a remarkable degree of genetic variability, largely as a result of the high error frequency intrinsic to reverse transcription. This results in antigenic variations that may serve to evade the host immune system. The genetic variability also contributes to development of drug-resistant strains of HIV. Despite the poor effectiveness of immune responses to the virus, it is important to characterize them for three reasons. First, the immune responses may be detrimental to the host, e.g., by stimulating uptake of opsonized virus into uninfected cells by Fc receptor–mediated endocytosis. Second, antibodies against HIV are diagnostic markers of HIV infection widely used for screening purposes. Third, the design of effective vaccines for immunization against HIV requires knowledge of the viral epitopes that are most likely to stimulate protective immunity.

The most immunogenic molecules on HIV appear to be the envelope glycoproteins, and high titers of anti-gp120 and anti-gp41 antibodies are present in most HIV-infected individuals. Other antibodies found frequently in patients' sera include antibodies to p24, reverse transcriptase, and *gag* and *pol* products. The effect of these antibodies on the clinical course of HIV infection is probably minimal. It is interesting that the anti-envelope antibodies are generally poor inhibitors of viral infectivity or cytopathic effects, supporting the hypothesis that the most immunogenic epitopes of the envelope glycoproteins are least important for the functions of these molecules. Furthermore, the antibodies are usually virus strain-specific, so that antibodies from one infected individual often do not recognize HIV isolated from other infected individuals. A region of the gp120 molecule, called the V3-loop, is one of the most antigenically variable components of the virus, and it may vary in HIV isolates taken from the same individual at different times. It is possible that the host immune response may work as a selective pressure that promotes survival of the most genetically variable viruses. Low titers of neutralizing antibodies that can inactivate HIV are present in HIV-infected patients, as are antibodies that can mediate antibody-dependent cell-mediated cytotoxicity (ADCC). These antibodies are usually specific for gp120. Whether there is a correlation between the titers of these antibodies and the clinical course remains controversial.

The role of T cell–mediated immune responses to HIV infection is also incompletely understood. Class I MHC–restricted CTLs specific for Env, Gag, and Pol proteins are present in HIV-infected individuals, and there is good evidence that CTL responses are important for the initial clearance of most of the virus during the early stages of infection. In addition, NK cell activity against HIV-infected targets is present in these patients. CD4⁺ T cell responses to HIV-derived peptides can also be detected early in HIV-infected individuals, and these responses are likely to promote humoral and CTL responses. These responses wane along with CD4⁺ T cell responses to other infections as the disease progresses.

Clinical Features of HIV Infection

Because of the complex biology of HIV, the clinical manifestations of infection are quite variable. Although initial infection may occur without accompanying symptoms, many patients experience an **acute HIV syndrome** within 2 to 6 weeks of exposure to the virus (Fig. 21–6). This syndrome is characterized by fever, headaches, sore throat with pharyngitis, generalized lymphadenopathy, and rashes. No aspect of this acute illness is specifically diagnostic for HIV infection. During this initial period after infection, the virus is replicating abundantly and is detectable in blood and cerebrospinal fluid. After the initial phase, a clinically **latent phase** begins, which may last up to 10 years. Although extracellular virus practically disappears from body fluids during this latent phase, and the majority of peripheral blood T cells do not harbor the virus, there is steady progression of the disease in lymphoid tissues, with increasing numbers of infected CD4⁺ T cells, macrophages, and follicular dendritic cells. Generalized lymphadenopathy may develop during this period, but the immune system remains competent at handling most infections. A heterogenous subset of patients develops a group of signs and symptoms that may persist for some time but do not fit the definition of clinical AIDS. These clinical features constitute **AIDS-related complex (ARC)** and include persistent fevers, night sweats, weight loss, diarrhea, inflammatory skin conditions, and generalized lymphadenopathy. ARC can persist for months or years before progression to AIDS. HIV-infected individuals who develop herpes zoster, oral candidal infection, and oral hairy leukoplakia are likely to succumb to full-blown AIDS relatively quickly.

The diagnosis of AIDS can be made based on a combination of laboratory evidence of HIV infection (discussed below) and the presence of many possible combinations of opportunistic infections, neoplasias, cachexia (HIV wasting syndrome), and CNS degeneration (AIDS encephalopathy). AIDS patients acquire numerous infections that can be

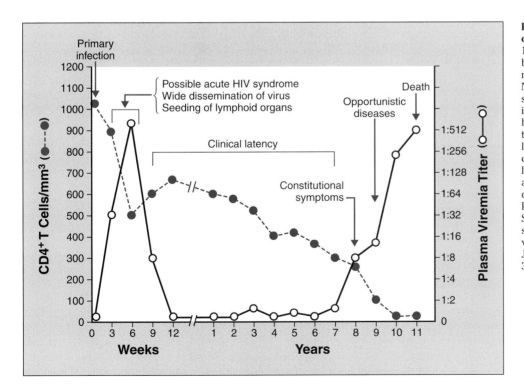

FIGURE 21–6. Typical course of HIV infection. After about 12 weeks post-infection, blood-borne virus (plasma viremia) is not detectable for many years. Nonetheless, CD4⁺ T cell counts steadily decline during this clinical latency period, probably because of active viral replication and T cell infection in lymph nodes. When CD4⁺ T cell counts drop below a critical level (about 200/mm³), there is a high risk of infection. (Reproduced with permission from Pantaleo, G., C. Graziosi, and A. S. Fauci. The immunopathogenesis of human immunodeficiency virus infection. New England Journal of Medicine 328:327–335, 1993.)

life-threatening, often with organisms that are not normally pathogenic for immunocompetent individuals. Pneumonia caused by the protozoan *Pneumocystis carinii* is the most commonly acquired opportunistic infection in AIDS, affecting up to 75 per cent of patients, and it is perhaps the most common cause of death in AIDS. Other protozoal organisms that frequently infect AIDS patients include *Cryptosporidium* and *Toxoplasma*. Bacteria that often cause infections in AIDS patients include *Mycobacterium* species such as *M. avium-intracellulare* and *M. kansasii, Nocardia,* and *Salmonella.* Fungal infections with *Candida, Cryptococcus neoformans, Coccidioides immitis,* and *Histoplasma capsulatum* are common, as are viral infections with cytomegalovirus, herpes simplex, and varicella-zoster. The inflammatory responses to these various organisms are often unlike those seen in immunocompetent individuals; this probably reflects the lack of T cells that would normally secrete cytokines that promote acute and chronic inflammation. For example, well-formed granulomas with activated macrophages are not seen in mycobacterial or fungally infected tissues in AIDS patients.

AIDS is also characterized by progressive weight loss (cachexia) and diarrhea with or without identifiable enteric infections. These symptoms are often called HIV wasting syndrome; the pathogenesis is not understood.

Various malignant neoplasms are frequently found in AIDS patients, and these represent another major cause of AIDS-related morbidity and mortality. This phenomenon is likely due to a failure of immunosurveillance against oncogenic virus infection (see Chapter 19). Up to 30 per cent of AIDS patients develop Kaposi's sarcoma, a mesenchymal tumor characterized histologically by vascular spaces and malignant spindle cells. AIDS-related Kaposi's tumors, unlike sporadic forms of this neoplasm, are highly aggressive and disseminated, involving skin, mucosa, lymph nodes, and multiple visceral organs. The cause of this tumor in the setting of AIDS may be related to a newly described herpesvirus, called Kaposi's sarcoma herpesvirus, which is consistently found in these tumors. There is some evidence that the incidence of Kaposi's sarcoma is declining. AIDS patients also develop malignant lymphomas at a much higher rate than immunocompetent individuals. Burkitt's lymphoma and other B cell tumors are most frequent, and they are sometimes positive for Epstein-Barr virus (EBV). Primary lymphomas of the CNS are also common in AIDS patients. Cervical carcinoma is also abnormally frequent in women with AIDS, and these tumors are usually positive for oncogenic strains of human papillomavirus.

The brain is a major site of HIV infection, and up to 66 per cent of AIDS patients suffer from a form of dementia called AIDS encephalopathy or AIDS dementia complex, characterized by memory loss and various other nonspecific neuropsychiatric disturbances. Macrophages are the cells most likely to be infected in the brain, although some evidence suggests direct infection of cerebrovascular endothelium and neurons. Neuronal damage is seen on pathologic examination of brains from AIDS autopsies, but the causes are not clear. It is possible that HIV-infected macrophages secrete cytokines that are toxic to neurons or that HIV interferes with neurotropic peptide factors or neurotransmitters.

A variety of clinical immunology laboratory techniques are used to diagnose and follow the progression of HIV infection. Antibodies to HIV proteins, including p24 and gp120, appear in the serum usually between 2 and 12 weeks after primary infection (see Fig. 21–6). Standard screening protocols for HIV utilize immunofluorescence or enzyme-linked immunoassays to detect these antibodies. Positive screening tests are often followed up by Western blot or radioimmunoassay determination of the presence of serum antibodies that bind to specific viral proteins. Viral antigens (usually p24) are present in the serum for up to 3 months after infection and then reappear years later when AIDS develops. These antigens can be detected using enzyme-linked immunoassays. Polymerase-chain reaction (PCR) assays are particularly sensitive in detecting the presence of viral genomes in cells or body fluids and may occasionally demonstrate HIV infection when other tests are negative. Reverse transcriptase-PCR is now being used as a very sensitive test to detect viral genomes or viral gene transcripts in peripheral blood. Culturing virus from infected individuals is possible but technically difficult. Diagnosis of asymptomatic infants born to HIV-infected mothers poses a special problem since maternal anti-HIV antibodies cross the placenta and will give false-positive results on screening tests. Neonatal diagnosis therefore requires viral antigen detection, viral culture, or PCR detection of viral genomes.

The peripheral CD4+ T cell count is the most commonly used laboratory test for assessing the progression of HIV disease (see Fig. 21–6). After an initial transient drop in the number of CD4+ T cells in the blood, there follows a steady decline over years. When the count drops below 200 cells/mm³, the risk of full-blown AIDS with opportunistic infections becomes high.

Transmission of HIV and Epidemiology of AIDS

The modes of transmission of HIV from one individual to another are the major determinants of the epidemiologic features of AIDS. The virus is transmitted by three major routes:

1. Intimate sexual contact is the most frequent mode of transmission, either between homosexual male partners or heterosexual couples. The virus is present in semen and gains access to the previously uninfected partner either through traumatized rectal mucosa or vaginal mucosa. Transmis-

sion from infected females to males may also occur.

2. Inoculation of a recipient with infected blood or blood products is the second most frequent mode of HIV transmission. Needles shared by intravenous drug abusers account for most cases of this form of transmission. Patients infected in these ways may then infect other individuals by sexual contact. With the advent of routine laboratory screening, transfusion of blood or blood products, in a clinical setting, accounts for a very small portion of HIV infections.

3. Mother to child transmission of HIV accounts for the majority of pediatric cases of AIDS. This occurs most frequently *in utero* or during childbirth, although transmission through breast milk is also possible.

Major groups at risk for developing AIDS in the United States include homosexual or bisexual males, intravenous drug abusers, heterosexual partners of members of other risk groups, and babies born from infected mothers. There is a small increased risk of infection among health care workers. In Africa, the vast majority of HIV infections occurs by sexual transmission from one heterosexual partner to another, with no identifiable risk factors.

Treatment of AIDS and Vaccine Development

Research efforts are aimed at developing reagents that interfere with various parts of the viral life cycle. Treatment of HIV infection and AIDS now includes the use of two classes of antiviral drugs, used in combination, that target viral molecules for which there are no human counterparts. The first type of drugs to be widely used are nucleotide analogs that inhibit reverse transcriptase activity. These drugs include azidothymidine (AZT), dideoxyinosine, and dideoxycytidine. When these drugs are used alone, they are often effective in significantly reducing plasma HIV RNA levels for several months to years. Usually they do not halt progression of HIV-induced disease, largely because of the evolution of virus with mutated forms of reverse transcriptase that are resistant to the drugs. More recently, effective viral protease inhibitors have been developed that block the processing of precursor proteins into mature viral capsid and core proteins. These protease inhibitors are now being used in combination with two different reverse-transcriptase inhibitors. This new triple-drug therapy has proved to be remarkably effective in reducing plasma viral RNA to undetectable levels in most patients treated for over 1 year. Whether or not resistance to this therapy will develop over longer time periods is not yet known. Two formidable problems associated with these new drug therapies, which may limit their effective use in many parts of the world, include very high expense and complicated administration schedules.

The individual infections experienced by AIDS patients are treated with the appropriate antibiotics and support measures. More aggressive antibiotic therapy is often required than for similar infections in less compromised hosts.

The development of an effective vaccine for immunoprophylaxis against HIV has become a major priority for biomedical research institutions worldwide. The task has been complicated by the genetic potential of the virus for great antigenic variability. Furthermore, although we know many of the viral gene products that induce humoral immune responses, these responses are ineffective in preventing disease. Perhaps vaccine development will require identification of viral epitopes and immunization methods that stimulate effective cell-mediated immunity. Vaccines effective in preventing SIV infection of macaques have already been developed. This is encouraging, because SIV is molecularly closely related to HIV and causes a disease similar to AIDS in macaques. Even after potentially effective vaccines are developed, there will remain enormous logistic and ethical problems in testing their effectiveness among populations at risk.

OTHER ACQUIRED IMMUNODEFICIENCIES

Overall, acquired immunodeficiencies caused by other factors besides HIV infection are still more common than AIDS. These immunodeficiency states fall into two general etiologic categories. First, immunosuppression may occur as a biologic complication of another disease process. Second, so-called iatrogenic immunodeficiencies may develop as complications of the therapy for other diseases.

Diseases in which immunodeficiency is a common complicating element include malnutrition, neoplasias, and infections. Protein-calorie malnutrition is extremely common in developing countries and is associated with impaired cellular and humoral immunity to microorganisms. Much of the morbidity and mortality that afflicts malnourished people is due to infections. The basis for the immunodeficiency is not well defined, but it is reasonable to assume that the global metabolic disturbances in these individuals, caused by deficient intake of protein, fat, vitamins, and minerals, will adversely affect the maturation and function of the cells of the immune system.

Patients with advanced, widespread cancers are often susceptible to infections because of impaired cell-mediated and humoral immune responses to a variety of organisms. Bone marrow tumors, including cancers metastatic to marrow and leukemias that arise in the marrow, may interfere with the growth and development of normal lymphocytes. Alternatively, tumors may produce substances that interfere with lymphocyte development or function, such as TGF-β. An example of malignancy-associated immunodeficiency is the impairment of T cell function commonly observed in patients with a type of lymphoma called Hodgkin's disease. This defect was first characterized as an inability to mount a DTH reaction upon dermal injection of various common antigens to which the patients were previously exposed, such as *Candida* or tetanus toxoid. Other *in vitro* measures of T cell function, such as proliferative responses to polyclonal activators, are also impaired in Hodgkin's disease. Such a generalized deficiency in DTH responses is called **anergy.** The basis for these T cell abnormalities is currently unknown.

Various types of infections lead to immunosuppression. Viruses other than HIV are known to impair immune responses and lead to complicating infections by other organisms. Examples include the measles virus and human T cell lymphotropic virus-1 (HTLV-1). Both viruses can infect lymphocytes, and this may be a basis for their immunosuppressive effects. Like HIV, HTLV-1 is a retrovirus with tropism for $CD4^+$ T cells; however, instead of killing helper T cells, it transforms them, producing an aggressive T cell malignancy called adult T cell leukemia/lymphoma (ATL). In ATL patients, typically there is severe immunosuppression with multiple opportunistic infections. Chronic infections with *Mycobacterium tuberculosis* and various fungi frequently result in anergy to many antigens. Chronic parasitic infections may also lead to immunosuppression. For example, African children with chronic malarial infections have depressed T cell function, and this may be important in the pathogenesis of EBV-associated malignancies (see Chapter 18, Box 18–2).

Iatrogenic immunosuppression is most often due to drug therapies that either kill or functionally inactivate lymphocytes. Some drugs are given intentionally to immunosuppress patients, either for treatment of inflammatory diseases or to prevent rejection of tissue allografts. The most commonly used immunosuppressive drugs are corticosteroids and cyclosporin A, discussed in Chapter 17. Various chemotherapeutic drugs are administered to cancer patients, and these drugs are usually cytotoxic to both mature and developing lymphocytes as well as to granulocyte and monocyte precursors. Thus, cancer chemotherapy is almost always accompanied by a period of immunosuppression and risk of infections. Radiation treatment of cancer patients carries the same risks.

One final form of acquired immunosuppression that should be mentioned results from the absence of a spleen, caused by surgical removal of the organ after trauma or for the treatment of certain hematologic diseases or as a result of infarction in sickle cell disease. Patients without spleens are more susceptible to infections by some organisms, particularly encapsulated bacteria such as *Streptococcus pneumoniae*. The spleen is required for the induction of protective humoral immune responses

to the thymus-independent capsular polysaccharide antigens of such organisms.

SUMMARY

Immunodeficiency diseases are caused by congenital or acquired defects in lymphocytes, phagocytes, and other mediators of specific and innate immunity. These diseases are associated with an increased susceptibility to infections, the nature and severity of which depend largely on which component of the immune system is abnormal and the extent of the abnormality. In addition, patients with immunodeficiencies often show an increased incidence of cancers and autoimmune diseases.

Congenital (or primary) immunodeficiencies may be due to defects in B or T lymphocytes or both. An example of a selective B cell deficiency is X-linked agammaglobulinemia, an inherited block in the maturation of pre–B cells to B lymphocytes that results in an absence of mature B cells and antibodies and is due to mutations in the B cell tyrosine kinase gene. Congenital disorders selectively affecting the production of one or a few Ig isotypes, including IgA, IgM, and various IgG subclasses, are usually due to a failure of mature B cells to switch to particular Ig heavy chain isotypes. One example is IgG and IgA deficiency with increased IgM (hyper-IgM syndrome), and this is due to mutations in the gene encoding the ligand for CD40. Common variable immunodeficiency is most often due to an intrinsic defect in the Ig secretory responses of mature B cells to antigenic stimulation.

The best-defined congenital T cell immunodeficiency is the DiGeorge syndrome, caused by a genetic abnormality in the development of the thymus. This results in a failure of T cell maturation. Many rare diseases are attributable to defects in genes encoding molecules required for T cell activation or antigen presentation to T cells.

Severe combined immunodeficiencies constitute a group of disorders with defects in the development or function of both B and T lymphocytes. Some cases are due to X-linked inherited deficiencies in the cytokine receptor common γ chain, and some are due to autosomal recessive deficiencies of enzymes, such as adenosine deaminase, that are involved in purine metabolism. Other cases are attributable to blocks in the development of antigen receptor-expressing lymphocytes from precursors in the bone marrow, including defects in antigen-receptor gene rearrangements. Deficiencies of B and T lymphocytes are also associated with diseases that affect multiple organ systems, such as the Wiskott-Aldrich syndrome and ataxia-telangiectasia.

AIDS is a severe T cell immunodeficiency caused by infection with HIV. This virus has a tropism for $CD4^+$ T lymphocytes, causing depletion of these cells mainly by direct lysis and possibly by other indirect mechanisms that lead to death, defective maturation, or abnormal function of uninfected T cells. The depletion of T cells results in greatly increased susceptibility to infection by a number of opportunistic microorganisms, including *Pneumocystis carinii,* mycobacteria, and various fungi and viruses. In addition, patients have an increased incidence of tumors, particularly Kaposi's sarcoma and EBV–associated B cell lymphomas, and they frequently develop an encephalopathy, the mechanism of which is not fully understood.

Acquired immunodeficiencies are also associated with malnutrition, disseminated cancers, and immunosuppressive therapy for transplant rejection or autoimmune diseases.

Selected Readings

Bloom, B. R. A perspective on AIDS vaccines. Science 272:1888–1890, 1996.

Conley, M. E. Primary immunodeficiencies: a flurry of new genes. Immunology Today 16:313–315, 1995.

Cournoyer, D., and C. T. Caskey. Gene therapy of the immune system. Annual Review of Immunology 11:297–329, 1993.

Curnutte, J. T. Chronic granulomatous disease: the solving of a clinical riddle at the molecular level. Clinical Immunology and Immunopathology. 67:S2–15, 1993.

Etzioni, A. Adhesion molecule deficiencies and their clinical significance. Cell Adhesion and Communication 2:257–260, 1994.

Fuleihan, R., N. Ramesh, and R. S. Geha. X-linked agammaglobulinemia and immunoglobulin deficiency with normal or elevated IgM: immunodeficiencies of B cell development and differentiation. Advances in Immunology 60:37–56, 1995.

Greene, W. C. Regulation of HIV-1 gene expression. Annual Review of Immunology 8:453–476, 1990.

Ochs, H. D., and R. J. Wedgwood. IgG subclass deficiencies. Annual Review of Medicine 38:325–340, 1987.

Pantaleo, G., and A. S. Fauci. New concepts in the immunopathogenesis of HIV infection. Annual Review of Immunology 13:487–512, 1995.

Rosen, F. S. (chairman). Primary immunodeficiency diseases: report of a WHO scientific group. Immunodeficiency Reviews 3:195–236, 1992.

Rotrosen, D., and J. I. Gallin. Disorders of phagocyte function. Annual Review of Immunology 5:127–150, 1987.

Sideras, P., and C. I. Edvard Smith. Molecular and cellular aspects of X-linked agammaglobulinemia. Advances in Immunology 59:135–223, 1995.

Steimle, V., W. Reith, and B. Mach. Major histocompatibility complex class II deficiency: a disease of gene regulation. Advances in Immunology 61:327–340, 1996.

Sugamura, K., H. Asao, M. Kondo, N. Tanaka, N. Ishii, K. Ohbo, M. Nakamura, and T. Takeshita. Interleukin-2 receptor gamma chain: its role in multiple cytokine receptor complexes and T cell development in XSCID. Annual Review of Immunology 14:179–206, 1996.

APPENDIX: PRINCIPAL FEATURES OF KNOWN CD MOLECULES

CD Designation	Common Synonym(s)	Molecular Structure	Main Cellular Expression	Known or Proposed Function(s)
CD1a*†	T6	49 kD; β_2 microglobulin–associated	Thymocytes, dendritic cells (including Langerhans cells)	Presentation of non-peptide antigens to some T cells
CD1b	—	45 kD; β_2 microglobulin–associated	Same as CD1a	Same as CD1a
CD1c	—	43 kD; β_2 microglobulin–associated	Same as CD1a; also B cells	Same as CD1a
CD2	T11; LFA-2; sheep red blood cell receptor	50 kD	T cells, NK cells	Adhesion molecule (binds LFA-3); T cell activation
CD3	T3; Leu-4	Composed of three chains, γ, δ, ϵ (see Chapter 7)	T cells	Associated with T cell antigen receptor; signal transduction as a result of antigen recognition by T cells
CD4	T4; Leu-3; L3T4 (mice)	55 kD	Class II MHC–restricted T cells	Adhesion molecule (binds to class II MHC); signal transduction
CD5	T1; Lyt-1	67 kD	T cells; B cell subset	Ligand for CD72; ? adhesion molecule
CD6	T12	100–130 kD	Subset of T cells; some B cells	Role in T cell activation
CD7	—	40 kD	Hematopoietic stem cells; subset of T cells	Signal transduction
CD8	T8; Leu-2; Lyt-2	Composed of two 34 kD chains; expressed as $\alpha\alpha$ or $\alpha\beta$ dimer	Class I MHC–restricted T cells	Adhesion (binds to class I MHC); signal transduction
CD9	—	24 kD	Pre-B and immature B cells; monocytes, platelets	? Role in platelet activation
CD10	CALLA; enkephalinase; neutral endopeptidose	100 kD	Immature and some mature B cells; lymphoid progenitors, granulocytes	Cell surface metallopeptidase
CD11a‡	LFA-1 α chain	180 kD; associates with CD18 to form LFA-1 integrin	Leukocytes	Adhesion (binds to ICAM-1, -2)
CD11b	Mac-1; CR3 (iC3b receptor) α chain	165 kD; associates with CD18 to form Mac-1 integrin	Granulocytes, monocytes, NK cells	Adhesion; phagocytosis of iC3b-coated (opsonized) particles
CD11c	p150,95; CR4 α chain	150 kD; associates with CD18 to form p150,95 integrin	Monocytes, granulocytes, NK cells	Adhesion; ? phagocytosis of iC3b-coated (opsonized) particles
CDw12§	—	? 90–120 kD	Monocytes, granulocytes	Phosphoprotein; no known function
CD13	—	150 kD	Monocytes, granulocytes	Aminopeptidase; ? role in oxidative burst
CD14	Mo2	55 kD; PI-linked	Monocytes	LPS receptor; ? role in oxidative burst
CD15	Lewis^x	Carbohydrate epitope	Granulocytes	Sialyl form is a ligand for selectins
CD16	FcγRIII	50–70 kD; PI-linked and transmembrane	NK cells, granulocytes, macrophages	Low-affinity Fcγ receptor: ADCC, activation of NK cells

APPENDIX: PRINCIPAL FEATURES OF KNOWN CD
MOLECULES *(Continued)*

CD Designation	Common Synonym(s)	Molecular Structure	Main Cellular Expression	Known or Proposed Function(s)
CDw17	—	Carbohydrate epitope (lactosylceramide)	Granulocytes, macrophages, platelets	?
CD18	β chain of LFA-1 family (β2 integrins)	95 kD; non-covalently linked to CD11a, CD11b, or CD11c	Leukocytes	See CD11a, CD11b, CD11c
CD19	B4	90 kD	Most B cells	Role in B cell activation
CD20	B1	Heterodimer: 35 and 37 kD chains	Most or all B cells	? Role in B cell activation or regulation; calcium ion channel
CD21	CR2; C3d receptor; B2	145 kD	Mature B cells	Role in B cell activation; receptor for C3d, Epstein-Barr virus
CD22	—	135 kD	B cells	Role in B cell activation
CD23	FcεRIIb	45–50 kD	Activated B cells, macrophages	Low-affinity Fcε receptor, induced by IL-4; function unknown
CD24	Heat-stable antigen	Heterodimer of 38 and 41 kD chains; PI-linked	B cells, granulocytes	? Role in costimulation of T cells
CD25	IL-2 receptor α chain; TAC; p55	55 kd	Activated T and B cells; activated macrophages	Complexes with IL-2Rβγc high-affinity IL-2 receptor; T cell growth
CD26	—	110 kD	Activated T and B cells; macrophages	Serine peptidase
CD27	—	Homodimer of 55 kD chains	Most T cells; ? some plasma cells	? Costimulation of T cells; member of TNF-R, Fas, CD40 family
CD28	Tp44	Homodimer of 44 kD chains	T cells (most CD4+, some CD8+ cells)	T cell receptor for costimulator molecule(s) B7-1, B7-2
CD29	β chain of VLA antigens (β1 integrins)	130 kD; non-covalently associated with VLA α chains (CD49)	Broad	Adhesion to extracellular matrix proteins, cell-cell adhesion (see CD49)
CD30	Ki-1	105 kD	Activated T and B cells; Reed-Sternberg cells in Hodgkin's disease	? Role in activation-induced cell death; member of TNF-R family
CD31	PECAM-1; platelet gpIIa	140 kD	Platelets; monocytes, granulocytes, B cells, endothelial cells, T cells	Role in leukocyte-endothelial adhesion
CD32	FcγRII	~40 kD	Macrophages, granulocytes, B cells, eosinophils	Fc receptor for aggregated IgG; role in phagocytosis, ADCC; feedback inhibition of B cells
CD33	—	67 kD	Monocytes, myeloid progenitor cells	?
CD34	—	90 kD	Precursors of hematopoietic cells; vascular endothelium	Ligand for L-selectin

Table continued on following page

APPENDIX: PRINCIPAL FEATURES OF KNOWN CD
MOLECULES *(Continued)*

CD Designation	Common Synonym(s)	Molecular Structure	Main Cellular Expression	Known or Proposed Function(s)
CD35	CR1; C3b receptor	Polymorphic; four forms are 190–280 kD	Granulocytes, monocytes, erythrocytes, B cells	Binding and phagocytosis of C3b-coated particles and immune complexes
CD36	Platelet gpIIIb	90 kD	Monocytes, platelets	? Platelet adhesion
CD37	—	Composed of two or three 40–52 kD chains	B cells, some T cells	?
CD38	T10	45 kD	Plasma cells, thymocytes, activated T cells	?
CD39	—	78 kD	Activated B cells, NK cells, some T cells	?
CD40	—	Heterodimer of 44 and 48 kD chains	B cells, macrophages, dendritic cells, endothelial cells, epithelial cells	Role in B cell and macrophage activation induced by T cell contact; receptor for T cell CD40 ligand; member of Fas/TNF-R family
CD41	gpIIb component of gpIIb/ IIIa complex (gpIIIa is CD61)	Complex of gpIIb heterodimer (120 and 23 kD) and gpIIIa (CD 61) (integrin)	Platelets	Platelet aggregation and activation: receptor for fibrinogen, fibronectin (binds to R-G-D sequence)
CD42a	Platelet gpIX	23 kD; forms complex with CD42b	Platelets, megakaryocytes	Platelet adhesion, binding to von Willebrand's factor
CD42b	Platelet gpIb	Dimer of 135 and 25 kD chains, forms complex with CD42a	See CD42a	See CD42a
CD43	Sialophorin	115 kD, highly sialylated	Leukocytes (except circulating B cells)	? Role in T cell activation
CD44	Pgp-1; Hermes	80–>100 kD, highly glycosylated	Leukocytes, erythrocytes	May function as homing receptor; receptor for matrix components (e.g., hyaluronate)
CD45	T200; leukocyte common antigen	Multiple isoforms, 180–220 kD	Leukocytes	Role in signal transduction (tyrosine phosphatase)
CD45R	Forms of CD45 with restricted cellular expression	CD45RO: 180 kD CD45RA: 220 kD CD45RB: 190, 205, and 220 kD isoforms	CD45RO: memory T cells CD45RA: naive T cells CD45RB: B cells, subset of T cells	See CD45
CD46	Membrane cofactor protein (MCP)	45–70 kD	Leukocytes; epithelial cells, fibroblasts	Regulation of complement activation; binds C3b and C4b
CD47	—	47–52 kD	Broad	Mediates neutrophil migration across epithelium
CD48	BLAST-1	41 kD; PI-linked	Leukocytes	?
CD49a	VLA α_1 chain	210 kD; associates with CD29 to form VLA-1 (β_1 integrin)	Activated T cells, monocytes; other connective tissue cells	Adhesion to collagen, laminin
CD49b	VLA α_2 chain; platelet gpIa	170 kD; associates with CD29 to form VLA-2 (β_1 integrin)	Platelets, activated T cells, monocytes, some B cells	Adhesion to extracellular matrix: receptor for collagen

APPENDIX: PRINCIPAL FEATURES OF KNOWN CD
MOLECULES (Continued)

CD Designation	Common Synonym(s)	Molecular Structure	Main Cellular Expression	Known or Proposed Function(s)
CD49c	VLA α_3 chain	Dimer of 130 and 25 kD; associates with CD29 to form VLA-3 (β_1 integrin)	T cells; some B cells, monocytes	Adhesion to fibronectin, laminin
CD49d	VLA α_4 chain	150 kD; associates with CD29 to form VLA-4 (β_1 integrin) or with β_7 integrin	T cells, monocytes, B cells	Peyer's patch homing receptor, binds to VCAM-1; adhesion to fibronectin
CD49e	VLA α_5 chain	Dimer of 135 and 25 kD; associates with CD29 to form VLA-5 (β_1 integrin)	T cells; few B cells and monocytes	Adhesion to fibronectin
CD49f	VLA α_6 chain	150 kD; associates with CD29 to form VLA-6 (β_1 integrin)	Platelets, megakaryocytes; activated T cells	Adhesion to extracellular matrix: receptor for laminin
CD50	ICAM-3	180–140 kD; ? PI-linked	Leukocytes	? Binds CD11aCD18
CD51	α chain of vitronectin receptor	140 kD heterodimer, associates with CD61	Platelets, activated endothelium, smooth muscle cells, leukocytes	Adhesion: receptor for vitronectin, fibrinogen, von Willebrand's factor (binds R-G-D sequence)
CD52	—	? 21–28 kD	Leukocytes	?
CD53	—	32–40 kD	Leukocytes, plasma cells	?
CD54	ICAM-1	80–114 kD	Broad; many activated cells (cytokine-inducible)	Adhesion: ligand for LFA-1, Mac-1
CD55	Decay-accelerating factor (DAF)	70 kD; PI-linked	Broad	Regulation of complement activation; binds C3b
CD56	Leu-19	Heterodimer of 135 and 220 kD chains	NK cells	Homotypic adhesion; isoform of neural cell adhesion molecule (N-CAM)
CD57	HNK-1, Leu-7	110 kD	NK cells, subset of T cells	?
CD58	LFA-3	55–70 kD; PI-linked or integral membrane protein	Broad	Adhesion: ligand for CD2
CD59	Membrane inhibitor of reactive lysis (MIRL)	18–20 kD; PI-linked	Broad	Regulation of complement (MAC) action
CD60	—	Carbohydrate epitope	Subset of T cells, platelets	?
CD61	β chain of vitronectin receptor; gpIIIa, component of gpIIb/gpIIIa complex	110 kD; associates with CD51 (α chain of vitronectin receptor) or CD41 (integrins)	Platelets, megakaryocytes, endothelial cells, leukocytes	See CD51, CD41
CD62E	E-selectin, ELAM-1	115 kD	Endothelial cells	Leukocyte-endothelial adhesion
CD62L	L-selectin, LAM-1	75–80 kD	T lymphocytes, other leukocytes	Leukocyte-endothelial adhesion; homing of naive T cells to peripheral lymph nodes

Table continued on following page

APPENDIX: PRINCIPAL FEATURES OF KNOWN CD
MOLECULES *(Continued)*

CD Designation	Common Synonym(s)	Molecular Structure	Main Cellular Expression	Known or Proposed Function(s)
CD62P	P-selectin, gmp140, PADGEM	130–150 kD	Platelets, endothelial cells	Leukocyte adhesion to endothelium, platelets
CD63	—	53 kD; present in platelet lysosomes, translocated to cell surface upon activation	Activated platelets; monocytes, macrophages	?
CD64	FcγRI	75 kD	Monocytes, macrophages	High-affinity Fcγ receptor: role in phagocytosis, ADCC, macrophage activation
CDw65	—	Carbohydrate epitope	Granulocytes	? Role in neutrophil activation
CD66	—	180–220 kD phosphorylated glycoprotein	Granulocytes	? Role in homotypic cell-cell adhesions (carcinoembryonic antigen, or CEA, is called CD66e)
CD67	—	100 kD; PI-linked	Granulocytes	?
CD68	—	110 kD, intracellular protein, weak surface expression	Monocytes, macrophages	?
CD69	—	Homodimer of 28–34 kD chains, phosphorylated glycoprotein	Activated B and T cells, macrophages, NK cells	?
CD70	—	?	Activated T and B cells	?
CD71	T9; transferrin receptor	95 kD homodimer	Activated T and B cells, macrophages, proliferating cells	Receptor for transferrin: role in iron metabolism, cell growth
CD72	Lyb-2 (mouse)	Heterodimer of 39 and 43 kD chains	B cells	Ligand for CD5; ? role in T cell–B cell interactions
CD73	—	69 kD; PI-linked	Subsets of T and B cells	Ecto-5′-nucleotidase, regulates nucleotide metabolism
CD74	Class II MHC invariant (γ) chain; I$_i$	Three protein species: 35, 41, and 53 kD	B cells, monocytes, macrophages; other class II$^+$ cells	Associates with newly synthesized class II MHC molecules
CDw75	—	53 kD	Mature B cells	?
CD76	—	Heterodimer of 67 and 85 kD chains	Mature B cells, subset of T cells	?
CD77	—	Carbohydrate epitope	Follicular center B cells	?
CDw78	Ba	?	B cells	?
CD79a	Igα, MB1	32–33 kD	Mature B cells	Component of B cell antigen receptor
CD79b	Igβ, B29	37–39 kD	Mature B cells	Component of B cell antigen receptor
CD80	B7-1	50–60 kD	Dendritic cells, activated B cells and macrophages	Costimulator for T lymphocyte activation; ligand for CD28 and CTLA-4
CD81	TAPA-1	26 kD	Broad	Associated with CD19 and CD21; ? role in B cell activation
CD82	—	50–53 kD	Broad	?

Appendix: Principal Features of Known CD Molecules *(Continued)*

CD Designation	Common Synonym(s)	Molecular Structure	Main Cellular Expression	Known or Proposed Function(s)
CD83	—	40–43 kD	Activated B and T cells	?
CDw84	—	73 kD	Monocytes, lymphocytes	?
CD85	—	120 kD	B cells, monocytes	?
CD86	B7-2	80 kD	B cells, monocytes	Costimulator for T lymphocyte activation; ligand for CD28 and CTLA-4
CD87	—	50–65 kD	Neutrophils, monocytes, endothelial cells	?
CD88	C5a receptor	40 kD	Neutrophils, macrophages, mast cells, eosinophils	Receptor for complement component; role in complement-induced inflammation
CD89	Fcα receptor	55–70 kD	Neutrophils, monocytes	IgA-dependent cytotoxicity
CDw90	Thy-1	25–35 kD; PI-linked	Thymocytes, peripheral T cells (mice), neurons (all species)	Marker for T cells; ? role in T cell activation
CD91	α2-macroglobulin receptor	600 kD	Macrophages and monocytes	?
CDw92	—	70 kD	Broad	?
CD93	—	118–129 kD	Neutrophils, monocytes, endothelial cells	?
CD94	—	Dimer of 43 kD units	NK cells	?
CD95	Fas, APO-1	42 kD	Multiple cell types	Role in activation-induced cell death
CD96	—	160, 180, 240 kD forms	T cells	?
CD97	—	74, 80, 89 kD forms	Broad	?
CD98	—	Dimer of 40 and 80 kD subunits	Broad	?
CD99	—	32 kD	Broad	?
CD100	—	150 kD	T cells, B cells, granulocytes, monocytes, NK cells	?
CDw101	—	Dimer of 140 kD subunits	Granulocytes, monocytes	?
CD102	ICAM-2	55–65 kD	Endothelial cells, monocytes, other leukocytes	Ligand for LFA-1 integrin
CD103	HML-1; αE integrin	Dimer of 150 and 25 kD subunits (integrin)	Some T lymphocytes, other cell types	? Role in T cell homing to mucosa; E cadherin binding
CD104	β4 integrin chain	205–220 kD	Lymphocytes, others	Adhesion
CD105	Endoglin	Dimer of 95 kD subunits	Endothelial cells, activated macrophages	? TGF-β receptor
CD106	VCAM-1	90–95 kD	Endothelial cells, macrophages, follicular dendritic cells, marrow stromal cells	Receptor for VLA-4 integrin; role in cell adhesion, lymphocyte activation, hematopoiesis
CD107a	LAMP-1	110 kD	Broad	Lysosomal protein of unknown function
CD107b	LAMP-2	120 kD	Broad	Lysomal protein of unknown function

Table continued on following page

APPENDIX: PRINCIPAL FEATURES OF KNOWN CD
MOLECULES *(Continued)*

CD Designation	Common Synonym(s)	Molecular Structure	Main Cellular Expression	Known or Proposed Function(s)
CDw108	—	75–83 kD	Broad	?
CDw109	—	170/150 kD	Endothelial cells, monocytes	?

In addition to the above, several cytokine receptors have been given CD designations. In this book, we refer to cytokine receptors by the more informative descriptive names, which identify them by their specificities. The following is a current list of the cytokine receptors that have been assigned CD numbers.

CD115	M-CSF (CSF-1) receptor
CDw116	GM-CSF receptor
CD117	c-Kit, stem cell factor receptor
CD118	IFN-α, β receptor
CDw119	IFN-γ receptor
CD120a	55 kD TNF receptor
CD120b	75 kD TNF receptor
CDw121a	Type 1 IL-1 receptor
CDw121b	Type 2 IL-1 receptor
CD122	IL-2 receptor β chain
CD123	IL-3 receptor α chain
CDw124	IL-4 receptor
CD125	IL-5 receptor α chain
CD126	IL-6 receptor
CDw127	IL-7 receptor
CDw128	IL-8 receptor
CDw130	130 kD signaling component of IL-6 receptor

This list has been compiled with the assistance of Drs. T. F. Tedder, Duke University School of Medicine, and S. Shaw, National Institutes of Health. The complete listing of CD molecules is published in *Leukocyte Typing V,* edited by S. Schlossman, L. Boumsell, W. Gilks, J. Harlan, T. Kishimoto, C. Morimoto, J. Ritz, S. Shaw, R. Silverstein, T. Springer, T. Tedder, and R. Todd, Oxford University Press, 1994. Additional details of individual CD molecules may be found in Barclay, A. N., M. L. Birkeland, M. H. Brown, A. D. Beyers, S. J. Davis, C. Somoza, and A. F. Williams, editors, *The Leukocyte Antigen Facts Book,* Academic Press, 1993.

* CD molecules to which reference has been made in the text of this book are indicated in boldface.

† The small letters affixed to some CD numbers refer to complex CD molecules that are encoded by multiple genes or that belong to families of structurally related proteins. For instance, CD1a, CD1b, and CD1c are structurally related but distinct forms of a β_2 microglobulin–associated nonpolymorphic protein.

‡ CD11a, CD11b, and CD11c are three α chains that can non-covalently associate with the same β chain (CD18) to form three different integrins, all of which are members of the "CD11CD18" family (also called the "LFA-1 family" or the "β_2 integrins").

§ Antibodies that have been submitted recently, or whose reactivity has not been fully confirmed, are said to identify putative CD molecules, indicated with a "w" (for "workshop") designation.

Abbreviations: ADCC, antibody-dependent cell-mediated cytotoxicity; GMP, granule membrane protein; GP, glycoprotein; HSA, heat-stable antigen; ICAM, intercellular adhesion molecule; Ig, immunoglobulin; IL, interleukin; kD, kilodalton; LFA, lymphocyte function–associated antigen; LPS, lipopolysaccharide; MAC, membrane attack complex; MHC, major histocompatibility complex; NK, natural killer; PI, phosphatidylinositol; TAC, T cell activation antigen; TGF, tumor growth factor; TNF, tumor necrosis factor; VCAM, vascular cell adhesion molecule; VLA, very late activation.

INDEX

Note: Page numbers in *italics* refer to illustrations. Page numbers followed by "t" indicate tables;
page numbers followed by "b" indicate boxed material.

A

O

O antigen, of blood, 371b
Oct-1, in interleukin-2 gene expression, 166
Off rate constant (k$_{off}$), of antibody-antigen interaction, 53b
    of peptide-MHC molecule binding, 105–106
OKT3 monoclonal antibody, in transplantation, 375–376
On rate constant (k$_{on}$), of antibody-antigen interaction, 53b
    of peptide-MHC molecule binding, 105–106
Oncofetal antigens, 394
Oncogenes, of retroviruses, 393
    tumor antigens encoded by, 389–390, 389t
Oncogenic viruses, tumor antigens encoded by, 390, 391b, 392–393
Oncostatin M, receptor for, 251b
Opsonins, 314, *314,* 332
Opsonization, 8, 55–57, *56*
    complement-mediated, 314, *314,* 332, 332t
    immunoglobulin G in, 55–57, *56*
Oral administration, of antigen, 209
Oral tolerance and, 245–246
Orthotopic transplantation, 363. See also *Transplantation.*
Osteoclasts, 24

P

p150,95 (CR4, CD11cCD18), 330t, 331
    deficiency of, 337
Papain cleavage, of immunoglobulin G, 50, *50*
Papillomavirus, tumors associated with, 390
Papovaviruses, genomes of, tumor antigen encoded by, 390
Paracrine action, of cytokines, 251
    of interleukin-2, 266
Parafollicular areas, of lymph nodes, *30, 31*
Parasites, anatomic sequestration of, 356
    antigen masking of, 356
    antigen variation of, 356–357, *357, 358*
    antigenic loss in, 357
    IgE-initiated response to, 301
    immune resistance of, 356
    immunity to, 353–357
        inhibition of, 357
        innate, 354
        specific, 354–356
        tissue injury with, 355–356
Paroxysmal nocturnal hemoglobinuria, 337
Passenger leukocytes, in rejection, 369
Passive cutaneous anaphylaxis, 299
Passive immunity, 7
Passive immunization, 226, 360
    in tetanus, 226, 344
    in tumor therapy, 403–405
Passive immunotherapy, with anti-tumor antibodies, 403–405
Pemphigus vulgaris, 431t
Pepsin proteolysis, of immunoglobulin G, 50, *50*
Peptide(s), immunodominant, 134–135
    MHC binding to. See *Peptide-MHC molecule complexes.*
    self, in T lymphocyte selection processes, 188–189
    signal (leader), in immunoglobulins, 74, *75, 76*

Peptide antigens. See also *Antigen(s); Antigen processing; Protein antigen(s).*
    altered, 168
    in helper T lymphocyte anergy, 218–219, *219*
    artificial generation of, 126, *126*
    endosome localization of, 124–125, *125*
    generation of, 125–127, *126*
    immunogenicity of, 134–136
    MHC molecule binding of. See *Peptide-MHC molecule complexes.*
    proteolytic generation of, 131–132
    T lymphocyte receptors for. See *T cell receptor(s).*
    T lymphocyte recognition of, 106–107, 116–117, *118,* 120, *121,* 121t
Peptide vaccines, in immunotherapy, 400–401
Peptide-binding region, of MHC molecules, 105–107
    class I, 102–104, *102, 103,* 106, Color Plate III
    class II, 104–105, *105,* 106, Color Plate III
Peptide-MHC molecule complexes, 105–106
    αβ T cell receptor binding of, 145–147, *146,* 147t
    cellular recycling of, 129
    determinant selection model of, 136
    double-positive thymocyte interactions with, *189,* 191–192
    formation of, 127–129, *128,* 130, *130,* 132–133, *133*
    in T lymphocyte antigen recognition, 106–107, 116–117, *118,* 120, *121,* 121t
    in T lymphocyte maturation, 185–192, *186, 189*
    life span of, 129
    physiologic significance of, 133–136, *135*
    presentation of. See *Antigen presentation; Antigen-presenting cells.*
Perforin (cytolysin), in cytolytic T lymphocyte–mediated cell lysis, 291, 293, *293*
Periarteriolar lymphoid sheaths, of spleen, 31
Peripheral lymphoid tissue, 27, *28, 29–32, 30, 32.* See also *Lymph nodes; Spleen.*
Peripheral tolerance, 407. See also *Tolerance.*
    to self-antigens, 407, *408,* 409–410, *409*
        failure of, 413–418, 414b, 415b, *416,* 417b
    in B lymphocytes, 412, 412t
    in T lymphocytes, 412t
Pernicious anemia, 431t, 433
Peyer's patches, 32, 234, 243, *243,* 244
PgP-1 (CD44 molecule), 236t, 465
    of T lymphocyte precursor, 174, 184
pH, in antigen processing, 126
Phagocytes. See also *Phagocytosis.*
    congenital disorders of, 448–449
    intracellular bacteria replication in, 346
    mononuclear. See *Mononuclear phagocytes.*
Phagocytic amebocytes, 6b
Phagocytosis, 289
    complement promotion of, 332, 332t, 333
    immunoglobulin G enhancement of, 55–57, *56*
    in antibody-mediated tissue injury, 426, *427*
    in bacterial infection, extracellular, 343, 344
    intracellular, 346
    of antibody-coated cells, 426, *427*

Phagocytosis *(Continued)*
    of immune complexes, 333
    resistance to, by bacteria, 346
        by parasites, 356
Pharyngeal tonsils, 244
Philadelphia chromosome, 92b
Phosphatidylinositol 4,5-bisphosphate, in T lymphocyte activation, 164, *164*
Phosphatidylinositol phospholipase Cγl (PI-PLC-γl), in B lymphocyte activation, 199
    in mast cell activation, 304–305, *305*
    in T lymphocyte activation, 164, *164*
Phospholipids, determinants (epitopes) of, 51
Phosphorylation, tyrosine, in B lymphocyte activation, 198, *199*
    in T lymphocyte activation, 161–164, *162,* 162b–163b
    of STAT proteins, 256b–257b
Pili, surface antigens of, variation in, 346
Plaque assay, hemolytic, for B lymphocyte activation, 197b
Plaque-forming cells, in B lymphocyte activation assay, 197b
Plasma cells, 21–22, *22*
    in rheumatoid arthritis, 434b
    mRNA of, 93
Plasma proteins, in acute phase response, 262b
Plasmacytoma, interleukin-6 effect on, 264
Plasmaphoresis, in immune-mediated diseases, 438
    in rejection prevention and treatment, 376
Plasmodium falciparum, immunity to, 354–356, 355b
    survival mechanisms of, 356–357
Platelet(s), 16t, *29*
    formation of, *276*
Platelet-activating factor, in delayed type hypersensitivity, 289
    in immediate hypersensitivity, 307, *308*
Platelet-derived growth factor, in delayed type hypersensitivity, 289
    receptor for, 144b
Platelet-endothelial adhesion molecule-1 (PECAM-1, CD31), in inflammation, 286
Pmell7, melanoma expression of, 393–394
Pneumocystis carinii pneumonia, in AIDS, 458
    in hyper-IgM syndrome, 443
pol gene, 393
    of HIV, 451, *451,* 453, *454*
Polyadenylation, in immunoglobulin heavy chain synthesis, 79
Polyarteritis nodosa, 429, 430t
Poly-C9 (complement component), 323–324
Polyclonal activators, 19
    in autoimmunity, 418, *418*
    of B lymphocytes, 210
Poly-Ig receptor, 58–59, 144b
Polymerase chain reaction, 109b–110b
    in MHC polymorphism study, 109
    in tissue typing, 374b
Polymorphism, 97–98
    of antibodies, 44b
    of major histocompatibility complex, 97, 101, 106, 107
    polymerase chain reaction study of, 109, 110b–111b
Polymorphonuclear leukocytes. See *Neutrophils.*
Polypeptide growth factors, in delayed type hypersensitivity, 289